Birkhäuser

Progress in Mathematical Physics

Volume 74

More information about this series at http://www.springer.com/series/4813

Michael Dütsch

From Classical Field Theory to Perturbative Quantum Field Theory

Michael Dütsch
University of Göttingen
Göttingen, Germany

ISSN 1544-9998 ISSN 2197-1846 (electronic)
Progress in Mathematical Physics
ISBN 978-3-030-04737-5 ISBN 978-3-030-04738-2 (eBook)
https://doi.org/10.1007/978-3-030-04738-2

Library of Congress Control Number: 2018968409

This book is published under the imprint Birkhäuser, www.birkhauser-science.com by the registered company Springer Nature Switzerland AG.
The registered company address is: Gewerbestrasse 11, 6330 Cham, Switzerland

Dedicated to the memory of Raymond Stora and Eberhard Zeidler. Besides H. Epstein and V. Glaser, Raymond Stora is one of the fathers of causal perturbation theory and he passionately worked on this topic until the end of his life [10]. *In innumerable discussions with Eberhard Zeidler I learned a lot about the mathematical aspects of quantum field theory and the "unity of mathematics and physics", and he strongly encouraged me to write this book.*

Contents

Foreword by Klaus Fredenhagen

Quantum field theory is a framework for the description of physics which explains a huge range of phenomena at quite different scales. Its main feature is the way, how infinitely many degrees of freedom are coupled in agreement with the principle of locality. Quantum field theory was especially successful in elementary particle physics, where the standard model yields a consistent picture, ranging from atomic physics up to the highest presently reachable energies. Its principles are also very fruitful for our understanding of condensed matter.

Quantum field theory has classical physics and nonrelativistic quantum mechanics as limit cases. But contrary to them, the mathematical status of quantum field theory is not fully clarified, and our confidence on QFT relies on the consistent picture, which evolves from different approaches, and, mainly, on the good, and in many instances excellent, agreement with experimental data.

Among the various approaches, formal perturbation theory is by far the most successful one. Formal perturbation theory, however, provides only formal power series as predictions. By resummation and truncation, numbers can be extracted from these formal series, which then can be compared with observations. The latter method is based on heuristic ideas and is, in general, not mathematically rigorous.

The computation of the terms of the formal power series is obstructed by various divergences, which occur in a direct application of the perturbative expansion. Here one can distinguish infrared (IR) divergences and ultraviolet (UV) divergences. UV-divergences are a consequence of the locality principle, which forces the inclusion of arbitrarily high energies in intermediate states. IR-divergences, on the other hand, are due to wrong ad hoc assumptions on the long distance behavior of interacting theories.

UV-divergences can be treated by the method of renormalization. This method was originally developed by Tomonaga, Schwinger, Feynman and Dyson, but had internal inconsistencies which could later be removed by the work of Bogoliubov, Parasiuk [11] and Hepp [98] (BPH). Important further steps were performed by Zimmermann [176], Epstein and Glaser [66], and Steinmann [155]. With the generalization to gauge theories by Faddeev [69], 't Hooft and Veltman [106], Becchi, Rouet and Stora [7, 8], the perturbative construction of quantum field theory was essentially complete.

A new interpretation of renormalization was developed by Wilson [171]. The main idea is that the theory at a given scale can be evaluated without knowing the theory at all scales. Thus introducing a UV-cutoff at a given scale yields a reasonable effective theory at this scale, provided the free parameters are given the correct dependence on the cutoff. In so-called renormalizable theories, the running of the free parameters can be obtained from perturbation theory.

The Wilsonian approach was mainly used in terms of the euclidean (imaginary time) path integral. But in this framework, important features of QFT are not directly visible, in particular, the noncommutative structure of quantum observables shows up only in different boundary values of analytically extended correlation functions. Moreover, when the theory is applied to more general spacetimes than the Minkowski space, in order to study effects of an external gravitational field, one finds that there is no generally covariant definition of the (real time) path integral, a difficulty arising from the non-existence of a generally covariant vacuum.

The correlation functions of the real time path integral are the vacuum expectation values of time-ordered products of interacting fields. According to the famous Lehmann–Symanzik–Zimmermann [124] (LSZ) relations, they are closely related to the S-matrix elements, as proven by Hepp [97] on the basis of the Haag–Ruelle scattering theory [92, 145]. One may therefore restrict oneself to computing these correlation functions. After splitting the action into a quadratic functional and the rest (the interaction), one obtains an expansion in terms of the familiar Feynman diagrams. They can be recursively computed by the method of BPH, and a closed expression was given by Zimmermann [176] in terms of the so-called Zimmermann-Forests.

In many cases, however, one is interested in other aspects of the theory. This holds in particular in cases, where there is no reasonable particle interpretation, e.g., on spacetimes without asymptotically flat regions. As proposed by Bogoliubov [12], one can define the interacting field as an operator-valued distribution on the Fock space of the free theory, provided one knows the time-ordered products of the interaction as operators on Fock space.

This approach (later called "causal perturbation theory") was rigorously elaborated by Epstein and Glaser [66]. They characterized the time-ordered products by certain axioms and proved that solutions exist and are uniquely determined, if the standard renormalization conditions are imposed. An essentially equivalent approach was developed by Steinmann [155] based on the so-called retarded products. Unfortunately, these extensions of the framework of perturbation theory did not find sufficient recognition (it is not even mentioned in most of the recent text books on QFT), perhaps due to problems in extending it to gauge theories and in relating it to the renormalization group in the spirit of Wilson. Actually, the absence of a regularization scale in causal perturbation theory is in apparent conflict with the Wilsonian interpretation of such a scale.

First steps for extending causal perturbation theory to gauge theory were performed by Scharf, Dütsch and collaborators [49, 148, 149]. In particular, these authors were able to show that massive nonabelian gauge fields can be consistently quantized, provided a real scalar field was added. The latter can be identified with the Higgs field. A remarkable feature of this approach is the absence of spontaneous breakdown of symmetry, which is crucial for the conventional formulation of the standard model. The model is nevertheless equivalent to the standard model, up to infrared problems which are open in both approaches.

Actually, this approach yields the definition of the S-matrix for gauge theories in the absence of IR-divergences, i.e., if all arising physical particles have non-zero masses. In the presence of massless particles (i.e., photons in the standard model) the S-matrix (in the usual sense) does not exist, and one has to construct other observables.

In the last two decades, a new effort was made to improve perturbative QFT, with the aim to construct the local observables for gauge theories and in the presence of external gravitational fields. It turned out that this required certain changes of the framework.

First of all, as previously observed by Dimock and Kay [34], one had to use the algebraic approach to quantum field theory [93] (also called Local Quantum Physics [94]). In this approach one first considers the algebra of local observables as an abstract algebra. States are then defined as linear functionals on the algebra yielding the expectation values. By the GNS construction, each state yields a representation of the algebra by operators on some Hilbert space. This approach is especially useful in the absence of a distinguished state (the typical situation on curved spacetimes) or, if the state space is a priori not well understood, as visible in IR-divergences for wrong choices.

The second ingredient is the use of structures of the associated classical field theory. Actually, QFT may be interpreted as a deformation quantization of classical field theory, as suggested by the relation of the labeling of Feynman diagrams by the number of loops and the expansion with respect to powers of $\hbar$.

The third ingredient is the choice of an appropriate framework of analysis which replaces the techniques for Hilbert space operators. Here, starting from the thesis of Radzikowski [140], the use of microlocal analysis turned out to be appropriate.

Based on these ingredients a new formalism of perturbative quantum field theory was developed, mainly by Brunetti, Dütsch, Fredenhagen, Hollands, Rejzner and Wald [24, 27, 48, 55, 74, 100, 101], and supported by advice from Stora [138, 158]. The formalism includes older approaches and provides a more complete solution. Nevertheless, it is less involved and allows explicit calculations. Moreover, it is mathematically rigorous. Clearly it is now the right time to give a coherent introduction to the field in order to make it easier accessible for newcomers.

There exists already a brief description of the framework by Katarzyna Rejzner [143], addressed to mathematicians. The present book by Michael Dütsch takes a more pragmatic point of view and is directed to physicists. So not only the general ideas are carefully developed, but many concrete calculations and other applications are included. In particular, certain parts of the literature are elaborated in more detail; moreover, some inconsistencies in the original papers are removed. While mathematically rigorous, an attempt is made to keep the used mathematics rather elementary and within standard knowledge of physicists.

The book should be helpful for master and Ph.D. students who want to understand QFT and want to become able to make concrete calculations, on the basis of a consistent framework. Also the experienced researcher will profit from reading this nice book.

Hamburg, July 2018 *Klaus Fredenhagen*

Preface

On the one hand, perturbative quantum field theory (pQFT) is very successful: The results obtained by computing the lowest orders of the perturbation series are in good agreement with experiments. In particular, for the magnetic moment of the electron the deviation of the g-factor from the Dirac value $g = 2$ is an effect of quantum electrodynamics (QED); the theoretical prediction and the experimental value for $(g - 2)$ agree with a spectacular accuracy.

On the other hand, pQFT appears to be somewhat mystic or magic. This relies not only on the fact that, for all known, physically relevant models in $d = 4$ dimensions, the convergence of the perturbation series is not under control; it relies also on the way pQFT is "derived" in many textbooks: Some of the arguments are only on a heuristic level; frequently rigorous treatments are hidden in the original literature, not intelligible for a non-expert.

Therefore, the aim of this book is to give a logically satisfactory route from the fundamental principles to the concrete applications of pQFT, which is well intelligible for students in mathematical physics on the master or Ph.D. level. This book is mainly written for the latter; it is made to be used as basis for an introduction to pQFT in a graduate-level course.

Let me explain the advantages of this book more explicitly and sketch the procedure. In conventional formulations of pQFT (see, e.g., [109, 134, 168]) one usually constructs the scattering matrix (S-matrix) as a formal power series in the coupling constant(s) κ. The coefficients of this series are obtained by integrating time-ordered products of the interaction over their spacetime arguments. Since the classical limit $\hbar \to 0$ of time-ordered products diverges, the quantization process is not very transparent in such an approach. In addition, many textbooks "derive" the rules for computing the S-matrix (i.e., the Feynman rules and their renormalization), by working at intermediate steps with ill-defined quantities.

In this book we present a novel way to pQFT [54, 55, 64], which avoids these drawbacks. The starting point is a perturbative formulation of classical field theory (clFT); quantization is then done by deformation quantization of the underlying free theory and by the principle that we want to maintain as much as possible from the classical structure. Similarly to Steinmann's first book [155], this procedure yields a formulation of pQFT in terms of *retarded products*, which are the coefficients of the perturbative expansion (in κ) of the interacting fields.

The main difference from Steinmann's work is that we work with localized interactions, i.e., the coupling constant κ is multiplied with a test function $\mathbb{M} \ni x \mapsto g(x)$ (here $\mathbb{M}$ denotes the Minkowski space) with compact support – this is similar to the Bogoliubov–Shirkov–Epstein–Glaser approach to pQFT (also called "causal perturbation theory") [12, 24, 66, 148, 158]. This switching of the interaction avoids infrared (IR) divergences; one first solves the ultraviolet (UV) problem, which is the problem of renormalization and amounts to the extension of distributions to coinciding points. To compute expectation values in scattering states, one performs the adiabatic limit $g(x) \to 1\ \forall x \in \mathbb{M}$ in a later step; in this limit one has to solve the IR-problem. However, as observed by Brunetti and Fredenhagen, the algebraic properties of interacting fields can be studied without really performing the adiabatic limit, i.e., without meeting IR-divergences, by the so-called "algebraic adiabatic limit" ([24, 52]; see Sect. 3.7). In particular the usual formulation of the renormalization group as a renormalization of the coupling constants can be obtained in this way ([27, 55, 57, 102]; see Sect. 3.8).

After translation of our retarded products into time-ordered products by Bogoliubov's formula ([12], see Sect. 3.3) the results agree with what one obtains in causal perturbation theory [66]. Finally, performing the adiabatic limit $g(x) \to 1$ for these time-ordered products, one obtains the same experimental predictions – mostly these are scattering experiments – as one finds in conventional textbooks on pQFT.

In our approach, Fock space is replaced by a description of (classical and quantum) fields as functionals on the classical configuration space; in case of a real scalar field the configuration space is $C^\infty(\mathbb{M}, \mathbb{R})$. Hence, technicalities of Fock space, such as domains of operators, do not appear, at the price of having to worry about wave front sets of distributions. However, a superficial knowledge of wave front sets, as given in App. A.3, suffices for an understanding of this book. In contrast to the description of quantum fields as Fock space operators, our (quantum) fields are "off-shell" fields, that is, they are not restricted by any field equation. This is a main reason why our description of quantum fields is more flexible, which is advantageous for various purposes: In particular for a discussion of the classical limit and for a regularization of the Feynman propagator and the definition of the pertinent regularized S-matrix (Sect. 3.9). Also more technical issues are strongly simplified, e.g., the formulation of the Action Ward Identity (Sect. 3.1.1) and the proof of the Main Theorem (Sect. 3.6.2), which is the key to the Stückelberg–Petermann renormalization group (Sect. 3.6). Finally, our off-shell field formalism bears a close resemblance to the functional integral approach to QFT, since also there one works with field configurations being not restricted by any field equation – see Sect. 3.9.5.

Proceeding in the spirit of this book's title, we treat *symmetries* by means of the Master Ward Identity (MWI, Chap. 4). Namely, the MWI is the straightforward generalization to pQFT of the most general identity valid in clFT due to the field equation and the fact that classical fields may be multiplied pointwise. Since the latter does not hold for quantum fields, the MWI is a highly nontrivial renormalization condition, which cannot always be satisfied.

To study a physically relevant model, we apply our formalism to QED, which requires a treatment of gauge or BRST symmetry (Chap. 5). The main problems appearing in the quantization of a gauge theory can be explained more clearly for QED than for the Standard Model of electroweak and strong interactions, since the latter is more complicated and has the additional difficulty that some of the spin-1 fields (also called "vector boson fields") are massive. For a treatment of massive spin-1 fields in the Epstein–Glaser framework we refer to the book [149] (or the enlarged edition [150]) or to the original papers [2, 49], for subsequent developments see [50, 59, 65]. In view of the generalization to massless non-Abelian gauge theories (in particular quantum chromodynamics) and to QFT on curved spacetimes, we give a *local* construction of the observables of QED – this seems to be the best one can do in a perturbative approach to such models (see Sect. 5.6).

The main subject of this book is pQFT and not classical field theory. Therefore, Chapters 1 and 2 about "Classical field theory" and "Deformation quantization of the free theory", respectively, are quite short; essentially only the material needed for the step from perturbative clFT to pQFT is presented. For example, we restrict the treatment of classical field theory to *local* interactions and the Poisson bracket is introduced only for the *free* theory – not for an interacting theory. An exception is that we include in Chapter 2 the possibility of quantizing with a Hadamard function instead of the Wightman two-point function – although the treatment of pQFT in Chapters 3, 4 and 5 is based on quantization with the Wightman two-point function. The modifications appearing in pQFT when quantizing with a Hadamard function are given in App. A.4.

The focus of this book is on the *physical* concepts – the mathematical tools are reduced to a minimum. For this reason, we consider only fields which are polynomials in the basic fields. The generalization of our formalism to non-polynomial fields is worked out in the literature (see Remark 1.3.8), but this requires more sophisticated mathematics, which is explained, e.g., in [143]. A reader more interested in the mathematical aspects of QFT is referred to the books [173–175] and [79].

The formalism developed in this book can be used to derive many structural results of pQFT in a transparent way, e.g., the "Main Theorem" and the pertinent version of the renormalization group (Sect. 3.6) or the "Quantum Action Principle" (Sect. 4.3). On the other hand, this formalism is also well suited for practical computations, as is explained in Sect. 3.5 ("Techniques to renormalize in practice") and by many examples and exercises.

Some proofs are only sketched – the complete versions can be found in the original literature. Some other proofs, which are helpful to make an important topic, such as the "Main Theorem", more intelligible for a non-expert reader, are given in greater detail than in the literature. Some statements can certainly be formulated in a shorter, more elegant way; however, since this book is written for non-experts, we prefer to give detailed and explicit formulas, to avoid misunderstandings. Some exercises are quite simple – the expert reader may skip them. However, the results of many exercises are used later on. Solutions are given for all exercises.

For the *expert reader* the scope of this book can be summarized as follows:

- Our approach to QFT is *formal perturbation theory*: The observables are constructed as *formal power series* in the coupling constant and in $\hbar$. Questions concerning the convergence of this series (or rather asymptotic convergence or Borel summability) are not touched.
- Quantization is done by *deformation of classical field algebras*. Since a direct quantization of the algebra of perturbative, interacting, classical fields is not yet solved, we apply deformation quantization to the underlying algebra of free fields.
- *Deformation quantization* [6] of the free theory is worked out not only for scalar fields (Chap. 2), also spinor fields, the photon field and Faddeev–Popov ghosts are treated (Sect. 5.1).
- Renormalization is done by a version of the Epstein–Glaser method [66]. So, to a large extent, the subject of this book is *causal perturbation theory for off-shell fields*, which has various advantages – as mentioned above.
- This book yields a perturbative construction of the net of algebras of observables ("perturbative algebraic QFT", Sect. 3.7), this net satisfies the Haag–Kastler axioms [93] of *algebraic QFT*, expect that there is no suitable norm available on these formal power series. For this construction it suffices to perform the algebraic adiabatic limit, which avoids IR-divergences. A main motivation for this construction is that it can be generalized to curved spacetimes [24, 100–103, 105]. However, in this book we remain in the Minkowski space.
- Such a local algebraic construction of pQFT is explicitly worked out for QED (Chap. 5): The observables are selected by means of the BRST transformation [7, 8, 164]. A construction of the pre-Hilbert space of physical states is given by a local version of the Kugo–Ojima formalism [118].
- To investigate the solvability of the Master Ward Identity (shortly introduced above), we prove a version of the Quantum Action Principle (QAP, Theorem 4.3.1). The main difference of our results to the original QAP of Lowenstein [125] and Lam [119] and its applications in "algebraic renormalization" (see, e.g., [135]) is that we work with an adiabatically switched off interaction (Sect. 4.5).
- The latter holds generally true: To make contact to traditional formulations of pQFT, we have to perform the adiabatic limit. For algebraic properties

of the theory, e.g., the renormalization group, the algebraic adiabatic limit suffices (Sect. 3.8). But, to compute expectation values in scattering states, or Wightman- or Green's functions, we have to perform the *weak adiabatic limit* (in the sense of Epstein–Glaser). We treat the latter only briefly, mainly under the simplifying assumption that all fields are massive (App. A.6).

Acknowledgement

There are two hidden co-authors: The first is Klaus Fredenhagen. To a very great extent this book is based on my common publications with Klaus and with some of his students and collaborators: Franz-Marc Boas, Ferdinand Brennecke, Romeo Brunetti, Kai Johannes Keller and Kasia Rejzner. Therefore, I am very pleased that this book starts with a preface written by Klaus. In addition, Klaus is also a kind of referee of this book, since he discussed very extensively a late version of the manuscript with me, suggesting a lot of improvements.

The second hidden co-author is Joseph C. Várilly: This book project started with a series of lectures – titled by the same phrase as this book – which I gave at the ZiF of the University of Bielefeld in Spring 2013. Joseph wrote up some 50 pages of notes from these lectures, which formed an early version of this book's manuscript; he also helped with a later version. The lecture series was initiated by José M. Gracia-Bondía, Philippe Blanchard and Joseph, too. They strongly encouraged me to write this book and my discussions with them have essentially shaped it.

I gave another series of lectures about the topic of this book at the University of Hannover in the summer semester 2017; I would like to thank Olaf Lechtenfeld and Elmar Schrohe for the invitation.

I profited a lot from enlightening discussions with Stefan Hollands, Raymond Stora and Eberhard Zeidler. In addition, I got very useful hints to improve the manuscript from Detlev Buchholz, Paweł Duch, Urs Schreiber and Stefan Waldmann.

During the writing of this book I was partly at the Max Planck Institute for Mathematics in the Sciences, Leipzig; I would like to thank the late Eberhard Zeidler and Jürgen Jost for the invitations. I also profited from an invitation to the Universidad de Costa Rica, and I thank the Vicerrectoría de Investigación for financial support. I am also grateful to Benjamin Bahr for financial support.

Göttingen, July 2018 *Michael Dütsch*

Introduction

Free quantum fields are well understood, but for all physically relevant models, interacting quantum field can only approximately be described. In this book, the approximation is perturbation theory: The interacting quantum fields are constructed as a "deformation" of the free ones. In addition, we quantize by the method of deformation quantization. Hence, the general scheme is as follows. There are two deformations, in the sense that a "field"[1]

$$F \quad \text{is deformed into a formal power series} \quad \sum_{n=0}^{\infty} F_n \, \lambda^n \quad \text{with} \quad F_0 = F,$$

where the F_n's are also "fields" and $\lambda \in \mathbb{R}$ is the "deformation parameter".

- One deformation, with parameter $\hbar$ (the Planck constant), is the quantization of the algebra of free classical fields ($\hbar = 0$), yielding the algebra of free quantum fields ($\hbar > 0$).
- The other deformation, with parameter κ (a coupling constant), is the deformation of an algebra of free fields ($\kappa = 0$) into the corresponding algebra of interacting fields ($\kappa \neq 0$), which applies to both the classical and the quantum theory.

Diagrammatically this can be described as follows:

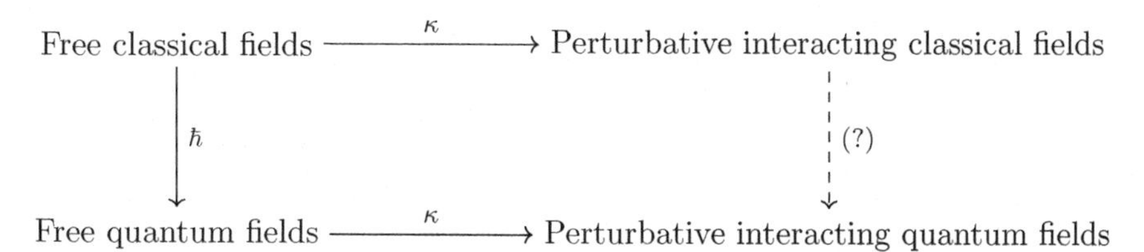

Unfortunately, the right vertical arrow, a putative direct quantization of interacting classical fields, is not yet solved; for recent progress in this direction, see

[1]The notion "field" is defined in Section 1.2.

[74, 95]. So, heuristically speaking, we quantize the perturbative interacting classical fields by an indirect route:

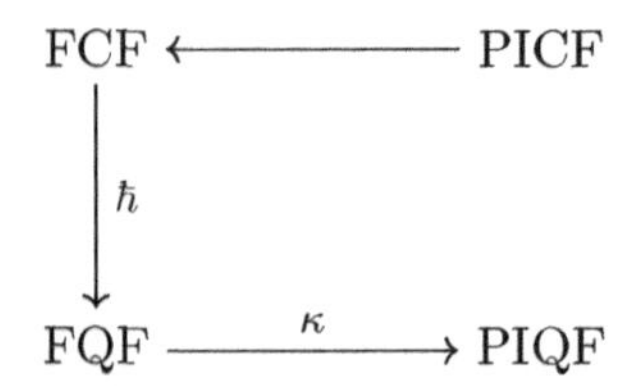

That is to say, our method is to quantize the underlying algebra of *free* fields.

Before we start to develop our formalism, we draw the reader's attention to App. A.1, in which some basic notations and conventions and a few mathematical preliminaries are given.

Physical units. We use the "Planck system of units"; for a detailed explanation of this system we refer to [173, App. A3]. We recall from this reference: "The Planck system is characterized by the fact that all the physical quantities are dimensionless and their numerical values coincide with the numerical values in natural SI units." In particular both c (the speed of light) and Planck's constant are equal to 1. Hence, we only write $\hbar$, when it has the meaning of a deformation parameter (as above); then $\hbar \in [0, \infty)$ is a switched value of Planck's constant, the physical value is $\hbar = 1$.

Chapter 1

Classical Field Theory

General references for the first part of this chapter are [54, 55].

1.1 Configuration space and the basic field φ

To keep the formulas as simple as possible, we consider only one basic field, which is a real scalar field on the d-dimensional Minkowski space $\mathbb{M}_d \equiv \mathbb{M}$; we assume $d > 2$. The pertinent free field may be massless, i.e., we allow $m \geq 0$. The generalization to more than one basic field and to more complicated fields is treated in some examples and exercises and in particular it is worked out for quantum electrodynamics in Chap. 5. The spacetime coordinates are $x = (x^\mu) \in \mathbb{M}$.

Most physicists understand by a classical field a solution of a given field equation. We use a different definition: Our "fields" are defined without reference to any field equation. Our definition of "configuration space" and "fields" (or "observables") in classical field theory (clFT) is motivated by analogy to classical mechanics.

Classical, non-relativistic mechanics (one point particle). The configurations are the trajectories

$$\vec{h} : \begin{cases} \mathbb{R} \longrightarrow \mathbb{R}^{d-1} \\ t \longmapsto \vec{h}(t) \ , \end{cases}$$

where $\vec{h}(t)$ is the position of the particle at time t. Assuming smoothness of the trajectories, the configuration space is $C^\infty(\mathbb{R}, \mathbb{R}^{d-1})$. The basic observable $\hat{\vec{x}}(t)$ – the position at time t – is the evaluation functional on the configuration space:

$$\hat{\vec{x}}(t) : \begin{cases} C^\infty(\mathbb{R}, \mathbb{R}^{d-1}) \longrightarrow \mathbb{R}^{d-1} \\ \vec{h} \longmapsto \vec{h}(t) \ . \end{cases}$$

M. Dütsch, *From Classical Field Theory to Perturbative Quantum Field Theory*,
Progress in Mathematical Physics 74, https://doi.org/10.1007/978-3-030-04738-2_1

Derivatives of $\hat{\vec{x}}(t)$, e.g., the velocity $\hat{\vec{v}}(t) = \frac{d}{dt}\hat{\vec{x}}(t) \equiv \hat{\vec{x}}'(t)$ at time t, are defined by

$$\frac{d^n}{dt^n}\hat{\vec{x}}(t) : \begin{cases} C^\infty(\mathbb{R}, \mathbb{R}^{d-1}) \longrightarrow \mathbb{R}^{d-1} \\ \vec{h} \longmapsto \frac{d^n}{dt^n}\vec{h}(t) \ . \end{cases}$$

A simple example for a non-linear functional is the kinetic energy at time t:

$$E_{\text{kin}}(t) \equiv \frac{m}{2}\,(\hat{\vec{x}}'(t))^2 : \begin{cases} C^\infty(\mathbb{R}, \mathbb{R}^{d-1}) \longrightarrow \mathbb{R} \\ \vec{h} \longmapsto \frac{m}{2}\left(\frac{d}{dt}\vec{h}(t)\right)^2 , \end{cases}$$

which can be written as

$$E_{\text{kin}}(t) = \int dt_1\, dt_2 \sum_{k_1,k_2=1}^{d-1} f_{k_1,k_2}(t_1,t_2)\, \hat{x}_{k_1}(t_1)\, \hat{x}_{k_2}(t_2)$$

with

$$f_{k_1,k_2}(t_1,t_2) := \frac{m}{2}\, \delta_{k_1,k_2}\, \delta'(t_1 - t)\, \delta'(t_2 - t) \ .$$

Generally, an observable is a function F of the map $\hat{\vec{x}} : \mathbb{R} \ni t \longmapsto \hat{\vec{x}}(t)$; $F(\hat{\vec{x}})$ is an $\mathbb{R}$-valued functional on the configuration space:

$$F(\hat{\vec{x}}) : \begin{cases} C^\infty(\mathbb{R}, \mathbb{R}^{d-1}) \longrightarrow \mathbb{R} \\ \vec{h} \longmapsto F(\hat{\vec{x}})(\vec{h}) := F(\vec{h}) , \end{cases}$$

that is, the number $F(\hat{\vec{x}})(\vec{h})$ is obtained by substituting $\vec{h}$ for $\hat{\vec{x}}$ in the formula for $F(\hat{\vec{x}})$. For example, the above formulas for $E_{\text{kin}}(t)$ can be understood in this way. If we restrict F to be a polynomial in $\hat{\vec{x}}$, it is of the form

$$F(\hat{\vec{x}}) = f_0 + \sum_{n=1}^{N} \int dt_1 \cdots dt_n \sum_{k_1 \dots k_n} \hat{x}_{k_1}(t_1) \cdots \hat{x}_{k_n}(t_n)\, f^n_{k_1 \dots k_n}(t_1, \dots, t_n) \ , \quad (1.1.1)$$

with $N < \infty$, $f_0 \in \mathbb{R}$ and $f^n_{k_1 \dots k_n}$ is an $\mathbb{R}$-valued distribution with compact support. The latter ensures existence of the integral in (1.1.1), when $h_{k_j} \in C^\infty(\mathbb{R}, \mathbb{R})$ is substituted for $\hat{x}_{k_j}$ for all $j = 1, \dots, n$.

Note: In contrast to $E_{\text{kin}}(t)$, the functionals (1.1.1) may be non-local in time, that is, the support of $f^n_{k_1 \dots k_n}$ does not need not to lie on the total diagonal $\{(t_1, \dots, t_n) \in \mathbb{R}^n \,|\, t_1 = \cdots = t_n\}$. Hence, values of $\vec{h}$ (and its derivatives) at different times may may be combined.

Classical field theory. Here, the configuration space is the space $\mathcal{C} := C^\infty(\mathbb{M}, \mathbb{R})$ of smooth functions. Since we are studying a model built from a real scalar field, we may use the word "field" instead of "observable". The *basic field* φ is simply the evaluation functional on the configuration space, explicitly:

$$\varphi : \mathbb{M} \ni x \longmapsto \varphi(x) \quad \text{where} \quad \varphi(x) : \begin{cases} \mathcal{C} \longrightarrow \mathbb{R} \\ h \longmapsto h(x) \ . \end{cases} \tag{1.1.2}$$

Partial derivatives $\partial^a \varphi$ are defined by

$$\partial^a \varphi(x) : \begin{cases} \mathcal{C} \longrightarrow \mathbb{R} \\ h \longmapsto \partial^a h(x) \ , \end{cases} \tag{1.1.3}$$

where $a \in \mathbb{N}^d$, see (A.1.23). A general field is a function F of the map φ (1.1.2); $F(\varphi)$ is a functional on $\mathcal{C}$:

$$F(\varphi) : \begin{cases} \mathcal{C} \longrightarrow \mathbb{C} \\ h \longmapsto F(\varphi)(h) := F(h) \ . \end{cases} \tag{1.1.4}$$

We point out that $F(\varphi)$ is $\mathbb{C}$-valued – for the following purpose: In view of QFT, we want to introduce a nontrivial "$*$-operation" by complex conjugation – see (1.2.7) below.

The first question to be asked is: *Which functionals $F(\varphi)$ are allowed?* Before answering this (in Def. 1.2.1 below), we study two simple examples.

Example 1.1.1. Let $x_1, x_2 \in \mathbb{M}$ and $a, b \in \mathbb{N}^d$ be given. The product of field derivatives

$$F_1(\varphi) := \partial^a \varphi(x_1)\, \partial^b \varphi(x_2) : h \longmapsto F_1(\varphi)(h) = \partial^a h(x_1)\, \partial^b h(x_2)$$

is well defined – even for $x_1 = x_2$, but F_1 is a non-linear functional of h.

The latter holds also for powers φ^n integrated out with $g \in \mathcal{D}(\mathbb{M}, \mathbb{C})$, where $n \in \mathbb{N}$:

$$F_2(\varphi) := \int dx\ \varphi^n(x)\, g(x) : h \longmapsto F_2(\varphi)(h) = \int dx\ \big(h(x)\big)^n\, g(x) \ .$$

In Sects. 1.2–1.4 we introduce the mathematical framework for our formulation of (perturbative) classical field theory, which will be used also in pQFT. Physics starts in Sect. 1.5: There we introduce the *free action* and *interactions* – with that the relation of our notion of a "classical field" to a "solution of a given field equation" will be clarified.

1.2 The space of fields $\mathcal{F}$

In this section we introduce the space of fields $\mathcal{F}$, which will be used throughout the whole book – also in QFT. In particular in view of deformation quantization, it is a main advantage of our approach that the fields of both the classical and the quantum theory are defined in terms of the *same* space of functionals. Classical and quantum fields differ only by the algebraic structures that we will introduce on this space: Main building blocks are the Poisson bracket for the classical algebra and the (non-commutative) star product for the quantum algebra.

A main point is that our fields are "off-shell fields", that is, they are not restricted by any field equation. For simplicity, we study only fields which are polynomials in the basic field φ (and its derivatives). Motivated by (1.1.1) we define:

Definition 1.2.1 (Set of fields). The *set of fields* $\mathcal{F}$ is defined as the totality of functionals $F \equiv F(\varphi)\colon \mathcal{C} \longrightarrow \mathbb{C}$ of the form

$$F(\varphi) = f_0 + \sum_{n=1}^{N} \int d^dx_1 \cdots d^dx_n \; \varphi(x_1) \cdots \varphi(x_n)\, f_n(x_1, \ldots, x_n)$$
$$=: \sum_{n=0}^{N} \langle f_n, \varphi^{\otimes n} \rangle \tag{1.2.1}$$

with $N < \infty$; the number $F(\varphi)(h)$ is obtained by substituting h for φ in this formula, according to (1.1.4). Here $f_0 \in \mathbb{C}$ is constant; but for $n \geq 1$, each expression f_n is a $\mathbb{C}$-valued distribution (i.e., $f_n \in \mathcal{D}'(\mathbb{M}^n, \mathbb{C})$) with compact support, which is

(i) *symmetric* in its arguments: $f_n(x_{\pi(1)}, \ldots, x_{\pi(n)}) = f_n(x_1, \ldots, x_n)$ for $\pi \in S_n$; and

(ii) whose *wave front set* satisfies the following property:

$$\mathrm{WF}(f_n) \subseteq \{ (x_1, \ldots, x_n; k_1, \ldots, k_n) \,\big|\, (k_1, \ldots, k_n) \notin \overline{V}_+^{\times n} \cup \overline{V}_-^{\times n} \}\,, \tag{1.2.2}$$

where (A.1.11) is used.

The set of distributions f_n of this kind will be denoted by $\mathcal{F}'(\mathbb{M}^n)$.

The condition that the distributions f_n have compact support implies convergence of the integral (1.2.1); hence, it is not needed that the configurations $h \in \mathcal{C}$ have any decay property.

The condition (i) is not necessary, however it avoids some redundancy of the set of allowed f_n's (because $\varphi(x_1) \cdots \varphi(x_n)$ is symmetric in $x_1, \ldots, x_n$) and it simplifies the expression for the functional derivatives given in (1.3.1).

A short introduction to wave front sets is given in App. A.3. The wave front set of a distribution is a characterization of its singularities by means of the Fourier transformation; the variables $k_1, \ldots, k_n \in \mathbb{M}$ appearing in (1.2.2) are the conjugate momenta of the spacetime variables $x_1, \ldots, x_n \in \mathbb{M}$. The purpose of the wave front set condition (ii) is to ensure the existence of pointwise products of distributions which appear in our definition of the Poisson bracket and, more generally, of the star product (see Sects. 1.8 and 2; both definitions will be given for the free theory).

One easily checks that $\mathcal{F}'(\mathbb{M}^n)$ is a *vector space*, hence, this holds also for $\mathcal{F}$: For $F = \sum_n \langle f_n, \varphi^{\otimes n}\rangle$, $G = \sum_n \langle g_n, \varphi^{\otimes n}\rangle \in \mathcal{F}$ and $\alpha \in \mathbb{C}$, the linear combination

$$F + \alpha G = \sum_n \langle (f_n + \alpha g_n), \varphi^{\otimes n}\rangle \; : \; h \longmapsto F(h) + \alpha\, G(h)$$

is also an element of $\mathcal{F}$.

Convergence in $\mathcal{F}$ is understood in the pointwise sense: For $F_n, F \in \mathcal{F}$ we define

$$\lim_{n\to\infty} F_n = F \quad \text{iff} \quad \lim_{n\to\infty} F_n(h) = F(h) \quad \forall h \in \mathcal{C}\ . \tag{1.2.3}$$

Example 1.2.2. A constant functional is $F = f_0 \in \mathbb{C}$, i.e., $F(h) = f_0$ for any h.

Example 1.2.3. If $\mathfrak{S}_n$ denotes symmetrization in $x_1, \dots, x_n$, we may consider

$$f_n(x_1, \dots, x_n) := \mathfrak{S}_n \int dx\; g(x)\, \partial^{a_1}\delta(x_1 - x) \cdots \partial^{a_n}\delta(x_n - x)\ , \quad g \in \mathcal{D}(\mathbb{M}, \mathbb{C})\ , \tag{1.2.4}$$

taking $f_k = 0$ for $k \neq n$. Then

$$F(\varphi) = (-1)^{\sum_j |a_j|} \int dx\, g(x)\, \partial^{a_1}\varphi(x) \cdots \partial^{a_n}\varphi(x)$$

in this case. That f_n (1.2.4) fulfills the wave front set condition (1.2.2) can be seen on general grounds, as explained after Remk. 1.2.6, or explicitly by a generalization of the argumentation given in the next example.

Example 1.2.4. In particular, the preceding example states that $\int dx\, g(x)\, \varphi^n(x) \in \mathcal{F}$ for $g \in \mathcal{D}(\mathbb{M})$. However, we point out that $\varphi^n(x) \notin \mathcal{F}$; for the following reason: Writing $\varphi^n(x)$ in the form (1.2.1) the pertinent $f_n(x_1, \dots, x_n) = \delta(x_1 - x) \cdots \delta(x_n - x)$ is a symmetric, x-dependent distribution with compact support, but it violates the wave front set condition (1.2.2). Namely, for $h \in \mathcal{D}(\mathbb{M}^n)$, $h(x, \dots, x) \neq 0$ we obtain $\widehat{hf_n}(k_1, \dots, k_n) \sim \int dx_1 \dots dx_n\, h(x_1, \dots, x_n)\, \delta(x_1 - x) \cdots \delta(x_n - x)\, e^{i(k_1x_1 + \cdots + k_nx_n)} = h(x, \dots, x)\, e^{i(k_1 + \cdots + k_n)x}$, which does not decay rapidly in any direction $(k_1, \dots, k_n)$.

In contrast, for $\int dx\, g(x)\, \varphi^n(x)$ we obtain $\widehat{hf_n}(k_1, \dots, k_n) \sim \widehat{l}(k_1 + \cdots + k_n)$ where $l(x) := g(x)\, h(x, \dots, x)$. Since $l \in \mathcal{D}(\mathbb{M}) \subset \mathcal{S}(\mathbb{M})$, we know that $\widehat{l} \in \mathcal{S}(\mathbb{M})$. Therefore, $\widehat{l}(k_1 + \cdots + k_n)$ decays rapidly for $(k_1, \dots, k_n) \in \overline{V}_+^{\times n} \cup \overline{V}_-^{\times n}$, that is, the wave front set condition (1.2.2) is satisfied.

Example 1.2.5. A typical example for a retarded product in perturbative clFT is

$$F(\varphi) = \int dx_1 dx_2 dx_3\; k(x_1) g(x_2) g(x_3)\, \partial_\mu\varphi(x_1)\, \partial^\mu p(x_1 - x_2)\, \varphi^2(x_2)\, p(x_2 - x_3)\, \varphi^3(x_3)\ ,$$

where $k, g \in \mathcal{D}(\mathbb{M})$ and p is the retarded propagator (see Definition 1.8.1) – for this example we only need that $p \in \mathcal{D}'(\mathbb{M})$. This functional is of the form (1.2.1) with $f_n = 0\ \forall n \neq 6$ and

$$f_6(x_1, \dots) = -\mathfrak{S}_6\Big(g(x_2)g(x_3)\, \partial_\mu^{x_1}\left(k(x_1)\partial^\mu p(x_1 - x_2)\right) \cdot \delta(x_2 - x_4)\, p(x_2 - x_3)\, \delta(x_3 - x_5)\delta(x_3 - x_6)\Big)\ .$$

The algebra of classical fields. We can form the pointwise product

$$F \cdot G \equiv FG : h \longmapsto F(h)G(h) , \tag{1.2.5}$$

which is commutative, we call it the "classical product". Written in terms of the expressions (1.2.1) for F and G it reads:

$$\sum_n \langle f_n, \varphi^{\otimes n}\rangle \cdot \sum_k \langle g_k, \varphi^{\otimes k}\rangle := \sum_{n,k} \langle f_n \otimes_{\mathrm{sym}} g_k, \varphi^{\otimes (n+k)}\rangle , \tag{1.2.6}$$

where $\otimes_{\mathrm{sym}}$ denotes the symmetrized tensor product. We also introduce the conjugation $F \longmapsto F^*$ with

$$F^* := \sum_n \langle \overline{f_n}, \varphi^{\otimes n}\rangle \equiv \sum_n \int dx_1 \cdots dx_n \; \varphi(x_1) \cdots \varphi(x_n) \, \overline{f_n(x_1, \ldots, x_n)} , \tag{1.2.7}$$

which we call the "$*$-operation". Obvious properties are

$$\begin{aligned} &F^*(h) = \overline{F(h)} \qquad \forall h \in \mathcal{C} \quad \text{and} \\ &(F + \alpha G)^* = F^* + \overline{\alpha}\, G^* , \quad (F \cdot G)^* = F^* \cdot G^* , \quad (F^*)^* = F . \end{aligned} \tag{1.2.8}$$

One can check that $f_n \otimes_{\mathrm{sym}} g_k$ and $\overline{f_n}$ also satisfy the wave front set condition (1.2.2), hence FG and F^* *lie also in* $\mathcal{F}$, so that the vector space of fields $\mathcal{F}$ becomes a *commutative* $*$*-algebra* (see App. A.1 for the definition of this notion). Moreover, this algebra is unital; the unit is the constant functional $F = 1$, that is, $1(h) = 1 \; \forall h \in \mathcal{C}$.

Remark 1.2.6. The wave front set condition (1.2.2) is a weak version of translation invariance. Indeed, note that the latter property means that f_n depends only on the relative coordinates,

$$f_n(x_1, \ldots, x_n) = \tilde{f}_n(x_1 - x_n, \ldots, x_{n-1} - x_n),$$

which has the following consequences:

$$\begin{aligned} &\Longrightarrow \operatorname{supp} \widehat{f}_n \subseteq \{\, (k_1, \ldots, k_n) \,|\, k_1 + \cdots + k_n = 0 \,\} \\ &\Longrightarrow \mathrm{WF}(f_n) \subseteq \mathbb{M}^n \times \{\, (k_1, \ldots, k_n) \,|\, k_1 + \cdots + k_n = 0 \,\} \\ &\Longrightarrow \mathrm{WF}(f_n) \cap \big(\mathbb{M}^n \times (\overline{V}_+^{\times n} \cup \overline{V}_-^{\times n})\big) = \emptyset. \end{aligned} \tag{1.2.9}$$

A stronger form of the wave front set condition (1.2.2) would be to ask that each $\mathrm{WF}(f_n)$ satisfy (1.2.9), i.e., that $k_1 + \cdots + k_n = 0$ for all $\big(x, (k_1, \ldots, k_n)\big) \in \mathrm{WF}(f_n)$ – this condition is used, e.g., in [54, 57], see also Remk. 1.3.6. However this stronger version does not work on a curved spacetime, since there the vectors $k_1, \ldots, k_n$ belong in general to different cotangent spaces, and hence cannot be added.

With this remark, the wave front set condition for Exap. 1.2.3 can be verified also in the following way: Let

$$\tilde{f}_n(x_1,\dots,x_n) := \mathfrak{S}_n \int dx\, \partial^{a_1}\delta(x_1-x)\cdots\partial^{a_n}\delta(x_n-x)\ .$$

Comparing the singularities of f_n and $\tilde{f}_n$, we see that the test function g appearing in f_n suppresses some singularities of $\tilde{f}_n$, hence $\mathrm{WF}(f_n)\subset\mathrm{WF}(\tilde{f}_n)$. Since $\tilde{f}_n$ is translation invariant, its wave front set fulfills (1.2.9). Therefore, the field $F(\varphi)$ studied in Exer. 1.2.3, is an element of $\mathcal{F}$.

1.3 Derivatives and support of functionals, local fields

The notions mentioned in the title of this section are frequently used in the traditional literature. Our point is that, by using the special form (1.2.1) of the functionals $F\in\mathcal{F}$, we can give simpler or more explicit definitions of these notions.

Functional derivative. We define directly the kth-order derivative for $k\in\mathbb{N}^*$.

Definition 1.3.1. The kth-order *functional derivative* of a field F (given by (1.2.1)) with respect to the basic fields $\varphi(y_j)$ is defined by

$$\frac{\delta^k F}{\delta\varphi(y_1)\cdots\delta\varphi(y_k)} \tag{1.3.1}$$
$$:= \sum_{n=k}^{N}\frac{n!}{(n-k)!}\int dx_1\cdots dx_{n-k}\,\varphi(x_1)\cdots\varphi(x_{n-k})\,f_n(y_1,\dots,y_k,x_1,\dots,x_{n-k})\ .$$

From the properties of the f_n, it follows that for all $h\in\mathcal{C}$, the functional derivative applied to h is a distribution, $\frac{\delta^k F}{\delta\varphi(y_1)\cdots\delta\varphi(y_k)}(h)\in\mathcal{D}'(\mathbb{M}^k,\mathbb{C})$, which has compact support and is symmetric in its arguments; the latter property can be written without application to h:

$$\frac{\delta^k F}{\delta\varphi(y_{\pi(1)})\cdots\delta\varphi(y_{\pi(k)})}=\frac{\delta^k F}{\delta\varphi(y_1)\cdots\delta\varphi(y_k)}\quad \forall\pi\in S_k\ .$$

The definition (1.3.1) uses the symmetry of f_n in its arguments: Due to this property it agrees with the usual definition of the functional derivative, e.g.,

$$\int dy_1dy_2\,\frac{\delta^2F}{\delta\varphi(y_1)\delta\varphi(y_2)}(h)\ g(y_1)g(y_2)=\frac{d^2}{d\varepsilon^2}\Big|_{\varepsilon=0}F(h+\varepsilon g)\quad\text{with}\quad h,g\in\mathcal{C} \tag{1.3.2}$$

for $k=2$. Note also the relation

$$\int dy\ \varphi(y)\frac{\delta}{\delta\varphi(y)}\sum_{n=0}^{N}\langle f_n,\varphi^{\otimes n}\rangle=\sum_{n=0}^{N}n\,\langle f_n,\varphi^{\otimes n}\rangle\ , \tag{1.3.3}$$

i.e., the operator $\int dy\ \varphi(y)\frac{\delta}{\delta\varphi(y)}$ counts the factors φ.

The functional derivative satisfies the *Leibniz rule*:

$$\frac{\delta(F\cdot G)}{\delta\varphi(x)}=\frac{\delta F}{\delta\varphi(x)}\cdot G+F\cdot\frac{\delta G}{\delta\varphi(x)}\qquad\forall\, F,G\in\mathcal{F}\;. \tag{1.3.4}$$

A simple way to prove this, is as follows:

$$\begin{aligned}\int dx\; g(x)\,\frac{\delta(F\cdot G)}{\delta\varphi(x)}(h)&=\frac{d}{d\varepsilon}\Big|_{\varepsilon=0}\big(F(h+\varepsilon g)\,G(h+\varepsilon g)\big)\\&=\frac{d}{d\varepsilon}\Big|_{\varepsilon=0}F(h+\varepsilon g)\,G(h)+F(h)\,\frac{d}{d\varepsilon}\Big|_{\varepsilon=0}G(h+\varepsilon g)\\&=\int dx\; g(x)\Big(\frac{\delta F}{\delta\varphi(x)}\cdot G+F\cdot\frac{\delta G}{\delta\varphi(x)}\Big)(h)\quad\forall\, h,g\in\mathcal{C}\;.\end{aligned}$$

Let a map $\beta:\mathcal{F}^{\otimes n}\longrightarrow\mathcal{F}$ be given. Since $D^kF:=\frac{\delta^k F}{\delta\varphi\cdots\delta\varphi}$ (1.3.1) is an $\mathcal{F}$-valued distribution – we write $D^kF\in\mathcal{D}'(\mathbb{M}^k,\mathcal{F})$ for that – the map β is well defined also on $D^{k_1}F_1\otimes\cdots\otimes D^{k_n}F_n$ by

$$\begin{aligned}&\langle\beta(D^{k_1}F_1\otimes\cdots\otimes D^{k_n}F_n)\,,\,g_1\otimes\cdots\otimes g_n\rangle\\&\qquad:=\beta\big(\langle D^{k_1}F_1,g_1\rangle\otimes\cdots\otimes\langle D^{k_n}F_n,g_n\rangle\big)\;,\qquad\forall g_l\in\mathcal{D}(\mathbb{M}^{k_l})\;;\end{aligned}$$

and similarly for maps $\beta:\mathcal{F}_1^{\otimes n}\longrightarrow\mathcal{F}$ if $\mathcal{F}_1$ is a subspace of $\mathcal{F}$ with $D^kF\in\mathcal{D}'(\mathbb{M}^k,\mathcal{F}_1)$ for $F\in\mathcal{F}_1$, in particular for the retarded products $\beta=R_{n-1,1}$ (1.7.2).

Example 1.3.2 (Complex scalar field)**.** A complex scalar field $\phi(x)$ and its $*$-conjugate field $\phi^*(x)$ are the evaluation functionals

$$\phi(x),\,\phi^*(x)\,:\begin{cases}C^\infty(\mathbb{M},\mathbb{C})\longrightarrow\mathbb{C}\\ \phi(x)(h):=h(x)\;,\quad\phi^*(x)(h):=\overline{h(x)}\;,\end{cases} \tag{1.3.5}$$

that is, the $*$-operation is defined by complex conjugation. The pertinent field space is the set of all functionals $F:C^\infty(\mathbb{M},\mathbb{C})\longrightarrow\mathbb{C}$ of the form

$$\begin{aligned}F&=f_{0,0}+\sum_{n=1}^{N}\sum_{l=0}^{n}\int dx_1\cdots dx_n\;\phi^*(x_1)\cdots\phi^*(x_l)\phi(x_{l+1})\cdots\phi(x_n)\,f_{n,l}(x_1,\ldots,x_n)\\&=:\sum_{n=0}^{N}\sum_{l=0}^{n}\big\langle f_{n,l}\,,\,(\phi^*)^{\otimes l}\otimes\phi^{\otimes(n-l)}\big\rangle\;,\quad N<\infty\;,\end{aligned} \tag{1.3.6}$$

where $f_{0,0}\in\mathbb{C}$ and $f_{n,l}\in\mathcal{F}'_{n,l}(\mathbb{M}^n)$. The distribution space $\mathcal{F}'_{n,l}(\mathbb{M}^n)$ is defined analogously to $\mathcal{F}'(\mathbb{M}^n)$ (Definition 1.2.1); the only difference is that "symmetry" is meant only with respect to permutations of the form $(1,\ldots,n)\longmapsto(\pi(1),\ldots\pi(l),\sigma(l+1),\ldots,\sigma(n))$ with $\pi\in S_l,\;\sigma\in S_{n-l}$.

The $*$-conjugate field is defined to be

$$F^*:=\sum_{n=0}^{N}\sum_{l=0}^{n}\big\langle\overline{f_{n,l}}\,,\,\phi^{\otimes l}\otimes(\phi^*)^{\otimes(n-l)}\big\rangle\;.$$

We emphasize that $\phi(x)$ and $\phi^*(x)$ behave as *independent* fields – this can be understood as the independence of $\mathrm{Re}\,\phi(x)$ and $\mathrm{Im}\,\phi(x)$ and becomes particularly clear in the definition of the functional derivative:

$$\frac{\delta^{k+j}F}{\delta\phi^*(y_1)\cdots\delta\phi^*(y_k)\delta\phi(z_1)\cdots\delta\phi(z_j)} := \sum_{n=k+j}^{N}\sum_{l=k}^{n-j}\frac{l!}{(l-k)!}\,\frac{(n-l)!}{(n-l-j)!}\int dx_1\cdots dx_{n-k-j}$$
$$\cdot\,\phi^*(x_1)\cdots\phi^*(x_{l-k})\phi(x_{l-k+1})\cdots\phi(x_{n-k-j})$$
$$\cdot\, f_{n,l}(y_1,\ldots,y_k,x_1,\ldots,x_{l-k},z_1,\ldots,z_j,x_{l-k+1},\ldots,x_{n-k-j})\,. \tag{1.3.7}$$

So, for the functional derivative with respect to ϕ, the field ϕ^* behaves as a constant; and the same for $\phi \leftrightarrow \phi^*$.

Support of a functional. We use the functional derivative to define the support of a functional.

Definition 1.3.3. The *support* of $F \in \mathcal{F}$ is defined as

$$\mathrm{supp}\,F := \mathrm{supp}\,\frac{\delta F}{\delta\varphi(\cdot)} \equiv \overline{\bigcup_{h\in\mathcal{C}} \mathrm{supp}\,\frac{\delta F}{\delta\varphi(\cdot)}(h)}\,;$$

on the r.h.s. we mean the support in the sense of distributions (see App. A.1). That is, $\mathrm{supp}\,F$ is the complement of the largest open subset of $\mathbb{M}$ on which *all* distributions $\{\frac{\delta F}{\delta\varphi(\cdot)}(h) \in \mathcal{D}'(\mathbb{M},\mathbb{C})\,\big|\, h\in\mathcal{C}\}$ vanish.

Hence $F(0) = f_0$ is irrelevant for the support of $F = \sum_{n=0}^{N}\langle f_n, \varphi^{\otimes n}\rangle$. The support $\mathrm{supp}\,F$ is given by the supports of the various f_n in the following way:

$$\mathrm{supp}\,F = \bigcup_{n=1}^{N} U_n \quad \text{where}$$
$$U_n := \{\, y\in\mathbb{M}\,\big|\,\exists(x_1,\ldots,x_{n-1})\in\mathbb{M}^{n-1} \text{ such that } (y,x_1,\ldots,x_{n-1})\in \mathrm{supp}\,f_n\,\}.$$

Since, for each f_n, the support $\mathrm{supp}\,f_n$ is bounded, all the sets U_n are bounded and, therefore, $\mathrm{supp}\,F$ *is bounded.* This is what is meant by saying that all fields $F\in\mathcal{F}$ are "localized".

From the Leibniz rule (1.3.4) we conclude that

$$\mathrm{supp}(F\cdot G) \subseteq (\mathrm{supp}\,F \cup \mathrm{supp}\,G) \qquad \forall F,G\in\mathcal{F}\,.$$

We point out that the stronger relation $\mathrm{supp}(f\cdot g) \subseteq (\mathrm{supp}\,f\cap\mathrm{supp}\,g)$, valid for functions $f,g:\mathbb{M}\longrightarrow\mathbb{C}$, does *not* hold for functionals $F,G\in\mathcal{F}$; a simple counter example is given for $F = c\in\mathbb{C}$ with $c\neq 0$ and $\mathrm{supp}\,G\neq\emptyset$, namely: $\mathrm{supp}(c\,G) = \mathrm{supp}\,G \neq \emptyset = (\mathrm{supp}\,F\cap\mathrm{supp}\,G)$.

Local fields. In clFT and QFT one mostly assumes that the interaction is *local* – our formulation of perturbative clFT and pQFT will use this assumption. To treat local interactions we introduce the space $\mathcal{F}_{\mathrm{loc}}$ of local fields, which is a subspace of $\mathcal{F}$. Local fields are linear combinations of fields of the form treated in Exap. 1.2.3. More explicitly we define:

Definition 1.3.4. The vector space $\mathcal{F}_{\mathrm{loc}}$ of *local fields* is defined as follows: Let $\mathcal{P}$ be the space of polynomials in the variables $\{\partial^a\varphi \,|\, a \in \mathbb{N}^d\}$ with real coefficients ("field polynomials"); then

$$\mathcal{F}_{\mathrm{loc}} := \Big\{\sum_{i=1}^{K} \int dx\; A_i(x)\, g_i(x) \,\Big|\, A_i \in \mathcal{P},\; g_i \in \mathcal{D}(\mathbb{M}),\; K < \infty\Big\}. \tag{1.3.8}$$

We abbreviate $A(g) := \int dx\, A(x)g(x)$. Note that any $A \in \mathcal{P}$ can be written as a finite sum

$$A = \sum_{n=0}^{N} \sum_{\substack{a_1,\dots,a_n \in \mathbb{N}^d \\ |a_j| \le M\ \forall j}} c_{a_1\cdots a_n}\, \partial^{a_1}\varphi \cdots \partial^{a_n}\varphi \quad \text{where} \quad c_{a_1\cdots a_n} \in \mathbb{R}\,,\quad N, M < \infty\,. \tag{1.3.9}$$

With this, $A(x)$ is the functional

$$A(x)(h) = \sum_{n} \sum_{a_1,\dots,a_n \in \mathbb{N}^d} c_{a_1\cdots a_n}\, \partial^{a_1}h(x) \cdots \partial^{a_n}h(x) \qquad \forall h \in \mathcal{C},\; x \in \mathbb{M}\,. \tag{1.3.10}$$

In the following exercise we derive some basic properties of local fields.

Exercise 1.3.5. Let $A(g) \in \mathcal{F}_{\mathrm{loc}}$. By $\frac{\partial A}{\partial(\partial^{\mu_1}\cdots\partial^{\mu_l}\varphi)}$ we mean the partial derivative of the polynomial $A \equiv A(\varphi, \partial^\mu\varphi, \dots, \partial^{\mu_1}\cdots\partial^{\mu_s}\varphi)$ with respect to the variable $(\partial^{\mu_1}\cdots\partial^{\mu_l}\varphi)$. Verify that

$$\begin{aligned} \frac{\delta A(g)}{\delta\varphi(y)} &= \sum_{l=0}^{s} (-1)^l\, \partial_y^{\mu_1} \cdots \partial_y^{\mu_l} \Big(g(y)\, \frac{\partial A}{\partial(\partial^{\mu_1}\cdots\partial^{\mu_l}\varphi)}(y) \Big) \\ &\equiv \sum_{a \in \mathbb{N}^d} (-1)^{|a|}\, \partial_y^{a} \Big(g(y)\, \frac{\partial A}{\partial(\partial^{a}\varphi)}(y) \Big), \end{aligned} \tag{1.3.11}$$

(the last expression is a shorthand notation for the second last) and the properties

$$\frac{\delta^2 A(g)}{\delta\varphi(x)\,\delta\varphi(y)} = 0 \quad \forall x \neq y \qquad \text{and} \qquad \operatorname{supp} A(g) \subseteq \operatorname{supp} g. \tag{1.3.12}$$

[*Solution*: Let $A = \partial^{a_1}\varphi \cdots \partial^{a_n}\varphi$. Inserting (1.2.4) into definition (1.3.1) we obtain

$$\frac{\delta A(g)}{\delta\varphi(y)} = \sum_{j=1}^{n} (-1)^{|a_j|}\, \partial_y^{a_j} \Big(g(y)\, \partial^{a_1}\varphi(y) \dots \hat{j} \dots \partial^{a_n}\varphi(y) \Big)\,,$$

where $\widehat{j}$ means that the jth factor is omitted. The r.h.s. can be written as

$$\sum_a (-1)^{|a|}\,\partial_y^a\Big(g(y)\,\frac{\partial A}{\partial(\partial^a\varphi)}(y)\Big) \quad \text{since} \quad \frac{\partial A}{\partial(\partial^a\varphi)} = \sum_{j=1}^{n} \delta_{a,a_j}\,\partial^{a_1}\varphi\ldots\widehat{j}\ldots\partial^{a_n}\varphi\ .$$

This yields the assertion (1.3.11). The latter implies immediately $\operatorname{supp} A(g) \subseteq \operatorname{supp} g$. A simple example for $\operatorname{supp} A(g) \neq \operatorname{supp} g$ is $A(g) = \int dx\; g(x)\,\partial^\mu\varphi(x)$ with $g|_{\mathcal{U}} = 1$ for some open set $\mathcal{U} \neq \emptyset$; namely, for this $A(g)$ we get $\frac{\delta A(g)}{\delta\varphi(y)} = -\partial^\mu g(y)$ and, hence, $\operatorname{supp} A(g) = \operatorname{supp} \partial^\mu g \neq \operatorname{supp} g$.
The statement about the second derivative follows from (1.3.1) and (1.2.4).]

From this exercise we know the following: For $F \in \mathcal{F}_{\text{loc}}$, the functional derivative $\frac{\delta F}{\delta\varphi(y)}$ is an $\mathcal{F}_{\text{loc}}$-valued distribution; and, in particular from (1.3.12), that any $F \in \mathcal{F}_{\text{loc}}$ automatically satisfies

$$\operatorname{supp}(\delta^2 F/\delta\varphi^2) \subseteq \Delta_2\ , \tag{1.3.13}$$

where we mean the support in the sense of distributions and Δ_n is the thin diagonal in $\mathbb{M}^n$ (A.1.12) – the property (1.3.13) is significant for locality.

Remark 1.3.6 (Characterization of locality). For $F = \sum_n \langle f_n, \varphi^{\otimes n}\rangle \in \mathcal{F}$, the condition (1.3.13) is only necessary but not sufficient for locality of F (1.3.8); this corrects an oversight in [55]. As explained below, the property (1.3.13) suffices to prove that each f_n is a linear combination of terms of the form (1.2.4) where the "coefficient" g is replaced by a distribution with compact support; but (1.3.13) does not imply that these coefficients are smooth.

However, if we define the space of fields $\mathcal{F}$ (Def. 1.2.1) by using the stronger version of the wave front set condition (1.2.2) mentioned after (1.2.9), that is,

$$\operatorname{WF}(f_n) \subseteq \Big\{\,(x_1,\ldots,x_n;k_1,\ldots,k_n)\,\Big|\,\sum_{j=1}^{n} k_j = 0\Big\} \quad \forall n \geq 1 \tag{1.3.14}$$

(in particular $\operatorname{WF}(f_1) = \emptyset$) instead of (1.2.2), then (1.3.13) is also sufficient for locality (1.3.8). To prove this statement we proceed in three steps, cf. [29].

(1) If $F = \sum_{n=0}^{N} \langle f_n, \varphi^{\otimes n}\rangle$ fulfills (1.3.13), then it holds that

$$\operatorname{supp} f_n \subseteq \Delta_n \quad \forall\, 2 \leq n \leq N\ . \tag{1.3.15}$$

This follows from

$$\frac{\delta^2 F}{\delta\varphi(y_1)\delta\varphi(y_2)} = \sum_{n=2}^{N} n(n-1)\int\Big(\prod_{j=1}^{n-2} dx_j\;\varphi(x_j)\Big)\,f_n(y_1,y_2,x_1,\ldots,x_{n-2})$$
$$\subseteq \Delta_2\ ,$$

taking also into account the symmetry of f_n in its arguments. As an incidental remark we note that (1.3.15) can equivalently be written as

$$\operatorname{supp}(\delta^n F/\delta\varphi^n) \subseteq \Delta_n \quad \forall n \geq 2\ . \tag{1.3.16}$$

(2) We are going to show that (1.3.15) and the compactness of the support of f_n imply that

$$f_n(x_1, \ldots, x_n) = \mathfrak{S}_n \sum_a \int dx\; t^n_a(x)\, \partial^a \delta(x_1 - x, \ldots, x_n - x) \tag{1.3.17}$$

for some $t^n_a \in \mathcal{D}'(\mathbb{M}, \mathbb{C})$ with compact support, where the sum over a is finite and $\mathfrak{S}_n$ denotes symmetrization in $x_1, \ldots, x_n$. For this purpose we introduce relative coordinates

$$\tilde{f}_n(\mathbf{y}, x_n) := f_n(x_1, \ldots, x_n)\ , \quad \mathbf{y} := (x_1 - x_n, \ldots, x_{n-1} - x_n)\ .$$

It suffices to consider test functions of the form $g(\mathbf{y}, x_n) = g_1(\mathbf{y})\, g_2(x_n)$:

$$\langle \tilde{f}_n, g \rangle = \langle \langle \tilde{f}_n(\mathbf{y}, x_n), g_1(\mathbf{y}) \rangle_{\mathbf{y}}, g_2(x_n) \rangle_{x_n}\ .$$

From (1.3.15) and (A.1.17) we conclude that

$$\langle \tilde{f}_n(\mathbf{y}, x_n), g_1(\mathbf{y}) \rangle_{\mathbf{y}} = \sum_a t^n_a(x_n)\, \langle \partial^a \delta(\mathbf{y}), g_1(\mathbf{y}) \rangle_{\mathbf{y}}$$

for some t^n_a, where the sum over a is finite. Since, for any fixed g_1, the map

$$\mathcal{D}(\mathbb{M}) \ni g_2 \longmapsto \langle \tilde{f}_n, g_1 \otimes g_2 \rangle = \sum_a (-1)^{|a|}\, \partial^a g_1(0)\, \langle t^n_a(x_n), g_2(x_n) \rangle_{x_n}$$

is linear and continuous, each t^n_a is a distribution. In addition, since f_n has compact support, this holds true also for each $t^n_a \in \mathcal{D}'(\mathbb{M}, \mathbb{C})$. Summing up we obtain

$$f_n(x_1, \ldots, x_n) = \sum_a \int dx\; \delta(x_n - x)\, t^n_a(x_n)\, \partial^a \delta(x_1 - x_n, \ldots, x_{n-1} - x_n)\ ,$$

which agrees with the assertion (1.3.17).

(3) Taking into account Exap. 1.2.3, it remains to verify that each t^n_a is smooth. It is only this step which uses any assumption about $\mathrm{WF}(f_n)$. Heuristically, it is clear that (1.3.14) implies smoothness of t^n_a, because (1.3.14) can be understood as the condition that "singularities may appear only in the dependence on the relative coordinates", as we see from (1.2.9). To show smoothness of t^n_a rigorously, we note that (1.3.14) implies that

$$\hat{f}_n(k) \sim \mathfrak{S}_n \sum_a (-ik)^a\, \hat{t}^n_a(k_1 + \cdots + k_n)\ , \quad k \equiv (k_1, \ldots, k_n)\ ,$$

(which is obtained by Fourier transformation of (1.3.17)) decays rapidly in all directions k except for $k_1 + \cdots + k_n = 0$. Hence, $\hat{t}^n_a(p)$ decays rapidly in all directions $p \in \mathbb{R}^d \setminus \{0\}$, that is, t^n_a is smooth for all n and a.

Obviously, the weaker wave front set condition (1.2.2) does not suffice for this conclusion. A simple example are linear functionals: $F(\varphi) = \int dx\, f(x)\,\varphi(x)$ with f a distribution with compact support. They trivially satisfy the condition (1.3.13): $\delta^2 F/\delta\varphi^2 = 0$. If f satisfies only the wave front set condition (1.2.2), it does not need to be smooth; but if it fulfills the stronger version (1.3.14), we have $\mathrm{WF}(f) = \emptyset$, that is, f is smooth. ⊟

To simplify the notations, we sometimes work with the integral kernel $A(x)$ instead of $A(g)$, and operations, which are defined on $\mathcal{F}_{\mathrm{loc}}$ or $\mathcal{F}$, are then applied to $A(x)$, although $A(x) \notin \mathcal{F}$ as explained in Exap. 1.2.4. Then, $A(x)$ has to be understood as $A(g)$, $g \in \mathcal{D}(\mathbb{M})$ arbitrary, in order that these operations are well defined. In particular, this yields the following definition of the functional derivative of $A(x)$, $A \in \mathcal{P}$:

$$\int dx\; g(x)\,\frac{\delta A(x)}{\delta\varphi(y)} := \frac{\delta A(g)}{\delta\varphi(y)} \quad \forall\, g \in \mathcal{D}(\mathbb{M})\ . \tag{1.3.18}$$

Inserting (1.3.11), we obtain the explicit formula

$$\frac{\delta A(x)}{\delta\varphi(y)} = \sum_{a\in\mathbb{N}^d} (\partial^a\delta)(x-y)\,\frac{\partial A}{\partial(\partial^a\varphi)}(x)\ . \tag{1.3.19}$$

By means of the Leibniz rule (1.3.4) for $\frac{\delta}{\delta\varphi(y)}\big(A_1(g_1)\cdots A_r(g_r)\big)$ we obtain the definition of the functional derivative of a *non-local* field *monomial*:

$$\frac{\delta\big(A_1(x_1)\cdots A_r(x_r)\big)}{\delta\varphi(y)} := \sum_{j=1}^{r} A_1(x_1)\cdots\frac{\delta A_j(x_j)}{\delta\varphi(y)}\cdots A_r(x_r)\ , \quad A_1,\dots,A_r \in \mathcal{P}\ . \tag{1.3.20}$$

The functional derivative of an arbitrary (local or non-local) field *polynomial* is obtained from these formulas by linearity.

Example 1.3.7. For $P(x_1,x_2,x_3) := \varphi^2(x_1)\,\Box\varphi(x_1)\,\varphi^5(x_2) + \partial^\mu\varphi\partial_\mu\varphi(x_3)$ we get

$$\begin{aligned}\frac{\delta P(x_1,x_2,x_3)}{\delta\varphi(y)} &= 2\,\delta(x_1-y)\,\varphi(x_1)\,\Box\varphi(x_1)\,\varphi^5(x_2) + (\Box\delta)(x_1-y)\,\varphi^2(x_1)\,\varphi^5(x_2)\\ &\quad + 5\,\delta(x_2-y)\,\varphi^2(x_1)\,\Box\varphi(x_1)\,\varphi^4(x_2) + 2\,(\partial^\mu\delta)(x_3-y)\,\partial_\mu\varphi(x_3)\ .\end{aligned}$$

Remark 1.3.8 (Non-polynomial fields)**.** One may go beyond Definition 1.2.1 to include *non-polynomial fields*, in the following way (see [22, 27, 29, 63]). One can redefine $\mathcal{F}$ to be the set of functionals $F\colon \mathcal{C} \longrightarrow \mathbb{R}$ which are infinitely differentiable and all of whose functional derivatives are compactly supported distributions and each $\mathrm{WF}(\delta^n F/\delta\varphi^n)$ satisfies (1.2.2). The definition of (continuous) differentiability used in infinite-dimensional calculus can be found, e.g., in [29, 129, 143].

Local functionals $\mathcal{F}_{\mathrm{loc}} \subset \mathcal{F}$ are then defined by the extra conditions:

(i) Additivity:

$$F(f+g+h) = F(f+g) - F(g) + F(g+h) \quad \text{if} \quad \operatorname{supp} f \cap \operatorname{supp} h = \emptyset\,; \tag{1.3.21}$$

(ii) $\mathrm{WF}(\delta^n F/\delta\varphi^n) \perp T\Delta_n$, where $T\Delta_n$ is the tangent bundle of the thin diagonal. In other words, if $(x,k) \in \mathrm{WF}(\delta^n F/\delta\varphi^n)$ and $v \in T_x\Delta_n$, then $\langle k, v\rangle = 0$.

One can show that the additivity condition (i) is equivalent to the condition (1.3.13) and that it is also equivalent to the seemingly stronger condition (1.3.16), see [27, Lemma 3.1] and [29, Proposition 1]. So the equivalence (1.3.13) $\Leftrightarrow$ (1.3.16), shown above for polynomial fields, holds true also for non-polynomial fields.

1.4 Balanced fields

There is a *problem of non-uniqueness* with the expression $F = \sum_{i=1}^K A_i(g_i)$ (1.3.8) for local fields: The pairs $(A_i, g_i)_{i=1}^K$ are not uniquely determined by F, since $\int dx\, \partial_\mu(A(x)g(x)) = 0$ for any $A \in \mathcal{P}$ and $g \in \mathcal{D}(\mathbb{M})$. To remove this non-uniqueness, we seek a subspace $\mathcal{P}_{\mathrm{bal}}$ of $\mathcal{P}$, called the space of "balanced fields", such that each element of $\mathcal{P}$ can be *uniquely* written as a finite sum of derivatives of balanced fields. For example, $\varphi\partial^\mu\varphi$ cannot be a balanced field, since $\varphi\partial^\mu\varphi = \partial^\mu(\frac{1}{2}\varphi^2)$.

In perturbative QFT, this result is crucial for the proof of the Action Ward Identity [55],[2] i.e., the statement that the retarded (or time-ordered) products can be renormalized so that they commute with partial derivatives with respect to the spacetime coordinates (see Sect. 3.1.1).

To find the space of balanced fields, note that any $A \in \mathcal{P}$ can be written as

$$A(x) = c + \sum_{n=1}^{N} p_n(\partial^1, \ldots, \partial^n)\,\varphi(x_1)\cdots\varphi(x_n)\Big|_{x_1=\cdots=x_n=x} \quad \text{with } c \in \mathbb{R},\ N < \infty \tag{1.4.1}$$

and $\partial^j := \left(\frac{\partial}{\partial x_j^\mu}\right)_{\mu=0,\ldots,d-1}$, for some polynomials p_n which are symmetric, that is, $p_n(\partial^1, \ldots, \partial^n) = p_n(\partial^{\pi(1)}, \ldots, \partial^{\pi(n)})$ for all $\pi \in S_n$.

We can introduce derivatives with respect to *relative* and *centre-of-mass* coordinates. We write $\partial^{ij} := \partial^i - \partial^j$ for the relative derivatives, and $\partial := (\partial^1 + \cdots + \partial^n)$ for the centre-of-mass derivative. Note that the application of $\partial_x \equiv \frac{\partial}{\partial x}$ to $A(x)$ is described on the r.h.s. of (1.4.1) by a centre-of-mass derivative ∂. One can then show that $p_n(\partial^1, \ldots, \partial^n)$ can be *uniquely* rewritten as

$$p_n(\partial^1, \ldots, \partial^n) = \sum_a \partial^a\, R_{na}(\partial^{1n}, \ldots, \partial^{n-1,n}) \tag{1.4.2}$$

for suitable polynomials R_{na}, where the sum is finite. Since p_n is symmetric, R_{na} has a corresponding permutation symmetry:

$$R_{na}(\partial^{\pi(1),\pi(n)}, \ldots, \partial^{\pi(n-1),\pi(n)}) = R_{na}(\partial^{1n}, \ldots, \partial^{n-1,n}) \quad \forall \pi \in S_n \tag{1.4.3}$$

where $\partial^{ji} = -\partial^{ij}$.

---

[2] Balanced fields have first been introduced in [31] for the purpose of non-equilibrium quantum field theory.

Definition 1.4.1 (Balanced fields)**.** Finite sums

$$B(x) = c + \sum_{n=1}^{N} B_n(x) \ , \quad \text{where} \quad N < \infty \ , \ c \in \mathbb{R}$$

and B_n is of the form

$$B_n(x) = P_n(\partial^{1n}, \dots, \partial^{n-1,n})\, \varphi(x_1) \cdots \varphi(x_n)\Big|_{x_1=\cdots=x_n=x} \ ,$$

for some polynomial P_n respecting the permutation symmetry (1.4.3), are called *balanced fields.* Obviously they form a vector space, which is denoted by $\mathcal{P}_{\text{bal}}$.

Example 1.4.2. Here are all $B \in \mathcal{P}_{\text{bal}}$ which are $\mathcal{L}_+^\uparrow$-invariant, with two φ-factors and two derivatives:

$$\begin{aligned} B &= c\,\partial_\mu^{12}\,\partial^{\mu,12}\varphi(x_1)\varphi(x_2)\Big|_{x_1=x_2=x} = c(\Box_{x_1} + \Box_{x_2} - 2\partial_\mu^{x_1}\,\partial_{x_2}^\mu)\varphi(x_1)\varphi(x_2)\Big|_{x_1=x_2=x} \\ &= 2c(\varphi\Box\varphi - \partial_\mu\varphi\,\partial^\mu\varphi), \quad c \in \mathbb{R} \ . \end{aligned}$$

By inserting (1.4.2) into (1.4.1) and by reordering the sums, we obtain

$$\begin{aligned} A(x) &= c + \sum_a \partial^a \sum_{n=1}^{N} R_{na}(\partial^{1n}, \dots, \partial^{n-1,n})\, \varphi(x_1) \cdots \varphi(x_n)\Big|_{x_1=\cdots=x_n=x} \\ &= \sum_a \partial^a B_a(x) \quad \text{where } B_a \in \mathcal{P}_{\text{bal}} \text{ and } B_a|_{\varphi=0} = 0 \ \forall a \neq 0. \end{aligned}$$

The point is that B_a is uniquely determined by these properties, due to the uniqueness of (1.4.2). Note that the constant field c is absorbed in the summand $a = 0$. This is part (a) of the following proposition [55, 57].

Proposition 1.4.3 (Crucial properties of balanced fields)**.** (a) *Every* $A \in \mathcal{P}$ *can* uniquely *be written as a finite sum of type*

$$A = \sum_{a \in \mathbb{N}^d} \partial^a B_a \ , \quad \textit{where} \quad B_a \in \mathcal{P}_{\text{bal}} \quad \textit{and} \quad B_a|_{\varphi=0} = 0 \ \forall a \neq 0 \ . \quad (1.4.4)$$

(b) *For each* $F \in \mathcal{F}_{\text{loc}}$*, there exists a* unique $f \in \mathcal{D}(\mathbb{M}, \mathcal{P}_{\text{bal}})$ *such that*[3]

$$F - F(0) = \int dx\, f(x) \quad \textit{and also} \quad f(x)|_{\varphi=0} = 0 \quad \forall x \in \mathbb{M} \ .$$

In part (b), $F(0) \in \mathbb{C}$ must be excluded, since there are infinitely many $\tilde{f} \in \mathcal{D}(\mathbb{M})$ fulfilling $F(0) = \int dx\, \tilde{f}(x)$.

[3] $f \in \mathcal{D}(\mathbb{M}, \mathcal{P}_{\text{bal}})$ means that f is of the form $f(x) = \sum_k h^{(k)}(x)\, B^{(k)}(x)$ with $h^{(k)} \in \mathcal{D}(\mathbb{M}) \equiv \mathcal{D}(\mathbb{M}, \mathbb{C})$ and $B^{(k)} \in \mathcal{P}_{\text{bal}}$, where the sum is finite.

Example 1.4.4. Let φ_1 and φ_2 be two different real scalar fields and let

$$A^\mu := \varphi_1^{l-1}\partial^\mu\varphi_1\,\varphi_2^k\ .$$

The unique representation claimed in part (a) reads

$$A^\mu = B_1^\mu + \partial^\mu B_2$$

with the balanced fields

$$B_1^\mu := \frac{k}{l+k}\big(\varphi_1^{l-1}(\partial^\mu\varphi_1)\,\varphi_2^k - \varphi_1^l\,\varphi_2^{k-1}\partial^\mu\varphi_2\big) \quad\text{and}\quad B_2 := \frac{1}{l+k}\varphi_1^l\varphi_2^k\ .$$

To derive this result, we write

$$A(x) = \partial_{x_1}\varphi_1(x_1)\cdots\varphi_1(x_l)\varphi_2(y_1)\cdots\varphi_2(y_k)\Big|_{x_r=y_s=x\ \forall r,s}\ .$$

The center-of-mass derivative is $\partial = l\,\partial_{x_1} + k\,\partial_{y_1}$. All non-vanishing relative derivatives of first order are equal to $\pm\partial^{\mathrm{rel}} := \pm(\partial_{x_1} - \partial_{y_1})$. The unique decomposition $\partial_{x_1} = \frac{1}{l+k}\,\partial + \frac{k}{l+k}\,\partial^{\mathrm{rel}}$ yields the assertion.

Proof of (a) $\Longrightarrow$ (b). Let $F = \sum_i A_i(h_i) \in \mathcal{F}_{\mathrm{loc}}$ with $A_i = \sum_{a_i}\partial^{a_i}B_{ia_i}$ where $B_{ia_i} \in \mathcal{P}_{\mathrm{bal}}$. Then, after integration by parts, we get

$$F = \int dx \sum_{i,a_i} B_{ia_i}(x)\,(-1)^{|a_i|}\,\partial^{a_i}h_i(x) = \int dx \sum_k B^{(k)}(x)\,h^{(k)}(x),$$

with $B^{(k)} \in \mathcal{P}_{\mathrm{bal}}$, $h^{(k)} \in \mathcal{D}(\mathbb{M})$. Since $\partial^a B^{(k)} \notin \mathcal{P}_{\mathrm{bal}}$ for $|a| > 0$, there is no further possibility of integration by parts, so we get uniqueness of $f(x) = \sum_k B^{(k)}(x)\,h^{(k)}(x)$. □

Exercise 1.4.5. In dimension $d = 1$, determine a basis of the vector space of all balanced fields with three factors φ and ≤ 3 derivatives.

[*Solution:* φ^3, $\varphi(\varphi\,\partial\partial\varphi - (\partial\varphi)^2)$ and $-\varphi^2\,\partial\partial\partial\varphi + 3\varphi(\partial\varphi)\,\partial\partial\varphi - 2(\partial\varphi)^3$.]

1.5 The field equation

General references for the remainder of Chap. 1 are [16, 29, 54].

Free action. To study the field equation, we start with the free action:

$$S_0 := \int dx\ L_0(x)\ , \quad\text{where}\quad L_0 := \frac{1}{2}(\partial^\mu\varphi\,\partial_\mu\varphi - m^2\varphi^2) \qquad (m \geq 0) \qquad (1.5.1)$$

is the free Lagrangian. However, $S_0 \notin \mathcal{F}$: This does not comply with Definition 1.2.1, since on writing S_0 in the form (1.2.1) the pertinent f_2 has non-compact support; or, comparing with the definition of local fields (1.3.8), the function g belonging to S_0 is $g(x) = 1$ for all x, which is not an element of $\mathcal{D}(\mathbb{M})$. This is not a shortcoming of our definitions, the problem is more fundamental: For $h \in \mathcal{C}$ the convergence of the integral $S_0(h) = \frac{1}{2}\int dx\, \big(\partial^\mu h(x)\,\partial_\mu h(x) - m^2(h(x))^2\big)$ is not under control for $\|x\| := \sqrt{x_0^2 + \cdots + x_{d-1}^2} \to \infty$, i.e., in the "IR-region". This fact is ignored in many textbooks. If one restricts the functional S_0 to $\mathcal{D}(\mathbb{M},\mathbb{R}) = \{\, h\colon \mathbb{M} \to \mathbb{R} \,|\, h \text{ is smooth and } \operatorname{supp} h \text{ is compact}\,\}$, there is the unsatisfactory feature that there does not exist any nontrivial solution of the Klein–Gordon equation $(\Box + m^2)\varphi(x) = 0$ which has compact support.[4]

In this book we avoid this problem in the following way: Although S_0 appears as an index of several quantities, it is (apart from very few exceptions, in which we understand S_0 as "generalized Lagrangian" in the sense of Remk. 1.5.1) never S_0 itself which appears in the definition of these quantities or in any formula, only the functional derivatives

$$\frac{\delta S_0}{\delta\varphi(x)} := -(\Box + m^2)\varphi(x) \ , \qquad \frac{\delta^2 S_0}{\delta\varphi(y)\,\delta\varphi(x)} := -(\Box + m^2)\delta(x-y) \tag{1.5.2}$$

are used – and these are well defined.

Remark 1.5.1 (Generalized Lagrangian). A rigorous definition of the free action and of non-localized interactions as, e.g., $\int dx\ \varphi^n(x) \quad (n \in \mathbb{N}^*)$, can be given by using the notion of a *generalized Lagrangian*, introduced in [27, Sect. 6.2], see also [29, Sect. 3]. By definition, a generalized Lagrangian is a map

$$\mathcal{L} : \mathcal{D}(\mathbb{M}) \longrightarrow \mathcal{F}_{\mathrm{loc}} \tag{1.5.3}$$

with the properties

$$\operatorname{supp}\mathcal{L}(f) \subseteq \operatorname{supp} f \quad \forall f \in \mathcal{D}(\mathbb{M}) \ , \quad \mathcal{L}(0) = 0$$

and the Additivity

$$\mathcal{L}(f+g+h) = \mathcal{L}(f+g) - \mathcal{L}(g) + \mathcal{L}(g+h) \quad \text{if} \quad \operatorname{supp} f \cap \operatorname{supp} h = \emptyset \ .$$

At first glance, the latter seems to agree with (1.3.21). However note that, in contrast to (1.3.21), the functions f, g and h appearing here are not field configurations; they play a different role: They are elements of $\mathcal{D}(\mathbb{M})$, which switch off the Lagrangian in the IR-region. We emphasize that the map (1.5.3) may be non-linear.

The action S given by the generalized Lagrangian $\mathcal{L}$ is defined as the *equivalence class of* $\mathcal{L}$, where two Lagrangians $\mathcal{L}_1$ and $\mathcal{L}_2$ are called equivalent, $\mathcal{L}_1 \sim \mathcal{L}_2$,

[4]The Fourier transform of a real, smooth solution of the Klein–Gordon equation is of the form

$$\widehat{f}(k_0,\vec{k}) = \delta(k_0^2 - \vec{k}^2 - m^2)\,\big(\theta(k_0)\,g(\vec{k}) + \theta(-k_0)\,\overline{g(-\vec{k})}\big)$$

for some $g \in \mathcal{D}'(\mathbb{R}^{d-1})$ with compact support. Since $\widehat{f}$ is singular, f cannot have compact support.

if and only if

$$\operatorname{supp}_x\Big(\frac{\delta \mathcal{L}_1(f)}{\delta\varphi(x)}-\frac{\delta \mathcal{L}_2(f)}{\delta\varphi(x)}\Big)\equiv \operatorname{supp}\big(\mathcal{L}_1(f)-\mathcal{L}_2(f)\big)\subseteq \bigcup_{\mu=0}^{d-1}\operatorname{supp}\,\partial_\mu f\quad \forall f\,,$$

that is, iff the pertinent field equations agree in the region in which f is constant. In particular, if $\mathcal{L}_1$ and $\mathcal{L}_2$ are of the form $\mathcal{L}_j(f)=\int dx\; L_j(x)\,f(x)$ for some $L_j\in\mathcal{P}$, then it holds that $\mathcal{L}_1\sim\mathcal{L}_2$ if $L_1-L_2=c+\sum_\mu\partial_\mu A^\mu$ for some $c\in\mathbb{C}$ and $A^\mu\in\mathcal{P}$. The free action of a real scalar field is then the equivalence class of the generalized Lagrangian $f\longmapsto\mathcal{L}_0(f):=\frac{1}{2}\int dx\;\big(\partial^\mu\varphi(x)\,\partial_\mu\varphi(x)-m^2\,\varphi^2(x)\big)\,f(x)$.

Localized and local interactions. As explained in the Preface, we work in pQFT with *localized* interactions S, that is $S\in\mathcal{F}$, to separate the UV-problem form the IR-problem. Therefore, we assume also in our treatment of clFT that the interactions are of this kind, i.e., they are switched off for $\|x\|$ large enough. Due to $S\in\mathcal{F}$, the problem discussed above for S_0 does not appear. However, the switching modifies the usual field equation.

We additionally assume that the interaction S is *local*: $S\in\mathcal{F}_{\mathrm{loc}}$, similarly to many textbooks.[5] More precisely we assume that S is of the simple form

$$S=\int dx\;\mathcal{L}_{\mathrm{int}}(x)\;;\quad\text{where}\quad \mathcal{L}_{\mathrm{int}}(x)=-\kappa\,g(x)\,L_{\mathrm{int}}(x),\tag{1.5.4}$$

is the interaction Lagrangian, with $L_{\mathrm{int}}\in\mathcal{P}_{\mathrm{bal}}$, $g\in\mathcal{D}(\mathbb{M})$ and so $S\in\mathcal{F}_{\mathrm{loc}}$. To make visible the dependence on κ and to simplify the notations, we sometimes write

$$\kappa\tilde{S}:=S\;,\quad \mathcal{L}(g):=S\;,\quad L(g):=\tilde{S}\;,\quad\text{that is,}\quad \mathcal{L}=\kappa L=-\kappa L_{\mathrm{int}}\in\kappa\mathcal{P}_{\mathrm{bal}}\;.$$

The *field equation* given by the total action S_0+S reads

$$\frac{\delta(S_0+S)}{\delta\varphi(x)}=0\;.$$

If L_{int} depends only on φ and its first derivatives $\partial^\mu\varphi$, this becomes the Euler–Lagrange equation (by using Exer. 1.3.5):

$$\frac{\partial(L_0+\mathcal{L}_{\mathrm{int}})}{\partial\varphi}(x)-\partial^\mu\,\frac{\partial(L_0+\mathcal{L}_{\mathrm{int}})}{\partial(\partial^\mu\varphi)}(x)=0\;.$$

For instance, if $L_{\mathrm{int}}=\varphi^k$, we get $(\square+m^2)\varphi(x)=-\kappa\,g(x)\,k\,\varphi^{k-1}(x)$.

Exercise 1.5.2 (Yang–Mills field equation in $d=4$ dimensions). The N vector fields $A_a\equiv(A_a^\mu(x))_{\mu=0,1,2,3}$, where $a\in\{1,\ldots,N\}$, constitute the evaluation functional

$$A(x):C^\infty(\mathbb{M},\mathbb{R}^{4N})\longrightarrow\mathbb{R}^{4N}\;:h\equiv(h_a^\mu)_{a=1,\ldots,N}^{\mu=0,1,2,3}\longmapsto A(x)(h)=h(x)\equiv(h_a^\mu(x))_{a=1,\ldots,N}^{\mu=0,1,2,3}\;.$$

[5] Naturally, one may also consider non-local interactions in our formalism – see Remk. 1.9.4: But local ones suffice, in this book, for the needs of pQFT.

Compute the field equation which follows from

$$L_0 + \mathcal{L}_{\text{int}} := -\frac{1}{4} F_a^{\mu\nu} F_{a\,\mu\nu} - \frac{\lambda}{2} (\partial_\nu A_a^\nu)^2 ,$$

where repeated indices are summed over (e.g., $\sum_a \big(\sum_\nu \partial_\nu A_a^\nu\big)^2$), with

$$F_a^{\mu\nu}(x) := \partial^\mu A_a^\nu(x) - \partial^\nu A_a^\mu(x) + \kappa\, g(x)\, f_{abc}\, A_b^\mu(x)\, A_c^\nu(x),$$

where $g \in \mathcal{D}(\mathbb{M})$ switches the coupling constant $\kappa \in \mathbb{R}$, $\lambda \in \mathbb{R}\backslash\{0\}$ is a constant parameter (the "gauge-fixing parameter") and $(f_{abc}) \in \mathbb{C}^{N\times N\times N}$ is a totally antisymmetric tensor which fulfills the Jacobi identity.[6] A gauge fixing term – we choose $-\frac{\lambda}{2}(\partial A_a)^2$ – needs to be added already for the free theory ($\kappa = 0$); the reason is the same as for the photon field, see Sect. 5.1.3.

[*Solution*: The Euler–Lagrange equation $\frac{\partial(L_0+\mathcal{L}_{\text{int}})}{\partial A_a^\nu}(x) = \partial^\mu \frac{\partial(L_0+\mathcal{L}_{\text{int}})}{\partial(\partial^\mu A_a^\nu)}(x)$ yields

$$\Box A_{\nu a}(x) - (1-\lambda)\,\partial_\nu \partial_\tau A_a^\tau(x) = \kappa f_{abc}\Big(g(x)\, A_b^\mu F_{\nu\mu c}(x) - \partial^\mu\big(g(x)\, A_{\mu b} A_{\nu c}(x)\big)\Big) .$$

This computation does not use the Jacobi identity. Note that the field equation contains a term which is $\sim (\kappa\, g(x))^2$, since $F_{\nu\mu c}(x)$ contains a term $\sim \kappa\, g(x)$.]

1.6 Retarded fields

Interacting fields. The interacting field $\varphi(x)$ to the interaction $S \in \mathcal{F}_{\text{loc}}$ (we will write $\varphi_S(x)$ for this field) should be defined such that it solves the field equation given by S. In our formalism, this can be realized as follows: Denote the space of smooth solutions of the field equation by

$$\mathcal{C}_{S_0+S} := \left\{ h \in \mathcal{C} \,\Big|\, \frac{\delta(S_0+S)}{\delta\varphi(x)}(h) = 0 \quad \forall x \in \mathbb{M} \right\} ;$$

in particular $\mathcal{C}_{S_0} := \{\, h \in \mathcal{C} \,|\, (\Box + m^2)h(x) = 0 \,\}$. Therewith, we define:

Definition 1.6.1. An *interacting field* belonging to a field $F \in \mathcal{F}_{\text{loc}}$ and an interaction S, is given by taking[7]

$$F_S := F\big|_{\mathcal{C}_{S_0+S}} \quad \text{or} \quad A_S(x) := A(x)\big|_{\mathcal{C}_{S_0+S}} \quad (A \in \mathcal{P}_{\text{bal}})$$

for the corresponding integral kernel.

Indeed, for $A = \varphi$, the field $\varphi_S(x)$ satisfies the field equation. An unrestricted field $F \in \mathcal{F}_{\text{loc}}$ is called an "off-shell field" and $F\big|_{\mathcal{C}_{S_0+S}}$ is an "on-shell field"; mostly

[6]The Jacobi identity is the relation

$$\sum_r f_{abr} f_{rcd} + [\text{cyclic permutations of } (a,b,c)] = 0 .$$

[7]In principle, this F could also be non-local – see Remk. 1.9.4; but in view of pQFT it suffices to study local F.

we will use these notions with respect to the *free* action, i.e., for the restriction to $\mathcal{C}_{S_0}$.[8]

For a scalar field the name "on-shell field" is motivated by the following: Restriction of the functionals $F \in \mathcal{F}$ to $\mathcal{C}_{S_0}$ means that the allowed configurations $h(x)$ solve the Klein–Gordon equation, that is, the support of $\hat{h}(p)$ lies on the mass shell

$$H_m := \{\, p \in \mathbb{M}_d \,\big|\, p^2 \equiv (p^0)^2 - \vec{p}^{\,2} = m^2 \,\} \,. \tag{1.6.1}$$

For an "off-shell field" the allowed configurations are not restricted this way, however note that $\operatorname{supp} \hat{h}$ may have a non-empty intersection with the mass shell H_m. Hence, "off-shell" does not mean "off H_m", it only means "not restricted to H_m". In addition, we point out that for a Dirac spinor field (introduced in Sect. 5.1.1) the condition that a configuration $h \in C^\infty(\mathbb{M}, \mathbb{C}^4)$ solves the Dirac equation $(i\partial\!\!\!/ - m)h(x) = 0$ is truly stronger than $\operatorname{supp} \hat{h} \subseteq H_m$.

Retarded wave operator. In perturbation theory, interacting fields are usually interpreted as functionals on $\mathcal{C}_{S_0}$; this corresponds to the interaction picture of QFT. For this purpose we need retarded wave operators (also called "retarded, classical Møller operators"), which map solutions of S_0 into solutions of $S_0 + S$. However, in pQFT there are good reasons to define the interacting fields as unrestricted functionals, i.e., on $\mathcal{C}$, see Remk. 3.6.5. Therefore, we define the retarded wave operators as off-shell operators, i.e., as maps from $\mathcal{C}$ to $\mathcal{C}$ (instead of $\mathcal{C}_{S_0} \longrightarrow \mathcal{C}_{S_0+S}$).

Definition 1.6.2. [16] A *retarded wave operator* is a family of maps $r_{S_0+S,S_0} : \mathcal{C} \longrightarrow \mathcal{C}$, one for each local interaction S, such that:

(i) $r_{S_0+S,S_0}(h)(x) = h(x)$ for x "sufficiently early";

(ii) $\dfrac{\delta(S_0 + S)}{\delta\varphi(x)} \circ r_{S_0+S,S_0} = \dfrac{\delta S_0}{\delta\varphi(x)}$.

Remark 1.6.3. If $h \in \mathcal{C}_{S_0}$, i.e., h is a solution of the free theory, then $r_{S_0+S,S_0}(h) \in \mathcal{C}_{S_0+S}$ is a solution of the interacting theory. Pictorially, the map $h \longmapsto r_{S_0+S,S_0}(h)$ is shown in Figure 1.1 in which time is advancing vertically: h and $r(h)$ coincide before the support of S and separate within that support. Propagation with respect to the free action is indicated by a straight line.

An arbitrary $h \in \mathcal{C}$ is mapped by r_{S_0+S,S_0} to that solution $f(x)$ of

$$-(\Box + m^2) f(x) + \frac{\delta S}{\delta\varphi(x)}(f) = g(x)\,, \quad \text{where} \quad g(x) := -(\Box + m^2) h(x)\,,$$

which agrees with $h(x)$ for x "sufficiently early".

[8]This is the usual terminology for "on-/off-shell field" in causal perturbation theory, which was introduced by Raymond Stora [159]. The notion "on-/off-shell field" appears, e.g., also in the book [32] about "Classical Mechanics" – essentially with the same meaning. In Sect. 3.4 we introduce and study the time-ordered and the retarded product of pQFT for on-shell fields. Further explanations about the notions "on-" and "off-shell" are given in App. A.7.

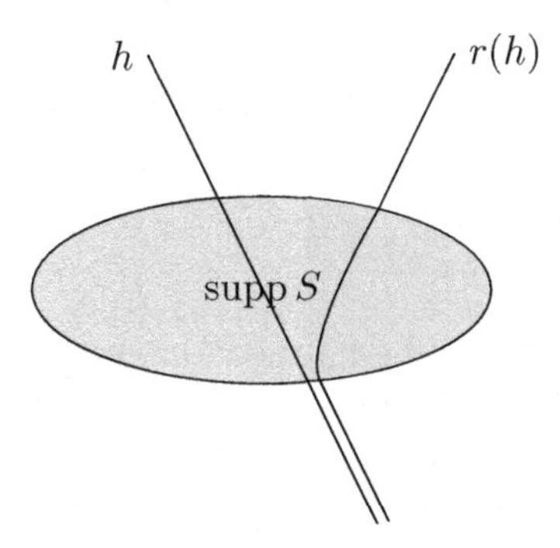

Figure 1.1: Effect of a retarded wave operator where time is advancing vertically. The interpretation given in Remark 1.6.3 can be made explicit by considering the case $d = 2$ and $m = 0$. Then, $h(x^0, x^1) := a\, f\big(x^0 \mp (x^1 - b)\big)$, where $a, b \in \mathbb{R}$ and $f \in C^\infty(\mathbb{R}, \mathbb{R})$, is a smooth solution of the wave equation, that is, $h \in \mathcal{C}_{S_0}$. Assuming that f is a smooth approximation of $\delta \in \mathcal{D}'(\mathbb{R})$, the straight line may be approximately the support of h and the curved line the support of $r(h)$.

In the framework of *perturbation theory*, i.e., in the sense of formal power series in S (or in the coupling constant κ, if you prefer), retarded wave operators always exist and are unique. A direct proof of this claim is given in Lemmas 1 and 2 of [16]. In this book we only give an implicit proof: In Proposition 1.9.1 we shall show that the perturbative expansion (1.7.3) of the retarded fields $F_S^{\mathrm{ret}} := F \circ r_{S_0+S,S_0}$ (1.6.2) exists and is unique for all $F, S \in \mathcal{F}_{\mathrm{loc}}$.

Remark 1.6.4. Since $r_{S_0+S,S_0} = \mathrm{Id} + \mathcal{O}(\kappa)$, the retarded wave operator is *invertible* in the sense of formal power series, see (A.1.3). Due to this, we can define the retarded wave operator connecting two interacting theories:

$$r_{S_0+S_1,S_0+S_2} := r_{S_0+S_1,S_0} \circ \big(r_{S_0+S_2,S_0}\big)^{-1} .$$

Obviously it satisfies the property (i) of Def. 1.6.2 and the analogon of property (ii):

$$\frac{\delta(S_0+S_1)}{\delta\varphi(x)} \circ r_{S_0+S_1,S_0+S_2} = \frac{\delta(S_0+S_2)}{\delta\varphi(x)} ;$$

and we also obtain

$$r_{S_0+S_1,S_0+S_2} \circ r_{S_0+S_2,S_0+S_3} = r_{S_0+S_1,S_0+S_3} .$$

Remark 1.6.5. Exact (i.e., non-perturbative) existence and uniqueness of retarded wave operators depends on the interaction S. For *non-localized interactions*, i.e., $S = -\kappa \int dx\, L_{\mathrm{int}}(x)$ with $L_{\mathrm{int}} \in \mathcal{P}_{\mathrm{bal}}$ (note that there is no function $g \in \mathcal{D}(\mathbb{M})$ switching off the interaction for $\|x\| \to \infty$), a lot is known about existence and uniqueness of exact solutions of nonlinear wave equations (see, e.g., [4, 141, 153]). A typical result is Jörgens' Theorem (from 1961), given, e.g., in Chapter 6 of [153]:

Theorem 1.6.6 (Smooth solutions of the φ^4-model). *Let $m > 0$, $\kappa > 0$ and $h, h_1 \in \mathcal{D}(\mathbb{R}^3)$ be given. Then the Cauchy problem*

$$(\Box + m^2) f(t, \vec{x}) + 4\kappa \left(f(t, \vec{x})\right)^3 = 0, \quad f(0, \vec{x}) = h(\vec{x}), \quad \partial_t f(0, \vec{x}) = h_1(\vec{x})$$

has a unique solution $f \in C^\infty(\mathbb{R}^4)$.

It seems that for the corresponding localized interaction, $S = -\kappa \int dx\, g(x)\, \varphi^4(x)$ with $g \in \mathcal{D}(\mathbb{R}^4)$ and $g(x) \geq 0$ for all x, the existence and uniqueness of solutions is even somewhat easier to prove. Hence, for this (localized) interaction S, it is likely that the "on-shell" retarded wave operator $r_{S_0+S,S_0}|_{\mathcal{C}_{S_0}}$ exists exactly and is unique.

Retarded fields. In perturbative clFT one usually deals with Taylor expansions with respect to κ; but, the fields to be expanded are not the interacting fields F_S (Definition 1.6.1) – one expands the so-called "retarded fields", which are defined in terms of the retarded wave operator.

Definition 1.6.7. Given $F \in \mathcal{F}_{\rm loc}$ and an interaction $S \in \mathcal{F}_{\rm loc}$, the *retarded field* $F_S^{\rm ret}$ is defined by

$$F_S^{\rm ret} := F \circ r_{S_0+S,S_0} : \mathcal{C} \longrightarrow \mathbb{C}, \tag{1.6.2}$$

and similarly $A_S^{\rm ret}(x) := A(x) \circ r_{S_0+S,S_0}$ for $A \in \mathcal{P}_{\rm bal}$.

Obviously, these two definitions are connected by

$$\int dx\, g(x) \big(A_S^{\rm ret}(x)\big) = A(g)_S^{\rm ret} \quad \text{for} \quad g \in \mathcal{D}(\mathbb{M}) , \tag{1.6.3}$$

which implies

$$\partial_x \big(A_S^{\rm ret}(x)\big) = (\partial_x A)_S^{\rm ret}(x), \tag{1.6.4}$$

by replacing $g \in \mathcal{D}(\mathbb{M})$ by ∂g and by performing integrations by parts.

These retarded (classical) fields *factorize* in the following sense: For $A, B \in \mathcal{P}_{\rm bal}$ we obtain

$$\begin{aligned}(AB)_S^{\rm ret}(x)(h) &= (AB)(x)(r_{S_0+S,S_0}(h)) = A(x)(r_{S_0+S,S_0}(h))\; B(x)(r_{S_0+S,S_0}(h)) \\ &= A_S^{\rm ret}(x)(h)\; B_S^{\rm ret}(x)(h),\end{aligned}$$

that is to say,

$$(AB)_S^{\rm ret}(x) = A_S^{\rm ret}(x) \cdot B_S^{\rm ret}(x) \tag{1.6.5}$$

as distributions $A_S^{\rm ret}(x)(h) : \mathcal{D}(\mathbb{M}) \ni g \longmapsto A(g)_S^{\rm ret}(h) \in \mathbb{C}$, $h \in \mathcal{C}$. This factorization property of classical field theory *cannot be maintained in the quantization,* since pointwise products of distributions in general need not exist.

Property (ii) of the retarded wave operator translates into the identity

$$(\Box_x + m^2)\varphi_S^{\rm ret}(x) = (\Box + m^2)\varphi(x) + \Big(\frac{\delta S}{\delta\varphi(x)}\Big)_S^{\rm ret} , \tag{1.6.6}$$

which we call the "off-shell field equation". Restricted to $\mathcal{C}_{S_0}$ the first term on the r.h.s. vanishes, that is, $\varphi_S^{\mathrm{ret}}(x)|_{\mathcal{C}_{S_0}}$ *solves the usual (interacting) field equation.* By means of the factorization (1.6.5) and (1.6.4), we can reach that there is only one unknown field in the differential equation (1.6.6), namely $\varphi_S^{\mathrm{ret}}(x)$. For example, for $S = -\kappa \int dx\, g(x)\, \varphi^k(x)$, we obtain

$$\Big(\frac{\delta S}{\delta\varphi(x)}\Big)_S^{\mathrm{ret}} = -\kappa\, g(x)\, k\, \big(\varphi^{k-1}\big)_S^{\mathrm{ret}}(x) = -\kappa\, g(x)\, k\, \big(\varphi_S^{\mathrm{ret}}(x)\big)^{k-1}\ ;$$

with that, (1.6.6) can be written as

$$(\square_x + m^2)\varphi_S^{\mathrm{ret}}(x) = (\square + m^2)\varphi(x) - \kappa\, g(x)\, k\, \big(\varphi_S^{\mathrm{ret}}(x)\big)^{k-1}\ . \tag{1.6.7}$$

1.7 Perturbative expansion of retarded fields

In this section we assume existence and uniqueness of the retarded wave operator r_{S_0+S,S_0}; in particular, uniqueness implies $r_{S_0,S_0} = \mathrm{Id}$.

The Taylor coefficient of a retarded field to nth order,

$$\frac{d^n}{d\kappa^n}\Big|_{\kappa=0} F_{\kappa\tilde S}^{\mathrm{ret}}\ ,\quad \text{depends linearly on } F \text{ and on } \tilde S^{\otimes n}. \tag{1.7.1}$$

Linearity in F follows immediately from $(F_1+\alpha F_2)_{\kappa\tilde S}^{\mathrm{ret}} = F_{1,\kappa\tilde S}^{\mathrm{ret}} + \alpha\, F_{2,\kappa\tilde S}^{\mathrm{ret}}$ for $\alpha\in\mathbb{C}$. Linearity in $\tilde S^{\otimes n}$ is obtained from (A.1.19)–(A.1.22) by writing

$$\frac{d^n}{d\kappa^n}\Big|_{\kappa=0} F_{\kappa\tilde S}^{\mathrm{ret}} = H_F^{(n)}(0)(\tilde S^{\otimes n})\ ,\quad \text{where}\quad H_F(S) := F_S^{\mathrm{ret}}\ .$$

The "(classical) retarded product" $R_{n,1}(\tilde S^{\otimes n}, F)$ is the nth Taylor coefficient of $F_{\kappa\tilde S}^{\mathrm{ret}}$ (1.7.1). The property that $H_F^{(n)}(0)$ is defined also for non-diagonal entries and, as a map $H_F^{(n)}(0) : \mathcal{F}_{\mathrm{loc}}^{\otimes n} \longrightarrow \mathcal{F}$, is totally symmetric and linear is required also for the (classical) retarded product:

$$R_{n,1}(\,\cdot\,, F) := H_F^{(n)}(0)\ .$$

So we define:

Definition 1.7.1 (Classical retarded product)**.** By the (*classical*) *retarded product* we mean a sequence of linear maps

$$R_{n,1} \equiv R \equiv R_{\mathrm{cl}} : \mathcal{F}_{\mathrm{loc}}^{\otimes n}\otimes\mathcal{F}_{\mathrm{loc}} \longrightarrow \mathcal{F}$$

(one for each $n\in\mathbb{N}$), which are symmetric in the first n arguments, given by

$$R_{n,1}(\tilde S^{\otimes n}, F) \equiv R(\tilde S^{\otimes n}, F) := \frac{d^n}{d\kappa^n}\Big|_{\kappa=0} F_{\kappa\tilde S}^{\mathrm{ret}}\ . \tag{1.7.2}$$

Due to linearity and symmetry (in the first n arguments), it suffices to define this on diagonal entries like $\tilde{S}^{\otimes n}$ – see (A.1.9). For $n = 0$, we obtain $R_{0,1}(F) = F$. The lower indices $n, 1$ indicate that $R_{n,1}$ has two kind of entries, the first n for the interaction and the last one for the pertinent field at $\kappa = 0$.

The perturbative expansion of $F^{\mathrm{ret}}_{\kappa\tilde{S}}$ (1.6.2) is symbolically written as

$$F^{\mathrm{ret}}_{\kappa\tilde{S}} \simeq \sum_{n=0}^{\infty} \frac{\kappa^n}{n!} R(\tilde{S}^{\otimes n}, F) = \sum_{n=0}^{\infty} \frac{1}{n!} R(S^{\otimes n}, F) \equiv R(e_{\otimes}^{S}, F) , \tag{1.7.3}$$

where $S = \kappa\tilde{S}$. A crucial point is that we understand the right-hand side of $\simeq$ as *formal* power series in κ, i.e., as an element of $\mathcal{F}[\![\kappa]\!]$ – see (A.1.1) for the notation. That is, no convergence is implied. In clFT there are a lot of examples for which the perturbation series (1.7.3) converges; but for all known, physically relevant QFT-models in $d = 4$ dimensions, the series diverges and, even worse, it is an open question whether an exact expression for F^{ret}_{S} (or for the S-matrix) exists.

To explain the last expression in (1.7.3) note that we have used the handy notation (A.1.6) for the first entry. There, we regard the (classical) retarded product as a map[9] $R\colon \mathcal{T}(\kappa\mathcal{F}_{\mathrm{loc}}) \otimes \mathcal{F}_{\mathrm{loc}} \longrightarrow \mathcal{F}[\![\kappa]\!]$, where (A.1.5) and (A.1.7) are used.

Defining $R\big(\tilde{S}^{\otimes n}, A(x)\big)$ for $A \in \mathcal{P}_{\mathrm{bal}}$ similarly to (1.7.2), i.e., $R\big(\tilde{S}^{\otimes n}, A(x)\big) := \frac{d^n}{d\kappa^n}\big|_{\kappa=0} A^{\mathrm{ret}}_{\kappa\tilde{S}}(x)$, the relations (1.6.3)–(1.6.4) translate into

$$\int dx\, g(x)\, R\big(S^{\otimes n}, A(x)\big) = R\big(S^{\otimes n}, A(g)\big) , \quad \forall\, g \in \mathcal{D}(\mathbb{M}),\ A \in \mathcal{P}_{\mathrm{bal}} , \tag{1.7.4}$$

and

$$\partial_x R\big(S^{\otimes n}, A(x)\big) = R\big(S^{\otimes n}, \partial_x A(x)\big) . \tag{1.7.5}$$

Also here, (1.7.4) implies (1.7.5).

The factorization (1.6.5) and the off-shell field equation (1.6.6) hold true also for the perturbative retarded fields:

$$R\big(e_{\otimes}^{S}, (AB)(x)\big) = R\big(e_{\otimes}^{S}, A(x)\big) \cdot R\big(e_{\otimes}^{S}, B(x)\big) , \quad \forall\, A, B \in \mathcal{P}_{\mathrm{bal}} , \tag{1.7.6}$$

and

$$(\Box_x + m^2) R\big(e_{\otimes}^{S}, \varphi(x)\big) = (\Box + m^2)\varphi(x) + R\Big(e_{\otimes}^{S}, \frac{\delta S}{\delta\varphi(x)}\Big) ; \tag{1.7.7}$$

for the field equation this is obvious, the factorization can be obtained by applying the Leibniz rule to $\frac{d^n}{d\kappa^n}\big|_{\kappa=0}\big(A^{\mathrm{ret}}_{\kappa\tilde{S}}(x)\, B^{\mathrm{ret}}_{\kappa\tilde{S}}(x)\big)$. We conclude that the perturbative

[9] So far, one could instead write the completion of the symmetric algebra $\overline{S(\kappa\mathcal{F}_{\mathrm{loc}})}$ in the place of $\mathcal{T}(\kappa\mathcal{F}_{\mathrm{loc}})$ (where the completion is understood in the same way as in $\mathcal{T}(\kappa\mathcal{F}_{\mathrm{loc}})$ (A.1.5)). However, later on we will also work with retarded products $R_{n,1}$ whose first n entries are non-diagonal (e.g., in the GLZ relation (1.10.3), which will play a crucial role), and for that it is more convenient to have $\mathcal{T}(\kappa\mathcal{F}_{\mathrm{loc}})$.

retarded field $R\big(e_\otimes^S,\varphi(x)\big)$ solves also the version (1.6.7) of the off-shell field equation. In addition, the defining property (i) of a retarded wave operator translates into

$$R\big(e_\otimes^S,A(x)\big)=A(x)\quad\text{for } x \text{ ``sufficiently early''},\ \forall\, A\in\mathcal{P}_{\mathrm{bal}},\tag{1.7.8}$$

since $A(\varphi)(x)\circ r_{S_0+S,S_0}(h)=A\big(r_{S_0+S,S_0}(h)\big)(x)=A\big(h\big)(x)=A(\varphi)(x)(h)$. The property (1.7.8) is what we mean by saying that "R is retarded".

1.8 The Poisson algebra of the free theory

The algebraic structure of the set of classical observables is usually that of a Poisson algebra, i.e., a commutative and associative algebra together with a second operation: A Poisson bracket, satisfying the Leibniz rule and the Jacobi identity. Hence, to complete our formulation of classical field theory, we need to introduce a Poisson bracket. We do this only for the free theory; this suffices for our purposes, since in perturbation theory the interacting fields are interpreted as functionals on $\mathcal{C}_{S_0}$. For the definition of the Poisson bracket for an interacting theory, by means of the Peierls bracket [133], and for a lot of further results about classical field theory in the present formalism, we refer to [16, 29, 54].

Retarded propagator. We will define the Poisson bracket by means of the retarded propagator belonging to the free action S_0.

Definition 1.8.1. Given the free field equation $(\Box+m^2)\varphi(x)=0$, the *retarded propagator* $\Delta_m^{\mathrm{ret}}\in\mathcal{D}'(\mathbb{R}^d)$ is defined as the fundamental solution to the equation

$$(\Box+m^2)\Delta_m^{\mathrm{ret}}(x)=-\delta(x),\quad\text{satisfying}\quad\operatorname{supp}\Delta_m^{\mathrm{ret}}\subseteq\overline{V}_+\,.\tag{1.8.1}$$

We point out that Δ_m^{ret} is uniquely determined by the conditions (1.8.1).

Exercise 1.8.2. Check that

$$\Delta_m^{\mathrm{ret}}(x)=\frac{1}{(2\pi)^d}\int d^dp\,\frac{e^{-ipx}}{p^2-m^2+ip^00}\,,\tag{1.8.2}$$

where px and p^2 are inner products with the Minkowski metric: $px:=p^0x^0-\vec{p}\cdot\vec{x}$. Verify also the relations

$$\Delta_m^{\mathrm{ret}}(\Lambda x)=\Delta_m^{\mathrm{ret}}(x)\quad\forall\Lambda\in\mathcal{L}_+^\uparrow\,,\quad\rho^{d-2}\,\Delta_{m/\rho}^{\mathrm{ret}}(\rho x)=\Delta_m^{\mathrm{ret}}(x)\quad\text{and}\quad\overline{\Delta_m^{\mathrm{ret}}(x)}=\Delta_m^{\mathrm{ret}}(x)\,.\tag{1.8.3}$$

[*Hint*: To be equipped with the Fourier transformation, consider Δ_m^{ret} as an element of $\mathcal{S}'(\mathbb{R}^d)$ and, to verify the differential equation, use (A.1.18). To check the support property, assume $m>0$ and perform the p^0-integration for $x^0<0$.]

[*Solution*: The statements (1.8.3) follow immediately from the Fourier representation (1.8.2).

To prove (1.8.2), let $f \in \mathcal{S}'(\mathbb{R}^d)$ be the r.h.s. of this assertion. We are going to verify that f fulfills the defining properties (1.8.1).
Differential equation:

$$\langle (\Box + m^2) f, g \rangle_x = \langle \mathfrak{F}((\Box + m^2) f), \mathfrak{F}^{-1} g \rangle_p = -\langle (2\pi)^{-d/2}, \mathfrak{F}^{-1} g \rangle_p = -\langle \delta, g \rangle_x$$

for all $g \in \mathcal{S}(\mathbb{R}^d)$.
Support property: Let $x \notin \overline{V}_+$. Due to the Lorentz invariance of f we may assume $x^0 < 0$. Considering the integral

$$\int dp_0 \, \frac{e^{-ip_0 x^0}}{p_0^2 - \omega_{\vec{p}}^2 + i\varepsilon \, \mathrm{sgn}\, p_0} \; , \qquad \omega_{\vec{p}} := \sqrt{m^2 + \vec{p}^2} > 0 \; , \; \varepsilon > 0 \; , \tag{1.8.4}$$

in the complex p_0-plane, the two poles are in the lower half-plane. Because of $|e^{-i\, p_0\, x^0}| = e^{\mathrm{Im}(p^0)\, x^0}$, the integrand decays sufficiently fast in the upper half-plane, to conclude that the integral (1.8.4) vanishes by a standard application of Cauchy's Theorem.]

The Fourier transformation (1.8.2) can be computed; for $d = 4$ it results

$$\Delta_m^{\mathrm{ret}\,(d=4)}(x) = \frac{-1}{2\pi} \, \theta(x^0) \Big(\delta(x^2) - \theta(x^2) \, \frac{m}{2\sqrt{x^2}} \, J_1(m\sqrt{x^2}) \Big) \tag{1.8.5}$$

(see, e.g., [148, Sect. 2.3]), where J_1 is the Bessel function of first kind and first order, which is analytic.

Due to the latter, $\Delta_m^{\mathrm{ret}\,(d=4)}(g)$ is smooth in $m \geq 0$ for all $g \in \mathcal{D}(\mathbb{R}^d)$; one can show that this holds in any dimension $d > 2$. Hence, the retarded propagator exists also for $m = 0$: $D^{\mathrm{ret}} := \Delta_{m=0}^{\mathrm{ret}}$ is the retarded Green's function of the wave operator (up to a sign). For $d = 4$, D^{ret} is given by the first term in (1.8.5). With that we obtain

$$\int d^4y \, D^{\mathrm{ret}}(x - y) \, g(y) = \frac{-1}{4\pi} \int d^3\vec{y} \; \frac{g(x^0 - |\vec{x} - \vec{y}| \, , \, \vec{y})}{|\vec{x} - \vec{y}|} \; , \qquad g \in \mathcal{D}(\mathbb{R}^4) \; ,$$

by using $\theta(x^0)\, \delta(x^2) = \frac{\delta(x^0 - |\vec{x}|)}{2|\vec{x}|}$, which exhibits future directed propagation with the speed of light.

Poisson bracket. The propagator of the Poisson bracket is the antisymmetric part of Δ_m^{ret}, which is the "commutator function" Δ_m (up to factor 2):

$$\Delta_m(x) := \Delta_m^{\mathrm{ret}}(x) - \Delta_m^{\mathrm{ret}}(-x). \tag{1.8.6}$$

From (1.8.1) we conclude that $(\Box + m^2)\Delta_m(x) = 0$ and $\mathrm{supp}\, \Delta_m \subseteq (\overline{V}_+ \cup \overline{V}_-)$. Inserting (1.8.2) and using (A.1.16) in the form

$$\frac{1}{p^2 - m^2 + ip^0 0} = \mathcal{P} \frac{1}{p^2 - m^2} - i\pi \; \mathrm{sgn}\, p^0 \, \delta(p^2 - m^2) \; ,$$

we obtain

$$\Delta_m(x) = -\frac{i}{(2\pi)^{d-1}} \int d^dp \; \mathrm{sgn}\, p^0 \, \delta(p^2 - m^2) \, e^{-ipx} \,. \tag{1.8.7}$$

Notice that $\Delta_m^{\mathrm{ret}}(x) = \Delta_m(x)\,\theta(x^0)$: A pointwise product which exists, although there is an overlapping singularity at $x = 0$! This is verified in Exer. A.3.4 by using Hörmander's criterion for the wave front sets (given in App. A.3).

Also the commutator function is well defined for $m = 0$; we write $D := \Delta_{m=0}$.

Definition 1.8.3. The *Poisson bracket* of fields, with respect to the free action S_0 (1.5.1), is the bilinear map $\mathcal{F} \times \mathcal{F} \longrightarrow \mathcal{F}$ given by

$$\{F, G\} := \int dx\, dy \; \frac{\delta F}{\delta\varphi(x)} \, \Delta_m(x-y) \, \frac{\delta G}{\delta\varphi(y)} \,. \tag{1.8.8}$$

Here F and G are off-shell fields, i.e., they are not restricted by any field equation, in particular they "do not know anything about the mass m". The information about the free action (and the mass m) lies in the propagator: Δ_m is a solution of the free field equation. We emphasize that the Poisson bracket of two local fields is in general non-local.

Example 1.8.4. Let

$$F := \int dx_1 dx_2 \; f(x_1,x_2)\, \varphi(x_1) \partial^\mu \varphi(x_2) = -\int dx_1 dx_2 \; \partial^\mu_{x_2} f(x_1,x_2)\, \varphi(x_1)\varphi(x_2) \in \mathcal{F}$$

and $G := \int dy_1 \; g(y_1)\, \varphi^k(y_1) \in \mathcal{F}_{\mathrm{loc}}$, where $f \in \mathcal{D}'(\mathbb{M}^2)$ has compact support and satisfies the wave front set property (1.2.2) and $g \in \mathcal{D}(\mathbb{M})$. Using

$$\frac{\delta F}{\delta\varphi(x)} = \int dx_1 \, \Big(f(x,x_1)\, \partial^\mu \varphi(x_1) - \partial^\mu_x f(x_1,x)\, \varphi(x_1) \Big)$$

and $\frac{\delta G}{\delta\varphi(y)} = k\, g(y)\, \varphi^{k-1}(y)$ we obtain

$$\{F, G\} = \int dx\, dy\, dx_1 \Big(f(x,x_1)\, \partial^\mu \varphi(x_1)\, \Delta_m(x-y) \\ + f(x_1,x)\, \varphi(x_1)\, \partial^\mu \Delta_m(x-y) \Big) \, k\, g(y)\, \varphi^{k-1}(y) \,.$$

Proposition 1.8.5 (Existence and properties of Poisson bracket, [51, 52]). *The Poisson bracket satisfies the following properties:*

(i) *The pointwise product of distributions in* (1.8.8) *exists, due to the wave front set properties of F and G. Moreover, $\{F, G\}$ again satisfies the wave front set property* (1.2.2), *hence $\{F, G\} \in \mathcal{F}$.*

(ii) *The following algebraic relations are verified:*

- *$\{F, G\}$ is* bilinear *in F and G;*
- *it is* skew-symmetric*: $\{G, F\} = -\{F, G\}$;*

- *the* Leibniz rule *holds:* $\{F, GH\} = \{F, G\}\, H + G\, \{F, H\}$;
- *the* Jacobi identity *holds:* $\{F, \{G, H\}\} + \{G, \{H, F\}\} + \{H, \{F, G\}\} = 0$.

Partial proof and comments. The first part of item (i) can be proved by applying a generalization of Thm. A.3.3 and by taking into account

$$\mathrm{WF}(\Delta_m) = \{\,(x, k)\,\big|\, x^2 = 0,\ k^2 = 0,\ x = \lambda k \text{ for some } \lambda \in \mathbb{R},\ k^0 \neq 0\,\}, \quad (1.8.9)$$

cf. (2.4.1).[10]

In item (ii), bilinearity is obvious, skew-symmetry follows from $\Delta_m(-x) = -\Delta_m(x)$, and the Leibniz rule from the analogous Leibniz rule for $\delta/\delta\varphi$. The Jacobi identity is a corollary of the associativity of the star-product introduced in Chap. 2; this is explained in Sect. 2.4.

If the propagator $\Delta \equiv \Delta_m$ of the Poisson bracket satisfies $\frac{\delta\Delta}{\delta\varphi} = 0$ – as it holds true for the Poisson bracket of the free theory treated here, a direct proof of the Jacobi identity is not a big deal [29]. Namely, using twice Def. 1.8.3, the Leibniz rule for $\delta/\delta\varphi$ and $\Delta(-z) = -\Delta(z)$, we obtain

$$\begin{aligned}
&\{F, \{G, H\}\} \\
&= \int \Big(\prod_{j=1,2} dx_j dy_j\Big) \frac{\delta F}{\delta\varphi(x_2)}\, \Delta(x_2 - y_2)\, \frac{\delta}{\delta\varphi(y_2)} \Big(\frac{\delta G}{\delta\varphi(x_1)}\, \Delta(x_1 - y_1)\, \frac{\delta H}{\delta\varphi(y_1)}\Big) \\
&= \int \Big(\prod_{j=1,2} dx_j dy_j\Big) \frac{\delta F}{\delta\varphi(x_2)}\, \Delta(x_2 - y_2) \Big(\frac{\delta^2 G}{\delta\varphi(y_2)\delta\varphi(x_1)}\, \Delta(x_1 - y_1)\, \frac{\delta H}{\delta\varphi(y_1)} \\
&\qquad + \frac{\delta G}{\delta\varphi(x_1)}\, \Delta(x_1 - y_1)\, \frac{\delta^2 H}{\delta\varphi(y_2)\delta\varphi(y_1)}\Big) \\
&= \int \Big(\prod_{j=1,2} dx_j dy_j\Big) \frac{\delta F}{\delta\varphi(x_2)}\, \Delta(x_2 - y_2) \\
&\qquad \cdot \Big(\frac{\delta G}{\delta\varphi(x_1)}\, \Delta(x_1 - y_1)\, \frac{\delta^2 H}{\delta\varphi(y_2)\delta\varphi(y_1)} - (G \leftrightarrow H)\Big)\,.
\end{aligned}$$

Summing over the three cyclic permutations of F, G, H we obviously get zero. □

Due to the Jacobi identity, the Poisson bracket is non-associative.

Exercise 1.8.6. Prove the identity

$$\{F, \{G, GH\}\} + \{G, \{H, F\}\}\, G = -\{G, F\}\, \{G, H\} + G\, \{H, \{G, F\}\}.$$

---

[10]That $\mathrm{WF}(\Delta_m)$ is contained in the set written on the right side can easily be obtained from formula (2.2.7) for $\mathrm{WF}(\Delta_m^+)$, by using $\Delta_m(x) = -i\big(\Delta_m^+(x) - \Delta_m^+(-x)\big)$ (A.2.7), the relation $\mathrm{WF}(t + s) \subseteq \mathrm{WF}(t) \cup \mathrm{WF}(s)$ and (A.3.4):

$$\mathrm{WF}(\Delta_m) \subseteq \big(\mathrm{WF}(\Delta_m^+(x)) \cup \mathrm{WF}(\Delta_m^+(-x))\big) = \mathrm{WF}(\Delta_m^+) \cup \big(-\mathrm{WF}(\Delta_m^+)\big)\,,$$

see App. A.3.

[*Solution*: By using twice the Leibniz rule, we obtain

$$\{F,\{G,GH\}\} = \{F,G\{G,H\}\} = \{F,G\}\{G,H\} + G\{F,\{G,H\}\}\ .$$

Applying the Jacobi identity to the last expression we get the assertion.]

Remark 1.8.7. For local or non-local field polynomials $P(x_1,\ldots,x_r)$, $Q(y_1,\ldots,y_s)$ we define the Poisson bracket similarly to (1.8.8):

$$\{P(x_1,\ldots,x_r),Q(y_1,\ldots,y_s)\} \\ := \int dx\,dy\ \frac{\delta P(x_1,\ldots,x_r)}{\delta\varphi(x)}\ \Delta_m(x-y)\ \frac{\delta Q(y_1,\ldots,y_s)}{\delta\varphi(y)}\ ,$$

where the functional derivative of $P(x_1,\ldots,x_r)$, $Q(y_1,\ldots,y_s)$ is defined in (1.3.18) and (1.3.20), see also Exap. 1.3.7.

We point out that

$$\{(\Box+m^2)\varphi(x)\,,\,F\} = \int dy\ (\Box_x+m^2)\Delta_m(x-y)\,\frac{\delta F}{\delta\varphi(y)} = 0\quad \forall F\in\mathcal{F}\ .$$

1.9 An explicit formula for the classical retarded product

From now on we do not assume non-perturbative existence and uniqueness of the retarded wave operator.

A *retarded product is now defined* to be a sequence $R\equiv(R_{n,1})_{n\in\mathbb{N}}$ of maps $R_{n,1}:\mathcal{F}_{\mathrm{loc}}^{\otimes n}\otimes\mathcal{F}_{\mathrm{loc}}\longrightarrow\mathcal{F}$, which are *linear, symmetric in the first* n *arguments*; and we define $R\big(S^{\otimes n},A(x)\big)$ for $A\in\mathcal{P}_{\mathrm{bal}}$ in terms of $R\big(S^{\otimes n},A(g)\big)$ by (1.7.4). We recall that this definition implies the relation (1.7.5). Additional defining properties of R are the *factorization* (1.7.6), the *off-shell field equation* (1.7.7) and that R is *retarded*, i.e., that it satisfies the relation (1.7.8).

We point out that the statements corresponding to the defining properties of a retarded wave operator (Def. 1.6.2), namely the off-shell field equation and that R is retarded (1.7.8), are contained in this definition.

In this section we derive an explicit formula for the classical retarded product for three purposes: Firstly, this formula shows existence and uniqueness of the classical retarded product; secondly, this formula will help us to find simple diagrammatic rules ("Feynman rules") to compute $R_{n,1}(\otimes_{j=1}^n F_j,F)$; and finally, we will derive from this formula basic structural properties of the classical retarded product (Sect. 1.10), which we want to maintain in the quantization (Sect. 3.1).

Proposition 1.9.1 (Existence, uniqueness and computation of retarded product, [54]). *The classical retarded product exists and is unique in the just defined sense.*

It can be computed in the following way: Let $S \in \mathcal{F}_{\mathrm{loc}}$ and define the (S-dependent) operator[11]

$$\mathcal{R}_S(x) \equiv \mathcal{R}(x) := -\int dy\, \frac{\delta S}{\delta\varphi(x)}\, \Delta^{\mathrm{ret}}(y-x)\, \frac{\delta}{\delta\varphi(y)} \quad \text{on } \mathcal{F}. \tag{1.9.1}$$

For all $F \in \mathcal{F}$, the pointwise product of distributions appearing in $\mathcal{R}(x)F$ exists and $\int dx\, \mathcal{R}(x)F$ lies again in $\mathcal{F}$.

Now, for $F \in \mathcal{F}_{\mathrm{loc}}$, the retarded product $R(S^{\otimes n}, F)$ is obtained by the formula

$$R_{n,1}(S^{\otimes n}, F) = n! \int_{x_1^0 \le \cdots \le x_n^0} dx_1 \cdots dx_n\, \mathcal{R}(x_1) \cdots \mathcal{R}(x_n)\, F\ . \tag{1.9.2}$$

Note that in general the r.h.s. of (1.9.2) is not equal to the simpler expression $\int dx_1 \cdots dx_n\, \mathcal{R}(x_1) \cdots \mathcal{R}(x_n)\, F$, because the operators $\mathcal{R}(x_k)_{k=1,\ldots,n}$ do not commute.

For instance, at level $n = 1$, the expression (1.9.2) yields:

$$R(S,F) = -\int dx\, dy\, \frac{\delta S}{\delta\varphi(x)}\, \Delta^{\mathrm{ret}}(y-x)\, \frac{\delta F}{\delta\varphi(y)}\ . \tag{1.9.3}$$

We see that

$$R(S,F) - R(F,S) = \{S,F\}\ , \tag{1.9.4}$$

which is essentially Peierls' own definition [133] of the Poisson bracket.

Suppose $S = \int dx_1\, L(x_1)g(x_1)$ and $F = \int dy_1\, P(y_1)f(y_1)$ where $L, P \in \mathcal{P}_{\mathrm{bal}}$; $g, f \in \mathcal{D}(\mathbb{M})$. Then we get

$$\begin{aligned} R(S,F) &= -\int dx_1 dy_1\; g(x_1)f(y_1) \int dx\, dy\, \frac{\delta L(x_1)}{\delta\varphi(x)}\, \Delta^{\mathrm{ret}}(y-x)\, \frac{\delta P(y_1)}{\delta\varphi(y)} \\ &= \int_{x_1^0 \le y_1^0} dx_1 dy_1\; g(x_1)f(y_1)\, \{L(x_1), P(y_1)\}\ , \end{aligned}$$

where we have used $\theta(-z^0)\, \Delta(z) = -\Delta^{\mathrm{ret}}(-z)$. More generally, (1.9.2) leads to

$$\begin{aligned} R(e_\otimes^S, F) = \sum_{n=0}^{\infty} \int dx_1 \cdots dx_n\, dy\; g(x_1) \cdots g(x_n)\, f(y)\, \theta(y^0 - x_n^0) \\ \cdot\, \theta(x_n^0 - x_{n-1}^0) \cdots \theta(x_2^0 - x_1^0)\ \{L(x_1), \{L(x_2), \ldots \{L(x_n), P(y)\} \ldots\}\}\ . \end{aligned} \tag{1.9.5}$$

Defining $R_{n,1}\big(L(x_1) \otimes \cdots, P(y)\big)$ by

$$\begin{aligned} \int dx_1 \cdots dx_n\, dy\; g(x_1) \cdots g(x_n)\, f(y) R_{n,1}\big(L(x_1) \otimes \cdots \otimes L(x_n), P(y)\big) \\ := R_{n,1}\big(L(g)^{\otimes n}, P(f)\big) \quad \forall g, f \in \mathcal{D}(\mathbb{M}) \end{aligned}$$

[11] Here we have dropped the mass label m on the propagators, for convenience.

in accordance with (1.7.4), the retarded product $R_{n,1}\big(L(x_1)\otimes\cdots,P(y)\big)$ is obtained from the iterated and time-ordered Poisson bracket in the second line of (1.9.5) by multiplication with $n!$ and symmetrization in $x_1,\dots,x_n$.

Example 1.9.2. The case $P=\varphi^2$, $L=\varphi^4$ of formula (1.9.5) gives, to order $n=2$, the expression

$$R_{2,1}\big(\varphi^4(x_1)\otimes\varphi^4(x_2),\varphi^2(y)\big)$$
$$=\theta(y^0-x_2^0)\,\theta(x_2^0-x_1^0)\,\{\varphi^4(x_1),\{\varphi^4(x_2),\varphi^2(y)\}\}+(x_1\leftrightarrow x_2)\;. \tag{1.9.6}$$

For the inner retarded Poisson bracket we obtain

$$\theta(y^0-x_2^0)\,\{\varphi^4(x_2),\varphi^2(y)\}=-8\,\varphi(y)\,\Delta^{\mathrm{ret}}(y-x_2)\,\varphi^3(x_2)\;.$$

For the second retarded Poisson bracket we get two terms:

$$\theta(y^0-x_2^0)\,\theta(x_2^0-x_1^0)\,\{\varphi^4(x_1),\varphi(y)\varphi^3(x_2)\}\,\Delta^{\mathrm{ret}}(y-x_2)$$
$$=-\,12\,\varphi(y)\,\Delta^{\mathrm{ret}}(y-x_2)\,\varphi^2(x_2)\,\Delta^{\mathrm{ret}}(x_2-x_1)\,\varphi^3(x_1)$$
$$-\,4\,\Delta^{\mathrm{ret}}(y-x_1)\,\varphi^3(x_1)\,\Delta^{\mathrm{ret}}(y-x_2)\,\varphi^3(x_2)\,\theta(x_2^0-x_1^0)\;.$$

Adding the $(x_1\leftrightarrow x_2)$-term and using $\theta(x_2^0-x_1^0)+\theta(x_1^0-x_2^0)=1$, we end up with

$$R_{2,1}\big(\varphi^4(x_1)\otimes\varphi^4(x_2),\varphi^2(y)\big)$$
$$=96\,\varphi(y)\,\Delta^{\mathrm{ret}}(y-x_2)\,\varphi^2(x_2)\,\Delta^{\mathrm{ret}}(x_2-x_1)\,\varphi^3(x_1)+(x_1\leftrightarrow x_2)$$
$$+\,32\,\Delta^{\mathrm{ret}}(y-x_1)\,\varphi^3(x_1)\,\Delta^{\mathrm{ret}}(y-x_2)\,\varphi^3(x_2)\;. \tag{1.9.7}$$

Each of these three terms may be visualized by a "Feynman diagram":

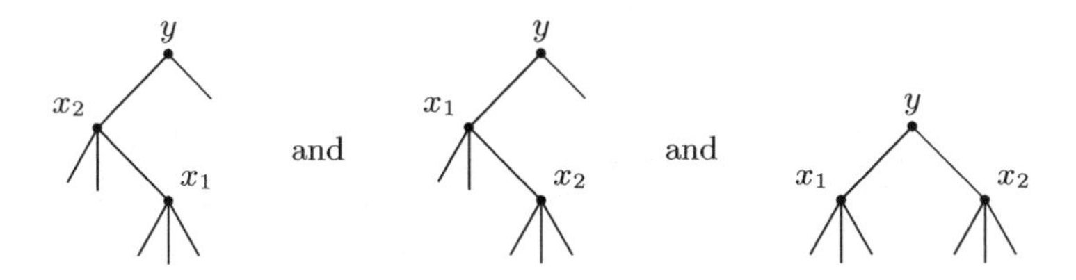

The large dots, labelled by x_1,x_2 and y, are called "vertices". An "inner line" connecting the vertices x_i [or y] and x_j symbolizes a propagator $\Delta^{\mathrm{ret}}(x_i-x_j)$ [or $\Delta^{\mathrm{ret}}(y-x_j)$, resp.], where x_i is the vertex which is nearer to the field vertex y – the inner lines are oriented in this sense. An "external line" starting at the vertex x_i stands for a factor $\varphi(x_i)$; and the same for y instead of x_i.

A nice feature of the result (1.9.7) is that the explicit time-ordering of the vertices, given by the θ-functions in (1.9.6), has disappeared; there is only an implicit time-ordering of the vertices due to $\operatorname{supp}\Delta^{\mathrm{ret}}\subset\overline{V}_+$. This holds generally – an illustration is part (b) of the following exercise.

Exercise 1.9.3.

(a) Derive

$$R(F_1 \otimes \cdots \otimes F_n, F) = \int_{x_1^0 \leq \cdots \leq x_n^0} dx_1 \cdots dx_n \sum_{\pi \in S_n} \mathcal{R}_{F_{\pi 1}}(x_1) \cdots \mathcal{R}_{F_{\pi n}}(x_n)\, F \quad (1.9.8)$$

by using (1.9.2), where $\mathcal{R}_F$ is defined in (1.9.1).

(b) For the particular case $n = 3$ show that formula (1.9.8) yields the following result for $R(F_1 \otimes F_2 \otimes F_3, F)$:

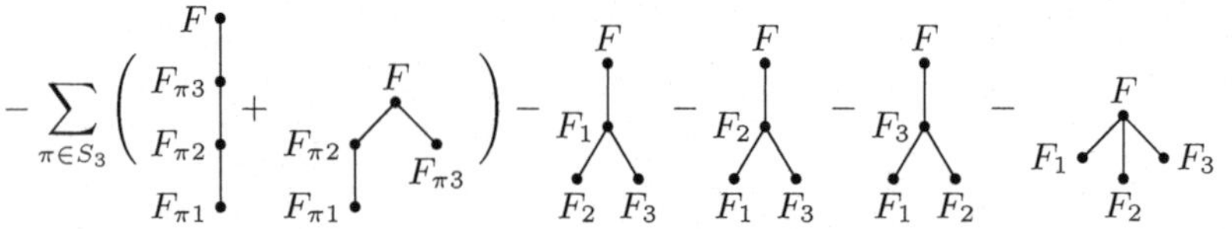

Here, an inner line connecting the vertices G and H, where G is nearer to the field vertex F ($G = F$ is possible), symbolizes the expression

$$\int dx\, dy\, \frac{\delta G}{\delta \varphi(y)}\, \Delta^{\mathrm{ret}}(y - x)\, \frac{\delta H}{\delta \varphi(x)}\ . \quad (1.9.9)$$

For example, the analytic expression for the second last diagram reads

$$\int dx_1\, dy_1\, dx_2\, dy_2\, dx_3\, dy_3\, \frac{\delta F}{\delta \varphi(y_3)}\, \Delta^{\mathrm{ret}}(y_3 - x_3)\, \frac{\delta^3 F_3}{\delta \varphi(x_3) \delta \varphi(y_1) \delta \varphi(y_2)}$$
$$\cdot\, \Delta^{\mathrm{ret}}(y_1 - x_1)\, \frac{\delta F_1}{\delta \varphi(x_1)}\, \Delta^{\mathrm{ret}}(y_2 - x_2)\, \frac{\delta F_2}{\delta \varphi(x_2)}\ . \quad (1.9.10)$$

[*Solution*: (a) We use (A.1.9) and (1.9.2): With $F_\lambda := \sum_{k=1}^n \lambda_k F_k$ we obtain

$$R(F_1 \otimes \cdots \otimes F_n, F) = \frac{1}{n!}\, \frac{\partial^n}{\partial \lambda_1 \cdots \partial \lambda_n}\, R(F_\lambda^{\otimes n}, F)$$
$$= \frac{\partial^n}{\partial \lambda_1 \cdots \partial \lambda_n} \int_{x_1^0 \leq \cdots \leq x_n^0} dx_1 \cdots dx_n\, \mathcal{R}_{F_\lambda}(x_1) \cdots \mathcal{R}_{F_\lambda}(x_n)\, F$$
$$= \int_{x_1^0 \leq \cdots \leq x_n^0} dx_1 \cdots dx_n \sum_{\pi \in S_n} \mathcal{R}_{F_{\pi 1}}(x_1) \cdots \mathcal{R}_{F_{\pi n}}(x_n)\, F\ ,$$

where ∂_{λ_k} is explained in Footn. 183.

(b) The contributions of (1.9.8) to the second last diagram can be written as

$$(-1)^3 \int_{x_1^0 \leq x_2^0} \prod_{j=1}^{3} (dx_j\, dy_j)\, \frac{\delta F}{\delta \varphi(y_3)}\, \Delta^{\mathrm{ret}}(y_3 - x_3)$$
$$\cdot \sum_{\pi \in S_2} \frac{\delta^3 F_3}{\delta \varphi(x_3) \delta \varphi(y_1) \delta \varphi(y_2)} \prod_{k=1,2} \Delta^{\mathrm{ret}}(y_k - x_k)\, \frac{\delta F_{\pi(k)}}{\delta \varphi(x_k)}\ .$$

Substituting $x_{\pi(k)}$ for x_k ($k = 1, 2$) and using that $\sum_{\pi \in S_2} \int_{x^0_{\pi(1)} \leq x^0_{\pi(2)}} dx_1\, dx_2 \cdots = \int dx_1\, dx_2 \ldots$ we get (1.9.10).
The terms of (1.9.8) contributing to the last diagram read

$$(-1)^3 \sum_{\pi \in S_3} \int_{x_1^0 \leq x_2^0 \leq x_3^0} \prod_{j=1}^{3} \Big(dx_j\, dy_j\, \frac{\delta F_{\pi(j)}}{\delta\varphi(x_j)}\, \Delta^{\mathrm{ret}}(y_j - x_j) \Big) \frac{\delta^3 F}{\delta\varphi(y_1)\delta\varphi(y_2)\delta\varphi(y_3)} \,.$$

Again, we may replace x_j by $x_{\pi(j)}$ ($j = 1, 2, 3$) and take into account that

$$\sum_{\pi \in S_3} \int_{x^0_{\pi(1)} \leq x^0_{\pi(2)} \leq x^0_{\pi(3)}} dx_1\, dx_2\, dx_3 \cdots = \int dx_1\, dx_2\, dx_3 \cdots \,;$$

it results

$$-\int \prod_{j=1}^{3} \Big(dx_j\, dy_j\, \frac{\delta F_j}{\delta\varphi(x_j)}\, \Delta^{\mathrm{ret}}(y_j - x_j) \Big) \frac{\delta^3 F}{\delta\varphi(y_1)\delta\varphi(y_2)\delta\varphi(y_3)} \,.$$

We symbolize the terms of (1.9.8) contributing to the second diagram by

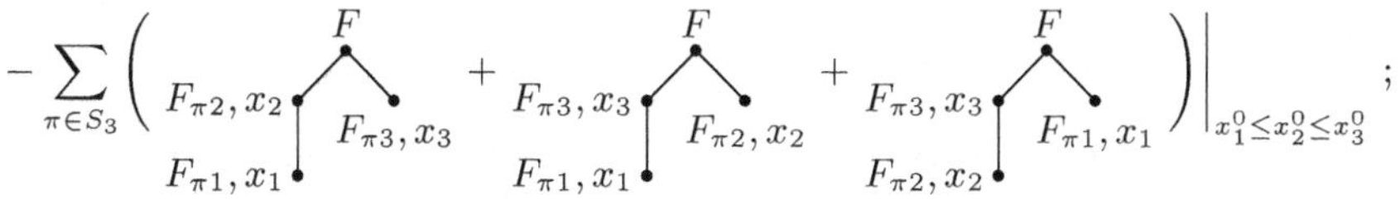

$- \sum_{\pi \in S_3} \Big($ [diagram: F; $F_{\pi 2}, x_2$; $F_{\pi 3}, x_3$; $F_{\pi 1}, x_1$] $+$ [diagram: F; $F_{\pi 3}, x_3$; $F_{\pi 2}, x_2$; $F_{\pi 1}, x_1$] $+$ [diagram: F; $F_{\pi 3}, x_3$; $F_{\pi 1}, x_1$; $F_{\pi 2}, x_2$] $\Big)\Big|_{x_1^0 \leq x_2^0 \leq x_3^0}$;

this sum gives indeed the asserted expression.
Finally, the computation of the term visualized by the first diagram is simple: The restriction $x_1^0 \leq x_2^0 \leq x_3^0$ can be omitted, since it follows from the support property of the retarded propagators.]

We meet here a modified version of the Feynman diagrams: The vertices symbolize *functionals* and, hence, there are no external lines. The diagrams contributing to $R(F_1 \otimes F_2 \otimes F_3, F)$ are precisely all (topologically distinct) connected tree diagrams[12] with vertices F_1, F_2, F_3 and F. This holds to all orders, as one can derive from formula (1.9.8).[13] That is, $R(F_1 \otimes \cdots \otimes F_n, F)$ can be obtained by the following "Feynman rules":

(1) sketch all (topologically distinct) *connected tree diagrams* which one can build from the vertices $F_1, \ldots, F_n$, F and n inner lines.

(2) From each diagram read off the pertinent analytic expression, as explained in (1.9.9)–(1.9.10). Then, $R(F_1 \otimes \cdots \otimes F_n, F)$ is $(-1)^n$ times the sum of all these expressions.

[12] By a "tree diagram" we mean a diagram which does not contain any closed line; more precisely, the set theoretic union of all (inner) lines and all vertices does not contain any closed line. A tree diagram may be connected or disconnected.

[13] In Sect. 3.1.7 we prove in a somewhat different way that *only connected tree diagrams* contribute to the classical retarded product.

Sketch proof of Proposition 1.9.1. The *existence* of $\mathcal{R}(x)F$ and the relation

$$\int dx\ \mathcal{R}(x)F \in \mathcal{F}$$

can be shown analogously to the proof of part (i) of Proposition 1.8.5.

Turning to the proof of (1.9.2), we denote the right-hand side of this equation by $R^0_{n,1}(S^{\otimes n}, F)$. We have to show that the sequence $R^0 = (R^0_{n,1})$ solves the defining properties for $R = (R_{n,1})$ given at the beginning of this section and that this solution is unique.

Obviously, $R^0(S^{\otimes n}, F)$ depends *linearly* on $S^{\otimes n} \otimes F$. The property

$$\partial_z R^0(e_\otimes^S, A(z)) = R^0(e_\otimes^S, \partial_z A(z))$$

follows immediately from the definition of R^0.

Next one shows that R^0 satisfies the *factorization* property (1.7.6). This can be done by using the Leibniz rule and taking careful account of integration regions $x_1^0 \leq \cdots \leq x_n^0$ – see the proof of [54, Propos. 3].

That R^0 is *retarded* (1.7.8), follows from

$$\mathcal{R}_S(x)A(z) = -\int dy\, \frac{\delta S}{\delta\varphi(x)}\, \Delta^{\mathrm{ret}}(y-x)\, \frac{\delta A(z)}{\delta\varphi(y)} = 0 \quad \text{if} \quad z \notin (\operatorname{supp} S + \overline{V}_+)\ ,$$

which relies on (1.3.19) and $\operatorname{supp}\Delta^{\mathrm{ret}} \subseteq \overline{V}_+$.

Turning to the *off-shell field equation* (1.7.7), that is

$$R^0\left(e_\otimes^S, \frac{\delta(S_0+S)}{\delta\varphi(x)}\right) \stackrel{?}{=} \frac{\delta S_0}{\delta\varphi(x)}\ , \tag{1.9.11}$$

we first find that

$$\begin{aligned}
\int dx\, \mathcal{R}(x)\, \frac{\delta S_0}{\delta\varphi(z)} &= -\int dx\, \frac{\delta S}{\delta\varphi(x)} \int dy\, \Delta^{\mathrm{ret}}(y-x)\, \frac{\delta^2 S_0}{\delta\varphi(y)\,\delta\varphi(z)} \\
&= +\int dx\, \frac{\delta S}{\delta\varphi(x)} \int dy\, (\square_y + m^2)\Delta^{\mathrm{ret}}(y-x)\, \delta(y-z) \\
&= -\int dx\, \frac{\delta S}{\delta\varphi(x)} \int dy\, \delta(y-x)\, \delta(y-z) = -\frac{\delta S}{\delta\varphi(z)}\,.
\end{aligned}$$

With that we indeed get the assertion (1.9.11): Starting with

$$\begin{aligned}
&R^0\left(e_\otimes^S, \frac{\delta S_0}{\delta\varphi(z)}\right) \\
&\quad = \frac{\delta S_0}{\delta\varphi(z)} + \sum_{n\geq 1} \int_{x_1^0\leq\cdots\leq x_{n-1}^0\leq x_n^0} dx_1 \cdots dx_n\, \mathcal{R}(x_1)\cdots\mathcal{R}(x_{n-1})\, \mathcal{R}(x_n)\, \frac{\delta S_0}{\delta\varphi(z)}\ ,
\end{aligned}$$

we also take into account that $\mathcal{R}(x_{n-1})\,\mathcal{R}(x_n)\,\frac{\delta S_0}{\delta\varphi(z)}$ vanishes for $x_{n-1}\notin(x_n+\overline{V}_-)$. Therefore, the restriction $x^0_{n-1}\le x^0_n$ on the domain of integration can be omitted and we obtain

$$\begin{aligned} R^0\Big(e_\otimes^S,\frac{\delta S_0}{\delta\varphi(z)}\Big) &= \frac{\delta S_0}{\delta\varphi(z)}-\sum_{n\ge 1}\int_{x_1^0\le\cdots\le x^0_{n-1}} dx_1\cdots dx_{n-1}\,\mathcal{R}(x_1)\cdots\mathcal{R}(x_{n-1})\,\frac{\delta S}{\delta\varphi(z)}\\ &=\frac{\delta S_0}{\delta\varphi(z)}-R^0\Big(e_\otimes^S,\frac{\delta S}{\delta\varphi(z)}\Big). \end{aligned}$$

Finally, to prove *uniqueness* of R^0, it suffices to prove uniqueness of $R^0(e_\otimes^S,\varphi(z))$; this follows from the factorization property (1.7.6). For simplicity, we now restrict to the case $S=-\kappa\int dx\;g(x)\,\varphi^k(x)$; with that the off-shell field equation (1.7.7) is equivalent to the simpler differential equation (1.6.7), due to the factorization (1.7.6). $R^0(e_\otimes^S,\varphi(z))$ solves (1.6.7), because it satisfies (1.7.6):

$$\begin{aligned} (\Box_z+m^2)R^0\big(e_\otimes^S,\varphi(z)\big) &\overset{(1.9.11)}{=} (\Box+m^2)\varphi(z)-k\kappa\,g(z)\,R^0\big(e_\otimes^S,\varphi^{k-1}(z)\big)\\ &\overset{(1.7.6)}{=} (\Box+m^2)\varphi(z)-k\kappa\,g(z)\Big(R^0\big(e_\otimes^S,\varphi(z)\big)\Big)^{k-1}. \end{aligned} \tag{1.9.12}$$

The point is now that perturbative, retarded solutions of (1.6.7) are unique. This can be seen as follows: Writing $\varphi_n(z):=\frac{1}{n!}\,R^0_{n,1}\big(\tilde S^{\otimes n},\varphi(z)\big)$, i.e., $R^0\big(e_\otimes^{\kappa\tilde S},\varphi(z)\big)=\sum_{n=0}^\infty\varphi_n(z)\,\kappa^n$, the field equation (1.9.12) to nth order reads

$$(\Box_z+m^2)\varphi_n(z)=-k\,g(z)\sum_{n_1+\cdots+n_{k-1}=n-1}\varphi_{n_1}(z)\cdots\varphi_{n_{k-1}}(z)\,. \tag{1.9.13}$$

Constructing the sequence (φ_n) by induction on n, we know the right-hand side. The boundary condition $\varphi_n(z)=0$ for z "sufficiently early" determines the solution of (1.9.13) uniquely: $\varphi_n(z)=-\int dx\;\Delta_m^{\mathrm{ret}}(z-x)\,f(x)$, where $f(z)$ denotes the right-hand side of (1.9.13).

By this proof of (1.9.2) we have shown that the *classical retarded product exists and is unique.* □

Remark 1.9.4 (Retarded products with non-local entries)**.** Definitions 1.6.2, 1.6.7 and 1.7.1 of the retarded wave operator r_{S_0+S,S_0}, the retarded field F_S^{ret} and the (classical) retarded product $R_{n,1}(S^{\otimes n},F)$, respectively, can directly be applied to arbitrary $S,F\in\mathcal{F}$; the retarded product to nth order is then a map $R_{n,1}\colon\mathcal{F}^{\otimes n}\otimes\mathcal{F}\longrightarrow\mathcal{F}$. But formula (1.9.2) holds true only for *local* entries $S,F\in\mathcal{F}_{\mathrm{loc}}$. An explicit expression for the computation of $R_{n,1}$ for non-local entries is given in [16, Eq. (2.21)], see also [54, Eqs. (40)–(42)].

1.10 Basic properties of the retarded product

In this section we point out some fundamental structural properties of the classical retarded product, which we will use as defining axioms for the retarded product (i.e., the interacting fields) of pQFT in Sect. 3.1.

(1) **Causality:** We first consider the expression (1.9.3) for $R(S,F)$. Due to

$$\operatorname{supp}\Delta^{\mathrm{ret}} \subseteq \overline{V}_+ \ ,$$

the integration over x is restricted to $x \in (\operatorname{supp}F + \overline{V}_-)$; therefore, the relation

$$R(S+H,F) = R(S,F) \quad \text{if} \quad \big(\operatorname{supp}F + \overline{V}_-\big) \cap \operatorname{supp}H = \emptyset \ .$$

holds true. Turning to higher orders we note that, e.g.,

$$\int dy\ \Delta^{\mathrm{ret}}(x_2-x_1)\,\Delta^{\mathrm{ret}}(y-x_2)\,\Delta^{\mathrm{ret}}(y-x_3)\,\frac{\delta F}{\delta\varphi(y)} = 0 \ \text{if} \ x_j \notin (\operatorname{supp}F+\overline{V}_-)$$

for at least one $j \in \{1,2,3\}$. In general we conclude from formula (1.9.2) for $R(S^{\otimes n},F)$, $\operatorname{supp}\Delta^{\mathrm{ret}} \subseteq \overline{V}_+$ and $\overline{V}_+ + \overline{V}_+ = \overline{V}_+$ that

$$R(e_\otimes^{S+H},F) = R(e_\otimes^S,F) \quad \text{if} \quad \big(\operatorname{supp}F + \overline{V}_-\big) \cap \operatorname{supp}H = \emptyset \ . \tag{1.10.1}$$

(2) **Field independence:** Applying $\delta/\delta\varphi(z)$ to the expression (1.9.3) for $R(S,F)$ we get

$$\frac{\delta}{\delta\varphi(z)}\,R(S,F) = R\Big(\frac{\delta S}{\delta\varphi(z)},F\Big) + R\Big(S,\frac{\delta F}{\delta\varphi(z)}\Big) \ .$$

Proceeding analogously for formula (1.9.2) for $R(S^{\otimes n},F)$, we obtain

$$\frac{\delta R(e_\otimes^S,F)}{\delta\varphi} = R\Big(e_\otimes^S \otimes \frac{\delta S}{\delta\varphi},F\Big) + R\Big(e_\otimes^S,\frac{\delta F}{\delta\varphi}\Big) \ ; \tag{1.10.2}$$

that is, the retarded product itself does not depend on φ, due to $\delta\Delta^{\mathrm{ret}}/\delta\varphi = 0$.

In a Taylor expansion of the retarded product $R_{n-1,1}(F_1 \otimes \cdots, F_n)$ with respect to φ, one may apply the relation (1.10.2) to the Taylor coefficients; in QFT this leads to the causal Wick expansion of retarded products [66], see Sect. 3.1.4.

(3) **The GLZ relation:** So called after Glaser, Lehmann and Zimmermann [84].

Proposition 1.10.1 ([54]). *Let $F,H,S \in \mathcal{F}_{\mathrm{loc}}$. Then the Poisson bracket of two interacting fields satisfies the relation:*

$$\begin{aligned}\Big\{R(e_\otimes^S,F),R(e_\otimes^S,H)\Big\} &= \frac{d}{d\lambda}\Big|_{\lambda=0}\Big(R(e_\otimes^{S+\lambda F},H) - R(e_\otimes^{S+\lambda H},F)\Big) \\ &= R(e_\otimes^S \otimes F,H) - R(e_\otimes^S \otimes H,F) \ ; \end{aligned} \tag{1.10.3}$$

see Footn. 183 *in App.* A.1 *for the definition of the derivative* $\frac{d}{d\lambda}$.

Note that both sides of the GLZ relation are manifestly antisymmetric under $F \leftrightarrow H$. For $S = 0$ the GLZ relation reduces to $\{F, H\} = R(F, H) - R(H, F)$, which is (1.9.4).

From Causality we know that

$$\operatorname{supp} R\big(e_\otimes^S \otimes A(x), B(y)\big) \subseteq \{\, (x,y) \in \mathbb{M}^2 \,\big|\, x \in (y + \overline{V}_-)\,\}\ ,$$
$$\operatorname{supp} R\big(e_\otimes^S \otimes B(y), A(x)\big) \subseteq \{\, (x,y) \in \mathbb{M}^2 \,\big|\, x \in (y + \overline{V}_+)\,\}$$

for $A, B \in \mathcal{P}_{\mathrm{bal}}$; hence, the GLZ relation implies

$$\operatorname{supp}\Big\{R\big(e_\otimes^S, A(x)\big), R\big(e_\otimes^S, B(y)\big)\Big\} \subseteq \{\, (x,y) \in \mathbb{M}^2 \,|\, (x-y)^2 \geq 0\,\}\ , \quad (1.10.4)$$

so in particular, the Poisson bracket *vanishes* when x and y are spacelike separated.

The GLZ relation, combined with causality, is the key for Steinmann's inductive construction of the retarded product of pQFT [155], which we will adopt in a different framework – see Sect. 3.2. The basic idea is the following: Assuming that the $R_{k,1}$ are known for $1 \leq k \leq n$, we can compute the left-hand side of (1.10.3) to order n in S, which is equal to the difference of two $R_{n+1,1}$ terms (on the right-hand side) with split supports, which allows one to extract the individual $R_{n+1,1}$.

The proof of the GLZ relation given in [54] uses Peierls' definition of the Poisson bracket for *interacting* theories. Since we have not introduced the Poisson bracket in this generality, we omit this proof; as a substitute we verify explicitly the GLZ relation to lowest orders (Exer. 1.10.3) and for particularly simple interacting fields (Exer. 1.10.2(c)).

We finish the chapter about Classical Field Theory with these exercises and some further ones.

Exercise 1.10.2. Let

$$S_1 := -\kappa \int dz\, g(z)\, \varphi(z), \quad S_2 := -\kappa \int dz\, g(z)\, \varphi^2(z), \quad S_3 := -\kappa \int dz\, g(z)\, \varphi^3(z).$$

(a) Compute $\varphi_{S_1}^{\mathrm{ret}}(x)$ non-perturbatively by solving the off-shell field equation (1.6.7).

(b) Compute $\varphi_{S_2}^{\mathrm{ret}}(x) = R\big(e_\otimes^{S_2}, \varphi(x)\big)$ perturbatively by using (1.9.2) or (1.9.5) or the "Feynman rules"; to control the result, check the off-shell field equation.

(c) For $S = S_1$ and $S = S_2$, compute $R\big(e_\otimes^S \otimes \varphi(y), \varphi(x)\big)$ perturbatively, and verify the GLZ relation $\{R\big(e_\otimes^S, \varphi(x)\big), R\big(e_\otimes^S, \varphi(y)\big)\} = \cdots$.

(d) Compute $\varphi_{S_3}^{\mathrm{ret}}(x) = R\big(e_\otimes^{S_3}, \varphi(x)\big) = \sum_{n=0}^\infty \varphi_n(x)\, \kappa^n$ perturbatively up to third order in κ, by solving the off-shell field equation recursively as explained in (1.9.13).

[*Solution:*

(a) $\varphi^{\rm ret}_{S_1}(x) = \varphi(x) + \kappa \int dy\, g(y)\, \Delta^{\rm ret}(x-y)$.

(b)
$$R\big(e_\otimes^{S_2}, \varphi(x)\big) = \varphi(x) + \sum_{n=1}^{\infty} \kappa^n 2^n \int dx_1 \cdots dx_n\, g(x_1)\cdots g(x_n)$$
$$\cdot\, \Delta^{\rm ret}(x-x_n)\, \Delta^{\rm ret}(x_n - x_{n-1}) \cdots \Delta^{\rm ret}(x_2 - x_1)\, \varphi(x_1).$$
To verify the off-shell field equation (1.6.7) for $\varphi^{\rm ret}_{S_2}(x) = R\big(e_\otimes^{S_2}, \varphi(x)\big)$, use $(\Box + m^2)\Delta^{\rm ret}_m(x) = -\delta(x)$.

(c) $R\big(e_\otimes^{S_1} \otimes \varphi(y), \varphi(x)\big) = -\Delta^{\rm ret}(x-y)$ and

$$R\big(e_\otimes^{S_2} \otimes \varphi(y), \varphi(x)\big) = -\Delta^{\rm ret}(x-y) - \sum_{n=1}^{\infty} \kappa^n 2^n \int dx_1 \cdots dx_n\, g(x_1)\cdots g(x_n)$$
$$\cdot\, \Delta^{\rm ret}(x-x_n)\, \Delta^{\rm ret}(x_n - x_{n-1}) \cdots \Delta^{\rm ret}(x_2 - x_1)\, \Delta^{\rm ret}(x_1 - y).$$

GLZ relation for S_1: Using the result of part (a), we get

$$\big\{R\big(e_\otimes^{S_1}, \varphi(x)\big), R\big(e_\otimes^{S_1}, \varphi(y)\big)\big\} = \{\varphi(x), \varphi(y)\} = \Delta(x-y)$$
$$= -\Delta^{\rm ret}(y-x) + \Delta^{\rm ret}(x-y) = R\big(e_\otimes^{S_1} \otimes \varphi(x), \varphi(y)\big) - R\big(e_\otimes^{S_1} \otimes \varphi(y), \varphi(x)\big)\ .$$

GLZ relation for S_2: By inserting the result of part (b), we obtain

$$\big\{R\big(e_\otimes^{S_2}, \varphi(x)\big), R\big(e_\otimes^{S_2}, \varphi(y)\big)\big\}$$
$$= \sum_{n=0}^{\infty} \sum_{l=0}^{n} \Big\{R\big(\tfrac{1}{(n-l)!} S_2^{\otimes n-l}, \varphi(x)\big), R\big(\tfrac{1}{l!} S_2^{\otimes l}, \varphi(y)\big)\Big\}$$
$$= \sum_{n=0}^{\infty} 2^n \kappa^n \int dx_1 \cdots dx_n\, g(x_1) \cdots g(x_n) \sum_{l=0}^{n} \Delta^{\rm ret}(x-x_n) \cdots \Delta^{\rm ret}(x_{l+2} - x_{l+1})$$
$$\cdot\, \{\varphi(x_{l+1}), \varphi(x_l)\}\, \Delta^{\rm ret}(y - x_1) \cdots \Delta^{\rm ret}(x_{l-1} - x_l)\ , \qquad x_0 := y\,,\ x_{n+1} := x\ .$$

Inserting $\{\varphi(x_{l+1}), \varphi(x_l)\} = \Delta^{\rm ret}(x_{l+1} - x_l) - \Delta^{\rm ret}(x_l - x_{l+1})$, most terms cancel in the sum over l; only the "boundary terms" $\sim \Delta^{\rm ret}(x-x_n) \cdots \Delta^{\rm ret}(x_1 - y)$ and $\sim -\Delta^{\rm ret}(y - x_1) \cdots \Delta^{\rm ret}(x_n - x)$ (coming from $l=0$ and $l=n$, resp.) remain, and these terms yield $R\big(e_\otimes^{S_2} \otimes \varphi(x), \varphi(y)\big) - R\big(e_\otimes^{S_2} \otimes \varphi(y), \varphi(x)\big)$.

(d) Starting with $\varphi_0 = \varphi$ the recursion formula

$$\varphi_n(x) = 3 \int dy\, \Delta^{\rm ret}(x-y)\, g(y) \sum_{n_1+n_2=n-1} \varphi_{n_1}(y)\, \varphi_{n_2}(y)$$

yields

$$\varphi_1(x) = 3 \int dy\, \Delta^{\rm ret}(x-y)\, g(y)\, \varphi^2(y)\ ,$$
$$\varphi_2(x) = 18 \int dy_1\, \Delta^{\rm ret}(x-y_1)\, g(y_1) \varphi(y_1) \int dy_2\, \Delta^{\rm ret}(y_1 - y_2)\, g(y_2)\, \varphi^2(y_2)\ ,$$

and

$$\varphi_3(x) = 3\int dy_1\, \Delta^{\mathrm{ret}}(x-y_1)\, g(y_1) \Big(36\, \varphi(y_1) \int dy_2\, \Delta^{\mathrm{ret}}(y_1-y_2)\, g(y_2)\varphi(y_2) \cdot \int dy_3\, \Delta^{\mathrm{ret}}(y_2-y_3)\, g(y_3)\, \varphi^2(y_3) + 9 \Big(\int dy_2\, \Delta^{\mathrm{ret}}(y_1-y_2)\, g(y_2)\, \varphi^2(y_2)\Big) \Big(\int dy_3\, \Delta^{\mathrm{ret}}(y_1-y_3)\, g(y_3)\, \varphi^2(y_3)\Big)\Big) .]$$

Exercise 1.10.3 (GLZ relation)**.** Verify the GLZ relation (1.10.3) to first and second order in S for arbitrary $F, H, S \in \mathcal{F}_{\mathrm{loc}}$, by using the "Feynman rules" for $R(F_1 \otimes \cdots \otimes F_n, F)$ explained in Sect. 1.9.

[*Hint*: To save time write down the various terms only diagrammatically – also for the Poisson bracket, e.g., with time advancing vertically the four terms on the r.h.s. of

$$\begin{aligned}\{R(S,F),H\} = \int dx_1\, dy_1\, dx_2\, dy_2 \cdot\Big(&- \frac{\delta F}{\delta\varphi(y_1)}\, \Delta^{\mathrm{ret}}(y_1-x_1)\, \frac{\delta^2 S}{\delta\varphi(x_1)\delta\varphi(y_2)} \Delta^{\mathrm{ret}}(y_2-x_2)\, \frac{\delta H}{\delta\varphi(x_2)} \\ &- \frac{\delta H}{\delta\varphi(x_1)}\, \Delta^{\mathrm{ret}}(y_1-x_1)\, \frac{\delta^2 F}{\delta\varphi(y_1)\delta\varphi(y_2)} \Delta^{\mathrm{ret}}(y_2-x_2)\, \frac{\delta S}{\delta\varphi(x_2)} \\ &+ \frac{\delta F}{\delta\varphi(y_1)}\, \Delta^{\mathrm{ret}}(y_1-x_1)\, \frac{\delta^2 S}{\delta\varphi(x_1)\delta\varphi(x_2)} \Delta^{\mathrm{ret}}(y_2-x_2)\, \frac{\delta H}{\delta\varphi(y_2)} \\ &+ \frac{\delta H}{\delta\varphi(y_1)}\, \Delta^{\mathrm{ret}}(y_1-x_1)\, \frac{\delta^2 F}{\delta\varphi(x_1)\delta\varphi(y_2)} \Delta^{\mathrm{ret}}(y_2-x_2)\, \frac{\delta S}{\delta\varphi(x_2)} \Big)\end{aligned}$$

are completely encoded in the diagrammatic expression

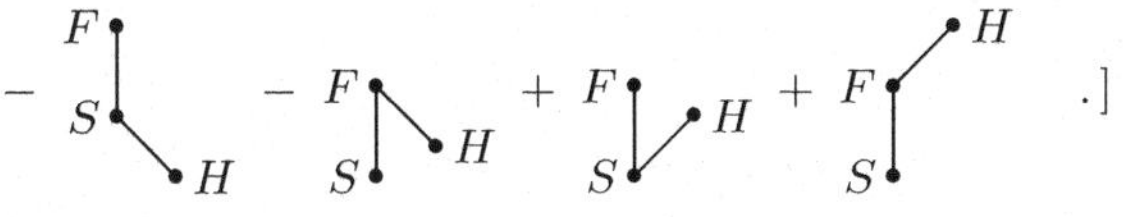

− (F, S, H) − (F, S, H) + (F, S, H) + (F, S, H) .]

[*Solution*: To first and second order in S the GLZ relation reads

$$\{R(S,F),H\} + \{F, R(S,H)\} = R(S\otimes F, H) - R(S\otimes H, F)$$

and

$$\frac{1}{2}\{R(S^{\otimes 2},F),H\} + \{R(S,F),R(S,H)\} + \frac{1}{2}\{F, R(S^{\otimes 2},H)\} = \frac{1}{2}\Big(R(S^{\otimes 2}\otimes F, H) + R(S^{\otimes 2}\otimes H, F)\Big),$$

respectively. Verification of the first-order equation: The l.h.s. is given by [the diagrams in the Hint] minus $[F \leftrightarrow H]$. The r.h.s. reads

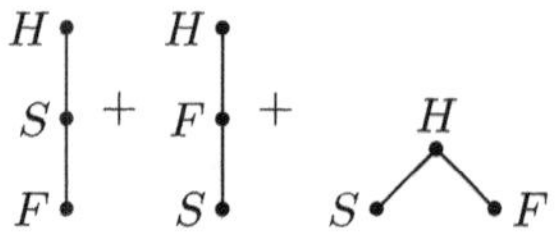

$$H \bullet\!\!-\!\!S \bullet\!\!-\!\!F \; + \; H \bullet\!\!-\!\!F \bullet\!\!-\!\!S \; + \; S \bullet\!\!\diagup^{H}\!\!\diagdown\!\bullet F \quad - \quad [F \leftrightarrow H] \, .$$

Sketching these diagrams, the left- and the right-hand side agree.
Verification of the second-order equation: For the l.h.s. we obtain 28 diagrams, 10 of them cancel out; for the r.h.s. we get 18 diagrams without any cancellation.]

Exercise 1.10.4 (*-structure). Prove the relation

$$R(e_\otimes^S, F)^* = R(e_\otimes^{S^*}, F^*) \qquad \forall S, F \in \mathcal{F}_{\mathrm{loc}} \, . \tag{1.10.5}$$

[*Solution*: From

$$\left(\frac{\delta G}{\delta\varphi(x)}\right)^* = \frac{\delta G^*}{\delta\varphi(x)} \quad \forall G \in \mathcal{F} \quad \text{and} \quad \overline{\Delta_m^{\mathrm{ret}}(x)} = \Delta_m^{\mathrm{ret}}(x)$$

we see that the operator $\mathcal{R}_S(x)$ (1.9.1) satisfies

$$\big(\mathcal{R}_S(x)\, G\big)^* = \mathcal{R}_{S^*}(x)\, G^* \quad \forall G \in \mathcal{F} \, , \quad \text{which implies} \quad R(S^{\otimes n}, F)^* = R\big((S^*)^{\otimes n}, F^*\big)$$

by means of (1.9.2).]

Exercise 1.10.5 (Field parity). The transformation $\varphi \longmapsto -\varphi$ can be defined as the linear map

$$\alpha\colon \begin{cases} \mathcal{F} \longrightarrow \mathcal{F} \\ (\alpha F)(h) = F(-h) \quad \text{for all} \quad h \in \mathcal{C} \, . \end{cases} \tag{1.10.6}$$

(a) Prove non-perturbatively the relation $\alpha(F_S^{\mathrm{ret}})\big|_{\mathcal{C}_{S_0}} = (\alpha F)_{\alpha S}^{\mathrm{ret}}\big|_{\mathcal{C}_{S_0}}$;

(b) and prove the corresponding perturbative statement without restriction to $\mathcal{C}_{S_0}$,

$$\alpha\big(R_{n,1}(S^{\otimes n}, F)\big) = R_{n,1}\big((\alpha S)^{\otimes n}, \alpha F\big)\big), \tag{1.10.7}$$

by using (1.9.2).

[*Hint for part* (a): Assume that $r_{S_0+S,S_0}\big|_{\mathcal{C}_{S_0}}$ exists and is unique, and use Remk. 1.6.3. A main step is to show that $r_{S_0+\alpha S,S_0}(-h_0) = -r_{S_0+S,S_0}(h_0)$ for $h_0 \in \mathcal{C}_{S_0}$.]

[*Solution*: From (1.3.1) we see that $\alpha\Big(\frac{\delta F}{\delta\varphi(x)}\Big) = -\frac{\delta(\alpha F)}{\delta\varphi(x)}$.
(a) Note that $h \in \mathcal{C}_{S_0+S} \Leftrightarrow 0 = \alpha\Big(\frac{\delta(S_0+S)}{\delta\varphi(x)}\Big)(-h) = -\frac{\delta(S_0+\alpha S)}{\delta\varphi(x)}(-h) \Leftrightarrow -h \in \mathcal{C}_{S_0+\alpha S}$, since $\alpha S_0 = S_0$. Let $h_0 \in \mathcal{C}_{S_0}$ and $h := r_{S_0+S,S_0}(h_0)$, hence $-h = r_{S_0+\alpha S,S_0}(-h_0)$. With this we obtain

$$\begin{aligned} \alpha(F_S^{\mathrm{ret}})(-h_0) &= F_S^{\mathrm{ret}}(h_0) = F(h) = (\alpha F)(-h) \\ &= (\alpha F) \circ r_{S_0+\alpha S,S_0}(-h_0) = (\alpha F)_{\alpha S}^{\mathrm{ret}}(-h_0) \, . \end{aligned}$$

(b) Applying α to (1.9.2), it commutes with the integrations over $x_1,\ldots,x_n$. The assertion follows from

$$\alpha(\mathcal{R}_S(x)\,G) = -\int dy\,\Big(-\frac{\delta(\alpha S)}{\delta\varphi(x)}\Big)\,\Delta^{\mathrm{ret}}(y-x)\,\Big(-\frac{\delta(\alpha G)}{\delta\varphi(y)}\Big) = \mathcal{R}_{\alpha S}(x)\,\alpha G\ , \quad \forall G\in\mathcal{F}\ .]$$

Exercise 1.10.6. Let $A,B\in\mathcal{P}_{\mathrm{bal}}$ and $S,H\in\mathcal{F}_{\mathrm{loc}}$. Prove the factorization property

$$\begin{aligned} R_{n+1,1}\big(S^{\otimes n}\otimes(AB)(x),H\big) = \sum_{l=0}^{n}\binom{n}{l}\Big(&R_{l,1}\big(S^{\otimes l},A(x)\big)\,R_{n-l+1,1}\big(S^{\otimes(n-l)}\otimes B(x),H\big)\\ &+R_{l+1,1}\big(S^{\otimes l}\otimes A(x),H\big)\,R_{n-l,1}\big(S^{\otimes(n-l)},B(x)\big)\Big). \end{aligned}$$

[*Hint*: Write the assertion as $R\big(e_\otimes^S\otimes(AB)(x),H\big)=\cdots$ and then use the GLZ relation, the factorization (1.6.5) and the Leibniz rule for the Poisson bracket.]

[*Solution*: Starting with the GLZ relation and applying twice the factorization we get

$$\begin{aligned} R\big(e_\otimes^S\otimes(AB)(x),H\big) &= R\big(e_\otimes^S\otimes H,(AB)(x)\big)+\Big\{R\big(e_\otimes^S,(AB)(x)\big)\,,\,R\big(e_\otimes^S,H\big)\Big\}\\ &=\frac{d}{d\lambda}\Big|_{\lambda=0}R\big(e_\otimes^{S+\lambda H},A(x)\big)\,R\big(e_\otimes^{S+\lambda H},B(x)\big)\\ &\quad+\Big\{R\big(e_\otimes^S,A(x)\big)\,R\big(e_\otimes^S,B(x)\big)\,,\,R\big(e_\otimes^S,H\big)\Big\}\ ; \end{aligned}$$

then we use the Leibniz rule for both the λ-derivative and the Poisson bracket, and finally we apply again the GLZ relation:

$$\begin{aligned} &=R\big(e_\otimes^S,A(x)\big)\Big(R\big(e_\otimes^S\otimes H,B(x)\big)+\Big\{R\big(e_\otimes^S,B(x)\big)\,,\,R\big(e_\otimes^S,H\big)\Big\}\Big)+(A\leftrightarrow B)\\ &=R\big(e_\otimes^S,A(x)\big)\,R\big(e_\otimes^S\otimes B(x),H\big)+(A\leftrightarrow B)\ .] \end{aligned}$$

Chapter 2

Deformation Quantization of the Free Theory

For this chapter, see the references [27, 51, 52, 55]. The basic reference for deformation quantization is due to Bayen, Flato, Fronsdal, Lichnerowicz and Sternheimer [6]. This seminal paper relies on much earlier work of John von Neumann [130] and on Gerstenhaber's algebraic deformation theory [83], and it gives an axiomatic formulation of the heuristic quantization formulas found by Weyl [169], Groenewold [89] and Moyal[126]. Further groundbreaking progress in the development of deformation quantization has been achieved (among others and in historical order) by De Wilde and Lecomte [33], Fedosov [70, 71] and Kontsevich [115, 116]. The book [165] gives a detailed introduction to deformation quantization. For the application to free fields we refer also to [19, 35, 36, 99, 166].

Deformation quantization is a way to quantize a classical mechanical system/ classical field theory to obtain a pertinent quantum mechanical system/quantum field theory. It focuses on the algebra of observables of a physical system: It provides rules for how to deform the commutative algebra of classical observables to a noncommutative algebra of quantum observables. Usually (and also in this book) deformation quantization refers to formal deformations, in the sense that the deformed product is a formal power series in Planck's constant $\hbar$.

Given a Poisson algebra $(\mathcal{A}, \{\cdot,\cdot\})$, a deformation quantization is a bilinear and associative product $\star$ on $\mathcal{A} \times \mathcal{A}$ ("star product") with values in $\mathcal{A}[\![\hbar]\!]$ (see (A.1.1)), subject to the following two axioms:

$$\begin{aligned} f \star g &= fg + O(\hbar), \\ [f,g]_\star &:= f \star g - g \star f = i\hbar\{f,g\} + O(\hbar^2) \end{aligned} \tag{2.0.1}$$

for all $f, g \in \mathcal{A}$. The star product corresponds to the operator product, if the quantum observables are represented by linear operators on a Hilbert space, as, e.g., in Schrödingers formulation of quantum mechanics. The second axiom in (2.0.1) is a precise formulation of the heuristic rule that under quantization the Poisson bracket is replaced by the commutator.

M. Dütsch, *From Classical Field Theory to Perturbative Quantum Field Theory*, Progress in Mathematical Physics 74, https://doi.org/10.1007/978-3-030-04738-2_2

2.1 Star products of fields

We seek a *deformation* $\star_\hbar : \mathcal{F} \times \mathcal{F} \longrightarrow \mathcal{F}[\![\hbar]\!]$ of the classical product of fields. The new product should be:

(a) *bilinear* in its arguments;

(b) *associative*; and moreover should satisfy:

(c) $F \star_\hbar G \to F \cdot G$ (the classical product) as $\hbar \to 0$;

(d) $(F \star_\hbar G - G \star_\hbar F)/i\hbar \to \{F, G\}$ (the Poisson bracket) as $\hbar \to 0$.

Since we have not introduced a topology on $\mathcal{F}[\![\hbar]\!]$, the "classical limit $\hbar \to 0$" is obtained by simply setting $\hbar := 0$, so that formal series in $\hbar$ are truncated at their $\hbar^0$-terms.

Further requirements on $\star_\hbar$ will emerge later on. For convenience, we abbreviate $\star_\hbar$ simply as $\star$ when no confusion is likely to arise.

We make the Ansatz that

$$\varphi(x) \star \varphi(y) = \varphi(x)\, \varphi(y) + \hbar\, H(x - y) \tag{2.1.1}$$

for some fixed distribution $H \in \mathcal{D}'(\mathbb{R}^d)$, which must fulfill certain properties, which we are now going to explain. Condition (d) determines the antisymmetric part of H:

$$\frac{1}{i}\big(H(x-y) - H(y-x)\big) = \{\varphi(x), \varphi(y)\} = \Delta_m(x - y). \tag{2.1.2}$$

With $F_0 := F|_{\mathcal{C}_{S_0}}$, we get $(\Box + m^2)\varphi_0 = 0$, i.e., φ_0 is a solution of the homogeneous Klein–Gordon equation. The following extra requirements on the star products may now be added:

(e) Requiring $(\Box_x + m^2)\big(\varphi(x) \star \varphi(y)\big)_0 = 0$, we ask that $H = H_m$ depends on m and that

$$(\Box_x + m^2) H_m(x - y) = 0. \tag{2.1.3}$$

Hence, the $\star$-product depends on m via H_m; to emphasize that, we sometimes write $\star_m$ instead of $\star$.

(f) *Lorentz invariance*: we require that

$$H_m(\Lambda z) = H_m(z) \quad \text{for all} \quad \Lambda \in \mathcal{L}_+^\uparrow. \tag{2.1.4}$$

To motivate the last two conditions on H_m, we first define $F \star G$ in terms of H_m for arbitrary $F, G \in \mathcal{F}[\![\hbar]\!]$.

Definition 2.1.1 (Star product, [51, 52, 55]). If the distribution H_m is known, the *general definition* of a star product $\star_\hbar : \mathcal{F}[\![\hbar]\!] \times \mathcal{F}[\![\hbar]\!] \longrightarrow \mathcal{F}[\![\hbar]\!]$ is given by

$$F \star_\hbar G := \sum_{n=0}^{\infty} \frac{\hbar^n}{n!} \int dx_1 \cdots dx_n\, dy_1 \cdots dy_n \qquad (2.1.5)$$
$$\cdot \frac{\delta^n F}{\delta\varphi(x_1)\cdots\delta\varphi(x_n)} \prod_{l=1}^{n} H_m(x_l - y_l)\, \frac{\delta^n G}{\delta\varphi(y_1)\cdots\delta\varphi(y_n)}\,.$$

For $1 \in \mathcal{F}$, i.e., $1(h) = 1\ \forall h \in \mathcal{C}$, we get $1 \star_\hbar F = F = F \star_\hbar 1$.

Substituting $\varphi(x)$ and $\varphi(y)$ for F and G, resp., formula (2.1.5) reduces to the Ansatz (2.1.1). In addition, it is obvious that this Definition of $F \star G$ satisfies the requirements bilinearity (a) and the first limit (c): In the limit $\lim_{\hbar\to 0} F \star_\hbar G$ only the $(n = 0)$-term survives and, for $F, G \sim \hbar^0$, this limit is equal to $F \cdot G$. Likewise, the second limit (d): In $(F \star_\hbar G - G \star_\hbar F)$ there is no term $\sim \hbar^0$, hence the limit $\lim_{\hbar\to 0}(F \star_\hbar G - G \star_\hbar F)/i\hbar$ exists and, for $F, G \sim \hbar^0$, it is equal to

$$\int dx\, dy\, \frac{\delta F}{\delta\varphi(x)}\, \frac{1}{i}\big(H(x-y) - H(y-x)\big)\, \frac{\delta G}{\delta\varphi(y)} = \{F\,,\,G\}\,,$$

due to (2.1.2). The associativity (b) is more difficult to prove; we defer this to Sect. 2.4.

The "integral" on the r.h.s. of (2.1.5) is finite for $\|x_j\| \to \infty$ or $\|y_j\| \to \infty$, because $\frac{\delta^n F}{\delta\varphi(x_1)\cdots\delta\varphi(x_n)}(h)$ and $\frac{\delta^n G}{\delta\varphi(y_1)\cdots\delta\varphi(y_n)}(h)$ ($h \in \mathcal{C}$ arbitrary) are distributions with compact support. But, the r.h.s. of (2.1.5) contains pointwise products of distributions, which may have overlapping singularities (cf. Thm. A.3.3)! In order that these pointwise products exist and that the result $F \star_\hbar G$ lies indeed in $\mathcal{F}[\![\hbar]\!]$, we need to restrict H_m further; see the requirement (g) below and Sect. 2.4.

Note also that this definition is given on $\mathcal{F}[\![\hbar]\!] \times \mathcal{F}[\![\hbar]\!]$ instead of only on $\mathcal{F} \times \mathcal{F}$. Since the elements of $\mathcal{F}$ are polynomials in φ, the following holds: Let $\mathcal{F}_\hbar$ be the space of fields which are *polynomials in* $\hbar$, more precisely

$$\mathcal{F}_\hbar := \Big\{ \sum_{s=0}^{S} F_s\, \hbar^s \,\Big|\, F_s \in \mathcal{F}\,,\ S < \infty \Big\}\,. \qquad (2.1.6)$$

We conclude that for $F, G \in \mathcal{F}_\hbar$ the sum over n in formula (2.1.5) for $F \star_\hbar G$ is *finite*[14] and, hence, $F \star_\hbar G \in \mathcal{F}_\hbar$; that is, the star product can be interpreted as a map $\star : \mathcal{F}_\hbar \times \mathcal{F}_\hbar \longrightarrow \mathcal{F}_\hbar$.

Analogously to our procedure for the Poisson bracket (Remk. 1.8.7), we adopt formula (2.1.5) to define $P(x_1, \ldots, x_r) \star_\hbar Q(y_1, \ldots, y_s)$ for (local or non-local) field polynomials $P(x_1, \ldots, x_r)$ and $Q(y_1, \ldots, y_s)$: We simply substitute the latter two for F and G, resp., in (2.1.5); see Exap. 2.1.2 below.

[14] Note that for $\sum_{s=0}^{\infty} F_s\, \hbar^s \in \mathcal{F}[\![\hbar]\!]$ the set of the degrees of the polynomials (in φ) F_s, $s \in \mathbb{N}$, may be unbounded.

Another way of writing (2.1.5), which is convenient for some proofs (in particular for the proof of the associativity of the star product), is the following:

$$F \star G = \mathcal{M} \circ e^{\hbar D_H}(F \otimes G) \quad \text{where} \quad D_H := \int dx\, dy\, H_m(x-y)\, \frac{\delta}{\delta\varphi(x)} \otimes \frac{\delta}{\delta\varphi(y)}\,. \tag{2.1.7}$$

In addition, $\mathcal{M}$ denotes the pointwise (or classical) multiplication (1.2.5), that is,

$$\mathcal{M}\colon \begin{cases} \mathcal{F} \otimes \mathcal{F} \longrightarrow \mathcal{F} \\ \mathcal{M}(F \otimes G)(h) := F(h)G(h)\ , \quad \forall h \in \mathcal{C}\ , \end{cases} \tag{2.1.8}$$

and

$$e^{\hbar D_H} := \mathrm{Id} \otimes \mathrm{Id} + \sum_{k=1}^{\infty} \frac{\hbar^k}{k!}\, (D_H)^k$$

is understood as formal power series in $\hbar$.

From the following example, we find a necessary condition for the existence of the r.h.s. of the definition (2.1.5).

(g) all powers

$$H_m(x-y)^l, \quad \text{for} \quad l = 2, 3, \ldots, \quad \text{must exist.} \tag{2.1.9}$$

Example 2.1.2. Let us compute the star product $\varphi(x_1)\,\varphi^2(x_2) \star \varphi^2(y)$. The general formula above indicates that the expansion in powers of $\hbar$ stops at order $O(\hbar^2)$. The h^0-term is the classical product. For the computation of the functional derivatives see (1.3.18)–(1.3.20). Altogether, we find five terms:

$$\begin{aligned} \varphi(x_1)\,\varphi^2(x_2) \star \varphi^2(y) &= \varphi(x_1)\,\varphi^2(x_2)\,\varphi^2(y) \\ &\quad + \hbar\big[2\varphi^2(x_2)\,H(x_1-y)\,\varphi(y) + 4\varphi(x_1)\,\varphi(x_2)\,H(x_2-y)\,\varphi(y)\big] \\ &\quad + \frac{\hbar^2}{2}\big[4\varphi(x_2)\,H(x_1-y)\,H(x_2-y) + 4\varphi(x_1)\,H(x_2-y)^2\big]. \end{aligned}$$

The appearance of $H(x_k - y)$ is called a "contraction of $\varphi(x_k)$ with $\varphi(y)$". Diagrammatically, the listed terms correspond to the following contraction schemes:

x_1, x_2, y $\quad$ x_1, x_2, y $\quad$ x_1, x_2, y $\quad$ x_1, x_2, y $\quad$ x_1, x_2, y

Comparing with Fock space calculations, we remark that if the classical, pointwise product is interpreted as the normally ordered product, then this computation arises from Wick's Theorem (A.5.33), see Exap. 2.6.4.

h) the last condition for H_m is

$$\overline{H_m(x)} = H_m(-x), \quad \text{which is equivalent to} \quad (F \star G)^* = G^* \star F^*\ , \tag{2.1.10}$$

due to $\big(\frac{\delta^n F}{\delta\varphi(x_1)\cdots\delta\varphi(x_n)}\big)^* = \frac{\delta^n F^*}{\delta\varphi(x_1)\cdots\delta\varphi(x_n)}$. A main motivation for this condition is given in Exer. 2.5.5.

Note that for $F := \int dx\, f(x)\, (\square + m^2)\varphi(x)$, where $f \in \mathcal{D}(\mathbb{M})$, it holds that

$$F \star_\hbar G = FG \quad \text{and} \quad [F, G]_{\star_\hbar} := F \star_\hbar G - G \star_\hbar F = 0 \quad \text{for all} \quad G \in \mathcal{F}[\![\hbar]\!]\,, \tag{2.1.11}$$

due to (2.1.3).

We also remark that the star product commutes with the field parity transformation α given in (1.10.6): By using $\alpha\Big(\frac{\delta^n F}{\delta\varphi(x_1)\cdots\delta\varphi(x_n)}\Big) = (-1)^n \frac{\delta^n(\alpha F)}{\delta\varphi(x_1)\cdots\delta\varphi(x_n)}$ we get

$$\alpha(F \star G) = (\alpha F) \star (\alpha G)\ , \quad \forall\, F, G \in \mathcal{F}[\![\hbar]\!]\ . \tag{2.1.12}$$

Exercise 2.1.3. Let $\varphi(f) := \int dx\, \varphi(x)\, f(x)$. Prove the relation

$$e^{\lambda\,\varphi(f)} \star_\hbar e^{\lambda\,\varphi(g)} = e^{\lambda\,\varphi(f+g)}\, e^{\lambda^2\hbar\,\langle f, Hg\rangle} \quad \text{for } f, g \in \mathcal{D}(\mathbb{M})\ ,$$

where

$$\langle f, Hg\rangle := \int dx\, dy\ f(x)\, H(x-y)\, g(y)\ , \tag{2.1.13}$$

which we understand in the sense of formal power series in λ; for instance, $e^{\lambda\,\varphi(f)} \equiv \sum_{n=0}^\infty \frac{\lambda^n}{n!}\, \varphi(f)^n$ with the pointwise product.

[*Solution*: Using the equation

$$\frac{\delta^n e^{\lambda\,\varphi(f)}}{\delta\varphi(x_1)\cdots\delta\varphi(x_n)} = \lambda^n\, f(x_1)\cdots f(x_n)\, e^{\lambda\,\varphi(f)}\ ,$$

the definition (2.1.5) gives

$$e^{\lambda\,\varphi(f)} \star_\hbar e^{\lambda\,\varphi(g)} = \sum_{n=0}^\infty \frac{\hbar^n \lambda^{2n}}{n!}\, e^{\lambda\,\varphi(f)}\, \langle f, Hg\rangle^n\, e^{\lambda\,\varphi(g)} = e^{\lambda\,\varphi(f+g)}\, e^{\lambda^2\hbar\,\langle f, Hg\rangle}\ .]$$

2.2 Solutions for the two-point function H_m

Wightman two-point function. The first candidate for a distribution which fulfills the conditions on H_m is the Wightman two-point function:

$$\begin{aligned} H_m(z) = \Delta_m^+(z) &:= \frac{1}{(2\pi)^{d-1}} \int d^dp\ \theta(p^0)\, \delta(p^2 - m^2)\, e^{-ipz} \\ &= \frac{1}{(2\pi)^{d-1}} \int d^{d-1}\vec{p}\ \frac{e^{-i(\omega_{\vec{p}} z^0 - \vec{p}\vec{z})}}{2\omega_{\vec{p}}}\ , \quad \omega_{\vec{p}} := \sqrt{\vec{p}^{\,2} + m^2} \end{aligned} \tag{2.2.1}$$

(cf. (A.5.1)), that is to say, the "positive frequencies of $i\Delta_m$" (since $i\widehat{\Delta_m}(p) \sim \operatorname{sgn} p^0\, \delta(p^2 - m^2)$). In other words, taking account of $\operatorname{sgn}(p^0) = \theta(p^0) - \theta(-p^0)$, we get

$$i\Delta_m(z) = \Delta_m^+(z) - \Delta_m^+(-z)\ , \tag{2.2.2}$$

which is the splitting of $i\Delta_m$ into positive ($\theta(p^0)$-part) and negative frequencies ($\theta(-p^0)$-part).[15]

The Fourier transformation in (2.2.1) can be computed; for $d=4$ the result reads

$$\Delta_m^{+\,(d=4)}(z) = \frac{-1}{4\pi^2(z^2 - iz^0 0)} + m^2\, f(m^2 z^2)\, \log\Big(-\frac{m^2}{4}\,(z^2 - iz^0 0)\Big) + m^2\, f_1(m^2 z^2), \tag{2.2.3}$$

with f and f_1 being certain analytic functions. f can be expressed in terms of the Bessel function J_1 of order 1, namely

$$f(x) := \frac{1}{8\pi^2\sqrt{x}}\, J_1(\sqrt{x}) = \sum_{k=0}^{\infty} C_k\, x^k\ , \quad C_k \in \mathbb{R}\ ; \tag{2.2.4}$$

and f_1 is given by a power series

$$f_1(x) := -\frac{1}{16\,\pi^2} \sum_{k=0}^{\infty} \Big(\psi(k+1) + \psi(k+2)\Big)\, \frac{(-x/4)^k}{k!(k+1)!}\ , \tag{2.2.5}$$

where the Psi-function (also called "Digamma function") is related to the Gamma-function by $\psi(x) := \Gamma'(x)\,/\,\Gamma(x)$. A proof of the result (2.2.3) is given, e.g., in [12, Sect. 15.1] or [148, Sect. 2.3]; a derivation of the corresponding formula for the Euclidean propagator is sketched in Exap. 3.9.2. Note that the first term in (2.2.3) is $D^+(z)$, the solution for $m=0$; it is computed in Exer. 2.2.2 below. The limit

$$D^+ := \lim_{m \downarrow 0} \Delta_m^+ \quad \text{exists in any dimension } d \geq 3, \tag{2.2.6}$$

for the usual (weak) convergence in $\mathcal{D}'(\mathbb{R}^d)$, see (A.1.15).

Verification of the requirements on $H_m = \Delta_m^+$: As already noted in (2.2.2), Δ_m^+ satisfies (2.1.2). Due to

$$\mathfrak{F}[(\square + m^2)\Delta_m^+](p) \sim (-p^2 + m^2)\,\theta(p^0)\,\delta(p^2 - m^2) = 0$$

(where $\mathfrak{F}$ denotes Fourier transformation), Δ_m^+ satisfies the homogeneous Klein–Gordon equation (2.1.3). The Lorentz invariance (2.1.4) and the relation (2.1.10) are obvious. The property (2.1.9) can be verified by using Hörmander's wave front set criterion (Thm. A.3.3) – see Exer. 2.2.1(b) below.[16] Since Δ_m^+ is the positive

---

[15] In [143] there is a mistake in the interpretation of its formula (5.5): The relation

$$\tfrac{i}{2}\,\Delta_m(z) = \Delta_m^+(z) - d_m(z)\ , \quad \text{where} \quad d_m(z) := \tfrac{1}{2}\,\big(\Delta_m^+(z) + \Delta_m^+(-z)\big)$$

is the symmetric part of Δ^+, is *not* the splitting into positive and negative frequencies.

[16] A possible alternative to the use of wave front sets is to employ *Riesz functions*, for which all complex powers exist: See [4], for instance.

frequency part of $i\Delta_m$, the wave front set of Δ_m^+ is the $(k^0>0)$-part of $\mathrm{WF}(\Delta_m)$ (given in (1.8.9)):

$$\mathrm{WF}(\Delta_m^+) = \{\, (x,k) \,|\, x^2=0,\ k^2=0,\ x=\lambda k \text{ for some } \lambda\in\mathbb{R},\ k^0>0 \,\}\,. \quad (2.2.7)$$

In particular note that the singular support of Δ_m^+ is the light cone $x^0=\pm|\vec{x}|$; and for $x=0$ the pertinent k-vectors of $\mathrm{WF}(\Delta_m^+)$ are

$$\mathrm{WF}(\Delta_m^+)\big|_{x=0} = \{\, \big(0\,,\,(|\vec{k}|,\vec{k})\big) \,\big|\, \vec{k}\in(\mathbb{R}^3\setminus\{0\}) \,\}\,.$$

We also point out that $\mathrm{WF}(\Delta_m^+)$ does not depend on m, which is a heuristic explanation for $\mathrm{WF}(D^+)=\mathrm{WF}(\Delta_m^+)$.

Exercise 2.2.1. (a) Verify explicitly that $\Delta_m^+(x)^2$ exists and that $\Delta_m^+(x)\Delta_m^+(-x)$ does not; by computing $\int d^dx\,\Delta_m^+(x)\Delta_m^+(\pm x)\,f(x)$, where $f\in\mathcal{D}(\mathbb{M})$ is arbitrary.

(b) Use the result (2.2.7) to show that each power $(\Delta_m^+)^l$, $l\in\mathbb{N}^*$, exists.
(Additional exercise for a Reader advanced in microlocal analysis: Derive formula (2.2.7).)

[*Solution*: (a) Changing the order of integrations the integral can be written as

$$\int d^dx\,\Delta_m^+(x)\Delta_m^+(\pm x)\,f(x) = \frac{1}{(2\pi)^{2d-2-d/2}}\int\frac{d^{d-1}\vec{p}}{2\omega_{\vec{p}}}\int\frac{d^{d-1}\vec{q}}{2\omega_{\vec{q}}}\,\hat{f}(\omega_{\vec{p}}\pm\omega_{\vec{q}},\vec{p}+\vec{q})\,.$$

For $\Delta_m^+(x)^2$ the integral exists, since $\hat{f}$ is fast decreasing. But for $\Delta_m^+(x)\Delta_m^+(-x)$ there is no decay in the directions $\vec{p}\approx-\vec{q}\,,\ |\vec{p}|\to\infty$.
(b) We proceed by induction on l and use Thm. A.3.3: Assuming that $(\Delta_m^+)^l$ exists and

$$\mathrm{WF}\big((\Delta^+)^l\big)\subset\mathbb{R}^d\times(\overline{V}_+\setminus\{0\})\,, \quad (2.2.8)$$

this theorem says that $(\Delta^+)^l(x)\Delta^+(x)$ exists and fulfills also (2.2.8), because $k_1,k_2\in\overline{V}_+\setminus\{0\}$ implies $(k_1+k_2)\in\overline{V}_+\setminus\{0\}$.
For a derivation of formula (2.2.7) for $\mathrm{WF}(\Delta_m^+)$ we refer to the literature, e.g., [21, Prop. 25] or [141, Vol. II, Thm. IX.48].]

From the definition (2.2.1), we see that Δ_m^+ satisfies the same *homogeneous scaling property* as Δ_m^{ret} (1.8.3):

$$\rho^{d-2}\Delta_{m/\rho}^+(\rho x) = \Delta_m^+(x). \quad (2.2.9)$$

An important property of Δ_m^+ is *positivity*:

$$\langle\overline{h},\Delta_m^+\,h\rangle = 2\pi\int d^dp\;\theta(p^0)\,\delta(p^2-m^2)\,|\hat{h}(p)|^2\geq 0\quad\forall h\in\mathcal{D}(\mathbb{M})\,, \quad (2.2.10)$$

where we use a notation introduced in (2.1.13). Positivity is necessary, in order that we can interpret the functional $\mathcal{F}\ni F\longmapsto F\big|_{\varphi=0}\in\mathbb{C}$ as (vacuum) state, as explained in Sect. 2.5, see in particular Definition 2.5.2.

A disadvantage of Δ_m^+ is that it is *not smooth* in the mass $m\geq 0$, due to the presence of the $\log(m^2)$ term in (2.2.3). However, smoothness can be reached using Hadamard functions instead.

Exercise 2.2.2 (Computation of the massless Wightman two-point function in $d=4$ dimensions). Compute the Fourier integral (2.2.1) for $D^+(x) \equiv D^{+\,(d=4)}(x)$ by using the following: It is well-known that D^+ is the limit of a function which is analytic in the forward tube $\mathbb{R}^4 - iV_+$. (This is, e.g., a consequence of the Wightman axioms [162].) Taking additionally into account the Lorentz invariance and $\operatorname{sing\,supp}(D^+) = \operatorname{sing\,supp}(\Delta_m^+) = \{x \in \mathbb{M} \,|\, x^2 = 0\}$ (see (2.2.7)), we conclude that D^+ is of the form

$$D^+(x) = \lim_{\varepsilon \downarrow 0} f(x^2 - ix^0\varepsilon) \;, \tag{2.2.11}$$

where $f(z)$ is analytic for $z \in \mathbb{C} \setminus \{0\}$. To determine the function f, it suffices to compute $D^+(x)$ for $x = (t, \vec{0})$.

[*Solution*: Since $(t - i\varepsilon, \vec{0}) \in \mathbb{R}^4 - iV_+$ for $\varepsilon > 0$, we proceed as follows, where $\omega := |\vec{p}|$:

$$\begin{aligned} D^+(t,\vec{0}) &= \lim_{\varepsilon\downarrow 0} D^+(t - i\varepsilon, \vec{0}) = \lim_{\varepsilon \downarrow 0} \frac{1}{(2\pi)^3} \int d^3\vec{p}\; \frac{e^{-i\omega(t-i\varepsilon)}}{2\omega} \\ &= \lim_{\varepsilon\downarrow 0} \frac{4\pi}{2(2\pi)^3} \int_0^\infty d\omega\; \omega\, e^{-i\omega(t-i\varepsilon)} = \lim_{\varepsilon\downarrow 0} \frac{-1}{4\pi^2}\, \frac{1}{(t^2 - it2\varepsilon)} \;. \end{aligned}$$

Writing the so-obtained expression in $\mathcal{L}_+^\uparrow$-invariant form in accordance with (2.2.11), we get the final result

$$D^+(x) = \frac{-1}{4\pi^2(x^2 - ix^0 0)} \;.] \tag{2.2.12}$$

Hadamard function. A second candidate for the propagator H_m of the star product is the so-called Hadamard function (or functions). These are needed for doing QFT on curved backgrounds, or QED with an external electromagnetic field. We shall use them in Minkowski spacetime. Our formulation of perturbative QFT given in the main text (in Chaps. 3–5) is based on quantization with the Wightman two-point function. The modifications appearing when quantizing with a Hadamard function are given in App. A.4.

To introduce the Hadamard function, let $\mu > 0$ be a mass parameter. For a given μ and with $d = 4$, the *Hadamard function* is the distribution H_m^μ defined by

$$H_m^\mu(z) := \Delta_m^+(z) - m^2\, f(m^2 z^2) \log(m^2/\mu^2) \;, \tag{2.2.13}$$

where f is the analytic function (2.2.4). Thus, the $\log(-\frac{m^2}{4}\,(z^2 - iz^0 0))$ factor in (2.2.3) is replaced by $\log(-\frac{\mu^2}{4}\,(z^2 - iz^0 0))$.

The difference $g(z) := H_m^\mu(z) - \Delta_m^+(z)$ is *even*, i.e., $g(-z) = g(z)$, so that H_m^μ also fulfills (2.1.2). Moreover, by using Bessel's differential equation one verifies $(\square + m^2) g = 0$; hence, (2.1.3) is satisfied, too. Since $g(\Lambda z) = g(z)$, we also get the Lorentz invariance (2.1.4), and $\overline{g(z)} = g(z) = g(-z)$ implies (2.1.10).

Since g is smooth in z, we find that $\mathrm{WF}(H_m^\mu) = \mathrm{WF}(\Delta_m^+)$ and thus $(H_m^\mu)^l$ also exists for each $l \in \mathbb{N}$. We obtain the nicer property that

$$H_m^\mu \text{ is smooth in } m \geq 0. \tag{2.2.14}$$

However, the scaling homogeneity is broken by a logarithmic term:

$$\rho^2 H^\mu_{m/\rho}(\rho y) - H^\mu_m(y) = 2m^2\, f(m^2 z^2) \log \rho \tag{2.2.15}$$

for $d = 4$. An even more severe disadvantage of quantization with a Hadamard function is that *positivity* (2.2.10) *may be violated*: From

$$\langle \overline{h}, H^\mu_m\, h\rangle = \langle \overline{h}, \Delta^+_m\, h\rangle - m^2 \log(m^2/\mu^2)\, f[\overline{h}, h]\ , \tag{2.2.16}$$

where

$$f[\overline{h}, h] := \int dx\, dy\; \overline{h}(x)\, f\big(m^2(x-y)^2\big)\, h(y)\ ,$$

we see that for each $h \in \mathcal{D}(\mathbb{M})$ with $f[\overline{h}, h] > 0$ [or $f[\overline{h}, h] < 0$] there exists a $\mu(h) \in (0, m)$ [or $\mu(h) \in (m, \infty)$] such that $\langle \overline{h}, H^\mu_m\, h\rangle < 0$ if and only if $0 < \mu < \mu(h)$ [or $\mu > \mu(h)$, respectively].

Finally, we get

$$\lim_{m\downarrow 0} H^\mu_m = D^+ \equiv \lim_{m\downarrow 0} \Delta^+_m \quad \text{in} \quad \mathcal{D}'(\mathbb{R}^d)\ .$$

On the matter of *uniqueness* of H^μ_m – see[17] [27, App. A], [114, Chap. III.3] and [63, Sect. IV.B] – we remark (without proof) that the listed properties determine H^μ_m, up to the parameter μ, in *even* dimensions d. More precisely, the conditions (2.1.2) to (2.1.4) and (2.1.9), (2.1.10), plus (2.2.14) and the condition that

$$\rho^{d-2} H^\mu_{m/\rho}(\rho z) \text{ is a polynomial in } \log \rho \tag{2.2.17}$$

as a replacement for (2.2.15), determine H^μ_m uniquely, up to the choice of $\mu > 0$.

In *odd* dimensions, there are crucial differences: Firstly, the Wightman two-point function Δ^+_m is smooth in the mass $m \geq 0$. There is also a one-parameter solution set of the just given conditions, which includes Δ^+_m. But, secondly, all solutions scale homogeneously, because no mass parameter μ enters here. (If one demands the stronger condition of smoothness in the squared mass m^2, not just in m, then there is a *unique* solution H_m – which is not equal to Δ^+_m.)

2.3 Equivalence of the several star products

Two products $\times_l \colon \mathcal{F} \times \mathcal{F} \longrightarrow \mathcal{F}$, where $l = 1, 2$, are called *equivalent*, if and only if there exists an algebra isomorphism $L\colon (\mathcal{F}, \times_1) \longrightarrow (\mathcal{F}, \times_2)$, that is, L is invertible, linear and intertwines the two products:

$$L(F \times_1 G) = (LF) \times_2 (LG)$$

(cf. App. A.1). We start with a preparatory exercise.

[17] The derivation of the general solution for the Hadamard function in even dimensions d, given in the first reference, contains a mistake, which is corrected in the other two references.

Exercise 2.3.1. Let us replace the distribution H_m in formula (2.1.5) by a smooth function $p \in C^\infty(\mathbb{R}^d)$, to get a "product with propagator p", denoted $F \star_p G$. With this replacement, the r.h.s. of (2.1.5) exists, because the distributions $\frac{\delta^n F}{\delta\varphi(x_1)\cdots\delta\varphi(x_n)}(h)$ and $\frac{\delta^n G}{\delta\varphi(y_1)\cdots\delta\varphi(y_n)}(h)$ ($h \in \mathcal{C}$ arbitrary) have compact support.

If $p(z) = p(-z)$, we immediately see that $\star_p$ is commutative. The aim of this exercise is to show that, in this case, $\star_p$ is equivalent to the classical product: Write

$$\Gamma_p := \frac{\hbar}{2} \int dx\, dy\, p(x-y)\, \frac{\delta^2}{\delta\varphi(x)\,\delta\varphi(y)} . \tag{2.3.1}$$

Verify that:

(i) The operator

$$e^{\Gamma_p} := \mathrm{Id} + \sum_{k=1}^{\infty} \frac{1}{k!} (\Gamma_p)^k \tag{2.3.2}$$

is invertible, with inverse $e^{-\Gamma_p}$; and

(ii) this operator intertwines the classical product with $\star_p$, i.e.,

$$F \star_p G = e^{\Gamma_p}\big((e^{-\Gamma_p} F)\cdot(e^{-\Gamma_p} G)\big). \tag{2.3.3}$$

[*Hint*: By using the notation $\frac{\hbar}{2}\langle p, \frac{\delta^2}{\delta\varphi^2}\rangle := \Gamma_p$ and (2.1.8), write the r.h.s. of (2.3.3) as

$$e^{\frac{\hbar}{2}\langle p, \frac{\delta^2}{\delta\varphi^2}\rangle} \circ \mathcal{M}\Big(e^{-\frac{\hbar}{2}\langle p, \frac{\delta^2}{\delta\varphi_1^2}\rangle} F(\varphi_1) \otimes e^{-\frac{\hbar}{2}\langle p, \frac{\delta^2}{\delta\varphi_2^2}\rangle} G(\varphi_2)\Big)\Big|_{\varphi_1=\varphi_2=\varphi} . \tag{2.3.4}$$

Then apply the Leibniz rule: $\frac{\delta}{\delta\varphi} \circ \mathcal{M} = \mathcal{M} \circ (\frac{\delta}{\delta\varphi_1} \otimes \mathrm{Id} + \mathrm{Id} \otimes \frac{\delta}{\delta\varphi_2})$.]

[*Solution*: (a) Due to $[\Gamma_p, (-\Gamma_p)] = 0$ we have $e^{\Gamma_p} e^{-\Gamma_p} = e^{\Gamma_p - \Gamma_p} = \mathrm{Id} = e^{-\Gamma_p} e^{\Gamma_p}$.
(b) The Leibniz rule and $p(-z) = p(z)$ imply

$$\Big\langle p, \frac{\delta^2}{\delta\varphi^2}\Big\rangle^n \circ \mathcal{M} = \mathcal{M} \circ \Big(\Big\langle p, \frac{\delta^2}{\delta\varphi_1^2}\Big\rangle \otimes \mathrm{Id} + \mathrm{Id} \otimes \Big\langle p, \frac{\delta^2}{\delta\varphi_2^2}\Big\rangle + 2\Big\langle p, \frac{\delta}{\delta\varphi_1} \otimes \frac{\delta}{\delta\varphi_2}\Big\rangle\Big)^n$$

for $n \in \mathbb{N}^*$. In addition, for $\Gamma_1 := \frac{\hbar}{2}\langle p, \frac{\delta^2}{\delta\varphi_1^2}\rangle$, $\Gamma_2 := \frac{\hbar}{2}\langle p, \frac{\delta^2}{\delta\varphi_1^2}\rangle$ and $\Gamma_{12} := \hbar\langle p, \frac{\delta}{\delta\varphi_1} \otimes \frac{\delta}{\delta\varphi_2}\rangle$ we may write

$$e^{\Gamma_1 \otimes \mathrm{Id} + \mathrm{Id} \otimes \Gamma_2 + \Gamma_{12}} = e^{\Gamma_{12}}\, e^{\Gamma_1 \otimes \mathrm{Id}}\, e^{\mathrm{Id} \otimes \Gamma_2} = e^{\Gamma_{12}}\,(e^{\Gamma_1} \otimes \mathrm{Id})\,(\mathrm{Id} \otimes e^{\Gamma_2}) = e^{\Gamma_{12}} \circ (e^{\Gamma_1} \otimes e^{\Gamma_2}) .$$

By using these auxiliary results, the expression (2.3.4) is equal to

$$\begin{aligned} &\mathcal{M} \circ e^{\hbar\langle p, \frac{\delta}{\delta\varphi_1} \otimes \frac{\delta}{\delta\varphi_2}\rangle} \\ &\qquad \circ \Big(e^{\frac{\hbar}{2}\langle p, \frac{\delta^2}{\delta\varphi_1^2}\rangle} \otimes e^{\frac{\hbar}{2}\langle p, \frac{\delta^2}{\delta\varphi_2^2}\rangle}\Big)\Big(e^{-\frac{\hbar}{2}\langle p, \frac{\delta^2}{\delta\varphi_1^2}\rangle} F(\varphi_1) \otimes e^{-\frac{\hbar}{2}\langle p, \frac{\delta^2}{\delta\varphi_2^2}\rangle} G(\varphi_2)\Big)\Big|_{\varphi_1=\varphi_2=\varphi} \\ &= \mathcal{M} \circ e^{\hbar\langle p, \frac{\delta}{\delta\varphi} \otimes \frac{\delta}{\delta\varphi}\rangle} (F \otimes G) = F \star_p G ; \end{aligned}$$

in the last step formula (2.1.7) is applied.]

To understand formula (2.3.3) diagrammatically first note that, for any $H \equiv H(\varphi) \in \mathcal{F}$, the term $e^{\Gamma_p} H$ is the sum over all possible contractions (with propagator p) of the entries φ in $H(\varphi)$. Applying this to the r.h.s. of (2.3.3), that is, to $H = (e^{-\Gamma_p} F) \cdot (e^{-\Gamma_p} G)$, we see that the operators $e^{-\Gamma_p}$ remove the tadpole diagrams, i.e., the contractions of F with F and of G with G, respectively. There remains the sum over all possible contractions of F with G, which is indeed equal to $F \star_p G$.

Due to property (d), a star product is non-commutative; hence, it cannot be equivalent to the classical product. However, the next exercise shows that the star products of Sect. 2.2 are equivalent to one another.

Exercise 2.3.2. Let $d = 4$. We write the notations

$$\Gamma := \hbar \int dx\, dy\, m^2\, f(m^2(x-y)^2)\, \frac{\delta^2}{\delta\varphi(x)\, \delta\varphi(y)} \quad \text{and} \quad r^\Gamma := \mathrm{Id} + \sum_{k=1}^{\infty} \frac{1}{k!} \big((\log r)\Gamma\big)^k, \tag{2.3.5}$$

where $r > 0$ and f is the analytic function (2.2.4), which appears in the definition of the Hadamard function (2.2.13). Verify that

$$F \star_{H_m^{\mu_2}} G = (\mu_2/\mu_1)^\Gamma \Big(\big((\mu_2/\mu_1)^{-\Gamma} F\big) \star_{H_m^{\mu_1}} \big((\mu_2/\mu_1)^{-\Gamma} G\big) \Big). \tag{2.3.6}$$

Thus $(\mu_2/\mu_1)^\Gamma$ is the operator intertwining the two star products. Similarly, the operator $(\mu/m)^\Gamma$ intertwines $\star_{\Delta_m^+}$ and $\star_{H_m^\mu}$, since $H_m^m = \Delta_m^+$.

[*Hint*: Use $H_m^{\mu_2}(z) - H_m^{\mu_1}(z) = 2m^2\, f(m^2 z^2) \log(\mu_2/\mu_1)$.]

[*Solution*: With the notations of the preceding exercise and $H_j := H_m^{\mu_j}$ we have

$$(\mu_2/\mu_1)^\Gamma = e^{\log\left(\frac{\mu_2}{\mu_1}\right)\Gamma} = e^{\frac{\hbar}{2}\left\langle H_2 - H_1\, ,\, \frac{\delta^2}{\delta\varphi^2}\right\rangle}.$$

With that and (2.1.7), the r.h.s. of (2.3.6) can be written as

$$e^{\frac{\hbar}{2}\langle H_2 - H_1\, ,\, \frac{\delta^2}{\delta\varphi^2}\rangle} \tag{2.3.7}$$
$$\circ\, \mathcal{M} \circ e^{\hbar\langle H_1, \frac{\delta}{\delta\varphi_1} \otimes \frac{\delta}{\delta\varphi_2}\rangle} \Big(e^{-\frac{\hbar}{2}\langle H_2 - H_1\, ,\, \frac{\delta^2}{\delta\varphi_1^2}\rangle} F(\varphi_1) \otimes e^{-\frac{\hbar}{2}\langle H_2 - H_1\, ,\, \frac{\delta^2}{\delta\varphi_2^2}\rangle} G(\varphi_2) \Big)\Big|_{\varphi_1 = \varphi_2 = \varphi}.$$

As shown in the solution of the preceding exercise, the Leibniz rule and $(H_2 - H_1)(-z) = (H_2 - H_1)(z)$ imply the relation

$$e^{\frac{\hbar}{2}\langle H_2 - H_1\, ,\, \frac{\delta^2}{\delta\varphi^2}\rangle} \circ \mathcal{M} = \mathcal{M} \circ e^{\hbar\langle H_2 - H_1\, ,\, \frac{\delta}{\delta\varphi_1} \otimes \frac{\delta}{\delta\varphi_2}\rangle} \circ \Big(e^{\frac{\hbar}{2}\langle H_2 - H_1\, ,\, \frac{\delta^2}{\delta\varphi_1^2}\rangle} \otimes e^{\frac{\hbar}{2}\langle H_2 - H_1\, ,\, \frac{\delta^2}{\delta\varphi_2^2}\rangle} \Big).$$

Inserting this relation into (2.3.7) we end up with

$$\mathcal{M} \circ e^{\hbar\langle (H_2 - H_1) + H_1\, ,\, \frac{\delta}{\delta\varphi_1} \otimes \frac{\delta}{\delta\varphi_2}\rangle} \Big(F(\varphi_1) \otimes G(\varphi_2) \Big)\Big|_{\varphi_1 = \varphi_2 = \varphi} = F \star_{H_2} G\, .]$$

Remark 2.3.3 (Time-ordered product of regular fields)**.** The distribution

$$\Delta_m^F(x) := \theta(x^0)\,\Delta_m^+(x) + \theta(-x^0)\,\Delta_m^+(-x) \tag{2.3.8}$$

is called the Feynman propagator. On the right-hand side of (2.3.8), both pointwise products of distributions exist, as shown in Exer. A.3.4. Due to $\Delta_m^+(x) = \Delta_m^+(-x)$ for $x^2 < 0$ (cf. the proof of Lemma 2.4.1), the definition (2.3.8) can be written in a manifestly Lorentz invariant way for $x \neq 0$:

$$\Delta_m^F(x) := \begin{cases} \Delta_m^+(x) & \text{for} \quad x \notin \overline{V}_- \\ \Delta_m^+(-x) & \text{for} \quad x \notin \overline{V}_+ \,. \end{cases} \tag{2.3.9}$$

Trying to introduce a "time-ordered product" by applying the definition of $\star_p : \mathcal{F} \times \mathcal{F} \longrightarrow \mathcal{F}$ (given in Exer. 2.3.1) to $p = \Delta_m^F$, we meet the problem that Δ_m^F is not smooth; the pointwise product of distributions appearing in $F \star_{\Delta_m^F} G$ does not exist for arbitrary $F, G \in \mathcal{F}$, because, e.g., powers of Δ_m^F do not exist.

A way out is to restrict $\star_{\Delta_m^F}$ to "regular fields" , which are defined as follows: The space $\mathcal{F}_{\text{reg}}$ of regular fields, is the subspace of $\mathcal{F}$ given by the additional condition that the several f_n in (1.2.1) are *smooth functions*, i.e., $f_n \in \mathcal{D}(\mathbb{M}^n)$ for all $n \geq 1$. If $F = \sum_n \langle f_n, \varphi^{\otimes n}\rangle \in \mathcal{F}_{\text{reg}}$ and $f_n \neq 0$ for some $n \geq 2$, then $F \notin \mathcal{F}_{\text{loc}}$; this is an immediate consequence of the first property in (1.3.12) of local fields.

For regular fields the "time-ordered product"

$$\tau_2 : \begin{cases} \mathcal{F}_{\text{reg}} \otimes \mathcal{F}_{\text{reg}} \longrightarrow \mathcal{F}_{\text{reg}} \\ \tau_2(F \otimes G) := F \star_{\Delta_m^F} G \,, \end{cases}$$

exists, because for $F, G \in \mathcal{F}_{\text{reg}}$ the nth functional derivatives $\frac{\delta^n F}{\delta\varphi(x_1)\cdots\delta\varphi(x_n)}$ and $\frac{\delta^n G}{\delta\varphi(y_1)\cdots\delta\varphi(y_n)}$, appearing in the general definition (2.1.5), depend smoothly on $(x_1, \ldots, x_n)$ and $(y_1, \ldots, y_n)$, respectively. The name "time-ordered product" is due to the property

$$F \star_{\Delta_m^F} G = \begin{cases} F \star_m G & \text{if} \quad \operatorname{supp} F \cap (\operatorname{supp} G + \overline{V}_-) = \emptyset \\ G \star_m F & \text{if} \quad \operatorname{supp} G \cap (\operatorname{supp} F + \overline{V}_-) = \emptyset \,, \end{cases}$$

which follows immediately from (2.3.9).

On account of $\Delta_m^F(-x) = \Delta_m^F(x)$, formula (2.3.3) applies, i.e., $\star_{\Delta^F}$ is equivalent to the classical product. Therefore, τ_2 is associative and we may introduce time-ordered products of higher orders: $\tau_n : \mathcal{F}_{\text{reg}}^{\otimes n} \longrightarrow \mathcal{F}_{\text{reg}}$ is defined by iteration of τ_2, which is equivalent to

$$\tau_n(F_1 \otimes \cdots \otimes F_n) := F_1 \star_{\Delta^F} \cdots \star_{\Delta^F} F_n = e^{\Gamma_{\Delta^F}}\Big((e^{-\Gamma_{\Delta^F}} F_1) \cdots (e^{-\Gamma_{\Delta^F}} F_n)\Big) \tag{2.3.10}$$

by using (2.3.3). So the algebraic structure of this time-ordered product τ_n is the same as that of the n-fold classical product.

However, the (renormalized) time-ordered product T_n relevant for perturbative QFT (introduced in Sect. 3.3) is a map $T_n \colon \mathcal{F}_{\mathrm{loc}}^{\otimes n} \longrightarrow \mathcal{F}$. Therefore, T_2 cannot be associative, T_n cannot be obtained as an iteration of T_2, and T_n cannot be equivalent to the n-fold classical product. In addition, for local entries $F_1, F_2 \in \mathcal{F}_{\mathrm{loc}}$, the expression $F_1 \star_{\Delta^F} F_2$ does not exist in general, for the same reason as explained above for $F_1, F_2 \in \mathcal{F}$ – renormalization is the removal of these divergences.

Nevertheless, a way to understand the renormalized time-ordered product T_n as an iteration of a binary product defined on a suitable domain, which is equivalent to the classical product (and hence associative and commutative), is given in [74].

2.4 Existence and properties of the star product

The existence of the general formula (2.1.5) for $F \star G$ amounts to the following problem (by using Definition 1.3.1 of the functional derivative): Writing $F = \sum_k \langle f_k, \varphi^{\otimes k}\rangle$ and $G = \sum_l \langle g_l, \varphi^{\otimes l}\rangle$ with f_k and g_l as in Definition 1.2.1, we have to show that

$$t(z_1,\ldots,z_{k+l-2n}) := \mathfrak{S} \int dx_1 \cdots dx_n \, dy_1 \cdots dy_n \; f_k(x_1,\ldots,x_n,z_1,\ldots,z_{k-n}) \\ \cdot \prod_{j=1}^{n} H(x_j - y_j)\, g_l(y_1,\ldots,y_n,z_{k-n+1},\ldots,z_{k+l-2n}) \tag{2.4.1}$$

($\mathfrak{S}$ denotes symmetrization in $z_1,\ldots,z_{k+l-2n}$), which contains pointwise products of distributions, exists in $\mathcal{D}'(\mathbb{M}^{k+l-2n})$. In addition, to confirm that $F \star G$ lies again in $\mathcal{F}[\![\hbar]\!]$, we have to verify that $t(z_1,\ldots,z_{k+l-2n})$ fulfills the wave front set condition (1.2.2). Both claims can be proved by a generalization of Hörmander's Theorem A.3.3, using the wave front set property (1.2.2) of both f_k and g_l and the analogous property (2.2.7) of H_m. For details we refer to [23].

It remains to prove the associativity (b); this is more involved, we follow [165, Sect. 6.2.4].

Proof. (Associativity of the $\star$-product) We will use the elegant way (2.1.7) of writing the $\star$-product. The strategy of the proof is to trace back the associativity of the $\star$-product to the associativity of the classical product $\mathcal{M}$ (2.1.8), which can be written as

$$\mathcal{M} \circ (\mathrm{id} \otimes \mathcal{M}) = \mathcal{M} \circ (\mathcal{M} \otimes \mathrm{id}) \; : \; \mathcal{F}^{\otimes 3} \longrightarrow \mathcal{F} \, . \tag{2.4.2}$$

Besides the differential operator $D_H : \mathcal{F}^{\otimes 2} \longrightarrow \mathcal{F}^{\otimes 2}$ (2.1.7), we will work with the related operators $D_H^{kl} : \mathcal{F}^{\otimes 3} \longrightarrow \mathcal{F}^{\otimes 3}$, $1 \le k < l \le 3$, which are defined by

$$D_H^{12} := D_H \otimes \mathrm{id} = \int dx\, dy\, H(x-y)\, \frac{\delta}{\delta\varphi(x)} \otimes \frac{\delta}{\delta\varphi(y)} \otimes \mathrm{id}$$

and analogously

$$D_H^{23} := \mathrm{id} \otimes D_H = \int dx\, dy\, H(x-y)\; \mathrm{id} \otimes \frac{\delta}{\delta\varphi(x)} \otimes \frac{\delta}{\delta\varphi(y)}$$

$$D_H^{13} := \int dx\, dy\, H(x-y)\; \frac{\delta}{\delta\varphi(x)} \otimes \mathrm{id} \otimes \frac{\delta}{\delta\varphi(y)}\;;$$

the upper indices (k,l) indicate at which position in the 3-fold tensor product the functional derivatives are. One easily verifies that the operators D_H^{kl} commute:

$$[D_H^{kl}, D_H^{rs}] = 0\;, \quad \forall 1 \leq k < l \leq 3\;,\; 1 \leq r < s \leq 3\;. \tag{2.4.3}$$

We verify the associativity by a direct calculation:

$$\begin{aligned} F \star (G \star J) &= \mathcal{M} \circ e^{\hbar D_H}\Big(F \otimes \big(\mathcal{M} \circ e^{\hbar D_H}(G \otimes J)\big)\Big) \\ &= \mathcal{M} \circ e^{\hbar D_H} \circ \big(\mathrm{id} \otimes \mathcal{M}\big) \circ \big(\mathrm{id} \otimes e^{\hbar D_H}\big)(F \otimes G \otimes J) \\ &= \mathcal{M} \circ e^{\hbar D_H} \circ \big(\mathrm{id} \otimes \mathcal{M}\big) \circ e^{\hbar D_H^{23}}(F \otimes G \otimes J)\;; \end{aligned} \tag{2.4.4}$$

in the second equality sign it is used that $\mathrm{id} \otimes (D_H)^k = (D_H^{23})^k$ for $k \in \mathbb{N}$.

Next we note that

$$\begin{aligned} &D_H \circ \big(\mathrm{id} \otimes \mathcal{M}\big)(K \otimes L \otimes P) = D_H(K \otimes LP) \\ &= \int dx\, dy\, H(x-y)\,\Big(\frac{\delta K}{\delta\varphi(x)} \otimes \big(\frac{\delta L}{\delta\varphi(y)}P + L\frac{\delta P}{\delta\varphi(y)}\big)\Big) \\ &= \big(\mathrm{id} \otimes \mathcal{M}\big) \circ \big(D_H^{12} + D_H^{13}\big)(K \otimes L \otimes P)\;. \end{aligned}$$

Iterated use of this identity gives $D_H^k \circ \big(\mathrm{id} \otimes \mathcal{M}\big) = \big(\mathrm{id} \otimes \mathcal{M}\big) \circ \big(D_H^{12} + D_H^{13}\big)^k$ for all $k \in \mathbb{N}$, which implies

$$e^{\hbar D_H} \circ \big(\mathrm{id} \otimes \mathcal{M}\big) = \big(\mathrm{id} \otimes \mathcal{M}\big) \circ e^{\hbar(D_H^{12} + D_H^{13})}\;.$$

Inserting the latter formula into (2.4.4) and using that the operators D_H^{kl} commute, we obtain

$$\begin{aligned} F \star (G \star J) &= \mathcal{M} \circ \big(\mathrm{id} \otimes \mathcal{M}\big) \circ e^{\hbar(D_H^{12} + D_H^{13})} \circ e^{\hbar D_H^{23}}(F \otimes G \otimes J) \\ &= \mathcal{M} \circ \big(\mathrm{id} \otimes \mathcal{M}\big) \circ e^{\hbar(D_H^{12} + D_H^{13} + D_H^{23})}(F \otimes G \otimes J)\;. \end{aligned} \tag{2.4.5}$$

By an analogous calculation one finds

$$(F \star G) \star J = \mathcal{M} \circ \big(\mathcal{M} \otimes \mathrm{id}\big) \circ e^{\hbar(D_H^{12} + D_H^{13} + D_H^{23})}(F \otimes G \otimes J)\;. \tag{2.4.6}$$

Due to associativity of the classical product (2.4.2), the expressions (2.4.5) and (2.4.6) agree. □

By using associativity of the $\star$-product, an alternative, very simple proof of the Jacobi identity for the Poisson bracket (part (ii) of Prop. 1.8.5) can be given.

Proof. (Jacobi identity for the Poisson bracket) Let $F, G, H \sim \hbar^0$. Since the $\star$-product is associative, its commutator

$$[F, G]_\star := F \star G - G \star F$$

fulfills the Jacobi identity:

$$[F, [G, H]_\star]_\star + [H, [F, G]_\star]_\star + [G, [H, F]_\star]_\star = 0 \, ; \tag{2.4.7}$$

this is obtained by writing out the 12 terms appearing in the Jacobi identity. From property (d) of the $\star$-product we conclude

$$[F, [G, H]_\star]_\star = -\hbar^2 \{F, \{G, H\}\} + \mathcal{O}(\hbar^3) \, . \tag{2.4.8}$$

Hence, the lowest non-vanishing order of the Jacobi identity for the $\star$-product is the Jacobi identity for the Poisson bracket. □

Lemma 2.4.1 ("Spacelike commutativity"). *If* $(x-y)^2 < 0$ *for all* $x \in \operatorname{supp} F$ *and* $y \in \operatorname{supp} G$, *then* $[F, G]_\star = 0$.

Proof. Notice that $H_m(z) - H_m(-z) = i\,\Delta_m(z) = 0$ for $z^2 < 0$; thus

$$\prod_{l=1}^n H_m(x_l - y_l) - \prod_{l=1}^n H_m(y_l - x_l) = 0 \quad \text{if} \quad (x_l - y_l)^2 < 0 \text{ for each } l.$$

The assertion follows by using (2.1.5). □

Remark 2.4.2. Notice that the space $\mathcal{F} \equiv \mathcal{F}[\![\hbar]\!]$ contains no reference to m, but the *algebra* $\mathcal{A}^{(m)} := (\mathcal{F}, \star_m)$ does indeed depend on m. Frequently, $\mathcal{A}^{(m)}$ is called the "algebra of observables".

Furthermore, if $\mathcal{O}$ is any open region in Minkowski spacetime, and if F and G belong to $\mathcal{F}(\mathcal{O}) := \{ F \in \mathcal{F} \mid \operatorname{supp} F \subset \mathcal{O} \}$, then $F \star G \in \mathcal{F}(\mathcal{O})$, too. This follows directly from $\operatorname{supp} F := \operatorname{supp} \delta F/\delta\varphi(\cdot)$ and the definition of the $\star$-product (2.1.5): Writing the latter as $F \star G := \sum_{n=0}^\infty \frac{\hbar^n}{n!} \langle F^{(n)}, H^{\otimes n} G^{(n)} \rangle$, we obtain

$$\begin{aligned} \operatorname{supp} \langle F^{(n)}, H^{\otimes n} G^{(n)} \rangle = \operatorname{supp}_x \Big(\Big\langle \Big(\frac{\delta F}{\delta\varphi(x)} \Big)^{(n)}, H^{\otimes n} G^{(n)} \Big\rangle \\ + \Big\langle F^{(n)}, H^{\otimes n} \Big(\frac{\delta G}{\delta\varphi(x)} \Big)^{(n)} \Big\rangle \Big) \\ \subseteq \operatorname{supp} F \cup \operatorname{supp} G \subset \mathcal{O} \, . \end{aligned}$$

Thus $\mathcal{A}^{(m)}(\mathcal{O}) := \big(\mathcal{F}(\mathcal{O}), \star_m \big)$ is also an algebra: Everything may be localized to the region $\mathcal{O}$. And $\mathcal{A}^{(m)}(\mathcal{O})$ is called the "algebra of observables localized in $\mathcal{O}$".

2.5 States

If the quantum observables are represented by linear operators on a Hilbert space (as, e.g., in Schrödingers formulation of quantum mechanics or in a Fock space description of QFT), one usually understands by a "state" a normalized vector of this Hilbert space. Here we do not assume the existence of such a representation and, hence, we generalize this definition of a state, by following the procedure in Algebraic QFT [94] (cf. Sect. 3.7).

In Sects. 2.5 and 2.6 we somewhat change the framework: Instead of dealing with formal power series in $\hbar$, we work with *polynomials in* $\hbar$. So we interpret the star product as a map $\star : \mathcal{F}_\hbar \times \mathcal{F}_\hbar \longrightarrow \mathcal{F}_\hbar$ (see (2.1.6)), where $\hbar > 0$ is a *fixed* parameter.

A *state* is a linear functional on the algebra of observables – for the model of a real scalar field this is the algebra $(\mathcal{F}_\hbar, \star)$ – with certain properties; in detail we define:

Definition 2.5.1 (State). A *state* ω on the algebra $\mathcal{A} \equiv \mathcal{A}_\hbar := (\mathcal{F}_\hbar, \star)$ is a map

$$\omega : \begin{cases} \mathcal{A} \longrightarrow \mathbb{C} \\ F \longmapsto \omega(F) \equiv \omega(F)_\hbar \ , \end{cases}$$

which itself may be a polynomial in $\hbar$, i.e., $\omega = \sum_{s=0}^{S} \omega_s \, \hbar^s$ for some $S < \infty$, and which is

$$\begin{aligned} &\text{(i) linear} \quad \omega(F + \alpha G) = \omega(F) + \alpha\, \omega(G) \ , \quad \forall F, G \in \mathcal{A} \ , \ \alpha \in \mathbb{C} \ , \\ &\text{(ii) real} \quad \omega(F^*)_\hbar = \overline{\omega(F)}_\hbar \ , \quad \forall F \in \mathcal{A} \ , \quad \forall \hbar > 0 \ , \\ &\text{(iii) positive} \quad \omega(F^* \star F)_\hbar \geq 0 \ , \quad \forall F \in \mathcal{A} \ , \quad \forall \hbar > 0 \ , \\ &\text{(iv) normalized} \quad \omega(1) = 1 \ , \end{aligned} \tag{2.5.1}$$

where $1 \in \mathcal{F}_\hbar$ is the functional $1(h) = 1 \ \forall h \in \mathcal{C}$.

In particular, if $F^* = F$ we have $\omega(F) \in \mathbb{R}$. Linearity implies that

$$\omega(F) \equiv \omega\Big(\sum_r F_r \, \hbar^r\Big) = \sum_{r,s} \omega_s(F_r) \, \hbar^{r+s} \ . \tag{2.5.2}$$

So $\omega(F) \equiv \omega(F)_\hbar$ is a polynomial in $\hbar$. We point out that this is a true polynomial, that is, the sum on the r.h.s. of (2.5.2) is an ordinary sum of complex numbers. Trying to generalize Definition 2.5.1 to $(\mathcal{F}[\![\hbar]\!], \star)$, the values $\omega(F)$ would be elements of $\mathbb{C}[\![\hbar]\!]$, that is, the sum on the r.h.s. of (2.5.2) would be a *formal sum*; hence, the property (iii) would require an appropriate definition of positivity in $\mathbb{C}[\![\hbar]\!]$ or $\mathbb{R}[\![\hbar]\!]$ – see Remk. 5.4.7. In this chapter – with regard to the Fock space representation of the on-shell fields (treated in Sect. 2.6) – we do not follow this path, however see (5.4.28).

A simple, but important, example is the *vacuum state*.

Definition 2.5.2 (Vacuum state)**.** When quantizing with the Wightman two-point function Δ_m^+, the *vacuum state* $\omega_0 \colon \mathcal{A} \longrightarrow \mathbb{C}$ is defined by setting $\omega_0(F) := f_0$, where $F = f_0 + \sum_{n\geq 1}\langle f_n, \varphi^{\otimes n}\rangle$.

Clearly, ω_0 is *linear* and *normalized.* Moreover, it is *real*: $\omega_0(F^*) = \overline{f_0} = \overline{\omega_0(F)}$. The verification of the *positivity* of ω_0 is more involved, because in the sum

$$\omega_0(F^* \star F) = \overline{f_0}\, f_0 + \sum_{n\geq 1} \omega_0\Big(\langle \overline{f_n}, \varphi^{\otimes n}\rangle \star \langle f_n, \varphi^{\otimes n}\rangle\Big)$$

there are in general non-vanishing contributions from the terms $n \geq 1$. A simple proof of the positivity can be given by using the representation of $\mathcal{A}|_{\mathcal{C}_{S_0}}$ in Fock space, treated in the next section. Hence, we postpone this verification to Remk. 2.6.6.

We point out that ω_0 is in general not positive, if we quantize with a Hadamard function:

$$\omega_0\big(\varphi(g)^* \star_{H^\mu_m} \varphi(g)\big) = \hbar\, \langle \overline{g}, H^\mu_m g\rangle \quad \text{may be } < 0$$

(where $\varphi(g) := \int dx\; \varphi(x)\, g(x)$ for $g \in \mathcal{D}(\mathbb{M})$), as noted in (2.2.16). However, for quantization with Δ_m^+ this problem does not appear, see (2.2.10).

Remark 2.5.3 (Coherent states)**.** Instead of evaluation at $\varphi = 0$ (vacuum state), we can also evaluate at $\varphi = h$, where h is a solution of the free field equation: $h \in \mathcal{C}_{S_0}$. This defines a "coherent state":[18]

$$\omega_h : \begin{cases} \mathcal{A} \longrightarrow \mathbb{C} \\ F \longmapsto \omega_h(F) := F(h) \end{cases} \qquad \text{for} \quad h \in \mathcal{C}_{S_0}\ . \tag{2.5.3}$$

Similarly to the vacuum state ω_0, it is obvious that ω_h is linear, normalized and real; e.g.,

$$\omega_h(F^*) = F^*(h) = \sum_n \langle \overline{f_n}, h^{\otimes n}\rangle = \sum_n \overline{\langle f_n, h^{\otimes n}\rangle} = \overline{\omega_h(F)}\ ,$$

since h is $\mathbb{R}$-valued. Again, the proof of positivity is more difficult and can be given by means of the Fock space representation – see Remk. 2.6.6; it is this proof which needs that h solves the free field equation.

Exercise 2.5.4. Compute

(a) the Wightman functions of the basic field

$$W_n(x_1, \ldots, x_n) := \omega_0\big(\varphi(x_1) \star_\hbar \cdots \star_\hbar \varphi(x_n)\big)\ ,$$

(b) and $\omega_0\big(\varphi^3(x_1) \star \varphi^3(x_2) \star \varphi^2(x_3)\big)$.

[18] Klaus Fredenhagen pointed this out to the author.

[*Solution*: (a) In the vacuum state only a term with all $\varphi(x_j)$ contracted does not vanish; hence $W_n = 0$ for n odd. For n even it results

$$W_n(x_1,\dots,x_n) = \hbar^{n/2} \sum \Delta^+(x_{i_1}-x_{i_2})\,\Delta^+(x_{i_3}-x_{i_4})\dots\Delta^+(x_{i_{n-1}}-x_{i_n}) ,$$

where the sum runs over all partitions of $\{1,\dots,n\}$ into $n/2$ disjoint pairs $\{i_1,i_2\}$, $\{i_3,i_4\}$, $\dots,\{i_{n-1},i_n\}$ with $i_{2k-1} < i_{2k}$.
(b) There is only one possibility to contract all $\varphi(x_j)$, namely

$$36\,\hbar^4\,\Delta^+(x_1-x_2)^2\,\Delta^+(x_1-x_3)\,\Delta^+(x_2-x_3) \text{ .]}$$

Exercise 2.5.5. Let $H \in \mathcal{F}_\hbar$ with $\omega_0(H^* \star H) = 1$ be given, where ω_0 is the vacuum state. Verify that

$$\omega_H : \begin{cases} \mathcal{A} \longrightarrow \mathbb{C} \\ F \longmapsto \omega_H(F) := \omega_0(H^* \star F \star H) , \end{cases}$$

is a state.

[*Solution*: Obviously, ω_H is linear and normalized. That ω_H is real and positive follows from (2.1.10):

$$\begin{aligned} \omega_H(F^*) &= \omega_0(H^* \star F^* \star H) = \omega_0\big((H^* \star F \star H)^*\big) = \overline{\omega_0(H^* \star F \star H)} = \overline{\omega_H(F)} \ ; \\ \omega_H(F^* \star F) &= \omega_0(H^* \star F^* \star F \star H) = \omega_0\big((F \star H)^* \star (F \star H)\big) \geq 0 \text{ .]} \end{aligned}$$

Vector states given by a Hilbert space representation. Let a nontrivial representation ρ of $\mathcal{A}$ in a (pre) Hilbert space $(\mathcal{H}, \langle\cdot,\cdot\rangle_{\mathcal{H}})$ be given. In detail we assume that a dense subspace D of $\mathcal{H}$ and a linear map

$$\rho : \mathcal{A} \longrightarrow \{L : D \to D \mid L \text{ is linear}\}$$

are given; in particular note that $\rho(F)D \subseteq D$, $\forall F \in \mathcal{A}$. In addition, ρ satisfies the relations

$$\rho(1) = \mathrm{Id} \ , \qquad \rho(F \star G) = \rho(F)\,\rho(G) \qquad \forall F, G \in \mathcal{A}$$

(where on the r.h.s. we have the operator product) and respects the $*$-operation:

$$\big\langle \psi_1, \rho(F^*)\psi_2 \big\rangle_{\mathcal{H}} = \big\langle \rho(F)\psi_1, \psi_2 \big\rangle_{\mathcal{H}} \qquad \forall \psi_1,\psi_2 \in D \ , \ \ F \in \mathcal{A} \ .$$

Moreover, we assume that the elements $\psi \equiv \psi_\hbar$ of D are polynomials in $\hbar$; Hilbert space positivity means that $\langle \psi_\hbar, \psi_\hbar \rangle \geq 0$ for all $\psi_\hbar \in D$ and $\hbar > 0$.

With these assumptions, any *normalized* vector $\psi \in D$ induces a state on $\mathcal{A}$ by the definition

$$\omega_\psi(F) := \big\langle \psi, \rho(F)\psi \big\rangle \ , \quad F \in \mathcal{A} \ . \tag{2.5.4}$$

Obviously, ω_ψ is linear and normalized; it is also real and positive, because

$$\begin{aligned} \omega_\psi(F^*) &= \big\langle \psi, \rho(F^*)\psi \big\rangle = \big\langle \rho(F)\psi, \psi \big\rangle = \overline{\omega_\psi(F)} \ , \\ \omega_\psi(F^* \star F) &= \big\langle \psi, \rho(F^*)\rho(F)\psi \big\rangle = \big\langle \rho(F)\psi, \rho(F)\psi \big\rangle \geq 0 \ . \end{aligned}$$

2.6 Connection to the algebra of Wick polynomials

We now consider the connection of the star product algebra with the algebra of Wick polynomials in Fock space (with the operator product). We assume that the reader is familiar with the latter algebra. A short introduction to the Fock space is given in App. A.5; to simplify the notation we write $\mathfrak{F}$ (instead of $\mathfrak{F}^+$) for the bosonic Fock space.

Comparing on a heuristic level the algebraic structure of Wick polynomials with the star product algebra, we observe the following.

(i) There is a crucial difference: The basic field in Fock space obeys the free field equation: $(\Box + m^2)\varphi^{\mathrm{op}}(x) = 0$; but our field $\varphi(x)$ is unrestricted.

(ii) The operator product in Fock space behaves as the star product: It is non-commutative and associative.

(iii) Let F^{op}, G^{op} be Wick polynomials integrated out with a test function. Neglecting questions of domains of Fock space operators, the map $F^{\mathrm{op}} \longmapsto F^{\mathrm{op}\,*}$ (here, "$*$" denotes the adjoint operator) corresponds to the $*$-operation in $(\mathcal{F}, \star)$: Both are involutions (see App. A.1), in particular, $(F^{\mathrm{op}}\, G^{\mathrm{op}})^* = G^{\mathrm{op}\,*}\, F^{\mathrm{op}\,*}$ corresponds to $(F \star G)^* = G^* \star F^*$.

(iv) The normally ordered product (A.5.21)–(A.5.22) behaves as the classical product: It is commutative and associative.

The aim of this section is to make these observations precise.

To take into account the difference (i), we will work in the following with the space of *on-shell fields* with respect to the free action S_0,

$$\mathcal{F}_0^{(m)} := \{\, F_0 := F\big|_{\mathfrak{C}_{S_0}} \,\big|\, F \in \mathcal{F} \,\}\,, \tag{2.6.1}$$

where $\mathfrak{C}_{S_0}$ is the space of solutions of the free field equation.

For $F, G \in \mathcal{F}$, we see that $F_0 = G_0$ if and only if each field monomial in $(F - G)$ has a factor $(\Box + m^2)\varphi(x_j)$ for some $1 \le j \le n$, that is, $(F - G)$ lies in

$$\mathcal{J}^{(m)} := \Big\{ \sum_{n=1}^{N} \int dx_1 \cdots dx_n\, \varphi(x_1) \cdots \varphi(x_n) \big(\Box_{x_1} + m^2\big) f_n(x_1, \ldots, x_n) \;\Big|\; f_n \in \mathcal{F}'(\mathbb{M}^n) \Big\}\,, \tag{2.6.2}$$

where $\mathcal{F}'(\mathbb{M}^n)$ is given in Definition 1.2.1. In particular we use that f_n is symmetric in its arguments. Obviously, $\mathcal{J}^{(m)}$ is an ideal (more precisely, a "2-sided ideal") of the algebra $(\mathcal{F}, \cdot)$, that is, $\mathcal{J}^{(m)}$ is a subalgebra of $(\mathcal{F}, \cdot)$ satisfying $\mathcal{F} \cdot \mathcal{J}^{(m)} = \mathcal{J}^{(m)} = \mathcal{J}^{(m)} \cdot \mathcal{F}$. We call $\mathcal{J}^{(m)}$ the "ideal of the free field equation". Due to these results, we can also describe the on-shell fields by using the quotient space $\mathcal{F}/\mathcal{J}^{(m)}$:

The map[19]

$$Q: \begin{cases} \mathcal{F}/\mathcal{J}^{(m)} \longrightarrow \mathcal{F}_0^{(m)} \\ F + \mathcal{J}^{(m)} \longmapsto F_0 \end{cases} \qquad \text{is an algebra isomorphism} \tag{2.6.3}$$

with respect to the classical product on both sides:

$$\begin{aligned} Q\big((F + \mathcal{J}^{(m)}) \cdot (G + \mathcal{J}^{(m)})\big) &= Q(FG + \mathcal{J}^{(m)}) = (FG)_0 = F_0 \cdot G_0 \\ &= Q(F + \mathcal{J}^{(m)}) \cdot Q(G + \mathcal{J}^{(m)}) \,. \end{aligned}$$

All these results remain true when we replace $\mathcal{F}$, $\mathcal{F}_0^{(m)}$ and $\mathcal{J}^{(m)}$ by $\mathcal{F}[\![\hbar]\!]$, $\mathcal{F}_0^{(m)}[\![\hbar]\!]$ and $\mathcal{J}^{(m)}[\![\hbar]\!]$, respectively. Note that

$$\mathcal{F}_0^{(m)}[\![\hbar]\!] := \Big\{ \sum_n F_n\, \hbar^n \,\Big|\, F_n \in \mathcal{F}_0^{(m)} \Big\} = \Big\{ F\big|_{\mathcal{C}_{S_0}} \,\Big|\, F \in \mathcal{F}[\![\hbar]\!] \Big\} =: \mathcal{F}[\![\hbar]\!]\big|_{\mathcal{C}_{S_0}} \,.$$

To introduce a star product on $\mathcal{F}_0^{(m)}[\![\hbar]\!] \simeq \mathcal{F}[\![\hbar]\!]/\mathcal{J}^{(m)}[\![\hbar]\!])$, we must check the following assertion.

Claim 2.6.1. *If $F \in \mathcal{F}[\![\hbar]\!]$ and $J \in \mathcal{J}^{(m)}[\![\hbar]\!]$, then $J \star F \in \mathcal{J}^{(m)}[\![\hbar]\!]$ and $F \star J \in \mathcal{J}^{(m)}[\![\hbar]\!]$.*

Proof. Any such J contains a factor $(\Box + m^2)\varphi(x_j)$, so that for the terms in $J \star F$ we get either

$$\int dx_i \cdots \int dy_1 \cdots (\Box_{x_j} + m^2)\varphi(x_j) \cdots \prod_{l=1}^{n} H(x_l - y_l)\, \frac{\delta^n F}{\delta\varphi(y_1) \cdots \delta\varphi(y_n)}$$

which lies in $\mathcal{J}^{(m)}$; or, if $(\Box + m^2)\varphi(x_j)$ is contracted,

$$\int dx_i \cdots \int dy_1 \cdots (\Box_{x_j} + m^2) H(x_j - y_j) \prod_{l \neq j} H(x_l - y_l)\, \frac{\delta^n F}{\delta\varphi(y_1) \cdots \delta\varphi(y_n)} \,,$$

which vanishes, since $(\Box_{x_j} + m^2) H(x_j - y_j) = 0$. □

Remark 2.6.2. An immediate consequence of this claim is that $\mathcal{J}^{(m)}$ is a *Poisson ideal* of $\mathcal{F}$ (cf. [29, Proposition 8]); to wit,

$$F \in \mathcal{F}\,,\ J \in \mathcal{J}^{(m)} \implies \{F, J\} \in \mathcal{J}^{(m)} \ \wedge\ \{J, F\} \in \mathcal{J}^{(m)} \,.$$

To see this, we recall that $\{F, J\} = -\{J, F\} = \lim_{\hbar \to 0}(F \star J - J \star F)/i\hbar$.

[19] The definition of an "algebra isomorphism" is given in App. A.1.

Due to Claim 2.6.1, the star product on $\mathcal{F}[\![\hbar]\!]$ induces a well-defined product on $\mathcal{F}_0^{(m)}[\![\hbar]\!] \simeq \mathcal{F}[\![\hbar]\!]/\mathcal{J}^{(m)}[\![\hbar]\!]$, also denoted by $\star\colon \mathcal{F}_0^{(m)}[\![\hbar]\!] \times \mathcal{F}_0^{(m)}[\![\hbar]\!] \to \mathcal{F}_0^{(m)}[\![\hbar]\!]$, given by

$$F_0 \star G_0 := (F \star G)_0 \quad \text{or} \quad (F + \mathcal{J}^{(m)}[\![\hbar]\!]) \star (G + \mathcal{J}^{(m)}[\![\hbar]\!]) := (F \star G) + \mathcal{J}^{(m)}[\![\hbar]\!] \,. \tag{2.6.4}$$

So Q is an algebra isomorphism also with respect to the star product.

For the remainder of this section we restrict the fields to be polynomials in $\hbar$, that is, we replace $\mathcal{F}[\![\hbar]\!]$ by $\mathcal{F}_\hbar$ (2.1.6). All equations are valid for all $\hbar > 0$, that is, they hold individually to each order of $\hbar$.

The following theorem gives the connection of our description of "fields" as functionals on the classical configuration space with the conventional description as Fock space operators.

Theorem 2.6.3 (Identification of on-shell fields with Fock space operators, [51, 52]). *Let $\varphi^{\mathrm{op}}(x)$ be the free, real scalar field (for a given mass m) on the Fock space $\mathfrak{F}$ and let $\mathcal{F}_{0,\hbar}^{(m)} := \mathcal{F}_\hbar\big|_{\mathcal{C}_{S_0}}$. Then the map*

$$\Phi\colon \mathcal{F}_{0,\hbar}^{(m)} \longrightarrow \Phi(\mathcal{F}_{0,\hbar}^{(m)}) \subset \{\textit{linear operators on } \mathfrak{F}\} \tag{2.6.5}$$

given by

$$\begin{aligned} F_0 &= \sum_{n=0}^{N} \int dx_1 \cdots dx_n \, \varphi_0(x_1) \cdots \varphi_0(x_n) \, f_n(x_1, \ldots, x_n) \\ &\longmapsto \Phi(F_0) = \sum_{n=0}^{N} \int dx_1 \cdots dx_n \, {:}\varphi^{\mathrm{op}}(x_1) \cdots \varphi^{\mathrm{op}}(x_n){:} \, f_n(x_1, \ldots, x_n) \end{aligned}$$

(where $:\!-\!:$ denotes normal ordering of Fock space operators (A.5.21)*) is an algebra isomorphism*

$$F_0 \star G_0 \longmapsto \Phi(F_0 \star G_0) = \Phi(F_0)\, \Phi(G_0)$$

for the star product on the left and the operator product on the right, which respects the $*$-operation:

$$\big\langle \psi_1, \Phi(F_0^*)\psi_2 \big\rangle_{\mathfrak{F}} = \big\langle \Phi(F_0)\psi_1, \psi_2 \big\rangle_{\mathfrak{F}} \qquad \forall F_0 \in \mathcal{F}_{0,\hbar}^{(m)} \tag{2.6.6}$$

and for all ψ_1, ψ_2 in the domain of $\Phi(F_0)$ or $\Phi(F_0^*)$, respectively. The same map Φ also gives an algebra isomorphism

$$F_0 \cdot G_0 \longmapsto \Phi(F_0 \cdot G_0) = {:}\Phi(F_0)\, \Phi(G_0){:}$$

with the classical product on the left and the normally ordered product on the right (A.5.22).

This theorem holds also for other kinds of fields and the pertinent Fock spaces; in particular for the complex scalar field (Example 1.3.2), gauge fields (Sect. 5.1.3) and also for fermionic fields as, e.g., Dirac spinors (Sect. 5.1.1) and Faddeev–Popov ghosts (Sect. 5.1.2). In Sect. 5.5 we will use this theorem for QED.

The *proof* relies on the observations

- that our definition of the classical product $F_0 \cdot G_0$ on $\mathcal{F}_{0,\hbar}^{(m)}$ (formula (1.2.6) restricted to $\mathcal{C}_{S_0}$) agrees under the bijection Φ with the definition of the normally ordered product $:\!\Phi(F_0)\,\Phi(G_0)\!:$ (A.5.23),
- and that our definition of the star product on $\mathcal{F}_{0,\hbar}^{(m)}$ (equations (2.1.5) and (2.6.4)) agrees under the bijection Φ with the version of Wick's Theorem given in (A.5.33).

This theorem yields a faithful representation of the algebra $\mathcal{A}_0^{(m)} := (\mathcal{F}_{0,\hbar}^{(m)}, \star_m)$ of on-shell fields – the "Fock space representation". That is, the algebra $\mathcal{A}_0^{(m)}$ describes uniquely the algebra of *smeared Wick products* on Fock space.[20]

Example 2.6.4. As an illustration of this theorem we verify

$$\Phi\big(\varphi_0(x_1)\,\varphi_0^2(x_2)\star\varphi_0^2(y)\big) \overset{?}{=} \Phi\big(\varphi_0(x_1)\,\varphi_0^2(x_2)\big)\,\Phi\big(\varphi_0^2(y)\big)\ .$$

By using the result of Exap. 2.1.2, we obtain for the l.h.s.:

$$\begin{aligned}
&\Phi\big(\varphi_0(x_1)\varphi_0^2(x_2)\varphi_0^2(y)\big)\\
&+\hbar\big[2\,\Phi\big(\varphi_0^2(x_2)\varphi_0(y)\big)\,\Delta_m^+(x_1-y)+4\,\Phi\big(\varphi_0(x_1)\varphi_0(x_2)\varphi_0(y)\big)\,\Delta_m^+(x_2-y)\big]\\
&+2\hbar^2\big[\Phi\big(\varphi_0(x_2)\big)\,\Delta_m^+(x_1-y)\,\Delta_m^+(x_2-y)+\Phi\big(\varphi_0(x_1)\big)\,\Delta_m^+(x_2-y)^2\big]\\
=\ &:\!\varphi^{\mathrm{op}}(x_1)\varphi^{\mathrm{op}}(x_2)^2\varphi^{\mathrm{op}}(y)^2\!:\\
&+\hbar\big[2:\!\varphi^{\mathrm{op}}(x_2)^2\varphi^{\mathrm{op}}(y)\!:\Delta_m^+(x_1-y)+4:\!\varphi^{\mathrm{op}}(x_1)\varphi^{\mathrm{op}}(x_2)\varphi^{\mathrm{op}}(y)\!:\Delta_m^+(x_2-y)\big]\\
&+2\hbar^2\big[\varphi^{\mathrm{op}}(x_2)\,\Delta_m^+(x_1-y)\,\Delta_m^+(x_2-y)+\varphi^{\mathrm{op}}(x_1)\,\Delta_m^+(x_2-y)^2\big]\ .
\end{aligned}$$

The r.h.s. is equal to

$$:\!\varphi^{\mathrm{op}}(x_1)\varphi^{\mathrm{op}}(x_2)^2\!:\ :\!\varphi^{\mathrm{op}}(y)^2\!:\ ;$$

by applying Wick's Theorem (A.5.33) we get exactly the same result as for the l.h.s..

Expectation values in states. Under this isomorphism, the state ω_0 corresponds to the *Fock vacuum state* $\langle\Omega\,|\cdot|\,\Omega\rangle$ (see (A.5.6) for the definition of $\Omega\in\mathfrak{F}$), that is,[21]

$$\omega_0(F_0) := \omega_0(F) = \langle\Omega\,\big|\,\Phi(F_0)\,\big|\,\Omega\rangle \quad \forall F\in\mathcal{F}_\hbar\ , \tag{2.6.7}$$

[20] In this way, we finesse the problem of having to worry about domains of operators on Fock space; instead, we have to prove nontrivial statements about wave front sets, e.g., that, for $F,G\in\mathcal{F}$, the product $F\star G$ again fulfills the wave front set property (1.2.2).

[21] Here and in the following we use the "bra-ket" notation , which is widespread in the physics literature: Let F^{op} be a linear operator on $\mathfrak{F}$, let $\phi\in D(F^{\mathrm{op}})\subseteq\mathfrak{F}$ and $\psi\in\mathfrak{F}$. Then we write

$$\langle\psi\,|\,F^{\mathrm{op}}\,|\,\phi\rangle := \langle\psi\,,\,F^{\mathrm{op}}\,\phi\rangle_{\mathfrak{F}}\ .$$

since

$$\omega_0\Big(\sum_{n=0}^{N}\int dx_1\cdots dx_n\,\varphi_0(x_1)\cdots\varphi_0(x_n)\,f_n(x_1,\ldots,x_n)\Big)=f_0$$
$$=\Big\langle\Omega\,\Big|\,\sum_{n=0}^{N}\int dx_1\cdots dx_n\,{:}\varphi^{\mathrm{op}}(x_1)\cdots\varphi^{\mathrm{op}}(x_n){:}\,f_n(x_1,\ldots,x_n)\,\Big|\,\Omega\Big\rangle\,.$$

It is well known (see, e.g., [162, Sect. 3]) that the Fock vacuum vector Ω is "cyclic", which means that the vectors[22]

$$\{\varphi^{\mathrm{op}}(f_1)\cdots\varphi^{\mathrm{op}}(f_n)\,\Omega=\Phi\big(\varphi_0(f_1)\star\cdots\star\varphi_0(f_n)\big)\,\Omega\;\big|\;f_1,\ldots,f_n\in\mathcal{D}(\mathbb{M})\,,\,n\in\mathbb{N}\}\tag{2.6.8}$$

generate a *dense subspace* D of $\mathfrak{F}$, where on the l.h.s. the operator product of the fields $\varphi^{\mathrm{op}}(f_j)$ is meant. One verifies the relation

$$\Phi(F_0)D\subseteq D\,,\qquad\forall F_0\in\mathcal{F}_{0,\hbar}^{(m)}\,,\tag{2.6.9}$$

due to which we may proceed as in (2.5.4): The functional $\omega:(\mathcal{F}_{0,\hbar}^{(m)},\star)\longrightarrow\mathbb{C}$ given by

$$\omega(F_0):=\frac{\langle\varphi^{\mathrm{op}}(f_1)\cdots\varphi^{\mathrm{op}}(f_n)\,\Omega\,|\,\Phi(F_0)\,|\,\varphi^{\mathrm{op}}(f_1)\cdots\varphi^{\mathrm{op}}(f_n)\Omega\rangle}{\langle\varphi^{\mathrm{op}}(f_1)\cdots\varphi^{\mathrm{op}}(f_n)\,\Omega\,|\,\varphi^{\mathrm{op}}(f_1)\cdots\varphi^{\mathrm{op}}(f_n)\Omega\rangle}\tag{2.6.10}$$

is a state on $(\mathcal{F}_{0,\hbar}^{(m)},\star)$. The Fock space matrix elements appearing here can be expressed in our functional formalism by using

$$\langle\varphi^{\mathrm{op}}(f_1)\cdots\varphi^{\mathrm{op}}(f_n)\,\Omega\,|\,\Phi(F_0)\,|\,\varphi^{\mathrm{op}}(g_1)\cdots\varphi^{\mathrm{op}}(g_k)\Omega\rangle$$
$$=\langle\Omega\,|\,\varphi^{\mathrm{op}}(\overline{f_n})\cdots\varphi^{\mathrm{op}}(\overline{f_1})\,\Phi(F_0)\,\varphi^{\mathrm{op}}(g_1)\cdots\varphi^{\mathrm{op}}(g_k)\,|\,\Omega\rangle$$
$$=\langle\Omega\,|\,\Phi\big(\varphi_0(\overline{f_n})\star\cdots\star\varphi_0(\overline{f_1})\star F_0\star\varphi_0(g_1)\star\cdots\star\varphi_0(g_k)\big)\,|\,\Omega\rangle$$
$$=\omega_0\Big(\varphi_0(\overline{f_n})\star\cdots\star\varphi_0(\overline{f_1})\star F_0\star\varphi_0(g_1)\star\cdots\star\varphi_0(g_k)\Big)\,,$$

where

$$\big\langle\varphi^{\mathrm{op}}(f)\psi_1\,,\,\psi_2\big\rangle_{\mathfrak{F}}=\big\langle\psi_1\,,\,\varphi^{\mathrm{op}}(\overline{f})\,\psi_2\big\rangle_{\mathfrak{F}}\,,\qquad\forall\psi_1,\psi_2\in D\,,$$

is taken into account, which follows from $\varphi^{\mathrm{op}}(\overline{f})=\Phi\big(\varphi_0(\overline{f})\big)=\Phi\big(\varphi_0(f)^*\big)$ and (2.6.6).

States with a fixed particle number are obtained, if we replace in (2.6.8) the operator product by the normally ordered product. In detail: According to (A.5.25), the Fock space vector

$$\frac{\psi}{\|\psi\|}\quad\text{with}\quad\psi\overset{\mathrm{Def}}{=}{:}\varphi^{\mathrm{op}}(f_1)\cdots\varphi^{\mathrm{op}}(f_n){:}\,\Omega=\Phi\big(\varphi_0(f_1)\cdots\varphi_0(f_n)\big)\,\Omega\tag{2.6.11}$$

[22] Analogously to $\varphi_0(f):=\int d^dx\;\varphi_0(x)f(x)$ we write $\varphi^{\mathrm{op}}(f):=\int d^dx\;\varphi^{\mathrm{op}}(x)f(x)$.

is a normalized eigenvector of the particle number operator (A.5.7) with eigenvalue n; therefore, it describes an *n-particle state*. Admitting all $n \in \mathbb{N}$ and all $f_1, \dots, f_n \in \mathcal{D}(\mathbb{M})$, the resulting set spans also a *dense subspace* of $\mathfrak{F}$. The expectation value of $\Phi(F_0)$, $F_0 \in \mathcal{F}_{0,\hbar}^{(m)}$, in such an n-particle state, can be expressed in our formalism by means of

$$\begin{aligned}\langle :\varphi^{\mathrm{op}}(f_1) \cdots \varphi^{\mathrm{op}}(f_n):\Omega \,|\, \Phi(F_0) \,|\, :\varphi^{\mathrm{op}}(f_1) \cdots \varphi^{\mathrm{op}}(f_n):\Omega\rangle \\ = \langle \Omega \,|\, \Phi\big(\varphi_0(\overline{f_1}) \cdots \varphi_0(\overline{f_n})\big) \cdot \Phi(F_0) \cdot \Phi\big(\varphi_0(f_1) \cdots \varphi_0(f_n)\big) \,|\, \Omega\rangle \\ = \omega_0\Big(\varphi_0(\overline{f_1}) \cdots \varphi_0(\overline{f_n}) \star F_0 \star \varphi_0(f_1) \cdots \varphi_0(f_n)\Big) \,.\end{aligned}$$

Exercise 2.6.5. Compute

$$\langle :\varphi^{\mathrm{op}}(f_1)\, \varphi^{\mathrm{op}}(f_2):\Omega \,|\, \Phi\Big(\varphi_0^3(x)\varphi_0(y)\, \Delta^{\mathrm{ret}}(y-x)\Big) \,|\, :\varphi^{\mathrm{op}}(f_3)\, \varphi^{\mathrm{op}}(f_4):\Omega\rangle$$

by using the functional formalism of this book. Up to constant prefactors, the field in the argument of Φ is the classical retarded product $R_{\mathrm{cl}}(\varphi^4(x), \varphi^2(y))$ restricted to $\mathcal{C}_{S_0}$, cf. (3.1.13).

[*Solution.* Since this is a straightforward computation of

$$\omega_0\Big(\varphi_0(\overline{f_1})\, \varphi_0(\overline{f_2}) \star \varphi_0^3(x)\varphi_0(y) \star \varphi_0(f_3)\, \varphi_0(f_4)\Big)\, \Delta^{\mathrm{ret}}(y-x) \,,$$

we only give the result:

$$\begin{aligned}6\hbar^4\, \Delta^{\mathrm{ret}}(y-x)\Big(&\Delta^+(y,f_4)\, \Delta^+(\overline{f_1},x)\, \Delta^+(\overline{f_2},x)\, \Delta^+(x,f_3) \\ &+ \Delta^+(y,f_3)\, \Delta^+(\overline{f_1},x)\, \Delta^+(\overline{f_2},x)\, \Delta^+(x,f_4) \\ &+ \Delta^+(\overline{f_1},y)\, \Delta^+(\overline{f_2},x)\, \Delta^+(x,f_3)\, \Delta^+(x,f_4) \\ &+ \Delta^+(\overline{f_2},y)\, \Delta^+(\overline{f_1},x)\, \Delta^+(x,f_3)\, \Delta^+(x,f_4)\Big) \,,\end{aligned}$$

where $\Delta^+(\overline{f},y) := \int dx\, \overline{f(x)}\, \Delta^+(x-y)$ and $\Delta^+(x,f)$ is defined analogously.]

Remark 2.6.6 (Positivity of the vacuum state and of coherent states).

Vacuum state: Using the Fock space representation of $(\mathcal{F}_{0,\hbar}^{(m)}, \star)$ and $\omega_0(G) = \omega_0(G_0)\ \forall G \in \mathcal{F}_\hbar$, the positivity of ω_0 can easily be verified:

$$\begin{aligned}\omega_0(F^* \star F) &= \omega_0\big((F^* \star F)_0\big) = \omega_0(F_0^* \star F_0) \\ &= \langle \Omega \,|\, \Phi(F_0^* \star F_0) \,|\, \Omega\rangle = \langle \Omega \,|\, \Phi(F_0^*)\, \Phi(F_0) \,|\, \Omega\rangle = \|\Phi(F_0)\,\Omega\|^2 \geq 0 \,. \quad (2.6.12)\end{aligned}$$

Coherent state: Given an $h \in \mathcal{C}_{S_0}$ we use the results of Exer. A.5.3: The vector $\psi_h \in \mathfrak{F}$, given in (A.5.39), satisfies the relation (A.5.40) and hence, a coherent state

(2.5.3) can be described in Fock space as follows:

$$\begin{aligned}\omega_h(F) = F(h) &= \sum_n \int dx_1 \cdots dx_n \; f_n(x_1,\ldots,x_n)\, h(x_1)\ldots h(x_n) \\ &= \sum_n \int dx_1 \cdots dx_n \; f_n(x_1,\ldots,x_n)\, \langle \psi_{\tilde{h}} \,|\, {:}\varphi^{\mathrm{op}}(x_1)\cdots\varphi^{\mathrm{op}}(x_n){:} \,|\, \psi_{\tilde{h}} \rangle \\ &= \langle \psi_{\tilde{h}} \,|\, \Phi(F_0) \,|\, \psi_{\tilde{h}} \rangle \quad \text{with} \quad \tilde{h}(x) := \hbar^{-1}\, h(x)\,. \end{aligned} \tag{2.6.13}$$

With that, positivity of ω_h can be shown analogously to the vacuum state (2.6.12):

$$\omega_h(F^* \star F) = \langle \psi_{\tilde{h}} \,|\, \Phi(F_0^* \star F_0) \,|\, \psi_{\tilde{h}} \rangle = \langle \psi_{\tilde{h}} \,|\, \Phi(F_0^*)\, \Phi(F_0) \,|\, \psi_{\tilde{h}} \rangle = \| \Phi(F_0)\, \psi_{\tilde{h}} \|^2 \geq 0\,.$$

Chapter 3

Perturbative Quantum Field Theory

Perturbation theory is the most widely used and, regarding the agreement with experimental data, most successful approach to QFT.

In this approach, one usually considers the interaction as a perturbation of the free theory. Consequently, physically relevant quantities, e.g., the S-matrix or Green functions (i.e., vacuum expectation values of time-ordered products of interacting fields), are power series in the coupling constant. For all physically relevant models in four-dimensional relativistic QFT, the convergence of these series is not under control – hence we understand them as *formal power series.*

Frequently, the formulas for the coefficients of these power series are "derived"

- by proceeding in analogy to the interaction picture of quantum mechanics (see, e.g., [109, Chap. 6-1] or Sect. 3.3.1);
- or from quantization by a path integral (also called "functional integral"), that is, the generalization of the path integral of quantum mechanics to QFT, see Sect. 3.9.5 and, e.g., [146] or [147] for the Euclidean theory.

We proceed in a different way: Similarly to the original formulation of causal perturbation theory (Epstein and Glaser [66]), we *define pQFT by axioms*; but in contrast to the latter, we give axioms for the retarded product (i.e., the interacting fields) and not for the time-ordered product, because we quantize the perturbative, interacting, classical fields. (An additional difference to [66] is that we work in the off-shell formalism: Our retarded product is a map from local off-shell fields to off-shell fields, cf. Sect. 3.4.) Our axioms are motivated by the principle that we want to maintain as much as possible from the structure of the perturbative expansion of the classical retarded fields [54], worked out in Chap. 1.

From now on we will always use the Wightman two-point function for the quantization. The modifications appearing when working with a Hadamard function are given in App. A.4.

M. Dütsch, *From Classical Field Theory to Perturbative Quantum Field Theory*,
Progress in Mathematical Physics 74, https://doi.org/10.1007/978-3-030-04738-2_3

3.1 Axioms for the retarded product

References for this section are the paper [55] and Steinmann's first book [155].

Recall that *quantization*, in the sense that we use the term here, means replacement of the classical product by the star product $(\cdot \mapsto \star)$ and the Poisson bracket by the star-commutator bracket, $\{\cdot,\cdot\} \mapsto (i\hbar)^{-1}[\cdot,\cdot]_{\star_\hbar}$, since we want to retain as much as possible from classical field theory.

What must we give up? First of all, the classical factorization (1.6.5), i.e.,

$$(AB)^{\mathrm{ret}}_S(x) = A^{\mathrm{ret}}_S(x) \cdot B^{\mathrm{ret}}_S(x) \tag{3.1.1}$$

for $A, B \in \mathcal{P}_{\mathrm{bal}}$ and with interaction S, cannot continue to hold, neither for the star product nor for the classical product on the r.h.s.; see the discussion about composite interacting fields in App. A.7. Secondly, certain *symmetries* of classical field theory cannot be maintained: These are the *anomalies* of the corresponding quantum field theory.

For any $F, S \in \mathcal{F}_{\mathrm{loc}}$ we wish to construct the *interacting (retarded) field*[23] F_S corresponding to F, as a formal power series in the interaction S (i.e., in the coupling constant κ), whose term of zeroth order is F. Compared with clFT, an essential new feature is that these interacting fields are also formal power series in $\hbar$, due to the quantization. Inspired by perturbative clFT, we make the Ansatz

$$F_{\kappa\tilde{S}} = \sum_{n=0}^{\infty} \frac{\kappa^n}{n!\,\hbar^n} R_{n,1}\big((\tilde{S})^{\otimes n}, F\big) =: R\Big(e_\otimes^{\kappa\tilde{S}/\hbar}, F\Big) \in \mathcal{F}[\![\hbar,\kappa]\!] \ ; \tag{3.1.2}$$

the shorthand notation on the right side uses Linearity of the maps $R_{n,1}$, which is the first axiom in the list below. The convention with respect to the factors $\hbar$ is motivated by the fact that $S/\hbar = \kappa\tilde{S}/\hbar$ is a dimensionless quantity. We shall streamline the notation a little, now writing mostly $\mathcal{F}$ for $\mathcal{F}[\![\hbar,\kappa]\!]$ and $\mathcal{F}_{\mathrm{loc}}$ for $\mathcal{F}_{\mathrm{loc}}[\![\hbar,\kappa]\!]$. Formal series in both parameters should henceforth be understood.

The maps $R_{n,1}$ introduced in (3.1.2) form the *retarded product* $R = (R_{n,1})_{n\in\mathbb{N}}$ (shortly "R-product"); sometimes we use the name "retarded product" also for an individual $R_{n,1}$. To define the retarded product, we impose axioms, which are motivated by their validity in classical field theory.

We divide the axioms into "basic axioms" and "renormalization conditions". As we will see, the former determine the R-products

$$R_{n,1}\big(B_1(x_1) \otimes \cdots \otimes B_n(x_n), B(x)\big) \ , \quad B_1,\ldots,B_n,B \in \mathcal{P}_{\mathrm{bal}} \ ,$$

(which are defined in terms of the R-products whose arguments are elements of $\mathcal{F}_{\mathrm{loc}}^{\otimes(n+1)}$ by adopting the relation (1.7.4) to pQFT, see (3.1.3) below) uniquely up

[23] We no longer write the superscript "ret" in the notation F^{ret}_S, but it should still be understood, and we omit the adjective "retarded". Remember that $S = \kappa\tilde{S}$.

to renormalization, that is, up to coinciding points (i.e., up to $x_i = x_j$ for some $i < j$ or $x_k = x$ for some k). The "renormalization conditions" restrict only the process of renormalization, that is, the extension of the R-products to coinciding points.

We use the following properties of the classical R-product as *basic axioms* for the R-product of QFT:

- From Definition 1.7.1 we adopt *Linearity* and *Symmetry* (in the first n entries) of the maps $R_{n,1} : \mathcal{F}_{\mathrm{loc}}^{\otimes(n+1)} \longrightarrow \mathcal{F}$ and the *Initial condition* $R_{0,1}(F) = F$. Besides these formal conditions,
- we require *Causality* (1.10.1) and the *GLZ relation* (1.10.3), which are profound properties of the classical retarded products – and at least Causality has a clear physical interpretation.

The solution of the basic axioms can be expressed by a closed formula, given in (3.1.69), which is valid only for non-coinciding points.

As *renormalization conditions* we use

- important properties of the classical R-product, namely *Field independence* (1.10.2), the *Off-shell field equation* (1.6.6) (or (1.7.7)) and the *Scaling and mass expansion*; the latter expresses two different properties of the R-products: "Smoothness in $m \geq 0$ modulo $\log m$" and "almost" homogeneous scaling under $(X, m) \mapsto (\rho X, \rho^{-1} m)$ (where $X := (x_1, \ldots, x_n, x)$), see Sect. 3.1.5;
- symmetries of the classical R-product, namely *Poincaré covariance*, *$*$-structure* and *Field parity* (see Exers. 1.10.4 and 1.10.5); in more complicated models *further symmetries* of the classical R-product, such as conservation of certain currents or gauge invariance belong also to the renormalization conditions;
- and the condition that *renormalization has to be done in each order of $\hbar$ individually*; which is motivated by the requirement that the classical limit exists and gives the classical R-product, see Sect. 3.1.6.

In the following Sects. 3.1.1–3.1.6 we work out the axioms and some direct consequences.

3.1.1 Basic axioms

The first axiom is

(a) **Linearity:** We require that

$$R_{n,1} \colon \mathcal{F}_{\mathrm{loc}}^{\otimes(n+1)} \longrightarrow \mathcal{F} \quad \text{be linear.}$$

Note that here both the arguments and the values of $R_{n,1}$ are *off-shell fields.* The latter is a significant difference to conventional formulations of pQFT, in which the arguments and the values of the retarded or time-ordered products are Fock space operators, i.e., on-shell fields (see Sect. 2.6).

The fact that the retarded products depend only on (local) *functionals* F, but not on how they are written as smeared field polynomials, $F = \int dx\, g(x)\, A(x)$, implies the *Action Ward Identity* (AWI), which is the relation $\partial_x R(\cdots A(x) \cdots) = R(\cdots (\partial A)(x) \cdots)$ for $A \in \mathcal{P}$. To explain this, we first define, for balanced fields $B_1, \ldots, B_n, B$ in $\mathcal{P}_{\text{bal}}$, the $\mathcal{F}$-valued distribution $R_{n,1}(B_1(x_1) \otimes \cdots \otimes B_n(x_n), B(x)) \in \mathcal{D}'(\mathbb{M}^{n+1}, \mathcal{F})$ by

$$\int dx_1 \cdots dx_n\, dx\; R_{n,1}(B_1(x_1) \otimes \cdots \otimes B_n(x_n), B(x))\; g_1(x_1) \cdots g_n(x_n)\, g(x)$$
$$:= R_{n,1}(B_1(g_1) \otimes \cdots \otimes B_n(g_n), B(g)) \;, \quad \forall\, g_1, \ldots, g_n, g \in \mathcal{D}(\mathbb{M}) \; ; \qquad (3.1.3)$$

this definition is motivated by the classical relation (1.7.4). Since the R-products depend only on (local) *functionals*, the r.h.s. of (3.1.3) is well defined. Next, for arbitrary field polynomials $A_1, \ldots, A_n, A \in \mathcal{P}$, we define the analogous $\mathcal{F}$-valued distributions $R_{n,1}(A_1(x_1) \otimes \cdots \otimes A_n(x_n), A(x))$ by first writing $A_i = \sum_{a_i} \partial^{a_i} B_{ia_i}$ where $B_{ia_i} \in \mathcal{P}_{\text{bal}}$ and setting

$$R_{n,1}(A_1(x_1) \otimes \cdots \otimes A_n(x_n), A(x)) \qquad (3.1.4)$$
$$:= \sum_{a_1, \ldots, a_n, a} \partial_{x_1}^{a_1} \cdots \partial_{x_n}^{a_n} \partial_x^a\, R_{n,1}(B_{1a_1}(x_1) \otimes \cdots \otimes B_{na_n}(x_n), B_a(x)) \;.$$

Here is where we make essential use of the *uniqueness* of the expressions of arbitrary fields A_i in terms of balanced fields B_{ia_i}, guaranteed by Proposition 1.4.3.

The definition (3.1.4) implies the *Action Ward Identity*, which is due to Raymond Stora [159, 160] and [55],[24]

$$\textbf{AWI:}\quad \partial_{x_l} R_{n,1}(\cdots \otimes A(x_l) \otimes \cdots) \qquad (3.1.5)$$
$$= R_{n,1}(\cdots \otimes \partial_{x_l} A(x_l) \otimes \cdots), \quad \forall A \in \mathcal{P}\;,\; 1 \le l \le n+1 \;.$$

A further simple consequence of these definitions is that *formula* (3.1.3) *actually holds for all* $B_1, \ldots, B_n, B \in \mathcal{P}$ (instead of only for $B_1, \ldots, B_n, B \in \mathcal{P}_{\text{bal}}$): Namely,

[24] Actually, for $A \in \mathcal{P}_{\text{bal}}$, the AWI (3.1.5) is a particular case of the definition (3.1.4) and for an arbitrary $A = \sum_a \partial^a B_a \in \mathcal{P}$ it can be obtained by using (3.1.4) twice:

$$\partial_x R(\cdots A(x) \cdots) = \sum_a \partial_x \partial_x^a R(\cdots B_a(x) \cdots) = R(\cdots \sum_a (\partial \partial^a B_a)(x) \cdots)$$
$$= R(\cdots (\partial A)(x) \cdots) \;.$$

for $A = \sum_a \partial^a B_a \in \mathcal{P}$, $B_a \in \mathcal{P}_{\mathrm{bal}}$ we obtain

$$\int dx\; g(x)\, R\big(\cdots A(x) \cdots\big) = \int dx\; g(x) \sum_a \partial_x^a R\big(\cdots B_a(x) \cdots\big)$$
$$= \int dx\; \sum_a (-1)^{|a|} (\partial^a g)(x) R\big(\cdots B_a(x) \cdots\big)$$
$$= R\big(\cdots \sum_a (-1)^{|a|} B_a(\partial^a g) \cdots\big) = R\big(\cdots A(g) \cdots\big) \,. \qquad (3.1.6)$$

Since we interpret $R_{n-1,1}$ sometimes as a map $R_{n-1,1}\colon \mathcal{F}_{\mathrm{loc}}^{\otimes n} \longrightarrow \mathcal{F}$ and sometimes as a map

$$R_{n-1,1}\colon \begin{cases} \mathcal{P}^{\otimes n} \longrightarrow \mathcal{D}'(\mathbb{M}^n, \mathcal{F}) \\ (A_1, \ldots, A_n) \longmapsto R_{n-1,1}\big(A_1(x_1) \otimes \cdots \otimes A_{n-1}(x_{n-1}), A_n(x_n)\big) \,, \end{cases} \qquad (3.1.7)$$

we shall often call it an "operation", and similarly for $J_{n-2,2}$ and $R^0_{n-1,1}$ introduced far below in (3.1.12) and in Prop. 3.2.2, respectively.

Remark 3.1.1. In this remark, we only assume that for a Lagrangian $g(x)\mathcal{L}(x)$ and a local field given by $g_1(x)A(x)$, where $g, g_1 \in \mathcal{D}(\mathbb{M})$ and $\mathcal{L}, A \in \mathcal{P}$, the pertinent R-product of order $(n,1)$ is obtained by

$$\int dx_1 \cdots dx_n\, dx\; g(x_1) \cdots g(x_n)\, g_1(x)\, R_{n,1}\big(\mathcal{L}(x_1) \otimes \cdots \otimes \mathcal{L}(x_n), A(x)\big) \,.$$

With this, the AWI is not only necessary, it is also *sufficient* for the dependence of the R-product only on (local) *functionals* F. Namely, writing F as

$$F = \int dx\; g(x)\, A(x) = \int dx\; \Big(g(x)\, A(x) + (\partial g_1)(x)\, A_1(x) + g_1(x)\, (\partial A_1)(x)\Big)$$

for an arbitrary pair $(g_1, A_1) \in \mathcal{D}(\mathbb{M}) \times \mathcal{P}$, this property of the R-product is equivalent to the relation

$$\int dx\; \Big((\partial g_1)(x)\, R\big(\cdots A_1(x) \cdots\big) + g_1(x)\, R\big(\cdots (\partial A_1)(x) \cdots\big)\Big) = 0 \,, \quad \forall g_1, A_1 \,,$$

which is an equivalent reformulation of the AWI.

The motivation for the AWI and the reason for the name "Action Ward Identity" is that physics depends only on the (inter)action S, and not on the choice of a corresponding Lagrangian $g(x)\mathcal{L}(x)$.

Our inductive construction of the retarded product (given in Sect. 3.2) is a *direct* construction of $R_{n,1}\big(A_1(x_1) \otimes \cdots \otimes A_n(x_n), A(x)\big)$ (for all $A_1, \ldots, A_n \in \mathcal{P}$, $n \in \mathbb{N}$); the retarded product as a sequence of maps $R_{n,1}\colon \mathcal{F}_{\mathrm{loc}}^{\otimes(n+1)} \longrightarrow \mathcal{F}$ is then obtained by (3.1.6). In this construction, the *AWI* (3.1.5) *plays the role of an additional axiom*; as it will turn out, it is a *renormalization condition*, i.e., it belongs to the axioms treated in Sects. 3.1.4–3.1.6.

Remark 3.1.2. If the entries F of the retarded products would be on-shell fields F_0 (2.6.1) (as this is the case when working in Fock space) and if $R_{n,1}$ would be linear as required by condition (a), then the AWI could not hold true, because there would be a contradiction to the field equation (1.7.7):

$$\begin{aligned} R\Big(e_\otimes^{(S)_0/\hbar}, (\Box + m^2)\varphi_0(x)\Big) &= 0 \qquad \text{versus} \\ (\Box_x + m^2) R\Big(e_\otimes^{(S)_0/\hbar}, \varphi_0(x)\Big) &= (\Box_x + m^2) R_{0,1}(\varphi_0(x)) \\ &\quad + R\Big(e_\otimes^{(S)_0/\hbar}, \Big(\frac{\delta S}{\delta\varphi(x)}\Big)_0(x)\Big) , \end{aligned}$$

where $(S)_0 := S\big|_{\mathcal{C}_{S_0}}$ is *not* the free action S_0. For the relation between R-products defined for on-shell fields and R-products defined for off-shell fields, see Sect. 3.4 and [161].

Remark 3.1.3. The main difference between our formalism and that of Steinmann [155] is that Steinmann works *only in the adiabatic limit*, i.e., he always sets $g_1(x) := \cdots g_n(x) := g(x) := 1$ for all x, in (3.1.3) – or more precisely in the corresponding formula for $B_1, \ldots, B_n, B \in \mathcal{P}$. Hence, to avoid infrared divergences, he assumes $m > 0$. In our formalism this assumption is not needed; the adiabatic switching by $g_1, \ldots, g \in \mathcal{D}(\mathbb{M})$ ensures IR-finiteness. This adiabatic switching is a main ingredient of causal perturbation theory (also called "Epstein–Glaser renormalization").

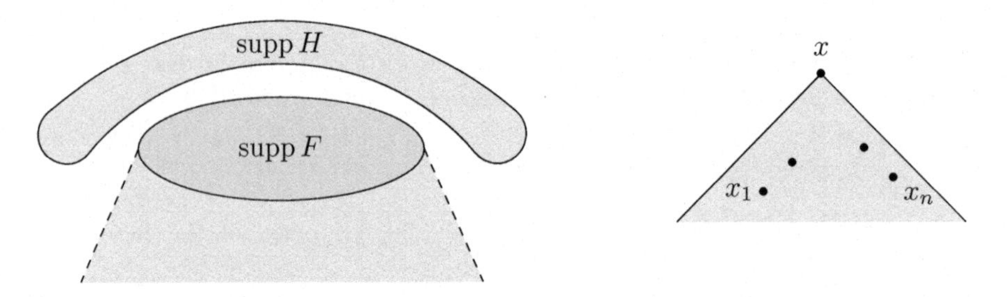

Figure 3.1: Visualizations of the two formulations of the Causality axiom

The further *basic axioms* needed to specify all $R_{n,1}$ now follow.

(b) **Symmetry:** $R_{n,1}(F_{\pi(1)} \otimes \cdots \otimes F_{\pi(n)}, F) = R_{n,1}(F_1 \otimes \cdots \otimes F_n, F)$ for all permutations $\pi \in S_n$.

Due to the axioms Linearity and Symmetry, the interacting field $R(e_\otimes^{F/\hbar}, G)$ can be interpreted as *generating functional* for the retarded products $R_{n,1}$: by using (A.1.10) we obtain

$$R_{n,1}(F_{j_1} \otimes \cdots \otimes F_{j_n}, G) = \frac{\hbar^n \partial^n}{\partial\lambda_{j_1} \cdots \partial\lambda_{j_n}}\bigg|_{\lambda_1 = \cdots = \lambda_J = 0} R\Big(e_\otimes^{\sum_{j=1}^J \lambda_j F_j/\hbar}, G\Big) \quad (3.1.8)$$

for all $n \in \mathbb{N}$, $1 \leq j_k \leq J$.

(c) **Initial condition:** $R_{0,1}(F) = F$.

(d) **Causality:** We require

$$R(e_\otimes^{(S+H)/\hbar}, F) = R(e_\otimes^{S/\hbar}, F) \quad \text{if} \quad (\operatorname{supp} F + \overline{V}_-) \cap \operatorname{supp} H = \emptyset \,, \tag{3.1.9}$$

see the left-hand part of Figure 3.1. This is equivalent to the set of conditions that, for each $n \in \mathbb{N}$:

$$\operatorname{supp} R_{n,1}\big(A_1(x_1) \otimes \cdots \otimes A_n(x_n), A(x)\big) \\ \subseteq \{\, (x_1, \ldots, x_n, x) \in \mathbb{M}^{n+1} \,|\, x_j \in x + \overline{V}_- \text{ for } j = 1, \ldots, n \} \tag{3.1.10}$$

for all $A_1, \ldots, A_n, A \in \mathcal{P}$, that is, all x_j must lie in the past of x; see the right-hand part of Figure 3.1.

Using the Linearity and Symmetry axioms, the equivalence of (3.1.9) and (3.1.10) can be shown as follows.

For (3.1.9) $\implies$ (3.1.10): We use

$$R_{n,1}\big(A_1(x_1) \otimes \cdots \otimes A_n(x_n), A(x)\big) \\ = \frac{\hbar^n \, \partial^n}{\partial\lambda_1 \ldots \partial\lambda_n}\bigg|_{\lambda_1 = \cdots = \lambda_n = 0} R\Big(e_\otimes^{\sum_{j=1}^n \lambda_j A_j(x_j)/\hbar}, A(x)\Big) \,,$$

see (3.1.8). Now if one of the x_j, say x_n, does not lie in $x + \overline{V}_-$, then (3.1.9) implies that the right-hand side may be modified as follows: In the argument of $R_{n,1}$ the sum $\sum_{j=1}^n \lambda_j A_j(x_j)$ may be reduced to $\sum_{j=1}^{n-1} \lambda_j A_j(x_j)$ and with that the derivative $\partial/\partial\lambda_n$ vanishes.

For (3.1.10) $\implies$ (3.1.9): Let $S = A_1(g_1)$, $H = A_2(g_2)$ and $F = A(g)$, with $A_1, A_2, A \in \mathcal{P}$ and $g_1, g_2, g \in \mathcal{D}(\mathbb{M})$; and let also $\operatorname{supp} g_2 \cap (\operatorname{supp} g + \overline{V}_-) = \emptyset$. Writing

$$R_{n,1}\big((S+H)^{\otimes n}, F\big) \\ = \sum_{k=0}^n \binom{n}{k} \int dx_1 \cdots dx_n \, dx \; g_1(x_1) \cdots g_1(x_k) \, g_2(x_{k+1}) \cdots g_2(x_n) \, g(x) \\ \cdot R_{n,1}\big(A_1(x_1) \otimes \cdots \otimes A_1(x_k) \otimes A_2(x_{k+1}) \otimes \cdots \otimes A_2(x_n), A(x)\big),$$

the assumption (3.1.10) implies that on the right side only the summand $k = n$ does not vanish; hence, the right-hand side is equal to $R_{n,1}\big(S^{\otimes n}, F\big)$.

Remark 3.1.4. An immediate consequence of the support property (3.1.10) is that

$$R_{n,1}\big(A_1(x_1) \otimes \cdots, A_{n+1}(x_{n+1})\big) = 0 \quad \text{if } n \geq 1 \text{ and some } A_j \text{ is a } C\text{-number},$$

i.e., $A_j(x) = c \in \mathbb{C}\ \forall x \in \mathbb{M}$, where $1 \leq j \leq n+1$. Namely, then we have $A_j(x_j) = A_j(x_j + a)$, $\forall a \in \mathbb{R}^d$, and choosing a suitably we can reach that $(x_1, \ldots, x_j + a, \ldots, x_{n+1})$ lies not in the set given on the r.h.s. of (3.1.10).

(e) **GLZ relation:** In the classical GLZ relation (1.10.3) we substitute $\{\cdot,\cdot\} \mapsto (i\hbar)^{-1}[\cdot,\cdot]_{\star_\hbar}$ and the first n arguments of $R_{n,1}$ get a factor $\hbar^{-1}$ according to (3.1.2). With that, we obtain the requirement that

$$\frac{1}{i\hbar}\big[R(e_\otimes^{S/\hbar},F),R(e_\otimes^{S/\hbar},H)\big]_{\star_\hbar} = R(e_\otimes^{S/\hbar}\otimes(F/\hbar),H) - R(e_\otimes^{S/\hbar}\otimes(H/\hbar),F) \tag{3.1.11a}$$

for all $S,H,F \in \mathcal{F}_{\mathrm{loc}}$. This is equivalent to the sequence of equations

$$\begin{aligned} R_{n-1,1}(G_1\otimes\cdots\otimes G_{n-2}\otimes F,H) - R_{n-1,1}(G_1\otimes\cdots\otimes G_{n-2}\otimes H,F)\\ = \hbar\, J_{n-2,2}(G_1\otimes\cdots\otimes G_{n-2},F\otimes H) \quad \forall G_1,\ldots,G_n,H,F\in\mathcal{F}_{\mathrm{loc}}\,; \end{aligned} \tag{3.1.11b}$$

where

$$\begin{aligned} J_{n-2,2}(G_1\otimes\cdots\otimes G_{n-2},F\otimes H) \\ := \frac{1}{i\hbar}\sum_{I\subseteq\{1,\ldots,n-2\}} \big[R_{|I|,1}(G_I,F),R_{|I^c|,1}(G_{I^c},H)\big]_{\star_\hbar} \end{aligned} \tag{3.1.12}$$

with $G_I := \bigotimes_{l\in I} G_l$ and $I^c := \{1,\ldots,n-2\}\setminus I$.

Exercise 3.1.5. Prove the equivalence of the relations (3.1.11a) and (3.1.11b).

[*Solution.* For (3.1.11a) $\implies$ (3.1.11b): In (3.1.11a) substitute $\sum_{j=1}^{n-2}\lambda_j G_j$ for S and apply the derivative $\frac{\partial^n}{\partial\lambda_1\cdots\partial\lambda_{n-2}}\big|_{\lambda_1=\cdots=\lambda_{n-2}=0}$ to the resulting equation.
For (3.1.11b) $\implies$ (3.1.11a): Set $G_j := S$ for all j. With that the r.h.s. of (3.1.12) is equal to

$$\frac{1}{i\hbar}\sum_{k=0}^{n-2}\binom{n-2}{k}\big[R_{k,1}(S^{\otimes k},F)\,,\,R_{n-2-k,1}(S^{\otimes(n-2-k)},H)\big]_\star\,;$$

insert this into (3.1.11b). Then apply $\sum_{n=2}^\infty \frac{1}{(n-2)!}\cdots$ to (3.1.11b).]

We point out that, in the basic axioms, information about the mass m appears only in the GLZ relation (3.1.11) – via the m-dependence of the propagator Δ_m^+ in the $\star$-product.

Remark 3.1.6 (Spacelike commutativity of interacting fields). Analogously to the relation (1.10.4) in classical field theory, causality (axiom (d)) and the GLZ-relation (axiom (e)) imply spacelike commutativity of the interacting fields:

$$[F_S,\,H_S]_\star = 0 \quad\text{if}\quad (x-y)^2<0 \quad\text{for all}\quad (x,y)\in\operatorname{supp}F\times\operatorname{supp}H.$$

The GLZ relation provides a splitting of the commutator $[A_S(x),B_s(y)]_\star$ (where $A,B\in\mathcal{P}_{\mathrm{bal}}$) into a "advanced" (i.e., $(x-y)\in\overline{V}_-$) and a "retarded part" ($(x-y)\in\overline{V}_+$).

3.1.2 A worked example

We now try to determine from the preceding axioms the value of

$$R_{1,1}\big(\varphi^4(x),\varphi^2(y)\big) =: R_{1,1}(x,y),$$

in dimension $d=4$.

The *initial condition* gives $R_{0,1}\big(\varphi^k(x)\big)=\varphi^k(x)\,,\ k=2,4$.

The *inductive step* $R_{0,1}\mapsto R_{1,1}$ is obtained as follows. First, the *GLZ relation* (3.1.11b) yields, according to the contraction patterns

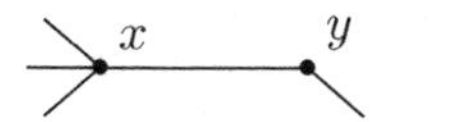

the result:

$$R_{1,1}(x,y)-R_{1,1}(y,x)=\hbar\,J_{0,2}(x,y):=\frac{1}{i}\,[\varphi^4(x),\varphi^2(y)]_\star$$
$$=\frac{4\cdot 2\hbar}{i}\,\varphi^3(x)\,\varphi(y)\,i\Delta_m(x-y)+\frac{4\cdot 3\cdot 2\hbar^2}{2i}\,\varphi^2(x)\big(\Delta_m^+(x-y)^2-\Delta_m^+(y-x)^2\big),$$

recalling that $\Delta_m^+(z)-\Delta_m^+(-z)=i\,\Delta_m(z)$ is the commutator function.

From Lemma 2.4.1 we know that

$$\operatorname{supp} J_{0,2}(x,y)\subseteq\{\,(x,y)\,\big|\,(x-y)^2\geq 0\,\}\,.$$

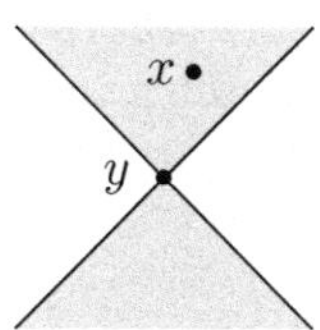

Causality gives

$$\operatorname{supp} R_{1,1}(x,y)\subseteq\{\,(x,y)\,\big|\,x\in y+\overline{V}_-\,\}\,;$$

so we seek a splitting of the form

$$\hbar\,J_{0,2}(x,y)=R_{1,1}(x,y)-R_{1,1}(y,x),$$

respecting these support properties:

supp $R_{1,1}(x,y)$

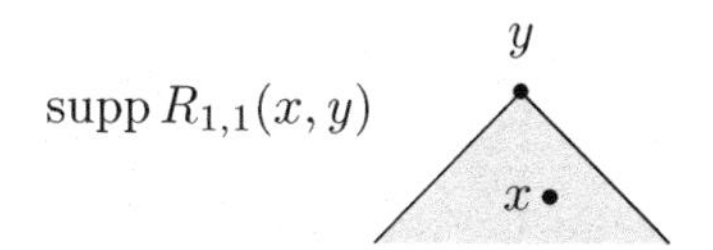

supp $R_{1,1}(y,x)$

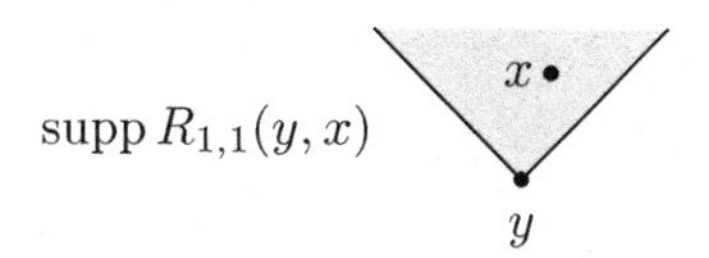

We can try a naïve splitting by setting

$$R_{1,1}(x,y):=\hbar J_{0,2}(x,y)\,\theta(y^0-x^0)$$
$$=\;-\,8\hbar\,\varphi^3(x)\,\varphi(y)\,\Delta_m^{\text{ret}}(y-x)+\frac{12}{i}\,\hbar^2\varphi^2(x)\,r^0(x-y)\,,\qquad(3.1.13)$$

on writing

$$r^0(x-y):=\big(\Delta_m^+(x-y)^2-\Delta_m^+(y-x)^2\big)\,\theta(y^0-x^0)\,.$$

The first term on the right-hand side certainly exists: $\Delta_m(z)\,\theta(z^0) = \Delta_m^{\rm ret}(z)$; but the second only exists for $x - y \neq 0$, so this second term requires renormalization, which is the extension of $r^0 \in \mathcal{D}'(\mathbb{R}^4 \setminus \{0\})$ to $r \in \mathcal{D}'(\mathbb{R}^4)$.

Definition 3.1.7 (Extension). An *extension* of a distribution $r^0 \in \mathcal{D}'(\mathbb{R}^k \setminus \{0\})$, defined away from the origin, is a distribution $r \in \mathcal{D}'(\mathbb{R}^k)$ such that $\langle r, f\rangle = \langle r^0, f\rangle$ for all $f \in \mathcal{D}(\mathbb{R}^k \setminus \{0\})$.

Due to the GLZ relation, the skew-symmetric part of r is already determined by $r(z) - r(-z) = \Delta_m^+(z)^2 - \Delta_m^+(-z)^2$, so we only need to extend the symmetrized distribution

$$s^0(z) := \tfrac{1}{2}\big[r^0(z) + r^0(-z)\big] \in \mathcal{D}'(\mathbb{R}^4 \setminus \{0\}).$$

Such an extension is non-unique: Given one extension $s(z)$, one can also use

$$s(z) + \sum_{|a|\ \rm even} C_a\, \partial^a \delta(z), \quad \text{with arbitrary constants} \quad C_a \in \mathbb{C}\ ;$$

cf. Theorem (A.1.17). For that reason, we shall need to add (later on) appropriate *renormalization conditions* to delimit the possible constants C_a that may appear.

To study the *classical limit* we require that the relation $s^0(z) \sim \hbar^0$ is maintained under renormalization, that is, $s(z) \sim \hbar^0$, which is a restriction of the constants C_a. This implies $r(z) \sim \hbar^0$. With that and with the interaction $S := \int \varphi^4(x)\, g(x)\, dx$, we get

$$\lim_{\hbar \to 0} \frac{1}{\hbar}\, R_{1,1}\big(S, \varphi^2(y)\big) = -8 \int dx\, g(x)\, \varphi^3(x)\, \Delta_m^{\rm ret}(y - x)\, \varphi(y) = R_{\rm cl}\big(S, \varphi^2(y)\big), \tag{3.1.14}$$

since the second term in (3.1.13) drops out on setting $\hbar \mapsto 0$. By $R_{\rm cl}$ we mean the classical retarded product, as defined in Chap. 1; the second equality sign results from (1.9.3).

3.1.3 Discussion of the GLZ relation

The l.h.s. of the GLZ relation (3.1.11a) is the commutator of two interacting fields, whose coefficients in the perturbative expansion are the fields $J_{n-2,2}(\cdots)$ (3.1.12). In this section we derive some basic properties of the operations $J_{n-2,2}$, which will be crucial for the inductive construction of the retarded product $(R_{n,1})_{n \in \mathbb{N}}$ – see Sect. 3.2.1.

In this section and later, we shall often write tensor products with commas, for instance:

$$R_{n-1,1}(G_1, \dots, G_{n-1}; F) \equiv R_{n-1,1}(G_1 \otimes \cdots \otimes G_{n-1}, F),$$

where the left-hand side can also be regarded as a *multilinear* function of n variables.

Assuming that the $R_{k,1}$ have been constructed for $k = 0, 1, \dots, n-2$, the functional

$$J_{n-2,2}(F_1, \dots, F_{n-2}; F, G) := \frac{1}{i\hbar} \sum_{I \subseteq \{1,\dots,n-2\}} \big[R_{|I|,1}(F_I, F), R_{|I^c|,1}(F_{I^c}, G)\big]_{\star_\hbar} \tag{3.1.15}$$

can be built. It is clear from the right-hand side that $J_{n-2,2}$ is symmetric in the first $(n-2)$ factors:

$$J_{n-2,2}(F_{\pi(1)}, \dots, F_{\pi(n-2)}; F, G) = J_{n-2,2}(F_1, \dots, F_{n-2}; F, G)\ , \quad \forall \pi \in S_{n-2}\ ;$$

and skew-symmetric in the last two factors:

$$J_{n-2,2}(F_1, \dots, F_{n-2}; G, F) = -J_{n-2,2}(F_1, \dots, F_{n-2}; F, G).$$

The GLZ relation (3.1.11b) and the causality axiom (3.1.10) – both stipulated for the as yet unknown $R_{n-1,1}$ operation – require that $J_{n-2,2}$ have the following additional two properties. Firstly, on combining the GLZ relation

$$\begin{aligned}\hbar\, J_{n-2,2}\big(A_1(x_1), \dots, A_{n-2}(x_{n-2}); A(y), B(z)\big) \\ = R_{n-1,1}\big(A_1(x_1), \dots, A_{n-2}(x_{n-2}), A(y); B(z)\big) \\ - R_{n-1,1}\big(A_1(x_1), \dots, A_{n-2}(x_{n-2}), B(z); A(y)\big)\end{aligned}$$

with the causality requirement on $R_{n-1,1}$, we get

$$\begin{aligned}\operatorname{supp} J_{n-2,2} \subset \mathbb{J}_n \quad \text{where} \quad \mathbb{J}_n := \Big\{ (x_1, \dots, x_{n-2}, y, z) \,\Big|\, \\ \{x_1, \dots, x_{n-2}, y\} \subset (z + \overline{V}_-) \quad \text{or} \quad \{x_1, \dots, x_{n-2}, z\} \subset (y + \overline{V}_-) \Big\},\end{aligned} \tag{3.1.16}$$

which in turn implies that $(y-z)^2 \geq 0$ in both cases:

Secondly, a further necessary condition for the GLZ relation (3.1.11b) is:

$$\begin{aligned}&J_{n-2,2}(F_1, \dots, F_{n-3}, H; F, G) + J_{n-2,2}(F_1, \dots, F_{n-3}, G; H, F) \\ &\quad + J_{n-2,2}(F_1, \dots, F_{n-3}, F; G, H) \\ &= \frac{1}{i\hbar}\Big(R_{n-1,1}(F_1, \dots, F_{n-3}, H, F; G) - R_{n-1,1}(F_1, \dots, F_{n-3}, H, G; F) \\ &\quad + R_{n-1,1}(F_1, \dots, F_{n-3}, G, H; F) - R_{n-1,1}(F_1, \dots, F_{n-3}, G, F; H) \\ &\quad + R_{n-1,1}(F_1, \dots, F_{n-3}, F, G; H) - R_{n-1,1}(F_1, \dots, F_{n-3}, F, H; G)\Big) = 0,\end{aligned}$$

since $R_{n-1,1}$ is symmetric in its first $(n-1)$ factors.

These two necessary conditions are indeed satisfied, as the next lemma shows.

Lemma 3.1.8 (Properties of $J_{n-2,2}$, [55, 155]). *Let operations* $R_{k,1}$*, for* $k = 0, 1, \ldots$ $\ldots, n-2$*, be given, satisfying the basic axioms* (a) *to* (e) *above. Then the operation* $J_{n-2,2}$ *defined by* (3.1.15) *fulfills both the Jacobi identity*

$$J_{n-2,2}(F_1, \ldots, F_{n-3}, H; F, G) + \mathrm{cyclic}(H, F, G) = 0 \tag{3.1.17}$$

and the support property $\mathrm{supp}\, J_{n-2,2} \subset \mathbb{J}_n$ (3.1.16).

Proof. Jacobi identity: we proceed by induction on n and start with the Jacobi identity of the commutator $[\cdot\,,\cdot] \equiv [\cdot\,,\cdot]_{\star_\hbar}$ and use again the notation $F_I := \otimes_{i\in I} F_i$, where $I \subset \{1, \ldots, n-3\}$. So we know

$$\sum_{I\sqcup M\sqcup L=\{1,\ldots,n-3\}} \big[[R(F_I; H), R(F_M; F)], R(F_L; G)\big] + \mathrm{cyclic}(H, F, G) = 0\ , \tag{3.1.18}$$

where $\sqcup$ means the disjoint union. Let $K := I \sqcup M$ be fixed. Summing the inner commutator over all decompositions of K into $I \sqcup M$, we obtain $i\hbar\, J_{|K|,2}(F_K; H, F)$, where $|K| \le n-3$, which splits into $i\big(R(F_K, H; F) - R(F_K, F; H)\big)$ due to the validity of the GLZ relation to lower orders. With that we obtain

$$\begin{aligned} 0 &= \sum_{K\sqcup L=\{1,\ldots,n-3\}} \big[\big(R(F_K, H; F) - R(F_K, F; H)\big), R(F_L; G)\big] + \mathrm{cycl}(H, F, G) \\ &= i\hbar\, J_{n-2,2}(F_1, \ldots, F_{n-3}, H; F, G) + \mathrm{cyclic}(H, F, G)\ . \end{aligned}$$

In the last equality sign we use that

$$\begin{aligned} &\sum_{K\sqcup L=\{1,\ldots,n-3\}} \big[R(F_K, H; F), R(F_L; G)\big] - \big[R(F_K, H; G), R(F_L; F)\big] \\ &\quad= \sum_{K\sqcup L=\{1,\ldots,n-3\}} \big[R(F_K, H; F), R(F_L; G)\big] + \big[R(F_K; F), R(F_L, H; G)\big] \\ &\quad= i\hbar\, J_{n-2,2}(F_1, \ldots, F_{n-3}, H; F, G)\ . \end{aligned}$$

Support property: for simplicity we write x_l for $A_l(x_l)$ and similarly y for $A(y)$, z for $B(z)$. By definition of J (3.1.12) and the support properties of R it follows that $J_{n-2,2}(x_1, \ldots, x_{n-2}; y, z)$ vanishes if one of the first $n-2$ arguments is not in $\{y, z\} + \overline{V}_-$.

It remains to show that it vanishes also for $(y-z)^2 < 0$. If one of the first $n-2$ arguments, e.g., x_1, is different from y and z, and is in $\{y, z\} + \overline{V}_-$, then J vanishes due to the Jacobi identity:

$$J(\ldots, x_1; y, z) = -J(\ldots, z; x_1, y) - J(\ldots, y; z, x_1) = 0\ ,$$

because $z \notin \{x_1, y\} + \overline{V}_-$ and $y \notin \{z, x_1\} + \overline{V}_-$.

If, on the other hand, all arguments x_i are sufficiently near to either y or z, then they are space-like to the other point. Hence, by Causality of the R-products of lower orders, all pairs of retarded products in the definition of J vanish up to one:

$$J(x_1,\ldots,x_{n-2};y,z)=\frac{1}{i\hbar}\big[R\big((x_k)_{k\in K};y\big),R\big((x_j)_{j\in K^c};z\big)\big]_\star\,, \tag{3.1.19}$$

where K is that subset of $\{1,\ldots,n-2\}$ for which $x_k\approx y\ \forall k\in K$ and $x_j\approx z\ \forall j\in K^c:=\{1,\ldots,n-2\}\setminus K$. By induction we know that the two retarded products fulfill the axiom (f) "Field independence" (given below) and, hence, the support property (3.1.21). Therefore, Lemma 2.4.1 can be applied: The commutator in (3.1.19) vanishes. □

3.1.4 Further axioms: Renormalization conditions

As we shall see later on (in Sect. 3.2.1), in the inductive step $R_{n-2,1}\mapsto R_{n-1,1}$ the basic axioms determine $R_{n-1,1}\big(A_1(x_1),\ldots,A_n(x_n)\big)$ *uniquely* on $\mathcal{D}(\mathbb{M}^n\setminus\Delta_n)$, where Δ_n is the thin diagonal (A.1.12). The further axioms restrict only the *extension* to $\mathcal{D}(\mathbb{M}^n)$ (which is called "renormalization") and, hence, are named "renormalization conditions". Similarly to the basic axioms, they are motivated by their validity in classical field theory. (Except for the last one, which is motivated by the classical limit.)

First, we require that the maps $R_{n-1,1}$ have no intrinsic dependence on the fields.

(f) **Field independence:**

$$\frac{\delta}{\delta\varphi(x)}R_{n-1,1}(F_1\otimes\cdots\otimes F_n)=\sum_{l=1}^n R_{n-1,1}\Big(F_1\otimes\cdots\otimes\frac{\delta F_l}{\delta\varphi(x)}\otimes\cdots\otimes F_n\Big). \tag{3.1.20}$$

In clFT Field independence relies on $\delta\Delta^{\mathrm{ret}}/\delta\varphi=0$, see (1.10.2). The "unrenormalized" R-product, given by (3.1.13) or generally by formula (3.1.69), contains also the propagator Δ^+. Since also $\delta\Delta^+/\delta\varphi=0$, axiom (f) is reasonable.

By using (1.3.19) and Linearity (axiom (a)), we find a nice consequence of Field independence:

$$\begin{aligned}\operatorname{supp}R\big(A_1(x_1),\ldots,A_{n-1}(x_{n-1});A_n(x_n)\big)&=\operatorname{supp}_x\frac{\delta}{\delta\varphi(x)}R\big(A_1(x_1),\ldots;A_n(x_n)\big)\\&=\operatorname{supp}_x\sum_{l=1}^n\sum_{a\in\mathbb{N}^d}(\partial^a\delta)(x_l-x)\,R\Big(A_1(x_1),\ldots,\frac{\partial A_l}{\partial(\partial^a\varphi)}(x_l),\ldots;A_n(x_n)\Big)\\&\subseteq\{x_1,\ldots,x_{n-1},x_n\}\end{aligned} \tag{3.1.21}$$

for all $A_1,\ldots,A_{n-1},A_n\in\mathcal{P}$, that is, the retarded products are localized at their arguments.

We are now going to derive a very important and useful application of Field independence, which will turn out to be an equivalent reformulation of this axiom. A retarded product $R_{n-1,1}(A_1(x_1),\dots;A_n(x_n))$ is an $\mathcal{F}$-valued distribution. Since (ordinary) $\mathbb{C}$-valued distributions are much more handy to work with, we are looking for a Taylor expansion of $R_{n-1,1}(\dots)$ in φ; the coefficients in this series are then the wanted $\mathbb{C}$-valued distributions. By means of Field independence, the latter can be expressed as vacuum expectation values of (in general) other R-products, as we will see.

As a preparation we remark that the defining formula (1.2.1) for a field,

$$\mathcal{F}\ni F=\sum_{n=0}^{N}\int dx_1\cdots dx_n\,\varphi(x_1)\cdots\varphi(x_n)\,f_n(x_1,\dots,x_n)\;, \tag{3.1.22}$$

implies the relation

$$n!\,f_n(x_1,\dots,x_n)=\omega_0\Big(\frac{\delta^n F}{\delta\varphi(x_1)\cdots\delta\varphi(x_n)}\Big)$$

according to (1.3.1). Hence, formula (3.1.22) can be interpreted as a (finite) Taylor expansion in φ with respect to $\varphi=0$. Applying this to $R_{n-1,1}(F_1\otimes\cdots\otimes F_n)$ and using the Field independence (3.1.20), we get

$$\begin{aligned}
&R_{n-1,1}(F_1\otimes\cdots\otimes F_n)\\
&\quad=\sum_{l=0}^{N}\frac{1}{l!}\int dx_1\cdots dx_l\;\omega_0\Big(\frac{\delta^l R_{n-1,1}(F_1\otimes\cdots\otimes F_n)}{\delta\varphi(x_1)\cdots\delta\varphi(x_l)}\Big)\varphi(x_1)\cdots\varphi(x_l)\\
&\quad\overset{(3.1.20)}{=}\sum_{l_1,\dots,l_n}\frac{1}{l_1!\dots l_n!}\int dx_{11}\cdots dx_{1l_1}\cdots dx_{n1}\cdots dx_{nl_n}\\
&\qquad\cdot\omega_0\Big(R_{n-1,1}\Big(\frac{\delta^{l_1}F_1}{\delta\varphi(x_{11})\cdots\delta\varphi(x_{1l_1})}\otimes\cdots\otimes\frac{\delta^{l_n}F_n}{\delta\varphi(x_{n1})\cdots\delta\varphi(x_{nl_n})}\Big)\Big)\\
&\qquad\cdot\varphi(x_{11})\cdots\varphi(x_{1l_1})\cdots\varphi(x_{n1})\cdots\varphi(x_{nl_n}).
\end{aligned}\tag{3.1.23}$$

One verifies (rather laboriously) that (3.1.23) provides a solution to (3.1.20); and so, the relations (3.1.20) and (3.1.23) are *equivalent.*

Remark 3.1.9. The expansion (3.1.23) is essentially the *causal Wick expansion* of Epstein and Glaser [66], see (3.1.24) below. For that, note that

$$\operatorname{supp}\frac{\delta^l F_k}{\delta\varphi(x_{k1})\cdots\delta\varphi(x_{kl})}\subseteq\Delta_l\,.$$

Explicitly: Letting $F_k=\int dx_k\;h_k(x_k)\,A_k(x_k)$ with $A_k\in\mathcal{P}$ a field monomial and $h_k\in\mathcal{D}(\mathbb{M})$, we integrate out the $\partial^a\delta(x_{k1}-x_k,\dots,x_{kl}-x_k)$ distributions coming from the functional derivative of $A_k(x_k)$, cf. (1.3.19). Removing the test functions h_k, we thereby arrive at the aforesaid *causal Wick expansion*: For *monomials*

$A_1, \ldots, A_n \in \mathcal{P}$ we obtain

$$R_{n-1,1}\big(A_1(x_1), \ldots, A_n(x_n)\big) \\ = \sum_{\underline{A}_l \subseteq A_l} \omega_0\Big(R_{n-1,1}\big(\underline{A}_1(x_1), \ldots, \underline{A}_n(x_n)\big)\Big)\, \overline{A}_1(x_1) \cdots \overline{A}_n(x_n), \tag{3.1.24}$$

where each *submonomial* $\underline{A}$ of a given monomial A and its *complementary submonomial* $\overline{A}$ are defined by[25]

$$\underline{A} := \frac{\partial^k A}{\partial(\partial^{a_1}\varphi) \cdots \partial(\partial^{a_k}\varphi)} \neq 0, \\ \overline{A} := C_{a_1 \ldots a_k}\, \partial^{a_1}\varphi \cdots \partial^{a_k}\varphi \quad \text{(no sum over } a_1, \ldots, a_k), \tag{3.1.25}$$

where each $C_{a_1 \ldots a_k}$ is a certain combinatorial factor and the range of the sum $\sum_{\underline{A} \subseteq A}$ are all $k \in \mathbb{N}$ and $a_1, \ldots, a_k \in \mathbb{N}^d$ which yield a non-vanishing $\underline{A}$. (For $k = 0$ we have $\underline{A} = A$ and $\overline{A} = 1$.)

The name "causal Wick expansion" is due to the fact that (3.1.24) is the "Wick expansion" (given below in (3.1.28)) for a product – the R-product – satisfying a causality relation.

We illustrate the causal Wick expansion (3.1.24) by a simple exercise.

Exercise 3.1.10. Derive the relation

$$R_{n,1}\big(\varphi^k(x_1), \ldots, \varphi^k(x_n); \varphi(x)\big) = \sum_{l_1, \ldots, l_n = 0}^{k-1} \binom{k}{l_1} \cdots \binom{k}{l_n} \\ \cdot \omega_0\Big(R_{n,1}\big(\varphi^{k-l_1}(x_1), \ldots, \varphi^{k-l_n}(x_n); \varphi(x)\big)\Big)\, \varphi^{l_1}(x_1) \cdots \varphi^{l_n}(x_n) \tag{3.1.26}$$

from formula (3.1.23). Note that in the sum the values $l_r = k$ are excluded.

[*Solution*: In formula (3.1.23) we substitute $\varphi^k(x_r)$ for F_r and $\varphi(x)$ for the last argument. Using

$$\frac{\delta^{l_r} \varphi^k(x_r)}{\delta\varphi(x_{r1}) \cdots \delta\varphi(x_{rl_r})} = \frac{k!}{(k-l_r)!}\, \delta(x_{r1} - x_r, \ldots, x_{rl_r} - x_r)\, \varphi^{k-l_r}(x_r)$$

and taking into account Remk. 3.1.4, we obtain the assertion (3.1.26).]

The point at issue is that $R_{n-1,1}\big(A_1(x_1), \ldots, A_n(x_n)\big)$ is *uniquely determined* by the vacuum expectation values

$$\omega_0\Big(R_{n-1,1}\big(\text{submonomials of } A_1(x_1), \ldots, A_n(x_n)\big)\Big)\,,$$

which are $\mathbb{C}$-valued distributions (also called "numerical distributions").

[25] The derivative $\frac{\partial A}{\partial(\partial^a \varphi)}$ is introduced in Exer. 1.3.5; $\underline{A}$ is defined by a k-fold derivative of this kind.

To verify that the expression on the right side of (3.1.23) lies in $\mathcal{F}$, we need to check the wave front condition (1.2.2); explicitly, that

$$\mathrm{WF}\Big(\omega_0\Big(R_{n-1,1}\Big(\frac{\delta^{l_1} F_1}{\delta\varphi(x_{11})\cdots\delta\varphi(x_{1l_1})}\otimes\cdots\otimes\frac{\delta^{l_n} F_n}{\delta\varphi(x_{n1})\cdots\delta\varphi(x_{nl_n})}\Big)\Big)\Big)$$
$$\cap\,(\mathbb{M}^L\times(\overline{V}_+^L\cup\overline{V}_-^L))=\emptyset,\quad\text{with}\quad L:=l_1+\cdots+l_n \tag{3.1.27}$$

holds true. According to Remk. 1.2.6, this follows from Translation covariance of the retarded products as required by axiom (h) below. More precisely, we will see that this axiom implies that the distributions $\omega_0(R_{n-1,1}(\ldots))$ (3.1.27) depend only on the relative coordinates and, hence, we may apply (1.2.9).

In turn, this result implies that the right-hand side of (3.1.24) exists. This corresponds to the famous Theorem 0 of Epstein and Glaser [66, Sect. 4] for causal perturbation theory in Fock space, which is originally due to Bogoliubov and Shirkov [12]. They say, in essence, that "the pointwise product of distributions $f_n(x_1-x_n,\ldots,x_{n-1}-x_n)$ $:A_1^{\mathrm{op}}(x_1)\ldots A_n^{\mathrm{op}}(x_n):$ exists on $\mathcal{D}(\mathbb{R}^{dn})$ for arbitrary $f_n\in\mathcal{D}'(\mathbb{R}^{d(n-1)})$ and $A_1,\ldots,A_n\in\mathcal{P}$, because the first factor depends only on the relative coordinates", cf. [24, Theorem 3.1].

Exercise 3.1.11. (a) By proceeding analogously to (3.1.23), prove

$$F_1\star\cdots\star F_n=\sum_{l_1,\ldots,l_n\in\mathbb{N}}\frac{1}{l_1!\cdots l_n!}\int dx_{11}\cdots dx_{1l_1}\cdots dx_{n1}\cdots dx_{nl_n}$$
$$\cdot\,\omega_0\Big(\frac{\delta^{l_1} F_1}{\delta\varphi(x_{11})\cdots\delta\varphi(x_{1l_1})}\star\cdots\star\frac{\delta^{l_n} F_n}{\delta\varphi(x_{n1})\cdots\delta\varphi(x_{nl_n})}\Big)$$
$$\cdot\,\varphi(x_{11})\cdots\varphi(x_{1l_1})\cdots\varphi(x_{n1})\cdots\varphi(x_{nl_n})\,. \tag{3.1.28}$$

This identity is called "Wick expansion", because it is the expansion of an n-fold star product in classical products, and the latter correspond (after restriction to $\mathcal{C}_0$) to normally ordered products (also called "Wick products") under the isomorphism Φ (2.6.5).

(b) Prove that the Field independence (3.1.20) implies the commutation relation:

$$\big[R(A_1(x_1),\ldots,A_n(x_n)),\,\varphi(y)\big]_\star$$
$$=i\hbar\sum_{k=1}^n\Delta_m(x_k-y)\sum_a(-1)^{|a|}\,\partial^a_{x_k}R\Big(A_1(x_1),\ldots,\frac{\partial A_k}{\partial(\partial^a\varphi)}(x_k),\ldots,A_n(x_n)\Big)\,.$$

[*Solution*: (a) Compared with the derivation of (3.1.23), the only new relation we need is the Leibniz rule

$$\frac{\delta}{\delta\varphi(x)}(F_1\star\cdots\star F_n)=\sum_{j=1}^n F_1\star\cdots\star\frac{\delta F_j}{\delta\varphi(x)}\star\cdots\star F_n\,,$$

which we prove by induction on n, using $F_1\star\cdots\star F_n=(F_1\star\cdots\star F_{n-1})\star F_n$ and Definition 2.1.1.

(b) We prove this equation by a straightforward computation of the l.h.s.: Starting with (2.1.5), we take into account (2.1.2). Then we apply the Field independence (3.1.20) and (1.3.19) and finally the AWI (3.1.5).]

The next supplementary axioms are the following.

(g) **$*$-structure and Field parity.**
- **$*$-structure:** This axiom reads[26]

$$R_{n-1,1}(F_1 \otimes \cdots \otimes F_n)^* = R_{n-1,1}(F_1^* \otimes \cdots \otimes F_n^*) \ . \tag{3.1.29}$$

If, as is usual, $S^* = S$, then this takes the form $R(e_\otimes^S, F)^* = R(e_\otimes^S, F^*)$.

For the classical retarded products, the validity of (3.1.29) relies on $\overline{\Delta_m^{\rm ret}(x)} = \Delta_m^{\rm ret}(x)$, as shown in Exercise 1.10.4. However, in QFT there appears also the propagator Δ_m^+, which has a different behaviour: $\overline{\Delta_m^+(x)} = \Delta_m^+(-x)$. So we motivate the $*$-structure axiom by the following exercise.

Exercise 3.1.12. Assume that $R_{k,1}$ satisfies the basic axioms for $0 \leq k \leq n-1$ and the $*$-structure axiom for $0 \leq k \leq n-2$. Prove that $R_{n-1,1}(\otimes_{l=1}^{n-1} G_l, H)$ fulfills the $*$-structure axiom if $\operatorname{supp} G_r \cap (\operatorname{supp} H + \overline{V}_+) = \emptyset$ for some $r \in \{1, \ldots, n-1\}$.

[*Solution*: Due to the axiom Symmetry we may assume that it is G_{n-1} which has the mentioned support property with respect to H. Writing $F := G_{n-1}$, the GLZ relation and Causality yield

$$R_{n-1,1}(G_1 \otimes \cdots \otimes G_{n-2} \otimes F, H) = \frac{1}{i} \sum_{I \subseteq \{1,\ldots,n-2\}} \big[R_{|I|,1}(G_I, F), R_{|I^c|,1}(G_{I^c}, H)\big]_\star \ .$$

With that, the assertion follows from

$$\Big(\frac{1}{i} \sum_I \big[R_{|I|,1}(G_I, F), R_{|I^c|,1}(G_{I^c}, H)\big]_\star\Big)^* = \frac{1}{i} \sum_I \big[R_{|I|,1}(G_I, F)^*, R_{|I^c|,1}(G_{I^c}, H)^*\big]_\star$$
$$= \frac{1}{i} \sum_I \big[R_{|I|,1}(G_I^*, F^*), R_{|I^c|,1}(G_{I^c}^*, H^*)\big]_\star = R_{n-1,1}(G_1^* \otimes \cdots \otimes G_{n-2}^* \otimes F^*, H^*) \ .]$$

(g) • **Field parity:** The Field parity axiom is the condition that the relation (1.10.7) for classical R-products holds also true for the R-products of pQFT:

$$\alpha \circ R_{n-1,1} = R_{n-1,1} \circ \alpha^{\otimes n} \ . \tag{3.1.30}$$

To make the relevance of this axiom more transparent, we give equivalent reformulations. For this purpose we introduce, for a field *monomial* A, the order $|A|$ of A in φ: For

$$A = c \prod_{l=1}^{L} \partial^{a_l} \varphi \quad \text{(where } c \in \mathbb{C}) \qquad \text{we define} \quad |A| := L \ ; \tag{3.1.31}$$

[26]For time-ordered products, this condition is much more complicated, see Sect. 3.3.

this can equivalently be written as

$$A(x)(\lambda h) = \lambda^{|A|}\, A(x)(h)\,,\ \forall h \in \mathcal{C}\,,\ x \in \mathbb{M}\,,\ \lambda > 0\,.$$

By means of

$$\alpha A = (-1)^{|A|}\, A\,, \tag{3.1.32}$$

the Field parity axiom (3.1.30) can equivalently be rewritten as

$$\alpha\Big(R_{n-1,1}\big(A_1(x_1),\ldots,A_n(x_n)\big)\Big) = (-1)^{\sum_{j=1}^n |A_j|}\, R_{n-1,1}\big(A_1(x_1),\ldots,A_n(x_n)\big) \tag{3.1.33}$$

for all monomials $A_1,\ldots,A_n \in \mathcal{P}$. Particularly useful is the following reformulation of the Field parity axiom: Applying ω_0 to (3.1.33) and using the obvious identity $\omega_0 \circ \alpha = \omega_0$, we get:

$$\omega_0\Big(R_{n-1,1}\big(A_1(x_1),\ldots,A_n(x_n)\big)\Big) = 0 \quad \text{if} \quad \sum_{j=1}^n |A_j| \quad \text{is odd,} \tag{3.1.34}$$

for all monomials $A_1,\ldots,A_n \in \mathcal{P}$. The condition (3.1.34) is not only necessary for the Field parity axiom (3.1.33), it is also sufficient; this follows by means of the causal Wick expansion (3.1.24) and $|\underline{A}_j| + |\overline{A}_j| = |A_j|$.

For the "unrenormalized" retarded product R^{unrenorm}, introduced below in (3.1.69), the validity of (3.1.34) is obvious: If $\sum_{j=1}^n |A_j|$ is odd, there is no term in the iterated commutator on the r.h.s. of (3.1.69) with all $\partial^a\varphi$ contracted, hence $\omega_0\big(R^{\text{unrenorm}}(A_1,\ldots,A_n)\big) = 0$.

Poincaré covariance. To formulate the axiom "Poincaré covariance", we introduce a (natural) linear action of the proper, orthochronous Poincaré group $\mathcal{P}_+^\uparrow$ on $\mathcal{F}$:

$$F \longmapsto \beta_{\Lambda,a} F := \sum_{n=0}^{N} \int dx_1 \cdots dx_n\, \varphi(\Lambda x_1 + a)\cdots\varphi(\Lambda x_n + a)\, f_n(x_1,\ldots,x_n)\,, \tag{3.1.35}$$

if F is given by (1.2.1) and $(\Lambda, a) \in \mathcal{P}_+^\uparrow$. In the usual way, this induces a linear action of $\mathcal{P}_+^\uparrow$ on $\{\,A(x)\,|\,A \in \mathcal{P}\,\}$, that is, $\beta_{\Lambda,a}\, A(x)$ is defined by

$$\int dx\ g(x)\,\beta_{\Lambda,a}\, A(x) := \beta_{\Lambda,a} \int dx\ g(x)\, A(x) \qquad \forall g \in \mathcal{D}(\mathbb{M})\,. \tag{3.1.36}$$

Example 3.1.13. For $A^{\mu\nu\lambda} := (\partial^\mu\partial^\nu\varphi)(\partial^\lambda\varphi)\varphi^{k-2}$, $k \geq 2$, this definition yields:

$$\begin{aligned}
&\int dx\ g(x)\,\beta_{\Lambda,a}\, A^{\mu\nu\lambda}(x) \\
&= -\beta_{\Lambda,a} \int dx_1 \cdots dx_k \int dx\ g(x)\, \partial^\mu_{x_1}\partial^\nu_{x_1}\partial^\lambda_{x_2}\delta(x_1 - x,\ldots,x_k - x)\,\varphi(x_1)\cdots\varphi(x_k) \\
&= \int dx\ g(x)\,\big(\partial^\mu_x\partial^\nu_x\varphi(\Lambda x + a)\big)\big(\partial^\lambda_x\varphi(\Lambda x + a)\big)\varphi^{k-2}(\Lambda x + a) \\
&= \int dx\ g(x)\,\Lambda_{\mu'}^{\ \mu}\Lambda_{\nu'}^{\ \nu}\Lambda_{\lambda'}^{\ \lambda}\, A^{\mu'\nu'\lambda'}(\Lambda x + a)\,,
\end{aligned}$$

that is, $A^{\mu\nu\lambda}(x)$ is a contravariant Lorentz tensor field of rank 3:

$$\begin{aligned}\beta_{\Lambda,a}\, A^{\mu\nu\lambda}(x) &= \Lambda_{\mu'}^{\ \mu}\Lambda_{\nu'}^{\ \nu}\Lambda_{\lambda'}^{\ \lambda}\, A^{\mu'\nu'\lambda'}(\Lambda x + a) \\ &= (\Lambda^{-1})^{\mu}{}_{\mu'}(\Lambda^{-1})^{\nu}{}_{\nu'}(\Lambda^{-1})^{\lambda}{}_{\lambda'}\, A^{\mu'\nu'\lambda'}(\Lambda x + a)\ .\end{aligned}$$

Obviously, $\mathcal{P}_+^\uparrow \ni (\Lambda, a) \longmapsto \beta_{\Lambda,a} \in \{\, L : \mathcal{F} \longrightarrow \mathcal{F} \,|\, L \text{ is linear} \,\}$ is a representation:

$$\beta_{\Lambda_2,a_2} \circ \beta_{\Lambda_1,a_1} = \beta_{(\Lambda_2,a_2)(\Lambda_1,a_1)}\ .$$

Note that $\beta_{\Lambda,a}$ commutes with the classical and the star product:

$$(\beta_{\Lambda,a}F)(\beta_{\Lambda,a}G) = \beta_{\Lambda,a}(FG)\ , \quad (\beta_{\Lambda,a}F) \star (\beta_{\Lambda,a}G) = \beta_{\Lambda,a}(F \star G) \tag{3.1.37}$$

for all $F, G \in \mathcal{F}$. For the classical product this is obvious; for the star product this follows from the definition (2.1.5) and $H_m\big((\Lambda x + a) - (\Lambda y + a)\big) = H_m(x - y)$ and

$$\beta_{\Lambda,a}\, \frac{\delta^n F}{\delta\varphi(x_1)\cdots\varphi(x_n)} = \frac{\delta^n(\beta_{\Lambda,a}F)}{\delta\varphi(\Lambda x_1 + a)\cdots\varphi(\Lambda x_n + a)}\ . \tag{3.1.38}$$

(h) **Poincaré covariance:** This axiom requires that $\beta_{\Lambda,a}$ *commutes with the retarded product*:

$$\beta_{\Lambda,a}R_{n-1,1}(F_1\otimes\cdots\otimes F_n) = R_{n-1,1}(\beta_{\Lambda,a}F_1\otimes\cdots\otimes\beta_{\Lambda,a}F_n) \quad \text{for} \quad (\Lambda, a) \in \mathcal{P}_+^\uparrow\ . \tag{3.1.39}$$

Considering only Translation covariance (i.e., $\Lambda = 0$) we get

$$\beta_a R_{n-1,1}\big(A_1(x_1)\otimes\cdots\otimes A_n(x_n)\big) = R_{n-1,1}\big(A_1(x_1+a)\otimes\cdots\otimes A_n(x_n+a)\big) \tag{3.1.40}$$

for all $a \in \mathbb{R}^d$ and all $A_1, \ldots, A_n \in \mathcal{P}$. Taking the vacuum expectation value $\omega_0(-)$ of this relation, we conclude that the $\mathbb{C}$-valued distributions

$$r_{n-1,1}(A_1, \ldots, A_n)(x_1 - x_n, \ldots, x_{n-1} - x_n) := \omega_0\Big(R_{n-1,1}\big(A_1(x_1), \ldots, A_n(x_n)\big)\Big) \tag{3.1.41}$$

depend *only* on the relative coordinates, since $\omega_0 \circ \beta_{\Lambda,a} = \omega_0$. By using the causal Wick expansion (3.1.24), one easily verifies that this property (3.1.41) is in fact *equivalent* to Translation covariance (3.1.40).

Remark 3.1.14. Using $\beta_a(F \star G) = (\beta_a F) \star (\beta_a G)$ and proceeding analogously to the derivation of (3.1.41), we conclude that the Wightman functions

$$w^{(m)}(x_1 - x_n, \ldots, x_{n-1} - x_n) := \omega_0\big(A_1(x_1) \star \cdots \star A_n(x_n)\big)\ ,\ A_1, \ldots, A_n \in \mathcal{P}\ , \tag{3.1.42}$$

depend *only* on the relative coordinates.

Exercise 3.1.15. Prove that the classical retarded product is Poincaré covariant, i.e., that $R_{\mathrm{cl}\,n-1,1}$ fulfills (3.1.39).

[*Solution*: The relation (3.1.38) and $\Delta^{\mathrm{ret}}(\Lambda z) = \Delta^{\mathrm{ret}}(z)$ imply

$$\beta_{\Lambda,a} \circ \mathcal{R}_S(x) = \mathcal{R}_{\beta_{\Lambda,a}S}(\Lambda x + a) \circ \beta_{\Lambda,a}$$

for the operator (1.9.1). Using this identity we apply $\beta_{\Lambda,a}$ to the expression (1.9.8) for $R(F_1 \otimes \cdots \otimes F_n, F)$. Taking into account that $x_k^0 \leq x_{k+1}^0$ is equivalent to $(\Lambda x_k + a)^0 \leq (\Lambda x_{k+1} + a)^0$, we obtain the assertion.]

(i) **Further symmetries** (such as conservation of certain currents or gauge symmetries, etc.): The discussion of these is postponed until we reach the Master Ward Identity, in Chaps. 4 and 5.

(j) **Off-shell field equation:** Similarly to (3.1.2) we write $A_S(x) := R(e_\otimes^{S/\hbar}, A(x))$ for $A \in \mathcal{P}$. We require that the classical off-shell field equation (1.6.6) (or (1.7.7)) holds true also for the interacting fields in pQFT:

$$(\Box + m^2)\varphi_S(x) = (\Box + m^2)\varphi(x) + \Big(\frac{\delta S}{\delta\varphi(x)}\Big)_S\,, \tag{3.1.43}$$

or, in integrated form,

$$\varphi_S(x) = \varphi(x) - \int dy\, \Delta_m^{\mathrm{ret}}(x-y)\Big(\frac{\delta S}{\delta\varphi(y)}\Big)_S \quad \text{for all} \quad S \in \mathcal{F}_{\mathrm{loc}}\,. \tag{3.1.44}$$

In contrast to classical field theory, the interacting field $\big(\frac{\delta S}{\delta\varphi}\big)_S$ cannot be expressed in terms of the basic interacting field φ_S, because the factorization (3.1.1) does not hold.

In clFT the off-shell field equation is essentially the defining property (ii) of a retarded wave operator (Definition 1.6.2); here it is *a renormalization condition*, as we will see in Sect. 3.2.

In view of the inductive construction of the retarded products given in Sect. 3.2, it is instructive to write the off-shell field equation in the form[27]

$$\begin{aligned} &R_{n,1}\big(F_1 \otimes \cdots \otimes F_n,\, \varphi(x)\big) \\ &\quad = -\hbar \int dy\, \Delta_m^{\mathrm{ret}}(x-y) \sum_{l=1}^{n} R_{n-1,1}\Big(F_1 \otimes \cdots \otimes \widehat{F_l} \otimes \cdots \otimes F_n,\, \frac{\delta F_l}{\delta\varphi(y)}\Big)\,, \end{aligned} \tag{3.1.45}$$

which is obtained by inserting $S := \sum_{j=1}^n \lambda_j F_j$ into (3.1.44) and applying the derivative $\frac{\partial^n}{\partial\lambda_1\cdots\partial\lambda_n}\big|_{\lambda_1=\cdots=\lambda_n=0}$.

[27] As usual, the notation $\widehat{-_k}$ indicates a missing item from a list.

Formula (3.1.45) may be illustrated by the diagram

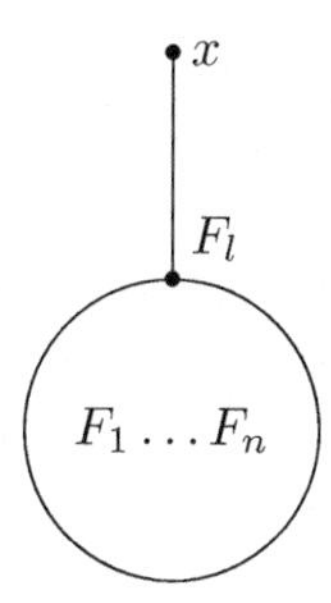

where time is advancing vertically. We see: Assuming that $R_{n-1,1}$ is constructed and fulfills all axioms, the off-shell field equation (3.1.45) determines $R_{n,1}\big(F_1 \otimes \cdots \otimes F_n,\, \varphi(x)\big)$ *uniquely* for all $F_1, \dots, F_n \in \mathcal{F}_{\mathrm{loc}}$. We easily verify that the so-constructed $R_{n,1}\big(F_1 \otimes \cdots \otimes F_n,\, \varphi(x)\big)$ satisfies the basic axioms (a) (Linearity), (b) (Symmetry), (d) (Causality) and the renormalization conditions (g) ($*$-structure and Field parity) and (h) (Poincaré covariance). The axiom (f) (Field independence) is verified in the following exercise.

Exercise 3.1.16 (Compatibility of the "Off-shell field equation" and the "Field independence"). Assume that $R_{n-1,1}$ fulfills the axiom (f) (Field independence). Prove that $R_{n,1}\big(\cdots,\, \varphi(x)\big)$, defined by the off-shell field equation (3.1.45), satisfies also the axiom (f).

[*Solution*: Due to $\frac{\delta\varphi(x)}{\delta\varphi(y)} = \delta(x-y)$ and Remk. 3.1.4 the assertion reads

$$\frac{\delta}{\delta\varphi(z)} R_{n,1}\big(F_1 \otimes \cdots \otimes F_n,\, \varphi(x)\big) = \sum_{j=1}^{n} R_{n,1}\Big(F_1 \otimes \cdots \otimes \frac{\delta F_j}{\delta\varphi(z)} \otimes \cdots \otimes F_n,\, \varphi(x)\Big)\,.$$

To compute the l.h.s. we use (3.1.45) and axiom (f) for $R_{n-1,1}$; this yields

$$\begin{aligned}\text{l.h.s.} = -\hbar \int dy\, \Delta_m^{\mathrm{ret}}(x-y) \sum_{l=1}^{n} \Big(& R_{n-1,1}\Big(F_1 \otimes \cdots \widehat{F_l} \cdots \otimes F_n,\, \frac{\delta^2 F_l}{\delta\varphi(z)\delta\varphi(y)}\Big) \\ & + \sum_{j=1,\, j\neq l}^{n} R_{n-1,1}\Big(F_1 \otimes \cdots \frac{\delta F_j}{\delta\varphi(z)} \cdots \widehat{F_l} \cdots \otimes F_n,\, \frac{\delta F_l}{\delta\varphi(y)}\Big)\Big)\,.\end{aligned}$$

For the r.h.s. of the assertion we obtain the same result by using (3.1.45) for $R_{n,1}\big(F_1 \otimes \cdots \frac{\delta F_j}{\delta\varphi(z)} \cdots \otimes F_n,\, \varphi(x)\big)$.]

3.1.5 The scaling and mass expansion

References for this section are [64, 101] and to some extent [55].

Mostly, the literature about causal perturbation theory (e.g., [24, 66, 148]) works with a condition for the scaling behaviour of the retarded products which concerns only the UV-region, that is, the asymptotic scaling behaviour of the distributions $r(A_1,\dots,A_n)(\mathbf{y})$ (3.1.41) (where $\mathbf{y} := (x_1 - x_n, \dots)$) for $\mathbf{y} \to \mathbf{0}$ – this is the "Scaling degree axiom" treated in Sect. 3.2.5. In view of the techniques to renormalize in practice explained in Sect. 3.5, we prefer to work with a clearly stronger version of this axiom, which takes into account that the *classical* retarded product $R^{(m)}_{n-1,1}$ (Def. 1.7.1) behaves homogeneously under the scaling $(X, m) \mapsto (\rho X, \rho^{-1} m)$, $\forall X := (x_1, \dots, x_n) \in \mathbb{M}^n$, $\forall m \geq 0$ and $\forall \rho > 0$ (where m denotes the mass of the retarded propagator, which is also the mass of the star product $\star_m$). This property cannot be maintained under quantization, but we may require that it is "almost" maintained (i.e., up to powers of $\log \rho$). In addition, since we want to work with distributions having good properties under scaling in X only, we require that the distributions $r^{(m)}(A_1,\dots,A_n)$ (3.1.41) can be expanded with respect to m in a "Taylor series modulo $\log m$"; with this, the first requirement amounts to the condition that the coefficients of this series are "almost" homogeneous (i.e., homogeneous up to logarithmic terms) with respect to $X \mapsto \rho X$.

Almost homogeneous scaling, mass dimension and scaling of fields. To work this idea out, we need some mathematical notions about scaling of distributions and fields.

Definition 3.1.17 ((Almost) homogeneous scaling)**.** We say that a distribution t in $\mathcal{D}'(\mathbb{R}^k)$ or in $\mathcal{D}'(\mathbb{R}^k \setminus \{0\})$ scales *almost homogeneously* with *degree* $D \in \mathbb{C}$ and *power* $N \in \mathbb{N}$ if and only if [28]

$$(\mathbb{E}_k + D)^{N+1} t(z_1, \dots, z_k) = 0 \quad \text{and} \tag{3.1.46a}$$

$$(\mathbb{E}_k + D)^{N} t(z_1, \dots, z_k) \neq 0 \; ; \tag{3.1.46b}$$

where $\mathbb{E}_k := \sum_{r=1}^k z_r \, \partial/\partial z_r$ is the Euler operator.

When $N = 0$, we say there is *homogeneous* scaling of degree D.

According to our definition, $t(z) = z^{-D} \log^N(z)$ is almost homogeneous with degree D and power N. The reason for the somewhat strange terminology with respect to the sign of D is that the scaling degree (introduced in Definition 3.2.5) of a distribution fulfilling (3.1.46a) is defined to be D rather than $(-D)$.[29]

As we explain below, the relation (3.1.46a) holds if and only if

$$0 = (\rho\,\partial_\rho)^{N+1} \big(\rho^D \, t(\rho z_1, \dots, \rho z_k)\big) = \frac{\partial^{N+1}}{\partial(\log \rho)^{N+1}} \big(\rho^D \, t(\rho z_1, \dots, \rho z_k)\big) \quad \forall \rho > 0 \; ; \tag{3.1.47}$$

and both parts of (3.1.46) hold if and only if $\rho^D \, t(\rho z_1, \dots, \rho z_k)$ is a polynomial in $\log \rho$, of degree N.

---

[28]The relations (3.1.46) have to be understood in the sense of distributions, that is, $f(z) = 0$ if and only if $\int dz \; f(z)\, g(z) = 0 \; \forall g \in \mathcal{D}(\mathbb{R}^k[\setminus\{0\}])$. In this section and later we shall write "$\mathcal{D}(\mathbb{R}^k[\setminus\{0\}])$" for "$\mathcal{D}(\mathbb{R}^k)$ or $\mathcal{D}(\mathbb{R}^k \setminus \{0\})$", and likewise for $\mathcal{D}'(\mathbb{R}^k[\setminus\{0\}])$.

[29]Other names appear in the literature: Distributions which scale almost homogeneously with degree D and power N are also called "associate homogeneous of scaling degree $(-D)$ and scaling order N" in [80, 131] or "log-homogeneous with bidegree $(-D, N)$" in [86].

To wit, computing the $(N+1)$th derivative in (3.1.47) at $\rho = 1$ we obtain (3.1.46a). That (3.1.46a) (or (3.1.47) at $\rho = 1$) is also *sufficient* for (3.1.47) can be seen as follows:

$$\begin{aligned} 0 &= \big\langle (\rho\,\partial_\rho)^{N+1}\big(\rho^D\, t(\rho z)\big), g(z)\big\rangle_z \\ &= \rho^D \big\langle \big((\mathbb{E}_k + D)^{N+1} t\big)(\rho z), g(z)\big\rangle_z \quad \forall \rho > 0\ ,\ g \in \mathcal{D}(\mathbb{R}^k[\setminus\{0\}]) \end{aligned}$$

is (by setting $\tilde{z} := \rho z$) equivalent to

$$0 = \big\langle (\mathbb{E}_k + D)^{N+1} t(\tilde{z}), \tilde{g}(\tilde{z})\big\rangle_{\tilde{z}} \quad \forall \tilde{g} \in \mathcal{D}(\mathbb{R}^k[\setminus\{0\}])\ . \qquad \boxminus$$

To define a scaling transformation on the field space $\mathcal{F}$, we need the notion of the "mass dimension" (also called "scaling dimension") of a monomial $A \in \mathcal{P}$.

Definition 3.1.18 (Mass dimension). The *mass dimension* of $\partial^a\varphi \in \mathcal{P}$ is defined by

$$\dim \partial^a \varphi := \frac{d-2}{2} + |a| \quad \text{for} \quad a \in \mathbb{N}^d\ .$$

For *monomials* $A_1, A_2 \in \mathcal{P}$, we agree that $\dim(A_1 A_2) := \dim A_1 + \dim A_2$.

Denote by $\mathcal{P}_j$ the vector space spanned by all monomials $A \in \mathcal{P}$ with $\dim A = j$. Write $\mathcal{P}_{\mathrm{hom}} := \bigcup_{j\in\mathbb{N}} \mathcal{P}_j$ (union, not direct sum) to denote the set of "homogeneous" polynomials, in this sense.

Remark 3.1.19. In practice, a way to determine the mass dimension of a basic field ϕ, e.g., a complex scalar field or a Dirac spinor field, is the following: Using that each monomial in the Lagrangian $\mathcal{L}_0$ of the free theory has mass dimension d, where one sets $\dim m^l := l$, and taking into account the relations $\dim \partial^a\phi = \dim \partial^a \phi^* = \dim \phi + |a|$ and the formula $\dim A_1 A_2 = \dim A_1 + \dim A_2$, one obtains the value of $\dim \phi$.

Now, the scaling of fields is defined as follows.

Definition 3.1.20 (Scaling transformation). The *scaling transformation* σ_ρ, for $\rho > 0$, is defined on the basic field $\varphi(x)$ by

$$\sigma_\rho\big(\varphi(x)\big) := \rho^{-\dim\varphi} \varphi(\rho^{-1}x)\ . \tag{3.1.48}$$

For general fields, $\sigma_\rho \colon \mathcal{F} \longrightarrow \mathcal{F}$ is given by

$$\begin{aligned} &\sigma_\rho\Big(\sum_{n=0}^{N} \langle f_n, \varphi^{\otimes n}\rangle\Big) \\ &\quad := \sum_{n=0}^{N} \rho^{-n\dim\varphi} \int dx_1 \cdots dx_n\, f_n(x_1,\ldots,x_n)\, \varphi(\rho^{-1}x_1)\cdots\varphi(\rho^{-1}x_n)\ . \end{aligned}$$

Heuristically speaking, "σ_ρ acts on each $\varphi(x_j)$ according to (3.1.48) and leaves everything else unchanged".

One easily verifies the following properties:

- σ_ρ is an algebra isomorphism of $(\mathcal{F}, \cdot)$; in particular,
$$\sigma_\rho(F_1 \cdot F_2) = \sigma_\rho(F_1) \cdot \sigma_\rho(F_2) \; ; \tag{3.1.49}$$
- the inverse σ_ρ^{-1} is given by $\sigma_\rho^{-1} = \sigma_{1/\rho}$.
- For $A \in \mathcal{P}$ we define $\sigma_\rho(A(x))$ by $\int dx\; g(x)\,\sigma_\rho(A(x)) := \sigma_\rho\big(\int dx\; g(x)\,A(x)\big)$ for all $g \in \mathcal{D}(\mathbb{M})$. This gives the frequently used relation
$$\rho^{\dim A}\,\sigma_\rho(A(\rho x)) = A(x)\;, \quad \forall A \in \mathcal{P}_{\mathrm{hom}}\;. \tag{3.1.50}$$
- $\omega_0 \circ \sigma_\rho = \omega_0$, a trivial but useful remark.

For $m > 0$ the physically relevant scaling is a simultaneous scaling of $X := (x_1, \ldots)$ and m: $(X, m) \mapsto (\rho X, \rho^{-1} m)$. The Wightman two-point function Δ_m^+ behaves homogeneously under that, as already noted in (2.2.9). In the next exercise we verify that this entails
$$\sigma_\rho(F_1 \star_{m/\rho} \cdots \star_{m/\rho} F_n) = \sigma_\rho F_1 \star_m \cdots \star_m \sigma_\rho F_n\;. \tag{3.1.51}$$
For $m = 0$, this relation simplifies to
$$\sigma_\rho(\sigma_\rho^{-1} F_1 \star_0 \cdots \star_0 \sigma_\rho^{-1} F_n) = F_1 \star_0 \cdots \star_0 F_n\;, \tag{3.1.52}$$
where $\star_0$ means the star product for $m = 0$.

Exercise 3.1.21.

(a) Verify the relation
$$\sigma_\rho\,\frac{\delta F}{\delta\varphi(x)} = \rho^{-1-d/2}\,\frac{\delta(\sigma_\rho F)}{\delta\varphi(\rho^{-1}x)} \tag{3.1.53}$$
and use this relation to prove (3.1.51).

(b) Let $A_1, \ldots, A_n \in \mathcal{P}$ be *monomials.* Show that the pertinent Wightman function (3.1.42) scales homogeneously under $(X,m) \mapsto (\rho X, m/\rho)$ (where $X := (x_1, \ldots, x_n)$) with degree $D := \sum_{l=1}^n \dim A_l$, i.e.,
$$\rho^D\, w^{(m/\rho)}\big(\rho(x_1 - x_n), \ldots, \rho(x_{n-1} - x_n)\big) = w^{(m)}(x_1 - x_n, \ldots, x_{n-1} - x_n)\;.$$

[*Solution*: (a) Formula (3.1.53) is obtained straightforwardly by combining Definitions 1.3.1 and 3.1.20 of the functional derivative and of σ_ρ, respectively. Computing $\sigma_\rho(F_1 \star_{m/\rho} F_2)$ by means of the definition of the star product (2.1.5) and using (3.1.53) and (2.2.9), we get $\sigma_\rho F_1 \star_m \sigma_\rho F_2$. The generalization to an n-fold star product is obtained by induction on n and by taking into account the associativity of the star product.
(b) By using $\omega_0 \circ \sigma_\rho^{-1} = \omega_0$, the relation (3.1.51) and finally $\sigma_\rho^{-1}(A(x)) = \rho^{\dim A}\, A(\rho x)$, we obtain
$$\begin{aligned} w^{(m)}(x_1 - x_n, \ldots, x_{n-1} - x_n) &= \omega_0\Big(\sigma_\rho^{-1}\big(A_1(x_1) \star_m \cdots \star_m A_n(x_n)\big)\Big) \\ &= \omega_0\Big(\sigma_\rho^{-1}(A_1(x_1)) \star_{m/\rho} \cdots \star_{m/\rho} \sigma_\rho^{-1}(A_n(x_n))\Big) \\ &= \rho^{\sum_l \dim A_l}\, w^{(m/\rho)}\big(\rho(x_1 - x_n), \ldots, \rho(x_{n-1} - x_n)\big)\;. \end{aligned}$$
]

The scaling and mass expansion for the classical retarded product. Next we investigate the scaling behaviour of the classical retarded product R_{cl}. Writing $\mathcal{R}_S^{(m)}(x)$ for the operator $\mathcal{R}_S(x)$ (1.9.1) and $R_{\text{cl}}^{(m)}$ for R_{cl}, where the upper index m stands for the mass of $\Delta^{\text{ret}} \equiv \Delta_m^{\text{ret}}$, we find

$$\sigma_\rho \circ \mathcal{R}^{(m/\rho)}_{\sigma_\rho^{-1}S}(x) = \rho^{-d}\,\mathcal{R}_S^{(m)}(\rho^{-1}x) \circ \sigma_\rho \,, \tag{3.1.54}$$

by proceeding analogously to part (a) of the preceding exercise and by taking into account $\rho^{d-2}\,\Delta^{\text{ret}}_{m/\rho}(\rho x) = \Delta^{\text{ret}}_m(x)$. By using (3.1.54), we apply σ_ρ to the expression (1.9.2) for $R^{(m/\rho)}\big((\sigma_\rho^{-1}S)^{\otimes n}, \sigma_\rho^{-1}F\big)$ and obtain

$$\begin{aligned} \sigma_\rho \circ R_{\text{cl}}^{(m/\rho)}\big((\sigma_\rho^{-1}S)^{\otimes n}, \sigma_\rho^{-1}F\big) &= R_{\text{cl}}^{(m)}\big(S^{\otimes n}, F\big) \\ \text{or} \quad \sigma_\rho\big((\sigma_\rho^{-1}F)^{(m/\rho)\,\text{ret}}_{\sigma_\rho^{-1}S}\big) &= F_S^{(m)\,\text{ret}} \end{aligned} \tag{3.1.55}$$

for all $S, F \in \mathcal{F}_{\text{loc}}$; that is, the *classical retarded fields scale homogeneously.* Working with (1.9.8) instead of (1.9.2), we obtain the scaling relation (3.1.55) for non-diagonal entries:

$$\sigma_\rho\, R_{\text{cl}}^{(m/\rho)}\big(\sigma_\rho^{-1}\,A_1(x_1) \otimes \cdots \otimes \sigma_\rho^{-1}\,A_n(x_n)\big) = R_{\text{cl}}^{(m)}\big(A_1(x_1) \otimes \cdots \otimes A_n(x_n)\big)\,. \tag{3.1.56}$$

Since R_{cl} is translation invariant (see Exer. 3.1.15), we may proceed similarly to (3.1.41): We define $r_{\text{cl}}^{(m)}(A_1, \ldots, A_n)(x_1 - x_n, \ldots)$ to be the vacuum expectation value of the r.h.s. of (3.1.56). Using $\omega_0 \circ \sigma_\rho = \omega_0$ and $\sigma_\rho^{-1}(A_l(x)) = \rho^{\dim A_l}\, A_l(\rho x)$, we conclude that

$$\begin{aligned} &\rho^D\, r_{\text{cl}}^{(m/\rho)}(A_1, \ldots, A_n)\big(\rho(x_1 - x_n), \ldots, \rho(x_{n-1} - x_n)\big) \\ &\quad = r_{\text{cl}}^{(m)}(A_1, \ldots, A_n)(x_1 - x_n, \ldots, x_{n-1} - x_n)\,, \quad D := \sum_{l=1}^n \dim A_l \in \mathbb{N}\,, \end{aligned} \tag{3.1.57}$$

for all $A_1, \ldots, A_n \in \mathcal{P}_{\text{hom}}$. The restriction to $\mathcal{P}_{\text{hom}}$ is necessary in order that $\dim A_l$ is well defined.

That D is a natural number can be seen as follows: Let $A_1, \ldots, A_n$ be monomials. Now, $r_{\text{cl}}(A_1, \ldots A_n)$ is non-vanishing only if $\sum_l |A_l|$ is even; this follows from Field parity, similarly to (3.1.34). From that and from

$$\dim A_l = |A_l| \cdot \tfrac{1}{2}(d-2) + [\text{number of derivatives in } A_l]\,,$$

we conclude that $D \in \mathbb{N}$.

The information contained in (3.1.56) is completely encoded in (3.1.57), for the following reason: Due to linearity of R_{cl} it suffices to consider *monomials* $A_1, \ldots, A_n$. Now, R_{cl} satisfies Field independence (1.10.2) and, hence, the causal Wick expansion (3.1.24). By the latter $R_{\text{cl}}(A_1, \ldots, A_n)$ is uniquely given in terms of the vacuum expectation values $r_{\rm cl}\big(\text{submonomials of } A_1,\dots,A_n\big)$. So, if the latter scale homogeneously, this holds also for $R_{\rm cl}\big(A_1,\dots,A_n\big)$. Or explicitly, (3.1.56) can be derived from (3.1.57) as follows: By using the causal Wick expansion (3.1.24) and the relation

$$\dim \underline{A} + \dim \overline{A} = \dim A \quad \text{for monomials } A, \tag{3.1.58}$$

which is a consequence of the definition (3.1.25), we obtain

$$\begin{aligned}\sigma_\rho\, R_{\rm cl}^{(m/\rho)}\big(\sigma_\rho^{-1}\, A_1(x_1)\otimes\cdots\big) &= \rho^{\sum_l \dim A_l}\, \sigma_\rho\, R_{\rm cl}^{(m/\rho)}\big(A_1(\rho x_1)\otimes\cdots\big)\\ &= \sum_{\underline{A}_l\subseteq A_l} \rho^{\sum_l \dim \underline{A}_l}\, r_{\rm cl}^{(m/\rho)}(\underline{A}_1,\dots)\big(\rho(x_1-x_n),\dots\big)\, \rho^{\dim \overline{A}_1}\, \sigma_\rho\big(\overline{A}_1(\rho x_1)\big)\cdots\\ &= \sum_{\underline{A}_l\subseteq A_l} r_{\rm cl}^{(m)}(\underline{A}_1,\dots)(x_1-x_n,\dots)\, \overline{A}_1(x_1)\cdots = R_{\rm cl}^{(m)}\big(A_1(x_1)\otimes\cdots\big)\ .\end{aligned} \tag{3.1.59}$$

As we have already seen in the example treated in Sect. 3.1.2, the main problem in pQFT is renormalization, i.e., the extension of distributions $u^0\in\mathcal{D}'(\mathbb{R}^k\setminus\{0\})$ to $u\in\mathcal{D}'(\mathbb{R}^k)$. If u^0 scales almost homogeneously under $\mathbb{R}^k\ni X\mapsto\rho X$, there exists an extension which maintains almost homogeneous scaling (Prop. 3.2.16 in Sect. 3.2.2) and a lot is known about the uniqueness and the construction of such an extension. However, in the scaling property (3.1.57) the mass is also scaled. In view of these facts, we are looking for a reformulation of (3.1.57) in terms of distributions which scale homogeneously under $X:=(x_1-x_n,\dots)\mapsto\rho X$ only, instead of $(X,m)\mapsto(\rho X,\rho^{-1}m)$. This can be done as follows: First we recall that $m\longmapsto\Delta_m^{\rm ret}(g)$ is smooth in $m\geq 0$, for each $g\in\mathcal{D}(\mathbb{R}^d)$. By (1.9.8), this implies that $R_{\rm cl}^{(m)}(F_1\otimes\cdots\otimes F_n)$ is smooth in $m\geq 0$ for all $F_1,\dots,F_n\in\mathcal{F}_{\rm loc}$; and hence this holds also for $m\longmapsto r_{\rm cl}^{(m)}(A_1,\dots,A_n)(g)$ for all $g\in\mathcal{D}(\mathbb{R}^{d(n-1)})$ and for all $A_1,\dots,A_n\in\mathcal{P}$; where we assume that $F_1,\dots,F_n$ or $A_1,\dots,A_n$, respectively, do not depend on m. Therefore, the following Taylor expansion is possible: For all $l,L\in\mathbb{N}$, there exist distributions u_l, $\mathfrak{r}_{L+1}^{(m)}\in\mathcal{D}'(\mathbb{R}^{d(n-1)})$ such that

$$r_{\rm cl}^{(m)}(A_1,\dots,A_n)(X)=\sum_{l=0}^{L} m^l\, u_l(X)+\mathfrak{r}_{L+1}^{(m)}(X)\quad\forall L\in\mathbb{N}, \tag{3.1.60}$$

where $X:=(x_1-x_n,\dots,x_{n-1}-x_n)$, and the following properties hold:

(a) The distributions $u_l(X)$ are independent of m and $u_0=r_{\rm cl}^{(m=0)}(A_1,\dots,A_n)$;

(b) the remainders $\mathfrak{r}_{L+1}^{(m)}$ fulfill

$$\lim_{m\downarrow 0}\Big(\frac{m}{M}\Big)^{-(L+1)+\varepsilon}\mathfrak{r}_{L+1}^{(m)}=0\qquad\forall\varepsilon>0\ ,$$

where $M>0$ is an arbitrary mass scale.

If $A_1, \ldots, A_n$ are monomials, the scaling behaviour (3.1.57) of $r_{\mathrm{cl}}^{(m)}(A_1, \ldots, A_n)$ translates into the following properties:

(c) The distributions $u_l(X)$ scale homogeneously in X with degree $D - l$;

(d) and the remainders $\mathfrak{r}_{L+1}^{(m)}(X)$ are homogeneous with degree D under the scaling $(X, m) \mapsto (\rho X, m/\rho)$, i.e., they fulfill (3.1.57).

The scaling and mass expansion: An axiom for pQFT. Trying to preserve this structure (i.e., the expansion (3.1.60) with the properties (a)–(d)) in the quantization, the following crucial complications appear:

- For *even* dimensions d, the Wightman two-point function $\Delta_m^+(z)$ is not smooth in m at $m = 0$, due to the term $m^2 f(m^2 z^2) \log\big(-\frac{m^2}{4}(z^2 - iz^0 0)\big)$ in (2.2.3). Therefore, the Taylor expansion (3.1.60) has to be generalized to an expansion of the form

$$r^{(m)}(A_1, \ldots, A_n)(X) = \sum_{l=0}^{L} m^l \sum_{p=0}^{P_l} \log^p\Big(\frac{m}{M}\Big)\ u_{l,p}(X) + \mathfrak{r}_{L+1}^{(m)}(X)\,, \quad L, P_l \in \mathbb{N}\,,$$

where $M > 0$ is a fixed mass scale.

- In general a homogeneous behaviour under $X \mapsto \rho X$ cannot be maintained in the renormalization; this is the "scaling anomaly" (also called "dilatation anomaly"). However, one can reach that the breaking of this symmetry is "weak", in the sense that the breaking term is $\sim \log \rho$. Or more general, if $u^0 \in \mathcal{D}'(\mathbb{R}^k \setminus \{0\})$ scales almost homogeneously with degree D and power N, it may happen that there is no extension $u \in \mathcal{D}'(\mathbb{R}^k)$ which scales also with D and N, but then there exists an extension which scales with D and $(N + 1)$, see Proposition 3.2.16.

  To give an example let us study the distribution r^0 appearing in (3.1.13) in the massless case: $r^0(z) := \big(D^+(z)^2 - D^+(-z)^2\big)\,\theta(z^0) \in \mathcal{D}'(\mathbb{R}^4 \setminus \{0\})$ scales homogeneously with degree 4. The best that exists is an extension $r \in \mathcal{D}'(\mathbb{R}^4)$ which scales almost homogeneously with degree 4 and power 1, see (3.2.119).

  Therefore, in the properties (c) and (d) above, we have to replace "homogeneous scaling" by "almost homogeneous scaling".

These facts and thoughts motivate the following axiom.

(k) **Scaling and mass expansion:** For all field *monomials* $A_1, \ldots, A_n \in \mathcal{P}$ which do not depend on m, the numerical distributions

$$\begin{aligned} r^{(m)}(A_1, \ldots, A_n)&(x_1 - x_n, \ldots, x_{n-1} - x_n) \\ &:= \omega_0\big(R_{n-1,1}^{(m)}\big(A_1(x_1) \otimes \cdots \otimes A_n(x_n)\big)\big) \end{aligned} \tag{3.1.61}$$

fulfill the scaling and mass expansion (shortly: The *Sm-expansion*) with degree $D := \sum_{l=1}^{n} \dim A_l$, where the following definition is used.

Definition 3.1.22. A distribution $f^{(m)} \in \mathcal{D}'(\mathbb{R}^k)$ or $f^{(m)} \in \mathcal{D}'(\mathbb{R}^k \setminus \{0\})$, depending on $m \geq 0$, fulfills the Sm-expansion with degree $D \in \mathbb{R}$ if and only if, for all $l, L \in \mathbb{N}$, there exist distributions $u_l^{(m)}$, $\mathfrak{r}_{L+1}^{(m)} \in \mathcal{D}'(\mathbb{R}^k[\setminus\{0\}])$ such that

$$f^{(m)}(X) = \sum_{l=0}^{L} m^l\, u_l^{(m)}(X) + \mathfrak{r}_{L+1}^{(m)}(X) \quad \forall L \in \mathbb{N},\ m > 0\ , \tag{3.1.62}$$

where $X := (x_1, \ldots, x_k)$, and the following properties hold true:

(a) The leading term $u_0 \equiv u_0^{(m)}$ is independent of m and it agrees with $f^{(m=0)}$: $u_0 = f^{(m=0)}$.

(b) For $l \geq 1$ the m-dependence of $u_l^{(m)}(X)$ is a polynomial in $\log \frac{m}{M}$, where $M > 0$ is a fixed mass scale. Explicitly, there exist m-independent distributions $u_{l,p} \in \mathcal{D}'(\mathbb{R}^k[\setminus\{0\}])$ such that

$$u_l^{(m)}(X) = \sum_{p=0}^{P_l} \log^p\Big(\frac{m}{M}\Big)\ u_{l,p}(X)\ , \quad P_l < \infty\ , \quad \forall m > 0\,. \tag{3.1.63}$$

(Of course, the distributions $u_{l,p}$ may depend on M.)

(c) $u_l^{(m)}(X)$ scales almost homogeneously in X with degree $D-l$ and, hence, this holds also for all $u_{l,p}\,,\ p = 0, 1, \ldots, P_l$ (3.1.63).

(d) $\mathfrak{r}_{L+1}^{(m)}(X)$ is almost homogeneous with degree D under the scaling $(X, m) \mapsto (\rho X, m/\rho)$.

(e) $\mathfrak{r}_{L+1}^{(m)}$ is smooth in m for $m > 0$ and

$$\lim_{m \downarrow 0} \Big(\frac{m}{M}\Big)^{-(L+1)+\varepsilon} \mathfrak{r}_{L+1}^{(m)} = 0 \qquad \forall \varepsilon > 0\,.$$

All properties are meant in the weak sense; e.g., property (e) holds for $\langle \mathfrak{r}_{L+1}^{(m)}, h\rangle$ for all $h \in \mathcal{D}(\mathbb{R}^k[\setminus\{0\}])$.

We point out that the defining properties of the Sm-expansion do not contain any statement about convergence of the limit $\lim_{L\to\infty} \mathfrak{r}_{L+1}^{(m)}$; the remainders are only restricted by the properties (d) and (e) and by

$$\mathfrak{r}_{L_1+1}^{(m)\,0}(X) - \mathfrak{r}_{L_2+1}^{(m)\,0}(X) = \sum_{l=L_1+1}^{L_2} m^l\, u_l^{(m)}(X)\,, \quad \forall 0 \leq L_1 < L_2\,, \tag{3.1.64}$$

which follows immediately from (3.1.62).

For the same reason as in (3.1.57), the degree $D = \sum_l \dim A_l$ is a natural number.

The name "scaling and mass expansion" refers to the following two possibilities to interpret (3.1.62)–(3.1.63): On the one hand it is an expansion in terms of m-independent, almost homogeneously scaling distributions $u_{l,p}$ and on the other hand it is a "Taylor expansion in the mass m modulo $\log m$".

Remark 3.1.23. For a *massless* model, i.e., for $m = 0$, the Sm-expansion axiom obviously reduces to the condition that the distributions $r(A_1, \ldots, A_n)$ (3.1.61) scale almost homogeneously with degree $D := \sum_{l=1}^{n} \dim A_l$ for all monomials $A_1, \ldots, A_n \in \mathcal{P}$.

Exercise 3.1.24. In $d = 4$ dimensions verify that the Wightman two-point function Δ_m^+ fulfills the Sm-expansion with degree $D = 2$, by using (2.2.3).

[*Solution*: Writing the analytic functions f, f_1 appearing in (2.2.3) as

$$f(y) = \sum_{k=0}^{\infty} a_k\, y^k \ , \quad f_1(y) = \sum_{k=0}^{\infty} b_k\, y^k \ , \quad a_k, b_k \in \mathbb{R} \ ,$$

and introducing the shorthand notation $Z := -(z^2 - iz^0 0)$, we obtain

$$\Delta_m^+(z) = u_0(z) + \sum_{l=1}^{L} \Big(u_{2l,0}(z) + u_{2l,1}(z) \log\Big(\frac{m}{M}\Big) \Big)\, m^{2l} + \mathfrak{r}^{(m)}_{2L+2}(z) \ ,$$

with

$$u_0(z) = \frac{1}{4\pi^2 Z} = D^+(z) \ , \quad u_{2l,0}(z) = \Big(a_{l-1} \log\Big(\frac{M^2 Z}{4}\Big) + b_{l-1} \Big)\, (z^2)^{l-1} \ ,$$

$$u_{2l,1}(z) = 2a_{l-1}\, (z^2)^{l-1} \ , \quad \mathfrak{r}^{(m)}_{2L+2}(z) = m^2 \sum_{l=L+1}^{\infty} \Big(a_{l-1} \log\Big(\frac{m^2 Z}{4}\Big) + b_{l-1} \Big)\, (m^2 z^2)^{l-1} \ ,$$

where D^+ denotes the massless two-point function (see Sect. 2.2). Note that only even powers of m appear, hence the remainder has the index $(2L + 2)$ instead of $(2L + 1)$.]

Example 3.1.25 (Sm-expansion axiom for $r(\varphi, \varphi)$ in $d \geq 3$ dimensions). For arbitrary $d \geq 3$, $\Delta_m^{+\,(d)}$ fulfills the Sm-expansion with degree $D = d - 2$. (If d is odd, $\Delta_m^{+\,(d)}$ is smooth in $m \geq 0$, hence the Sm-expansion is simply the Taylor expansion.) Proceeding analogously to the example in Sect. 3.1.2, we obtain

$$r^{(m)}(\varphi, \varphi)(x_1 - x_2) = \frac{1}{i}\, \omega_0\Big([\varphi(x_1, \varphi(x_2)]_\star\, \theta(x_2^0 - x_1^0) \Big) = \frac{\hbar}{i} \Big(\Delta_m^+(X) - \Delta_m^+(-X) \Big)\, \theta(-X^0)$$

(where $X := x_1 - x_2$). We conclude that $r(\varphi, \varphi)$ fulfills the Sm-expansion with degree $D = d - 2$. Since $\dim \varphi = \frac{d-2}{2}$, the axiom (k) is satisfied.

The following lemma gives basic properties of distributions fulfilling the Sm-expansion.

Lemma 3.1.26. *If $f^{(m)} \equiv f_1^{(m)} \in \mathcal{D}'(\mathbb{R}^k[\setminus\{0\}])$ and $f_2^{(m)} \in \mathcal{D}'(\mathbb{R}^q[\setminus\{0\}])$ satisfy the Sm-expansion with degree $D \equiv D_1$ or D_2, respectively, then the following statements hold true:*

(0) *If $k = q$ and $D_1 = D_2$, then every linear combination $\alpha f_1^{(m)} + \beta f_2^{(m)}$, $(\alpha, \beta) \in \mathbb{C}^2$ arbitrary, fulfills also the Sm-expansion with degree D_1.*

(1) *$f^{(m)}$ is smooth in m for $m > 0$ and $\lim_{m\downarrow 0} f^{(m)} = u_0 = f^{(m=0)}$.*

(2) *$f^{(m)}(X)$ is almost homogeneous with degree D under the scaling $(X, m) \mapsto (\rho X, m/\rho)$.*

(3) *$\partial_X^a f^{(m)}(X)$ (where $a \in \mathbb{N}^k$) fulfills the Sm-expansion with degree $D + |a|$.*

(4) *We assume that the product of distributions $f_1^{(m)}(X_1) f_2^{(m)}(X_2)$, which may be a (partly) pointwise product[30], exists. Then, $f_1^{(m)}(X_1)\, f_2^{(m)}(X_2)$ fulfills also the Sm-expansion with degree $(D_1 + D_2)$.*

(5) *The Sm-expansion is unique, i.e., if we know that a given $f^{(m)}$ has such an expansion, then the "coefficients" $u_l^{(m)}$ (and, hence, also the "remainders" $\mathfrak{r}_{L+1}^{(m)}$) are uniquely determined.*

(6) *The scaling behaviour for $X \to 0$ of the remainder $\mathfrak{r}_{L+1}^{(m)}(X)$ is bounded by*

$$\lim_{\rho\downarrow 0} \rho^{D-(L+1)+\varepsilon}\, \mathfrak{r}_{L+1}^{(m)}(\rho X) = 0 \quad \forall \varepsilon > 0\,;$$

or, in terms of the scaling degree introduced later in Definition 3.2.5: $\mathrm{sd}(\mathfrak{r}_{L+1}^{(m)}) \leq D - (L + 1)$.

Proof. Part (0) is obvious.

Part (1) follows immediately from (3.1.62) and properties (a), (b) and (e).

Part (2): We have to show that $m^l\, u_l^{(m)}(X)$ has the asserted scaling property. This can be done by using (3.1.46a):

$$\begin{aligned}(X\,\partial_X + D - m\,\partial_m)^N\, m^l\, u_l^{(m)}(X) &= m^l\,(X\,\partial_X + (D-l) - m\,\partial_m)^N\, u_l^{(m)}(X)\\ &= m^l \sum_{k=0}^{N} \binom{N}{k} (X\,\partial_X + D - l)^k (-m\,\partial_m)^{N-k}\, u_l^{(m)}(X)\,,\end{aligned}$$

where $X\,\partial_X := \sum_{i=1}^k x_i \partial_{x_i}$. Now, choosing N sufficiently large, at least one of the operators $(X\,\partial_X + D - l)^k$ or $(-m\,\partial_m)^{N-k}$ yields zero when applied to $u_l^{(m)}(X)$, due to properties (c) and (b), respectively.

Part (3): We show that $\partial_X^a u_l^{(m)}(X)$ and $\partial_X^a \mathfrak{r}_{L+1}^{(m)}(X)$ satisfy the properties (a)–(e) with degree $D + |a|$. To verify (d) let $N \in \mathbb{N}$ be such that $(X\,\partial_X + D - m\,\partial_m)^N\, \mathfrak{r}_{L+1}^{(m)}(X) = 0$. It follows that

$$0 = \partial_X^a\,(X\,\partial_X + D - m\,\partial_m)^N\, \mathfrak{r}_{L+1}^{(m)}(X) = (X\,\partial_X + D + |a| - m\,\partial_m)^N\, \partial_X^a\, \mathfrak{r}_{L+1}^{(m)}(X).$$

[30] More precisely: Let $X_1 = (x_1, \ldots, x_k) \in \mathbb{R}^k$ and $X_2 = (x_{k+1}, \ldots, x_{k+q}) \in \mathbb{R}^q$, respectively. Then, the set $\{x_1, \ldots, x_{k+q}\}$ may be linearly dependent.

(c) can be shown analogously. To verify (a),(b) and (e) we use that these properties hold for $\langle g^{(m)}, h\rangle$, where $g^{(m)} = u_l^{(m)}$ or $g^{(m)} = \mathfrak{r}_{L+1}^{(m)}$, for all $h \in \mathcal{D}(\mathbb{R}^k[\setminus\{0\}])$. Hence, they hold also for $(-1)^{|a|}\langle g^{(m)}, \partial^a h\rangle = \langle \partial^a g^{(m)}, h\rangle\ \forall h$.

Part (4): By a straightforward calculation we obtain

$$f_1^{(m)}(X_1)\, f_2^{(m)}(X_2) = \sum_{l=0}^{L} m^l\, u_l^{(m)}(X_1, X_2) + \mathfrak{r}_{L+1}^{(m)}(X_1, X_2)\,,$$

where

$$u_l^{(m)}(X_1, X_2) := \sum_{k=0}^{l} u_{1,k}^{(m)}(X_1)\, u_{2,l-k}^{(m)}(X_2), \quad (0 \le l \le L)$$

$$\begin{aligned}\mathfrak{r}_{L+1}^{(m)}(X_1, X_2) := {}& \mathfrak{r}_{1,L+1}^{(m)}(X_1)\, \mathfrak{r}_{2,L+1}^{(m)}(X_2) + \mathfrak{r}_{1,L+1}^{(m)}(X_1) \sum_{l=0}^{L} m^l\, u_{2,l}^{(m)}(X_2) \\ & + \Big(\sum_{l=0}^{L} m^l\, u_{1,l}^{(m)}(X_1)\Big)\, \mathfrak{r}_{2,L+1}^{(m)}(X_2) \\ & + \sum_{l=L+1}^{2L} m^l \sum_{k=l-L}^{L} u_{1,k}^{(m)}(X_1)\, u_{2,l-k}^{(m)}(X_2)\,.\end{aligned}$$

With that, it is an easy task to verify that $u_l^{(m)}(X_1, X_2)$ and $\mathfrak{r}_{L+1}^{(m)}(X_1, X_2)$ satisfy the properties (a)–(e) with degree $D = D_1 + D_2$, by using that $u_{j,l}^{(m)}$ and $\mathfrak{r}_{j,L+1}^{(m)}$ fulfill these properties with degree D_j (where $j = 1, 2$).

Part (5): The determination of u_0 is given in *part* (1). For $l \ge 1$ we assume that $u_k^{(m)}$ is known for $k < l$ and we determine the coefficients $u_{l,p}$ of $u_l^{(m)}$ (3.1.63) as follows: For $\mathbb{N} \ni P > P_l$ the limit

$$\lim_{m \downarrow 0}\Big(f^{(m)}(X) - \sum_{k=0}^{l-1} m^k\, u_k^{(m)}(X)\Big)\, m^{-l} \log^{-P}\Big(\frac{m}{M}\Big) \tag{3.1.65}$$

gives zero, for $P = P_l$ it gives u_{l,P_l} and for $P < P_l$ it diverges. Since P_l is unknown, we start with a P which is sufficiently high that the limit exists, if it vanishes we lower P by 1 etc. Having determined P_l and u_{l,P_l} in this way, we compute

$$\begin{aligned}&\lim_{m \downarrow 0}\Big(f^{(m)}(X) - \sum_{k=0}^{l-1} m^k\, u_k^{(m)}(X) - m^l \log^{P_l}\Big(\frac{m}{M}\Big)\ u_{l,P_l}(X)\Big)\, m^{-l} \log^{-(P_l-1)}\Big(\frac{m}{M}\Big) \\ &\quad = u_{l,P_l-1}\ ;\end{aligned}$$

and so on.

Part (6): From property (e) we know that the distribution

$$t^{(m)}(X) := m^{-(L+1)}\, \mathfrak{r}^{(m)}_{L+1}(X) \quad \text{fulfills} \quad \lim_{m\downarrow 0}\Big(\frac{m}{M}\Big)^{\varepsilon}\, t^{(m)} = 0 \qquad \forall \varepsilon > 0\,.$$

From (d) we conclude that

$$\rho^{D-(L+1)}\, t^{(m)}(\rho X) = t^{(\rho m)}(X) + \sum_{k=1}^{N} l_k^{(\rho m)}(X)\, \log^k \rho \qquad \forall \rho > 0$$

with some $l_k^{(m)} \in \mathcal{D}'(\mathbb{R}^k[\setminus\{0\}])$. Multiplying the latter equation by $(\rho m/M)^{\varepsilon}$ and performing the limit $m\downarrow 0$, we conclude that

$$\lim_{m\downarrow 0}\Big(\frac{m}{M}\Big)^{\varepsilon}\, l_k^{(m)} = 0 \qquad \forall \varepsilon > 0\,,\ k = 1,\dots,N\,.$$

It follows that

$$\lim_{\rho\downarrow 0} \rho^{D-(L+1)+\varepsilon}\, \mathfrak{r}^{(m)}_{L+1}(\rho X) = m^{L+1}\Big(\lim_{\rho\downarrow 0}\rho^{\varepsilon}\, t^{(\rho m)}(X)$$

$$+ \sum_{k=1}^{N}\big(\lim_{\rho\downarrow 0}\rho^{\varepsilon/2}\, l_k^{(\rho m)}(X)\big)\,\big(\lim_{\rho\downarrow 0}\rho^{\varepsilon/2}\log^k\rho\big)\Big) = 0 \quad \forall\varepsilon > 0\,. \quad \square$$

From *part* (1) of the lemma and linearity of $R^{(m)}_{n-1,1}$, we see that the axiom (k) implies *continuity* of the functions $0 \le m \longmapsto r^{(m)}(A_1,\dots,A_n)(g)$ for all $g\in\mathcal{D}(\mathbb{R}^{d(n-1)})$ and for all $A_1,\dots,A_n \in \mathcal{P}$. Due to the causal Wick expansion (3.1.24), this continuity holds also for $0\le m \longmapsto R^{(m)}(F_1\otimes\cdots\otimes F_n)$ for all $F_1,\dots,F_n\in\mathcal{F}_{\mathrm{loc}}$.

Because of *part* (2) of the lemma, the Sm-expansion axiom implies that the distributions $r^{(m)}(A_1,\dots,A_n)(X)$ (3.1.61) are almost homogeneous with degree $D := \sum_l \dim A_l$ under $(X,m)\mapsto(\rho X,\rho^{-1}m)$, for all monomials $A_1,\dots,A_n$. Proceeding similarly to (3.1.59), we use the causal Wick expansion (3.1.24) to obtain

$$\begin{aligned}
&(\rho\partial_\rho)^{N+1}\Big(\sigma_\rho\, R^{(m/\rho)}\big(\sigma_\rho^{-1}A_1(x_1)\otimes\cdots\otimes\sigma_\rho^{-1}A_n(x_n)\big)\Big)\\
&\quad = \sum_{\underline{A}_l\subseteq A_l}(\rho\partial_\rho)^{N+1}\Big(\rho^{\sum_l \dim \underline{A}_l}\, r^{(m/\rho)}(\underline{A}_1,\dots,\underline{A}_n)(\rho(x_1-x_n),\dots)\Big)\\
&\qquad\qquad \cdot\overline{A}_1(x_1)\cdots\overline{A}_n(x_n)\\
&\quad = 0
\end{aligned} \tag{3.1.66}$$

for $N\in\mathbb{N}$ sufficiently large. Using linearity of the retarded product, this relation can equivalently be written as the following property: For all $F_1,\dots,F_n\in\mathcal{F}_{\mathrm{loc}}$, there exist m-dependent functionals $L_1^{(m)},\dots,L_N^{(m)}\in\mathcal{F}$, which depend also on $(F_1,\dots,F_n)$, such that

$$\sigma_\rho R^{(m/\rho)}_{n-1,1}(\sigma_\rho^{-1}F_1\otimes\cdots\otimes\sigma_\rho^{-1}F_n) = R^{(m)}_{n-1,1}(F_1\otimes\cdots\otimes F_n) + \sum_{k=1}^{N} L_k^{(m)}\log^k\rho \tag{3.1.67}$$

for some $N < \infty$. Or in other words: The ρ-dependence of the left-hand side of (3.1.67) is a polynomial in $\log \rho$. This relation gives the scaling behaviour of the retarded product. For $m = 0$ it is a property of $R^{(m=0)}$, since in that case we have $R^{(m=0)}$ on both sides of (3.1.67); but for $m > 0$ it relates retarded products for different masses.

Comparing with the scaling behaviour of the star product, given in (3.1.51) and (3.1.52), we see that, for the retarded product, this relation can only be maintained up to powers of $\log \rho$.

Summarizing roughly, the Sm-expansion axiom requires two different properties for the retarded product: "Smoothness in $m \geq 0$ modulo $\log m$" and almost homogeneous scaling under $(X, m) \mapsto (\rho X, \rho^{-1} m)$.

3.1.6 The classical limit

An important additional reference for this section is [51].

Since we describe classical and quantum fields by the *same* space $\mathcal{F}$ of functionals, the requirement that "the limit $\hbar \to 0$ of an interacting field of pQFT gives the corresponding classical, interacting field" makes directly sense. To fulfill this condition, we need to control the number of factors $\hbar$ appearing in an R-product.

We start with the "unrenormalized" R-product: Let

$$\check{\mathbb{M}}^{n+1} := \{\, (x_1, \ldots, x_n, x) \in \mathbb{M}^{n+1} \,\big|\, x \neq x_l \neq x_j \neq x \ \forall 1 \leq l < j \leq n \,\} \,. \quad (3.1.68)$$

On $\mathcal{D}(\check{\mathbb{M}}^{n+1})$ there is a closed formula for the retarded product $R_{n,1}$, which is obtained from the classical formula (1.9.5) by the replacement $\{\cdot,\cdot\} \mapsto i^{-1}[\cdot,\cdot]_{\star_\hbar}$:[31]

$$\begin{aligned} &R_{n,1}\big(A_1(x_1), \ldots, A_n(x_n); A(x)\big) \qquad (3.1.69) \\ &\quad = \sum_{\pi \in S_n} \theta(x^0 - x^0_{\pi(n)})\, \theta(x^0_{\pi(n)} - x^0_{\pi(n-1)}) \cdots \theta(x^0_{\pi(2)} - x^0_{\pi(1)}) \\ &\qquad\quad \cdot \frac{1}{i^n}\, [A_{\pi(1)}(x_{\pi(1)}), [A_{\pi(2)}(x_{\pi(2)}), \ldots [A_{\pi(n)}(x_{\pi(n)}), A(x)]_\star \ldots]_\star]_\star \,. \end{aligned}$$

This iterated, retarded commutator gives the *"unrenormalized"* retarded product (analogously to the "unrenormalized" time-ordered product, which results from the Feynman rules, see Remks. 3.3.3 and 3.3.4).

The commutators are taken with respect to the star product; hence, each commutator gives at least one factor of $\hbar$. We conclude that (3.1.69) is of order $\mathcal{O}(\hbar^n)$ if $A_1, \ldots, A_n, A \sim \hbar^0$.

Our final renormalization condition ensures that this property be maintained in the extension to coinciding points:[32]

[31] In contrast to (2.0.1), the commutator does not get a factor $\hbar^{-1}$ here, as can be seen from the GLZ relation (3.1.11a).

[32] The extension to partial diagonals (e.g., $x_1 = x_2$ with $x_3, \ldots, x_n, x$ arbitrary) is the renormalization of "subdiagrams", which is accomplished in earlier steps of the inductive construction of the retarded products.

(l) **$\hbar$-dependence:** Renormalization is done in each order of $\hbar$ individually and, proceeding this way, $0 \in \mathcal{D}'(\mathbb{M}^{n+1} \setminus \Delta_{n+1}, \mathcal{F})$ is extended by $0 \in \mathcal{D}'(\mathbb{M}^{n+1}, \mathcal{F})$.

For later purpose we give an equivalent reformulation of this axiom: Using the notation (3.1.31), the condition is that for all monomials $A_1, \ldots, A_{n+1}$ it holds that[33]

$$r_{n,1}(A_1, \ldots, A_{n+1}) \sim \hbar^{\sum_{j=1}^{n+1} |A_j|/2} \quad \text{if} \quad A_1, \ldots, A_{n+1} \sim \hbar^0 \,. \tag{3.1.70}$$

Due to the field parity axiom (3.1.34) solely integer powers of $\hbar$ appear. The equivalence of the condition (3.1.70) to the axiom (l) is due to the fact that the vacuum expectation value of the "unrenormalized" retarded product (3.1.69) fulfills the condition (3.1.70) (because each term contributing to this vacuum expectation value has $\sum_{j=1}^{n+1} |A_j|/2$ contractions and each contraction gives precisely one factor $\hbar$).

From this axiom we conclude that

$$R_{n,1}(F_1 \otimes \cdots \otimes F_n, F) = \mathcal{O}(\hbar^n) \quad \text{if} \quad F_1, \ldots, F_n, F \sim \hbar^0.$$

Thus, if $S, F \sim \hbar^0$, then $\lim_{\hbar \to 0} R(e_\otimes^{S/\hbar}, F)$ exists. This limit fulfills the classical version of the axioms (i.e., the axioms for $\hbar \to 0$). These have a unique solution, namely $R_{\mathrm{cl}}(e_\otimes^S, F)$. This will follow immediately from the construction of the retarded products given in Sect. 3.2.

Summing up, the classical limit exists and gives the classical retarded product:

$$\lim_{\hbar \to 0} R(e_\otimes^{S/\hbar}, F) = R_{\mathrm{cl}}(e_\otimes^S, F) \quad \text{if} \quad S, F \sim \hbar^0 \,. \tag{3.1.71}$$

For an illustration, see (3.1.14).

3.1.7 (Feynman) diagrams

In Sect. 1.9 we have seen that a classical $R_{n,1}$-product can be understood as the sum over all connected tree diagrams, and that each inner line stands for a propagator Δ^{ret}. For a $R_{n,1}$-product of pQFT the following three complications appear: (i) also loop diagrams[34] contribute and (ii) there are two kind of propagators, Δ^{ret} and Δ^+, being symbolized by the inner lines. Finally (iii), if different diagrams (that is, different contraction schemes) contribute to the same coefficient $r(\underline{A}_1, \ldots, \underline{A}_n)$ in the causal Wick expansion (3.1.24), it is unclear to which diagram a finite renormalization term of $r(\underline{A}_1, \ldots, \underline{A}_n)$ (i.e., a local term being undetermined in the extension of $r(\underline{A}_1, \ldots, \underline{A}_n)$ from $\mathcal{D}'(\mathbb{R}^{d(n-1)} \setminus \{0\})$ to $\mathcal{D}'(\mathbb{R}^{d(n-1)})$) belongs.

[33] We use the convention that $0 \in \mathcal{D}'(\mathbb{R}^{dn})$ satisfies: $0 \sim \hbar^r \ \forall r \in \mathbb{R}, r > 0$.

[34] A (Feynman) diagram that is not a tree diagram (see Footn. 12) is called a "loop diagram". That is, the defining criterion for a loop diagram is that it contains at least one closed line. A loop diagram may be connected or disconnected.

To explain what we understand by a (Feynman) diagram for a $R_{n,1}$-product of pQFT, we start with the formula for the unrenormalized $R_{n,1}$-product (3.1.69), in which solely the propagator Δ^+ appears. Writing each commutator in (3.1.69) as $[A, B] = A \star B - B \star A$, we get $n! \cdot 2^n$ terms, where the symmetrization in $(x_1, \dots, x_n)$ is also taken into account. For example, for $x \neq x_1 \neq x_2 \neq x$ we obtain

$$\begin{aligned} R_{2,1}\big(\varphi^3(x_1), \varphi^3(x_2); \varphi^3(x)\big) &= -\hbar^4\, 108\, \varphi(x_1) \\ &\cdot\Big\{\theta(x^0 - x_2^0)\,\theta(x_2^0 - x_1^0)\,\Big(\Delta^+(x_2 - x)^2 - \Delta^+(x - x_2)^2\Big) \\ &\quad \cdot \Big(\Delta^+(x_1 - x_2)\Delta^+(x_1 - x) - \Delta^+(x_2 - x_1)\Delta^+(x - x_1)\Big) \\ &+\theta(x^0 - x_1^0)\,\theta(x_1^0 - x_2^0)\,\Big(\Delta^+(x_1 - x) - \Delta^+(x - x_1)\Big) \\ &\quad \cdot \Big(\Delta^+(x_2 - x_1)\Delta^+(x_2 - x)^2 - \Delta^+(x_1 - x_2)\Delta^+(x - x_2)^2\Big)\Big\} + \cdots , \end{aligned} \tag{3.1.72}$$

where the dots stand for terms coming from other contraction schemes. The $2! \cdot 2^2 = 8$ terms written on the r.h.s. of (3.1.72) could be visualized by eight diagrams which differ by the orientation of the inner lines[35] (due to $\Delta^+(z) \neq \Delta^+(-z)$) and by the time-ordering of all vertices (given by the θ-functions). However, to reduce the number of diagrams, we represent these eight terms by *one* diagram, which has *only* the time-ordering $x_1^0 \leq x^0$ and $x_2^0 \leq x^0$ (due to the Causality axiom), namely the diagram given in Figure 3.2.

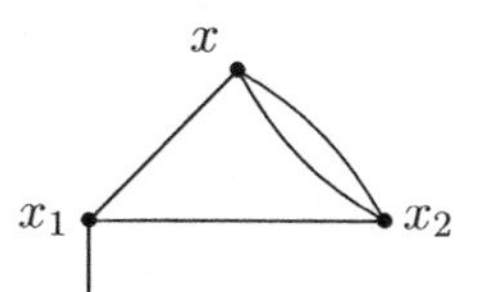

Figure 3.2: This diagram symbolizes the sum of all terms which contribute to $R_{2,1}\big(\varphi^3(x_1), \varphi^3(x_2); \varphi^3(x)\big)$ and which are $\sim \varphi(x_1)\,\Delta^+(\pm(x_1 - x_2))\Delta^+(\pm(x_1 - x))\,\Delta^+(\pm(x_2 - x))^2$, where $x \neq x_1 \neq x_2 \neq x$ is assumed. These are the eight terms written out in (3.1.72).

Generally speaking, by a (Feynman) diagram we symbolize a *contraction scheme for the unrenormalized $R_{n,1}$-product*, that is, for the iterated, retarded and symmetrized commutator (3.1.69). Such a diagram has only the time-ordering $x_j^0 \leq x^0$ $\forall j = 1, \dots, n$, given by the causality axiom.

For some diagrams, some of the pertinent $n! \cdot 2^n$ terms can be summarized by using

$$\Delta^{\text{ret}}(z) = -i\big(\Delta^+(z) - \Delta^+(-z)\big)\,\theta(z^0)\ ; \tag{3.1.73}$$

[35] The terms "vertex", "inner line" and "external line" are explained in Exap. 1.9.2.

a simple example is the first diagram in Sect. 3.1.2, an example belonging to a loop diagram appears in the second last line of (3.1.72).

We are now going to prove two basic properties of the diagrams belonging to a retarded product.

Claim 3.1.27. *Only* connected *diagrams contribute to the unrenormalized retarded product* (3.1.69).

Proof. Considering the causal Wick expansion for $R_{n,1}(A_1,\ldots,A)$ (3.1.24), the appearing contraction schemes are encoded in the vacuum expectation values $r_{n,1}(\underline{A}_1,\ldots,\underline{A})$ (3.1.41) (where $A_1,\ldots,A$ are monomials and $\underline{A}_l \subseteq A_l$, $\underline{A} \subseteq A$); they are visualized by vacuum diagrams, i.e., diagrams without external legs. Let us assume that a disconnected vacuum diagram contributes. By Translation covariance, it depends only on the variables $(x_j - x_i)$ with x_j and x_i in the *same* connected component (where x_i stands also for the field vertex x). Hence, the supports of different connected components can be translated relatively to each other without changing the analytic expression. Therefore, the terms corresponding to such a diagram would violate the causality axiom. □

Claim 3.1.28. *Let $L, A \in \mathcal{P}$ with $L, A \sim \hbar^0$. On $\mathcal{D}(\check{\mathbb{M}}^{n+1})$* (3.1.68) *the classical R-product $R_{\mathrm{cl}\,n,1}\big(\otimes_{j=1}^n L(x_j), A(x)\big)$ is precisely the contribution of all* (*connected*) tree *diagrams to $R_{n,1}\big(\otimes_{j=1}^n L(x_j)/\hbar, A(x)\big)$* (*given by* (3.1.69)). *This statement is valid also for the renormalized R-product* (*i.e., on $\mathcal{D}(\mathbb{M}^{n+1})$*).

Proof. On $\mathcal{D}(\check{\mathbb{M}}^{n+1})$: The diagrams belonging to $R_{n,1}\big(\otimes_{j=1}^n L(x_j)/\hbar, A(x)\big)$ have n interaction vertices, each having a factor $\hbar^{-1}$; and one field vertex, which is $\sim \hbar^0$. Each inner line has a factor $\hbar$, due to the definition of the $\star$-product (2.1.5). Since all diagrams are connected, each diagram has at least n inner lines. The connected tree diagrams are precisely the connected diagrams with n inner lines. Counting the factors of $\hbar$ for each diagram contributing to $R_{n,1}\big(\otimes_{j=1}^n L(x_j)/\hbar, A(x)\big)$, we conclude that the limit $\hbar \to 0$, which gives $R_{\mathrm{cl}\,n,1}\big(\otimes_{j=1}^n L(x_j), A(x)\big)$ (3.1.71), selects precisely all connected diagrams with n inner lines, i.e., all connected tree diagrams.

To prove the validity of the Claim on $\mathcal{D}(\mathbb{M}^{n+1})$, first note that the contribution of any (connected) tree diagram is well defined on $\mathcal{D}(\mathbb{M}^{n+1})$, since it is given by tensor products of distributions (pointwise products of distributions do not appear) – renormalization concerns only loop diagrams. Hence, $R_{\mathrm{tree}\,n,1} = R_{\mathrm{cl}\,n,1}$ holds also on $\mathcal{D}(\mathbb{M}^{n+1})$ and, under renormalization, the relation

$$R^{\mathrm{unrenorm}} = R_{\mathrm{cl}} + R^{\mathrm{unrenorm}}_{\mathrm{loop}} \quad \text{goes over into} \quad R = R_{\mathrm{cl}} + R^{\mathrm{renormalized}}_{\mathrm{loop}} . \qquad \square$$

A simple illustration for the argumentation in this Proof is the example in Sect. 3.1.2.

Taking into account the Feynman rules for R_{cl}, given after Exer. 1.9.3, we obtain the following statement for any connected tree diagram contributing to the unrenormalized R-product $R_{n,1}\big(\otimes_{j=1}^n L(x_j)/\hbar, A(x)\big)$: The pertinent $n!\cdot 2^n$ terms of the iterated, retarded commutator (3.1.69) can be summarized by *one* term in which all propagators are Δ^{ret}-distributions, by using the relation (3.1.73). Having done this, all $\theta(y^0-z^0)$ (where $y,z\in\{x,x_1,\dots,x_n\}$) appearing in (3.1.69) have vanished or can be omitted, since there is an implicit chain-like time-ordering of the vertices along each branch of the tree, due $\mathrm{supp}\,\Delta^{\mathrm{ret}}\subseteq\overline{V}_+$. See Exap. 1.9.2.

3.2 Construction of the retarded product

We continue to follow Steinmann [155] and the papers [55, 64].

To construct the general solution $(R_{n,1})_{n\in\mathbb{N}}$ of the axioms, we proceed by induction on n. Axiom (c) gives the initial value: $R_{0,1}(F)=F$. The inductive step is done in two substeps:

- First the construction of $R_{n-1,1}\big(A_1(x_1),\dots;A_n(x_n)\big)$, $A_1,\dots,A_n\in\mathcal{P}$, off the thin diagonal Δ_n (A.1.12), i.e., on $\mathcal{D}(\mathbb{M}^n\setminus\Delta_n)$. This part is uniquely determined by the basic axioms.
- Then the extension to Δ_n, i.e., to a distribution in $\mathcal{D}'(\mathbb{M}^n,\mathcal{F})$. This second part is non-unique, but it is strongly restricted by the renormalization conditions.

3.2.1 Inductive step, off the thin diagonal Δ_n

Let $R_{l,1}$, for $l=0,1,\dots,n-2$, be constructed already such that they satisfy all axioms (a)–(l) – this is the inductive assumption which is made in the following two propositions. Now construct the functionals

$$J_{n-2,2}\big(A_1(x_1),\dots,A_{n-2}(x_{n-2});A_{n-1}(x_{n-1}),A_n(x_n)\big)$$

according to the recipe (3.1.15).

Proposition 3.2.1 (Uniqueness, [55, 155]). *If some operation $R_{n-1,1}$ fulfilling the basic axioms exists, then*

(a) *it is uniquely determined on $\mathcal{D}(\mathbb{M}^n\setminus\Delta_n)$; and*

(b) *if the extension of the* totally symmetric part

$$S_n\big(A_1(x_1),\dots,A_n(x_n)\big) := \frac{1}{n}\sum_{k=1}^n R_{n-1,1}\big(A_1(x_1),\dots,\widehat{A_k(x_k)},\dots,A_n(x_n);A_k(x_k)\big) \tag{3.2.1}$$

from $\mathcal{D}(\mathbb{M}^n\setminus\Delta_n)$ to $\mathcal{D}(\mathbb{M}^n)$ is constructed for all $A_1,\dots,A_n\in\mathcal{P}$, then the operation $R_{n-1,1}$ is completely determined (see Footn. 27 for the notation).

As a consequence of this proposition, two solutions of the basic axioms for the operation $R_{n-1,1}$ evaluated at $(A_1(x_1),\dots;A_n(x_n))$ differ by an element of $\mathcal{D}'(\mathbb{M}^n,\mathcal{F})$, which has support on Δ_n and is invariant under permutations of the pairs $(A_1,x_1),\dots,(A_n,x_n)$.

Proof. Ad (a): Consider the following open subsets of $\mathbb{M}^n$:

$$\mathcal{M}_0 := \{\,(x_1,\dots,x_n)\,\big|\,\exists j\in\{1,\dots,n-1\}\text{ with } x_j\notin(x_n+\overline{V}_-)\,\},$$

$$\mathcal{M}_k := \{\,(x_1,\dots,x_n)\,\big|\,x_k\notin(x_n+\overline{V}_+)\,\},\quad \text{for}\quad k=1,\dots,n-1. \tag{3.2.2}$$

First we check that $\{\mathcal{M}_0,\mathcal{M}_1,\dots,\mathcal{M}_{n-1}\}$ is an open cover (i.e., a cover by open sets) of $\mathbb{M}^n\setminus\Delta_n$, that is,

$$\bigcup_{k=0}^{n-1}\mathcal{M}_k=\mathbb{M}^n\setminus\Delta_n\ .$$

(Figure 3.3 illustrates the case $n=2$.) The relation $\bigcup_k\mathcal{M}_k\subseteq\mathbb{M}^n\setminus\Delta_n$ is obvious. To prove "$\supseteq$", let $x:=(x_1,\dots,x_n)\notin\Delta_n$. Then there exists a $j\in\{1,\dots,n-1\}$ with $x_j\neq x_n$. If $x_j\notin(x_n+\overline{V}_-)$ we have $x\in\mathcal{M}_0$; and if $x_j\notin(x_n+\overline{V}_+)$ we have $x\in\mathcal{M}_j$.

Now we are going to show that $R_{n-1,1}\big(A_1(x_1),\dots;A_n(x_n)\big)$ is uniquely determined on each $\mathcal{M}_k$, $0\le k\le n-1$: For $(x_1,\dots,x_n)\in\mathcal{M}_0$, we get, due to causality (3.1.10):

$$R_{n-1,1}\big(A_1(x_1),\dots;A_n(x_n)\big)=0. \tag{3.2.3a}$$

For $(x_1,\dots,x_n)\in\mathcal{M}_k$ with $k>0$, causality gives

$$R_{n-1,1}\big(A_1(x_1),\dots,\widehat{A_k(x_k)},\dots,A_n(x_n);A_k(x_k)\big)=0.$$

The GLZ relation then implies that

$$\begin{aligned}&R_{n-1,1}\big(A_1(x_1),\dots;A_n(x_n)\big) \qquad (3.2.3b)\\ &\quad=\hbar\,J_{n-2,2}\big(A_1(x_1),\dots,\widehat{A_k(x_k)},\dots,A_{n-1}(x_{n-1});A_k(x_k),A_n(x_n)\big)\ .\end{aligned}$$

Ad (b): The GLZ relation also shows that symmetrizing $J_{n-2,2}$ with respect to the first $(n-1)$ arguments we obtain

$$\begin{aligned}&\frac{\hbar}{n}\sum_{k=1}^{n-1}J_{n-2,2}\big(A_1(x_1),\dots,\widehat{A_k(x_k)},\dots,A_{n-1}(x_{n-1});A_k(x_k),A_n(x_n)\big)\\ &\quad=\frac{1}{n}\sum_{k=1}^{n}\Big(R_{n-1,1}\big(A_1(x_1),\dots,A_{n-1}(x_{n-1});A_n(x_n)\big)\\ &\qquad\quad-R_{n-1,1}\big(A_1(x_1),\dots,\widehat{A_k(x_k)},\dots,A_n(x_n);A_k(x_k)\big)\Big)\\ &\quad=R_{n-1,1}\big(A_1(x_1),\dots;A_n(x_n)\big)-S_n\big(A_1(x_1),\dots,A_n(x_n)\big). \qquad (3.2.4)\end{aligned}$$

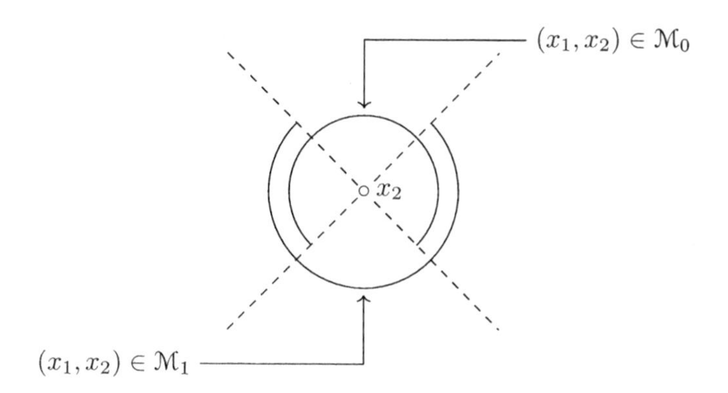

Figure 3.3: The cover of $\mathbb{M}^2 \setminus \Delta_2$. For a fixed x_2 (in the middle of the figure), the position of x_1 is sketched.

Note that in the second and third line the contributions coming from $k = n$ cancel. We conclude: Once $S_n(A_1(x_1), \ldots, A_n(x_n))$ is extended from $\mathcal{D}(\mathbb{M}^n \setminus \Delta_n)$ to $\mathcal{D}(\mathbb{M}^n)$, the distribution $R_{n-1,1}(A_1(x_1), \ldots; A_n(x_n)) \in \mathcal{D}'(\mathbb{M}^n, \mathcal{F})$ is completely determined. □

Proposition 3.2.2 (Existence, [55, 155]). *Under the inductive assumption, there exists an operation*

$$R^0_{n-1,1} \colon \mathcal{P}^{\otimes n} \longrightarrow \mathcal{D}'(\mathbb{M}^n \setminus \Delta_n, \mathcal{F}) ,$$

fulfilling all *the aforesaid axioms, i.e., the basic axioms and all renormalization conditions.*

Proof. First we check that the partial recipes (3.2.3) for $R_{n-1,1}$ on the several $\mathcal{M}_k$ coincide on the overlaps $\mathcal{M}_k \cap \mathcal{M}_l$, for $0 \leq k < l \leq n-1$. Then, by the sheaf theorem on distributions, these assemble to give a well-defined operation $R^0_{n-1,1}$ on the union $\mathbb{M}^n \setminus \Delta_n$. This compatibility check uses the support property of $J_{n-2,2}$ and the Jacobi identity given in Lemma 3.1.8.

In detail: First let $(x_1, \ldots, x_n) \in \mathcal{M}_0 \cap \mathcal{M}_l$. Due to (3.2.3a) and (3.2.3b) we have to show that[36] $J(\ldots; x_l, x_n) = 0$. If $(\ldots, x_l, x_n) \in \operatorname{supp} J_{n-2,2} \subseteq \mathbb{J}_n$ (3.1.16), we conclude from $x_l \notin (x_n + \overline{V}_+)$ that $\{x_1, \ldots, x_{n-1}\} \subset (x_n + \overline{V}_-)$, which contradicts $(x_1, \ldots, x_n) \in \mathcal{M}_0$.

Now let $(x_1, \ldots, x_n) \in \mathcal{M}_k \cap \mathcal{M}_l$. Then we have to show that

$$J_{n-2,2}(\ldots, x_l; x_k, x_n) = J_{n-2,2}(\ldots, x_k; x_l, x_n) .$$

By the Jacobi identity the left minus the right side of this equation is equal to $J_{n-2,2}(\ldots, x_n; x_k, x_l)$, which vanishes since x_n is neither in the past of x_k nor x_l.

It remains to verify the axioms for $R^0_{n-1,1}$:

[36] To simplify the notations we write x_r for $A_r(x_r)$ in the arguments of $R^0_{n-1,1}$ and $J_{n-2,2}$.

- *Causality* follows from (3.2.3a).
- *Symmetry* is obvious from the construction.
- To verify the *GLZ relation*, we have to show that

$$R^0_{n-1,1}(\dots,x,y;z) - R^0_{n-1,1}(\dots,x,z;y) = \hbar\, J_{n-2,2}(\dots,x;y,z) \qquad (3.2.5)$$

holds whenever $(x,y,z) \notin \Delta_3$. If $y \neq z$, then $y \notin z + \overline{V}_+$ or $y \notin z + \overline{V}_-$. In the latter case we have $(\dots,x,y,z) \in \mathcal{M}_0$ and $(\dots,x,z,y) \in \mathcal{M}_{n-1}$, hence $R^0(\dots,x,y;z) = 0$ and

$$R^0(\dots,x,z;y) = \hbar\, J_{n-2,2}(\dots,x;z,y) = -\hbar\, J_{n-2,2}(\dots,x;y,z)\ .$$

The case $y \notin z + \overline{V}_+$ is analogous.

So it remains to treat the case when y is arbitrary close to z and $x \neq y \wedge x \neq z$, i.e., when $x \notin (y + \overline{V}_\epsilon \cup z + \overline{V}_{\epsilon'})$ for some pair $(\varepsilon,\varepsilon')$, where $\epsilon,\epsilon' \in \{+,-\}$. In the case $\epsilon = \epsilon' = -$, all terms in (3.2.5) vanish (by construction of $R^\circ_{n-1,1}$ and due to the support property of J). In the case $\epsilon = -$ and $\epsilon' = +$, we analogously find $R^\circ(\dots,x,z;y) = 0$ and $J(\dots,z;x,y) = 0$. Hence, by the Jacobi identity the assertion (3.2.5) becomes $R^\circ(\dots,x,y;z) = \hbar\, J(\dots,y;x,z)$ which is the construction (3.2.3b) of $R^\circ_{n-1,1}(\dots,x,y;z)$ since $(\dots,x,y,z) \in \mathcal{M}_{n-2}$. The case $\epsilon = +$ and $\epsilon' = -$ is analogous. Finally, for $\epsilon = \epsilon' = +$ we have $R^\circ(\dots,x,y;z) = \hbar\, J(\dots,y;x,z)$ and $R^\circ(\dots,x,z;y) = \hbar\, J(\dots,z;x,y)$ by means of (3.2.3b). The assertion (3.2.5) follows then by the Jacobi identity.
- The *renormalization conditions* (axioms (f)–(l) and the AWI) go over from $R_{0,1},\dots,R_{n-2,1}$ to $J_{n-2,2}$ via (3.1.15). Due to (3.2.3), they hold also for $R^0_{n-1,1}$. This can be verified straightforwardly; only the Sm-expansion axiom is more involved.

  To prove the latter, we have to show that

$$j_{n-2,2}(A_1,\dots;A_{n-1},A_n)(x_1-x_n,\dots) = \frac{1}{i\hbar} \qquad (3.2.6)$$
$$\cdot\,\omega_0\Big(\sum_{I\subseteq\{1,\dots,n-2\}} \big[R(\otimes_{i\in I}A_i(x_i),A_{n-1}(x_{n-1})),R(\otimes_{j\in I^c}A_j(x_j),A_n(x_n))\big]\Big)$$

fulfills the Sm-expansion with degree $D = \sum_{l=1}^n \dim A_l$, where $A_1,\dots,A_n$ are field monomials. By means of (3.2.3) this property goes then over to $r^0_{n-1,1}(A_1,\dots,A_n) \in \mathcal{D}'(\mathbb{R}^{d(n-1)}\setminus\{0\})$, which is defined similarly to (3.1.41).

  Inserting the causal Wick expansion (3.1.24) into (3.2.6) we obtain a linear combination of products of the form

$$r(\otimes_{i\in I}\underline{A}_i,\underline{A}_{n-1})(\dots,x_i-x_{n-1},\dots)\,r(\otimes_{j\in I^c}\underline{A}_j,\underline{A}_n)(\dots,x_j-x_n,\dots)$$
$$\cdot\,\omega_0\Big(\big[\big(\prod_{i\in I}\overline{A}_i(x_i)\big)\,\overline{A}_{n-1}(x_{n-1}),\big(\prod_{j\in I^c}\overline{A}_j(x_j)\big)\,\overline{A}_n(x_n)\big]\Big)\,. \qquad (3.2.7)$$

The $\omega_0(\dots)$-factor is, if it does not vanish, a linear combination of terms of the form

$$\prod_{k=1}^{K} \partial^{a_k} \Delta_m^+(x_{i_k} - x_{j_k}) - \prod_{k=1}^{K} (-1)^{|a_k|} \partial^{a_k} \Delta_m^+(x_{j_k} - x_{i_k}) \tag{3.2.8}$$

with

$$K(d-2) + \sum_{k=1}^{K} |a_k| = \sum_{l=1}^{n} \dim \overline{A}_l \,,$$

where $i_k \in I \cup \{n-1\}$ and $j_k \in I^c \cup \{n\}$. By induction $r(\otimes_{i \in I} \underline{A}_i, \underline{A}_{n-1})$ and $r(\otimes_{j \in I^c} \underline{A}_j, \underline{A}_n)$ fulfill the Sm-expansion with degree $D_I := \dim \underline{A}_{n-1} + \sum_{i \in I} \dim \underline{A}_i$ and $D_{I^c} := \dim \underline{A}_n + \sum_{j \in I^c} \dim \underline{A}_j$, respectively; in addition $\partial^{a_k} \Delta_m^+$ satisfies this expansion with degree $D_k := d - 2 + |a_k|$ (due to *part (3)* of Lemma 3.1.26). By means of *parts (4) and (0)* of the same lemma, we conclude that (3.2.7) fulfills the Sm-expansion with degree

$$D_I + D_{I^c} + \sum_{k=1}^{K} D_k = \sum_{l=1}^{n} \dim A_l \,, \tag{3.2.9}$$

where we use the relation (3.1.58). It follows that $J_{n-2,2}^{(m)}$ fulfills the Sm-expansion axiom. □

The cover $\bigcup_{k=0}^{n-1} \mathcal{M}_k = \mathbb{M}^n \setminus \Delta_n$ is well suited for the proofs of the two propositions. However, for practical computations one usually works with a somewhat modified cover: Let $\Delta_{j=n}$ $(1 \le j \le n-1)$ be the partial diagonal

$$\Delta_{j=n} := \{\, (x_1, \dots, x_n) \,|\, x_j = x_n \,\} \quad \text{and note that} \quad \Delta_{j=n}^c := \mathbb{M}^n \setminus \Delta_{j=n}$$

is an open set. We consider the open cover

$$\bigcup_{j=1}^{n-1} \Delta_{j=n}^c = \mathbb{M}^n \setminus \Delta_n \,.$$

Due to

$$\operatorname{supp} J\big(\dots; A_j(x_j), A_n(x_n)\big) \subset \{\, (x_1, \dots, x_n) \,|\, (x_j - x_n)^2 \ge 0 \,\} \,,$$

the overlapping singularities of $J\big(\dots; A_j(x_j), A_n(x_n)\big)$ and $\theta(x_n^0 - x_j^0)$ are on $\Delta_{j=n}$. Therefore, the product

$$\begin{aligned} &R^0\big(\dots A_j(x_j) \dots; A_n(x_n)\big) \\ &\quad := \hbar\, J\big(\dots; A_j(x_j), A_n(x_n)\big)\, \theta(x_n^0 - x_j^0) \quad \text{exists on } \mathcal{D}(\Delta_{j=n}^c). \end{aligned} \tag{3.2.10}$$

One easily verifies that on each $\mathcal{D}(\Delta^{\rm c}_{j=n})$ $(j=1,\dots,n-1)$ this formula agrees with the $R^0_{n-1,1}\big(\dots;A_n(x_n)\big)\in\mathcal{D}'(\mathbb{M}^n\setminus\Delta_n,\mathcal{F})$ constructed by (3.2.3) (Proposition 3.2.2); therefore, the partial recipes (3.2.10) for $R^0_{n-1,1}\big(\dots;A_n(x_n)\big)$ coincide on the overlaps $\mathcal{D}(\Delta^{\rm c}_{j=n})\cap\mathcal{D}(\Delta^{\rm c}_{k=n})$ for all $1\le j<k\le n-1$.

There are a few cases in which the product of distributions on the r.h.s. of (3.2.10) exists on $\mathcal{D}(\mathbb{M}^n)$, since it amounts to

$$\Delta^{\rm ret}_m(z)\,\theta(z^0)=\Delta^{\rm ret}_m(z)\ ,\quad \Delta^{\rm ret}_m(-z)\,\theta(z^0)=0\quad\text{or}\quad \Delta_m(z)\,\theta(z^0)=\Delta^{\rm ret}_m(z)\ . \tag{3.2.11}$$

Then formula (3.2.10) yields the renormalized retarded product on $\mathcal{D}(\mathbb{M}^n)$, that is, $R\big(A_1(x_1),\dots;A_n(x_n)\big)\in\mathcal{D}'(\mathbb{M}^n,\mathcal{F})$. Examples are given in Exer. 3.2.3.

To express $R^0\big(A_1(x_1),\dots;A_n(x_n)\big)$ on $\mathcal{D}(\mathbb{M}^n\setminus\Delta_n)$ by *one* formula, we introduce a partition of unity: Let $f_j\in C^\infty(\mathbb{M}^n\setminus\Delta_n,\mathbb{R})$ such that

$$1=\sum_{j=1}^{n-1}f_j(x)\quad\forall x\in\mathbb{M}^n\setminus\Delta_n\quad\text{and}\quad \operatorname{supp}f_j\subseteq\Delta^{\rm c}_{j=n}\ . \tag{3.2.12}$$

With that and by using (3.2.10), we may write

$$\begin{aligned}&R^0\big(\dots A_j(x_j)\dots;A_n(x_n)\big) \qquad\qquad (3.2.13)\\ &\quad:=\hbar\sum_{j=1}^{n-1}f_j(x_1,\dots,x_n)\,J\big(\dots;A_j(x_j),A_n(x_n)\big)\,\theta(x_n^0-x_j^0)\end{aligned}$$

on $\mathcal{D}(\mathbb{M}^n\setminus\Delta_n)$.

Exercise 3.2.3. Let S_1 and S_2 be the same interactions as in Exer. 1.10.2. Compute

(a) $R\big(e_\otimes^{S_1/\hbar},\varphi(x)\big)$ to all orders in κ,

(b) $R\big(e_\otimes^{S_2/\hbar},\varphi(x)\big)$ up to second order in κ,

from the axioms for retarded products. Explain the agreement with the results of Exercise 1.10.2.

[*Solution*: (a) $R_{0,1}\big(\varphi(x)\big)=\varphi(x)$. Formula (3.2.10) gives

$$\frac{1}{\hbar}R_{1,1}\big(S_1;\varphi(x)\big)=\frac{-\kappa}{i\hbar}\int dy\;g(y)\,[\varphi(y),\varphi(x)]_\star\,\theta(x^0-y^0)=\kappa\int dy\;g(y)\,\Delta^{\rm ret}(x-y)\ . \tag{3.2.14}$$

Since $R_{1,1}(S_1;\varphi(x))=\omega_0\big(R_{1,1}(S_1;\varphi(x))\big)$ and also $R_{1,1}(S_1;S_1)=\omega_0\big(R_{1,1}(S_1;S_1)\big)$, and since $[\omega_0(F),G]_\star=0\ \forall F,G\in\mathcal{F}$, the commutators contributing to $J_{1,2}\big(S_1;S_1,\varphi(x)\big)$ vanish. Hence, $R_{2,1}\big(S_1^{\otimes 2};\varphi(x)\big)=0$ and, for the same reason, $R_{n,1}\big(S_1^{\otimes n};\varphi(x)\big)=0$ for all $n\ge 2$.

(b) Analogously to (3.2.14) we get

$$\begin{aligned}&\frac{1}{\hbar}R_{1,1}(S_2;\varphi(x))=2\kappa\int dy\,g(y)\,\Delta^{\rm ret}(x-y)\,\varphi(y)\ , \qquad (3.2.15)\\ &R_{1,1}(\varphi^2(y_2);\varphi^2(y_1))=4\hbar\,\varphi(y_1)\,\Delta^{\rm ret}(y_1-y_2)\,\varphi(y_2)+\omega_0\big(R_{1,1}(\varphi^2(y_2);\varphi^2(y_1))\big)\ .\end{aligned}$$

By using (3.2.10) for $j = 1$ and the relations (3.2.11) and $\Delta(z) = \Delta^{\mathrm{ret}}(z) - \Delta^{\mathrm{ret}}(-z)$, we obtain

$$R_{2,1}\big(\varphi^2(y_2), \varphi^2(y_1); \varphi(x)\big) = 4\hbar^2 \Big(\Delta^{\mathrm{ret}}(y_1 - y_2)\big(\Delta(x - y_1)\varphi(y_2) + \Delta(x - y_2)\varphi(y_1)\big) + \Delta^{\mathrm{ret}}(x - y_2)\Delta(y_2 - y_1)\varphi(y_1)\Big)\,\theta(x^0 - y_1^0)$$
$$= 4\hbar^2 \Delta^{\mathrm{ret}}(x - y_1)\Delta^{\mathrm{ret}}(y_1 - y_2)\varphi(y_2) + (y_1 \leftrightarrow y_2)\ .$$

Note that the latter result is manifestly symmetric under $y_1 \leftrightarrow y_2$, although this does not hold for the procedure. We end up with

$$\frac{1}{\hbar^2}\, R_{2,1}\big(S_2^{\otimes 2}; \varphi(x)\big) = 8\kappa^2 \int dy_1 dy_2\ g(y_1)g(y_2)\,\Delta^{\mathrm{ret}}(x - y_1)\Delta^{\mathrm{ret}}(y_1 - y_2)\varphi(y_2)\ .$$

(a) and (b): Since only (connected) tree diagrams contribute to $R_{n,1}\big(S_j^{\otimes n}; \varphi(x)\big)$ ($j = 1, 2$), Claim 3.1.28 implies that $\frac{1}{\hbar^n}\, R_{n,1}\big(S_j^{\otimes n}; \varphi(x)\big) = R_{\mathrm{cl}\,n,1}\big(S_j^{\otimes n}; \varphi(x)\big)$.]

There are also cases involving loop diagrams, in which the use of a partition of unity (3.2.13) can be avoided – the following exercise gives an example.

Exercise 3.2.4 (Triangle diagram). In $d = 4$ dimensions, compute

$$r^0(x - z, y - z) := r^0_{2,1}(\varphi^2, \varphi^2, \varphi^2)(x - z, y - z) \in \mathcal{D}'\big(\mathbb{R}^8 \setminus \{0\}\big)\ ,$$

where the notation (3.1.41) is used.

[*Hint*: Compute first r^0 on $\mathcal{D}\big(\mathbb{R}^8 \setminus (\mathbb{R}^4 \times \{0\})\big)$ by using (3.2.10). Then rewrite this result such that it is manifestly symmetrical under $x \leftrightarrow y$. Explain, why the so-obtained result gives r^0 on $\mathcal{D}(\mathbb{R}^8 \setminus \{0\})$.]

[*Solution.* Restricting to $\mathcal{D}\big(\mathbb{R}^8 \setminus (\mathbb{R}^4 \times \{0\})\big)$ we have $y \neq z$. Now, $R(\varphi^2(x); \varphi^2(y)) - \omega_0\big(R(\varphi^2(x); \varphi^2(y))\big)$ is given in (3.2.15). With that we get

$$\omega_0\Big(J_{1,2}\big(\varphi^2(x); \varphi^2(y), \varphi^2(z)\big)\Big)$$
$$= \frac{8\hbar}{i}\bigg(\Big(\Delta^+(y - x)\Delta^+(y - z) - \Delta^+(x - y)\Delta^+(z - y)\Big)\Delta^{\mathrm{ret}}(z - x) + \Big(\Delta^+(x - z)\Delta^+(y - z) - \Delta^+(z - x)\Delta^+(z - y)\Big)\Delta^{\mathrm{ret}}(y - x)\bigg)\ .$$

Hence, only one diagram contributes – the triangle diagram

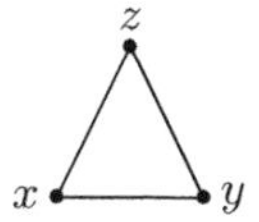

(3.2.16)

Next we use in both terms the identity

$$\Delta^+(u)\Delta^+(v) - \Delta^+(-u)\Delta^+(-v)$$
$$= i\Big(\Delta^+(u)\big(\Delta^{\mathrm{ret}}(v) - \Delta^{\mathrm{ret}}(-v)\big) + \big(\Delta^{\mathrm{ret}}(u) - \Delta^{\mathrm{ret}}(-u)\big)\Delta^+(-v)\Big)\ .$$

Since $y \neq z$, we may multiply with $\theta(z^0 - y^0)$, according to (3.2.10). Using (3.2.11) we get

$$r^0(x-z,y-z) = -8\hbar^2\Big(\Delta^+(y-x)\,\Delta^{\rm ret}(z-y)\,\Delta^{\rm ret}(z-x) \tag{3.2.17}$$
$$+\Delta^+(z-y)\,\Delta^{\rm ret}(z-x)\,\Delta^{\rm ret}(x-y)+\Delta^+(x-z)\,\Delta^{\rm ret}(z-y)\,\Delta^{\rm ret}(y-x)\Big)\,.$$

To rewrite this result in $(x \leftrightarrow y)$-symmetric form, we add

$$\begin{aligned}
0 = 4i\hbar^2 &\Big(\Big(\Delta^{\rm ret}(y-x)-\Delta^{\rm ret}(x-y)\Big)\,\Delta^{\rm ret}(z-y)\,\Delta^{\rm ret}(z-x)\\
&+\Big(\Delta^{\rm ret}(z-y)-\Delta^{\rm ret}(y-z)\Big)\,\Delta^{\rm ret}(z-x)\,\Delta^{\rm ret}(x-y)\\
&+\Big(\Delta^{\rm ret}(x-z)-\Delta^{\rm ret}(z-x)\Big)\,\Delta^{\rm ret}(z-y)\,\Delta^{\rm ret}(y-x)\Big)\\
= 4\hbar^2 &\Big(\Big(\Delta^{+}(y-x)-\Delta^{+}(x-y)\Big)\,\Delta^{\rm ret}(z-y)\,\Delta^{\rm ret}(z-x)\\
&+\Big(\Delta^{+}(z-y)-\Delta^{+}(y-z)\Big)\,\Delta^{\rm ret}(z-x)\,\Delta^{\rm ret}(x-y)\\
&+\Big(\Delta^{+}(x-z)-\Delta^{+}(z-x)\Big)\,\Delta^{\rm ret}(z-y)\,\Delta^{\rm ret}(y-x)\Big)\ ;
\end{aligned} \tag{3.2.18}$$

in the first equality sign we have taken into account that $\operatorname{supp}\Delta^{\rm ret} \subseteq \overline{V}_+$ and $y \neq z$. With the notation $\Delta_s^+(u) := \frac{1}{2}\big(\Delta^+(u)+\Delta^+(-u)\big)$, the sum of (3.2.17) and (3.2.18) reads

$$r^0(x-z,y-z) = -8\hbar^2\Big(\Delta_s^+(x-y)\,\Delta^{\rm ret}(z-y)\,\Delta^{\rm ret}(z-x) \tag{3.2.19}$$
$$+\Delta_s^+(y-z)\,\Delta^{\rm ret}(z-x)\,\Delta^{\rm ret}(x-y)+\Delta_s^+(x-z)\,\Delta^{\rm ret}(z-y)\,\Delta^{\rm ret}(y-x)\Big)\,.$$

This formula for r^0 is manifestly symmetrical under $x \leftrightarrow y$ and, due to our procedure, it is valid on $\mathcal{D}\big(\mathbb{R}^8 \setminus (\mathbb{R}^4 \times \{0\})\big)$.
Now we repeat the computation with the roles of x and y reversed. This yields the same result (3.2.19) for r^0, but now valid on $\mathcal{D}\big(\mathbb{R}^8 \setminus (\{0\}\times\mathbb{R}^4)\big)$. Since $\big(\mathbb{R}^8 \setminus (\mathbb{R}^4 \times \{0\})\big) \cup \big(\mathbb{R}^8 \setminus (\{0\}\times\mathbb{R}^4)\big) = \mathbb{R}^8 \setminus \{0\}$ is an open cover, we may construct $r^0 \in \mathcal{D}'(\mathbb{R}^8 \setminus \{0\})$ by means of a partition of unity, as explained in (3.2.12)–(3.2.13). Inserting (3.2.19) into both $r^0\big|_{\mathcal{D}(\mathbb{R}^8\setminus(\mathbb{R}^4\times\{0\}))}$ and $r^0\big|_{\mathcal{D}(\mathbb{R}^8\setminus(\{0\}\times\mathbb{R}^4))}$, we see that (3.2.19) is valid on $\mathcal{D}(\mathbb{R}^8 \setminus \{0\})$.]

3.2.2 The extension to the thin diagonal Δ_n

As in the inductive Epstein–Glaser construction of time-ordered products [24, 66, 158] (explained shortly in Sect. 3.3), the extension to Δ_n corresponds to what is called "renormalization" in conventional approaches. For this reason, this extension is named "Epstein–Glaser renormalization" (or only "renormalization").

Due to Proposition 3.2.1, we need only extend the symmetric part S_n given by (3.2.1). More precisely, if $\mathfrak{S}$ denotes the operation of symmetrization, it is necessary to extend $S_n^0 := \mathfrak{S}R_{n-1,1}^0$ from $\mathcal{D}'(\mathbb{M}^n \setminus \Delta_n, \mathcal{F})$ to $\mathcal{D}'(\mathbb{M}^n, \mathcal{F})$.

By the causal Wick expansion (3.1.24), the operations $H = R^0_{n-1,1}, R_{n-1,1}, S^0_n, S_n$ are uniquely determined by their respective vacuum expectation values (VEV)[37]

$$h(A_1,\dots,A_n)(x_1-x_n,\dots,x_{n-1}-x_n) := \omega_0\Big(H\big(A_1(x_1),\dots,A_n(x_n)\big)\Big), \quad (3.2.20)$$

for $A_1,\dots,A_n \in \mathcal{P}$, which are translation-invariant numerical distributions. The task, then, is to extend $s^0(A_1,\dots,A_n) \in \mathcal{D}'(\mathbb{R}^{d(n-1)} \setminus \{0\})$ to a distribution $s(A_1,\dots,A_n)$ in $\mathcal{D}'(\mathbb{R}^{d(n-1)})$.

Scaling degree. The existence and uniqueness of solutions of the just given task can be described in terms of Steinmann's scaling degree of $s^0 \equiv s^0(A_1,\dots,A_n)$ [24, 155], which – heuristically speaking – is a measure for the strength of the singularity of $s^0(x)$ at $x=0$.

Definition 3.2.5. The *scaling degree* (with respect to the origin) of a distribution $t \in \mathcal{D}'(\mathbb{R}^k)$ or $t \in \mathcal{D}'(\mathbb{R}^k \setminus \{0\})$ is given by[38]

$$\mathrm{sd}(t) := \inf\{\, r \in \mathbb{R} \;\big|\; \lim_{\rho\downarrow 0} \rho^r\, t(\rho x) = 0\,\}\ ,$$

where $\inf\emptyset := \infty$ and $\inf\mathbb{R} := -\infty$.

If t scales homogeneously, with degree $D \in \mathbb{C}$, then $\mathrm{sd}(t) = \mathrm{Re}\, D$. Note that this holds also when t merely scales *almost* homogeneously with degree D, since then $\rho^D\, t(\rho x)$ is a polynomial in $\log\rho$.

Example 3.2.6. If $\delta_{(k)}$ denotes the Dirac delta supported at the origin of $\mathbb{R}^k$ and $a \in \mathbb{N}^k$, then[39]

$$\mathrm{sd}\big(\partial^a\delta_{(k)}(x)\big) = k + |a|, \quad \text{since} \quad \partial^a\delta_{(k)}(\rho x) = \rho^{-k-|a|}\,\partial^a\delta_{(k)}(x).$$

If $t \in \mathcal{D}'(\mathbb{R}^k)$ with $0 \notin \mathrm{supp}\, t$, then $\mathrm{sd}(t) = -\infty$, because for each $g \in \mathcal{D}(\mathbb{R}^k)$ there exist a $\rho_g > 0$ such that $\mathrm{supp}\, t(\rho\,\cdot\,) \cap \mathrm{supp}\, g = \emptyset$ for all $0 < \rho < \rho_g$.

For $t \in \mathcal{D}'(\mathbb{R}^k)$, the relation $\mathrm{sd}(t) < \infty$ always holds; but for $t \in \mathcal{D}'(\mathbb{R}^k \setminus \{0\})$ the value $\mathrm{sd}(t) = \infty$ is possible, a one-dimensional example is $t(x) = \theta(x)\, e^{1/x}$, since

$$\lim_{\rho\downarrow 0} \rho^r \int_0^\infty dx\; e^{1/(\rho x)}\, h(x) \quad \text{diverges} \quad \forall r \in \mathbb{R}$$

and for a suitable $h \in \mathcal{D}(\mathbb{R}\setminus\{0\})$ – details are worked out in the following exercise.

[37] The convention implicitly adopted here is that each VEV is denoted by a lowercase letter when the operation is given by the corresponding uppercase letter. We will use this convention also when the arguments are elements of $\mathcal{F}_{\mathrm{loc}}$, explicitly:

$$h(F_1,\dots,F_n) := \omega_0\Big(H(F_1,\dots,F_n)\Big) \qquad \forall F_1,\dots,F_n \in \mathcal{F}_{\mathrm{loc}}\ .$$

[38] The defining condition for the set $\{\cdots\}$ has to be understood in the sense of distributions: $\lim_{\rho\downarrow 0}\rho^r \int d^kx\; t(\rho x)\, g(x) = 0$ for all $g \in \mathcal{D}(\mathbb{R}^k[\setminus\{0\}])$.

[39] One could write $\delta^{(k)}$ instead; but this risks confusion with the standard notation for the kth derivative of the one-dimensional δ.

Exercise 3.2.7.

(a) Prove $\mathrm{sd}\big(\theta(x)\,e^{1/x}\big)=\infty$.

(b) Compute the scaling degree

(i) of a continuous function $f\colon \mathbb{R}^k \longrightarrow \mathbb{C}$ with $f(0)\neq 0$, considered as an element of $\mathcal{D}'(\mathbb{R}^k)$;

(ii) of the Wightman two-point function Δ_m^+ (2.2.1) and of the Feynman propagator Δ_m^F (2.3.8) in d dimensions.

(c) Prove the following basic properties of the scaling degree for $a\in\mathbb{N}^k$:

(iii) $\mathrm{sd}(\partial^a t)\leq \mathrm{sd}(t)+|a|$;

(iv) $\mathrm{sd}(x^a\,t)\leq \mathrm{sd}(t)-|a|$;

(v) $\mathrm{sd}(t_1\otimes t_2)=\mathrm{sd}(t_1)+\mathrm{sd}(t_2)$, where $\otimes$ denotes the tensor product of distributions.

The properties (iii) and (iv) are precise formulations of the well-known heuristic statements that "in general differentiation increases the strength of the singularities" and that "multiplication with x^a makes the distribution less singular at $x=0$". For example, for $f\in\mathcal{D}'(\mathbb{R})$, given by

$$f(x):=\begin{cases} x & \text{for} \quad x\geq 0 \\ 0 & \text{for} \quad x<0 \end{cases}, \quad \text{we obtain} \quad \mathrm{sd}(f)=-1\ , \quad \mathrm{sd}(f')=\mathrm{sd}(\theta)=0\ ,$$

and $\mathrm{sd}(f^{(2+k)})=\mathrm{sd}(\delta^{(k)})=1+k\ \ (k\in\mathbb{N})$.

[*Solution*: (a) Let $h\in\mathcal{D}(\mathbb{R}\setminus\{0\})$ be an approximation of $\theta(2-x)\theta(x-1)$. Then

$$\left|\rho^r\int_0^\infty dx\ e^{1/(\rho x)}\,h(x)\right|\approx\left|\rho^r\int_1^2 dx\ e^{1/(\rho x)}\right|\geq \rho^r\,e^{1/(2\rho)}\to\infty \quad \text{for} \quad \rho\downarrow 0\ \forall r\in\mathbb{R}\ .$$

(b) (i) $\mathrm{sd}(f)=0$.
(ii) By using (2.2.9) we obtain

$$\lim_{\rho\downarrow 0}\rho^{d-2}\,\Delta_m^+(\rho x)=\lim_{\rho\downarrow 0}\Delta_{\rho m}^+(x)=D^+(x)\ ,$$

hence $\mathrm{sd}(\Delta_m^+)=d-2$. This calculation holds also for Δ_m^F, i.e., we get the same result for $\mathrm{sd}(\Delta_m^F)$.
(c) (iii) and (iv): Let $D:=\mathrm{sd}(t)$. By using $(\partial^a t)(\rho x)=\rho^{-|a|}\,\partial_x^a\,t(\rho x)$ and $(x^a t)(\rho x)=\rho^{|a|}\,x^a\,t(\rho x)$ we obtain

$$\lim_{\rho\downarrow 0}\rho^{D+|a|+\varepsilon}\int dx\ (\partial^a t)(\rho x)\,h(x)=(-1)^{|a|}\lim_{\rho\downarrow 0}\rho^{D+\varepsilon}\int dx\ t(\rho x)\,(\partial^a h)(x)=0\ ,$$

$$\lim_{\rho\downarrow 0}\rho^{D-|a|+\varepsilon}\int dx\ (x^a t)(\rho x)\,h(x)=\lim_{\rho\downarrow 0}\rho^{D+\varepsilon}\int dx\ t(\rho x)\,(x^a h)(x)=0\ ,$$

for all $h\in\mathcal{D}(\mathbb{R}^k)$ and for all $\varepsilon>0$. This proves the assertions.
(v) *Proof of* $\mathrm{sd}(t_1\otimes t_2)\leq\mathrm{sd}(t_1)+\mathrm{sd}(t_2)$: It suffices to consider test functions of the form $h(x_1,x_2)=h_1(x_1)\,h_2(x_2)$ with $h_1,h_2\in\mathcal{D}(\mathbb{R}^k)$. The assertion follows from the factorization

$$\langle (t_1\otimes t_2)(\rho x_1,\rho x_2),h_1(x_1)h_2(x_2)\rangle_{(x_1,x_2)}=\langle t_1(\rho x_1),h_1(x_1)\rangle_{x_1}\cdot\langle t_2(\rho x_2),h_2(x_2)\rangle_{x_2}\ .$$

Proof of $\mathrm{sd}(t_1 \otimes t_2) \geq \mathrm{sd}(t_1) + \mathrm{sd}(t_2)$: Given $\varepsilon > 0$, there exist $h_1, h_2 \in \mathcal{D}(\mathbb{R}^k)$ such that $\rho^{\mathrm{sd}(t_j)-\varepsilon/2} \langle t_j(\rho x_j), h_j(x_j)\rangle_{x_j}$ diverges for $\rho \downarrow 0$, where $j = 1, 2$. Hence, this holds also for $\rho^{\mathrm{sd}(t_1)+\mathrm{sd}(t_2)-\varepsilon} \langle (t_1 \otimes t_2)(\rho x_1, \rho x_2), h_1(x_1)h_2(x_2)\rangle_{(x_1,x_2)}$.]

Existence and uniqueness of extension. From the definition of the scaling degree, it follows immediately that any extension $t \in \mathcal{D}'(\mathbb{R}^k)$ of a given $t^0 \in \mathcal{D}'(\mathbb{R}^k \setminus \{0\})$ obeys $\mathrm{sd}(t) \geq \mathrm{sd}(t^0)$. We are looking for extensions which do not increase the scaling degree. The following fundamental theorem is taken from [24], the proof given there relies on earlier work of Epstein and Glaser [66] and of Hörmander [107]. Techniques which are similar or related to the ones of this theorem and of Proposition 3.2.16 have been developed also in the older work of Güttinger and Rieckers [90, 91].

Theorem 3.2.8. *Let* $t^0 \in \mathcal{D}'(\mathbb{R}^k \setminus \{0\})$. *Then:*

(a) *If* $\mathrm{sd}(t^0) < k$, *there is a* unique *extension* $t \in \mathcal{D}'(\mathbb{R}^k)$ *fulfilling the condition* $\mathrm{sd}(t) = \mathrm{sd}(t^0)$.

(b) *If* $k \leq \mathrm{sd}(t^0) < \infty$, *there are* several *extensions* $t \in \mathcal{D}'(\mathbb{R}^k)$ *satisfying the condition* $\mathrm{sd}(t) = \mathrm{sd}(t^0)$. *In this case, given a particular solution* t_0, *the general solution is of the form*

$$t = t_0 + \sum_{|a| \leq \mathrm{sd}(t^0)-k} C_a\, \partial^a \delta_{(k)} \quad \textit{with} \quad C_a \in \mathbb{C}. \tag{3.2.21}$$

In case (b), the addition of a term $\sum_a C_a\, \partial^a \delta_{(k)}$ is called a "finite renormalization".

Outline of the proof. (See also [73].)

Ad (a): Let $\chi \in C^\infty(\mathbb{R}^k)$ be such that $0 \leq \chi(x) \leq 1$, $\chi(x) = 0$ for $|x| \leq 1$ and $\chi(x) = 1$ for $|x| \geq 2$.

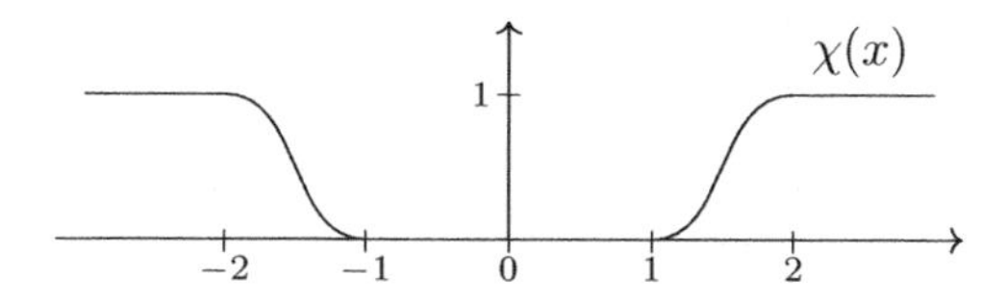

If $h \in \mathcal{D}(\mathbb{R}^k)$, then $\chi_\rho h \colon x \longmapsto \chi(\rho x)h(x)$ lies in $\mathcal{D}(\mathbb{R}^k \setminus \{0\})$ for any $\rho > 0$, hence $\langle t^0, \chi_\rho h\rangle$ exists. For $h_1 \in \mathcal{D}(\mathbb{R}^k \setminus \{0\})$ it holds that $\chi_\rho h_1 = h_1$ for ρ sufficiently large. Therefore, assuming that the limit

$$\langle t, h\rangle := \lim_{\rho \to \infty} \langle t^0, \chi_\rho h\rangle \qquad \forall h \in \mathcal{D}(\mathbb{R}^k)\,, \tag{3.2.22}$$

exists and defines a distribution $t \in \mathcal{D}'(\mathbb{R}^k)$, this t is an extension of t^0. These assumptions can indeed be proved. In addition one shows that $\mathrm{sd}(t) = \mathrm{sd}(t^0)$. We call (3.2.22) the *direct extension* of t^0. Independence of χ follows from the uniqueness.

For that uniqueness, let t_1 and t_2 be extensions of t^0 with the same scaling degree. Then $\mathrm{supp}(t_1 - t_2) \subseteq \{0\}$, so that $t_1 - t_2 = \sum_a C_a\, \partial^a \delta_{(k)}$ for some coefficients C_a (Theorem (A.1.17)). However, since $\mathrm{sd}(t_1 - t_2) \leq \max\{\mathrm{sd}(t_1), \mathrm{sd}(t_2)\} = \mathrm{sd}(t^0) < k$, whereas $\mathrm{sd}(\partial^a \delta_{(k)}) = k + |a|$, we see that $C_a = 0$ for all a.

Ad (b): *General solution* (3.2.21). Again, two solutions t_1 and t_2 differ by $t_1 - t_2 = \sum_a C_a\, \partial^a \delta_{(k)}$ for some $C_a \in \mathbb{C}$. However, now we have $\mathrm{sd}(t_1 - t_2) \leq \mathrm{sd}(t^0) \geq k$. Hence, we obtain $C_a = 0$ only for $|a| > \mathrm{sd}(t^0) - k$.

Existence of t: Let $\omega := \mathrm{sd}(t^0) - k$ be the *singular order*[40] of t^0. Note that ω need not be an integer; let $\lfloor \omega \rfloor$ be its integer part. Introduce the subspace of test functions

$$\mathcal{D}_\omega := \mathcal{D}_\omega(\mathbb{R}^k) := \{\, h \in \mathcal{D}(\mathbb{R}^k) \,|\, \partial^a h(0) = 0 \text{ for } |a| \leq \lfloor \omega \rfloor \,\} \,. \tag{3.2.23}$$

This subspace clearly includes $\mathcal{D}(\mathbb{R}^k \setminus \{0\})$.

Claim 3.2.9. *t^0 has a unique extension t_ω to $\mathcal{D}'_\omega$ satisfying* $\mathrm{sd}(t_\omega) = \mathrm{sd}(t^0)$.

⟦*Proof of the claim*: Any $h \in \mathcal{D}_\omega$ can be written as

$$h(x) = \sum_{|a| = \lfloor \omega \rfloor + 1} x^a g_a(x) \quad \text{with} \quad g_a \in \mathcal{D}(\mathbb{R}^k) \,. \tag{3.2.24}$$

This decomposition of h is non-unique in general, however, we will see that this does not matter. Then we can define

$$\langle t_\omega, h \rangle := \sum_{|a| = \lfloor \omega \rfloor + 1} \langle \overline{x^a t^0}, g_a \rangle, \tag{3.2.25}$$

where the overline denotes the direct extension of $x^a t^0$, which is guaranteed by part (a), since $\mathrm{sd}(x^a t^0) \leq \mathrm{sd}(t^0) - (\lfloor \omega \rfloor + 1) < k$. Inserting (3.2.22) we obtain

$$\begin{aligned} \langle t_\omega, h \rangle &= \lim_{\rho \to \infty} \sum_{|a| = \lfloor \omega \rfloor + 1} \langle x^a t^0, \chi_\rho g_a \rangle \\ &= \lim_{\rho \to \infty} \langle t^0, \chi_\rho \sum_{|a| = \lfloor \omega \rfloor + 1} x^a g_a \rangle = \lim_{\rho \to \infty} \langle t^0, \chi_\rho h \rangle \,, \end{aligned} \tag{3.2.26}$$

from which we see that the definition of $t_\omega \in \mathcal{D}'_\omega$ (3.2.25) is independent of the choice of the decomposition (3.2.24). Due to the similarity of (3.2.26) and (3.2.22), we call t_ω the "direct extension of t^0 to $\mathcal{D}'_\omega$."

Uniqueness of t_ω follows from the fact that $\sum_{|a| \leq \omega} C_a\, \partial^a \delta_{(k)}$ vanishes on $\mathcal{D}_\omega$.⟧

Each projector $W \colon \mathcal{D}(\mathbb{R}^k) \longrightarrow \mathcal{D}_\omega$ – by which is meant a linear, continuous and idempotent map with[41] range $\mathcal{D}_\omega$ – defines an extension $t^W \in \mathcal{D}'(\mathbb{R}^k)$ (called "W-extension") by

$$\langle t^W, h \rangle := \langle t_\omega, W h \rangle \,, \tag{3.2.27}$$

[40]This is the same as the power-counting degree in momentum space.

[41]Since $\mathcal{D}_\omega$ has finite codimension in $\mathcal{D}(\mathbb{R}^k)$, such continuous projectors exist, but are not unique.

where the right-hand pairing denotes duality on $\mathcal{D}_\omega$. Since $Wh = h$ for $h \in \mathcal{D}(\mathbb{R}^k \setminus \{0\})$, the relations

$$\langle t^W, h\rangle = \langle t_\omega, Wh\rangle = \langle t_\omega, h\rangle = \langle t^0, h\rangle$$

show that t^W is indeed an extension of t^0.

It is somewhat harder to prove that $\mathrm{sd}(t^W) = \mathrm{sd}(t^0)$ for any such W; we omit this.

Any set of functions $w_a \in \mathcal{D}(\mathbb{R}^k)$, one for each $a \in \mathbb{N}^k$ such that $|a| \le \lfloor\omega\rfloor$, satisfying

$$\partial^b w_a(0) = \delta^b_a \quad \forall b \in \mathbb{N}^k\ ,\ |b| \le \lfloor\omega\rfloor \tag{3.2.28}$$

(Kronecker delta, not Dirac!) defines such a projector W by[42]

$$Wh(x) := h(x) - \sum_{|a|\le\lfloor\omega\rfloor} \partial^a h(0)\, w_a(x). \tag{3.2.29}$$

Except for the continuity this can easily be verified: Since $\partial^b(Wh)(0) = 0\ \forall |b| \le \lfloor\omega\rfloor$ and $\forall h \in \mathcal{D}(\mathbb{R}^k)$, we have $\mathrm{Ran}\, W \subseteq \mathcal{D}_\omega$; taking additionally into account $Wh_1 = h_1\ \forall h_1 \in \mathcal{D}_\omega$, we conclude $W^2 = W$ and also $\mathcal{D}_\omega \subseteq \mathrm{Ran}\, W$.

For the functions w_a one can, for instance, take

$$w_a(x) := \frac{x^a}{a!}\, w(x) \qquad (|a| \le \lfloor\omega\rfloor) \tag{3.2.30}$$

where $w \in \mathcal{D}(\mathbb{R}^k)$ satisfies $w(0) = 1$ and $\partial^b w(0) = 0\ \forall 1 \le |b| \le \lfloor\omega\rfloor$. □

Example 3.2.10 (Exer. 3.2.4 continued). The distribution $r^0 := r^0_{2,1}(\varphi^2, \varphi^2, \varphi^2) \in \mathcal{D}'(\mathbb{R}^8 \setminus \{0\})$, computed in Exer. 3.2.4, has $\mathrm{sd}(r^0) = 6 < 8$. Therefore, we are in case (a) of the theorem and the direct extension (3.2.22) applies. In this sense, one writes for the extension $r \in \mathcal{D}'(\mathbb{R}^8)$ the same explicit expression (3.2.19) as for r^0. Although the result (3.2.19) involves a pointwise product of distributions (which is equivalent to the observation that the corresponding diagram contains a loop), this result is well defined in $\mathcal{D}'(\mathbb{R}^8)$ by the direct extension.

Example 3.2.11. [43] Let

$$t^0(x) := \theta(x)\, x^{-l} \in \mathcal{D}'(\mathbb{R} \setminus \{0\}), \quad l = 1, 2, \ldots$$

Obviously $\mathrm{sd}(t^0) = l$, $\omega = l - 1$ and $h \in \mathcal{D}(\mathbb{R})$ lies in $\mathcal{D}_\omega$ if and only if $\partial^a h(0) = 0$ for all $0 \le a \le l - 1$. Let $w(x)$ be as in (3.2.30) and let W be the pertinent projector (3.2.29). Writing any $h \in \mathcal{D}(\mathbb{R})$ as

$$h(x) = Wh(x) + \sum_{|a|\le\lfloor\omega\rfloor} \partial^a h(0)\, \frac{x^a}{a!}\, w(x)$$

we see that

$$\mathcal{D}(\mathbb{R}) = \mathcal{D}_\omega \oplus \Big[\Big\{\frac{x^a}{a!}\, w(x)\,\Big|\, 0 \le a \le l-1\Big\}\Big],$$

[42] Actually, there are circumstances under which not all w_a need to be test functions: See the explanations below about the central solution and the "improved Epstein–Glaser renormalization" treated in Lemma 3.2.19.

[43] This is a generalization of an example in [73].

where $[-]$ denotes the linear span. We find that

$$\langle t^W, h\rangle = \lim_{\varepsilon\downarrow 0}\int_\varepsilon^\infty dx\,\frac{1}{x^l}\Big(h(x) - w(x)\sum_{0\le a\le l-1}\frac{x^a}{a!}\,\partial^a h(0)\Big).$$

For $l = 1$ we obtain, with integration by parts,

$$\begin{aligned}\langle t^W, h\rangle &= \lim_{\varepsilon\downarrow 0}\Big(-\log\varepsilon\,\big(h(\varepsilon) - w(\varepsilon)h(0)\big) - \int_\varepsilon^\infty dx\,\log x\,\big(h'(x) - w'(x)h(0)\big)\Big)\\ &= -\int_{-\infty}^\infty dx\,\theta(x)\,\log x\,h'(x) + \Big(\int_0^\infty dx\,\log x\,w'(x)\Big)\,\langle\delta, h\rangle\end{aligned}$$

by using that $\theta(x)\log x \in \mathcal{D}'(\mathbb{R})$. Hence,

$$t^W(x) = \frac{d}{dx}\big(\theta(x)\log x\big) + \Big(\int_0^\infty dy\,\log y\,w'(y)\Big)\,\delta(x).$$

Exercise 3.2.12. For $l = 2$ derive analogously that

$$\begin{aligned}t^W(x) = -\frac{d^2}{dx^2}\big(\theta(x)\log x\big) &+ \Big(\int_0^\infty dy\,\log y\,w''(y)\Big)\,\delta(x)\\ &- \Big(\int_0^\infty dy\,\log y\,\big(yw''(y) + 2w'(y)\big)\Big)\,\delta'(x).\end{aligned}$$

[*Solution*: Similarly to the preceding example, we perform integrations by parts. Taking into account that $w(\varepsilon) = 1,\ w'(\varepsilon) = 0$ for $\varepsilon > 0$ sufficiently small, we get

$$\begin{aligned}\langle t^W, h\rangle &= \lim_{\varepsilon\downarrow 0}\int_\varepsilon^\infty dx\,\frac{1}{x^2}\Big(h(x) - w(x)\,\big(h(0) + x\,h'(0)\big)\Big)\\ &= \lim_{\varepsilon\downarrow 0}\int_\varepsilon^\infty dx\,\frac{1}{x}\Big(h'(x) - w'(x)\,h(0) - \big(xw'(x) + w(x)\big)\,h'(0)\Big)\\ &= \lim_{\varepsilon\downarrow 0}\Big(\int_\varepsilon^\infty dx\,\frac{h'(x)}{x} + h(0)\int_0^\infty dy\,\log y\,w''(y)\\ &\qquad + h'(0)\Big[\log\varepsilon + \int_0^\infty dy\,\log y\,\big(yw''(y) + 2w'(y)\big)\Big]\Big)\end{aligned}$$

By using

$$-\Big\langle\frac{d^2}{dx^2}\big(\theta(x)\log x\big), h(x)\Big\rangle = \lim_{\varepsilon\downarrow 0}\Big(h'(0)\,\log\varepsilon + \int_\varepsilon^\infty dx\,\frac{h'(x)}{x}\Big)$$

we obtain the assertion.]

Central solution. If all fields are massive, the infrared behavior of t^0 is harmless and, for the existence of (3.2.27), it is not necessary that $Wh(x)$ decays as $|x| \to \infty$. In that case, one may choose $w(x) := 1$ for all $x \in \mathbb{R}^k$ in (3.2.30); this is the *"central solution"* $t^c \in \mathcal{D}'(\mathbb{R}^k)$ of Epstein and Glaser [66]:

$$\langle t^c, h\rangle := \langle t_\omega, W_1 h\rangle \quad\text{with}\quad (W_1 h)(x) := h(x) - \sum_{|a|\le\omega}\partial^a h(0)\,\frac{x^a}{a!}\,. \tag{3.2.31}$$

Remark 3.2.13 (Central solution in momentum space). Epstein and Glaser define the central solution in momentum space, by proceeding as follows: First note that Theorem 3.2.8 holds also for the extension from $\mathcal{S}'(\mathbb{R}^k \setminus \{0\})$ to $\mathcal{S}'(\mathbb{R}^k)$, hence we may use Fourier transformation. In [66] it is proved that in the massive case the Fourier transformation $\widehat{t^W}(p)$ of the W-extension (3.2.27) is analytic in a neighbourhood of $p=0$; hence, this holds for any extension (3.2.21). Relying on this result, Epstein–Glaser define the central solution t^c by the conditions

$$\mathrm{sd}(t^c) = \mathrm{sd}(t^0) \quad \text{and} \quad \partial_p^a \widehat{t^c}(0) = 0 \qquad \forall |a| \leq \mathrm{sd}(t_0) - k \equiv \omega\ , \tag{3.2.32}$$

which determines t^c uniquely, due to part (b) of Theorem 3.2.8.

To show the equivalence of the x- and p-space definitions of t^c, we verify that $t^c := t_\omega \circ W_1$ (3.2.31) fulfills (3.2.32). $\mathrm{sd}(t^c) = \mathrm{sd}(t^0)$ can be shown analogously to $\mathrm{sd}(t^W) = \mathrm{sd}(t^0)$ (we have omitted this in the above sketch proof of Theorem 3.2.8); to check the second condition in (3.2.32) we proceed as follows. Due to analyticity of $\hat{t}^c(p)$ in a neighbourhood of $p=0$, we may use $\partial^a \delta(p)$ as "test function": For $|a| \leq \omega$ we obtain

$$\langle \hat{t}^c, \partial^a \delta \rangle_p = \langle t^c, \widehat{\partial^a \delta} \rangle_x = \frac{(-i)^{|a|}}{(2\pi)^{k/2}} \langle t^c, x^a \rangle_x = \frac{(-i)^{|a|}}{(2\pi)^{k/2}} \langle t_\omega, W_1\, x^a \rangle_x = 0\ ,$$

since $W_1\, x^a = 0$.

From (3.2.32) we see that the central solution can be obtained from any extension $t \in \mathcal{S}'(\mathbb{R}^k)$ with $\mathrm{sd}(t) = \mathrm{sd}(t^0)$ by Taylor subtraction in momentum space:

$$\widehat{t^c}(p) = \hat{t}(p) - \sum_{|a| \leq \lfloor \omega \rfloor} \frac{p^a}{a!}\, \partial^a \hat{t}(0)\ . \tag{3.2.33}$$

This corresponds to "BPHZ subtraction" at $p=0$.[44]

If there exists an extension $t^{(m)} \in \mathcal{S}'(\mathbb{R}^k)$ which scales almost homogeneously with degree $D \in \mathbb{N}$ and power N (cf. Proposition 3.2.16 below), i.e.,

$$(\rho \partial_\rho)^{N+1} \Big(\rho^{D-k}\, \hat{t}^{(m/\rho)}(p/\rho) \Big) = 0, \tag{3.2.34}$$

then this holds also for the central solution $t^{(m)\,c}$ with the same degree and the same power. This follows from (3.2.33), because the scaling relation (3.2.34) implies that $(\rho \partial_\rho)^{N+1} \Big(\rho^{D-k}\, (p/\rho)^a\, (\partial^a \hat{t}^{(m/\rho)})(0) \Big) = 0$.

But, it is well known that the limit $\lim_{m \downarrow 0} t^{(m)\,c}$ diverges in general; hence, the central solution is in conflict with the Sm-expansion axiom.

We give an example for the existence of the central solution, which belongs to electrostatics and is taken from [139].

[44] BPHZ renormalization is a famous renormalization method which is due to Bogoliubov, Parasiuk [11] (see also [12]), Hepp [98] and Zimmermann [176].

Example 3.2.14 (Self-energy of a classical point charge)**.** The electrostatic potential $\Phi \in \mathcal{D}'(\mathbb{R}^3)$ of a classical point charge at rest, sitting at the origin, is determined by

$$\Delta\Phi(\vec{x}) = -4\pi q\,\delta(\vec{x}) \;, \quad \text{which gives} \quad \Phi(\vec{x}) = \frac{q}{|\vec{x}|} \in \mathcal{D}'(\mathbb{R}^3) \;,$$

where q is the charge of the point particle. The pertinent electric field

$$\vec{E}(\vec{x}) = -\vec{\nabla}\Phi(\vec{x}) = \frac{q\,\vec{x}}{|\vec{x}|^3} \quad \text{exists in} \quad \mathcal{D}'(\mathbb{R}^3,\mathbb{R}^3) \;;$$

but the energy density

$$U^0(\vec{x}) = \frac{1}{4\pi}\,\vec{E}^2(\vec{x}) = \frac{q^2}{4\pi\,|\vec{x}|^4} \quad \text{exists only in} \quad \mathcal{D}'(\mathbb{R}^3 \setminus \{0\}) \;.$$

The singular order reads $\omega(U^0) = 4 - 3 = 1$. We can extend U^0 to a distribution $U \in \mathcal{D}'(\mathbb{R}^3)$ by means of the W-extension (3.2.27). Working with a projector W of the form (3.2.29) with w_a fulfilling (3.2.30), the general solution reads

$$\langle U,h\rangle = \frac{q^2}{4\pi}\int d^3x\,\frac{1}{|\vec{x}|^4}\Big(h(x) - w(\vec{x})\big(h(0) + \vec{x}\cdot(\vec{\nabla}h)(0)\big)\Big) + C_0\,h(0) - \sum_{k=1,2,3} C_k\partial_k h(0) \;,$$

where $w(\vec{x}) = 1$ for all $\vec{x}$ in some neighbourhood of $\vec{0}$. To compute the self-energy E of the point charge we would like to choose $h(\vec{x}) = 1\ \forall\vec{x}$. This is indeed possible since $1/|\vec{x}|^4$ decays sufficiently fast for $|\vec{x}| \to \infty$. For the same reason we may choose $w(\vec{x}) = 1\ \forall\vec{x}$, i.e., the central solution exists. With that we obtain

$$E = \langle U,1\rangle = C_0 \;.$$

We may determine C_0 by the requirement that the self-energy E is given by the mass m of the point charge: $C_0 = mc^2$.

Universality of W-extension. The central solution preserves most symmetries of t^0. For example: If t^0 is Lorentz covariant, the central solution manifestly maintains this symmetry. For a W-extension t^W as in (3.2.27) and (3.2.29), Lorentz covariance is at least not manifest, since there does not exist any Lorentz covariant $w_a \in \mathcal{D}(\mathbb{R}^k)$. On the other hand, Lorentz covariance can always be maintained in the extension, as explained in Sect. 3.2.7. From this point of view, the following lemma is somewhat surprising: It states that every extension t having $\mathrm{sd}(t) = \mathrm{sd}(t^0)$ – in particular any Lorentz covariant extension fulfilling this condition – is a W-extension (3.2.27) with the projector W given in terms of a family of functions (w_a) (3.2.28) by (3.2.29).

Lemma 3.2.15 ([55])**.** *The following two maps are* surjective*:*

(a) $\{w_a\}_{|a|\le\lfloor\omega\rfloor} \longmapsto W = \mathrm{Id} - \sum_{|a|\le\omega} w_a\,(-1)^{|a|}\partial^a\delta$ *, from the set of families of test functions* $\{w_a\}_{|a|\le\lfloor\omega\rfloor}$ *with the above property* (3.2.28)*, to the set of continuous projectors* W *from* $\mathcal{D}(\mathbb{R}^k)$ *onto* $\mathcal{D}_\omega$*; and*

(b) $W \longmapsto t^W = t_\omega \circ W$, *from the latter set to the set of extensions* t *of the given* t^0 *satisfying* $\mathrm{sd}(t) = \mathrm{sd}(t^0)$.

Proof. We follow [55, App. B].

Ad (a): Let a continuous projectors W from $\mathcal{D} := \mathcal{D}(\mathbb{R}^k)$ onto $\mathcal{D}_\omega$ be given. We consider the pertinent decomposition

$$\mathcal{D} = \mathcal{D}_\omega \oplus \mathcal{E} \quad \text{where} \quad \mathcal{E} := \mathrm{Ran}(\mathrm{Id} - W) \; . \tag{3.2.35}$$

First we show that there is a basis $(w_a)_{|a|\leq\omega}$ of $\mathcal{E}$ satisfying (3.2.28). The decomposition (3.2.35) induces a decomposition of the dual space: Splitting any $f \in \mathcal{D}'$ into

$$\begin{aligned} &f = (f \circ W) + f \circ (\mathrm{Id} - W) \quad \text{we get the decomposition} \quad \mathcal{D}' = \mathcal{D}'_\omega \oplus \mathcal{D}^\perp_\omega \quad \text{where} \\ &\mathcal{D}'_\omega := \{ f \in \mathcal{D}' \,|\, \langle f, h \rangle = 0 \;\; \forall h \in \mathcal{E} \} \; , \\ &\mathcal{D}^\perp_\omega := \{ f \in \mathcal{D}' \,|\, \langle f, h \rangle = 0 \;\; \forall h \in \mathcal{D}_\omega \} \; . \end{aligned} \tag{3.2.36}$$

A basis of $\mathcal{D}^\perp_\omega$ is given by $(\partial^a \delta)_{|a|\leq\omega}$. We define $\big((-1)^{|a|} w_a\big)_{|a|\leq\omega}$ to be the dual basis in $\mathcal{E}$, that is, $(-1)^{|a|} w_a \in \mathcal{E}$ is determined by the condition $\langle \partial^b \delta, (-1)^{|a|} w_a \rangle = \delta^b_a \;\forall |b| \leq \omega$, which is equivalent to (3.2.28).

It remains to show that the so-constructed functions w_a fulfill (3.2.29). For any $h \in \mathcal{D}$, the function $(\mathrm{Id} - W)h$ can be expanded in terms of the basis $(w_a)_{|a|\leq\omega}$ of $\mathcal{E}$: $(\mathrm{Id} - W)h(x) = \sum_a c_a \, w_a(x)$ for some $c_a \in \mathbb{C}$. Since $Wh \in \mathcal{D}_\omega$, we obtain

$$\partial^b h(0) = \partial^b (\mathrm{Id} - W)h(0) = c_b \; , \quad \forall |b| \leq \omega \; .$$

Hence,

$$Wh(x) = h(x) - (\mathrm{Id} - W)h(x) = h(x) - \sum_{|a|\leq\omega} (\partial^a h)(0) \, w_a(x) \; .$$

Ad (b): Let an extension t of t^0 with $\mathrm{sd}(t) = \mathrm{sd}(t^0)$ be given. We show that there exists a complementary space $\mathcal{E}$ of $\mathcal{D}_\omega$ in $\mathcal{D}$ fulfilling $t|_\mathcal{E} = 0$. Then, defining W to be the projector $W : \mathcal{D} \longrightarrow \mathcal{D}_\omega$ with $\mathrm{Ran}(\mathrm{Id} - W) = \mathcal{E}$, we indeed obtain

$$\langle t, h \rangle = \langle t, Wh \rangle + \langle t, (\mathrm{Id} - W)h \rangle = \langle t, Wh \rangle = \langle t_\omega, Wh \rangle = \langle t^W, h \rangle \; ,$$

by using $t|_\mathcal{E} = 0$ and $t|_{\mathcal{D}_\omega} = t_\omega$, which follows from the uniqueness of t_ω.

To prove the existence of $\mathcal{E}$, let $\mathcal{E}_1$ be an arbitrary complementary space of $\mathcal{D}_\omega$ in $\mathcal{D}$ and W_1 the pertinent projector $W_1 : \mathcal{D} \longrightarrow \mathcal{D}_\omega$, that is, $\mathrm{Ran}(\mathrm{Id} - W_1) = \mathcal{E}_1$. Now we choose a $g \in \mathcal{D}_\omega$ with $\langle t, g \rangle = 1$ and set

$$\mathcal{E} := \{ k - \langle t, k \rangle \, g \,|\, k \in \mathcal{E}_1 \} \; . \tag{3.2.37}$$

Obviously $\mathcal{E}$ is a vector space and it holds that $t|_{\mathcal{E}} = 0$. To show that $\mathcal{D} = \mathcal{D}_\omega + \mathcal{E}$ we split any $h \in \mathcal{D}$ into $h = h_1 + h_2$, $h_1 \in \mathcal{E}_1$, $h_2 \in \mathcal{D}_\omega$. With that we obtain

$$\begin{aligned} &h = (h_1 - \langle t, h_1 \rangle\, g) + (h_2 + \langle t, h_1 \rangle\, g) \quad \text{where} \\ &(h_1 - \langle t, h_1 \rangle\, g) \in \mathcal{E}\ ,\ (h_2 + \langle t, h_1 \rangle\, g) \in \mathcal{D}_\omega\ . \end{aligned}$$

It remains to verify that $\mathcal{E} \cap \mathcal{D}_\omega = \{0\}$. Let $l \in \mathcal{E} \cap \mathcal{D}_\omega$. Then, on the one hand we have $l = k - \langle t, k \rangle\, g$ for some $k \in \mathcal{E}_1$, and on the other hand we get $l = W_1 l = W_1 k - \langle t, k \rangle\, W_1 g = -\langle t, k \rangle\, g$. We find $k = 0$ and hence $l = 0$. □

Almost homogeneous extension. In our construction of the retarded product, we will trace back the renormalization (or extension) problem to direct extensions (3.2.22) and the extension of almost homogeneously scaling distributions t^0. In the latter case, an arbitrary extension t satisfying $\mathrm{sd}(t) = \mathrm{sd}(t^0)$ does not maintain almost homogeneous scaling; e.g., a term $C_a\, \partial^a \delta_{(k)}$ with $|a| < \mathrm{sd}(t^0) - k$ (which may be added according to (3.2.21)) violates this property. However, there exist extensions which preserve almost homogeneous scaling, but in some cases it cannot be avoided that the power N (see Definition 3.1.17) is increased by 1. This is essentially the statement of the following proposition.

Proposition 3.2.16 (Extension of almost homogeneously scaling distributions, [55, 86, 101, 107, 131]). *Let $t^0 \in \mathcal{D}'(\mathbb{R}^k \setminus \{0\})$ scale almost homogeneously with degree $D \in \mathbb{C}$ and power $N_0 \in \mathbb{N}$. Then there exists an extension $t \in \mathcal{D}'(\mathbb{R}^k)$ which scales also almost homogeneously with degree D and power $N \geq N_0$:*

(i) *If $D \notin \mathbb{N} + k$, then t is unique and $N = N_0$;*

(ii) *if $D \in \mathbb{N} + k$, then t is non-unique and $N = N_0$ or $N = N_0 + 1$. In this case, given a particular solution t_0, the general solution is of the form*

$$t = t_0 + \sum_{|a|=D-k} C_a\, \partial^a \delta_{(k)} \quad \textit{with arbitrary} \quad C_a \in \mathbb{C}. \tag{3.2.38}$$

In addition, there exists a homogeneous polynomial

$$p(\partial) = \sum_{|a|=D-k} \tilde{C}_a\, \partial^a\ ,\quad \tilde{C}_a \in \mathbb{C}\ , \tag{3.2.39}$$

such that

$$\big(x\partial_x + D\big)^{N_0+1} t(x) = p(\partial_x)\delta_{(k)}(x)\ ,\quad \textit{where} \quad x\partial_x := \sum_{r=1}^{k} x_r \partial_{x_r}\ , \tag{3.2.40}$$

for all *solutions t. So $p(\partial)$ is universal in the sense that it depends only on t^0, but not on t. Exercise 3.2.18(c) and Remark 3.5.15 give formulas to compute $p(\partial)$.*

For the subcase $\mathrm{Re}\, D\ (= \mathrm{sd}(t^0)) < k$ *of case* (i), *the unique t agrees with the direct extension of t^0* (3.2.22).

The last sentence is due to the following: For $\operatorname{Re} D < k$ the unique t of the proposition fulfills $\operatorname{sd}(t) = \operatorname{Re} D = \operatorname{sd}(t^0) < k$, hence it agrees with the unique extension of part (a) of Theorem 3.2.8, which is the direct extension.

Concerning the statement (3.2.39)–(3.2.40), note that universality of $p(\partial)$ follows immediately from (3.2.38), since $(x\partial_x + D)\sum_{|a|=D-k} C_a\, \partial^a \delta_{(k)}(x) = 0$. In particular in the case $N_0 = 0$, the breaking of homogeneous scaling is the *same* local term for all almost homogeneous extensions t: $(x\partial_x + D)t(x) = p(\partial)\delta(x)$.

Example 3.2.17. Recall that the massless retarded propagator $D^{\mathrm{ret}} \in \mathcal{D}'(\mathbb{R}^d)$ scales homogeneously with degree $d-2$. Let $Q(\partial)$ be a polynomial in the partial derivatives $\partial_0, \dots, \partial_{d-1}$, which is homogeneous of degree q. All homogeneous extensions of

$$Q(\partial)\, D^{\mathrm{ret}}(x) \in \mathcal{D}'(\mathbb{R}^d \setminus \{0\}) \quad \text{are} \quad Q(\partial)\, D^{\mathrm{ret}}(x) + \sum_{|a|=q-2} C_a\, \partial^a \delta_{(d)}(x) \in \mathcal{D}'(\mathbb{R}^d)\ ,$$

where the constants $C_a \in \mathbb{C}$ are arbitrary. Note: For $q \geq 2$ we are in case (ii) of the proposition, in which the direct extension is not (directly) applicable; nevertheless, homogeneous scaling is maintained.

We prove Proposition 3.2.16 in the following exercise, by following [55, 101].

Exercise 3.2.18. Again we set $x\partial_x := \sum_{r=1}^k x_r \partial_{x_r}$. Let t^0, D and N_0 be given as in the proposition and let t_1 be an arbitrary extension of t^0 with $\operatorname{sd}(t_1) = \operatorname{sd}(t^0) = \operatorname{Re} D$, which exists by Theorem 3.2.8.

(a) Prove that

$$(x\partial_x + D)^{N_0+1} t_1(x) = \sum_{|a| \leq \operatorname{Re} D - k} C_a\, \partial^a \delta_{(k)}(x) \tag{3.2.41}$$

for some constant numbers $C_a \in \mathbb{C}$; this relation will be used in the following steps.

(b) In case (i) show that

$$t := t_1 - \sum_{|a| \leq \operatorname{Re} D - k} \frac{C_a}{(D - k - |a|)^{N_0+1}}\, \partial^a \delta_{(k)} \tag{3.2.42}$$

is an extension with the wanted properties, where the constants C_a are given by (3.2.41).

(c) In case (ii) the subtraction of the $\partial^a\delta$-terms in (3.2.42) does not work for $|a| = D-k$. However, show that

$$t := t_1 - \sum_{|a| < D-k} \frac{C_a}{(D - k - |a|)^{N_0+1}}\, \partial^a \delta_{(k)}\ . \tag{3.2.43}$$

fulfills the assertion. In addition, verify the claim (3.2.39)–(3.2.40) and show that $p(\partial)$ can be obtained from t_1 by the formula

$$p(\partial_x)\delta_{(k)}(x) = \frac{1}{\prod_{j=0}^{D-k-1}(j - D + k)} \prod_{l=0}^{D-k-1} (x\partial_x + l + k)\, (x\partial_x + D)^{N_0+1} t_1(x)\ . \tag{3.2.44}$$

(d) Prove uniqueness in case (i) and formula (3.2.38) for case (ii).

[*Solution*: (a) Since t^0 scales almost homogeneously with power N_0, the support of the left side of (3.2.41) must be contained in $\{0\}$; therefore, it is of the form $\sum_a C_a\, \partial^a \delta_{(k)}(x)$ where the sum is finite. The upper bound on $|a|$ follows from $\mathrm{sd}\big((x\partial_x + D)^{N_0+1}\, t_1\big) \leq \mathrm{sd}(t_1) = \mathrm{Re}\, D$.
(b) Due to (3.2.21), the distribution t (3.2.42) is an extension of t^0. By using

$$(x\partial_x + D)\, \partial^a \delta_{(k)}(x) = (D - k - |a|)\, \partial^a \delta_{(k)}(x)\ , \tag{3.2.45}$$

we verify $(x\partial_x + D)^{N_0+1}\, t = 0$.
(c) Again, t (3.2.43) is obviously an extension. Now we get

$$(x\partial_x + D)^{N_0+1}\, t(x) = \sum_{|a|=D-k} C_a \partial^a \delta_{(k)}(x)\ . \tag{3.2.46}$$

However, applying the operator $(x\partial_x + D)$ once more we get zero, i.e., t scales almost homogeneously with power $\leq N_0 + 1$.
The claim (3.2.39)–(3.2.40) can be verified as follows: Relation (3.2.46) says that it is satisfied for the particular solution t constructed in (3.2.43); this holds then for all solutions t, due to the universality of $p(\partial)$ explained above.
The idea behind formula (3.2.44) is as follows: $p(\partial_x)\delta_{(k)}(x)$ is equal to the right side of (3.2.46), which can be obtained from t_1 by selecting from (3.2.41) the terms with $|a| = D - k$. The operator

$$\frac{1}{\prod_{j=0}^{D-k-1}(j - D + k)} \prod_{l=0}^{D-k-1} (x\partial_x + l + k)$$

does precisely this selection, since $x\partial_x \partial^a \delta_{(k)} = (-k - |a|)\, \partial^a \delta_{(k)}$.
(d) Since $\partial^a \delta_{(k)}$ scales homogeneously with degree $(k + |a|)$, the terms $C_a\, \partial^a \delta_{(k)}$ in the formula for the general extension (3.2.21), must fulfill $|a| = D - k$, otherwise almost homogeneous scaling is broken. In case (i) there is no a solving this condition.]

In contrast to this indirect construction (3.2.42)–(3.2.43) of an almost homogeneous extension, direct constructive proofs of the proposition are given in [105, 131].[45] In both references the basic idea is to introduce spherical coordinates; with that the problem reduces to a one-dimensional extension in the radial coordinate. The disadvantage of this method is that in general the so-obtained extension breaks Lorentz covariance.

Another constructive proof of part (b) of the proposition is given by the so-called "improved Epstein–Glaser renormalization", which is due to of [85, 122], see also [86, 131].

Lemma 3.2.19 (Improved W-extension)**.** *Assume that a given $t^0 \in \mathcal{D}'(\mathbb{R}^k \setminus \{0\})$ scales almost homogeneously with degree $D \in \mathbb{N} + k$ and power $N \in \mathbb{N}$. Let $\omega := D - k$ and let t_ω be the unique extension of t^0 to $\mathcal{D}'_\omega$ satisfying $\mathrm{sd}(t_\omega) = \mathrm{sd}(t^0)$ (see Claim 3.2.9). Then, the following statements hold true:*

[45] Formula (186) in [105] contains an error: The extension given by that formula does not scale almost homogeneously. This can be corrected by using the "improved Epstein–Glaser renormalization" given in Lemma 3.2.19.

(a) *t_ω scales also almost homogeneously with degree D and power N.*

(b) *The "improved W-extension"*

$$t := t_\omega(x) \circ \Big(\mathrm{Id} - \sum_{|a|<\omega} \frac{(-1)^{|a|}}{a!}\, x^a\, \partial^a \delta - w(x) \sum_{|a|=\omega} \frac{(-1)^{|a|}}{a!}\, x^a\, \partial^a \delta\Big)\ , \quad (3.2.47)$$

where $w \in \mathcal{D}(\mathbb{R}^k)$ is identically 1 in some neighbourhood[46] *of 0, is an extension of t^0 to $\mathcal{D}'(\mathbb{R}^k)$ which scales also almost homogeneously with degree D, but the power may be increased to $N+1$.*

Comparing t (3.2.47) with the W-extension t^W (3.2.27) given by $w_a(x) := \frac{x^a}{a!}\, w(x)$ (3.2.30), where w is as assumed in the lemma, we see: Similarly to the definition of the central solution (3.2.31), the test function $w(x)$ is replaced by 1, but here this is done only for the ($|a| < \omega$)-terms. Unfortunately also formula (3.2.47) is not manifestly Lorentz covariant, since there does not exist a Lorentz invariant $w \in \mathcal{D}(\mathbb{R}^k)$.

Motivated by (3.2.38) one might naively think that a W-extension maintaining almost homogeneous scaling contains only W-subtraction terms with $|a| = \omega$ (i.e., the sum in (3.2.29) is restricted to $|a| = \omega$); but this is wrong, as we see explicitly from (3.2.47).

Proof. First we note that for any distribution $f \in \mathcal{D}'(\mathbb{R}^k)$ (or $\mathcal{D}'(\mathbb{R}^k \setminus \{0\})$ or $\mathcal{D}'_\omega(\mathbb{R}^k)$) it holds that

$$\big\langle (x\partial_x + \tilde{D})^{\tilde{N}} f(x)\, ,\, h(x) \big\rangle_x = (\rho\partial_\rho)^{\tilde{N}} \rho^{\tilde{D}-k} \langle f\, ,\, h_\rho \rangle \big|_{\rho=1} \quad \text{where} \quad h_\rho(x) := h(x/\rho) \quad (3.2.48)$$

and $\tilde{D} \in \mathbb{C}\ ,\ \tilde{N} \in \mathbb{N}^*$.

(a) By using (3.2.48) and (3.2.24)–(3.2.25), we obtain

$$\begin{aligned} \big\langle (x\partial_x + D)^{N+1} t_\omega(x)\, ,\, h(x) \big\rangle_x &= \sum_{|a|=\omega+1} (\rho\partial_\rho)^{N+1} \rho^{D-k-|a|} \langle t_\omega\, ,\, x^a g_{a\,\rho} \rangle \big|_{\rho=1} \\ &= \sum_{|a|=\omega+1} (\rho\partial_\rho)^{N+1} \rho^{-1} \langle \overline{x^a t^0}\, ,\, g_{a\,\rho} \rangle \big|_{\rho=1} = 0\ . \end{aligned}$$

In the last step we use that $x^a t^0$ scales almost homogeneously with degree $D - |a| = k-1$ and power N; and that this property goes over to the direct extension $\overline{x^a t^0}$, as we know from Proposition 3.2.16.

(b) *First part: t exists in $\mathcal{D}'(\mathbb{R}^k)$ and is an extension of t^0.* To prove this we compute $(t - t^W)$, where t^W is as just before:

$$t - t^W = - \sum_{|a|<\omega} \frac{(-1)^{|a|}}{a!} \langle t_\omega, (1 - w(x))\, x^a \rangle\ \partial^a \delta\ .$$

[46] Actually, in order that $w_a(x) = \frac{x^a}{a!}\, w(x)$, where $|a| = \omega$, fulfills (3.2.28), it suffices that $w(0) = 1$. The lemma holds also for this weaker assumption, see [85].

We have to show that the expression on the r.h.s. is well defined: Since $(1-w(x))$ vanishes in a neighbourhood of $x=0$, we may write

$$\langle t_\omega, (1-w(x))\, x^a\rangle = \int d^kx\; t^0(x)\,(1-w(x))\,x^a\ .$$

This "integral" is IR-finite, because $|t^0(x)\,x^a| \sim |x|^{-D+|a|}$ for $|x|\to\infty$ (up to powers of $\log x$) and $-D+|a| < -D+\omega = -k$. Therefore, $\langle t_\omega, (1-w(x))\,x^a\rangle$ exists for $|a|<\omega$ and, hence, $(t-t^W)$ is a linear combination of derivatives of the δ-distribution.

Second part: t scales almost homogeneously with degree D and power $\leq N+1$. To verify this we first use (3.2.48) and $\partial^a h_\rho(0) = \rho^{-|a|}\partial^a h(0)$:

$$\begin{aligned}
\big\langle (x\partial_x + D)^{N+1} t(x)\,,\, h(x)\big\rangle_x &= (\rho\partial_\rho)^{N+1}\rho^\omega\,\langle t\,,\, h_\rho\rangle\big|_{\rho=1}\\
&= (\rho\partial_\rho)^{N+1}\rho^\omega\,\Big\langle t_\omega(x), \Big(h_\rho(x) - \sum_{|a|<\omega}\frac{x^a}{a!\,\rho^{|a|}}\,\partial^a h(0)\\
&\qquad\qquad - w(x)\sum_{|a|=\omega}\frac{x^a}{a!\,\rho^{|a|}}\,\partial^a h(0)\Big)\Big\rangle_x\Big|_{\rho=1}\\
&= (\rho\partial_\rho)^{N+1}\Big\langle \rho^D t_\omega(\rho y), \Big(h(y) - \sum_{|a|<\omega}\frac{y^a}{a!}\,\partial^a h(0)\\
&\qquad\qquad - w(\rho y)\sum_{|a|=\omega}\frac{y^a}{a!}\,\partial^a h(0)\Big)\Big\rangle_y\Big|_{\rho=1}\ .
\end{aligned}$$

Since $(\rho\partial_\rho)^{N+1}(\rho^D t_\omega(\rho y)) = 0$, there is only a contribution if at least one derivative $\rho\partial_\rho$ acts on the test function, that is, on $w(\rho y)$. Hence, we get

$$\begin{aligned}
&\langle (x\partial_x + D)^{N+1} t(x)\,,\, h(x)\rangle_x \qquad\qquad (3.2.49)\\
&= -\sum_{|a|=\omega}\sum_{k=1}^{N+1}\binom{N+1}{k}\Big\langle (\rho\partial_\rho)^{N+1-k}\big(\rho^D t_\omega(\rho y)\big)\,,\,(\rho\partial_\rho)^k w(\rho y)\,y^a\Big\rangle_y\Big|_{\rho=1}\frac{\partial^a h(0)}{a!}\ .
\end{aligned}$$

We point out: $\langle t_\omega(\rho y), w(\rho y)\,y^a\rangle_y$ does not exist, since $w(\rho y)\,y^a \notin \mathcal{D}_\omega$; but (3.2.49) exists, since $(\rho\partial_\rho)^k w(\rho y)\,y^a$ vanishes in a neighbourhood of $y=0$ for $k\geq 1$. We conclude that $(x\partial_x + D)^{N+1}t(x) = \sum_{|a|=\omega} C_a\,\partial^a\delta(x)$ for some coefficients $C_a\in\mathbb{C}$. This implies $(x\partial_x + D)^{N+2}t(x) = 0$. □

Some techniques to compute the almost homogeneous extension t of Proposition 3.2.16 in practice are:

- *Differential renormalization* – explained in Sect. 3.5.1.
- *Analytic regularization* – see Sect. 3.5.2, which includes "dimensional regularization in position space" [63] – a method which is always applicable.
- A *renormalization method relying on the Källén–Lehmann representation* (explained in Sect. 3.2.8); however, this method is restricted to 2-point functions, i.e., to the case $k=d$.

3.2.3 Maintaining scaling and mass expansion in the extension

Due to Proposition 3.2.2, the distributions[47] $r^{(m)\,0}(A_1,\ldots,A_n)\in\mathcal{D}'(\mathbb{R}^{d(n-1)}\setminus\{0\})$ fulfill the Sm-expansion (Definition 3.1.22) with degree $D:=\sum_{k=1}^n \dim A_k$ for all *monomials* $A_1,\ldots,A_n\in\mathcal{P}$. It follows that this holds also for the totally symmetric part

$$s_n^{(m)\,0}(X):=s^0(A_1,\ldots,A_n)(x_1-x_n,\ldots)$$
$$=\frac{1}{n}\sum_{k=1}^n r^0(A_1,\ldots,\widehat{A_k},\ldots,A_k)(x_1-x_k,\ldots)\;, \tag{3.2.50}$$

that is,

$$s_n^{(m)\,0}(X)=u_0^0(X)+\sum_{l=1}^L m^l\,u_l^{(m)\,0}(X)+\mathfrak{r}_{L+1}^{(m)\,0}(X)\quad\text{with}$$
$$u_l^{(m)\,0}(X)=\sum_{p=0}^{P_l}\log^p\Big(\frac{m}{M}\Big)\,u_{l,p}^0(X)\;,\quad\forall L\in\mathbb{N}\;. \tag{3.2.51}$$

The next step in our inductive construction of the retarded product $(R_{n,1})_{n\in\mathbb{N}}$, is to extend $s_n^{(m)\,0}\in\mathcal{D}'(\mathbb{R}^{d(n-1)}\setminus\{0\})$ to a distribution $s_n^{(m)}$ in $\mathcal{D}'(\mathbb{R}^{d(n-1)})$ such that the Sm-expansion (with degree D) is preserved.

- For $m=0$ Remk. 3.1.23 applies; hence, we only have to find an almost homogeneous extension $s_n(X)=u_0(X)\in\mathcal{D}'(\mathbb{R}^{d(n-1)})$ of $s_n^0(X)=u_0^0(X)\in\mathcal{D}'(\mathbb{R}^{d(n-1)}\setminus\{0\})$ – this is the problem treated in Proposition 3.2.16 and Sect. 3.5.
- For $m>0$ we extend each distribution u_0^0, $u_{l,p}^0$, $\mathfrak{r}_{L+1}^{(m)\,0}\in\mathcal{D}'(\mathbb{R}^{d(n-1)}\setminus\{0\})$ individually.

    Due to *part* (6) of Lemma 3.1.26, the remainders

    $$\mathfrak{r}_{L+1}^{(m)\,0}\quad\text{with}\quad L\geq L_0:=D-d(n-1)$$

    may be extended by the direct extension (3.2.22).

    The distributions $u_{l,p}^0$ $(l\geq 1)$ and u_0^0 $(l=0)$ scale almost homogeneously in X with degrees $(D-l)$. Thus, by Proposition 3.2.16, there exist extensions $u_{l,p}\in\mathcal{D}'(\mathbb{R}^{d(n-1)})$ and $u_0\in\mathcal{D}'(\mathbb{R}^{d(n-1)})$, respectively, which scale almost homogeneously with the same degree as the corresponding $u^0_{\ldots}$-distributions. For $l>L_0$ the almost homogeneous extension is unique and agrees with the direct extension (3.2.22). For $0\leq l\leq L_0$ the extension needs a mass scale $M_1>0$; we choose M_1 independent of m, such that $\partial_m u_{l,p}=0$ and $\partial_m u_0=0$. One may choose $M_1=M$.

[47] Here and in what follows notations of type (3.2.20) are used.

We have to maintain the relation (3.1.64), that is,

$$\mathfrak{r}^{(m)\,0}_{L_1+1}(X)=\mathfrak{r}^{(m)\,0}_{L_2+1}(X)+\sum_{l=L_1+1}^{L_2} m^l\sum_{p=0}^{P_l}\log^p\Big(\frac{m}{M}\Big)\,u^0_{l,p}(X)\,,\quad 0\le L_1<L_2\,. \tag{3.2.52}$$

For $L_1 \geq L_0$ the extensions indeed satisfy this relation, because all distributions appearing in (3.2.52) are extended by the unique direct extension. For $L_1 < L_0$ we fulfill (3.2.52) for the extensions by *defining* the extension of $\mathfrak{r}^{(m)\,0}_{L_1+1}$ by

$$\mathfrak{r}^{(m)}_{L_1+1}(X):=\mathfrak{r}^{(m)}_{L_0+1}(X)+\sum_{l=L_1+1}^{L_0} m^l\sum_{p=0}^{P_l}\log^p\Big(\frac{m}{M}\Big)\,u_{l,p}(X)\ \text{ for } 0\le L_1<L_0\,. \tag{3.2.53}$$

This definition gives indeed an extension of $\mathfrak{r}^{(m)\,0}_{L_1+1}$, because (3.2.53) holds true when restricted to $\mathcal{D}'(\mathbb{R}^{d(n-1)}\setminus\{0\})$, as we know from (3.2.52).

An extension $s^{(m)}_n\in\mathcal{D}'(\mathbb{R}^{d(n-1)})$ of $s^{(m)\,0}_n$, which fulfills the Sm-expansion (with the same degree D as $s^{(m)\,0}_n$), is obtained by inserting the constructed extensions of the various distributions into (3.2.51); it does not matter which L we use, since the extensions fulfill (3.2.52).

The Sm-expansion is very helpful for practical computations: Choosing $L=L_0\equiv D-d(n-1)$ it reduces the main problem – the extension $\mathcal{D}'(\mathbb{R}^{d(n-1)}\setminus\{0\})\ni s^{(m)\,0}_n\mapsto s^{(m)}_n\in\mathcal{D}'(\mathbb{R}^{d(n-1)})$ – to a minimal set of almost homogeneously scaling distributions (namely $\{u^0_{l,p}\,|\,0\le l\le L_0,\,0\le p\le P_l\}$); the direct extension of the remainder gives no computational work. Examples are given in Sects. 3.5.1 and 3.5.2. An alternative method to reduce Epstein–Glaser renormalization of a massive model to the extension $\mathcal{D}'(\mathbb{R}^k\setminus\{0\})\to\mathcal{D}'(\mathbb{R}^k)$ of almost homogeneously scaling distributions, is proposed in [132].

3.2.4 Completing the inductive step

In this section we complete the inductive step in the construction of $(R_{n,1})_{n\in\mathbb{N}}$: We construct an extension $S_n\in\mathcal{D}'(\mathbb{M}^n,\mathcal{F})$ of $S^0_n\in\mathcal{D}'(\mathbb{M}\setminus\Delta_n,\mathcal{F})$ such that all axioms are satisfied.

Step 1. In the construction of the preceding section, which preserves the Sm-expansion, the extension of $s^0_n(A_1,\dots,A_n)\equiv s^{(m)\,0}_n(A_1,\dots,A_n)$ is done for *monomials* $A_1,\dots,A_n$, only. On the other hand, to fulfill the AWI[48] we intend to con-

[48] The upper index (m) of $s^{(m)\,0}_n$ and $s^{(m)}_n$ is omitted in the following. The proof that the AWI can be maintained in process of renormalization was first given in [55]; essentially, we follow the construction in that reference.

struct the retarded products by using the definition (3.1.4), hence, we want to extend $s_n^0(A_1,\ldots,A_n)$ only for[49] *balanced* $A_1,\ldots,A_n$. To give a construction which preserves the Sm-expansion and uses the definition (3.1.4), we proceed as follows: First note that, due to linearity of the map $\otimes_{i=1}^n A_i \longmapsto s_n^0(A_1,\ldots,A_n)$ and *part (0)* of Lemma 3.1.26, the Sm-expansion holds for $s_n^0(A_1,\ldots,A_n)$ for all $A_1,\ldots,A_n \in \mathcal{P}_{\mathrm{hom}}$ (and not only for field *monomials*). Using this, we extend $s_n^0(A_1,\ldots,A_n)(x_1-x_n,\ldots,x_{n-1}-x_n)$ for all $A_1,\ldots,A_n \in \mathcal{P}_{\mathrm{bal}} \cap \mathcal{P}_{\mathrm{hom}}$ by means of the construction of the preceding section. This gives distributions

$$\tilde{s}_n(A_1,\ldots,A_n)(x_1-x_n,\ldots,x_{n-1}-x_n) \in \mathcal{D}'(\mathbb{R}^{d(n-1)})$$

which fulfill the Sm-expansion with degree $D := \sum_{k=1}^n \dim A_k$.

There is the slight issue that the resulting $\tilde{s}_n$ might not be symmetric in $(A_1,x_1),\ldots,(A_n,x_n)$; so we simply symmetrize it:

$$\begin{aligned} &s_n(A_1,\ldots,A_n)(x_1-x_n,\ldots,x_{n-1}-x_n) \\ &:= \frac{1}{n!}\sum_{\pi\in S_n} \tilde{s}_n(A_{\pi(1)},\ldots,A_{\pi(n)})(x_{\pi(1)}-x_{\pi(n)},\ldots,x_{\pi(n-1)}-x_{\pi(n)}). \end{aligned}$$

This is also an extension of $s_n^0(A_1,\ldots,A_n)$, since $\tilde{s}_n(A_{\pi(1)},\ldots)(x_{\pi(1)}-x_{\pi(n)},\ldots)$ is an extension of $s_n^0(A_{\pi(1)},\ldots)(x_{\pi(1)}-x_{\pi(n)},\ldots) = s_n^0(A_1,\ldots)(x_1-x_n,\ldots)$. In addition, the Sm-expansion is maintained under this symmetrization.

Step 2. $R_{n-1,1}$ of *arbitrary* field polynomials is defined in terms of $R_{n-1,1}$ of *balanced* fields by (3.1.4). This definition goes over to the symmetric part S_n and to its VEV s_n. In detail: We write arbitrary polynomials A_i $(1\le i\le n)$ as $A_i = \sum_{a_i} \partial^{a_i} B_{ia_i}$ with $B_{ia_i} \in \mathcal{P}_{\mathrm{bal}}$ by using Proposition 1.4.3. From the definition of balanced fields (Definition 1.4.1) we see that every $B \in \mathcal{P}_{\mathrm{bal}}$ can uniquely be written as a finite sum $B = \sum_l B^l$ with $B^l \in \mathcal{P}_{\mathrm{bal}} \cap \mathcal{P}_l$, where $\mathcal{P}_l$ is introduced in Definition 3.1.18. With that we can uniquely write

$$A_i = \sum_{a_i,l_i} \partial^{a_i} B_{ia_i}^{l_i} \quad \text{with} \quad B_{ia_i}^{l_i} \in \mathcal{P}_{\mathrm{bal}} \cap \mathcal{P}_{l_i}\ . \tag{3.2.54}$$

Using that the extension $s_n(B_{1a_1}^{l_1},\ldots,B_{na_n}^{l_n})$ is already constructed in Step 1, we then define

$$\begin{aligned} &s_n(A_1,\ldots,A_n)(x_1-x_n,\ldots) \\ &:= \sum_{a_1,l_1,\ldots,a_n,l_n} \partial_{x_1}^{a_1}\cdots\partial_{x_n}^{a_n} s_n(B_{1a_1}^{l_1},\ldots,B_{na_n}^{l_n})(x_1-x_n,\ldots)\,. \end{aligned} \tag{3.2.55}$$

This expression is well-defined, since one can always take derivatives of distributions.

[49]Note that monomials are in general not balanced fields; but, balanced fields A with a fixed order $|A|$ in φ (3.1.31) and a fixed number of derivatives are in $\mathcal{P}_{\mathrm{hom}}$ – see, e.g., Exap. 1.4.2 and Exer. 1.4.5.

Remark 3.2.20. Let $A_1,\ldots,A_n \in \mathcal{P}$ arbitrary. Formula (3.2.55) holds also for $s_n^0(A_1,\ldots,A_n) \in \mathcal{D}'(\mathbb{R}^{d(n-1)} \setminus \{0\})$, since S^0 (constructed in Sect. (3.2.1)) fulfills the AWI. Due to that the following holds: $s_n(A_1,\ldots,A_n)$ constructed by (3.2.55) is an extension of $s_n^0(A_1,\ldots,A_n)$ for all $A_1,\ldots,A_n \in \mathcal{P}$; because, by restricting (3.2.55) to $\mathcal{D}(\mathbb{R}^{d(n-1)} \setminus \{0\})$, we see that $s_n(A_1,\ldots)$ is an extension of $\sum_{a_1,l_1\ldots} \partial_{x_1}^{a_1} \cdots s_n^0(B_{1a_1}^{l_1},\ldots)$, which agrees with $s_n^0(A_1,\ldots)$.

This remark is only a consistency check for our construction of the retarded product; however, it contains also the basic idea of differential renormalization, which we will study in Sect. 3.5.1.

For completeness, we have to verify, on the level of the extensions, that the Sm-expansion holds true for all *monomials* $A_1,\ldots,A_n$ (and not only for $A_1,\ldots,A_n \in \mathcal{P}_{\mathrm{bal}} \cap \mathcal{P}_{\mathrm{hom}}$) – as formulated in our inductive assumption (axiom (k)). The problem is that in general the elements of $\mathcal{P}_{\mathrm{bal}} \cap \mathcal{P}_{\mathrm{hom}}$ are not monomials, see Exap. 1.4.2. For this purpose we consider the definition (3.2.55) in the case that $A_1,\ldots,A_n$ are monomials. Then there is no sum over l_i in (3.2.54) and in (3.2.55) and we have $\dim B_{ia_i} + |a_i| = \dim A_i$ for all (i,a_i). From *part (3)* of Lemma 3.1.26 we know that each summand in (3.2.55) fulfills the Sm-expansion with degree

$$\sum_{i=1}^n \dim B_{ia_i} + \sum_{i=1}^n |a_i| = \sum_{i=1}^n \dim A_i\,,$$

hence, this holds also for $s_n^{(m)}(A_1,\ldots,A_n)$ (by *part (0)* of that lemma).

Step 3. To construct the $\mathcal{F}$-valued distribution $S_n\big(A_1(x_1),\ldots\big)$ from the VEVs $s_n = \omega_0\big(S_n(\text{subpolynomials of } A_1(x_1),\ldots)\big)$ we use the causal Wick expansion (3.1.23):

$$\begin{aligned} S_n\big(A_1(x_1),\ldots,A_n(x_n)\big) := \sum_{l_1,\ldots,l_n} \frac{1}{l_1!\ldots l_n!} \int dx_{11}\cdots dx_{1l_1}\cdots dx_{n1}\cdots dx_{nl_n} \\ \cdot s_n\bigg(\frac{\delta^{l_1} A_1(x_1)}{\delta\varphi(x_{11})\cdots\delta\varphi(x_{1l_1})},\ldots \frac{\delta^{l_n} A_n(x_n)}{\delta\varphi(x_{n1})\cdots\delta\varphi(x_{nl_n})}\bigg) \\ \cdot \varphi(x_{11})\cdots\varphi(x_{1l_1})\cdots\varphi(x_{n1})\cdots\varphi(x_{nl_n}). \qquad (3.2.56) \end{aligned}$$

By construction, S_n is a symmetric extension of $S_n^0 := \mathfrak{S}R_{n-1,1}^0$, in addition $S_n\big(A_1(x_1),\ldots,A_n(x_n)\big)$ is linear in each $A_j(x_j)$ and fulfills the Sm-expansion, Translation covariance and Field independence; concerning this last property we recall that the expansion (3.1.23) satisfies the Field independence axiom (3.1.20).

The AWI for S_n (3.2.56) can be verified straightforwardly by using that (3.2.55) implies the AWI for s_n (this is analogous to the step from (3.1.4) to (3.1.5), see Footn. 24) and the relation

$$\partial_y \frac{\delta A(y)}{\delta\varphi(x)} = \frac{\delta(\partial A)(y)}{\delta\varphi(x)}\,.$$

Similarly to (3.1.24), the causal Wick expansion (3.2.56) can be written as

$$S_n\big(A_1(x_1),\ldots,A_n(x_n)\big)=\sum_{\underline{A}_l\subseteq A_l} s_n\big(\underline{A}_1,\ldots,\underline{A}_n\big)(x_1-x_n,\ldots)\,\overline{A}_1(x_1)\cdots\overline{A}_n(x_n)\ , \tag{3.2.57}$$

if $A_1,\ldots,A_n$ are *monomials*.

For any symmetric extension S_n of S^0_n, the corresponding $R_{n-1,1}$ obtained by (3.2.4) satisfies Causality and the GLZ relation; for the following reasons: *Causality* cannot get lost in the extension of $R^0_{n-1,1}\big(A_1(x_1),\ldots\big)$ from $\mathcal{D}(\mathbb{M}^n\setminus\Delta_n)$ to $\mathcal{D}(\mathbb{M}^n)$, because the thin diagonal is contained in the set on the r.h.s. of (3.1.10). Also the *GLZ relation* cannot get lost, because the extension $R^0_{n-1,1}\mapsto R_{n-1,1}$ concerns only the symmetric part S_n of $R_{n-1,1}$, and the contributions of S_n to the GLZ relation (3.1.11b) drop out.

The properties which we have just found for S_n, hold then also for $R_{n-1,1}$.

Step 4. How, then, may one fulfill the remaining renormalization conditions?

(i) For the *$\hbar$-dependence* (axiom (l)): in Step 1, the extension $s^0_n(A_1,\ldots,A_n)\mapsto s_n(A_1,\ldots,A_n)$ (where $A_1,\ldots,A_n\in\mathcal{P}_{\mathrm{bal}}\cap\mathcal{P}_{\mathrm{hom}}$) is done in each order of $\hbar$ individually.

(ii) For the *$*$-structure* (axiom (g)): if $\widetilde{S}_n(F_1,\ldots,F_n)$ fulfills the aforementioned properties, this is also true for[50]

$$S_n(F_1,\ldots,F_n):=\tfrac{1}{2}\big(\widetilde{S}_n(F_1,\ldots,F_n)+\widetilde{S}_n(F_1^*,\ldots,F_n^*)^*\big), \tag{3.2.58}$$

which moreover satisfies $S_n(F_1^*,\ldots,F_n^*)=S_n(F_1,\ldots,F_n)^*$. This S_n is an extension of S^0_n, because S^0_n is $*$-symmetric.

The *Field parity* axiom (3.1.34) can trivially be fulfilled: For any n-tupel $(A_1,\ldots,A_n)$ of field monomials for which $\sum_{j=1}^n|A_j|$ is odd, we extend

$$\begin{aligned}&s^0_n(A_1,\ldots,A_n)=0\in\mathcal{D}'(\mathbb{R}^{d(n-1)}\setminus\{0\})\quad\text{by}\\&s_n(A_1,\ldots,A_n):=0\in\mathcal{D}'(\mathbb{R}^{d(n-1)})\ .\end{aligned}$$

(iii) How more complicated *symmetries*, in particular Lorentz covariance, current conservation and gauge symmetries, can be maintained in the extension $s^0_n\mapsto s_n$, is explained in Sect. 3.2.7 and Chaps. 4 and 5.

(iv) It remains the *Off-shell field equation* (axiom (j)). Due to the causal Wick expansion (3.1.23)–(3.1.24), the Field equation (3.1.45) is *equivalent* to

$$\begin{aligned}r_{n-1,1}\big(F_1,\ldots,F_{n-1};\varphi(g)\big)&=-\hbar\int dx\,g(x)\int dy\;\Delta^{\mathrm{ret}}(x-y)\\&\cdot\sum_{k=1}^{n-1}r_{n-2,1}\big(F_1,\ldots\widehat{F_k}\ldots,F_{n-1};\frac{\delta F_k}{\delta\varphi(y)}\big)\ ,\quad\forall F_1,\ldots,F_{n-1}\in\mathcal{F}_{\mathrm{loc}}\end{aligned} \tag{3.2.59}$$

[50]Recall that if $F=\langle f_n,\varphi^{\otimes n}\rangle$, then $F^*:=\langle\bar{f}_n,\varphi^{\otimes n}\rangle$.

and for all $n \geq 2$, $g \in \mathcal{D}(\mathbb{M})$. The expression on the r.h.s. gives an extension of $r^0_{n-1,1}(F_1,\dots,F_{n-1};\varphi(g))$, because the Field equation holds outside the thin diagonal. It is compatible with all other axioms and respects the AWI; this can be verified straightforwardly by using the inductive assumption (cf. Exer. 3.1.16 and the sentences directly before that exercise) – except for the GLZ relation. Concerning the latter, we proceed as follows: From (3.2.59) and the inductively known $j_{n-2,2}\big(F_1,\dots;F_{n-1},\varphi(g)\big)$ we obtain $r_{n-1,1}\big(F_1,\dots,\varphi(g),\dots;F_{n-1}\big)$ by using the GLZ relation and the symmetry in the first $(n-1)$ factors:

$$\begin{aligned} r_{n-1,1}\big(F_1,\dots,F_{n-2},\varphi(g);F_{n-1}\big) &= -\hbar\int dx\; g(x)\int dy\;\Delta^{\mathrm{ret}}(y-x) \\ &\quad\cdot\Big(r_{n-2,1}\big(F_1,\dots,F_{n-2};\frac{\delta F_{n-1}}{\delta\varphi(y)}\big) \\ &\quad+\sum_{k=1}^{n-2} r_{n-2,1}\big(F_1,\dots,\frac{\delta F_k}{\delta\varphi(y)},\dots,F_{n-2};F_{n-1}\big)\Big)\;, \end{aligned} \tag{3.2.60}$$

this computation is worked out in Exer. 3.2.21. The two terms on the r.h.s. may be illustrated by the diagrams

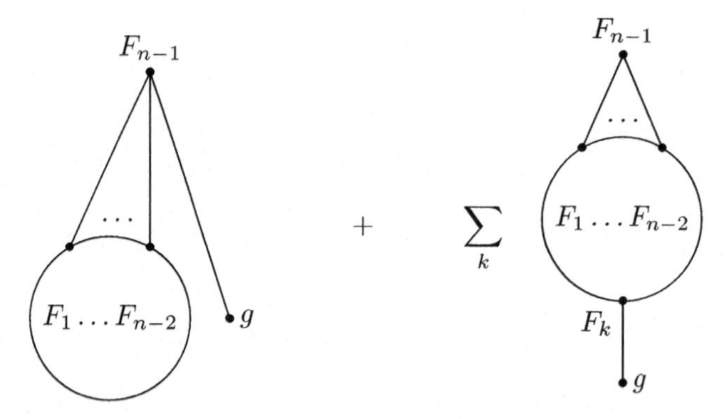

where time is advancing vertically. Since, $r^0_{n-1,1}\big(F_1,\dots,\varphi(g);F_{n-1}\big)$ satisfies the GLZ relation and, hence, (3.2.60) for $\mathrm{supp}\,g\cap\big(\bigcap_{j=1}^{n-1}\mathrm{supp}\,F_j\big)=\emptyset$, formula (3.2.60) gives an extension of $r^0_{n-1,1}\big(F_1,\dots,\varphi(g);F_{n-1}\big)$; which also satisfies all axioms and the AWI, as one verifies straightforwardly. Computing $s_n\big(\dots,\varphi(g)\big)$ by taking the VEV of (3.2.1) and by inserting (3.2.59) and (3.2.60), $s_n\big(\dots,\varphi(g)\big)$ is uniquely determined in terms of the inductively known $r_{n-2,1}$.

So, in order to fulfill the Field equation, we modify the Step 1 as follows: $s_n(A_1,\dots,A_{n-1},\varphi(g))$, $A_1,\dots,A_{n-1}\in\mathcal{P}_{\mathrm{bal}}\cap\mathcal{P}_{\mathrm{hom}}$, is uniquely given by the Field equation in the just described way and fulfills the required properties. However, the construction of $s_n(A_1,\dots,A_n)$ remains unchanged if

$A_1, \ldots, A_n \in \mathcal{P}_{\mathrm{bal}} \cap \mathcal{P}_{\mathrm{hom}}$ and $|A_j| \geq 2\ \forall 1 \leq j \leq n$. (If at least one factor A_j is a C-number, i.e., $|A_j| = 0$, the retarded product vanishes and hence also s_n, see Remk. 3.1.4.) Finally Steps 2 and 3 are done as before.

Exercise 3.2.21. For $n = 3$ derive formula (3.2.60) by proceeding as described.

[*Solution*: Applying the GLZ relation, we have to compute the r.h.s. of

$$r\big(F_1, \varphi(g); F_2\big) = r\big(F_1, F_2; \varphi(g)\big) + i\,\omega_0\Big(\big[R(F_1; F_2), \varphi(g)\big] + \big[F_2, R\big(F_1; \varphi(g)\big)\big]\Big) .$$

By means of (3.2.59), the first term is equal to

$$-\hbar \int dx\, dy\; g(x)\, \Delta^{\mathrm{ret}}(x-y) \Big(r\big(F_1; \frac{\delta F_2}{\delta\varphi(y)}\big) + r\big(F_2; \frac{\delta F_1}{\delta\varphi(y)}\big)\Big) .$$

For the second term we get

$$\hbar \int dx\, dy\; g(x)\Big(r\big(\frac{\delta F_1}{\delta\varphi(y)}; F_2\big) + r\big(F_1; \frac{\delta F_2}{\delta\varphi(y)}\big)\Big)\big(\Delta^{\mathrm{ret}}(x-y) - \Delta^{\mathrm{ret}}(y-x)\big)$$

by using Field independence and $i\big(\Delta^+(z) - \Delta^+(-z)\big) = \Delta^{\mathrm{ret}}(-z) - \Delta^{\mathrm{ret}}(z)$. To compute the last term we use the Field equation,

$$R\big(F_1; \varphi(g)\big) = -\hbar \int dx\, dy\; g(x)\, \Delta^{\mathrm{ret}}(x-y)\, \frac{\delta F_1}{\delta\varphi(y)} ,$$

and the GLZ relation

$$i\,\omega_0\Big(\big[F_2, \frac{\delta F_1}{\delta\varphi(y)}\big]\Big) = -r\big(F_2; \frac{\delta F_1}{\delta\varphi(y)}\big) + r\big(\frac{\delta F_1}{\delta\varphi(y)}; F_2\big) .$$

Inserting all these results we get the assertion

$$r\big(F_1, \varphi(g); F_2\big) = -\hbar \int dx\, dy\; g(x)\, \Delta^{\mathrm{ret}}(y-x) \Big(r\big(F_1; \frac{\delta F_2}{\delta\varphi(y)}\big) + r\big(\frac{\delta F_1}{\delta\varphi(y)}; F_2\big)\Big) .]$$

Remark 3.2.22. Similarly to the equivalence of the Off-shell field equation in terms of $R(\cdots)$ (3.1.45) and $r(\cdots)$ (3.2.59), the validity of (3.2.60) for all $F_1, \ldots, F_{n-1} \in \mathcal{F}_{\mathrm{loc}}$ is equivalent to the same equation for $R(\cdots)$ in place of $r(\cdots)$, that is,

$$\begin{aligned} R_{n-1,1}\big(F_1, \ldots, F_{n-2}, \varphi(g); F_{n-1}\big) &= -\hbar \int dx\, g(x) \int dy\; \Delta^{\mathrm{ret}}(y-x) \\ &\cdot \Big(R_{n-2,1}\big(F_1, \ldots, F_{n-2}; \frac{\delta F_{n-1}}{\delta\varphi(y)}\big) \\ &\quad + \sum_{k=1}^{n-2} R_{n-2,1}\big(F_1, \ldots, \frac{\delta F_k}{\delta\varphi(y)}, \ldots, F_{n-2}; F_{n-1}\big)\Big) , \end{aligned} \tag{3.2.61}$$

for all $F_j \in \mathcal{F}_{\mathrm{loc}}$. Again, this equivalence relies on the causal Wick expansion (3.1.23)–(3.1.24). Alternatively, (3.2.61) can be obtained from (3.1.45) by using the GLZ relation, analogously to the step from (3.2.59) to (3.2.60); for example, in the solution of Exer. 3.2.21 we may throughout replace $r(\cdots)$ by $R(\cdots)$ if we omit ω_0.

3.2.5 Scaling degree axiom versus Sm-expansion axiom

In the literature (e.g.,[24, 66, 148]) the axiom (k) – the Sm-expansion – is frequently replaced by the following weaker axiom:

Scaling degree: For all $A_1,\dots,A_n\in\mathcal{P}_{\mathrm{hom}}$ one requires

$$\mathrm{sd}\, r_{n-1,1}(A_1,\dots,A_n)(x_1-x_n,\dots)\le\sum_{j=1}^{n}\dim A_j\,. \tag{3.2.62}$$

We give the precise formulation of the relation to the Sm-expansion axiom:

Corollary 3.2.23.

(a) *Every* $f^{(m)}\in\mathcal{D}'(\mathbb{R}^k[\setminus\{0\}])$ *fulfilling the Sm-expansion with degree* $D\in\mathbb{R}$*, satisfies* $\mathrm{sd}(f^{(m)})\le D$.

(b) *It follows: Every solution* $(R^{(m)}_{n-1,1})_{n\in\mathbb{N}^*}$ *of the axioms given in Sect.* 3.1 *fulfills the scaling degree property* (3.2.62).

Proof. (a) By definition of the Sm-expansion we may write $f^{(m)}(X)=u_0(X)+\mathfrak{r}_1^{(m)}(X)$, where u_0 scales almost homogeneously with degree D, hence $\mathrm{sd}(u_0)=D$. From *part* (6) of Lemma 3.1.26 we know $\mathrm{sd}(\mathfrak{r}_1^{(m)})\le D-1$. We conclude

$$\mathrm{sd}(f^{(m)})\le\max\{\mathrm{sd}(u_0),\,\mathrm{sd}(\mathfrak{r}_1^{(m)})\}=D\,.$$

(b) By the Sm-expansion axiom the distributions $r^{(m)}_{n-1,1}(A_1,\dots,A_n)$ fulfill the Sm-expansion with degree $D:=\sum_{j=1}^n\dim A_j$ for all *monomials* $A_1,\dots,A_n$. As explained in Step 1 above, this holds then even for all $A_1,\dots,A_n\in\mathcal{P}_{\mathrm{hom}}$. With that, the claim follows by means of part (a). □

If one uses the scaling degree axiom (3.2.62) instead of the axiom (k) in the inductive construction of the retarded product, this weaker requirement plays also the role of a renormalization condition: Below in Remk. 3.2.25 we verify that the scaling degree of $s_n^0(A_1,\dots,A_n)\in\mathcal{D}'(\mathbb{R}^{d(n-1)}\setminus\{0\})$ is bounded by $\sum_j\dim A_j$ for all $A_1,\dots,A_n\in\mathcal{P}_{\mathrm{hom}}$. The axiom (3.2.62) translates then into the requirement, that the extension $s_n(A_1,\dots,A_n)\in\mathcal{D}'(\mathbb{R}^{d(n-1)})$ satisfies also this bound for all $A_j\in\mathcal{P}_{\mathrm{hom}}$.

Since, up to very few exceptions, it holds that $\mathrm{sd}\,s_n^0(A_1,\dots,A_n)=\sum_j\dim A_j$, this axiom can be understood as the requirement that renormalization may not make the retarded products "more singular" (in the UV-region). Substituting (3.2.62) for the axiom (k), the set of allowed retarded products gets truly bigger. An example for this is given in Remk. 3.5.17.

Remark 3.2.24 (Scaling degree axiom for arbitrary field polynomials)**.** We are going to show that the scaling degree condition (3.2.62) can equivalently be formulated as follows: For a *polynomial*

$$A=\sum_l A_l\ ,\quad \text{where}\quad A_l\in\mathcal{P}_{\text{hom}}\ \forall l\ ,\quad \text{we define}\quad \dim A:=\max_l \dim A_l\ . \tag{3.2.63}$$

With that the condition reads

$$\operatorname{sd} r_{n-1,1}(A_1,\ldots,A_n)(x_1-x_n,\ldots)\le\sum_{j=1}^n \dim A_j\ ,\quad \forall A_1,\ldots,A_n\in\mathcal{P}\ . \tag{3.2.64}$$

That (3.2.64) implies (3.2.62) is trivial. To prove "(3.2.62) $\Rightarrow$ (3.2.64)" let $A_j=\sum_{l_j}A_{jl_j}$ with $A_{jl_j}\in\mathcal{P}_{\text{hom}}$; we use linearity of $A_1\otimes\cdots\otimes A_n\longmapsto r_{n-1,1}(A_1,\ldots,A_n)$ and $\operatorname{sd}(t_1+t_2)\le\max\{\operatorname{sd}(t_1),\operatorname{sd}(t_2)\}$:

$$\begin{aligned}\operatorname{sd} r(A_1,\ldots,A_n)&=\operatorname{sd}\sum_{l_1,\ldots,l_n} r(A_{1l_1},\ldots,A_{nl_n})\le\max_{l_1,\ldots,l_n}\operatorname{sd} r(A_{1l_1},\ldots,A_{nl_n})\\&\le\max_{l_1,\ldots,l_n}\sum_{j=1}^n\dim A_{jl_j}=\sum_{j=1}^n\dim A_j\ .\end{aligned}$$

Remark 3.2.25 (Proof of Scaling degree axiom for S_n^0)**.** In this remark we prove the following statement: *Let $(R_{l,1})_{l=0,1,\ldots,n-2}$ be constructed such that they satisfy the basic axioms and the axiom Scaling degree* (3.2.62) *and let $R^0_{n-1,1}(A_1(x_1),\ldots)\in\mathcal{D}'(\mathbb{M}^n\setminus\Delta_n,\mathcal{F})$ and its totally symmetric part $S_n^0(A_1(x_1),\ldots)\in\mathcal{D}'(\mathbb{M}^n\setminus\Delta_n,\mathcal{F})$* (3.2.1) *be constructed (for all $A_1,\ldots\in\mathcal{P}$) by means of the basic axioms, as explained in Propositions* 3.2.1 *and* 3.2.2. *Then, S_n^0 fulfills also the axiom Scaling degree* (3.2.62).

Proof. Similarly to the proof of the Sm-expansion axiom for S_n^0 (Proposition 3.2.2), the nontrivial step is to show that $j_{n-2,2}(A_1,\ldots;A_{n-1},A_n)(x_1-x_n,\ldots)\in\mathcal{D}'(\mathbb{R}^{d(n-1)})$ (3.2.6) fulfills $\operatorname{sd}\big(j_{n-2,2}(A_1,\ldots)\big)\le\sum_{j=1}^n\dim A_j$ for all $A_1,\ldots,A_n\in\mathcal{P}_{\text{hom}}$. From (3.2.7)–(3.2.9) we see that it suffices to prove the following:[51] if $r_1(x_1-x_l,\ldots,x_{l-1}-x_l)\in\mathcal{D}'(\mathbb{R}^{d(l-1)})$ and $r_2(y_1-y_p,\ldots,y_{p-1}-y_p)\in\mathcal{D}'(\mathbb{R}^{d(p-1)})$, where $l,p\ge 1$ and $l+p=n$, have scaling degree $D_1:=\operatorname{sd}(r_1)<\infty$ and $D_2:=\operatorname{sd}(r_2)<\infty$, respectively, then the scaling degree of

$$r_1(x_1-x_l,\ldots,x_{l-1}-x_l)\,r_2(y_1-y_p,\ldots,y_{p-1}-y_p)\prod_{k=1}^K(\partial^{a_k}\Delta_m^+)(x_{i_k}-y_{j_k}) \tag{3.2.65}$$

[51] The corresponding proof for the inductive construction of the time-ordered product $T\equiv(T_n)_{n\in\mathbb{N}^*}$ based on the Scaling degree axiom, i.e., that $T_n^0(\cdots)\in\mathcal{D}'(\mathbb{M}^n\setminus\Delta_n,\mathcal{F})$ (given in (3.3.17)–(3.3.18)) fulfills the Scaling degree axiom, has to master precisely the same problem as stated here in (3.2.65)–(3.2.66). That proof can be found in [66, Sect. 6.1], [148, Chap. 4.3] or [24, Lemma 6.6]. However, the first two references work with different definitions of the scaling degree, in addition they both proceed in momentum space. And the last reference needs a refinement of Steinmann's definition of the scaling degree (Definition 3.2.5), because it treats the problem on curved spacetimes – the latter makes the problem mathematically more demanding.

(which exists in $\mathcal{D}'(\mathbb{R}^{d(n-1)})$) is bounded by

$$\mathrm{sd}\Big(r_1\, r_2 \prod_k (\partial^{a_k}\Delta^+_m)\Big) \le D_1 + D_2 + K(d-2) + \sum_{k=1}^{K} |a_k| \ , \tag{3.2.66}$$

where $i_k \in \{1,\dots,l\}$ and $j_k \in \{1,\dots,p\}$. Introducing the variables

$$\mathbf{u} = (u_i)_{i=1}^{l-1}\ ,\ u_i := x_i - x_l\ ,\quad \mathbf{v} = (v_j)_{j=1}^{p-1}\ ,\ v_j := y_j - y_p\ ,\quad z := x_l - y_p\ ,$$

and setting $u_l := 0$ and $v_p := 0$, we may write the distribution (3.2.65) as

$$f(\mathbf{u},\mathbf{v},z) := r_1(\mathbf{u})\, r_2(\mathbf{v}) \prod_{k=1}^{K} (\partial^{a_k}\Delta^+_m)(u_{i_k} - v_{j_k} + z)\ . \tag{3.2.67}$$

The difficulty is that $f(\mathbf{u},\mathbf{v},z)$ contains pointwise products of distributions. As mentioned at the beginning of Sect. 2.4, using wave front set arguments one can prove that such pointwise products exist.

For an arbitrary $h \in \mathcal{D}(\mathbb{R}^{d(n-1)})$ and an arbitrary $\varepsilon > 0$ we have to show that

$$\lim_{\rho\downarrow 0} \rho^{D_1+D_2+K(d-2)+\sum_{k=1}^K |a_k|+\varepsilon}\, \langle f(\rho\mathbf{u},\rho\mathbf{v},\rho z), h(\mathbf{u},\mathbf{v},z)\rangle_{\mathbf{u},\mathbf{v},z} = 0\ .$$

For this purpose we write the l.h.s. as

$$\lim_{\rho\downarrow 0} \rho^{D_1+D_2+\varepsilon}\, \langle r_1(\rho\mathbf{u})\, r_2(\rho\mathbf{v}), \tilde{h}_{\rho m}(\mathbf{u},\mathbf{v})\rangle_{\mathbf{u},\mathbf{v}} \tag{3.2.68}$$

with

$$\begin{aligned}\tilde{h}_{\rho m}(\mathbf{u},\mathbf{v}) &:= \int dz\ \prod_{k=1}^{K} \Big(\rho^{d-2+|a_k|}(\partial^{a_k}\Delta^+_m)\big(\rho(u_{i_k} - v_{j_k} + z)\big)\Big)\, h(\mathbf{u},\mathbf{v},z)\\ &= \int dz\ \prod_{k=1}^{K} \Big((\partial^{a_k}\Delta^+_{\rho m})(u_{i_k} - v_{j_k} + z)\Big)\, h(\mathbf{u},\mathbf{v},z)\ ,\end{aligned}$$

where (2.2.9) is used. Because h has compact support, this holds also for $\tilde{h}_m$. As we explain below, $\tilde{h}_m$ is also smooth; hence $\tilde{h}_m \in \mathcal{D}(\mathbb{R}^{d(n-2)})$. We point out that this holds true for all $m \ge 0$; in particular also for $m = 0$.

From the Sm-expansion of Δ^+_m (see Exer. 3.1.24 for the case of $d = 4$ dimensions) we know that[52]

$$\partial^a \Delta^+_{\rho m}(z) = \partial^a D^+(z) + \mathcal{O}(\rho m)\ ,$$

where $\mathcal{O}(\rho m)$ includes terms $\sim \rho m \log^p(\frac{\rho m}{M})$ with $p \in \mathbb{N}$. Inserting this into the definition of $\tilde{h}_{\rho m}$ we obtain

$$\tilde{h}_{\rho m} = \tilde{h}_0 + \mathcal{O}(\rho m)\ .$$

[52] In even dimensions the relation is stronger: $\partial^a \Delta^+_{\rho m}(z) = \partial^a D^+(z) + \mathcal{O}(\rho^2 m^2)$.

Therefore, considering the limit (3.2.68), the leading term comes from the massless two point function, that is, from $\tilde{h}_0$:

$$(3.2.68) = \lim_{\rho\downarrow 0} \rho^{D_1+D_2+\varepsilon} \langle r_1(\rho\mathbf{u})\, r_2(\rho\mathbf{v}), \tilde{h}_0(\mathbf{u},\mathbf{v})\rangle_{\mathbf{u},\mathbf{v}} \,. \tag{3.2.69}$$

Because $\tilde{h}_0 \in \mathcal{D}(\mathbb{R}^{d(n-2)})$, we may apply the tensor product rule, $\mathrm{sd}(r_1 \otimes r_2) = \mathrm{sd}(r_1) + \mathrm{sd}(r_2)$ (Exer. 3.2.7(c)(v)), to conclude that the limit (3.2.69) vanishes indeed.

To show that $\tilde{h}_m$ is smooth for all $m \geq 0$, we verify that

$$\widehat{\tilde{h}_m}(\mathbf{p},\mathbf{q}) \sim \int d\mathbf{u}\, d\mathbf{v}\; e^{i(\mathbf{up}+\mathbf{vq})}\, \tilde{h}_m(\mathbf{u},\mathbf{v})$$

decays rapidly in all directions. To simplify the notations we do this for a typical example:

$$l = 2 = p\,, \quad k = 3\,, \quad i_1 = i_2 = i_3 = 1\,, \quad j_1 = j_2 = 1\,,\ j_3 = 2\,.$$

Omitting constant prefactors we get

$$\widehat{\tilde{h}_m}(\mathbf{p},\mathbf{q}) \sim \int dk_1 dk_2 dk_3\, \prod_{j=1}^{3} k_j^{a_j}\, \theta(k_j^0)\, \delta(k_j^2 - m^2) \tag{3.2.70}$$

$$\cdot\, \hat{h}\big((p_1 - k_1 - k_2 - k_3, p_2), (q_1 + k_1 + k_2, q_2 + k_3), -k_1 - k_2 - k_3\big)\,.$$

Since $h \in \mathcal{D}(\mathbb{R}^{5d}) \subset \mathcal{S}(\mathbb{R}^{5d})$, we have $\hat{h} \in \mathcal{S}(\mathbb{R}^{5d})$. In addition, if, e.g., $\|k_3\| \to \infty$ or $\|k_1 + k_2\| \to \infty$, then $\|-k_1 - k_2 - k_3\| \to \infty$, due to the factors $\theta(k_j^0)\, \delta(k_j^2 - m^2)$. These facts imply that $\widehat{\tilde{h}_m}$ decays rapidly in all directions. Namely, for example, if $\|q_1\| \to \infty$ and we want to keep $\|q_1 + k_1 + k_2\|$ finite, then $\|k_1 + k_2\| \to \infty$ and, hence, $\|-k_1 - k_2 - k_3\| \to \infty$. □

3.2.6 The general solution of the axioms

We are going to study the general solution of the inductive step $R_{n-2,1} \to R_{n-1,1}$; the result depends on the precise formulation of the Sm-expansion axiom.

Stronger version of the Sm-expansion axiom. The first part of the inductive step – the "off Δ_n" part – is uniquely determined by the basic axioms; and, as explained in Sect. 3.2.4, the second part – the extension to Δ_n – amounts to the extension $s_n^0(A_1, \ldots, A_n) \mapsto s_n(A_1, \ldots, A_n)$ for all $A_1, \ldots, A_n \in \mathcal{P}_{\mathrm{bal}} \cap \mathcal{P}_{\mathrm{hom}}$. By the Sm-expansion axiom, the extension $s_n^0 \mapsto s_n$ is reduced to the almost homogeneous extension of the coefficients $u_l^{(m)\,0}$ (defined in (3.2.51)), where the extensions $u_l^{(m)}$ must also be polynomials in $\log \frac{m}{M}$. The latter extension is restricted by some further renormalization conditions, as explained in Step 4 of Sect. 3.2.4; but in general it is still non-unique. It can be further restricted by the following supplement to the Sm-expansion axiom, which forbids that in the extension $u_l^{(m)\,0} \mapsto u_l^{(m)}$ *new* powers of $\log \frac{m}{M}$ emerge.

(k′) **Addition to Sm-expansion axiom:** In each inductive step and for all $A_1,\dots$ $\dots,A_n\in\mathcal{P}_{\text{bal}}\cap\mathcal{P}_{\text{hom}}$, do the extension $s_n^0(A_1,\dots,A_n)\mapsto s_n(A_1,\dots,A_n)$ as described in Sect. 3.2.3, that is, extend the coefficients u_0^0, $u_{l,p}^0\in\mathcal{D}'(\mathbb{R}^{d(n-1)}\setminus\{0\})$ in the Sm-expansion of $s_n^{(m)\,0}$ (3.2.51) by almost homogeneous and *m-independent* distributions u_0, $u_{l,p}\in\mathcal{D}'(\mathbb{R}^{d(n-1)})$. In particular, we extend zero by zero, that is, if $u_{l_0,p_0}^0=0$ we set $u_{l_0,p_0}:=0$. The general solution for u_0, $u_{l,p}$ is given by (3.2.38), where the coefficients C_a are *m-independent.*

The axiom (k) completed by (k′) is called the "stronger version of the Sm-expansion axiom". Using the additional prescription (k′), we obtain a particular subset $\mathcal{B}_{\text{part}}$ of the set $\mathcal{B}$ of all retarded products $R\equiv(R_{n,1})_{n\in\mathbb{N}}$ solving the axioms without (k′). Since the Sm-expansion of each $s_n^{(m)\,0}$ (3.2.51) is unique, this definition of $\mathcal{B}_{\text{part}}$ is unique.

Remark 3.2.26 (Even spacetime dimensions)**.** We claim: If d is even and if we require the addition (k′), then solely *even* powers of m appear in the Sm-expansion of $r(A_1,\dots,A_n)$ for all monomials $A_1,\dots,A_n$ not depending on m.

This can be seen as follows: From Exer. 3.1.24 we know that the claim holds true for Δ_m^+ in $d=4$ dimensions; and one verifies analogously that this result is correct for all even values of d, see [55, App. A]. Using that, and following the inductive construction of the R-products, we inductively conclude from (3.2.6)–(3.2.8) that the claim holds for $j_{n-2,2}$; hence, it holds also for $r_{n-1,1}^0$ and for s_n^0. Finally, due to the addition (k′), it is maintained in the extension $s_n^0\mapsto s_n$.

The most general solution of the axioms without (k′). The set $\mathcal{B}$ (defined shortly above Remk. 3.2.26) is obtained by adding – in each inductive step – to a particular solution $s^{(m)}(A_1,\dots,A_n)(x_1-x_n,\dots)$, $A_1,\dots,A_n\in\mathcal{P}_{\text{bal}}\cap\mathcal{P}_{\text{hom}}$, a polynomial in derivatives of the delta distribution which fulfills the Sm-expansion:

$$\sum_{l,p} m^l\,\log^p\Big(\frac{m}{M}\Big)\;\mathfrak{S}_n\sum_a C_{l,p,a}(A_1,\dots,A_n)\;\partial^a\delta(x_1-x_n,\dots,x_{n-1}-x_n)\,. \quad (3.2.71)$$

Here, $\mathfrak{S}_n$ denotes symmetrization w.r.t. permutations of the pairs $(A_1,x_1),\dots$ $\dots,(A_n,x_n)$, and the sum runs over $l\in\mathbb{N}$, $p\in\mathbb{N}$ and $a\in\mathbb{N}^{d(n-1)}$, with the restrictions

$$|a|+l=D-d(n-1)\,,\quad p\le P_l\ \text{ for some } P_l<\infty \text{ and (for } l=0)\ P_0=0 \quad (3.2.72)$$

(the latter is due to $\partial_m u_0=0$); the numbers $C_{l,p,a}(A_1,\dots,A_n)\in\mathbb{C}$ do not depend on m. In addition (3.2.71) has to be Lorentz covariant; the coefficients $C_{l,p,a}(A_1,\dots,A_n)$ are also restricted by further axioms as, e.g., the $*$-structure: $C_{l,p,a}(A_1^*,\dots,A_n^*)=\overline{C_{l,p,a}(A_1,\dots,A_n)}$.

Or, to compare with the above prescription (k′) we may say: To obtain the set $\mathcal{B}$, the general extension of the coefficients $u_l^{(m)\,0}\in\mathcal{D}'(\mathbb{R}^{d(n-1)}\setminus\{0\})$ (3.2.51) (we do not mean $u^0_{l,p}$!) in the Sm-expansion of s^0_n reads

$$u_l^{(m)} = [\text{particular extension}] + \sum_{|a|=D-l-d(n-1)} P_a\Big(\log\frac{m}{M}\Big)\,\partial^a\delta_{(d(n-1))} \quad \text{for } l\geq 1\ ,$$

where $P_a(\log\frac{m}{M})$ is an arbitrary polynomial in $\log\frac{m}{M}$ which respects the other axioms. Also in the case $u_l^{(m)\,0}=0$, the axiom (k) (without (k′)) allows arbitrary polynomials $P_a(\log\frac{m}{M})$; however, in that case, usually Lorentz covariance requires $u_l^{(m)}=0$.

The difference between the sets $\mathcal{B}_{\text{part}}$ and $\mathcal{B}$ is illustrated in Exap. 3.5.4 and Remk. 3.5.17; it plays also a deciding role for the mass renormalization of the φ^4-model in $d=4$ dimensions, see Exap. 3.8.2 and Sect. 3.8.2.

Remark 3.2.27 (Odd spacetime dimensions). In our construction of the particular solutions $\mathcal{B}_{\text{part}}$ of the axioms, the factors $\log\frac{m}{M}$ come solely from the $(\log\frac{m}{M})$-term in $\Delta_m^{+\,(d)}$, i.e., the appearance of these factors is due to the (unique) "off Δ_n part" of the inductive step (given in Sect. 3.2.1). For d odd, $\Delta_m^{+\,(d)}$ does not contain any $(\log\frac{m}{M})$-term – it is smooth in $m\geq 0$. Hence, the *stronger version of the Sm-expansion axiom reduces to a Taylor expansion in* m, i.e., $P_l=0$ $\forall l$ in (3.2.51). Or: For d odd the stronger version of the Sm-expansion axiom is equivalent to the conditions that the retarded products $R^{(m)}$ scale almost homogeneously under $(X,m)\mapsto(\rho X,m/\rho)$ (by which we mean (3.1.67)) and that the maps $0\leq m\longmapsto R^{(m)}_{n-1,1}(F_1\otimes\cdots\otimes F_n)$ be smooth, for all $F_1,\ldots,F_n\in\mathcal{F}_{\text{loc}}$ and for all n.

When quantizing with a Hadamard function, this remark applies to all dimensions $d>2$, see App. A.4 and [55].

Counting the indeterminate parameters before the adiabatic limit. In the last part of this section we investigate the number of indeterminate parameters in the inductive step $R_{n-1,1}\big(L(g)^{\otimes(n-1)},F\big)\to R_{n,1}\big(L(g)^{\otimes n},F\big)$ as a function of n. Obviously, this function depends on the interaction L. If it is unbounded, one needs infinitely many additional informations to fix the model uniquely to all orders of the perturbation series. In particular, if the number of indeterminate parameters having observable consequences is unbounded in n, one would need infinitely many experiments to determine the observable predictions of the model completely. An important example for this unsatisfactory situation is *spin-2 gauge theory* (see, e.g., [149, Chap. 5]) – this is a main reason why a perturbative approach to quantum gravity is conceptually problematic. A way out is to interpret perturbative quantum gravity as an "effective" theory, that is, it is used only to model physical situations in which the effects of quantum gravity are small and, hence, the higher orders of the perturbation series may be neglected, see [28] and [143, Chap. 8]. However, we point out, that even for such "non-renormalizable" models, causal perturbation theory produces a *finite* result for the renormalized R-product to every fixed order in the coupling constant and in $\hbar$.

Essentially we follow [55, Sect. 4.1]. The interacting fields

$$A_{\kappa L(g)}(x) = \sum_{n=0}^{\infty} \frac{\kappa^n}{n!} R_{n,1}\big((L(g))^{\otimes n}; A(x)\big) \ , \quad L, A \in \mathcal{P},\ g \in \mathcal{D}(\mathbb{M}) \ , \tag{3.2.73}$$

are left with an indefiniteness coming from the extension of the symmetric part $S^0_{n+1}\big(L(x_1), \ldots, L(x_n); A(x)\big) \in \mathcal{D}'(\mathbb{M}^{n+1} \setminus \Delta_{n+1})$ to $\mathcal{D}'(\mathbb{M}^{n+1})$ in each inductive step. Let $N(L, A, n) \in \mathbb{N}$ be the number of indeterminate parameters in $R_{n,1}\big((L(g))^{\otimes n}; A(x)\big)$ coming from the inductive step $(n-1, 1) \to (n, 1)$, i.e., the number of independent constants $C_{l,p,a}(\ldots)$ in (3.2.71). This number depends on the choice of the renormalization conditions. In the following we presume only Translation covariance, Field independence (axiom (f)) and the stronger version of the Sm-expansion (axioms (k) and (k′)), in particular Lorentz covariance and the AWI will not be used. The latter makes an essential difference, as the following example shows: If we require the AWI, the retarded product

$$r(\partial^\mu\varphi, \partial^\nu\varphi)(x-y) = \partial^\mu_x\partial^\nu_y\, r(\varphi,\varphi)(x-y) = \hbar\,(\partial^\mu\partial^\nu\Delta^{\mathrm{ret}})(x-y)$$

is unique; in contrast, if we omit the AWI but require Lorentz covariance, we may add a term $C\, g^{\mu\nu}\, \delta(x-y)$ (where $C \in \mathbb{C}$ is an arbitrary constant), and if we omit also Lorentz covariance a term $C^{\mu\nu}\, \delta(x-y)$ is admitted for any constant tensor $C^{\mu\nu}$.

In the following proposition it suffices to study the case that L and A are *monomials*, due to linearity of $R_{n,1}$ (axiom (a)).

Proposition 3.2.28 (Renormalizability by power counting, [55]). *Let $L, A \in \mathcal{P}$ be monomials and let $N(L, A, \cdot)$ be the function*

$$N(L,A,\cdot)\colon \begin{cases} \mathbb{N} \longrightarrow \mathbb{N} \\ n \longmapsto N(L,A,n) \ . \end{cases}$$

With the just mentioned renormalization conditions the following statements[53] *hold true for all monomials $A \in \mathcal{P} \setminus \mathbb{C}$.*

(a) *$N(L, A, \cdot)$ is bounded if and only if* $\dim L \le d$.

(b) *The number of non-vanishing values of $N(L, A, \cdot)$ is* finite *if and only if* $\dim L < d$.

An interaction L fulfilling (a) is called "renormalizable by power counting". An important example is $L = \kappa\varphi^4$ in $d = 4$ dimensions. In the (algebraic) adiabatic limit and with the usual renormalization conditions (also Lorentz covariance), this model has 4 indeterminate parameters, each being a formal power series in κ and $\hbar$; see Exap. 3.8.2 in Sect. 3.8.1.

[53] The case $A \in \mathbb{C}$ is excluded, because in that case it holds that $R_{n,1}\big((L(g))^{\otimes n}; A(x)\big) = 0\ \forall L(g)$ and for all $n \ge 1$, see Remk. 3.1.4.

If $\dim L < d$ (case (b) of the proposition), only the lowest orders of the perturbation series (3.2.73) require "nontrivial" renormalization, all higher orders can be renormalized by the direct extension (3.2.22), that is, they are unique; such an L is called "super-renormalizable".

In the literature (also in causal perturbation theory [66, 148, 149]) the counting of indeterminate parameters is usually done in terms of the S-matrix *in the adiabatic limit* (i.e., for $\sum_n \frac{\kappa^n}{n!} \lim_{g\to 1} T_n\big((L(g))^{\otimes n}\big)$ where T_n is the time-ordered product introduced in Sect. 3.3), and the corresponding version of the proposition can be proved rather easily, see, e.g., [12, Sect. 28.1]. It does not make an essential difference that we count in terms of retarded products. But, since we do not perform the adiabatic limit, our discussion is more involved.

Proof. We only treat the case "d is even", which is more difficult than the case "d is odd", as explained in Remk. 3.2.27.

Using the causal Wick expansion (3.1.24), the indefiniteness of the R-product $R_{n,1}\big((L(g))^{\otimes n}; A(x)\big)$ is precisely the indefiniteness of all C-number distributions $r_{n,1}(U_1, \ldots, U_n; U)$ with $U_1, \ldots, U_n \subseteq L$ and $U \subseteq A$. We will use the relations

$$0 \leq \dim U_j \leq \dim L \qquad \text{and} \qquad 0 \leq \dim U \leq \dim A \;,$$

which follow from (3.1.58). From (3.2.71) we know that the indefiniteness of $r_{n,1}(U_1, \ldots, U_n; U)(x_1 - x, \ldots)$ is of the form

$$\sum_{l=0}^{\lfloor \omega_n/2 \rfloor} m^{2l} \sum_{p=0}^{P^n_{2l}} \log^p\Big(\frac{m}{M}\Big)\; \mathfrak{S}_{n+1} \sum_{|a|=\omega_n - 2l} C^n_{l,p,a}(U_1, \ldots, U_n, U) \cdot \partial^a \delta(x_1 - x, \ldots, x_n - x)\,, \tag{3.2.74}$$

where $\mathfrak{S}_{n+1}$ denotes symmetrization in $(U_1, x_1), \ldots, (U_n, x_n), (U, x)$ and

$$\omega_n \equiv \omega_n(U_1, \ldots, U_n, U) = \sum_{j=1}^n \dim U_j + \dim U - dn\;. \tag{3.2.75}$$

Here we use that for "d is even" only even powers of m appear in (3.2.71), as explained in Remk. 3.2.26.

$N(L, A, n)$ is the number of independent parameters

$$C^n_{l,p,a}(U_1, \ldots, U_n, U)\;,\; U_1, \ldots, U_n \subseteq L\;,\; U \subseteq A\;,\; l \leq \frac{\omega_n}{2}\;,\; p \leq P^n_{2l}\;,\; |a| = \omega_n - 2l\;, \tag{3.2.76}$$

after performing the symmetrization $\mathfrak{S}_{n+1}$.

Next we are going to show that the stronger version of the Sm-expansion implies the bound

$$P^n_{2l} \leq \frac{2l}{d-2} \leq \frac{\omega_n}{d-2}\;. \tag{3.2.77}$$

Namely, in the stronger version of the Sm-expansion all factors $\log \frac{m}{M}$ come from the ($\log \frac{m}{M}$)-term in $\Delta_m^{+(d)}$ (cf. Remk. 3.2.27), which contains a factor m^{d-2} according to formula (2.2.3) (case $d=4$) or formula (A.17) in [55] (case $d=4+2k$, $k\in\mathbb{N}^*$). Therefore, in the Sm-expansion of any distribution $r(A_1,\ldots,A_{n+1})$ (3.1.61), each factor $\log \frac{m}{M}$ is accompanied by at least $(d-2)$ factors m; that is, the power of m $(=2l)$ and the pertinent powers of $\log \frac{m}{M}$ $(=p)$ fulfill the relation $2l\geq p(d-2)$. This must hold also for (3.2.74).

The main point of the proof is that ω_n is bounded by

$$\omega_n\equiv\omega_n(U_1,\ldots,U_n,U)\leq \dim A+n(\dim L-d)\;,\quad \forall U_j\subseteq L\;,\;U\subseteq A\;,\;n\in\mathbb{N}\;. \tag{3.2.78}$$

With that the only non-obvious statement of the proposition is that for an interaction L with $\dim L=d$ the function $N(L,A,\cdot)$ is bounded for any fixed A. The boundedness of $n\longmapsto\omega_n$ *alone* (given by (3.2.78)) does not suffice to prove this, we have to take into account also the following arguments:

- The number of n-tuples $(U_1,\ldots,U_n)$ with $U_j\subseteq L$ $\forall j$, is increasing with n. But due to the restriction $\omega_n(U_1,\ldots,U_n,U)\geq 0$ and the symmetry in $U_1,\ldots,U_n$, only n-tuples of the form

$$(U_1,\ldots,U_s,L,\ldots,L)\quad\text{with}\quad U_1,\ldots,U_s\subseteq L\;,\quad s:=\lfloor\min\{n,\tfrac{\dim A}{\dim\varphi}\}\rfloor \tag{3.2.79}$$

can be relevant. For $n\geq\frac{\dim A}{\dim\varphi}$ the number of such n-tupels does not depend on n.

- The number of indices $a\in\mathbb{N}^{dn}$ with $|a|=\omega$, for some fixed $\omega>0$, is also increasing with n. But here, the symmetrization $\mathfrak{S}_n$ saves the matter: Taking into account also (3.2.79), we conclude that only indices $a=(a_1,\ldots,a_s,a_{s+1},\ldots,a_n)$ (where $a_j\in\mathbb{N}^d$ and $|a|=\sum_{j=1}^n|a_j|=\omega$) appear which are *symmetrical in* $a_{s+1},\ldots,a_n$. The number of such indices becomes independent of n if n is big enough.

  To give an example let s be fixed and $d=2$; we consider the set of all indices $(a_{s+1},\ldots,a_n)$ with $\sum_{j=s+1}^n|a_j|=2$. Then, for all $n\geq s+2$ this set has only 6 elements:
  $(a_{s+1},\ldots,a_n)=\mathfrak{S}_{n-s}\left(\binom{2}{0},\binom{0}{0},\ldots\right)$, $\mathfrak{S}_{n-s}\left(\binom{0}{2},\binom{0}{0},\ldots\right)$, $\mathfrak{S}_{n-s}\left(\binom{1}{1},\binom{0}{0},\ldots\right)$
  or
  $\mathfrak{S}_{n-s}\left(\binom{1}{0},\binom{1}{0},\binom{0}{0},\ldots\right)$, $\mathfrak{S}_{n-s}\left(\binom{0}{1},\binom{0}{1},\binom{0}{0},\ldots\right)$, $\mathfrak{S}_{n-s}\left(\binom{1}{0},\binom{0}{1},\binom{0}{0},\ldots\right)$,
  where $\mathfrak{S}_{n-s}$ denotes symmetrization in $a_{s+1},\ldots,a_n$.

Since, in addition, there are n-independent bounds for the ranges of l and p, we conclude from (3.2.76) and (3.2.78) that the function $N(L,A,\cdot)$ is indeed bounded. □

If we omit the addition (k′) to the Sm-expansion axiom, the statement (a) of the proposition does not hold, because P^n_{2l} could be unbounded as a function of n.

3.2.7 Maintaining symmetries in the extension of distributions

Mostly the hard (and not always solvable) part of the renormalization problem is to preserve the symmetries of the initial (i.e., unrenormalized) expression.

The procedure in this section is limited to symmetries with respect to compact groups and some further groups, the most important of which is the Lorentz group. But it does not apply to scaling transformations, the conservation of certain currents and gauge invariance. To treat also the latter symmetries, a general study of symmetries is given in Chap. 4, in terms of the "Master Ward Identity".

We investigate the question whether the symmetry with respect to a given group G can be maintained in the process of renormalization. Or in mathematical terms: Given a $t^0 \in \mathcal{D}'(\mathbb{R}^k \setminus \{0\})$ which is symmetric with respect to G and scales almost homogeneously with degree $D \in \mathbb{N}$ under $x \mapsto \rho x$ (3.1.46), does there exist an extension $t \in \mathcal{D}'(\mathbb{R}^k)$ which is also G-symmetric and scales also almost homogeneously with degree D? We follow [55, App. D], which is based on lecture notes of K. Fredenhagen [73].[54]

Existence of a symmetric extension. Let V be a representation of a group G defined on both $\mathcal{D}'(\mathbb{R}^k)$ and $\mathcal{D}'(\mathbb{R}^k \setminus \{0\})$, i.e.,

$$V(g)\,\mathcal{D}'(\mathbb{R}^k) \subseteq \mathcal{D}'(\mathbb{R}^k) \quad \text{and} \quad V(g)\,\mathcal{D}'(\mathbb{R}^k \setminus \{0\}) \subseteq \mathcal{D}'(\mathbb{R}^k \setminus \{0\}) \quad \forall g \in G\ ,$$

which fulfills the following assumptions:

(a) The given $t^0 \in \mathcal{D}'(\mathbb{R}^k \setminus \{0\})$ is invariant under V: $V(g)t^0 = t^0\ ,\ \forall g \in G$.

(b) If $t \in \mathcal{D}'(\mathbb{R}^k)$ is an arbitrary extension of t^0, then $V(g)t$ is an extension of $V(g)t^0 = t^0$.

(c) Almost homogeneous scaling with degree D (3.1.46) is preserved under $V(g)$, $\forall g \in G$.

Remark 3.2.29. Frequently, and in particular in the case of Lorentz covariance, the representation V is given in the following way: There is a corresponding representation W of G on $\mathcal{D}(\mathbb{R}^k)$ under which $\mathcal{D}(\mathbb{R}^k \setminus \{0\})$ and t^0 are invariant, by the latter we mean

$$\langle t^0, W(g)h\rangle = \langle t^0, h\rangle \qquad \forall h \in \mathcal{D}(\mathbb{R}^k \setminus \{0\})\ ,\ g \in G\ .$$

Then, $V := W^T$ is the transposed representation of G on $\mathcal{D}'(\mathbb{R}^k)$ or $\mathcal{D}'(\mathbb{R}^k \setminus \{0\})$, respectively, that is, it is defined by

$$\langle V(g)s, h\rangle := \langle s, W(g^{-1})h\rangle\ , \quad \forall s \in \mathcal{D}'(\mathbb{R}^k[\setminus\{0\}])\ ,\ h \in \mathcal{D}(\mathbb{R}^k[\setminus\{0\}])\ ,\ g \in G\ . \tag{3.2.80}$$

[54] The later paper [3] by Bahns and Wrochna develops an alternative method to address this issue.

With that, (a) and (b) need not to be assumed, they can be derived: This is obvious for (a); and (b) is obtained as follows: If t is an extension of t^0 and if $h \in \mathcal{D}(\mathbb{R}^k \setminus \{0\})$, we know that $W(g^{-1})h \in \mathcal{D}(\mathbb{R}^k \setminus \{0\})$ and, hence,

$$\langle V(g)t, h\rangle = \langle t, W(g^{-1})h\rangle = \langle t^0, W(g^{-1})h\rangle = \langle t^0, h\rangle \quad \forall g \in G\,. \tag{3.2.81}$$

To show the existence of a G-invariant and almost homogeneous extension of the given t^0, we start with an arbitrary extension $t \in \mathcal{D}'(\mathbb{R}^k)$ of t^0 which scales almost homogeneously with degree D. From the assumptions (b), (c) and from (3.2.38) we conclude that[55]

$$l(g) := V(g)t - t \tag{3.2.82}$$

satisfies

$$l(g) \in \tilde{\mathcal{D}}^{\perp}_{\omega}(\mathbb{R}^k) := \Big\{ \sum_{|a|=\omega} C_a \partial^a \delta_{(k)} \;\Big|\; C_a \in \mathbb{C} \Big\} \tag{3.2.83}$$

where $\omega := D - k$.[56] For any $\tilde{l} \in \tilde{\mathcal{D}}^{\perp}_{\omega}(\mathbb{R}^k)$ we find

$$V(g)\tilde{l} = \big(V(g)(t + \tilde{l}) - V(g)t\big) \in \tilde{\mathcal{D}}^{\perp}_{\omega}(\mathbb{R}^k)\,,$$

since this is the difference of two almost homogeneous extensions of t^0. Hence,

$$G \ni g \longrightarrow \pi(g) := V(g)\Big|_{\tilde{\mathcal{D}}^{\perp}_{\omega}(\mathbb{R}^k)} \tag{3.2.84}$$

is a sub-representation of V.

We are searching an $l_0 \in \tilde{\mathcal{D}}^{\perp}_{\omega}(\mathbb{R}^k)$ such that $t + l_0$ is invariant,

$$V(g)(t + l_0) = t + l(g) + V(g)l_0 \overset{!}{=} t + l_0\,, \tag{3.2.85}$$

i.e., l_0 must fulfill

$$l(g) = l_0 - \pi(g)l_0\,, \qquad \forall g \in G\,. \tag{3.2.86}$$

From (3.2.82) it follows that $l(g)$ has the property

$$l(g_1g_2) = \pi(g_1)l(g_2) + l(g_1)\,, \quad \forall g_1, g_2 \in G\,, \tag{3.2.87}$$

because

$$l(g_1g_2) = V(g_1g_2)t - t = V(g_1)\big(V(g_2)t - t\big) + V(g_1)t - t = \pi(g_1)l(g_2) + l(g_1)\,.$$

A solution $G \ni g \longmapsto l(g) \in \tilde{\mathcal{D}}^{\perp}_{\omega}(\mathbb{R}^k)$ of (3.2.87) is called a "cocycle". If $l(g)$ is of the form (3.2.86) for some $l_0 \in \tilde{\mathcal{D}}^{\perp}_{\omega}(\mathbb{R}^k)$, it is called a "coboundary"; and such

[55] In the notation $l(g)$ we suppress that $l(g)$ depends on t, since t is fixed in the following.

[56] In contrast to the definition of $\mathcal{D}_{\omega}(\mathbb{R}^k)$ (3.2.23) and $\mathcal{D}^{\perp}_{\omega}(\mathbb{R}^k)$ (3.2.36), the sum over a in the definition of $\tilde{\mathcal{D}}^{\perp}_{\omega}(\mathbb{R}^k)$ runs only over $|a| = \omega$ (and not $|a| \leq \omega$); to mark this difference we write $\tilde{\mathcal{D}}^{\perp}_{\omega}(\mathbb{R}^k)$ instead of $\mathcal{D}^{\perp}_{\omega}(\mathbb{R}^k)$.

an $l(g)$ solves automatically the cocycle equation (3.2.87), as one verifies easily. The space of the cocycles modulo the coboundaries is called the cohomology of the group with respect to the representation π. Summing up, *the invariance of* t^0 *with respect to the representation* V *of* G *can be maintained in the extension if the cohomology of* G *with respect to* π (3.2.84) *is trivial, i.e., if every cocycle is a coboundary.*

We are now going to show that *this supposition holds true if all finite-dimensional representations of* G *are completely reducible.* For this purpose we consider the restriction of $V(g)$ to the space $(\mathbb{C}\cdot t)\oplus\tilde{\mathfrak{D}}^{\perp}_{\omega}(\mathbb{R}^k)$. From (3.2.83) we see that this is a finite-dimensional representation of G, which may be identified with the matrix representation

$$g \longrightarrow \overline{\pi}(g)=\begin{pmatrix} 1 & 0 \\ l(g) & \pi(g) \end{pmatrix}, \tag{3.2.88}$$

since

$$V(g)(ct+d)=ct+\big(c\,l(g)+\pi(g)\,d\big)\quad \forall c\in\mathbb{C}\,,\ d\in\tilde{\mathfrak{D}}^{\perp}_{\omega}(\mathbb{R}^k)\,,$$

where we identify $(ct+d)$ with $\binom{c}{d}$. Note that all elements of $(\mathbb{C}\cdot t)\oplus\tilde{\mathfrak{D}}^{\perp}_{\omega}(\mathbb{R}^k)$ scale almost homogeneously with degree D. Because $\tilde{\mathfrak{D}}^{\perp}_{\omega}(\mathbb{R}^k)$ is an invariant subspace for $\overline{\pi}$ and because we presuppose that $\overline{\pi}$ is completely reducible, there exists a one-dimensional subspace U which is complementary to $\tilde{\mathfrak{D}}^{\perp}_{\omega}(\mathbb{R}^k)$,

$$(\mathbb{C}\cdot t)\oplus\tilde{\mathfrak{D}}^{\perp}_{\omega}(\mathbb{R}^k)=U\oplus\tilde{\mathfrak{D}}^{\perp}_{\omega}(\mathbb{R}^k)\,,\quad \text{and which is invariant:}\quad \overline{\pi}(g)\,U\subseteq U\ \ \forall g\in G\,.$$

Such a subspace is of the form $U=\mathbb{C}\cdot\binom{1}{l_0}$ for some $l_0\in\tilde{\mathfrak{D}}^{\perp}_{\omega}(\mathbb{R}^k)$. That is, there exists such an l_0 with

$$\overline{\pi}(g)\begin{pmatrix} 1 \\ l_0 \end{pmatrix}=\begin{pmatrix} 1 \\ l(g)+\pi(g)l_0 \end{pmatrix}\in\mathbb{C}\cdot\begin{pmatrix} 1 \\ l_0 \end{pmatrix}. \tag{3.2.89}$$

Hence, $\overline{\pi}\big|_U=\mathrm{Id}$, which means that l_0 solves (3.2.86). ⊟

For the Lorentz group $\mathcal{L}^{\uparrow}_{+}$ all finite-dimensional representations are completely reducible and, hence, the Lorentz invariance can be maintained in perturbative renormalization.[57]

However, for the scaling transformations one has to consider the representations of $(\mathbb{R}_+,\,\cdot\,)$, where "$\cdot$" denotes multiplication. They are not always completely reducible. An example for a reducible but not completely reducible representation is

$$\mathbb{R}_+\ni\rho\longmapsto\begin{pmatrix} 1 & 0 \\ \ln\rho & 1 \end{pmatrix}; \tag{3.2.90}$$

because this representation has only one nontrivial invariant subspace, namely $\mathbb{C}\cdot\binom{0}{1}$. The existence of such representations can be understood as the reason for the breaking of homogeneous scaling in perturbative renormalization.

[57] Related proofs for the *existence* of a Lorentz covariant extension are given in [138], [43, App. A] and, e.g., [18, 105]. Historically, the first one was [138]; the next one was [43, App. A], whose authors did not know the paper [138], since the latter was published only much later.

Construction of a symmetric extension. There remains the question: Given $l(g)$, how to find the solution l_0 of the equation (3.2.86) (if it exists). For *compact groups*[58] this can be done in the following way: We set

$$l_0 := \int_G dg\; l(g)\ , \tag{3.2.91}$$

where dg is the uniquely determined measure on G which has norm 1 and is invariant under left- and right-translations (Haar measure). To verify that (3.2.91) solves (3.2.86) we use the cocycle equation (3.2.87):

$$l_0 - \pi(g)l_0 = \int_G dh\,\big(l(h) - \pi(g)l(h)\big) = \int_G dh\,\big(l(h) - l(gh) + l(g)\big)\ . \tag{3.2.92}$$

Due to translation invariance of the Haar measure, the integrals over the first two terms cancel, and since the measure is normalized we indeed obtain (3.2.86).

Example 3.2.30 ($*$-structure). A very simple example for the method (3.2.91) is the symmetrization (3.2.58). In detail: In terms of the coefficients

$$s(x_1 - x_n, \ldots) := s_n\big(\underline{A}_1, \ldots, \underline{A}_n\big)(x_1 - x_n, \ldots)$$

appearing in the causal Wick expansion of $S_n\big(A_1(x_1), \ldots\big)$ (3.2.57), the $*$-structure condition $S_n\big(A_1(x_1), \ldots\big)^* = S_n\big(A_1(x_1), \ldots\big)$ reads $\overline{s} = s$, where we assume that the monomials A_j are real: $A_j^* = A_j$ (which implies $\underline{A}_j^* = \underline{A}_j$ and $\overline{A}_j^* = \overline{A}_j$, $\forall \underline{A}_j \subseteq A_j$, see (3.1.25)) for all j – this assumption does not restrict generality.

For the coefficients $s^0 \in \mathcal{D}'(\mathbb{R}^{d(n-1)} \setminus \{0\})$ of $S_n^0(A_1(x_1), \ldots) \in \mathcal{D}'(\mathbb{M}^n \setminus \Delta_n, \mathcal{F})$ we inductively know that $\overline{s^0} = s^0$. Let u be an arbitrary, almost homogeneous extension of s^0. The symmetry group is $G = \big(\{-1, 1\}, \cdot\big)$ where "$\cdot$" denotes multiplication. The representation V is given by

$$V(-1)t = \overline{t}\ , \quad V(1)t = t\ , \quad \forall t \in \mathcal{D}'(\mathbb{R}^{d(n-1)}[\setminus\{0\}])\ ;$$

obviously V satisfies the assumptions (a), (b) and (c). Following our procedure, we obtain

$$l(1) = 0\ , \quad l(-1) = \overline{u} - u \in \tilde{\mathcal{D}}_\omega^\perp(\mathbb{R}^{d(n-1)})\ .$$

The integral (3.2.91) is a finite sum, explicitly

$$l_0 = \frac{1}{2}\big(l(1) + l(-1)\big) = \frac{1}{2}\left(\overline{u} - u\right)\ ,$$

and the pertinent symmetric extension reads

$$s = u + l_0 = \frac{1}{2}\left(u + \overline{u}\right)\ .$$

This formula is the symmetrization (3.2.58) in terms of the coefficients $s \equiv s_n(\underline{A}_1, \cdots)$.

[58] A compact group G is a topological group (that is, G is also a topological space such that the group operations $(g_1, g_2) \longmapsto g_1 g_2$ and $g \longmapsto g^{-1}$ are continuous) whose topology is compact (i.e., every open cover of G contains a finite cover). Compact groups are a generalization of finite groups with the discrete topology.

Lorentz invariance. For the Lorentz group the method (3.2.91) fails, since it is non-compact – however, see [66].

To construct a Lorentz invariant extension, we give here a method which has some resemblance with [155] and [18]: The strategy is to construct a projector P on the space $E_{\rm inv}$ of all Lorentz invariant vectors in the representation $\overline{\pi}$ (3.2.88). Then, with t being the above used arbitrary extension of t^0 (with the mentioned scaling property), the definition

$$t_{\rm inv} := Pt \;, \tag{3.2.93}$$

yields a $\mathcal{L}_+^\uparrow$-invariant distribution. And it is also an extension of t^0 with the required scaling property, if P is such that $(t_{\rm inv} - t) \in \tilde{\mathfrak{D}}_\omega^\perp(\mathbb{R}^k)$.

To construct P we search an operator $A\colon (\mathbb{C}\cdot t)\oplus\tilde{\mathfrak{D}}_\omega^\perp(\mathbb{R}^k) \longrightarrow (\mathbb{C}\cdot t)\oplus\tilde{\mathfrak{D}}_\omega^\perp(\mathbb{R}^k)$ such that $A^{-1}(0) = E_{\rm inv}$ and, writing A as a matrix (with the identification used also in (3.2.88)), the upper right coefficient of A is $= 0$ and A is diagonalizable. Assuming that we have found such an operator we proceed as follows: Let $a_0 = 0,\, a_1, \ldots, a_l$ (where $a_r \neq a_s$ for $r < s$) be the eigenvalues of A and let $E_s := (A - a_s\,{\rm Id})^{-1}(0)$ be the corresponding eigenspaces; note $E_0 = E_{\rm inv}$. Since A is diagonalizable we may write

$$(\mathbb{C}\cdot t) \oplus \tilde{\mathfrak{D}}_\omega^\perp(\mathbb{R}^k) = E_{\rm inv} \oplus \bigoplus_{s=1}^{l} E_s \;.$$

For $r, s \neq 0,\ e_s \in E_s$ and $e_0 \in E_{\rm inv}$ we have

$$\frac{a_r\,{\rm Id} - A}{a_r}\, e_s = \frac{a_r - a_s}{a_r}\, e_s \;(= 0 \text{ for } r = s) \quad \text{and} \quad \frac{a_s\,{\rm Id} - A}{a_s}\, e_0 = e_0 \;;$$

from these relations we conclude that the operator

$$P := \frac{a_1\,{\rm Id} - A}{a_1} \cdot \frac{a_2\,{\rm Id} - A}{a_2} \cdots \frac{a_l\,{\rm Id} - A}{a_l} \tag{3.2.94}$$

is a projector on $E_{\rm inv}$. Namely, write any $v \in (\mathbb{C}\cdot t) \oplus \tilde{\mathfrak{D}}_\omega^\perp(\mathbb{R}^k)$ as $v = \sum_{s=0}^{l} e_s$ with $e_s \in E_s$ for all $0 \leq s \leq l$; then $Pv = e_0$.

To verify $(t_{\rm inv} - t) \in \tilde{\mathfrak{D}}_\omega^\perp(\mathbb{R}^k)$ we write

$$t_{\rm inv} = Pt = P(t + l_0) - Pl_0 = t + l_0 - Pl_0 \;, \tag{3.2.95}$$

where $l_0 \in \tilde{\mathfrak{D}}_\omega^\perp(\mathbb{R}^k)$ is chosen such that it fulfills (3.2.86), i.e., $t + l_0$ is invariant (3.2.85). We see from (3.2.94) that the upper right coefficient of the matrix P is $= 0$, since this holds for A. Hence,

$$Pl_0 = \begin{pmatrix} * & 0 \\ * & * \end{pmatrix} \begin{pmatrix} 0 \\ l_0 \end{pmatrix} = \begin{pmatrix} 0 \\ * \end{pmatrix} \in \tilde{\mathfrak{D}}_\omega^\perp(\mathbb{R}^k) \;,$$

and this gives the assertion.

To find A we use the following general fact: For a quadratic Casimir operator[59] C (in any representation $\tilde{\pi}$ of a group G) the invariant vectors (with respect to G) $E_{\rm inv}$ are a subspace of its kernel $\tilde{\pi}(C)^{-1}(0)$, because $\tilde{\pi}(X^a)\,E_{\rm inv}=0$ for all a (see Footn. 59 for the notation). In case of the Lorentz group $\mathcal{L}_+^\uparrow$, we will find a Casimir operator C_0 with $E_{\rm inv}=\tilde{\pi}(C_0)^{-1}(0)$ in each finite-dimensional representation $\tilde{\pi}$.

For the representation $\overline{\pi}$ (3.2.88), the matrix $\overline{\pi}(C_0)$ is a lower triangle matrix, since $\overline{\pi}(X^a)$ is a lower triangle matrix for all a. In addition, $\overline{\pi}(C_0)$ is diagonalizable; this follows by means of Schur's lemma[60] from the facts that $\overline{\pi}(C_0)$ commutes with all $\overline{\pi}(X^a)$ and that $\overline{\pi}$ is completely reducible. Summing up, $A:=\overline{\pi}(C_0)$ fulfills the above made assumptions.

To determine a Casimir operator C_0 of the Lorentz group with $\tilde{\pi}(C_0)^{-1}(0)=E_{\rm inv}$ in each finite-dimensional representation $\tilde{\pi}$, we use the well-known result that there are two quadratic Casimirs,

$$C_0=\vec{L}^2-\vec{M}^2 \quad \text{and} \quad C_1=\vec{L}\cdot\vec{M}\ , \tag{3.2.96}$$

where $\vec{L}$ denotes the infinitesimal rotations and $\vec{M}$ the infinitesimal Lorentz boosts. For the representation on $\mathcal{D}(\mathbb{R}^4)$ it holds that

$$\vec{L}=\frac{1}{i}\vec{x}\times\vec{\partial}\ , \qquad \vec{M}=\frac{1}{i}(x^0\vec{\partial}-\vec{x}\,\partial^0)\ . \tag{3.2.97}$$

Since the representation $\tilde{\pi}$ is completely reducible, we can investigate the question whether the Casimirs (3.2.96) satisfy $\tilde{\pi}(C_j)^{-1}(0)\overset{?}{=}E_{\rm inv}$ by studying the irreducible finite-dimensional representations of the Lorentz group $\mathcal{L}_+^\uparrow$. The latter are those irreducible finite-dimensional representations of $\mathrm{SL}(2,\mathbb{C})$ which represent the matrix $(-\mathbf{1})$ by Id. The irreducible finite-dimensional representations of $\mathrm{SL}(2,\mathbb{C})$ are indexed by two spin quantum numbers $j_1,j_2\in\frac{1}{2}\mathbb{N}$ and have the form

$$\pi_{j_1j_2}(A)\big(\xi^{\otimes 2j_1}\otimes\eta^{\otimes 2j_2}\big)=(A\xi)^{\otimes 2j_1}\otimes((A^*)^{-1}\eta)^{\otimes 2j_2} \tag{3.2.98}$$

with $\xi,\eta\in\mathbb{C}^2$, $A\in\mathrm{SL}(2,\mathbb{C})$ and $A^*:=\overline{A^T}$. For $j_1+j_2\in\mathbb{N}$ this yields a representation of the Lorentz group, since $\pi_{j_1j_2}(-\mathbf{1})=(-1)^{2(j_1+j_2)}\,\mathrm{Id}$. We denote by $\vec{L}_i,\vec{M}_i$ the representations of $\vec{L}$ and $\vec{M}$ on the left $(i=1)$ and the right factor

[59]It suffices here to consider Casimir operators which are quadratic. By a quadratic Casimir C for a Lie group G, we mean an expression which is quadratic in the basis vectors (X^a) of the pertinent Lie algebra $\mathcal{G}$, $C=\sum_{a,b}c_{a,b}X^aX^b$ for some $c_{a,b}\in\mathbb{C}$, and which commutes with all X^a. For example, the quadratic Casimirs of the Lorentz group are given in (3.2.96). For a representation $\tilde{\pi}$ of $\mathcal{G}$, $\tilde{\pi}(C)$ is defined to be the linear operator $\tilde{\pi}(C)=\sum_{a,b}c_{a,b}\,\tilde{\pi}(X^a)\tilde{\pi}(X^b)$.

[60]We mean here the following version of Schur's lemma: Let $\rho\colon G\longrightarrow V$ be an irreducible representation of a group G with $n:=\dim V<\infty$ and let $L\in\mathbb{C}^{n\times n}$ commute with all $\rho(g)$:

$$L\,\rho(g)=\rho(g)\,L \qquad \forall g\in G\ .$$

Then there exists a $\lambda\in\mathbb{C}$ such that $L=\lambda\,\mathbf{1}_{n\times n}$.

($i = 2$), respectively, i.e., $\vec{L} = \vec{L}_1 \otimes \mathrm{Id} + \mathrm{Id} \otimes \vec{L}_2$ and analogously for $\vec{M}$. In the fundamental representation of SL(2, $\mathbb{C}$) (i.e., $(j_1, j_2) = (\frac{1}{2}, 0)$) we have

$$\vec{L}_1 = \frac{1}{2}\vec{\sigma} \, , \qquad \vec{M}_1 = \frac{i}{2}\vec{\sigma} \, , \tag{3.2.99}$$

and in the conjugated representation (i.e., $(j_1, j_2) = (0, \frac{1}{2})$) it holds that

$$\vec{L}_2 = \frac{1}{2}\vec{\sigma} \, , \qquad \vec{M}_2 = -\frac{i}{2}\vec{\sigma} \, , \tag{3.2.100}$$

where $\vec{\sigma} = (\sigma_1, \sigma_2, \sigma_3)$ denotes the Pauli matrices. So we have

$$\vec{M}_1 = i\vec{L}_1 \, , \qquad \vec{M}_2 = -i\vec{L}_2 \, , \tag{3.2.101}$$

and the validity of these relations goes over to all representations π_{j_1,j_2}, where $j_1, j_2 \in \frac{1}{2}\mathbb{N}$ (3.2.98). With that we obtain for the Casimir operators

$$\begin{aligned} C_0 &= (\vec{L}_1 \otimes \mathrm{Id} + \mathrm{Id} \otimes \vec{L}_2)^2 - (\vec{M}_1 \otimes \mathrm{Id} + \mathrm{Id} \otimes \vec{M}_2)^2 \\ &= 2(\vec{L}_1^2 \otimes \mathrm{Id} + \mathrm{Id} \otimes \vec{L}_2^2) = 2\left(j_1(j_1+1) + j_2(j_2+1)\right)(\mathrm{Id} \otimes \mathrm{Id}) \end{aligned} \tag{3.2.102}$$

and

$$\begin{aligned} C_1 &= (\vec{L}_1 \otimes \mathrm{Id} + \mathrm{Id} \otimes \vec{L}_2) \cdot (\vec{M}_1 \otimes \mathrm{Id} + \mathrm{Id} \otimes \vec{M}_2) \\ &= i(\vec{L}_1^2 \otimes \mathrm{Id} - \mathrm{Id} \otimes \vec{L}_2^2) = i\left(j_1(j_1+1) - j_2(j_2+1)\right)(\mathrm{Id} \otimes \mathrm{Id}) \, . \end{aligned} \tag{3.2.103}$$

We find indeed that C_0 vanishes on the trivial representation $j_1 = j_2 = 0$ only, that is, on E_{inv} only. However, the kernel of C_1 is much bigger. Hence, $A = \overline{\pi}(C_0)$.

As an example let us consider a Lorentz invariant distribution $t^0 \in \mathcal{D}'(\mathbb{R}^4 \setminus \{0\})$ which scales almost homogeneously with degree $D = 6$. Let t_ω be the unique extension of t^0 to $\mathcal{D}'_\omega$ (3.2.23), where $\omega = D - 4 = 2$; t_ω is also Lorentz invariant since it is a direct extension (3.2.22) of t^0. An extension t of t^0 to $\mathcal{D}'(\mathbb{R}^4)$ which maintains the almost homogeneous scaling with degree $D = 6$, is obtained by the improved Epstein–Glaser renormalization, see Lemma 3.2.19:

$$t = t_\omega \circ W \, , \quad W := \mathrm{Id} - \sum_{|a|<\omega} \frac{(-1)^{|a|}}{a!} x^a \, \partial^a \delta - w(x) \sum_{|a|=\omega} \frac{(-1)^{|a|}}{a!} x^a \, \partial^a \delta \, . \tag{3.2.104}$$

Next we determine the eigenvalues of C_0 restricted to $(\mathbb{C} \cdot t) \oplus \tilde{\mathcal{D}}_2^\perp$, where $\tilde{\mathcal{D}}_2^\perp := \{\sum_{|a|=2} C_a \partial^a \delta \,|\, C_a \in \mathbb{C}\} \subset \mathcal{D}'(\mathbb{R}^4)$. Using (3.2.102), this can be done by finding the irreducible subrepresentations of the representation of the Lorentz group on $(\mathbb{C} \cdot t) \oplus \tilde{\mathcal{D}}_2^\perp$. From (3.2.104) we see that these are the representations[61]

$$(j_1, j_2) = (0, 0), \left(\frac{1}{2}, \frac{1}{2}\right), (1, 1) \, .$$

[61]Note that δ and $\Box\delta$ transform with the representation $(j_1, j_2) = (0, 0)$, $\partial^\mu \delta$ with $(j_1, j_2) = (\frac{1}{2}, \frac{1}{2})$ and $(\partial^\mu \partial^\nu - \frac{1}{4} g^{\mu\nu} \Box)\delta$ with $(j_1, j_2) = (1, 1)$.

Hence, the eigenvalues of C_0 restricted to $(\mathbb{C}\cdot t)\oplus\tilde{\mathfrak{D}}_2^\perp$ are $2\big(j_1(j_1+1)+j_2(j_2+1)\big) = 0, 3, 8$. By means of (3.2.93)–(3.2.94) we obtain the following formula for a Lorentz invariant extension of t^0:

$$t_{\mathrm{inv}} := P(t_\omega \circ W) \quad \text{with} \quad P := \Big(\mathrm{Id} - \frac{1}{3}\,\overline{\pi}(C_0)\Big)\Big(\mathrm{Id} - \frac{1}{8}\,\overline{\pi}(C_0)\Big) \,. \tag{3.2.105}$$

Now, $\overline{\pi}(C_0)$ (3.2.88) is completely given by $l(C_0)$ (3.2.82) and $\pi(C_0)$ (3.2.84). The latter two can explicitly be computed by means of

$$C_0 = -\frac{1}{2}(x_\mu\partial_\nu - x_\nu\partial_\mu)(x^\mu\partial^\nu - x^\nu\partial^\mu) = (x^\mu\partial_\mu)^2 + 2x^\mu\partial_\mu - x^2\Box \,, \tag{3.2.106}$$

which results from (3.2.96) and (3.2.97).

3.2.8 Explicit computation of an interacting field – renormalization of two-point functions using the Källén–Lehmann representation

In this section we explicitly compute the interacting field

$$\varphi_S(x) \quad \text{for} \quad S = -\kappa \int dx\, g(x)\, \varphi^4(x) \quad \text{and } m = 0, \text{ in } d = 4 \text{ dimensions,}$$

up to second order in κ:

$$\begin{aligned} \varphi_S(x) = \varphi(x) &- \frac{\kappa}{\hbar}\int dx_1\, g(x_1)\, R_{1,1}\big(\varphi^4(x_1), \varphi(x)\big) \\ &+ \frac{\kappa^2}{2\,\hbar^2}\int dx_1\, dx_2\, g(x_1)\, g(x_2)\, R_{2,1}\big(\varphi^4(x_1)\otimes\varphi^4(x_2), \varphi(x)\big) + O(\kappa^3). \end{aligned} \tag{3.2.107}$$

The appearing retarded products can most easily be computed by means of the Off-shell field equation axiom (3.1.45) (with $\Delta^{\mathrm{ret}}_{m=0} = D^{\mathrm{ret}}$) and the derivative formula

$$\frac{\delta\varphi^4(x_1)}{\delta\varphi(y)} = 4\,\varphi^3(x_1)\,\delta(x_1 - y) \,.$$

To first order we obtain

$$R_{1,1}\big(\varphi^4(x_1), \varphi(x)\big) = -4\hbar\, D^{\mathrm{ret}}(x - x_1)\,\varphi^3(x_1) \,. \tag{3.2.108}$$

To second order, the Off-shell field equation yields

$$\begin{aligned} &R_{2,1}\big(\varphi^4(x_1)\otimes\varphi^4(x_2), \varphi(x)\big) \\ &\quad = -4\hbar\, D^{\mathrm{ret}}(x - x_1)\, R_{1,1}\big(\varphi^4(x_2), \varphi^3(x_1)\big) + (x_1 \leftrightarrow x_2) \,. \end{aligned} \tag{3.2.109}$$

It remains to compute $R(\varphi^4, \varphi^3)$; this requires non-direct extensions of distributions. The corresponding $J_{0,2}(\varphi^4 \otimes \varphi^3)$ reads:

$$\begin{aligned} J_{0,2}\big(\varphi^4(x_2) \otimes \varphi^3(x_1)\big) &= \frac{1}{i\,\hbar}\,[\varphi^4(x_2), \varphi^3(x_1)] \\ &= 4 \cdot 3\,\varphi^3(x_2)\,D(x_2 - x_1)\,\varphi^2(x_1) \\ &+ \frac{(4\cdot 3)\,(3\cdot 2)\,\hbar}{2!\,i}\,\varphi^2(x_2)\Big(D^+(x_2-x_1)^2 - D^+(x_1-x_2)^2\Big)\,\varphi(x_1) \\ &+ \frac{(4\cdot 3\cdot 2)\,(3\cdot 2\cdot 1)\,\hbar^2}{3!\,i}\,\varphi(x_2)\Big(D^+(x_2-x_1)^3 - D^+(x_1-x_2)^3\Big)\,. \end{aligned} \tag{3.2.110}$$

So the (Feynman) diagrams contributing to the unrenormalized retarded product $R_{2,1}\big(\varphi^4(x_1) \otimes \varphi^4(x_2), \varphi(x)\big)$ are

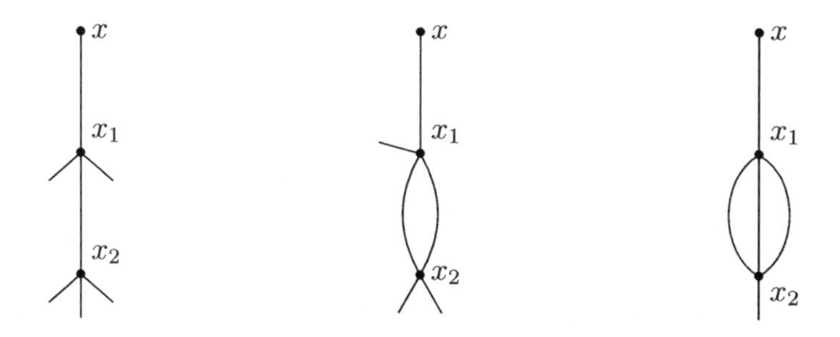

plus the three diagrams obtained by mutual exchange of x_1 with x_2.

We have to extend

$$R^0_{1,1}\big(\varphi^4(x_2), \varphi^3(x_1)\big) = \hbar\,J_{0,2}\big(\varphi^4(x_2) \otimes \varphi^3(x_1)\big)\,\theta(x_1^0 - x_2^0), \tag{3.2.111}$$

or at any rate its symmetric part $S_2(-)$ (3.2.1), from $\mathcal{D}'(\mathbb{M}^2 \setminus \Delta_2, \mathcal{F})$ to $\mathcal{D}'(\mathbb{M}^2, \mathcal{F})$. For the method we shall use, the transition to the symmetric part does not simplify the problem, hence we extend $R^0_{1,1}$ itself. For the first term in (3.2.110), we can use the direct extension and obtain

$$12\,\varphi^3(x_2)\,D(x_2 - x_1)\,\theta(x_1^0 - x_2^0)\,\varphi^2(x_1) = -12\,\varphi^3(x_2)\,D^{\rm ret}(x_1 - x_2)\,\varphi^2(x_1).$$

This extension preserves homogeneous scaling and, hence, fulfills the Sm-expansion axiom (see Remk. 3.1.23), since $\rho^4\,D^{\rm ret}(\rho y) = D^{\rm ret}(y)$.

For the second and third term, we write

$$j_k(y) := -i\,\big(D^+(y)^k - D^+(-y)^k\big), \quad y := x_2 - x_1 \quad \text{for} \quad k = 2, 3,$$

and obtain

$$\mathrm{sd}\big(j_k(y)\,\theta(-y^0)\big) = 2k \geq d = 4.$$

Hence, the direct extension does not apply.

Following [55, App. C] (which is based on lecture notes of K. Fredenhagen [73]), the idea is to use the *Källén–Lehmann spectral representation* [111, 123]: Generally, the latter is a representation of the two-point function of any quantum field (which may be an interacting field) as a "weighted sum" of free propagators with different masses m; more precisely, this sum is an integral over m^2 with a measure $f(m^2)\,dm^2$ on $[0,\infty)$, where f is a suitable weight function or distribution. In the following we will need the Källén–Lehmann representation only for the fields φ^2 and φ^3 of the massless, free theory: Below we prove that there exist weight functions f_2 and f_3 such that

$$\begin{aligned}\omega_0\big(\varphi^k(x_2)\star\varphi^k(x_1)\big) &= \hbar^k\,k!\,D^+(x_2-x_1)^k\\ &= -\,i\hbar^k\,k!\int_0^\infty dm^2\,f_k(m^2)\,\Delta_m^+(x_2-x_1)\end{aligned}\tag{3.2.112}$$

for $k=2,3$. The proof will be given by computing explicitly f_2 and f_3. Inserting (3.2.112) into the definition of j_k, we get

$$j_k(y)=-i\int_0^\infty dm^2\,f_k(m^2)\,\Delta_m(y)\tag{3.2.113}$$

Considering $j_k(y)\,\theta(-y^0)$, we are tempted to use $\Delta_m(y)\,\theta(-y^0)=-\Delta_m^{\rm ret}(-y)$ in the integrand. However, a trick will be needed to ensure convergence of the resulting m^2-integration.

To compute the weight functions f_2 and f_3 (3.2.112), we first derive the Källén–Lehmann representation for $\Delta_{m_1}^+(y)\,D^+(y)\ ,\ m_1\geq 0$. We write

$$\begin{aligned}\Delta_{m_1}^+(y)\,D^+(y)=\frac{1}{(2\pi)^6}\int d^4p_1\,d^4p_2\,\theta(p_1^0)\,\theta(p_2^0)\,\delta(p_1^2-m_1^2)\,\delta(p_2^2)\,e^{-i(p_1+p_2)y}\\ \cdot\int_0^\infty dm^2\int d^4q\,\theta(q^0)\,\delta(q^2-m^2)\,\delta(p_1+p_2-q).\end{aligned}$$

Note that the integrals in the second line (which we have added) are equal to 1, since $(p_1+p_2)^2\geq 0$ and $p_1^0+p_2^0\geq 0$. After replacing $e^{-i(p_1+p_2)y}$ by e^{-iqy}, we use that the integral

$$I_{m_1}(q):=\frac{1}{(2\pi)^3}\int d^4p_1\,d^4p_2\,\theta(p_1^0)\,\theta(p_2^0)\,\delta(p_1^2-m_1^2)\,\delta(p_2^2)\,\delta(p_1+p_2-q)\ ,\quad q\in\overline{V}_+\ ,$$

is Lorentz invariant: $I_{m_1}(\Lambda q)=I_{m_1}(q)\ ,\ \forall\Lambda\in\mathcal{L}_+^\uparrow$; therefore it depends only on q^2 and we may set[62]

$$\tilde f_{m_1}(m^2)=I_{m_1}(q)|_{q^2=m^2,\,q^0\geq 0}\ .$$

Changing the order of integration, we obtain

$$\Delta_{m_1}^+(y)\,D^+(y)=\int_0^\infty dm^2\,\tilde f_{m_1}(m^2)\,\Delta_m^+(y)\ .\tag{3.2.114}$$

[62] To distinguish the two families of weight-functions $\tilde f_{m_1}$ and f_k, we write a tilde over f_{m_1}.

To compute the integral for $\tilde{f}_{m_1}(m^2)$, we may choose $q = (m, \vec{0})$ and first perform the integrations over the energies p_i^0 and evaluate the momentum conservation $\vec{p}_2 = -\vec{p}_1$. The remaining integral over $\vec{p}_1$ reads, after performing the angular integration and setting $p := |\vec{p}_1|$,

$$\tilde{f}_{m_1}(m^2) = \frac{4\pi}{4(2\pi)^3} \int_0^\infty \frac{dp\, p^2}{p\,\sqrt{p^2 + m_1^2}}\, \delta\big(\sqrt{p^2 + m_1^2} + p - m\big)\ .$$

The argument of the δ-function vanishes at

$$m - p = \sqrt{p^2 + m_1^2}\ , \quad \text{i.e.,} \quad p = \frac{1}{2m}\,(m^2 - m_1^2)\ ,$$

provided that $m \geq m_1$. From this, we obtain

$$\tilde{f}_{m_1}(m^2) = \frac{1}{4(2\pi)^2}\,\frac{1}{m^2}\,(m^2 - m_1^2)\,\theta(m^2 - m_1^2)\ . \tag{3.2.115}$$

Using this result, we can easily compute the Källén–Lehmann representation of j_2 and j_3:

$$j_2(y) = \frac{1}{4(2\pi)^2} \int_0^\infty dm^2\, \Delta_m(y)$$

and, taking into account $D^+(\pm y)^2 = \frac{1}{4(2\pi)^2} \int_0^\infty dm_1^2\, \Delta_{m_1}^+(\pm y)$ (3.2.114), we get

$$\begin{aligned} j_3(y) &= \frac{-i}{4(2\pi)^2} \int_0^\infty dm_1^2\, \Big(\Delta_{m_1}^+(y)\, D^+(y) - \Delta_{m_1}^+(-y)\, D^+(-y)\Big) \\ &= \frac{-i}{16(2\pi)^4} \int_0^\infty dm^2 \int_0^{m^2} dm_1^2\, \frac{1}{m^2}\,(m^2 - m_1^2)(\Delta_m^+(y) - \Delta_m^+(-y)) \\ &= \frac{1}{32(2\pi)^4} \int_0^\infty dm^2\, m^2\, \Delta_m(y). \end{aligned}$$

Exercise 3.2.31. Compute the weight function $f_{m_1,m_2}(m^2)$ in the Källén–Lehmann representation

$$\Delta_{m_1}^+(y)\,\Delta_{m_2}^+(y) = \int_0^\infty dm^2\, f_{m_1,m_2}(m^2)\,\Delta_m^+(y)$$

in the d-dimensional Minkowski space (for arbitrary $d \geq 3$).

[*Solution*: Proceeding analogously to the above derivation of (3.2.114)–(3.2.115), we obtain

$$f_{m_1,m_2}(m^2) = \frac{|S^{d-2}|}{4 \cdot 2^{d-3}\,(2\pi)^{d-1}}\,\theta(m - m_1 - m_2)\, m^{2-d}\Big((m^2 - m_1^2 - m_2^2)^2 - 4m_1^2 m_2^2\Big)^{(d-3)/2}$$

where $|S^d| = \frac{2\pi^{(d+1)/2}}{\Gamma(\frac{1}{2}(d+1))}$ is the surface of the unit sphere S^d in $(d+1)$ dimensions; details are given in App. A of [60].]

Now, if we replace $\Delta_m(y)$ in (3.2.113) by $-\Delta_m^{\rm ret}(-y)$, then the m^2-integral becomes UV-divergent. Using $f_k(m^2) \sim m^{2k-4}$, an indication of this different behaviour is that in momentum space the integral

$$\int_0^\infty dm^2\, m^{2k-4}\, \hat{\Delta}_m(p) \;\text{ exists, but }\; \int_0^\infty dm^2\, m^{2k-4}\, \hat{\Delta}_m^{\rm ret}(-p) \;\text{ diverges,}$$

where $k = 2, 3$. However, for $y \neq 0$ (or more correctly, on $\mathcal{D}(\mathbb{R}^4 \setminus \{0\})$) we have $-\Delta_m^{\rm ret}(-y) = \Delta_m(y)$ if $y^0 \le 0$, and $-\Delta_m^{\rm ret}(-y) = 0$ if $y^0 \ge 0$. Hence we may write

$$r_k^0(y) = j_k(y)\,\theta(-y^0) = i\int_0^\infty dm^2\, f_k(m^2)\, \Delta_m^{\rm ret}(-y) \in \mathcal{D}'(\mathbb{R}^4 \setminus \{0\}). \tag{3.2.116}$$

Note that r_k^0 scales homogeneously: $\rho^{2k}\, r_k^0(\rho y) = r_k^0(y)$.

For $k = 2$, let $\mu > 0$ be an arbitrary mass scale; we define

$$r_{2\,\mu}(y) := \frac{1}{4(2\pi)^2}\,(\Box_y - \mu^2)\int_0^\infty dm^2\, \frac{\Delta_m^{\rm ret}(-y)}{m^2+\mu^2}\,, \tag{3.2.117}$$

which exists as a distribution in $\mathcal{D}'(\mathbb{R}^4)$, since the m^2-integral is finite. For $y \neq 0$ the operator $(\Box_y - \mu^2)$ may be applied before the m^2-integration and we obtain $r_2^0(y)$ (3.2.116), i.e., $r_{2\,\mu}$ is an extension of r_2^0. To avoid an IR-divergence of the m^2-integral, we need to introduce a scale $\mu > 0$, which breaks homogeneous scaling, due to

$$\rho^4 r_{2\,\mu}(\rho y) = r_{2\,\rho\mu}(y).$$

We still have to check the Sm-expansion axiom, i.e., that $r_{2\,\mu}$ scales almost homogeneously with degree $D = 4$. According to Proposition 3.2.16, the power of the almost homogeneous scaling should be $N = 1$, i.e.,

$$(\rho\,\partial_\rho + 4)^2\, r_{2\,\mu}(\rho y) = 0. \tag{3.2.118}$$

For this purpose, we compute

$$\begin{aligned}
\rho^4 r_{2\,\mu}(\rho y) - r_{2\,\mu}(y) &= r_{2\,\rho\mu}(y) - r_{2\,\mu}(y) \\
&= \frac{1}{4(2\pi)^2}\Big(\big((\Box_y - \mu^2) + (1-\rho^2)\mu^2\big)\int dm^2\, \frac{\Delta_m^{\rm ret}(-y)}{m^2+\rho^2\mu^2} \\
&\qquad - (\Box_y - \mu^2)\int dm^2\, \frac{\Delta_m^{\rm ret}(-y)}{m^2+\mu^2}\Big) \\
&= \frac{1}{4(2\pi)^2}\Big[\int dm^2\, (\Box_y - \mu^2)\,\Delta_m^{\rm ret}(-y)\Big(\frac{1}{m^2+\rho^2\mu^2} - \frac{1}{m^2+\mu^2}\Big) \\
&\qquad + (1-\rho^2)\mu^2\int dm^2\, \frac{\Delta_m^{\rm ret}(-y)}{m^2+\rho^2\mu^2}\Big] \\
&= \frac{-1}{4(2\pi)^2}\,\delta(y)\int dm^2\Big(\frac{1}{m^2+\rho^2\mu^2} - \frac{1}{m^2+\mu^2}\Big) \\
&= \frac{1}{2(2\pi)^2}\,\log\rho\;\delta(y).
\end{aligned} \tag{3.2.119}$$

Hence, (3.2.118) is indeed satisfied for all $\mu > 0$. So we obtain explicitly the breaking of homogeneous scaling by computing only elementary integrals, i.e., without really computing the integral (3.2.117) – this is a main advantage of the renormalization method used here. As a byproduct, the calculation (3.2.119) shows explicitly that the choice of $\mu > 0$ is precisely the choice of the indeterminate parameter C in the general solution $r_2(y) + C\,\delta(y)$.

In the case $k = 3$ we proceed analogously to $k = 2$: By the same arguments as above, one verifies that

$$r_{3\,\mu_1,\mu_2}(y) := \frac{-1}{32\,(2\pi)^4}(\Box_y - \mu_1^2)(\Box_y - \mu_2^2)\int_0^\infty dm^2\;\frac{m^2\,\Delta_m^{\mathrm{ret}}(-y)}{(m^2+\mu_1^2)(m^2+\mu_2^2)}$$

is an extension of $r_3^0(y)$ (3.2.116), where $\mu_1, \mu_2 > 0$ are arbitrary mass scales.

Exercise 3.2.32. Setting $\mu_1 = \mu_2 =: \mu$ in the formula for $r_{3\,\mu_1,\mu_2}$, show that the breaking of homogeneous scaling is

$$\rho^6 r_{3\,\mu,\mu}(\rho y) - r_{3\,\mu,\mu}(y) = \frac{1}{32\,(2\pi)^4}\Big(-\Box_y\delta(y)\,2\log\rho + \delta(y)\,\mu^2(\rho^2-1)\Big). \tag{3.2.120}$$

[*Solution*: The l.h.s. of (3.2.120) is equal to $(r_{3\,\rho\mu,\rho\mu} - r_{3\,\mu,\rho\mu}) + (r_{3\,\mu,\rho\mu} - r_{3\,\mu,\mu})$; to compute each of these two brackets we use the method (3.2.119).]

We see that $r_{3\,\mu,\mu}$ violates almost homogeneous scaling with degree 6; and by a generalization of Exer. 3.2.32, one finds that this holds true even for all $r_{3\,\mu_1,\mu_2}$ with $(\mu_1,\mu_2) \in \mathbb{R}_+ \times \mathbb{R}_+$. However, from the result (3.2.120) we read off that

$$r_{3\,C}(y) := r_{3\,\mu,\mu}(y) - \frac{1}{32\,(2\pi)^4}\,\mu^2\,\delta(y) + C\,\Box\delta(y), \tag{3.2.121}$$

where $C \in \mathbb{R}$ is arbitrary, fulfills our requirements. The prescription (3.2.121) is the most general solution which is additionally Lorentz-invariant and respects axiom (g) ($*$-structure). For the breaking of homogeneous scaling, we obtain

$$\rho^6 r_{3\,C}(\rho y) - r_{3\,C}(y) = \frac{-1}{16\,(2\pi)^4}\,\log\rho\;\Box\delta(y). \tag{3.2.122}$$

A general feature of massless models shows up in the scaling results (3.2.119) and (3.2.122): The breaking of homogeneous scaling is independent of the normalization (i.e., of the choice of μ in (3.2.117) and C in (3.2.121)), because the undetermined polynomial $\sum_{|a|=D-k} C_a\,\partial^a\delta_{(k)}(y)$ scales homogeneously.

Remark 3.2.33. The expressions r_2 and r_3 differ from $r(\varphi^2,\varphi^2)$ and $r(\varphi^3,\varphi^3)$ (3.1.41) by combinatorial factors and powers of $\hbar$: Using (2.1.5) we find

$$r(\varphi^2,\varphi^2)(y) = 2!\,\hbar^2\,r_2(y), \quad r(\varphi^3,\varphi^3)(y) = 3!\,\hbar^3\,r_3(y). \tag{3.2.123}$$

With $y = x_2 - x_1$, the corresponding diagrams are the "fish diagram"

x_2 x_1 and the "setting sun diagram" x_2 x_1 . (3.2.124)

Summing up, we obtain

$$R_{1,1}\big(\varphi^4(x_2),\,\varphi^3(x_1)\big) = -12\,\hbar\,\varphi^3(x_2)\,D^{\mathrm{ret}}(x_1-x_2)\,\varphi^2(x_1) \\ +36\,\hbar^2\,\varphi^2(x_2)\,r_{2\,\mu}(x_2-x_1)\,\varphi(x_1)+24\,\hbar^3\,\varphi(x_2)\,r_{3\,C}(x_2-x_1)\,. \quad (3.2.125)$$

Inserting this result into (3.2.109), we end up with

$$R_{2,1}\big(\varphi^4(x_1)\otimes\varphi^4(x_2),\,\varphi(x)\big) = D^{\mathrm{ret}}(x-x_1)\Big(48\,\hbar^2\,D^{\mathrm{ret}}(x_1-x_2)\,\varphi^3(x_2)\,\varphi^2(x_1) \\ -144\,\hbar^3\,r_{2\,\mu}(x_2-x_1)\,\varphi^2(x_2)\,\varphi(x_1)-96\,\hbar^4\,r_{3\,C}(x_2-x_1)\,\varphi(x_2)\Big)+(x_1\leftrightarrow x_2)\,. \quad (3.2.126)$$

The non-uniqueness lies in the choice of the parameters $\mu>0$ in $r_{2\,\mu}$ and $C\in\mathbb{R}$ in $r_{3\,C}$.

If we do not require the Off-shell field equation axiom, formula (3.2.126) still gives a solution of all other axioms. But, in this case $R_{2,1}(\varphi^4\otimes\varphi^4,\varphi)$ contains a further non-uniqueness: We may add a term

$$C_1\,\hbar^4\,\delta(x_1-x,x_2-x)\,\varphi(x)$$

to (3.2.126), because $\big(D^{\mathrm{ret}}(x-x_1)\,r_{3\,C}(x_2-x_1)\big)$ scales almost homogeneously with degree 8. The $*$-structure axiom restricts the number C_1 to be real.

Exercise 3.2.34. Let $d=4$, $m_1,m_2>0$ and

$$r^{(m_1,m_2)\,0}(y) := \frac{\hbar^2}{i}\,\big(\Delta^+_{m_1}(y)\,\Delta^+_{m_2}(y)-\Delta^+_{m_1}(-y)\,\Delta^+_{m_2}(-y)\big)\,\theta(-y^0)\in\mathcal{D}'(\mathbb{R}^4\setminus\{0\})\,.$$

Compute all extensions $r^{(m_1,m_2)}\in\mathcal{D}'(\mathbb{R}^4)$ which maintain the scaling relations

$$\rho^4\,r^{(m_1/\rho,m_2/\rho)\,0}(\rho y)=r^{(m_1,m_2)\,0}(y)\quad\text{and}\quad \mathrm{sd}(r^{(m_1,m_2)\,0})=4\,, \quad (3.2.127)$$

by using the Källén–Lehmann representation derived in Exer. 3.2.31.

[*Solution*: Similarly to (3.2.116) we obtain

$$r^{(m_1,m_2)\,0}(y)=\frac{-\hbar^2}{4(2\pi)^2}\int_{(m_1+m_2)^2}^{\infty}dm^2\;\Delta^{\mathrm{ret}}_m(-y)\,\sqrt{\Big(1-\frac{m_1^2}{m^2}-\frac{m_2^2}{m^2}\Big)^2-4\,\frac{m_1^2m_2^2}{m^4}}$$

in $\mathcal{D}'(\mathbb{R}^4\setminus\{0\})$. An extension to $\mathcal{D}'(\mathbb{R}^4)$ is obtained analogously to (3.2.117), however, it is not necessary to introduce a mass scale $\mu>0$, since the m^2-integral starts at $(m_1+m_2)^2>0$:

$$r^{(m_1,m_2)}(y):=\frac{\hbar^2}{4(2\pi)^2}\,\square_y\int_{(m_1+m_2)^2}^{\infty}dm^2\;\frac{\Delta^{\mathrm{ret}}_m(-y)}{m^2}\,\sqrt{\big(1-\frac{m_1^2}{m^2}-\frac{m_2^2}{m^2}\big)^2-4\,\frac{m_1^2m_2^2}{m^4}}\,.$$

Obviously, this extension fulfills the required scaling relations (3.2.127). The most general extension with these properties is obtained by adding a term $C(\frac{m_1}{m_2})\,\delta(y)$, where $C(\frac{m_1}{m_2})$ is an arbitrary function of $\frac{m_1}{m_2}$.

Note that the limit $m_1,m_2\to 0$, with $\frac{m_1}{m_2}=$ constant, of $r^{(m_1,m_2)}(y)+C(\frac{m_1}{m_2})\,\delta(y)$ diverges, due to an IR-divergence of the m^2-integral. Therefore, $r^{(m_1,m_2)}(y)+C(\frac{m_1}{m_2})\,\delta(y)$ violates the Sm-expansion axiom for any $C(\frac{m_1}{m_2})$.]

Remark 3.2.35. The observation made in this exercise is of general validity (cf. [65, Remk. 6.1]): For $r^{(m)}(A_1,\ldots,A_n)(X) \in \mathcal{D}'(\mathbb{R}^{d(n-1)})$ with $\sum_j \dim A_j \geq d(n-1)$ (where $A_1,\ldots,A_n$ are monomials) the Sm-expansion axiom is not compatible with *homogeneous* scaling under $(X,m) \mapsto (\rho X, m/\rho)$ for the following reason: If $r^{(m)}(A_1,\ldots,A_n)$ would have both properties, the distribution $r^{(m=0)}(A_1,\ldots,A_n)$ would be well defined in $\mathcal{D}'(\mathbb{R}^{d(n-1)})$ and it would scale homogeneously under $X \mapsto \rho X$; in general this is impossible, as we know from the proof of Proposition 3.2.16 (Exer. 3.2.18).

3.3 The time-ordered product

References for this section are [12, 24, 66, 131, 148] and [55, App. E].

Mostly, pQFT is formulated in terms of time-ordered products. Compared with the retarded products, they have the advantage that they are symmetric in all their arguments – this simplifies their computation. Since our main techniques to renormalize – differential renormalization and analytic regularization – work for both retarded and time-ordered products,[63] we first introduce time-ordered products and then explain these methods (Sect. 3.5).

3.3.1 Heuristic explanation of the physical relevance of the time-ordered product

The retarded products are the coefficients in the Taylor expansion of the interacting fields with respect to the coupling constant κ; analogously, the time-ordered products of the interaction are the coefficients in the Taylor expansion (also with respect to κ) of the S-matrix. On a very heuristic level, the latter statement can be explained by analogy to the interaction picture of quantum mechanics (cf. [77, Sect. 2.4]): Let $\kappa H_{\mathrm{int}}(t)$ and $\psi(t) \equiv \psi(t,\vec{x})$ be the interaction Hamiltonian and the wave function, respectively, at time t, both in the interaction picture. Introducing the time evolution operator $U(t,t_0)$ by

$$U(t,t_0)\,\psi(t_0) := \psi(t)\ , \quad \forall t, t_0 \in \mathbb{R}\ , \ \forall \psi\ ,$$

the time evolution equation in the interaction picture ("Schwinger-Tomonaga equation"), which is equivalent to the Schrödinger equation in the Schrödinger picture, can be written as

$$i\hbar\,\frac{\partial}{\partial t} U(t,t_0) = \kappa\, H_{\mathrm{int}}(t)\, U(t,t_0)\ .$$

[63] This hold also for the renormalization method relying on the Källén–Lehmann representation explained in the preceding section.

This differential equation can be solved perturbatively by the "Dyson series":

$$
\begin{aligned}
U(t,t_0) &= \mathrm{Id} + \sum_{n=1}^{\infty} \frac{(-i\kappa)^n}{\hbar^n} \int_{t_0\le t_1\le\cdots\le t_n\le t} dt_1\cdots dt_n\; H_{\mathrm{int}}(t_n)\cdots H_{\mathrm{int}}(t_1) \\
&= \mathrm{Id} + \sum_{n=1}^{\infty} \frac{(-i\kappa)^n}{n!\,\hbar^n} \int_{[t_0,t]^{\times n}} dt_1\cdots dt_n\; T_n\big(H_{\mathrm{int}}(t_1)\otimes\cdots\otimes H_{\mathrm{int}}(t_n)\big)\;,
\end{aligned}
\tag{3.3.1}
$$

where the time-ordering map T_n is defined by

$$
T_n\big(H_{\mathrm{int}}(t_1)\otimes\cdots\otimes H_{\mathrm{int}}(t_n)\big) := H_{\mathrm{int}}(t_{\pi(1)})\cdots H_{\mathrm{int}}(t_{\pi(n)}) \tag{3.3.2}
$$

whenever $t_{\pi(1)}\ge\cdots\ge t_{\pi(n)}$ for a permutation $\pi\in S_n$. The S-matrix is obtained by the double limit

$$
\mathbf{S} := \lim_{\substack{t\to\infty \\ s\to-\infty}} U(t,s)\;.
$$

Proceeding *heuristically*, this formula for the S-matrix can be translated into our formalism for pQFT in the following way: By using the formal[64] relation

$$
H_{\mathrm{int}}(t) = -\int d\vec{x}\; L_{\mathrm{int}}(t,\vec{x})\;,
$$

we get for the S-matrix the following heuristic expression, which should be a functional on the (classical) configuration space $\mathcal{C}$:

$$
\mathbf{S} = 1 + \sum_{n=1}^{\infty} \frac{(i\kappa)^n}{n!\,\hbar^n} \int_{\mathbb{R}^{dn}} dx_1\cdots dx_n\; T_n\big(L_{\mathrm{int}}(x_1)\otimes\cdots\otimes L_{\mathrm{int}}(x_n)\big)\;, \tag{3.3.3}
$$

where the time-ordered product of n factors, T_n, is given by

$$
T_n\big(L_{\mathrm{int}}(x_1)\otimes\cdots\otimes L_{\mathrm{int}}(x_n)\big) := L_{\mathrm{int}}(x_{\pi(1)})\star_m\cdots\star_m L_{\mathrm{int}}(x_{\pi(n)}) \tag{3.3.4}
$$

whenever $x^0_{\pi(1)}\ge\cdots\ge x^0_{\pi(n)}$ for a $\pi\in S_n$. Note that the operator product appearing on the r.h.s. of (3.3.2) translates into the star product, because it is the star product which corresponds to the product of Fock space operators (Theorem 2.6.3).

But formula (3.3.3) is ill defined due to three kinds of divergences; the first two can be avoided by the Epstein–Glaser construction [66] (also called "causal perturbation theory"). More in detail:

- *UV-divergences*: Since $[L_{\mathrm{int}}(x_i), L_{\mathrm{int}}(x_j)]_\star = 0$ for $(x_i-x_j)^2<0$, the naive "definition" (3.3.4) of the time-ordered product works if $x_i\neq x_j$ $\forall 1\le i< j\le n$ (i.e., (3.3.4) is well defined as an $\mathcal{F}$-valued distribution on $\mathcal{D}(\check{\mathbb{M}}^n)$, see

[64]We ignore divergences of the integral appearing for $|\vec{x}|\to\infty$.

(3.1.68) for the definition of $\check{\mathbb{M}}^n$), and it is even Lorentz covariant – cf. Lemma 3.3.2 below. But formula (3.3.4) is in general divergent for coinciding points, $x_i = x_j$ for some $i < j$ (i.e., (3.3.4) is ill defined on $\mathcal{D}(\mathbb{M}^n)$). We solve this UV-problem analogously to our construction of the R-product $R \equiv (R_{n,1})_{n\in\mathbb{N}}$: We construct the sequence $(T_n)_{n\in\mathbb{N}^*}$ by induction on n and take into account translation invariance. In each inductive step UV-divergences can be avoided by a rigorous extension of the relevant distributions from $\mathcal{D}'(\mathbb{R}^{d(n-1)} \setminus \{0\})$ to $\mathcal{D}'(\mathbb{R}^{d(n-1)})$; that is, the results and techniques explained in Sects. 3.2.2 and 3.5 solve the UV-problem ("renormalization") also for the time-ordered product.

- *IR-divergences*: The integral in (3.3.3) is in general divergent for $\|x_j\| \to \infty$ for some j. As we do in the construction of the R-product, we multiply the coupling constant κ by a spacetime-dependent test function $g(x)$ with compact support. Having solved the UV-problem as explained above, the integral in (3.3.3) becomes then the application of the distribution $T_n\big(L_{\text{int}}(x_1)\otimes\cdots\otimes L_{\text{int}}(x_n)\big) \in \mathcal{D}'(\mathbb{M}^n, \mathcal{F})$ to the test function $g(x_1)\cdots g(x_n) \in \mathcal{D}(\mathbb{M}^n)$.
- *Divergence of the series* : Generally the sum over n in (3.3.3) is divergent. Our way out is to interpret this sum as a *formal* power series, similarly to the perturbation series for the interacting fields (3.1.2).

Having "cured" formula (3.3.3) as sketched, $\mathbf{S} \equiv \mathbf{S}(g)$ is a well-defined element of $\mathcal{F}[\![\kappa]\!]$.

3.3.2 Axioms for the time-ordered product and its inductive construction

Basic axioms. Similarly to the retarded product $R \equiv (R_{n,1})_{n\in\mathbb{N}}$, we define the time-ordered product $T \equiv (T_n)_{n\in\mathbb{N}^*}$ by basic axioms and additional axioms called "renormalization conditions".

Definition 3.3.1 (Basic axioms for the time-ordered product). The *time-ordered product* (shortly "T-product"), at the nth order, is a map

$$T_n^{(m)} \equiv T_n : \mathcal{F}_{\text{loc}}^{\otimes n} \longrightarrow \mathcal{F}\ ,$$

which satisfies:

(i) **Linearity:** T_n is *linear*;

(ii) **Initial condition:** $T_1(F) = F$ for any $F \in \mathcal{F}_{\text{loc}}$;

(iii) **Symmetry:** T_n is *symmetric* in all its arguments,

$$T_n(F_{\pi(1)}\otimes\cdots\otimes F_{\pi(n)}) = T_n(F_1\otimes\cdots\otimes F_n) \quad \forall \pi \in S_n,\ F_1,\ldots,F_n \in \mathcal{F}_{\text{loc}}\,;$$

(iv) **Causality.** For all $A_1,\dots,A_n\in\mathcal{P}$, T_n fulfills the causal factorization:

$$T_n\big(A_1(x_1),\dots,A_n(x_n)\big)=T_k\big(A_{\pi(1)}(x_{\pi(1)}),\dots,A_{\pi(k)}(x_{\pi(k)})\big)\\ \star_m\, T_{n-k}\big(A_{\pi(k+1)}(x_{\pi(k+1)}),\dots,A_{\pi(n)}(x_{\pi(n)})\big) \tag{3.3.5}$$

whenever $\{x_{\pi(1)},\dots,x_{\pi(k)}\}\cap\big(\{x_{\pi(k+1)},\dots,x_{\pi(n)}\}+\overline{V}_-\big)=\emptyset$ for a permutation $\pi\in S_n$;

and the renormalization conditions (axioms (v)–(x)) given below.

The T-products $T_n\big(A_1(x_1),\dots,A_n(x_n)\big)$ with $A_1,\dots,A_n\in\mathcal{P}$, appearing in the Causality axiom, are defined analogously to (3.1.3)–(3.1.4). Similarly to Remk. 3.1.1, the dependence of the T-products only on (local) *functionals* is equivalent to the AWI:

$$\partial_{x_l}T(\cdots\otimes A(x_l)\otimes\cdots)=T(\cdots\otimes\partial_{x_l}A(x_l)\otimes\cdots),\quad A\in\mathcal{P}\,. \tag{3.3.6}$$

The four properties listed in Definition 3.3.1 are the *basic axioms* for time-ordered products. Notice that the mass m (i.e., the information about the free field equation) appears only in the Causality axioms.

The Causality axiom is a rigorous and explicitly Lorentz covariant formulation of "time-ordering"; it is very restrictive. A first simple consequence is that

$$T_n\big(A_1(x_1),\dots,A_{n-1}(x_{n-1}),c\big)=c\cdot T_{n-1}\big(A_1(x_1),\dots,A_{n-1}(x_{n-1})\big) \tag{3.3.7}$$

for all $c\in\mathbb{C}$, $A_1,\dots,A_{n-1}\in\mathcal{P}$; this result differs significantly from the corresponding statement for retarded products given in Remk. 3.1.4. To derive (3.3.7) we write c as $c=A(x)$ where $x\in\mathbb{M}$ is arbitrary. Choosing x suitably, the causal factorization (3.3.5) yields $T_n\big(A_1(x_1),\dots,c\big)=c\star T_{n-1}\big(A_1(x_1),\dots\big)=c\cdot T_{n-1}\big(A_1(x_1),\dots\big)$.

An important consequence of the Causality axiom is that the two factors on the right-hand side of (3.3.5) commute under the $\star$-product if $(x_{\pi(l)}-x_{\pi(j)})^2<0$ $\forall 1\le l\le k<j\le n$; this property is called *spacelike commutativity.*

It is remarkable that the axioms Causality and Initial condition determine $T_n^{(m)}\big(A_1(x_1),\dots,A_n(x_n)\big)$ uniquely if $x_l\neq x_j$ $\forall l<j$. In detail and in rigorous terms, the statement reads as follows:

Lemma 3.3.2 (Consequences of Causality). *Let* $\check{\mathbb{M}}^n:=\{\,(x_1,\dots,x_n)\in\mathbb{M}^n\,\big|\,x_l\neq x_j\ \forall 1\le l<j\le n\,\}$.

(a) *Roughly speaking, the time-ordered product* $T_n(A_1(x_1),\dots,A_n(x_n))$ *can be expressed as a multiple $\star$-product on* $\mathcal{D}(\check{\mathbb{M}}^n)$*. This can be done as follows: For any permutation* $\pi\in S_n$ *the set*[65]

$$U_\pi:=\Big\{(x_1,\dots,x_n)\,\big|\,x_{\pi(j)}\notin\big(\{x_{\pi(j+1)},\dots,x_{\pi(n)}\}+\overline{V}_-\big)\ \forall 1\le j\le n-1\Big\}$$

[65] In the definition of the set U_π we take into account that $(x_1\notin x_2+\overline{V}_-)\wedge(x_2\notin x_3+\overline{V}_-)$ does not imply $x_1\notin x_3+\overline{V}_-$.

is an open subset of $\check{\mathbb{M}}^n$ *and* $\{U_\pi \,|\, \pi \in S_n\}$ *is an open cover of* $\check{\mathbb{M}}^n$*:*

$$\bigcup_{\pi \in S_n} U_\pi = \check{\mathbb{M}}^n \ .$$

On $\mathcal{D}(U_\pi)$ *it holds that*

$$T_n^{(m)}\big(A_1(x_1) \otimes \cdots \otimes A_n(x_n)\big) = A_{\pi(1)}(x_{\pi(1)}) \star_m \cdots \star_m A_{\pi(n)}(x_{\pi(n)}) \quad (3.3.8)$$

for all $A_1, \ldots, A_n \in \mathcal{P}$*. This formula is consistent in the sense that*

$$A_{\pi(1)}(x_{\pi(1)}) \star_m \cdots \star_m A_{\pi(n)}(x_{\pi(n)}) = A_{\sigma(1)}(x_{\sigma(1)}) \star_m \cdots \star_m A_{\sigma(n)}(x_{\sigma(n)}) \quad (3.3.9)$$

on $\mathcal{D}(U_\pi) \cap \mathcal{D}(U_\sigma)$ *for all* $\pi, \sigma \in S_n$*.*

(b) *On* $\mathcal{D}(\check{\mathbb{M}}^n)$ *the time-ordered product* $T_n^{(m)}$ *agrees with the n-fold product* $\star_{\Delta_m^F}$*, where* $\star_{\Delta_m^F}$ *is defined by replacing* H_m *in formula (2.1.5) by the Feynman propagator* Δ_m^F *(2.3.8), analogously to the procedure in Remk. 2.3.3:*

$$T_n^{(m)}\big(A_1(x_1) \otimes \cdots \otimes A_n(x_n)\big) = A_1(x_1) \star_{\Delta_m^F} \cdots \star_{\Delta_m^F} A_n(x_n) \quad \text{on } \mathcal{D}(\check{\mathbb{M}}^n) \quad (3.3.10)$$

for all $A_1, \ldots, A_n \in \mathcal{P}$*.*

The r.h.s. of (3.3.10) is called the *"unrenormalized time-ordered product"*. According to (3.3.10) and Linearity, it is defined on the space

$$\big[\{\, F_1 \otimes \cdots \otimes F_n \in \mathcal{F}_{\mathrm{loc}}^{\otimes n} \,\big|\, \operatorname{supp} F_l \cap \operatorname{supp} F_j = \emptyset \ \forall 1 \le l < j \le n \}\big] \quad (3.3.11)$$

where $[-]$ denotes the linear span. The problem that the choice $p = \Delta_m^F$ does not fit into the definition of $\star_p : \mathcal{F} \times \mathcal{F} \longrightarrow \mathcal{F}$ (given in Exer. 2.3.1) because Δ_m^F is not smooth, is circumvented here by the restriction of $F_1 \star_{\Delta^F} \cdots \star_{\Delta^F} F_n$ to the space (3.3.11); this differs from the restriction to $\mathcal{F}_{\mathrm{reg}}^{\otimes n}$ used in the definition of τ_n (2.3.10) in Remk. 2.3.3.

Proof. Ad (a): It is quite easy to verify the geometrical statements about the sets U_π. Formula (3.3.8) is obtained by iterated use of the causal factorization (3.3.5), taking into account also the axiom (ii) (Initial condition):

$$\begin{aligned} T_n\big(A_1(x_1) \otimes \cdots \otimes A_n(x_n)\big) &= A_{\pi 1}(x_{\pi 1}) \star T_{n-1}\big(A_{\pi 2}(x_{\pi 2}) \otimes \cdots \otimes A_{\pi n}(x_{\pi n})\big) \\ &= A_{\pi 1}(x_{\pi 1}) \star A_{\pi 2}(x_{\pi 2})) \star T_{n-2}\big(A_{\pi 3}(x_{\pi 3}) \otimes \cdots \otimes A_{\pi n}(x_{\pi n})\big) = \cdots \ , \end{aligned}$$

where we write πj for $\pi(j)$ and $\star$ for $\star_m$. The relation (3.3.9) follows immediately from the proof of part (b): On $\mathcal{D}(U_\pi) \cap \mathcal{D}(U_\sigma)$ both sides of (3.3.9) are equal to $A_1(x_1) \star_{\Delta^F} \cdots \star_{\Delta^F} A_n(x_n)$.

Ad (b): Due to the commutativity of the product $\star_{\Delta_m^F}$ and due to $\Delta_m^F(x) = \Delta_m^+(x)$ for $x \notin \overline{V}_-$ (see (2.3.9)), the r.h.s. of (3.3.10) agrees with the r.h.s. of

(3.3.8) on $\mathcal{D}(U_\pi)$ for all $\pi \in S_n$; namely for $(x_1,\dots,x_n) \in U_\pi$ we obtain

$$\begin{aligned} A_1(x_1) \star_{\Delta^F} \cdots \star_{\Delta^F} A_n(x_n) &= A_{\pi 1}(x_{\pi 1}) \star_{\Delta^F} \cdots \star_{\Delta^F} A_{\pi n}(x_{\pi n}) \\ &= A_{\pi 1}(x_{\pi 1}) \star \Big(A_{\pi 2}(x_{\pi 2}) \star_{\Delta^F} \cdots \star_{\Delta^F} A_{\pi n}(x_{\pi n}) \Big) \\ &= A_{\pi 1}(x_{\pi 1}) \star A_{\pi 2}(x_{\pi 2}) \star \Big(A_{\pi 3}(x_{\pi 3}) \star_{\Delta^F} \cdots \star_{\Delta^F} A_{\pi n}(x_{\pi n}) \Big) = \cdots . \quad \square \end{aligned}$$

Remark 3.3.3 (UV-divergences and UV-regularization). Trying to interpret the unrenormalized T-product (3.3.10) as a distribution in $\mathcal{D}'(\mathbb{M}^n, \mathcal{F})$ (or the unrenormalized R-product (3.1.69) as a distribution in $\mathcal{D}'(\mathbb{M}^{n+1}, \mathcal{F})$) one meets the well-known UV-divergences of perturbative QFT. For example, for $d = 4$ the powers $\Delta_m^F(x_i - x_j)^k$, $k \geq 2$, exist on $\mathcal{D}(\check{\mathbb{M}}^n)$ (because $\Delta_m^F(z) = \Delta_m^+(z)$ or $\Delta_m^F(z) = \Delta_m^+(-z)$ for $z \neq 0$, due to (2.3.9)), but they do not exist for coinciding points, i.e., on $\mathcal{D}(\mathbb{M}^n)$.

In terms of the explicit formula (A.3.5) for $\mathrm{WF}(\Delta_m^F)$ and Hörmander's criterion (Theorem A.3.3) the existence of $(\Delta_m^F)^k$ can be discussed as follows:

- The wave front set of Δ_m^F restricted to $\mathcal{D}(\mathbb{R}^d \setminus \{0\})$ is given by the second set in the set union on the r.h.s. of (A.3.5). From this we see that $\mathrm{WF}(\Delta_m^F|_{\mathcal{D}(\mathbb{R}^d \setminus \{0\})}) \oplus \mathrm{WF}(\Delta_m^F|_{\mathcal{D}(\mathbb{R}^d \setminus \{0\})}) = \mathrm{WF}(\Delta_m^F|_{\mathcal{D}(\mathbb{R}^d \setminus \{0\})})$.
  Proceeding by induction on k we conclude that $(\Delta_m^F)^k|_{\mathcal{D}(\mathbb{R}^d \setminus \{0\})}$ exists and $\mathrm{WF}(\Delta_m^F)^k|_{\mathcal{D}(\mathbb{R}^d \setminus \{0\})} = \mathrm{WF}(\Delta_m^F|_{\mathcal{D}(\mathbb{R}^d \setminus \{0\})})$, cf. Exer. 2.2.1(b).
- But, for $x = 0$ the Hörmander criterion is not fulfilled, already for $k = 2$, because $\big(\{0\} \times (\mathbb{R}^d \setminus \{0\})\big) \subset \mathrm{WF}(\Delta_m^F)$.

A widespread method to overcome the UV-divergences is to regularize the unrenormalized T-product (3.3.10) by an UV-cutoff Λ. Let $T_\Lambda\big(\otimes_{j=1}^n A_j(x_j)\big)$, $A_j \in \mathcal{P}$, be a regularized T-product, that is, $T_\Lambda\big(\otimes_{j=1}^n A_j(x_j)\big)$ exists in $\mathcal{D}'(\mathbb{M}^n, \mathcal{F})$ and[66]

$$\lim_{\Lambda \to \infty} T_\Lambda\big(\otimes_{j=1}^n A_j(x_j)\big) = T\big(\otimes_{j=1}^n A_j(x_j)\big) \quad \text{in} \quad \mathcal{D}'(\check{\mathbb{M}}^n, \mathcal{F}) .$$

But, on $\mathcal{D}(\mathbb{M}^n)$ the cutoff cannot be removed directly, more explicitly, the limit $\lim_{\Lambda \to \infty} T_\Lambda\big(\otimes_{j=1}^n A_j(x_j)\big)$ diverges in general. Therefore, one adds suitable, local "counter terms" $N_\Lambda\big(\otimes_{j=1}^n A_j(x_j)\big) \in \mathcal{D}'(\mathbb{M}^n, \mathcal{F})$ (which diverge for $\Lambda \to \infty$), such that the limit

$$\lim_{\Lambda \to \infty} (T_\Lambda + N_\Lambda)\big(\otimes_{j=1}^n A_j(x_j)\big) \quad \text{exists in} \quad \mathcal{D}'(\mathbb{M}^n, \mathcal{F}) . \tag{3.3.12}$$

[66] According to (A.1.15) and (1.2.3) convergence in $\mathcal{D}'(\Omega, \mathcal{F})$ (where $\Omega \subseteq \mathbb{M}^n$ is open) is defined by

$$\lim_{n \to \infty} T_n = T \quad \text{iff} \quad \lim_{n \to \infty} \langle T_n, g\rangle(h) = \langle T, g\rangle(h) \quad \forall g \in \mathcal{D}(\Omega) \ , \ h \in \mathcal{C} ,$$

with $T_n, T \in \mathcal{D}'(\Omega, \mathcal{F})$.

The counter terms are called "local" because they are localized on (partial) diagonals:

$$\operatorname{supp} N_\Lambda\big(\otimes_{j=1}^n A_j(x_j)\big) \subseteq \{(x_1,\dots,x_n)\in\mathbb{M}^n \,\big|\, x_j = x_k \text{ for at least one pair } (j,k)\ ,\ j<k\}\cdot$$

In the framework of the inductive Epstein–Glaser construction of the T-product (or R-product), we develop the following UV-regularization methods:

- An analytic regularization to explicitly compute the extension of almost homogeneously scaling distributions, i.e., to solve the extension problem of Proposition 3.2.16 in practice; this method is given in Sect. 3.5.2.
- Regularization of the Feynman propagator: an UV-regularization of the unrenormalized T-product (3.3.10) is obtained by replacing $\star_{\Delta_m^F}$ by $\star_{p_\Lambda}$ (introduced in Exer. 2.3.1), where $(p_\Lambda)_{\Lambda>0}$ is a family of smooth functions with $\lim_{\Lambda\to\infty} p_\Lambda = \Delta_m^F$ in an appropriate sense. This is the starting point for Sect. 3.9. For this regularization method, Theorem 3.9.4 shows that the addition of local counter terms (3.3.12) can consistently be done to obtain a time-ordered product $T\equiv (T_n)_{n\in\mathbb{N}^*}$ solving our axioms.

However, we emphasize: *Epstein–Glaser renormalization is well defined without any regularization or divergent counter terms.* We introduce these devices only as a method for practical computation of the extension of distributions (see Sect. 3.5.2 about analytic regularization), or to be able to mimic Wilson's renormalization group (see Sect. 3.9).

Remark 3.3.4 (Feynman diagrams for time-ordered products). Similarly to the R-product (Sect. 3.1.7), we understand by a (Feynman) diagram for a T-product $T_n\big(\otimes_{j=1}^n A_j(x_j)\big)$ a contraction scheme for the corresponding unrenormalized expression (3.3.10). For example, the diagram

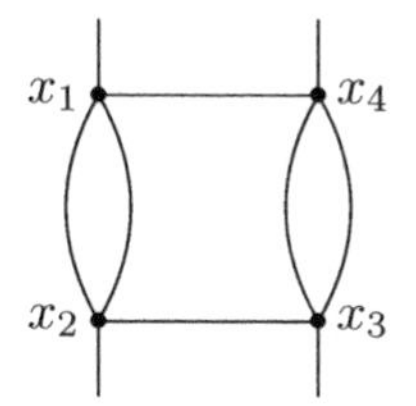

symbolizes the sum of all terms of $T_4\big(\otimes_{j=1}^4 \varphi^4(x_j)\big)|_{\mathcal{D}(\check{\mathbb{M}}^4)}$ which are

$$\sim \Delta^F(x_1-x_2)^2\,\Delta^F(x_2-x_3)\,\Delta^F(x_3-x_4)^2\,\Delta^F(x_1-x_4)\,\varphi(x_1)\,\varphi(x_2)\,\varphi(x_3)\,\varphi(x_4)\ .$$

Inner lines are not oriented, since $\Delta^F(-z) = \Delta^F(z)$. A diagram Γ_1 is called a "subdiagram" of the diagram Γ if and only if all vertices and all (inner and external) lines of Γ_1 are also vertices and lines of Γ; we write

$$\Gamma_1 \subseteq \Gamma$$

for this. An example is given by the fish diagram and the setting sun diagram (introduced in (3.2.124), however here understood as diagrams for the unrenormalized T-product): The former is a subdiagram of the latter.

The "Feynman rules" are rules by which one can read off from a diagram the corresponding analytical, "unrenormalized" expression (inclusive of all signs and combinatorial factors) contributing to the r.h.s. of (3.3.10).

Example 3.3.5. From part (b) of the lemma we conclude that

$$T_2\big(\varphi(x_1)\otimes\varphi^2(x_2)\big)=\varphi(x_1)\,\varphi^2(x_2)+2\,\hbar\,\Delta_m^F(x_1-x_2)\,\varphi(x_2)\quad\text{on}\quad\mathcal{D}(\check{\mathbb{M}}^2)\,. \tag{3.3.13}$$

For this particular example, the product $\star_{\Delta_m^F}$ (i.e., the r.h.s.) is well defined on the whole space $\mathcal{D}(\mathbb{M}^2)$ and gives a solution of the basic axioms.

Properties of the Feynman propagator. Motivated by (3.3.10), we derive some basic properties of the Feynman propagator in the following exercise. The wave front set of Δ_m^F is computed in Exer. A.3.5.

Exercise 3.3.6. Derive from the definition of Δ_m^F (2.3.8) the relations

$$\Delta_m^F(x)=i\Delta_m^{\text{ret}}(x)+\Delta_m^+(-x)=\frac{i}{(2\pi)^d}\int d^dp\,\frac{e^{-ipx}}{p^2-m^2+i0} \tag{3.3.14}$$

and

$$(\square+m^2)\Delta_m^F(x)=-i\,\delta(x).$$

[*Solution*: Starting with (2.3.8) we get

$$\Delta_m^F(x)=\theta(x^0)\big(\Delta_m^+(x)-\Delta^+(-x)\big)+\Delta_m^+(-x)\big(\theta(x^0)+\theta(-x^0)\big)=i\Delta_m^{\text{ret}}(x)+\Delta_m^+(-x)\,,$$

which implies immediately $(\square+m^2)\Delta_m^F(x)=i\,(\square+m^2)\Delta_m^{\text{ret}}(x)=-i\,\delta(x)$. Inserting the momentum-space representations of Δ_m^{ret} (1.8.2) and Δ_m^+ (2.2.1) and using (A.1.16) we obtain

$$\begin{aligned}\Delta_m^F(x)&=\frac{i}{(2\pi)^d}\int d^dp\;e^{-ipx}\Big(P\Big(\frac{1}{p^2-m^2}\Big)-i\pi\,\text{sgn}(p^0)\,\delta(p^2-m^2)\\&\qquad\qquad-i2\pi\,\theta(-p^0)\,\delta(p^2-m^2)\Big)\\&=\frac{i}{(2\pi)^d}\int d^dp\;e^{-ipx}\Big(P\Big(\frac{1}{p^2-m^2}\Big)-i\pi\,\delta(p^2-m^2)\Big)\,,\end{aligned}$$

which agrees with the r.h.s. of (3.3.14).]

Inductive construction of the time-ordered product. According to Lemma 3.3.2, the basic axioms determine T_n uniquely on $\mathcal{D}(\check{\mathbb{M}}^n)$. An even stronger statement holds true: Given $T_1,\ldots,T_{n-1}$, the basic axioms determine T_n uniquely on the larger space $\mathcal{D}(\mathbb{M}^n\setminus\Delta_n)$. For this inductive construction of $T_n^0:=T_n|_{\mathcal{D}(\mathbb{M}^n\setminus\Delta_n)}$ we refer to [138], more details are given in unpublished notes of Raymond Stora[67] from

[67] Actually, Raymond Stora told us that this method was first proposed by H. Epstein.

1993 [158], and we also refer to [24, 131]. The basic idea is to work with an open cover $\{C_I \,|\, I \text{ as explained below}\}$ of $\mathbb{M}^n \setminus \Delta_n$, which is such that on each $\mathcal{D}(C_I)$ the time-ordered product T_n^0 is uniquely determined by causal factorization (3.3.5) in terms of time-ordered products of lower orders. More explicitly: For each $I \subset \{1, \ldots, n\}$ with $I \neq \emptyset$ and $I \neq \{1, \ldots, n\}$ let

$$C_I := \{(x_1, \ldots, x_n) \in \mathbb{M}^n \,|\, x_i \notin (x_j + \overline{V}_-) \; \forall i \in I,\, j \in I^c\} \tag{3.3.15}$$

(where $I^c := \{1, \ldots, n\} \setminus I$), which is an open subset of $\mathbb{M}^n \setminus \Delta_n$. One verifies that

$$\bigcup_I C_I = \mathbb{M}^n \setminus \Delta_n \,. \tag{3.3.16}$$

On $\mathcal{D}(C_I)$ the Causality axiom determines T_n^0 uniquely in terms of $T_1, \ldots, T_{n-1}$:

$$T_n^0(\otimes_{j=1}^n A_j(x_j)) := T_{|I|}(\otimes_{j\in I} A_j(x_j)) \star T_{|I^c|}(\otimes_{k\in I^c} A_k(x_k)) \quad \text{on} \quad \mathcal{D}(C_I)\,. \tag{3.3.17}$$

However, if $C_{I_1} \cap C_{I_2} \neq \emptyset$, the prescription (3.3.17) gives two different definitions of T_n^0 on $\mathcal{D}(C_{I_1} \cap C_{I_2})$; one checks compatibility, that is, that these two definitions are equal. This construction is quite analogous to the inductive construction of $R_{n-1,1}^0$ given in Sect. 3.2.1; similarly to (3.2.13) it has the disadvantage that in general a closed formula for T_n^0 needs a partition of unity:[68]

$$T_n^0(\otimes_{j=1}^n A_j(x_j)) = \sum_I f_I(x_1, \ldots, x_n) \cdot [\text{r.h.s. of (3.3.17)}]\,, \tag{3.3.18}$$

where

$$f_I \in C^\infty(\mathbb{M}^n \setminus \Delta_n, \mathbb{R})\,, \quad 1 = \sum_I f_I(x) \;\forall x \in \mathbb{M}^n \setminus \Delta_n \quad \text{and} \quad \operatorname{supp} f_I \subseteq C_I\,. \tag{3.3.19}$$

We point out that this result for T_n^0 does not depend on the choice of the family $\{f_I\}$, due to the compatibility of (3.3.17) on $\mathcal{D}(C_{I_1} \cap C_{I_2})$. The use of a partition of unity can be avoided by working with the distribution splitting method of Epstein and Glaser [66, 148] (see also Exer. 3.5.7), on the price of a more complicated combinatorics.

Similarly to the inductive construction of $R_{n-1,1} \in \mathcal{D}'(\mathbb{M}^n)$ (Sects. 3.2.2–3.2.7), renormalization is then the extension $\mathcal{D}'(\mathbb{M}^n \setminus \Delta_n) \ni T_n^0 \to T_n \in \mathcal{D}'(\mathbb{M}^n)$ which is restricted by further axioms – the renormalization conditions for the time-ordered product. The latter correspond 1 : 1 to the renormalization conditions for the retarded product (axioms (f)–(l) in Sect. 3.1).

[68] An example, for which a partition of unity cannot be avoided, is given in Remk. 3.5.20. For this example, a suitable partition of unity is explicitly constructed, by a method which seems to be of general applicability.

Renormalization conditions. These additional axioms for the T-product read:

(v) **Field independence:** $\delta T_n/\delta\varphi = 0$, or more explicitly

$$\frac{\delta T_n(F^{\otimes n})}{\delta\varphi(x)} = n\, T_n\Big(\frac{\delta F}{\delta\varphi(x)} \otimes F^{\otimes(n-1)}\Big)\ .$$

Similarly to the R-product, this axiom is equivalent to the requirement that T_n satisfies the *causal Wick expansion* (3.1.23) or (3.1.24).

(vi) **$*$-structure and field parity:** Field parity is the condition

$$\alpha\circ T_n = T_n\circ\alpha^{\otimes n}\ ,$$

where α is defined in (1.10.6); the formulation of the $*$-structure condition is given below in (3.3.29), since the necessary tools are not yet introduced.

(vii) **Poincaré covariance:** $\beta_{\Lambda,a}\circ T_n = T_n\circ\beta_{\Lambda,a}^{\otimes n}\quad\forall(\Lambda,a)\in\mathcal{P}_+^\uparrow$.

(viii) **Off-shell field equation:**

$$\begin{aligned} T_n\big(\varphi(g)\otimes F_1\otimes\cdots\otimes F_{n-1}\big) &= \varphi(g)\ T_{n-1}\big(F_1\otimes\cdots\otimes F_{n-1}\big) \\ &\quad + \hbar\int dx\,dy\ g(x)\,\Delta_m^F(x-y)\,\frac{\delta}{\delta\varphi(y)}T_{n-1}\big(F_1\otimes\cdots\otimes F_{n-1}\big)\ , \end{aligned} \tag{3.3.20}$$

where $g\in\mathcal{D}(\mathbb{M})$.

Note that the r.h.s. agrees with $\varphi(g)\star_{\Delta^F} T_{n-1}\big(F_1\otimes\cdots\otimes F_{n-1}\big)$; however, in contrast to (3.3.10), there is no restriction of the domain here – this $\star_{\Delta^F}$-product exists for all g and for all $F_1,\ldots,F_{n-1}$, since there is at most one contraction.

(ix) **Sm-expansion:** For all *monomials* $A_1,\ldots,A_n\in\mathcal{P}$, the distributions

$$t^{(m)}(A_1,\ldots,A_n)(x_1-x_n,\ldots,x_{n-1}-x_n) := \omega_0\big(T_n^{(m)}\big(A_1(x_1)\otimes\cdots\otimes A_n(x_n)\big)\big)$$

fulfill the Sm-expansion with degree $D:=\sum_{k=1}^n \dim A_k$.

Scaling degree: Similarly to the R-product (see Sect. 3.2.5), the Scaling degree axiom,

$$\mathrm{sd}\, t(A_1,\ldots,A_n)(x_1-x_n,\ldots)\leq\sum_{j=1}^n \dim A_j\quad\forall A_1,\ldots,A_n\in\mathcal{P}_{\mathrm{hom}}\ , \tag{3.3.21}$$

is a less restrictive substitute for the Sm-expansion axiom.

(x) **$\hbar$-dependence:**

$$t(A_1,\ldots,A_n)\sim\hbar^{\sum_{j=1}^n |A_j|/2}$$

for all monomials $A_1,\ldots,A_n$ which fulfill $A_j\sim\hbar^0\ \forall j$ (see (3.1.31) for the definition of $|A|$).

Note that the corresponding "unrenormalized" expression, that is $\omega_0\big(A_1(x_1)\star_{\Delta_m^F}\cdots\star_{\Delta_m^F} A_n(x_n)\big)$, fulfills the condition (x) – this is similar to the $\hbar$-dependence axiom for the R-product.

Example 3.3.7. Let $d = 4$. The aim of this example is to compute $T_3(\varphi \otimes \varphi^2 \otimes \varphi)$ on $\mathcal{D}(\mathbb{M}^3)$ without requiring the axiom (viii) (Off-shell field equation). Applying (3.3.10) we obtain the restriction of this time-ordered product to $\mathcal{D}(\check{\mathbb{M}}^3)$:

$$\begin{aligned} T_3\big(\varphi(x_1) \otimes \varphi^2(x_2) \otimes \varphi(x_3)\big) = \varphi(x_1)\, \varphi^2(x_2)\, \varphi(x_3) + \hbar\Big(2\, \Delta_m^F(x_1 - x_2)\, \varphi(x_2)\, \varphi(x_3) \\ + 2\, \varphi(x_1)\, \varphi(x_2)\, \Delta_m^F(x_2 - x_3) + \Delta_m^F(x_1 - x_3)\, \varphi^2(x_2)\Big) \\ + 2\,\hbar^2\, \Delta_m^F(x_1 - x_2)\, \Delta_m^F(x_2 - x_3)\ . \end{aligned} \tag{3.3.22}$$

To show that this formula holds even on $\mathcal{D}(\mathbb{M}^3 \setminus \Delta_3)$ we verify relation (3.3.17) for formula (3.3.22) for all I. E.g., for $I = \{2, 3\}$ it should hold

$$[\text{r.h.s. of}(3.3.22)] \stackrel{?}{=} T_2\big(\varphi^2(x_2) \otimes \varphi(x_3)\big) \star T_1\big(\varphi(x_1)\big)\ . \tag{3.3.23}$$

To verify this, we use formula (3.3.13) for $T_2(\varphi^2 \otimes \varphi)$; this is allowed because, taking into account also the relevant renormalization conditions, the solution for $T_2(\varphi^2 \otimes \varphi)$ is unique and it is given by (3.3.13). The r.h.s. of (3.3.23) is equal to

$$\begin{aligned} &\Big(\varphi^2(x_2)\, \varphi(x_3) + 2\,\hbar\, \Delta^F(x_2 - x_3)\, \varphi(x_2)\Big) \star \varphi(x_1) \\ &\quad = \varphi^2(x_2)\, \varphi(x_3)\, \varphi(x_1) + \hbar\Big(\varphi^2(x_2)\, \Delta^+(x_3 - x_1) + 2\, \varphi(x_2)\, \varphi(x_3)\, \Delta^+(x_2 - x_1) \\ &\qquad + 2\, \Delta^F(x_2 - x_3)\, \varphi(x_2)\, \varphi(x_1)\Big) + 2\,\hbar^2\, \Delta^F(x_2 - x_3)\, \Delta^+(x_2 - x_1)\ . \end{aligned} \tag{3.3.24}$$

Now, for $(x_1, x_2, x_3) \in C_{\{2,3\}}$ we have $\Delta^+(x_j - x_1) = \Delta^F(x_1 - x_j)$ for $j = 2, 3$, due to (2.3.8). With that (3.3.24) agrees with the r.h.s. of (3.3.22).

The extension to $\mathcal{D}(\mathbb{M}^3)$ can be done by the direct extension, since the distributions $\Delta^F(x_j - x_k) \in \mathcal{D}'(\mathbb{R}^8 \setminus \{0\})$ (where $j < k$) and $\Delta^F(x_1 - x_2)\, \Delta^F(x_2 - x_3) \in \mathcal{D}'(\mathbb{R}^8 \setminus \{0\})$ have scaling degree $2(< 8)$ and $4(< 8)$, respectively. We conclude that the solution for $T_3(\varphi \otimes \varphi^2 \otimes \varphi) \in \mathcal{D}'(\mathbb{M}^3)$ is unique and that it is given by formula (3.3.22).

S-matrix as generating functional of the time-ordered product. The main motivation to study the S-matrix is certainly its striking physical relevance for the description of scattering experiments in elementary particle physics, cf. Sect. 3.3.1. However, there is also a mathematical motivation: Many properties of the T-product can be formulated more elegantly in terms of its generating functional, which is the S-matrix.

Definition 3.3.8. The *S-matrix* is defined as

$$\mathbf{S}(F) := 1 + \sum_{n=1}^{\infty} \frac{i^n}{n!\, \hbar^n}\, T_n(F^{\otimes n}) \equiv T(e_{\otimes}^{iF/\hbar})\ . \tag{3.3.25}$$

In the last expression we use $T_0(c) := c$ for $c \in \mathbb{C}$, and we interpret T as a (linear and totally symmetric) map

$$T : \mathcal{T}(\kappa\mathcal{F}_{\mathrm{loc}}) \longrightarrow \mathcal{F}[\![\kappa]\!]\ ; \tag{3.3.26}$$

for the notation see (A.1.5)–(A.1.7).

It follows that

$$\frac{i^n}{\hbar^n}\,T_n(F^{\otimes n}) = \mathbf{S}^{(n)}(0)(F^{\otimes n}) = \frac{d^n}{d\kappa^n}\Big|_{\kappa=0} \mathbf{S}(\kappa F)\ ;$$

see (A.1.19)–(A.1.22) for the definition of the nth derivative of $\mathbf{S} : \kappa\mathcal{F}_{\mathrm{loc}} \longrightarrow \mathcal{F}[\![\kappa]\!]$ at 0.

Remark 3.3.9. Let $F, G \sim \hbar^0$. According to (3.1.71), $G_F = R(e_\otimes^{F/\hbar}, G)$ is a formal power series in $\hbar$. But, $\mathbf{S}(F) = T(e_\otimes^{iF/\hbar})$ is a formal *Laurent series* in $\hbar$, because

$$T_n(F^{\otimes n}) = F\cdot F\cdot\,\cdots\,\cdot F + \mathcal{O}(\hbar)\ ,$$

(classical product in the first term) due to axiom (x) and the causal Wick expansion (axiom (v)). It is only the connected part of the S-matrix, more precisely $\hbar\, T^c(e_\otimes^{iF/\hbar})$ (see Definition 4.4.1), which is a formal power series in $\hbar$, as explained after (4.4.4).

The causality axiom for the time-ordered product (3.3.5) can equivalently be expressed in terms of the S-matrix by the condition:

$$\mathbf{S}(H+G+F) = \mathbf{S}(H+G)\star\mathbf{S}(G)^{\star-1}\star\mathbf{S}(G+F)$$
$$\text{if } \operatorname{supp} H\cap(\operatorname{supp} F+\overline{V}_-)=\emptyset \tag{3.3.27}$$

where the notation ${}^{\star-1}$ denotes the inverse with respect to the $\star$-product; or by the seemingly weaker condition[69] obtained from (3.3.27) by setting $G=0$:

$$\mathbf{S}(H+F) = \mathbf{S}(H)\star\mathbf{S}(F) \quad\text{if}\quad \operatorname{supp} H\cap(\operatorname{supp} F+\overline{V}_-)=\emptyset. \tag{3.3.28}$$

That this last condition is equivalent to the causal factorization of the time-ordered product (3.3.5) is verified in the following exercise.

Exercise 3.3.10 (Equivalence of two Causality conditions)**.** Let maps $T_n : \mathcal{F}_{\mathrm{loc}}^{\otimes n} \longrightarrow \mathcal{F}$ be given, satisfying the axioms (i) Linearity and (iii) Symmetry, and let $\mathbf{S}$ be defined in terms of these maps by (3.3.25). Prove that (3.3.5) $\Leftrightarrow$ (3.3.28).

[*Hint*: Use the techniques of the proof of (3.1.9) $\Leftrightarrow$ (3.1.10).]

[*Solution.* (3.3.5) $\Longrightarrow$ (3.3.28): Let $H=A(h)$, $F=B(g)$, with $A,B\in\mathcal{P}$, $h,g\in\mathcal{D}(\mathbb{M})$ and $\operatorname{supp} h\cap(\operatorname{supp} g+\overline{V}_-)=\emptyset$. Using the axioms Symmetry and Causality (3.3.5), we get

$$\begin{aligned}\mathbf{S}\big(A(h)+B(g)\big) &= \sum_{n=0}^\infty \frac{i^n}{n!\,\hbar^n}\sum_{k=0}^n\binom{n}{k}\int dx_1\cdots dx_n\; h(x_1)\ldots h(x_k)\,g(x_{k+1})\cdot\cdots\cdot g(x_n)\\ &\qquad\cdot T_n\big(A(x_1),\ldots,A(x_k),B(x_{k+1}),\ldots,B(x_n)\big)\\ &=\sum_{n=0}^\infty\sum_{k=0}^n \frac{i^k}{k!\,\hbar^k}\,T_k\big(A(h)^{\otimes k}\big)\star\frac{i^{n-k}}{(n-k)!\,\hbar^{n-k}}\,T_{n-k}\big(B(g)^{\otimes(n-k)}\big)\\ &=\mathbf{S}\big(A(h)\big)\star\mathbf{S}\big(B(g)\big)\ .\end{aligned}$$

[69] In a non-perturbative framework, (3.3.28) is in general a truly weaker condition than (3.3.27).

(3.3.28) $\Longrightarrow$ (3.3.5): Let $\{x_1,\dots,x_k\}\cap\big(\{x_{k+1},\dots,x_n\}+\overline{V}_-\big)=\emptyset$. By applying (3.3.28) we obtain

$$\begin{aligned}T_n\big(\otimes_{j=1}^n A_j(x_j)\big) &= \frac{\hbar^n\,\partial^n}{i^n\,\partial\lambda_1\dots\partial\lambda_n}\Big|_{\lambda_1=\dots=\lambda_n=0}\mathbf{S}\Big(\sum_{j=1}^n\lambda_jA_j(x_j)\Big)\\ &= \frac{\hbar^k\,\partial^k}{i^k\,\partial\lambda_1\dots\partial\lambda_k}\Big|_{\lambda_1=\dots=\lambda_k=0}\mathbf{S}\Big(\sum_{j=1}^k\lambda_jA_j(x_j)\Big)\\ &\qquad\star\frac{\hbar^{n-k}\,\partial^{n-k}}{i^{n-k}\,\partial\lambda_{k+1}\dots\partial\lambda_n}\Big|_{\lambda_{k+1}=\dots=\lambda_n=0}\mathbf{S}\Big(\sum_{l=k+1}^n\lambda_lA_l(x_l)\Big)\\ &= T_k\big(\otimes_{j=1}^kA_j(x_j)\big)\star T_{n-k}\big(\otimes_{l=k+1}^nA_l(x_l)\big)\ .]\end{aligned}$$

To complete the proof of the equivalence of the three causality conditions (3.3.5), (3.3.27) and (3.3.28), it remains to show that (3.3.5) (or (3.3.28)) implies (3.3.27). This proof is more involved, we give it in Exer. 3.3.16 by following Epstein and Glaser [66, Sect. 8.1].[70] Alternatively, a direct proof that (3.3.28) is indeed equivalent to (3.3.27) in the framework of perturbation theory, can be given by using the method of [27, App. B].

We point out that the formulation (3.3.27) or (3.3.28) of the causality condition does not have a direct physical foundation, because switching on and off the interaction is unphysical.

By means of the S-matrix, the $*$-structure axiom for the time-ordered product can be formulated as

$$\mathbf{S}(F)^*=\mathbf{S}(F^*)^{\star-1},\quad\text{for all}\quad F\in\mathcal{F}_{\text{loc}}\,. \tag{3.3.29}$$

For a real interaction, $F=F^*$, this condition requires that $\mathbf{S}(F)$ is *unitary*. The validity of the axiom (3.3.29) for the unrenormalized T-product is verified in Exer. 3.3.13.

3.3.3 Connection between the T- and the R-product

Let an S-matrix $\mathbf{S}$ be given. One may then define a sequence of maps

$$R_{n,1}\colon\mathcal{F}_{\text{loc}}^{\otimes(n+1)}\longrightarrow\mathcal{F}\quad\text{by linearity, symmetry in the first } n \text{ arguments,}$$

and the Bogoliubov formula:

$$R(e_\otimes^{F/\hbar},G)\equiv\sum_{n=0}^\infty\frac{1}{n!\,\hbar^n}R_{n,1}(F^{\otimes n},G):=\frac{\hbar}{i}\frac{d}{d\lambda}\Big|_{\lambda=0}\mathbf{S}(F)^{\star-1}\star\mathbf{S}(F+\lambda G). \tag{3.3.30}$$

We are now going to show that the so-defined maps $R_{n,1}$ fulfill the axioms for the R-product if $\mathbf{S}$ satisfies the axioms for the T-product.

[70] Since this proof uses some tools, which are not yet introduced, we postpone it till the end of this section.

Exercise 3.3.11.

(a) Derive the causality relation (axiom (d), formula (3.1.9)) for the maps $R_{n,1}$ defined in (3.3.30), from the causality of the time-ordered product in the form (3.3.27).

(b) The *relative S-matrix* is defined by

$$\mathbf{S}_K(F) := \mathbf{S}(K)^{\star -1} \star \mathbf{S}(K+F) \quad \text{for} \quad K, F \in \mathcal{F}_{\text{loc}}\,. \tag{3.3.31}$$

Prove the causality relation (3.3.27) for $\mathbf{S}_K$ in place of $\mathbf{S}$, and derive from this relation spacelike commutativity of $\mathbf{S}_K$, i.e., the property

$$[\mathbf{S}_K(F), \mathbf{S}_K(H)]_\star = 0 \quad \text{if} \quad (x-y)^2 < 0 \quad \text{for all} \quad (x,y) \in \operatorname{supp} F \times \operatorname{supp} H.$$

[*Solution*: (a) For $\operatorname{supp} H \cap (\operatorname{supp} G + \overline{V}_-) = \emptyset$ we obtain

$$\begin{aligned} R\big(e_\otimes^{(F+H)/\hbar}, G\big) &= \frac{\hbar}{i}\frac{d}{d\lambda}\Big|_{\lambda=0} \mathbf{S}(F+H)^{\star -1} \star \mathbf{S}(F+H+\lambda G) \\ &= \frac{\hbar}{i}\frac{d}{d\lambda}\Big|_{\lambda=0} \mathbf{S}(F+H)^{\star -1} \star \Big(\mathbf{S}(F+H) \star \mathbf{S}(F)^{\star -1} \star \mathbf{S}(F+\lambda G)\Big) \\ &= R\big(e_\otimes^{F/\hbar}, G\big)\,. \end{aligned}$$

(b) Let $\operatorname{supp} F \cap (\operatorname{supp} H + \overline{V}_-) = \emptyset$. By using (3.3.31) and (3.3.27) we get

$$\begin{aligned} &\mathbf{S}_K(G+F) \star \mathbf{S}_K(G)^{\star -1} \star \mathbf{S}_K(G+H) \\ &\quad = \mathbf{S}(K)^{\star -1} \star \mathbf{S}(K+G+F) \star \mathbf{S}(K+G)^{\star -1} \star \mathbf{S}(K) \star \mathbf{S}(K)^{\star -1} \star \mathbf{S}(K+G+H) \\ &\quad = \mathbf{S}(K)^{\star -1} \star \mathbf{S}(K+G+F+H) = \mathbf{S}_K(G+F+H)\,. \end{aligned}$$

Setting $G := 0$ in this relation, we obtain the following: If $\operatorname{supp} F$ and $\operatorname{supp} H$ are spacelike separated it holds that

$$\mathbf{S}_K(F) \star \mathbf{S}_K(H) = \mathbf{S}_K(F+H) = \mathbf{S}_K(H) \star \mathbf{S}_K(F)\,.]$$

The maps $R_{n,1}$ defined by (3.3.30) fulfill also the GLZ relation (axiom (e), formula (3.1.11a)). The proof of that uses *only* the definition (3.3.30) and Linearity and Symmetry of the T-product – further axioms for the T-product (in particular Causality) are not needed. The proof can be given as follows [48].

Proof. Due to $\frac{d}{d\lambda}\big|_{\lambda=0}\mathbf{S}(F+\lambda G)^{\star -1} \star \mathbf{S}(F+\lambda G) = 0$, we may write

$$R\big(e_\otimes^{F/\hbar}, G\big) = -\frac{\hbar}{i}\frac{d}{d\lambda}\Big|_{\lambda=0} \mathbf{S}(F+\lambda G)^{\star -1} \star \mathbf{S}(F) \tag{3.3.32}$$

and with that we obtain

$$R\big(e_\otimes^{F/\hbar}, G\big) \star R\big(e_\otimes^{F/\hbar}, H\big) = \hbar^2 \frac{\partial^2}{\partial\lambda_1\,\partial\lambda_2}\Big|_{\lambda_1=0=\lambda_2} \mathbf{S}(F+\lambda_1 G)^{\star -1} \star \mathbf{S}(F+\lambda_2 H). \tag{3.3.33}$$

Next we note that

$$R\big(e_\otimes^{F/\hbar}\otimes(G/\hbar),H\big)=\frac{\hbar}{i}\,\frac{\partial^2}{\partial\lambda_1\,\partial\lambda_2}\Big|_{\lambda_1=0=\lambda_2}\mathbf{S}(F+\lambda_1 G)^{\star-1}\star\mathbf{S}(F+\lambda_1 G+\lambda_2 H).$$

Hence,

$$\begin{aligned}
&R\big(e_\otimes^{F/\hbar}\otimes(G/\hbar),H\big)-R\big(e_\otimes^{F/\hbar}\otimes(H/\hbar),G\big)\\
&=\frac{\hbar}{i}\,\frac{\partial^2}{\partial\lambda_1\,\partial\lambda_2}\Big|_{\lambda_1=0=\lambda_2}\Big(\mathbf{S}(F+\lambda_1 G)^{\star-1}\star\mathbf{S}(F+\lambda_1 G+\lambda_2 H)\\
&\qquad\qquad-\mathbf{S}(F+\lambda_2 H)^{\star-1}\star\mathbf{S}(F+\lambda_1 G+\lambda_2 H)\Big)\\
&=\frac{\hbar}{i}\,\frac{\partial^2}{\partial\lambda_1\,\partial\lambda_2}\Big|_{\lambda_1=0=\lambda_2}\Big(\mathbf{S}(F+\lambda_1 G)^{\star-1}\star\mathbf{S}(F+\lambda_2 H)\\
&\qquad\qquad-\mathbf{S}(F+\lambda_2 H)^{\star-1}\star\mathbf{S}(F+\lambda_1 G)\Big)\\
&=\frac{1}{i\hbar}\Big[R\big(e_\otimes^{F/\hbar},G\big),\,R\big(e_\otimes^{F/\hbar},H\big)\Big]_\star\,,
\end{aligned}$$

where (3.3.33) is used in the last step. □

In addition, one verifies easily that the maps $R_{n,1}$ defined by (3.3.30) fulfill the renormalization conditions, axioms (f)–(l), if $\mathbf{S}$ satisfies the corresponding conditions – only the verification of the Off-shell field equation (axiom (j)) is somewhat more involved, see Exer. 3.3.15 below. We conclude that $R:=(R_{n,1})_{n\in\mathbb{N}}$ is indeed a retarded product.

Among the axioms for the T-product there is none corresponding to the GLZ relation. The latter can be interpreted as an integrability condition[71] for the "vector fields" $G\longmapsto R(e_\otimes^F,G)$, which ensures the existence of a "potential" $\mathbf{S}(F)$ from which R can be recovered by (3.3.30).

To write (3.3.30) as a relation expressing $R_{n,1}$ in terms of time-ordered products of different orders, we need to express $\mathbf{S}(F)^{\star-1}$ in terms of time-ordered products $T_l(F^{\otimes l})$ for $l\in\mathbb{N}^*$. For this purpose we define the *antichronological product* $\overline{T}\equiv(\overline{T}_n)_{n\in\mathbb{N}^*}$ to be a sequence of linear maps $\overline{T}_n\colon\mathcal{F}_{\mathrm{loc}}^{\otimes n}\longrightarrow\mathcal{F}$ which are symmetric in all arguments and given by

$$\overline{T}\big(e_\otimes^{-iF/\hbar}\big)\equiv 1+\sum_{n=1}^\infty\frac{(-i)^n}{n!\,\hbar^n}\,\overline{T}_n(F^{\otimes n}):=T\big(e_\otimes^{iF/\hbar}\big)^{\star-1}\equiv\mathbf{S}(F)^{\star-1}\,. \tag{3.3.34}$$

[71] According to an (unpublished) note of R. Brunetti, K. Fredenhagen and the author, one can regard $\omega_F\colon G\longmapsto iR(e_\otimes^F,G)$ as a principal connection on a trivial unitary-group bundle over $\mathcal{F}_{\mathrm{loc}}$, whose curvature vanishes if and only if the GLZ relation holds true.

As the term "antichronological" indicates, the $\overline{T}_n$ satisfy (3.3.5), with the $\overline{T}$-products on the right-hand side in reverse order:

$$\begin{aligned}&\overline{T}_n\big(A_1(x_1),\ldots,A_n(x_n)\big) \qquad (3.3.35)\\ &=\overline{T}_{n-k}\big(A_{\pi(k+1)}(x_{\pi(k+1)}),\ldots,A_{\pi n}(x_{\pi n})\big)\star_m \overline{T}_k\big(A_{\pi 1}(x_{\pi 1}),\ldots,A_{\pi k}(x_{\pi k})\big)\end{aligned}$$

whenever $\{x_{\pi 1},\ldots,x_{\pi k}\}\cap\big(\{x_{\pi(k+1)},\ldots,x_{\pi n}\}+\overline{V}_-\big)=\emptyset$ for a permutation $\pi\in S_n$. This "anticausal" factorization follows by taking the inverse $\star^{-1}$ of the causality relation (3.3.28):

$$\mathbf{S}(H+F)^{\star-1}=\mathbf{S}(F)^{\star-1}\star\mathbf{S}(H)^{\star-1} \quad \text{if} \quad \operatorname{supp} H\cap(\operatorname{supp} F+\overline{V}_-)=\emptyset\ . \qquad (3.3.36)$$

Analogously to (3.3.28) $\Leftrightarrow$ (3.3.5) (Exer. 3.3.10) one proves (3.3.36) $\Leftrightarrow$ (3.3.35). With that we conclude that (3.3.5) $\Leftrightarrow$ (3.3.35).

Note that $\overline{T}_n$ is *uniquely* determined by $T_1,\ldots,T_n$. More precisely, we get the following relation.

Exercise 3.3.12. Prove the formula

$$\overline{T}_n(F^{\otimes n})=\sum_{r=1}^{n}(-1)^{n-r}\sum_{\substack{n_1+\cdots+n_r=n\\ n_j\geq 1}}\frac{n!}{n_1!\cdots n_r!}\,T_{n_1}(F^{\otimes n_1})\star\cdots\star T_{n_r}(F^{\otimes n_r}). \qquad (3.3.37)$$

[*Solution*: Writing $1+\tau(F/\hbar):=T\big(e_\otimes^{iF/\hbar}\big)$, the formula for the geometric series (which we use here as a purely algebraic relation for formal power series) gives

$$\overline{T}\big(e_\otimes^{-iF/\hbar}\big)=1+\sum_{r=1}^{\infty}\Big(-\tau(F/\hbar)\Big)^{\star r}\ ,$$

where the exponent $\star r$ denotes the rth power with respect to the $\star$-product. Selecting from this equation the term of nth order in F we get the assertion (3.3.37).]

Exercise 3.3.13 (Unrenormalized antichronological product). Let $\check{\mathbb{M}}^n$ be defined as in Lemma 3.3.2. The "anti-Feynman propagator" is defined by

$$\Delta_m^{AF}(x):=\theta(x^0)\,\Delta_m^+(-x)+\theta(-x^0)\,\Delta_m^+(x)\ . \qquad (3.3.38)$$

Due to $\overline{\Delta_m^+(x)}=\Delta_m^+(-x)$, it is related to the Feynman propagator by

$$\Delta_m^{AF}(x)=\overline{\Delta_m^F(x)} \quad \text{and obviously it is symmetric:} \quad \Delta_m^{AF}(-x)=\Delta_m^{AF}(x)\ .$$

(a) Prove that

$$\overline{T}_n^{(m)}\big(A_1(x_1)\otimes\cdots\otimes A_n(x_n)\big)=A_1(x_1)\star_{\Delta_m^{AF}}\cdots\star_{\Delta_m^{AF}}A_n(x_n) \quad \text{on } \mathcal{D}(\check{\mathbb{M}}^n), \qquad (3.3.39)$$

where the r.h.s. is the n-fold product with propagator Δ_m^{AF}, which is commutative and associative for the same reasons as for $\star_{\Delta_m^F}$. From the analogy to (3.3.10) we see that the unrenormalized $\overline{T}$-product can be computed by the Feynman rules with Δ_m^{AF} in place of Δ_m^F.

(b) Verify the $*$-structure axiom (3.3.29) for the unrenormalized T-product, that is,

$$T_n^{(m)}\big(A_1(x_1)\otimes\cdots\otimes A_n(x_n)\big)^* = \overline{T}_n^{(m)}\big(A_1^*(x_1)\otimes\cdots\otimes A_n^*(x_n)\big) \quad \text{on } \mathcal{D}(\check{\mathbb{M}}^n). \tag{3.3.40}$$

[*Solution*: (a) We proceed analogously to the proof of Lemma 3.3.2: Due to anticausal factorization (3.3.35), $\overline{T}_n$ fulfills (3.3.8) on $\mathcal{D}(U_\pi)$ with the factors on the r.h.s. in reverse order. This n-fold $\star$-product can be rewritten as an n-fold $\star_{\Delta_m^{AF}}$-product by using $\Delta_m^+(x) = \Delta_m^{AF}(x)$ for $x \notin \overline{V}_+$, cf. (2.3.9).
(b) Analogously to (2.1.10) we obtain

$$\big(A_1(x_1)\star_{\Delta_m^F} A_2(x_2)\big)^* = A_1^*(x_1)\star_{\Delta_m^{AF}} A_2^*(x_2) \quad \text{on } \mathcal{D}(\check{\mathbb{M}}^2),$$

since $\overline{\Delta_m^F(x)} = \Delta_m^{AF}(x)$. By iteration we generalize this formula to the n-fold $\star_{\Delta_m^F}$- or $\star_{\Delta_m^{AF}}$-product. This yields the assertion, due to (3.3.10) and (3.3.39).]

Now we are ready for the proposition giving the connection between the T- and the R-product [55].

Proposition 3.3.14 (Duality of T- and R-product).

(a) *The Bogoliubov formula* (3.3.30) *has the following property: Given the finite family* $\{T_1,\dots,T_{n+1}\}$ *fulfilling the axioms* (i) (*Linearity*) *and* (iii) (*Symmetry*), *it determines uniquely the finite family* $\{R_{0,1},\dots,R_{n,1}\}$, *which is required to fulfill the axioms* (a) (*Linearity*) *and* (b) (*Symmetry*); *and vice versa.*

(b) *Given* $T=(T_n)_{n\in\mathbb{N}^*}$ *solving the axioms for the* T*-product, then the corresponding* $R=(R_{n,1})_{n\in\mathbb{N}}$ *constructed according to part* (a), *fulfills the axioms for the* R*-product.*

(c) *The same statement as part* (b), *but with the roles of* T *and* R *reversed.*

So, in this sense, the axioms for the R-product and those for the T- product are equivalent.

Proof. Ad (a): Expanding formula (3.3.30) and using Linearity and Symmetry, we get

$$\begin{aligned}&R_{n,1}(F_1\otimes\cdots\otimes F_n,G) - i^n\,T_{n+1}(F_1\otimes\cdots\otimes F_n\otimes G)\\ &\quad = i^n \sum_{\emptyset\neq I\subseteq\{1,\dots,n\}} (-1)^{|I|}\,\overline{T}_{|I|}(F_I)\star T_{|I^c|+1}(F_{I^c}\otimes G)\,,\end{aligned} \tag{3.3.41}$$

where $F_J := \otimes_{j\in J}F_j$ for $J\subseteq\{1,\dots,n\}$. The second term on the left-hand side corresponds to the case $I=\emptyset$. From (3.3.41) we directly see that $\{T_1,\dots,T_{n+1}\}$ determines $\{R_{0,1},\dots,R_{n,1}\}$ uniquely. To verify the other direction of the claim, we proceed by induction on n. So, given $\{R_{0,1},\dots,R_{n,1}\}$, the family $\{T_1,\dots,T_n\}$ is known inductively and, hence, this holds also for the right-hand side of (3.3.41), and in this way we obtain T_{n+1}.

Ad (b): This is shown above.

Ad (c): We give the verification of the axioms Symmetry and Causality, for which the assertion is far from being obvious.

Symmetry[72]: Obviously, $T_{n+1}(F_1 \otimes \cdots \otimes F_n \otimes G)$ constructed inductively by means of (3.3.41), is symmetrical in $F_1, \ldots, F_n$. Hence, it suffices to study

$$\begin{aligned} & i^n\Big(T_{n+1}(F_1,\ldots,F_{n-1},H,G) - T_{n+1}(F_1,\ldots,F_{n-1},G,H)\Big) \qquad (3.3.42) \\ & = R_{n,1}(F_1,\ldots,F_{n-1},H;G) - R_{n,1}(F_1,\ldots,F_{n-1},G;H) \\ & \quad - i^n \sum_{I\subseteq\{1,\ldots,n-1\}} (-1)^{|I|+1}\Big(\overline{T}(F_I,H)\star T(F_{I^c},G) - \overline{T}(F_I,G)\star T(F_{I^c},H)\Big)\,. \end{aligned}$$

The GLZ-relation says that

$$R_{n,1}(F_1,\ldots,H;G) - R_{n,1}(F_1,\ldots,G;H) = \frac{1}{i} \sum_{I\subseteq\{1,\ldots,n-1\}} \Big[R(F_I;H), R(F_{I^c};G\Big]_\star\,. \tag{3.3.43}$$

From (3.3.33), which is a consequence of the Bogoliubov formula (3.3.30), we get

$$\begin{aligned} \Big[R(e_\otimes^{F/\hbar},H), R(e_\otimes^{F/\hbar},G)\Big]_\star &= \overline{T}(e_\otimes^{-iF/\hbar}\otimes H)\star T(e_\otimes^{iF/\hbar}\otimes G) \\ &\quad - \overline{T}(e_\otimes^{-iF/\hbar}\otimes G)\star T(e_\otimes^{iF/\hbar}\otimes H)\,. \end{aligned}$$

Expanding this relation, we see that (3.3.43) is equal to minus the term in the last line of (3.3.42), hence, the r.h.s. of (3.3.42) vanishes.

Causality: Since we have verified the axiom Symmetry, it suffices to consider the following configuration. Let $P \subseteq \{1,\ldots,n\}$ with $P \neq \emptyset$, $P^c := \{1,\ldots,n\} \setminus P$ and assume that $x_r \notin (x_s + \overline{V}_-)$ $\forall r \in P$, $s \in P^c \cup \{n+1\}$. To show the pertinent causal factorization (3.3.5) of T_{n+1}, we start with Causality of $R_{n,1}$: $R_{n,1}\big(A_1(x_1),\ldots,A_n(x_n);A_{n+1}(x_{n+1})\big) = 0$. With that, the Bogoliubov formula (3.3.41) yields

$$\begin{aligned} T_{n+1}(A_1(x_1),\ldots,A_{n+1}(x_{n+1})) &= - \sum_{\emptyset\neq I\subseteq\{1,\ldots,n\}} (-1)^{|I|}\,\overline{T}(I)\star T(I^c,n+1) \\ = - \sum_{\substack{K\subseteq P^c,\,J\subseteq P\\ K\cup J\neq\emptyset}} &(-1)^{|K|}\,\overline{T}(K)\star(-1)^{|J|}\,\overline{T}(J)\star T(P\setminus J)\star T(P^c\setminus K,n+1)\,, \end{aligned} \tag{3.3.44}$$

[72]The inductive construction of $i^n\,T_{n+1}$ as $R_{n,1}$ minus the r.h.s. of (3.3.41) agrees essentially with the procedure in the distribution splitting method of Epstein and Glaser [66, 148] (see also Exer. 3.5.7). However, in that method the GLZ-relation for the R-product is not taken into account. Therefore, it may happen that the so-obtained T_{n+1} does not fulfill the axiom Symmetry; hence, Epstein and Glaser symmetrize it. This is allowed, because this symmetrization amounts to the addition of a local term, i.e., a term with support $\subseteq \Delta_{n+1}$.

where the shorthand notation $T(I) := T_{|I|}(\otimes_{i\in I} A_i(x_i))$ and the (anti)causal factorizations (3.3.5) and (3.3.35) are also used. Now we take into account that $\overline{T}(e_\otimes^{-iF/\hbar}) \star T(e_\otimes^{iF/\hbar}) = 1$ implies that

$$\sum_{J\subseteq P} (-1)^{|J|}\, \overline{T}(J) \star T(P\setminus J) = 0 \quad \text{if} \quad P \neq \emptyset \,. \tag{3.3.45}$$

Due to that, the terms with $K \neq \emptyset$ cancel out in (3.3.44), because for these terms $J = \emptyset$ is admitted; there remain the $(K = \emptyset)$-terms:

$$-\Big(\sum_{J\subseteq P,\, J\neq\emptyset} (-1)^{|J|}\, \overline{T}(J) \star T(P\setminus J)\Big) \star T(P^{\mathrm{c}}, n+1) = T(P) \star T(P^{\mathrm{c}}, n+1)\,,$$

by using once more (3.3.45). □

Due to (3.3.32), we may express $R_{n,1}$ in terms of $\{T_1, \ldots, T_{n+1}\}$ also by the formula

$$R_{n,1}(F_1\otimes\cdots\otimes F_n, G) = i^n \sum_{I\subseteq\{1,\ldots,n\}} (-1)^{|I|}\, \overline{T}_{|I|+1}(F_I\otimes G) \star T_{|I^{\mathrm{c}}|}(F_{I^{\mathrm{c}}})\,. \tag{3.3.46}$$

Similarly to (3.3.41), this is an equivalent reformulation of the Bogoliubov formula (3.3.30).

Exercise 3.3.15. Given a time-ordered product $T \equiv (T_n)_{n\in\mathbb{N}^*}$ satisfying all axioms – in particular the Off-shell field equation (axiom (viii)), prove that the pertinent retarded product $R \equiv (R_{n,1})_{n\in\mathbb{N}}$ defined by the Bogoliubov formula (3.3.30) fulfills also the Off-shell field equation (axiom (j)).

[*Solution*: By using the Bogoliubov formula (3.3.41) and axiom (viii) we get

$$\begin{aligned} R_{n,1}(F_1\otimes\cdots\otimes F_n, \varphi(x)) = i^n &\sum_{I\subseteq\{1,\ldots,n\}} (-1)^{|I|}\, \overline{T}(F_I) \star \Big(T(F_{I^{\mathrm{c}}})\,\varphi(x) \\ &+ \hbar \int dy\; \Delta^F(x-y)\, \frac{\delta T(F_{I^{\mathrm{c}}})}{\delta\varphi(y)}\Big)\,. \end{aligned} \tag{3.3.47}$$

For the first term we obtain

$$\begin{aligned} \overline{T}(F_I) \star \Big(T(F_{I^{\mathrm{c}}})\,\varphi(x)\Big) = &\Big(\overline{T}(F_I) \star T(F_{I^{\mathrm{c}}})\Big)\,\varphi(x) \\ &+ \hbar \int dy\; \Delta^+(y-x)\, \frac{\delta \overline{T}(F_I)}{\delta\varphi(y)} \star T(F_{I^{\mathrm{c}}})\,. \end{aligned}$$

Now we apply $i^{n-1}\sum_{I\subseteq\{1,\ldots,n\}}(-1)^{|I|}\ldots$ to the two terms on the r.h.s.. The first one vanishes, due to (3.3.45). For the second one we use the Field independence (axiom (v)):

Omitting $\hbar \int dy\, \Delta^+(y-x)$ we get

$$
\begin{aligned}
i^{n-1} & \sum_{\emptyset \neq I \subseteq \{1,\ldots,n\}} \sum_{l \in I} (-1)^{|I|}\, \overline{T}\Big(F_{I\setminus\{l\}} \otimes \frac{\delta F_l}{\delta\varphi(y)}\Big) \star T(F_{I^c}) \\
&= -\sum_{l=1}^{n} i^{n-1} \sum_{I \subseteq \{1,\ldots\widehat{l}\ldots,n\}} (-1)^{|I|}\, \overline{T}\Big(F_I \otimes \frac{\delta F_l}{\delta\varphi(y)}\Big) \star T(F_{I^c}) \\
&= -\sum_{l=1}^{n} R\Big(F_1 \otimes \cdots \widehat{F_l} \cdots \otimes F_n, \frac{\delta F_l}{\delta\varphi(y)}\Big) .
\end{aligned}
$$

where we have rearranged the sums and used (3.3.46). For the second term in (3.3.47) we proceed analogously: Omitting $i\hbar \int dy\, \Delta^F(x-y)$ we obtain

$$
\begin{aligned}
i^{n-1} & \sum_{\substack{I \subset \{1,\ldots,n\} \\ I^c \neq \emptyset}} \sum_{l \in I^c} (-1)^{|I|}\, \overline{T}(F_I) \star T\Big(F_{I^c\setminus\{l\}} \otimes \frac{\delta F_l}{\delta\varphi(y)}\Big) \\
&= \sum_{l=1}^{n} i^{n-1} \sum_{I \subseteq \{1,\ldots\widehat{l}\ldots,n\}} (-1)^{|I|}\, \overline{T}(F_I) \star T\Big(F_{I^c} \otimes \frac{\delta F_l}{\delta\varphi(y)}\Big) \\
&= \sum_{l=1}^{n} R\Big(F_1 \otimes \cdots \widehat{F_l} \cdots \otimes F_n, \frac{\delta F_l}{\delta\varphi(y)}\Big) ,
\end{aligned}
$$

where we have used (3.3.41) in place of (3.3.46). Inserting these results into (3.3.47), we end up with

$$
\begin{aligned}
R_{n,1}(F_1 \otimes \cdots \otimes F_n, \varphi(x)) = \hbar \int dy\, & \big(i\Delta^F(x-y) - i\Delta^+(y-x)\big) \\
& \cdot \sum_{l=1}^{n} R\Big(F_1 \otimes \cdots \widehat{F_l} \cdots \otimes F_n, \frac{\delta F_l}{\delta\varphi(y)}\Big) .
\end{aligned}
$$

Due to (3.3.14), this formula agrees with the assertion (3.1.45).]

Finally, we complete the proof of the equivalence of the three versions (3.3.5), (3.3.27) and (3.3.28) of the Causality axiom. We use that the causal factorization of the T-product (3.3.5) implies the anticausal factorization of the $\overline{T}$-product (3.3.35), as explained after (3.3.36).

Exercise 3.3.16 (Generalized retarded product and equivalence of Causality conditions). In this exercise we prove that (3.3.5) and (3.3.35) imply the (seemingly stronger) causality condition (3.3.27) by following [66, Sect. 8.1]: We make the same assumptions as in Exer. 3.3.10. In addition let $\overline{T}$ be defined by (3.3.34) and let

$$
\mathbf{S}_G(F) := \overline{T}\big(e_\otimes^{-iG/\hbar}\big) \star T\big(e_\otimes^{i(G+F)/\hbar}\big) \tag{3.3.48}
$$

be the relative S-matrix (3.3.31). For $n, l \in \mathbb{N}$ we define the "generalized retarded product" $R_{n,l} : \mathcal{F}_{\mathrm{loc}}^{\otimes(n+l)} \longrightarrow \mathcal{F}$ by linearity, symmetry in the first n and in the last l factors,

$$
R_{n,l}(G_{\pi(1)}, \ldots, G_{\pi(n)}; F_{\sigma(1)}, \ldots, F_{\sigma(l)}) = R_{n,l}(G_1, \ldots, G_n; F_1, \ldots, F_l)
$$

for all $\pi \in S_n$, $\sigma \in S_l$, and by

$$\mathbf{S}_G(F) =: \sum_{n,l=0}^{\infty} \frac{i^l}{n!\,l!\,\hbar^{n+l}} R_{n,l}(G^{\otimes n}; F^{\otimes l}) \equiv R\big(e_\otimes^{G/\hbar}; e_\otimes^{iF/\hbar}\big) \quad \forall G, F \in \mathcal{F}_{\mathrm{loc}} \,. \tag{3.3.49}$$

In particular we have $R_{0,0} = 1$ and for $l = 1$ the generalized retarded products $(R_{n,l})_{n\in\mathbb{N}}$ agree with the retarded products $(R_{n,1})_{n\in\mathbb{N}}$ which are the coefficients of the interacting fields; this follows immediately from the Bogoliubov formula (3.3.30).

(a) Prove that (3.3.5) and (3.3.35) imply the support property

$$\begin{aligned} &\operatorname{supp} R_{n,l} \\ &\subseteq \{\, (y_1,\ldots,y_n; x_1,\ldots,x_l) \in \mathbb{M}^{n+l} \,\big|\, y_j \in (\{x_1,\ldots,x_l\} + \overline{V}_-),\ \forall 1 \le j \le n \,\} \,. \end{aligned} \tag{3.3.50}$$

(b) Show that the relation (3.3.50) is equivalent to the assertion (3.3.27).

[*Solution*: (a) Analogously to (3.3.30) $\Leftrightarrow$ (3.3.41), the definitions (3.3.48) and (3.3.49) are equivalent to

$$R_{n,l}\big(\otimes_{j=1}^n G_j; \otimes_{r=1}^l F_r\big) = i^n \sum_{I\subseteq\{1,\ldots,n\}} (-1)^{|I|}\, \overline{T}_{|I|}(G_I) \star T_{|I^c|+l}\big(G_{I^c} \otimes (\otimes_{r=1}^l F_r)\big) \tag{3.3.51}$$

by taking into account linearity and symmetry of $R_{n,l}$. Now, we replace G_j by $A_j(y_j)$ and F_r by $B_r(x_r)$, where A_j, $B_r \in \mathcal{P}$. To simplify the notation we write y_j for $A_j(y_j)$ and x_r for $B_r(x_r)$; in addition let $Y := (y_1,\ldots,y_n)$ and $X := (x_1,\ldots,x_l)$. With that formula (3.3.51) can be written as

$$R(Y;X) = i^n \sum_{I\subseteq Y} (-1)^{|I|}\, \overline{T}(I) \star T(I^c, X) \,.$$

If (Y, X) does not lie in the set given on the r.h.s. of (3.3.50), there are subsets P and Q of Y such that

$$P \neq \emptyset \ \wedge\ P \cup Q = Y \ \wedge\ P \cap Q = \emptyset \ \wedge\ P \cap (X + \overline{V}_-) = \emptyset \ \wedge\ Q \subseteq (X + \overline{V}_-) \,.$$

Due to the (anticausal) factorization of T (3.3.5) and $\overline{T}$ (3.3.35) the following equations hold true in a neighbourhood of such a point $(Y, X) = (P \cup Q, X)$:

$$\begin{aligned} R(Y,X) &= i^n \sum_{J\subseteq P,\, K\subseteq Q} (-1)^{|J|+|K|}\, \overline{T}(J\cup K) \star T(P\setminus J,\, Q\setminus K,\, X) \\ &= i^n \sum_{J\subseteq P,\, K\subseteq Q} (-1)^{|J|+|K|}\, \overline{T}(K) \star \overline{T}(J) \star T(P\setminus J) \star T(Q\setminus K,\, X) = 0 \,. \end{aligned}$$

In the last step we apply (3.3.45), which is allowed since $P \neq \emptyset$.
(b) Similarly to (3.1.9) $\Leftrightarrow$ (3.1.10), the support property (3.3.50) is equivalent to:

$$\begin{aligned} &\mathbf{S}_{G+H}(F) \equiv R\big(e_\otimes^{(G+H)/\hbar}; e_\otimes^{iF/\hbar}\big) = R\big(e_\otimes^{G/\hbar}; e_\otimes^{iF/\hbar}\big) \equiv \mathbf{S}_G(F) \\ &\text{if} \quad \operatorname{supp} H \cap (\operatorname{supp} F + \overline{V}_-) = \emptyset \,. \end{aligned}$$

By multiplying this relation with $\mathbf{S}(G+H) \star \cdots$ from the left side, we see that it is equivalent to (3.3.27).]

Comparison with Exer. 3.3.11(a): In the particular case $l = 1$, the solution of part (a) (of the exercise here) is an alternative way to derive the Causality of the R-product (defined by the Bogoliubov formula (3.3.41)) from the Causality of the T-product, which is more direct since it uses Causality of the T-product in the original form (3.3.5) and (3.3.35) instead of (3.3.27). The statement shown in Exer. 3.3.11(a), is part of the equivalence verified in part (b) of the exercise here.

3.4 The time-ordered (or retarded) product for on-shell fields

This section is based on references [54] and [20], see also [161].

In the vast majority of the literature (in particular in the work of Epstein and Glaser [66]), the T- (or R-) product is a map from local Wick polynomials to Wick polynomials. Roughly speaking, this corresponds to a map from local on-shell fields to on-shell fields in our formalism, due to Theorem 2.6.3. The aim of this section is to define a corresponding on-shell T-product in our formalism, which we denote by $T^{\mathrm{on}} \equiv (T_n^{\mathrm{on}})_{n\in\mathbb{N}^*}$; the definition will be given in terms of our off-shell version $T_n\colon \mathcal{F}_{\mathrm{loc}}^{\otimes n} \longrightarrow \mathcal{F}$ (introduced in Sect. 3.3.2) and it will be such that the axioms for $T \equiv (T_n)_{n\in\mathbb{N}^*}$ translate into "nice" properties for T^{on} – essentially these properties agree with the axioms of Epstein and Glaser used to define their on-shell T-product. By proceeding in a completely analogous way, one can define also an on-shell R-product $R^{\mathrm{on}} \equiv (R_{n,1}^{\mathrm{on}})_{n\in\mathbb{N}}$ in terms of our off-shell R-product (introduced in Sect. 3.1); the Bogoliubov formula (3.3.30) (or (3.3.41)) gives also the connection between T^{on} and R^{on}.

3.4.1 Definition of the on-shell T-product

A crucial point in the definition of T^{on} are the Remks. 3.1.2 and 3.1.1: Requiring Linearity and the Field equation for T^{on}, the AWI cannot hold true and, hence, T^{on} cannot depend only on (local) *functionals*! Therefore, we define the arguments of T^{on} to be (local) on-shell field *polynomials*.

To define the latter, we introduce the ideal of $\mathcal{P}$ generated by the free field equation:

$$\mathcal{J}_{\mathcal{P}} = \Big\{ \sum_{a\in\mathbb{N}^d} B_a \partial^a (\Box + m^2)\varphi \,\Big|\, B_a \in \mathcal{P} \Big\} . \tag{3.4.1}$$

Indeed, $\mathcal{J}_{\mathcal{P}}$ is a 2-sided ideal in $\mathcal{P}$: $\mathcal{P}\mathcal{J}_{\mathcal{P}} = \mathcal{J}_{\mathcal{P}} = \mathcal{J}_{\mathcal{P}}\mathcal{P}$. Therefore, the quotient of the commutative algebra $\mathcal{P}$ by the ideal $\mathcal{J}_{\mathcal{P}}$,

$$\mathcal{P}_0 := \frac{\mathcal{P}}{\mathcal{J}_{\mathcal{P}}} \tag{3.4.2}$$

is also a commutative algebra – the algebra of "local on-shell field polynomials" – and the canonical surjection

$$\pi : \begin{cases} \mathcal{P} \longrightarrow \mathcal{P}_0 \\ A \longmapsto \pi A := A + \mathcal{J}_{\mathcal{P}} \end{cases} \tag{3.4.3}$$

is an algebra homomorphism, i.e., it is linear and commutes with multiplication: $\pi(A_1 A_2) = (\pi A_1)(\pi A_2)$, $\forall A_1, A_2 \in \mathcal{P}$. Derivatives in $\mathcal{P}_0$ are defined as follows: For $\pi A \in \mathcal{P}_0$ choose a $B \in \mathcal{P}$ such that $\pi B = \pi A$. Then,

$$\partial^\mu(\pi A) := \pi(\partial^\mu B) \tag{3.4.4}$$

is well defined, since $\pi B_1 = \pi B_2$ implies $(\partial^\mu B_1 - \partial^\mu B_2) \in \partial^\mu \mathcal{J}_{\mathcal{P}} \subset \mathcal{J}_{\mathcal{P}}$. Note that the definition (3.4.4) can also be written as

$$\partial^\mu(\pi A) := \pi(\partial^\mu A) \quad \forall A \in \mathcal{P} .$$

In particular,

$$(\Box + m^2)\pi\varphi = \pi\big((\Box + m^2)\varphi\big) = 0 \in \mathcal{P}_0 ,$$

because $(\Box + m^2)\varphi \in \mathcal{J}_{\mathcal{P}}$.

In analogy to $F_0 := F|_{\mathcal{C}_0}$ $\forall F \in \mathcal{F}$ (2.6.1), we introduce some notations: For $A, A_1, \ldots, A_n \in \mathcal{P}$ we set

$$A_0(x) := A(x)\big|_{\mathcal{C}_0} , \quad \text{or equivalently} \quad A_0(g) := A(g)_0 \quad \forall g \in \mathcal{D}(\mathbb{M}) , \tag{3.4.5}$$

and

$$T_n\big(A_1(x_1), \ldots, A_n(x_n)\big)_0 := T_n\big(A_1(x_1), \ldots, A_n(x_n)\big)\big|_{\mathcal{C}_0} .$$

Since

$$\mathcal{J}_{\mathcal{P}} = \{A \in \mathcal{P} \,|\, A_0(x) = 0 \;\; \forall x \in \mathbb{M}\} , \tag{3.4.6}$$

the map

$$Q_{\mathcal{P}} : \begin{cases} \mathcal{P}_0 \longrightarrow \{A_0 : \mathbb{M} \ni x \mapsto A_0(x) \,|\, A \in \mathcal{P}\} \\ \pi A \longmapsto B_0 \quad \text{with} \quad B \in \pi A \equiv A + \mathcal{J}_{\mathcal{P}} \end{cases}$$

is a well-defined algebra isomorphism, where the product on the r.h.s. is the pointwise product w.r.t. both the dependence on x and the dependence on $h \in \mathcal{C}_{S_0}$, that is,

$$(A_{1,0}A_{2,0})(x)(h) := A_{1,0}(x)(h) \cdot A_{2,0}(x)(h) = (A_1 A_2)_0(x)(h) \quad \forall x \in \mathbb{M} , \; h \in \mathcal{C}_{S_0} ,$$

the second equality is obvious when taking into account (1.3.10). Hence, we indeed obtain

$$Q_{\mathcal{P}}(\pi A_1 \cdot \pi A_2) = Q_{\mathcal{P}}\big(\pi(A_1 A_2)\big) = (A_1 A_2)_0 = A_{1,0}\, A_{2,0} = Q_{\mathcal{P}}(\pi A_1)\, Q_{\mathcal{P}}(\pi A_2) .$$

So, we may identify πA with A_0 for all $A \in \mathcal{P}$. The map $Q_{\mathcal{P}}$ is a version of the algebra isomorphism $Q : \mathcal{F}/\mathcal{J} \longrightarrow \mathcal{F}_0 := \mathcal{F}|_{\mathcal{C}_0}$ given in (2.6.3) – the version for local field polynomials.

The main idea of this section is to define

$$T_n^{\mathrm{on}} : \mathcal{P}_0^{\otimes n} \longrightarrow \mathcal{D}'(\mathbb{M}^n, \mathcal{F}_0) \qquad \forall n \in \mathbb{N}^* \tag{3.4.7}$$

– cf. (3.1.7) – in terms of T_n by

$$\begin{aligned} T_n^{\mathrm{on}}\big((A_{1,0}(x_1), \ldots, A_{n,0}(x_n)\big) &\equiv T_n^{\mathrm{on}}(\pi A_1 \otimes \cdots \otimes \pi A_n)(x_1, \ldots, x_n) \\ &:= T_n\big(\xi \circ \pi(A_1)(x_1), \ldots, \xi \circ \pi(A_n)(x_n)\big)_0 , \end{aligned} \tag{3.4.8}$$

where ξ is an algebra homomorphism[73]

$$\xi : \mathcal{P}_0 \longrightarrow \mathcal{P} , \tag{3.4.9}$$

which chooses a representative of the equivalence class $A + \mathcal{J}_{\mathcal{P}}$, that is, $\xi(A + \mathcal{J}_{\mathcal{P}}) \in A + \mathcal{J}_{\mathcal{P}}$ or equivalently $\pi \circ \xi = \mathrm{Id}$. In order that the defining axioms for T translate into "nice" properties for T^{on}, we require that ξ satisfies certain additional conditions.

In detail, for the *model of one real scalar field* the map $\xi : \mathcal{P}_0 \longrightarrow \mathcal{P}$ is defined by the following axioms:[74]

(a) $\pi\xi = \mathrm{Id}$.

(b) ξ is an algebra homomorphism.

(c) The Lorentz transformations commute with $\xi\pi$: $b_\Lambda \circ \xi\pi = \xi\pi \circ b_\Lambda \ \forall \Lambda \in \mathcal{L}_+^\uparrow$, where b_Λ is the action of $\mathcal{L}_+^\uparrow$ on $\mathcal{P}$, that is, it is defined by $(b_\Lambda A)(\Lambda x) := \beta_\Lambda A(x) \ \forall A \in \mathcal{P}$, with β_Λ given in (3.1.35)–(3.1.36).

(d) $\xi\pi(\mathcal{P}_1) \subseteq \mathcal{P}_1$, where $\mathcal{P}_1 \subset \mathcal{P}$ is the subspace of fields linear in φ and its partial derivatives.

(e) $\xi\pi$ does not increase the mass dimension of the fields: $\dim \xi\pi(A) \leq \dim A$.

Due to (a), $\xi\pi : \mathcal{P} \longrightarrow \mathcal{P}$ is a *projection*: $(\xi\pi)(\xi\pi) = \xi\pi$. From the linearity of ξ and the condition (a) we conclude that

$$\mathrm{Ker}\, \xi\pi = \mathrm{Ker}\, \pi = \mathcal{J}_{\mathcal{P}} \tag{3.4.10}$$

and, hence,

$$\xi\pi\Box\partial^a\varphi = -m^2\xi\pi\partial^a\varphi \qquad \forall a \in \mathbb{N}^d . \tag{3.4.11}$$

Since $\xi(A + \mathcal{J}_{\mathcal{P}}) \in A + \mathcal{J}_{\mathcal{P}}$, we know

$$\xi\pi(A) - A \in \mathcal{J}_{\mathcal{P}} , \quad \text{that is,} \quad \xi\pi(A)_0 = A_0 \qquad \forall A \in \mathcal{P} . \tag{3.4.12}$$

[73] In references [54] and [20] this map is called σ; we use a different Greek letter, because σ, or more precisely σ_ρ, is used in this book to denote a scaling transformation.

[74] To shorten the notations we write $\xi\pi$ and $\pi\xi$ for $\xi \circ \pi$ and $\pi \circ \xi$, respectively. We recall the definition of the mass dimension of a field polynomial (3.2.63).

We are now searching the most general solution ξ of the above axioms; that is, the most general map $\xi\pi$, because ξ is uniquely determined by $\xi\pi$. Due to (b), $\xi\pi$ is an algebra homomorphism; hence, it holds that

$$\xi\pi(c) = c \quad \forall c \in \mathbb{R} \subset \mathcal{P} \tag{3.4.13}$$

and it suffices to determine $\xi\pi(\partial^a\varphi)$ for all $a \in \mathbb{N}^d$:

$$\xi\pi(A) = \sum_n \sum_{a_1,\dots,a_n} c_{a_1\cdots a_n}\, \xi\pi(\partial^{a_1}\varphi)\cdots\xi\pi(\partial^{a_n}\varphi)\ ,$$

when A is given by (1.3.9). From (3.4.12) and (d) we conclude that $\xi\pi(\partial^a\varphi)$ takes the form

$$\xi\pi(\partial^a\varphi) = \partial^a\varphi + \sum_{|b|\le|a|-2} c^a_b\, \partial^b(\Box + m^2)\varphi \quad \text{with constants} \quad c^a_b \in \mathbb{R}\ ; \tag{3.4.14}$$

the upper bound on $|b|$ comes from the condition (e): $\dim(\xi\pi\partial^a\varphi) \le \dim(\partial^a\varphi)$. We immediately get

$$\xi\pi(\varphi) = \varphi \quad \text{and} \quad \xi\pi(\partial^\mu\varphi) = \partial^\mu\varphi\ . \tag{3.4.15}$$

We are now going to determine the coefficients c^a_b for $|a| = 2, 3$, by proceeding by induction on $|a|$. Using additionally Lorentz covariance (c) and invariance w.r.t. permutations of the Lorentz indices, the ansatz (3.4.14) yields

$$\begin{aligned}
\xi\pi(\partial^\mu\partial^\nu\varphi) &= \partial^\mu\partial^\nu\varphi + c\, g^{\mu\nu}\,(\Box + m^2)\varphi\ ,\\
\xi\pi(\partial^\mu\partial^\nu\partial^\lambda\varphi) &= \partial^\mu\partial^\nu\partial^\lambda\varphi + c_1\,(g^{\mu\nu}\partial^\lambda + g^{\mu\lambda}\partial^\nu + g^{\nu\lambda}\partial^\mu)(\Box + m^2)\varphi
\end{aligned}$$

with unknown coefficients $c, c_1 \in \mathbb{R}$. To determine the latter we contract with $g_{\mu\nu}$ and use $g_{\mu\nu}g^{\mu\nu} = d$:

$$\begin{aligned}
\xi\pi(\Box\varphi) &= \Box\varphi + cd\,(\Box + m^2)\varphi\\
\xi\pi(\partial^\lambda\Box\varphi) &= \partial^\lambda\Box\varphi + c_1(d+2)\,\partial^\lambda(\Box + m^2)\varphi
\end{aligned}$$

On the other hand we know from (3.4.11) and the lower $|a|$ results (3.4.15) that

$$\begin{aligned}
\xi\pi(\Box\varphi) &= -m^2\,\xi\pi(\varphi) = -m^2\,\varphi\\
\xi\pi(\partial^\lambda\Box\varphi) &= -m^2\,\xi\pi(\partial^\lambda\varphi) = -m^2\,\partial^\lambda\varphi\ .
\end{aligned}$$

Hence, we end up with

$$\begin{aligned}
\xi\pi(\partial^\mu\partial^\nu\varphi) &= \partial^\mu\partial^\nu\varphi - \frac{g^{\mu\nu}}{d}\,(\Box + m^2)\varphi\ , \qquad (3.4.16)\\
\xi\pi(\partial^\mu\partial^\nu\partial^\lambda\varphi) &= \partial^\mu\partial^\nu\partial^\lambda\varphi - \frac{1}{d+2}\,(g^{\mu\nu}\partial^\lambda + g^{\mu\lambda}\partial^\nu + g^{\nu\lambda}\partial^\mu)(\Box + m^2)\varphi\ .
\end{aligned}$$

In [54, Sect. 3.2 and App. A] it is shown that in the four-dimensional Minkowski space $\xi\pi(\partial^a\varphi)$ is uniquely determined for all $a \in \mathbb{N}^d$, that is, the above axioms have a unique solution ξ. However, this solution for $\xi\pi(\partial^a\varphi)$ is given there only by induction on $|a|$. In [20] this recurrence relation is solved and a fully explicit expression for $\xi\pi(\partial^a\varphi)$ in d-dimensions, d arbitrary, is given.

Exercise 3.4.1. For all $n \geq 2$, prove the relations

$$\omega_0\big(F_1 \star \cdots \star \xi\pi(A)(x) \star \cdots \star F_{n-1}\big) = \omega_0\big(F_1 \star \cdots \star A(x) \star \cdots \star F_{n-1}\big) \tag{3.4.17}$$

(with $A \in \mathcal{P}$ and $F_1, \ldots, F_{n-1} \in \mathcal{F}$), and

$$\omega_0\Big(T_n\big(\xi\pi(A_1)(x_1), \ldots, \xi\pi(A_n)(x_n)\big)\Big) = \omega_0\Big(T_n\big(A_1(x_1), \ldots, A_n(x_n)\big)\Big) \quad \text{on } \mathcal{D}(\check{\mathbb{M}}^n) \tag{3.4.18}$$

(with $A_1, \ldots, A_n \in \mathcal{P}$), where $\check{\mathbb{M}}^n$ is defined in Lemma 3.3.2. Using the obvious definition $\omega_0(F_0) := \omega_0(F)\ \forall F \in \mathcal{F}$ (2.6.7), the result (3.4.18) can equivalently be written as

$$\omega_0\Big(T_n^{\mathrm{on}}(\pi A_1 \otimes \cdots \otimes \pi A_n)(x_1, \ldots, x_n)\Big) = \omega_0\Big(T_n\big(B_1(x_1), \ldots, B_n(x_n)\big)\Big) \quad \text{on } \mathcal{D}(\check{\mathbb{M}}^n)$$

for any $B_1, \ldots B_n \in \mathcal{P}$ satisfying $\pi B_j = \pi A_j\ \forall 1 \leq j \leq n$.

[*Solution*: The assertion (3.4.17) follows from $\xi\pi(A) - A \in \mathcal{J}_\mathcal{P}$ (3.4.12) and

$$\omega_0\big(F_1 \star \cdots \star B(x) \star \cdots \star F_{n-1}\big) = 0 \qquad \forall B \in \mathcal{J}_\mathcal{P}\ .$$

The latter relation is obtained as follows: Recall that $B = \sum_a B_a \partial^a(\square + m^2)\varphi$ with some $B_a \in \mathcal{P}$ (3.4.1). Since solely those terms contribute to the vacuum state in which all $\partial^b\varphi$ are contracted, $\partial^a(\square + m^2)\varphi(x)$ must be contracted; however $\omega_0\big(\partial^a(\square+m^2)\varphi(x) \star \partial^b\varphi(y)\big) = 0$ and $\omega_0\big(\partial^b\varphi(y) \star \partial^a(\square+m^2)\varphi(x)\big) = 0$ for all $a, b \in \mathbb{N}^d$, due to $(\square+m^2)\Delta^+ = 0$. To derive the second assertion (3.4.18), we first mention that iterated use of (3.4.17) yields

$$\omega_0\big(\xi\pi(A_1)(x_1) \star \cdots \star \xi\pi(A_n)(x_n)\big) = \omega_0\big(A_1(x_1) \star \cdots \star A_n(x_n)\big)\ .$$

From this relation the assertion follows by means of Lemma 3.3.2(a).]

Other models. The definition of T^{on} (3.4.7)–(3.4.8) is of general validity; however, for other models the axioms defining the map ξ need to be modified and supplemented.

The model of *one complex scalar field* ϕ (introduced in Exap. 1.3.2) can be viewed as the model of two real scalar fields, φ_1 and φ_2, given by the real and imaginary part of ϕ, that is, $\phi = \varphi_1 + i\,\varphi_2$. $\mathcal{P}$, $\mathcal{P}_1$ and $\mathcal{J}_\mathcal{P}$ are modified: $\mathcal{P}$ is the complex $*$-algebra generated by $\partial^a\phi$ and $\partial^a\phi^*$ ($a \in \mathbb{N}^d$). $\mathcal{P}_1$ is the subspace of fields linear in ϕ and ϕ^* and their partial derivatives; and $\mathcal{J}_\mathcal{P}$ is the ideal in $\mathcal{P}$ generated by $(\square + m^2)\phi$ and $(\square + m^2)\phi^*$, hence $\mathcal{J}_\mathcal{P}^* = \mathcal{J}_\mathcal{P}$ and $\pi(A^*) = \pi(A)^*$. In addition, the mass dimension is given by $\dim \partial^a\phi := \dim \partial^a\phi^* := \frac{d-2}{2} + |a|$. With these explanations, the axioms (a)–(e) are well defined. The axiom

(iia) $\xi\pi(A^*) = \xi\pi(A)^*\ ,\ \forall A \in \mathcal{P}\ ,$

is added in order that ξ is a $*$-algebra homomorphism.

It follows $\xi\pi(\phi) = \phi$ and $\xi\pi(\phi^*) = \phi^*$. Namely, (d) and (e) imply $\xi\pi(\phi) = a\,\phi + b\,\phi^*$ with unknown $a, b \in \mathbb{C}$. From $(\xi\pi(\phi) - \phi) \in \mathcal{J}_{\mathcal{P}}$ we conclude $a = 1$, $b = 0$.

It is not known whether ξ is uniquely determined by these axioms. The most obvious solution is obtained from the (unique) map ξ_{real} of the real scalar field (treated above) by setting

$$\xi\pi(\partial^a\phi) = \xi_{\mathrm{real}}\pi_{\mathrm{real}}(\partial^a\varphi_1) + i\,\xi_{\mathrm{real}}\pi_{\mathrm{real}}(\partial^a\varphi_2)$$
$$= \partial^a\phi + \sum_{|b|\le|a|-2} c_b^a\,\partial^b(\square + m^2)\phi\ ,$$

where the coefficients $c_a^b \in \mathbb{R}$ are the ones defined in (3.4.14). It follows

$$\xi\pi(\partial^a\phi^*) = \partial^a\phi^* + \sum_{|b|\le|a|-2} c_b^a\,\partial^b(\square + m^2)\phi^*\ .$$

For the model of *one Dirac spinor field* and the model of N *gauge fields* $(A_a^\mu)_{a=1,\dots,N}$ in Feynman gauge (see Sect. 5.1.3), the axioms defining the map ξ and their solution are worked out in [20], for arbitrary values of the spacetime dimension d. In the Dirac spinor case the solution is unique.

3.4.2 Properties of the on-shell T-product

Basic axioms. The basic axioms for $T_n\colon \mathcal{F}_{\mathrm{loc}}^{\otimes n} \longrightarrow \mathcal{F}$ translate directly into essentially the same properties for T_n^{on} (3.4.7)–(3.4.8):

(i) **Linearity:** $T_n^{\mathrm{on}} : \mathcal{P}_0^{\otimes n} \longrightarrow \mathcal{D}'(\mathbb{M}^n, \mathcal{F}_0)$ is linear, because ξ, T_n and $\mathcal{F} \ni F \longmapsto F_0 \in \mathcal{F}_0$ are linear.

(ii) **Initial condition:**

$$T_1^{\mathrm{on}}\big(A_0(x)\big) = T_1\big(\xi\pi(A)(x)\big)_0 = \xi\pi(A)_0(x) = A_0(x)\ ,\quad \forall A \in \mathcal{P}\ ,$$

by using (3.4.12) in the last step.

Combining the definition of T_n^{on} (3.4.8) with the Symmetry axiom (iii) and the Causality axiom (iv) for T_n, respectively, we straightforwardly obtain:

(iii) **Symmetry:** For all $\tau \in S_n$ and $A_1, \dots, A_n \in \mathcal{P}$ it holds that

$$T_n^{\mathrm{on}}\big(A_{\tau 1,0}(x_{\tau 1}), \dots, A_{\tau n,0}(x_{\tau n})\big) = T_n^{\mathrm{on}}\big(A_{1,0}(x_1), \dots, A_{n,0}(x_n)\big)\ . \tag{3.4.19}$$

(iv) **Causality:** T_n^{on} satisfies causal factorization, that is, for all $A_{1,0}, \dots, A_{n,0}$ it holds that

$$T_n^{\mathrm{on}}\big(A_{1,0}(x_1), \dots, A_{n,0}(x_n)\big)$$
$$= T_k^{\mathrm{on}}\big(A_{1,0}(x_1), \dots, A_{k,0}(x_k)\big) \star T_{n-k}^{\mathrm{on}}\big(A_{k+1,0}(x_{k+1}), \dots, A_{n,0}(x_n)\big) \tag{3.4.20}$$

whenever $\{x_1, \dots, x_k\} \cap \big(\{x_{k+1}, \dots, x_n\} + \overline{V}_-\big) = \emptyset$.

Renormalization conditions. We turn to the discussion of the renormalization conditions:

(AWI) **Action Ward identity:** The term violating the AWI can be written as

$$\partial_x^\mu T_n^{\rm on}\big(A_0(x),\dots\big) - T_n^{\rm on}\big(\partial^\mu A_0(x),\dots\big) = T_n\big([\partial^\mu,\xi\pi](A)(x),\dots\big)_0\ ,$$

by using the AWI for T_n. The commutator $[\partial^\mu,\xi\pi]$ is in general nonvanishing, e.g., $[\partial^\mu,\xi\pi](\partial^\nu\varphi) = (g^{\mu\nu}/d)\,(\Box+m^2)\varphi$ by using the results (3.4.15)–(3.4.16). From (3.4.12) we get $\partial^\mu\xi\pi(A) - \partial^\mu A \in \partial^\mu\mathcal{J}_\mathcal{P} \subset \mathcal{J}_\mathcal{P}$ and $\xi\pi(\partial^\mu A) - \partial^\mu A \in \mathcal{J}_\mathcal{P}$, hence, $[\partial^\mu,\xi\pi](A) \in \mathcal{J}_\mathcal{P}$.

In the particular case $A = \partial^a\varphi$ the failure of the AWI can also be written as

$$\begin{aligned}\partial_x^a T_n^{\rm on}\big(\varphi_0(x),\dots\big) - T_n^{\rm on}\big(\partial^a\varphi_0(x),\dots\big) &= T_n\big([\partial^a,\xi\pi](\varphi)(x),\dots\big)_0\\ &= -\sum_{|b|\le|a|-2} c^a_b\,\partial^b(\Box+m^2)\,T_n\big(\varphi(x),\dots\big)_0\\ &= -\sum_{|b|\le|a|-2} c^a_b\,\partial^b(\Box+m^2)\,T_n^{\rm on}\big(\varphi_0(x),\dots\big)\ ,\end{aligned}$$

by using again the AWI for T_n, the result $\xi\pi(\varphi)=\varphi$ and (3.4.14).

So far, the $*$-operation (1.2.7), the field parity transformation α (1.10.6), and the action of the Poincaré group $\beta_{\Lambda,a}$ (3.1.35) are defined only on $\mathcal{F}$; however, in an obvious way they induce corresponding maps $*$, α, $\beta_{\Lambda,a}:\mathcal{F}_0\longrightarrow\mathcal{F}_0$:

$$(F_0)^* := (F^*)_0\ ,\quad \alpha(F_0) := (\alpha F)_0\ ,\quad \beta_{\Lambda,a}(F_0) := (\beta_{\Lambda,a}F)_0\ ,\quad \forall F\in\mathcal{F}\,. \tag{3.4.21}$$

Concerning the vacuum state ω_0, we recall from (2.6.7) that $\omega_0(F_0) := \omega_0(F)$ for all $F\in\mathcal{F}$.

(via) **Field parity:** The transformation $\alpha:\mathcal{F}\longrightarrow\mathcal{F}$ (defined in (1.10.6)) induces a linear map $\alpha:\mathcal{P}\longrightarrow\mathcal{P}$ which may be given by

$$(\alpha A)(x)(h) := A(x)(-h)\quad \forall h\in\mathcal{C}\ ,\ \ A\in\mathcal{P}\ ,$$

cf. (3.1.32). Writing A as in (1.3.9) we get

$$\begin{aligned}\alpha\big(\xi\pi(A)\big) &= \sum_n\sum_{a_1,\dots,a_n} c_{a_1\cdots a_n}\,\alpha\big(\xi\pi(\partial^{a_1}\varphi)\cdots\xi\pi(\partial^{a_n}\varphi)\big)\\ &= \sum_n\sum_{a_1,\dots,a_n} c_{a_1\cdots a_n}\,(-1)^n\,\xi\pi(\partial^{a_1}\varphi)\cdots\xi\pi(\partial^{a_n}\varphi)\\ &= \xi\pi(\alpha A)\ ,\end{aligned}$$

by using the defining properties (b) and (d) of ξ. Due to $\alpha\mathcal{J}_\mathcal{P} = \mathcal{J}_\mathcal{P}$, α is well defined on $\mathcal{P}_0$:

$$\alpha(\pi A) = \alpha A + \alpha\mathcal{J}_\mathcal{P} = \alpha A + \mathcal{J}_\mathcal{P} = \pi(\alpha A)\ ,\quad \text{that is,}\quad \alpha\circ\pi = \pi\circ\alpha\ ; \tag{3.4.22}$$

hence we get $\alpha \circ \xi = \xi \circ \alpha$. Using these results and $\alpha \circ T_n = T_n \circ \alpha^{\otimes n}$ we obtain

$$\begin{aligned}\alpha \circ T_n^{\mathrm{on}}(\pi A_1 \otimes \cdots \otimes \pi A_n)(x_1, \ldots, x_n) &= T_n\big(\xi\alpha\pi(A_1)(x_1), \ldots, \xi\alpha\pi(A_n)(x_n)\big)_0 \\ &= T_n^{\mathrm{on}}(\alpha\pi A_1 \otimes \cdots \otimes \alpha\pi A_n)(x_1, \ldots, x_n)\ ,\end{aligned}$$

that is,

$$\alpha \circ T_n^{\mathrm{on}} = T_n^{\mathrm{on}} \circ \alpha^{\otimes n}\ . \tag{3.4.23}$$

(vii) **Poincaré covariance:** Analogously to (3.4.22), the relation $b_\Lambda \mathcal{J}_{\mathcal{P}} = \mathcal{J}_{\mathcal{P}}$ implies $b_\Lambda \circ \pi = \pi \circ b_\Lambda$ for all $\Lambda \in \mathcal{L}_+^\uparrow$. Taking into account also the defining property (c) of ξ, we get

$$\beta_{\Lambda,a}\, \xi\pi(A)(x) = \big(b_\Lambda\, \xi\pi(A)\big)(\Lambda x + a) = \big(\xi \circ b_\Lambda(\pi A)\big)(\Lambda x + a)\ .$$

Using additionally $\beta_{\Lambda,a} \circ T_n = T_n \circ \beta_{\Lambda,a}^{\otimes n}$ we conclude that

$$\begin{aligned}\beta_{\Lambda,a} \circ T_n^{\mathrm{on}}(\pi A_1 \otimes \cdots \otimes \pi A_n)(x_1, \ldots, x_n) &= T_n\big(\xi \circ b_\Lambda(\pi A_1)\big)(\Lambda x_1 + a), \ldots\big)_0 \\ &= T_n^{\mathrm{on}}\big(b_\Lambda(\pi A_1) \otimes \cdots \otimes b_\Lambda(\pi A_n)\big)(\Lambda x_1 + a, \ldots, \Lambda x_n + a)\end{aligned} \tag{3.4.24}$$

for all $(\Lambda, a) \in \mathcal{P}_+^\uparrow$ and $A_1, \ldots, A_n \in \mathcal{P}$.

(x) **$\hbar$-dependence:** By using Translation covariance we may set

$$\begin{aligned}&t^{\mathrm{on}}(A_{1,0}, \ldots, A_{n,0})(x_1 - x_n, \ldots) \\ &\quad := \omega_0\Big(T_n^{\mathrm{on}}\big(A_{1,0}(x_1), \ldots, A_{n,0}(x_n)\big)\Big) \in \mathcal{D}'(\mathbb{R}^{d(n-1)})\ ,\end{aligned} \tag{3.4.25}$$

in analogy to (3.2.20). Inserting the definition of T_n^{on} into (3.4.25), we get

$$t^{\mathrm{on}}(A_{1,0}, \ldots, A_{n,0}) = t\big(\xi\pi(A_1), \ldots, \xi\pi(A_n)\big)\ . \tag{3.4.26}$$

From the defining properties (b) and (d) of ξ we conclude that, for any monomial $A \in \mathcal{P}$,

$$\xi\pi(A) = \sum_k B_k \quad \text{where all } B_k \text{ are monomials satisfying} \quad |B_k| = |A|\ .$$

With these preparations, linearity of $(A_1, \ldots, A_n) \longmapsto t(A_1, \ldots, A_n)$ and the $\hbar$-dependence axiom for T_n imply

$$t^{\mathrm{on}}(A_{1,0}, \ldots, A_{n,0}) \sim \hbar^{\sum_{j=1}^n |A_j|/2} \tag{3.4.27}$$

for all monomials $A_1, \ldots, A_n$ fulfilling $A_j \sim \hbar^0\ \forall j$.

To discuss the $*$-structure property of T^{on} and the Bogoliubov formula (3.3.41) for $(T^{\mathrm{on}}, R^{\mathrm{on}})$, we introduce the *on-shell S-matrix and its inverse* (w.r.t. the star product) and the *on-shell antichronological product* $\overline{T}^{\mathrm{on}} \equiv (\overline{T}_n^{\mathrm{on}})_{n\in\mathbb{N}^*}$. We first define the on-shell S-matrix as the generating functional of T_n^{on}:

$$\mathbf{S}^{\mathrm{on}}(A_0, g) \tag{3.4.28}$$
$$:= 1 + \sum_{n=1}^{\infty} \frac{i^n}{n!\,\hbar^n} \int dx_1 \cdots dx_n \; T_n^{\mathrm{on}}\big(A_0(x_1), \ldots, A_0(x_n)\big)\, g(x_1)\cdots g(x_n) \;;$$

similarly to T_n^{on}, $\mathbf{S}^{\mathrm{on}}$ cannot depend only on the local functional $A_0(g) \equiv A(g)_0$ (3.4.5). Inserting the definition of T_n^{on} (3.4.8), we obtain the relation to the off-shell S-matrix (Definition 3.3.8):

$$\mathbf{S}^{\mathrm{on}}(A_0, g) = \mathbf{S}\Big(\big(\xi\pi(A)\big)(g)\Big)_0 \;. \tag{3.4.29}$$

We now define the on-shell antichronological product $\overline{T}_n^{\mathrm{on}}$ to be a linear map $\overline{T}_n^{\mathrm{on}} : \mathcal{P}_0^{\otimes n} \longrightarrow \mathcal{D}'(\mathbb{M}^n, \mathcal{F}_0)$ which is symmetric in all arguments (similarly to (3.4.19)) and is given by

$$1 + \sum_{n=1}^{\infty} \frac{(-i)^n}{n!\,\hbar^n} \int dx_1 \cdots dx_n \; \overline{T}_n^{\mathrm{on}}\big(A_0(x_1), \ldots, A_0(x_n)\big)\, g(x_1)\cdots g(x_n)$$
$$:= \mathbf{S}^{\mathrm{on}}(A_0, g)^{\star -1} \tag{3.4.30}$$

for all $g \in \mathcal{D}(\mathbb{M})$, where the exponent ${}^{\star -1}$ denotes the inverse with respect to the star product in $\mathcal{F}_0$. We recall that the latter is defined by $F_0 \star G_0 := (F \star G)_0$ (see (2.6.4)). Using this definition and the relation (3.4.29), one verifies easily that

$$\mathbf{S}^{\mathrm{on}}(A_0, g)^{\star -1} = \Big(\mathbf{S}\Big(\big(\xi\pi(A)\big)(g)\Big)^{\star -1}\Big)_0 \,, \tag{3.4.31}$$

where on the r.h.s. ${}^{\star -1}$ denotes the inverse with respect to the star product in $\mathcal{F}$. Considering (3.4.31) to nth order in g and taking into account the definition of $\overline{T}_n$ (3.3.34), we conclude that $\overline{T}_n^{\mathrm{on}}$ is related to $\overline{T}_n$ in precisely the same way as T_n^{on} to T_n:

$$\overline{T}_n^{\mathrm{on}}\big(A_0(x_1), \ldots, A_0(x_n)\big) = \overline{T}_n\big(\xi\pi(A_1)(x_1), \ldots, \xi\pi(A_n)(x_n)\big)_0 \;. \tag{3.4.32}$$

Formula (3.3.37), which expresses $\overline{T}_n$ in terms of $\{T_k \,|\, 1 \leq k \leq n\}$, translates into

$$\overline{T}_n^{\mathrm{on}}\big(A_{1,0}(x_1), \ldots, A_{n,0}(x_n)\big) = \sum_{r=1}^{n} (-1)^{n-r} \tag{3.4.33}$$
$$\cdot \sum_{\substack{I_1 \sqcup \cdots \sqcup I_r = \{1,\ldots,n\} \\ |I_j| \geq 1}} T_{|I_1|}^{\mathrm{on}}\big(\otimes_{j_1 \in I_1} A_{j_1,0}(x_{j_1})\big) \star \cdots \star T_{|I_r|}^{\mathrm{on}}\big(\otimes_{j_r \in I_r} A_{j_r,0}(x_{j_r})\big) \,,$$

where "$\sqcup$" means the disjoint union, by using first (3.4.32), then (3.3.37) and $(F\star G)_0 = F_0\star G_0$, and finally again (3.4.32). Alternatively, the relation (3.4.33) can be derived directly from (3.4.30) by the usual inversion of a formal power series (cf. (A.1.4)), that is by proceeding as in Exer. 3.3.12.

(vib) **$*$-structure:** Since the coefficients of the polynomial $\xi\pi(A)\in\mathcal{P}$ are real, it holds that $(\xi\pi(A))(g)^* = (\xi\pi(A))(\overline{g})$. With this and the relations (3.4.31) and (3.4.29), the axiom $\mathbf{S}(F)^{\star-1} = \mathbf{S}(F^*)^*$ translates into

$$\mathbf{S}^{\mathrm{on}}(A_0,g)^{\star-1} = \Big(\mathbf{S}\Big(\big(\xi\pi(A)\big)(\overline{g})\Big)^*\Big)_0 = \Big(\mathbf{S}\Big(\big(\xi\pi(A)\big)(\overline{g})\Big)_0\Big)^* = \mathbf{S}^{\mathrm{on}}(A_0,\overline{g})^* \tag{3.4.34}$$

(for all $A\in\mathcal{P}$, $g\in\mathcal{D}(\mathbb{M})$), which can equivalently be written as

$$\overline{T}_n^{\mathrm{on}}\big(A_{1,0}(x_1),\ldots,A_{n,0}(x_n)\big) = T_n^{\mathrm{on}}\big(A_{1,0}(x_1),\ldots,A_{n,0}(x_n)\big)^* \tag{3.4.35}$$

for all $A_1,\ldots,A_n\in\mathcal{P}$ and all $n\in\mathbb{N}^*$.

In view of the discussion of the renormalization conditions Field independence and Field equation, we define the functional derivative in $\mathcal{F}_0$ in the same way as in $\mathcal{F}$ (Definition 1.3.1): Writing $\mathcal{F}_0\in F_0 = \sum_{n=0}^N\langle f_n,\varphi_0^{\otimes n}\rangle$ with f_n as in Definition 1.2.1 we define[75]

$$\frac{\delta F_0}{\delta\varphi_0(x)} := \sum_{n=1}^N n\int dy_1\cdots dy_{n-1}\; f_n(x,y_1,\ldots,y_{n-1})\,\varphi_0(y_1)\cdots\varphi_0(y_{n-1})\,. \tag{3.4.36}$$

An immediate and important consequence is

$$\frac{\delta F_0}{\delta\varphi_0(x)} = \Big(\frac{\delta F}{\delta\varphi(x)}\Big)_0\,. \tag{3.4.37}$$

Defining $\frac{\delta A_0(x)}{\delta\varphi_0(y)}$ analogously to (1.3.18), $\int dx\; g(x)\,\frac{\delta A_0(x)}{\delta\varphi_0(y)} := \frac{\delta A_0(g)}{\delta\varphi_0(y)}$ for all $g\in\mathcal{D}(\mathbb{M})$, the relation (3.4.37) yields

$$\frac{\delta A_0(x)}{\delta\varphi_0(y)} = \Big(\frac{\delta A(x)}{\delta\varphi(y)}\Big)_0 = \sum_{a\in\mathbb{N}^d}(\partial^a\delta)(x-y)\,\Big(\frac{\partial A}{\partial(\partial^a\varphi)}\Big)_0(x)\,, \tag{3.4.38}$$

the second equality is obtained by inserting (1.3.19).

[75]Note that it is a priori less obvious how to define the functional derivative in $\mathcal{F}/\mathcal{J}$. This is a main reason why we define T_n^{on} by (3.4.7)–(3.4.8) instead of $T_n^{\mathrm{on}}:\mathcal{P}_0^{\otimes n}\longrightarrow\mathcal{D}'(\mathbb{M}^n,\mathcal{F}/\mathcal{J})$ with

$$\langle T_n^{\mathrm{on}}((A_{1,0}(x_1),\ldots),g(x_1,\ldots,x_n)\rangle := \langle T_n(\xi\pi(A_1)(x_1),\ldots),g(x_1,\ldots,x_n)\rangle + \mathcal{J}\quad\forall g\in\mathcal{D}(\mathbb{M}^n)\,.$$

(v) **Field independence:** Using (3.4.37) and Field independence of T_n we get

$$\frac{\delta}{\delta\varphi_0(y)}\, T_n^{\mathrm{on}}\big(A_{1,0}(x_1),\ldots,A_{n,0}(x_n)\big)$$
$$= \sum_{k=1}^{n} T_n\Big(\xi\pi(A_1)(x_1),\ldots,\frac{\delta}{\delta\varphi(y)}\xi\pi(A_k)(x_k),\ldots\Big)_0 \,. \tag{3.4.39}$$

The problem is that in general $\frac{\delta}{\delta\varphi(y)}$ does not commute with $\xi\pi$, more precisely,

$$\frac{\delta}{\delta\varphi(y)}\xi\pi(A)(x) \neq \sum_{a\in\mathbb{N}^d} (\partial^a\delta)(x-y)\;\xi\pi\Big(\frac{\partial A}{\partial(\partial^a\varphi)}\Big)(x)\;,$$

by taking into account (1.3.19).

However, if A is a polynomial in φ and $\partial^\mu\varphi$ only, that is,

$$\frac{\partial A}{\partial(\partial^a\varphi)} = 0 \qquad \forall |a|\geq 2\;, \tag{3.4.40}$$

then we know from (3.4.15) that

$$\xi\pi\Big(\frac{\partial A}{\partial(\partial^a\varphi)}\Big) = \frac{\partial A}{\partial(\partial^a\varphi)} \qquad \forall |a|\leq 1 \tag{3.4.41}$$

and, hence,

$$\frac{\delta}{\delta\varphi(y)}\xi\pi(A)(x) = \frac{\delta A(x)}{\delta\varphi(y)} = \sum_{|a|\leq 1}(\partial^a\delta)(x-y)\,\frac{\partial A}{\partial(\partial^a\varphi)}(x)$$
$$= \sum_{|a|\leq 1}(\partial^a\delta)(x-y)\;\xi\pi\Big(\frac{\partial A}{\partial(\partial^a\varphi)}\Big)(x)\;. \tag{3.4.42}$$

Assuming that $A_1,\ldots,A_n$ *satisfy* (3.4.40), we may insert (3.4.42) into the r.h.s. of (3.4.39):

$$\frac{\delta}{\delta\varphi_0(y)}\, T_n^{\mathrm{on}}\big(A_{1,0}(x_1),\ldots,A_{n,0}(x_n)\big)$$
$$= \sum_{k=1}^{n}\sum_{|a|\leq 1}(\partial^a\delta)(x_k-y)\,T_n^{\mathrm{on}}\Big(A_{1,0}(x_1),\ldots,\Big(\frac{\partial A_k}{\partial(\partial^a\varphi)}\Big)_0(x_k),\ldots\Big)$$
$$= \sum_{k=1}^{n} T_n^{\mathrm{on}}\Big(A_{1,0}(x_1),\ldots,\frac{\delta A_{k,0}(x_k)}{\delta\varphi_0(y)},\ldots\Big)\;, \tag{3.4.43}$$

by applying (3.4.38) in the last equality sign. So, under this assumption, $T_n^{\mathrm{on}}\big(A_{1,0}(x_1),\ldots\big)$ satisfied Field independence and, hence, one can derive the causal Wick expansion by proceeding similarly to (3.1.23)–(3.1.24). We give a somewhat shorter derivation: On the r.h.s. of (3.4.8), we first use $\xi\pi(A_j) = A_j$

for all $1 \leq j \leq n$ and then we insert the causal Wick expansion (3.1.24) for $T_n\big(A_1(x_1), \ldots, A_n(x_n)\big)$. This yields

$$T_n^{\text{on}}\big(A_{1,0}(x_1), \ldots, A_{n,0}(x_n)\big) \tag{3.4.44}$$
$$= \sum_{\underline{A_l} \subseteq A_l} t(\underline{A}_1, \ldots, \underline{A}_n)(x_1 - x_n, \ldots)\, \overline{A}_{1,0}(x_1) \cdots \overline{A}_{n,0}(x_n)\ ,$$

where $\overline{A}_0 := \big(\overline{A}\big)_0$ for all monomials $A = A_1, \ldots, A_n$. Understanding A_0 as monomial in $\{\partial^a \varphi_0 \,|\, a \in \mathbb{N}^d\}$, the submonomials $\underline{(A_0)}$ and $\overline{(A_0)}$ can be defined analogously to (3.1.25) by throughout replacing A by A_0 and $\partial^a\varphi$ by $\partial^a\varphi_0$. With this and with the assumption (3.4.40), it obviously holds

$$\underline{A}_0 := \big(\underline{A}\big)_0 = \underline{(A_0)} \quad \text{and} \quad \overline{A}_0 := \big(\overline{A}\big)_0 = \overline{(A_0)}\ .$$

Having clarified this and using that our assumption implies also $\xi\pi(\underline{A}_j) = \underline{A}_j$, the relation (3.4.26) yields

$$t(\underline{A}_1, \ldots, \underline{A}_n) = t^{\text{on}}(\underline{A}_{1,0}, \ldots, \underline{A}_{n,0})\ ; \tag{3.4.45}$$

that is, we indeed get the causal Wick expansion for $T_n^{\text{on}}\big(A_{1,0}(x_1), \ldots\big)$.

We point out: If $A_1, \ldots, A_n$ contain higher derivatives of φ (more precisely, not all of them fulfill (3.4.40)), then it may be that $T^{\text{on}}(A_{1,0}(x_1), \ldots)$ violates Field independence (3.4.43) and the causal Wick expansion (3.4.44)–(3.4.45). Epstein–Glaser [66] give an axiomatic definition of an on-shell T-product,[76] $T_n^{\text{EG}} : \mathcal{P}_0^{\otimes n} \longrightarrow \mathcal{D}'(\mathbb{M}^n, \mathcal{F}_0)$, in which they require the causal Wick expansion for T_n^{EG} as a renormalization condition. So, it may happen that the renormalization prescriptions for T^{on} and T^{EG} are non-compatible! If one is interested only in an on-shell T-product, the procedure of Epstein–Glaser has the crucial advantage that $T_n^{\text{EG}}\big(A_{1,0}(x_1), \ldots, A_{1,0}(x_1)\big)$ is completely determined by the VEVs $\{t^{\text{EG}}(\underline{A}_{1,0}, \ldots, \underline{A}_{n,0})(x_1 - x_n, \ldots) \,|\, \underline{A}_{l,0} \subseteq A_{l,0}\}$ (where $t^{\text{EG}}(\cdots) := \omega_0\big(T_n^{\text{EG}}(\cdots)\big)$ similarly to (3.4.25)) and, hence, formulas (3.4.27) and (3.4.48) (with t^{EG} in place of t^{on}) can directly be used as formulation of the axioms $\hbar$-dependence and Scaling degree for T^{EG}.

(viii) **Field equation:** Writing (3.4.14) as

$$\xi\pi(\partial^a\varphi) =: \chi(\partial^a)\varphi\ , \quad \text{that is,} \quad \chi(\partial^a) := \partial^a + \sum_{|b| \leq |a|-2} c_b^a\, \partial^b(\Box + m^2)\ ,$$

the Off-shell field equation for T (3.3.20) translates into the following relation for T^{on} (3.4.8), by using also the AWI for T (in the first step) and (3.4.37)

[76] More precisely, by T^{EG} we mean here the translation of Epstein–Glaser's T-product, which is defined in Fock space, into our functional formalism. This translation is given by the algebra isomorphism Φ of Theorem 2.6.3, see (A.7.12).

(in the last step):

$$
\begin{aligned}
T_n^{\text{on}}\big(\partial^a\varphi_0(x), A_{1,0}(x_1),\ldots,A_{n-1,0}(x_{n-1})\big) &= \chi(\partial_x^a)\, T_n\big(\varphi(x), \xi\pi(A_1)(x_1),\ldots\big)_0 \\
&= \big(\chi(\partial^a)\varphi\big)_0(x)\; T_{n-1}\big(\xi\pi(A_1)(x_1),\ldots\big)_0 \\
&\quad + \hbar \int dy\; \chi(\partial_x^a)\Delta^F(x-y)\,\Big(\frac{\delta}{\delta\varphi(y)} T_{n-1}\big(\xi\pi(A_1)(x_1),\ldots\big)\Big)_0 \\
&= \big(\chi(\partial^a)\varphi\big)_0(x)\; T_{n-1}^{\text{on}}\big(A_{1,0}(x_1),\ldots,A_{n-1,0}(x_{n-1})\big) \\
&\quad + \hbar \int dy\; \chi(\partial_x^a)\Delta^F(x-y)\,\frac{\delta}{\delta\varphi_0(y)} T_{n-1}^{\text{on}}\big(A_{1,0}(x_1),\ldots,A_{n-1,0}(x_{n-1})\big)\;,
\end{aligned}
\tag{3.4.46}
$$

that is, the Field equation holds true also for T^{on}. If $A_1,\ldots,A_{n-1}$ satisfy (3.4.40), Field independence of T_{n-1}^{on} (3.4.43) can be used to shift the functional derivative $\frac{\delta}{\delta\varphi_0(y)}T_{n-1}^{\text{on}}(\cdots)$ (in the last term on the r.h.s. of (3.4.46)) to the arguments of T_{n-1}^{on}.

(ix) **Scaling degree:** In general, the Sm-expansion axiom does not translate into a simple property of t^{on} (3.4.25)–(3.4.26). Namely, assuming that A is a field monomial, the problem is that in general $\xi\pi(A)$ is a sum of monomials which have different mass dimensions, that is, $\xi\pi(A)\notin\mathcal{P}_{\text{hom}}$, see the examples given in (3.4.16). (Of course, this problem does not appear if A is a monomial in φ and $\partial^\mu\varphi$ only, because then $\xi\pi(A)=A$ due to (3.4.15).) For this reason we study here the axiom Scaling degree (3.3.21).

In the following we work with the mass dimension of field *polynomials* and the Scaling degree axiom for field *polynomials*; both are given in Remk. 3.2.24. Defining the mass dimension in $\mathcal{P}_0$ by

$$
\dim(\pi A)\equiv\dim(A_0):=\dim\big(\xi\pi(A)\big)\;,
\tag{3.4.47}
$$

the defining property (e) of ξ says $\dim(A_0)\le\dim A$. In particular, for $0\neq A\in\mathcal{J}_{\mathcal{P}}$, we have $\xi\pi(A)=0$ (3.4.10) and, hence, $\dim A_0=0<\dim A$.

By using (3.4.26) and the Scaling degree axiom for $t\big(\xi\pi(A_1),\ldots\big)$, we obtain

$$
\text{sd}\; t^{\text{on}}(A_{1,0},\ldots,A_{n,0})(x_1-x_n,\ldots)\le\sum_{j=1}^n\dim(A_{j,0})\;\Big(\le\sum_{j=1}^n\dim A_j\Big)\;.
\tag{3.4.48}
$$

On-shell *R*-product. In complete analogy to (3.4.7)–(3.4.8), we define the on-shell R-product $R^{\text{on}}\equiv(R_{n,1}^{\text{on}})_{n\in\mathbb{N}}$ in terms of the off-shell R-product:

$$
R_{n,1}^{\text{on}}\;:\;\begin{cases}\mathcal{P}_0^{\otimes n+1}\longrightarrow\mathcal{D}'(\mathbb{M}^{n+1},\mathcal{F}_0)\\ R_{n,1}^{\text{on}}\big(A_{1,0}(x_1),\ldots,A_{n,0}(x_n);A_0(x)\big)\\ \qquad:=R_{n,1}\big(\xi\pi(A_1)(x_1),\ldots,\xi\pi(A_n)(x_n);\xi\pi(A)(x)\big)_0\;,\end{cases}
\tag{3.4.49}
$$

for all $A_1,\ldots,A_n,A\in\mathcal{P}$ and all $n\in\mathbb{N}$.

$R^{\rm on}$ is related to $T^{\rm on}$ by the Bogoliubov formula (3.3.41). Namely, inserting (3.3.41) into the r.h.s. of (3.4.49), using $(F\star G)_0 = F_0 \star G_0$ and finally expressing $\overline{T}_{|I|}$ and $T_{|I^c|+1}$ by $T^{\rm on}_{|I|}$ and $T^{\rm on}_{|I^c|+1}$ by means of (3.4.32) and (3.4.8), respectively, we obtain

$$R^{\rm on}_{n,1}\big(A_{1,0}(x_1),\ldots,A_{n,0}(x_n);A_0(x)\big) = i^n \sum_{I\subseteq\{1,\ldots,n\}} (-1)^{|I|}$$
$$\cdot\, \overline{T}^{\rm on}_{|I|}\Big(\otimes_{i\in I}A_{i,0}(x_i)\Big) \star T^{\rm on}_{|I^c|+1}\Big(\big(\otimes_{j\in I^c}A_{j,0}(x_j)\big)\otimes A_0(x)\Big)\,. \tag{3.4.50}$$

Analogously to the above procedure for $(T, T^{\rm on})$, the basic axioms and most renormalization conditions for R (given in Sect. 3.1) can be translated into corresponding nice properties of $R^{\rm on}$. Alternatively, one may proceed analogously to Proposition 3.3.14(b): The above-found properties of $T^{\rm on}$ translate into corresponding properties of $R^{\rm on}$ by means of the Bogoliubov formula (3.4.50).

3.5 Techniques to renormalize in practice

In the traditional literature renormalization is usually done in *momentum space*. Since causal perturbation theory is an x-space construction of the R- (or T-) product,[77] and in view of the generalization to curved spacetimes, the topic of this section are two important *x-space* techniques.

Apart from the maintenance of symmetries (investigated in Sect. 3.2.7 and Chap. 4), we have reduced renormalization to the following mathematical problem by means of the Sm-expansion axiom:

Problem 3.5.1. *Given a distribution $t^0 \in \mathcal{D}'(\mathbb{R}^k\setminus\{0\})$ which scales almost homogeneously with degree $D\in\mathbb{C}$, find an extension $t\in\mathcal{D}'(\mathbb{R}^k)$ which scales also almost homogeneously with degree D.*

In Proposition 3.2.16 we have answered the existence and uniqueness of solutions t. A first method to explicitly compute t in practice, is given in Sect. 3.2.8, here we give two further techniques, which are of more general applicability.

3.5.1 Differential renormalization

Differential renormalization is due to [78, 120]. It was first applied to causal perturbation theory in [45, 139]; see in particular [86].

The method. Differential renormalization is a way to trace back the case $k \le \operatorname{Re} D < \infty$ to the simple case $\operatorname{Re} D < k$, in which the solution of Problem 3.5.1 is

[77] In the original work of Epstein and Glaser [66], renormalization is a distribution splitting, which can be interpreted as an extension problem in the sense of Theorem 3.2.8, see Exer. 3.5.7. This distribution splitting can be solved in momentum space by a dispersion integral [66] – this is a handy method for concrete computations, as demonstrated in [148].

unique and obtained by the direct extension (3.2.22), that is, the extension is given by the same formula as the non-extended distribution. The idea is to write t^0 as a derivative of a distribution $f^0 \in \mathcal{D}'(\mathbb{R}^k \setminus \{0\})$ which scales almost homogeneously with degree $D - l < k$, where $l \in \mathbb{N}^*$; more precisely

$$t^0 = \mathfrak{D} f^0 \quad \text{with} \quad \mathfrak{D} = \sum_{|a|=l} C_a \partial^a \ , \quad C_a \in \mathbb{C} \ , \quad \text{and} \quad (x\partial_x + D - l)^N f^0 = 0 \tag{3.5.1}$$

for $N \in \mathbb{N}$ sufficiently large, where $x\partial_x := \sum_{r=1}^k x_r \partial_{x_r}$. Let $f \in \mathcal{D}'(\mathbb{R}^k)$ be the direct extension of f^0; from Proposition 3.2.16 we know that f scales also almost homogeneously with the same degree $D - l$ and the same power. The term

$$t := \mathfrak{D} f \tag{3.5.2}$$

exists in $\mathcal{D}'(\mathbb{R}^k)$, since distributions are ∞-times differentiable. The so-constructed t is an extension of t^0, because for $h \in \mathcal{D}(\mathbb{R}^k \setminus \{0\})$ we have $\mathfrak{D}h \in \mathcal{D}(\mathbb{R}^k \setminus \{0\})$ and with that we get

$$\langle t, h \rangle = \langle \mathfrak{D} f, h \rangle = (-1)^l \langle f, \mathfrak{D} h \rangle = (-1)^l \langle f^0, \mathfrak{D} h \rangle = \langle \mathfrak{D} f^0, h \rangle = \langle t^0, h \rangle \ . \tag{3.5.3}$$

In addition, t scales almost homogeneously with degree D, since

$$(x\partial_x + D)^N t = (x\partial_x + D)^N \mathfrak{D} f = \mathfrak{D} (x\partial_x + D - l)^N f = 0 \ . \tag{3.5.4}$$

The non-uniqueness of t shows up in the non-uniqueness of f^0: One may add to f^0 a distribution $g^0 \in \mathcal{D}'(\mathbb{R}^k \setminus \{0\})$ which scales almost homogeneously with degree $D - l$ and with $\mathfrak{D} g^0 = 0$ on $\mathcal{D}(\mathbb{R}^k \setminus \{0\})$. The direct extension g of g^0 (which scales also almost homogeneously with degree $D - l$) fulfills $\operatorname{supp} \mathfrak{D} g \subseteq \{0\}$, that is, the addition of g^0 can change t (3.5.2) only by a local term. For example let $D = k$ and $\mathfrak{D} = \square$ (the wave operator in the k-dimensional Minkowski space). Then, g is of the form: $g = \alpha D^{\mathrm{ret}} +$[solution g_1 of $\square g_1 = 0$ which scales almost homogeneously with degree $k - 2$] ($\alpha \in \mathbb{C}$), and this yields $t_{\mathrm{new}} := \mathfrak{D}(f + g) = t_{\mathrm{old}} - \alpha\, \delta$.

Trying to solve Problem 3.5.1 by means of differential renormalization, the difficult step is to find a distribution $f^0 \in \mathcal{D}'(\mathbb{R}^k \setminus \{0\})$ satisfying the conditions (3.5.1). A general method to solve this problem is not known. For the extension from $\mathcal{D}'(\mathbb{M}_4 \setminus \{0\})$ to $\mathcal{D}'(\mathbb{M}_4)$ the following formulas are helpful:

Lemma 3.5.2. *In $\mathcal{D}'(\mathbb{M}_4 \setminus \{0\})$ it holds that*

$$\frac{1}{(x^2 - i0)^2} = -\square_x \frac{\log(-M^2(x^2 - i0))}{4\,(x^2 - i0)} \ , \tag{3.5.5}$$

where $M > 0$ is an arbitrary mass (which is needed in order that the argument of the logarithm is dimensionless), and [122]

$$\square_x (x^2 - i0)^\alpha = 4\alpha(\alpha + 1)\,(x^2 - i0)^{\alpha - 1} \quad \forall \alpha \in \mathbb{C} \quad (\textit{also valid in } \mathcal{D}'(\mathbb{M}_4 \setminus \{0\})) \tag{3.5.6}$$

and the same formulas for $(x^2 - ix^0 0)$ in place of $(x^2 - i0)$.

These formulas can be verified by straightforward computation of the derivatives. As we see from (A.2.10), the negative, integer powers of $(x^2 - i0)$ and $(x^2 - ix^0 0)$ are relevant for the renormalization of $D^F(x)^n$ and $r^0(\varphi^n, \varphi^n)(x) \sim \big(D^+(x)^n - D^+(-x)^n\big)\,\theta(-x^0)$, respectively; for the latter problem see Exer. 3.5.7.

Remark 3.5.3. [27, App. C] Formula (3.5.5) and its generalization to $\mathcal{D}'(\mathbb{M}_d \setminus \{0\})$, where $d \geq 4$ is even, can be derived systematically by proceeding in the following way: Setting $X := -(x^2 - i0)$ we search a distribution $F(X)$ such that

$$\square_x F(X) = X^{-\frac{d}{2}} \,. \tag{3.5.7}$$

Due to

$$\begin{aligned}\square_x F(X) &= \partial^x_\mu\big(-2x^\mu F'(X)\big) = -2d\,F'(X) - 4X\,F''(X)\\ &= \frac{-4}{X^{\frac{d}{2}-1}}\,\frac{d}{dX}\big(X^{\frac{d}{2}}\,F'(X)\big)\,,\end{aligned}$$

a first integration can easily be done:

$$F'(X) = -\frac{\log(M^2X) + a}{4\,X^{\frac{d}{2}}} \tag{3.5.8}$$

where $M > 0$ and $a \in \mathbb{C}$ are arbitrary, the choice of M and a is the choice of the integration constant. For the second integration we make the ansatz

$$F(X) = -C\,\frac{\log(M^2X)}{X^{\frac{d}{2}-1}}\,. \quad \text{Comparing} \quad F'(X) = -C\,\frac{1 + (1-\frac{d}{2})\,\log(M^2X)}{X^{\frac{d}{2}}}$$

with (3.5.8), we see that (3.5.7) is solved if we choose $C = \frac{1}{4-2d}$ and $a = 4C$. Summing up, we have derived

$$\frac{1}{4-2d}\,\square_x\,\frac{\log(-M^2(x^2 - i0))}{(x^2 - i0)^{\frac{d}{2}-1}} = \frac{1}{(x^2 - i0)^{\frac{d}{2}}} \quad \text{in } \mathcal{D}'(\mathbb{M}_d \setminus \{0\}). \qquad \boxminus \tag{3.5.9}$$

Examples and applications. The first example is an application of differential renormalization by means of Lemma 3.5.2, and it is also an application of the Sm-expansion axiom – we use the stronger version of this axiom, i.e., the addition (k′) given in Sect. 3.2.6.

Example 3.5.4 (Setting sun diagram for $m \geq 0$ in $d = 4$ dimensions). Cf. [64, Sect. 6]. The setting sun diagram is given in (3.2.124). For a change we work with time-ordered products. We have to extend

$$t^{m\,0}_{\text{ss}}(x) = \Delta^F_m(x)^3 \in \mathcal{D}'(\mathbb{M}_4 \setminus \{0\})\,, \quad x := x_1 - x_2\,, \tag{3.5.10}$$

where Δ^F_m is the Feynman propagator (2.3.8). We omit a combinatorial prefactor and factors $\hbar$, which are contained in the term belonging to the setting sun diagram. Using Exer. 3.1.24, the Sm-expansion of Δ^F_m can be written as

$$\Delta^F_m(x) = \frac{a}{X} + m^2\Big(\big(a_0\,\log(\frac{M^2X}{4}) + b_0\big) + 2a_0\,\log\frac{m}{M}\Big) + \mathfrak{R}^{(m)}_4(x)\,, \quad X := -(x^2 - i0)\,, \tag{3.5.11}$$

where

$$a = \frac{1}{4\pi^2}\,, \quad \text{and} \quad a_0 = \frac{1}{16\pi^2}\,, \quad b_0 = \frac{2\gamma - 1}{16\,\pi^2}$$

($\gamma = 0,5772\ldots$ is the Euler–Mascheroni constant) are the same numbers as in that exercise, and $M > 0$ is a mass scale. Inserting (3.5.11) into (3.5.10) we obtain the Sm-expansion (3.1.62) of $t_{\rm ss}^{m\,0}$; in order that the direct extension can be applied to the remainder we choose it such that it has scaling degree $= 2$:

$$t_{\rm ss}^{m\,0}(x) = u_0^0(x) + m^2\Big(u_{2,0}^0(x) + u_{2,1}^0(x)\,\log\frac{m}{M}\Big) + \mathfrak{r}_4^{(m)\,0}(x)\,, \tag{3.5.12}$$

where

$$u_0^0(x) = \frac{a^3}{X^3}\,, \qquad u_{2,0}^0(x) = \frac{3\,a^2\,(a_0\,\log(\frac{M^2X}{4}) + b_0)}{X^2}\,, \qquad u_{2,1}^0(x) = \frac{6\,a^2\,a_0}{X^2}\,,$$

$$\mathfrak{r}_4^{(m)}(x) = 3\Delta_m^F(x)^2\,\mathfrak{R}_4^{(m)}(x) + 3m^4\Big(a_0\,\log(\frac{m^2X}{4}) + b_0\Big)^2\frac{a}{X} + m^6\Big(a_0\,\log(\frac{m^2X}{4}) + b_0\Big)^3.$$

The property that only even powers of m appear, goes over from (3.5.11) to (3.5.12).

The non-direct, almost homogeneous extensions of $u_0^0(x)$, $u_{2,0}^0$ and $u_{2,1}^0$ can be computed by using differential renormalization – as renormalization mass scale we use the same mass M as in the Sm-expansion (3.5.12):

$$\begin{aligned}
u_0(x) &= a^3\,\Box_x\Box_x\Big(\overline{\frac{-\log(M^2X)}{32\,X}}\Big) + C\,\Box_x\delta(x),\\
u_{2,0}(x) &= 3\,a^2\,\Big[a_0\,\Box_x\Big(\overline{\frac{\log^2(M^2X) + 2\,\log(M^2X)}{8\,X}}\Big)\\
&\qquad + (b_0 - a_0\,\log 4)\,\Box_x\Big(\overline{\frac{\log(M^2X)}{4\,X}}\Big)\Big] + C_0\,\delta(x)\,,\\
u_{2,1}(x) &= 6\,a^2\,a_0\,\Box_x\Big(\overline{\frac{\log(M^2X)}{4\,X}}\Big) + C_1\,\delta(x)\,,
\end{aligned} \tag{3.5.13}$$

where C, C_0, $C_1 \in \mathbb{C}$ are arbitrary constants, in particular they do not depend on m. These formulas have to be understood as follows: For $x \neq 0$ (i.e., on $\mathcal{D}(\mathbb{M}_4 \setminus \{0\})$) the derivatives can straightforwardly be computed by means of formulas (3.5.5)–(3.5.6) and we obtain the corresponding $u_{\ldots}^0$-distributions. In addition, the expressions in $(\ldots)$-brackets scale almost homogeneously with degree 2. Hence, the assumptions (3.5.1) are satisfied. The overline denotes the direct extension. So formulas (3.5.13) are examples for (3.5.2). Homogeneous scaling of $u_{2,1}^0$ and u_0^0 is broken, and the power of the almost homogeneous scaling of $u_{2,0}^0$ is increased from 1 to 2.

We end up with the following result for the renormalized setting sun diagram:

$$t_{\rm ss}^{m,M}(x) = u_0(x) + m^2\Big(u_{2,0}(x) + u_{2,1}(x)\,\log\frac{m}{M}\Big) + \mathfrak{r}_4^{(m)}(x) \in \mathcal{D}'(\mathbb{M}_4)\,, \tag{3.5.14}$$

where $\mathfrak{r}_4^{(m)}$ is the direct extension of $\mathfrak{r}_4^{(m)\,0}$.

We point out that, with the renormalization prescription used here, which is the strong version of the Sm-expansion axiom, homogeneous scaling of the unrenormalized distribution,

$$\rho^6\,t_{\rm ss}^{m/\rho\,0}(\rho x) = t_{\rm ss}^{m\,0}(x)\,,$$

is broken by terms $\sim \log\rho$ and terms $\sim \log^2\rho$:

$$\rho^6\, t_{\rm ss}^{m/\rho,\,M}(\rho x) - t_{\rm ss}^{m,\,M}(x) = K_1\, m^2\, \delta(x)\, \log^2\rho$$
$$+ \Big(K_2\,\Box\delta(x) + m^2\,\big((K_3 - C_1) + K_4\,\log\frac{m}{M}\big)\,\delta(x)\Big)\,\log\rho\ , \qquad (3.5.15)$$

with numbers K_j which can be computed – this is done in Exer. 3.5.5; the contribution of the renormalization constant C_1 (introduced in (3.5.13)) is written explicitly.

Note that choosing $M := m$ for both, the mass scale of the Sm-expansion and the renormalization mass scale, homogeneous scaling is preserved, because then the mass scale is also scaled:

$$\rho^6\, t_{\rm ss}^{m/\rho,\,m/\rho}(\rho x) = t_{\rm ss}^{m,\,m}(x)\ . \qquad (3.5.16)$$

But this choice *violates the Sm-expansion axiom*, because $t_{\rm ss}^{m,m}$ diverges for $m \downarrow 0$ – this is a further example for Remk. 3.2.35.

The non-uniqueness of (3.5.14) reads: $\big(C\,\Box_x + m^2\,(C_0 + C_1\,\log\frac{m}{M})\big)\,\delta(x)$. If we would work with the weaker version of the Sm-expansion axiom (axiom (k) without (k′)), the term $C\,\Box_x\delta(x)$ would not be changed, but $(C_0 + C_1\,\log\frac{m}{M})$ would be replaced by an arbitrary polynomial in $\log\frac{m}{M}$.

Exercise 3.5.5 (Breaking of homogeneous scaling for differential renormalization).

(a) Verify the following: The above results (3.5.13) for $u_{2,1} \equiv u_{2,1}^M$ and $u_0 \equiv u_0^M$ have the property that a change of M amounts to a change of the indeterminate parameter C_1 or C, respectively. In addition, derive the following results for the breaking of homogeneous scaling:

$$\rho^4\, u_{2,1}^M(\rho x) - u_{2,1}^M(x) = -12\,i\,\pi^2\, a^2 a_0\, \log\rho\; \delta(x)\,,$$
$$\rho^6\, u_0^M(\rho x) - u_0^M(x) = a^3\,\frac{i\pi^2}{4}\,\log\rho\;\Box\delta(x)\,. \qquad (3.5.17)$$

(b) Compute $\rho^6\, t_{\rm ss}^{m/\rho,\,M}(\rho x) - t_{\rm ss}^{m,\,M}(x)$, that is, the numbers K_j introduced on the r.h.s. of (3.5.15). In particular verify the cancellation of the nonlocal terms coming from the scaling of $m^2\, u_{2,0}$ and $m^2\,\log\frac{m}{M}\, u_{2,1}$, respectively.

[*Solution*: (a) The key to this exercise is the formula

$$\Box\left(\overline{\frac{1}{X}}\right) = 4\pi^2\,\Box D^F(x) = -4i\pi^2\,\delta(x)\ , \qquad (3.5.18)$$

which follows from (A.2.10) and (A.2.9). With that we get

$$u_{2,1}^{M_2}(x) - u_{2,1}^{M_1}(x) = 3\,a^2 a_0\,\log\frac{M_2}{M_1}\,\Box\left(\overline{\frac{1}{X}}\right) = -12\,i\,\pi^2\,a^2 a_0\,\log\frac{M_2}{M_1}\,\delta(x)$$
$$u_0^{M_2}(x) - u_0^{M_1}(x) = -\frac{a^3}{16}\,\log\frac{M_2}{M_1}\,\Box\Box\left(\overline{\frac{1}{X}}\right) = a^3\,\frac{i\pi^2}{4}\,\log\frac{M_2}{M_1}\,\Box\delta(x)\ .$$

The scaling relations (3.5.17) are obtained by taking into account $\rho^D\, u_{\dots}^M(\rho x) - u_{\dots}^M(x) = u_{\dots}^{\rho M}(x) - u_{\dots}^M(x)$, where $D = 4$ or $D = 6$, respectively.
(b) From the scaling behaviour of u_0 (3.5.17) we immediately get

$$K_2 = a^3\,\frac{i\pi^2}{4} = \frac{i}{\pi^4\, 2^8}\ .$$

In addition we obtain

$$\log \frac{m}{\rho M}\, \rho^4\, u_{2,1}(\rho x) - \log \frac{m}{M}\, u_{2,1}(x) = \log \frac{m}{M}\, \big(\rho^4\, u_{2,1}(\rho x) - u_{2,1}(x)\big) - \log \rho\, \rho^4\, u_{2,1}(\rho x)$$

$$= \log \frac{m}{M}\, (-12) i\pi^2 a^2 a_0\, \log \rho\, \delta(x) - \log \rho\, 6a^2 a_0 \Big[\Box_x \Big(\overline{\frac{\log(M^2 X)}{4X}}\Big) + 2\log \rho\, \Box_x \Big(\overline{\frac{1}{4X}}\Big)\Big] \tag{3.5.19}$$

and

$$\rho^4\, u_{2,0}(\rho x) - u_{2,0}(x) = 3a^2\big(a_0(1-\log 4) + b_0\big)(-2i)\,\pi^2 \log \rho\, \delta(x)$$
$$+3a^2 a_0 \Big[4\, \log^2 \rho\, \Box_x \Big(\overline{\frac{1}{8X}}\Big) + 4\, \log \rho\, \Box_x \Big(\overline{\frac{\log(M^2 X)}{8\,X}}\Big)\Big]\,. \tag{3.5.20}$$

Now we insert (3.5.18), if this is not yet done. In the sum of (3.5.19) and (3.5.20) the nonlocal terms, i.e., the terms $\sim \Box_x \Big(\overline{\frac{\log(M^2 X)}{X}}\Big)$, cancel indeed; and we can read off

$$K_4 = (-12) i\,\pi^2\, a^2 a_0 = \frac{-3i}{2^6\,\pi^4}\,, \quad K_1 = i\pi^2 a^2 a_0\, (6\cdot 2 - 3\cdot \tfrac{4}{2}) = \frac{3i}{2^7\,\pi^4}\,,$$
$$K_3 = -6i\pi^2\, a^2\big(a_0(1-\log 4) + b_0\big) = \frac{3i}{2^6\,\pi^4}\,(\log 2 - \gamma)\ .]$$

If in the Problem 3.5.1 the given distribution t^0 is of the form

$$t^0(x) = P(x)\, t^0_1(x) \in \mathcal{D}'(\mathbb{R}^k \setminus \{0\})\,, \quad P(x) = \sum_{|a|=p} C_a x^a \quad \text{for some} \quad p \in \mathbb{N}^*\,,$$

and if a distribution f^0 is known for t^0_1, i.e., f^0 fulfills the conditions (3.5.1) with t^0_1 in place of t^0 in particular $\mathfrak{D} f^0 = t^0_1$, then we can proceed as follows: With f being the direct extension of f^0, the distribution

$$t(x) := P(x)\, \mathfrak{D} f(x) \tag{3.5.21}$$

is well defined in $\mathcal{D}'(\mathbb{R}^k)$ and it solves the Problem 3.5.1 for t^0. The solution of the following exercise illustrates this procedure. A further example is given in the solution of Exer. 5.1.28.

Exercise 3.5.6. Compute the most general extension $t^{\mu\nu} \in \mathcal{D}'(\mathbb{M}_4)$ of

$$t^{\mu\nu\,0}(x) := \partial^\mu \Delta^F_m(x)\, \partial^\nu \Delta^F_m(x) \in \mathcal{D}'(\mathbb{M}_4 \setminus \{0\})\,,$$

which satisfies Lorentz covariance and the stronger version of the Sm-expansion axiom.

[*Solution*: Using (3.5.11) we get the Sm-expansion

$$\partial^\mu \Delta^F_m(x) = \frac{2a\, x^\mu}{X^2} - m^2 \frac{2a_0\, x^\mu}{X} + \partial^\mu \mathfrak{R}^{(m)}_4(x)\,, \quad X := -(x^2 - i0)\,, \quad \mathrm{sd}(\partial^\mu \mathfrak{R}^{(m)}_4) = -1\,,$$

in which $\log \frac{m}{M}$ appears only in the remainder $\partial^\mu \mathfrak{R}^{(m)}_4$; the latter holds also for the Sm-expansion of $t^{\mu\nu\,0}$:

$$t^{\mu\nu\,0}(x) = \frac{4a^2\, x^\mu x^\nu}{X^4} - m^2 \frac{8aa_0\, x^\mu x^\nu}{X^3} + \mathfrak{r}^{\mu\nu\,0}_m(x)\,,$$

where $\mathrm{sd}(\mathfrak{t}_m^{\mu\nu\,0}) = 2 < 4$. By means of Lemma 3.5.2 we obtain the following relations in $\mathcal{D}'(\mathbb{M}_4 \setminus \{0\})$:

$$\frac{x^\mu x^\nu}{X^3} = -\frac{x^\mu x^\nu}{8} \Box_x \Box_x \Big(\frac{\log(M^2 X)}{4\,X}\Big) ,$$
$$\frac{x^\mu x^\nu}{X^4} = \frac{x^\mu x^\nu}{192} \Box_x \Box_x \Box_x \Big(\frac{\log(M^2 X)}{4\,X}\Big) .$$

The version (3.5.21) of differential renormalization yields the extension

$$t^{\mu\nu}(x) = \Big(\frac{a^2\, x^\mu x^\nu}{48} \Box_x \Box_x \Box_x + m^2\, a a_0\, x^\mu x^\nu\, \Box_x \Box_x\Big) \overline{\Big(\frac{\log(M^2 X)}{4\,X}\Big)} + \mathfrak{t}_m^{\mu\nu}(x)$$
$$+ \big(C_0\, g^{\mu\nu}\, \Box_x + C_1 \partial_x^\mu \partial_x^\nu + m^2\, C_2\, g^{\mu\nu}\big)\, \delta(x) , \quad a = 4a_0 = \frac{1}{4\pi^2} . \qquad (3.5.22)$$

of $t^{\mu\nu\,0}$, where $\mathfrak{t}_m^{\mu\nu}$ is the direct extension of $\mathfrak{t}_m^{\mu\nu\,0}$ and $C_0, C_1, C_2 \in \mathbb{C}$ are arbitrary. This is the most general solution fulfilling the required axioms.]

Exercise 3.5.7 (Differential renormalization for the distribution splitting method). The distribution splitting method of Epstein and Glaser [66] is a construction of the time-ordered product $(T_n)_{n\in\mathbb{N}^*}$ by induction on n. In the inductive step from n to $(n+1)$, renormalization amounts to the following problem for the coefficients in the causal Wick expansion: Let $j \in \mathcal{D}'(\mathbb{R}^{dn})$ be given with $\mathrm{supp}\, j \subseteq \big((\overline{V}_+)^{\times n} \cup (\overline{V}_-)^{\times n}\big)$ and let j scale almost homogeneously with degree D. Find a splitting

$$j = r - a , \quad r, a \in \mathcal{D}'(\mathbb{R}^{dn}) , \quad \text{such that} \quad \mathrm{supp}\, r \subseteq (\overline{V}_-)^{\times n} , \ \mathrm{supp}\, a \subseteq (\overline{V}_+)^{\times n} \qquad (3.5.23)$$

and that r and a scale also almost homogeneously with degree D.

For simplicity we study only the inductive step from $n = 1$ to 2. For this value of n, the splitting problem (3.5.23) agrees with the renormalization problem in our construction of the retarded product $R_{1,1}$ from $R_{0,1}$, since the latter can be interpreted as the splitting problem $\hbar\, j_{0,2}(y) = r_{1,1}(y) - r_{1,1}(-y)$ for a given $j_{0,2}$ with $\mathrm{supp}\, j_{0,2} \subseteq (\overline{V}_+ \cup \overline{V}_-)$ and an $r_{1,1}$ being searched for with $\mathrm{supp}\, r_{1,1} \subseteq \overline{V}_-$, see Sect. 3.1.2. However, we point out that for higher n, the splitting problem resulting from the GLZ relation and Causality (axioms (e) and (d)), explicitly $\hbar\, j_{n-1,2}(y_1, \ldots, y_{n-1}; y_n) = r_{n,1}(y_1, \ldots, y_n) - r_{n-1,1}(y_1 - y_n, \ldots, y_{n-1} - y_n, -y_n)$, differs from the splitting problem (3.5.23) – the support properties are different.

(a) Show that
$$r^0(y) := j(y)\, \theta(-y^0) \quad \text{and} \quad a^0 := r^0 - j$$
solve the splitting problem (3.5.23) in $\mathcal{D}'(\mathbb{R}^d \setminus \{0\})$.

With that the splitting problem is traced back to finding an extension $r \in \mathcal{D}'(\mathbb{R}^d)$ of r^0 which preserves the almost homogeneous scaling. The pertinent $a \in \mathcal{D}'(\mathbb{R}^d)$ is then obtained by $a := r - j$.

(b) Assume we have found a $J \in \mathcal{D}'(\mathbb{R}^d)$ which scales almost homogeneously with degree
$$D - l < d \quad \text{and with} \quad \mathrm{supp}\, J \subset (\overline{V}_+ \cup \overline{V}_-)$$
and
$$j = \mathfrak{D}\, J \quad \text{on} \quad \mathcal{D}(\mathbb{R}^d \setminus \{0\}) , \quad \text{where} \quad \mathfrak{D} = \sum_{|b|=l} C_b \partial^b$$

for some numbers $C_b \in \mathbb{C}$. Prove that

$$r(y) := \mathfrak{D}_y\big(\overline{J(y)\,\theta(-y^0)}\big) \tag{3.5.24}$$

exists in $\mathcal{D}'(\mathbb{R}^d)$ and is an extension of r^0 with the required scaling property. (The overline denotes the direct extension.)

(c) To give an explicit example, let $d=4$, $a \in \mathbb{N}$ and

$$j(y) := \Big(\frac{\log^a(-M^2Y)}{Y^2} - (y \mapsto -y)\Big)\,, \quad Y := y^2 - iy^0 0\,,$$

where $M > 0$ is a mass scale. The cases $a = 0$ and $a = 1$ appear in the Sm-expansion of the setting sun diagram, see Exap. 3.5.4 and [55, App. B]. Verify that $\operatorname{supp} j \subset (\overline{V}_+ \cup \overline{V}_-)$.

(d) For the example of part (c), we have $D = 4$; hence, the most simple Lorentz-invariant choice for $\mathfrak{D}$ is $\mathfrak{D} = \square$. With that, compute a possible J for the particular value $a = 2$, by using the ansatz

$$J(y) = \frac{P\big(\log(-M^2Y)\big)}{Y} - (y \mapsto -y)\,,$$

where $P(x)$ is a polynomial in x.

[*Solution*: (a) The pointwise product $j(y)\,\theta(-y^0)$ exists in $\mathcal{D}'(\mathbb{R}^d \setminus \{0\})$ because the overlapping singularities of j and $\theta(-y^0)$ are in $(\overline{V}_+ \cup \overline{V}_-) \cap \{y \,|\, y^0 = 0\} = \{0\}$. The asserted support and scaling properties of r^0 and a^0 can easily be verified.
(b) The pointwise product $J(y)\,\theta(-y^0)$ exists in $\mathcal{D}'(\mathbb{R}^d \setminus \{0\})$ for the same reason as for $j(y)\,\theta(-y^0)$. Obviously, $J(y)\,\theta(-y^0)$ scales also almost homogeneously with degree $D-l<d$; hence, the direct extension can be applied and the so-constructed $r(y)$ (3.5.24) exists in $\mathcal{D}'(\mathbb{R}^d)$.
In $\mathcal{D}'(\mathbb{R}^d \setminus \{0\})$ we have

$$r^0(y) = \theta(-y^0)\;\mathfrak{D}J(y) = \mathfrak{D}\big(\theta(-y^0)J(y)\big)\,,$$

since $\operatorname{supp} \partial^b\theta(-y^0) \subseteq \{y \,|\, y^0 = 0\}$ for $|b| \geq 1$. With $f^0(y) := \theta(-y^0)J(y)$, $t^0 := r^0$ and $t := r$ the relations (3.5.1) and (3.5.2) are satisfied, and from (3.5.3) and (3.5.4) we know that r is an extension of r^0 and scales almost homogeneously with degree D.
(c) For $y^2 < 0$ we have $\frac{\log^a(-M^2Y)}{Y^2} = \frac{\log^a(M^2\,|y^2|)}{(y^2)^2}$ which is even in y; hence, $j(y) = 0$.
(d) The given ansatz for J scales almost homogeneously with degree $2 = D - l$ (since $l = 2$) and it fulfills $\operatorname{supp} J \subset (\overline{V}_+ \cup \overline{V}_-)$ for the same reason as for j, see the solution of part (c). r^0 scales almost homogeneously with power $N = a = 2$; hence for the extension r this power is $\leq N+1 = 3$ (Prop. 3.2.16). Therefore, it suffices to consider polynomials P which are of degree ≤ 3. Taking this into account when inserting the given ansatz for J into $j(y) = \square J(y)$ (where $y \neq 0$) and using

$$\square_y \frac{\log^a(-M^2Y)}{Y} = \frac{-4a\,\log^{a-1}(-M^2Y)) + 4a(a-1)\,\log^{a-2}(-M^2Y)}{Y^2} \quad \text{for} \quad y \neq 0$$

and $a \in \mathbb{N}$, we obtain

$$J = \frac{-\log^3(-M^2Y) - 3\,\log^2(-M^2Y) - 6\,\log(-M^2Y)}{12\,Y} - (y \mapsto -y)\,.]$$

Extension of distributions scaling homogeneously with a non-integer degree. Finally we give an application of differential renormalization, which will be important in the next section (Sect. 3.5.2): If, in Problem 3.5.1, $t^0 \in \mathcal{D}'(\mathbb{R}^k \setminus \{0\})$ scales *homogeneously with a non-integer degree* D, more precisely

$$\sum_{r=1}^{k} x_r \, \partial_r t^0 = -D\, t^0 \ , \quad D \in \mathbb{C} \setminus (k + \mathbb{N}) \ , \tag{3.5.25}$$

where $x := (x_1, \ldots, x_k)$ and $\partial_r := \partial_{x_r}$, then the unique, homogeneous extension $t \in \mathcal{D}'(\mathbb{R}^k)$ (see Prop. 3.2.16) can be computed by means of the following lemma and differential renormalization [63].[78]

Lemma 3.5.8 ([63]). *The scaling relation* (3.5.25) *implies the formula*

$$t^0 = \frac{1}{\prod_{j=0}^{n-1}(k + j - D)} \sum_{r_1 \ldots r_n} \partial_{r_1} \ldots \partial_{r_n} \Big(x_{r_1} \cdots x_{r_n} \, t^0 \Big) \ , \quad \forall n \in \mathbb{N} \ . \tag{3.5.26}$$

Proof. We prove this by induction on n. The case $n = 1$ is the scaling relation (3.5.25) written in the form

$$\sum_{r=1}^{k} \partial_r (x_r \, t^0) = (k - D)\, t^0 \ . \tag{3.5.27}$$

Assuming (3.5.26) to hold true for $n \leq l$, we take into account that $x_{r_1} \cdots x_{r_l} \, t^0$ is homogeneous with degree $(D - l)$ (i.e., it satisfies (3.5.27) correspondingly modified) and obtain

$$\begin{aligned} \sum_{r_1 \ldots r_{l+1}} \partial_{r_1} \ldots \partial_{r_{l+1}} \Big(x_{r_1} \cdots x_{r_{l+1}} \, t^0 \Big) &= \sum_{r_1 \ldots r_l} \partial_{r_1} \ldots \partial_{r_l} \Big(\sum_r \partial_r \, x_r \, \Big(x_{r_1} \cdots x_{r_l} \, t^0 \Big) \Big) \\ &= (k - (D - l)) \sum_{r_1 \ldots r_l} \partial_{r_1} \ldots \partial_{r_l} \Big(x_{r_1} \cdots x_{r_l} \, t^0 \Big) \\ &= (k + l - D) \Big(\prod_{j=0}^{l-1} (k + j - D) \Big) \, t^0 \ , \end{aligned}$$

which is (3.5.26) for $n = l + 1$. □

As we see from the proof, this lemma holds also for distributions $f \in \mathcal{D}'(\mathbb{R}^k)$ fulfilling the scaling relation (3.5.25).

---

[78] For $\mathrm{Re}\, D < k$ we know from Prop. 3.2.16 that t can be computed by the direct extension. Nevertheless, we include the case $\mathrm{Re}\, D < k$, because Lemma 3.5.8 holds also in this case. Note that the extension formula (3.5.30), which we are going to derive, reduces to the direct extension for $\mathrm{Re}\, D < k$.

To construct the unique, homogeneous extension $t \in \mathcal{D}'(\mathbb{R}^k)$ of $t^0 \in \mathcal{D}'(\mathbb{R}^k \setminus \{0\})$, let $\omega \in \mathbb{Z}$ be the minimal integer fulfilling

$$\omega > \operatorname{Re} D - k - 1 \ . \tag{3.5.28}$$

With that we get

$$\operatorname{sd}\big(x_{r_1} \cdots x_{r_{\omega+1}}\, t^0\big) \le \operatorname{Re} D - (\omega+1) < k \ , \tag{3.5.29}$$

by using Exer. 3.2.7(c). Hence, $x_{r_1} \cdots x_{r_{\omega+1}}\, t^0$ can be uniquely extended to a homogeneous distribution $\overline{x_{r_1} \cdots x_{r_{\omega+1}}\, t^0} \in \mathcal{D}'(\mathbb{R}^k)$ by the direct extension. Using (3.5.26) with $n = \omega + 1$ and differential renormalization, the solution t is obtained by

$$t = \frac{1}{\prod_{j=0}^{\omega}(k+j-D)} \sum_{r_1 \ldots r_{\omega+1}} \partial_{r_1} \ldots \partial_{r_{\omega+1}} \Big(\overline{x_{r_1} \cdots x_{r_{\omega+1}}\, t^0}\Big) \ . \tag{3.5.30}$$

It is now clear, why the assumption $D \notin k + \mathbb{N}$ is needed.

The extension method (3.5.30) can be generalized to a large class of *almost* homogeneously scaling distributions, provided we still have $D \in \mathbb{C} \setminus (k + \mathbb{N})$ – we explain this in Exap. 3.5.16.

The Sm-expansion and differential renormalization can be used to compute lowest-order diagrams of QED, this is done in Sect. 5.1.7.

3.5.2 Analytic regularization

Analytic regularization was introduced by Speer [154] in the context of BPHZ renormalization (i.e., in momentum space), and first applied to x-space Epstein–Glaser renormalization by Hollands [105]. Essentially, we follow [63].

The method. For massless 1-loop diagrams[79] (or more generally for massless, primitive divergent diagrams, see Exap. 3.5.14), the distribution $t^0 \in \mathcal{D}'(\mathbb{R}^k \setminus \{0\})$ of Problem 3.5.1 scales *homogeneously*, because $D^F \equiv \Delta^F_{m=0}$ scales homogeneously. But even for these diagrams, there is a snag to formula (3.5.30): The degree D of the homogeneous scaling is an *integer*, $D \in \mathbb{N}$, as explained in Sect. 3.1.5. Hence, for $D \ge k$, formula (3.5.30) cannot be applied.

The basic idea of analytic regularization to solve Problem 3.5.1 is the following: Given a distribution $t^0 \in \mathcal{D}'(\mathbb{R}^k \setminus \{0\})$ which scales almost homogeneously with a degree $D \in k + \mathbb{N}$ and with power $N \in \mathbb{N}$, one introduces a ζ-dependent "regularized" distribution $t^{\zeta\,0} \in \mathcal{D}'(\mathbb{R}^k \setminus \{0\})$ (where $\zeta \in \mathbb{C} \setminus \{0\}$ and $|\zeta|$ sufficiently

[79] A "n-loop diagram" is a loop diagram containing n independent loops (i.e., n independent closed lines, see Footn. 34). For example, the setting sun diagram (given in (3.2.124)) is a 2-loop diagram, although it contains three loops: , and ; but the union of two of them contains the third as a subset.

small), such that $\lim_{\zeta\to 0} t^{\zeta\,0} = t^0$ (in $\mathcal{D}'(\mathbb{R}^k \setminus \{0\})$) and that $t^{\zeta\,0}$ scales almost homogeneously with a *non-integer degree* D_ζ (which is of the form $D_\zeta = D + D_1\,\zeta$ for some constant $D_1 \in \mathbb{C} \setminus \{0\}$), and with the same power N as t^0.

From Prop. 3.2.16 we know that $t^{\zeta\,0}$ has a unique extension $t^\zeta \in \mathcal{D}'(\mathbb{R}^k)$ which scales almost homogeneously with the same degree D_ζ and the same power N. The explicit computation of t^ζ is mostly much simpler than the computation of a solution of the original extension task (Problem 3.5.1) – this is the gain of the regularization. Namely, for $N = 0$, t^ζ can be computed by formula (3.5.30), and for $N \geq 1$ a further development of that formula does mostly the job, as explained in Exap. 3.5.16.

For the resulting extension t^ζ one then removes the regularization, i.e., one performs the limit $\zeta \to 0$. In order that this limit exists in $\mathcal{D}'(\mathbb{R}^k)$, one has to subtract a suitable local term s^ζ, that is, s^ζ is a linear combination of derivatives of the δ-distribution, which diverges for $\zeta \to 0$. The subtraction term s^ζ should be such that the limit $t := \lim_{\zeta\to 0}(t^\zeta - s^\zeta)$ is an almost homogeneous extension $t \in \mathcal{D}'(\mathbb{R}^k)$ of t^0.

An example for an analytic regularization is obtained by *dimensional regularization of the Feynman propagator*; this is worked out in [63]. We give here a related example for an analytic regularization.

Example 3.5.9. For a Lorentz covariant $t^0(y_1, \ldots, y_l) \in \mathcal{D}'(\mathbb{R}^{dl} \setminus \{0\})$ a simple way to obtain a Lorentz covariant regularization $t^{\zeta\,0} \in \mathcal{D}'(\mathbb{R}^{dl} \setminus \{0\})$ is

$$t^{\zeta\,0}(y_1, \ldots, y_l) := t^0(y_1, \ldots, y_l)\,(M^{2J} Y_{j_1} \ldots Y_{j_J})^\zeta\,, \quad J \geq 1\,, \quad Y_j := \pm(y_j^2 + i\,s(y_j)\,0)\,, \tag{3.5.31}$$

where $s(y) \in \{1, -1, \operatorname{sgn} y^0, -\operatorname{sgn} y^0\}$ and $M > 0$ is a mass scale. The latter and the addition $i\,s(y_j)\,0$ are needed in order that

$$(M^{2J} Y_{j_1} \ldots Y_{j_J})^\zeta = \exp\big(\zeta\,\log(M^{2J} Y_{j_1} \ldots Y_{j_J})\big) = 1 + \sum_{p=1}^{\infty} \zeta^p\,\frac{\log^p(M^{2J} Y_{j_1} \ldots Y_{j_J})}{p!} \tag{3.5.32}$$

is well defined in $\mathcal{D}'(\mathbb{R}^{dl} \setminus \{0\})$. If t^0 is a time-ordered product, it is convenient to choose $Y_j := -(y_j^2 - i\,0)$, as in (3.5.11). Usually, the indices $j_1, \ldots, j_J \in \{1, \ldots, l\}$ are chosen such that symmetries of t^0 of type $t^0(y_{\pi(1)}, \ldots, y_{\pi(l)}) = t^0(y_1, \ldots, y_l)$, for some permutation $\pi \in S_l$, hold true also for $t^{\zeta\,0}$. Due to

$$(y\partial_y + D - 2J\zeta)^K\, t^{\zeta\,0}(y_1, \ldots, y_l) = (M^{2J} Y_{j_1} \ldots Y_{j_J})^\zeta\,(y\partial_y + D)^K\, t^0(y_1, \ldots, y_l)\,,$$

where $K \in \mathbb{N}^*$ and $y\partial_y := \sum_{r=1}^{l} y_r^\mu \partial_{y_r^\mu}$, the regularized distribution $t^{\zeta\,0}$ scales almost homogeneously with degree

$$D_\zeta = D - 2J\zeta \tag{3.5.33}$$

and with the same power as t^0.

In the general part of this section, we give a refinement of [63, Sect. IV.A]: We will generally define the notion of "analytic regularization" and, based on this definition, we will give a prescription for the choice of the subtraction term s^ζ, for which the limit $t := \lim_{\zeta\to 0}(t^\zeta - s^\zeta)$ has the required properties.

For this purpose we first recall Lemma 3.2.19(a): Let $t^0 \in \mathcal{D}'(\mathbb{R}^k \setminus \{0\})$ be a distribution which scales almost homogeneously with degree $D \in k + \mathbb{N}$ and power N, let $\omega := D - k$ and let $\mathcal{D}_\omega(\mathbb{R}^k)$ be defined as in (3.2.23). With that, the direct extension t_ω of t^0 to $\mathcal{D}'_\omega(\mathbb{R}^k)$, defined by (3.2.25), scales also almost homogeneously with degree D and power N. In addition, t_ω is the only extension to $\mathcal{D}'_\omega(\mathbb{R}^k)$ which maintains almost homogeneous scaling with degree D, due to the uniqueness of t_ω in Claim 3.2.9.

Definition 3.5.10 ((Analytic) regularization). [63] Let $t^0 \in \mathcal{D}'(\mathbb{R}^k \setminus \{0\})$ be a distribution which scales almost homogeneously with degree $D \in k+\mathbb{N}$ and power N, let ω, $\mathcal{D}_\omega(\mathbb{R}^k)$ and t_ω be as just mentioned. A family of distributions $\{t^\zeta\}_{\zeta \in \Omega \setminus \{0\}}$, $t^\zeta \in \mathcal{D}'(\mathbb{R}^k)$, with $\Omega \subseteq \mathbb{C}$ a neighbourhood of the origin, is called a *regularization* of t^0, if

$$\lim_{\zeta \to 0} \langle t^\zeta, h \rangle = \langle t_\omega, h \rangle \qquad \forall h \in \mathcal{D}_\omega(\mathbb{R}^k) , \tag{3.5.34}$$

and if t^ζ scales almost homogeneously with a degree $D_\zeta \notin k + \mathbb{N}$ (for $\zeta \in \Omega \setminus \{0\}$) and the same power N as t^0, where D_ζ is of the form $D_\zeta = D + D_1 \zeta$ for some constant $D_1 \in \mathbb{C} \setminus \{0\}$.[80] The regularization $\{t^\zeta\}$ is called *analytic*, if for all $h \in \mathcal{D}(\mathbb{R}^k)$ the map

$$\Omega \setminus \{0\} \ni \zeta \longmapsto \langle t^\zeta, h \rangle \tag{3.5.35}$$

is analytic with a pole of finite order at the origin.

Example 3.5.11. We continue the Exap. 3.5.9. Let t^ζ be the unique extension of $t^{\zeta\,0}$ (given in (3.5.31)) to $\mathcal{D}'(\mathbb{R}^{dl})$ which scales almost homogeneously with degree D_ζ (3.5.33) and the same power as $t^{\zeta\,0}$ and t^0. The pertinent family $\{t^\zeta\}$ is indeed a regularization of the given $t^0 \in \mathcal{D}'(\mathbb{R}^{dl} \setminus \{0\})$ in the sense of Definition 3.5.10: Namely, by using the functions χ_ρ (3.2.22) and (3.2.26) and that $\lim_{\zeta \to 0} t^{\zeta\,0} = t^0$ in $\mathcal{D}'(\mathbb{R}^{dl} \setminus \{0\})$, we get

$$\begin{aligned} \langle t_\omega, h \rangle &= \lim_{\rho \to \infty} \langle t^0, \chi_\rho\, h \rangle = \lim_{\rho \to \infty} \lim_{\zeta \to 0} \langle t^{\zeta\,0}, \chi_\rho\, h \rangle \\ &= \lim_{\zeta \to 0} \lim_{\rho \to \infty} \langle t^{\zeta\,0}, \chi_\rho\, h \rangle = \lim_{\zeta \to 0} \langle t^\zeta_\omega, h \rangle = \lim_{\zeta \to 0} \langle t^\zeta, h \rangle , \quad \forall h \in \mathcal{D}_\omega(\mathbb{R}^{dl}) , \end{aligned}$$

where t^ζ_ω is the direct extension of $t^{\zeta\,0}$ to $\mathcal{D}'_\omega(\mathbb{R}^{dl})$ and we use that $t^\zeta_\omega = t^\zeta|_{\mathcal{D}_\omega(\mathbb{R}^{dl})}$ (which is due to the uniqueness of t^ζ_ω). The crucial step is that the order of the limits may be reversed. For the limits at hand this is an allowed operation.

In addition, in all applications of (3.5.31) we will study, the explicit formula for the extension $t^{\zeta\,0} \mapsto t^\zeta$ will show that this regularization is analytic.

Let t^0 and t_ω be as in Definition 3.5.10, let $\{t^\zeta\}$ be a regularization of t^0 and let t be an almost homogeneous extension of t^0. From Lemma 3.2.15 we know that there exist functions $w_a \in \mathcal{D}(\mathbb{R}^k)$, $a \in \mathbb{N}^k$ with $|a| \leq \omega$, satisfying

[80] The formalism can be generalized to an analytic dependence of D_ζ on ζ: $D_\zeta = D + \sum_{s=1}^\infty D_s\, \zeta^s$, $D_s \in \mathbb{C}$ with $D_1 \neq 0$ – the latter condition is crucial. However, the regularization (3.5.31) is of type $D_\zeta = D + D_1\, \zeta$. So, for simplicity, we work with this special type of ζ-dependence.

$\partial^b w_a(0) = \delta^b_a \ \forall b \in \mathbb{N}^k$, such that t can be written as

$$\langle t, h\rangle = \langle t_\omega, Wh\rangle \quad \text{where} \quad Wh(x) := h(x) - \sum_{|a|\leq\omega} \partial^a h(0)\, w_a(x)\ , \quad \forall h \in \mathcal{D}(\mathbb{R}^k)\ . \tag{3.5.36}$$

Since $Wh \in \mathcal{D}_\omega(\mathbb{R}^k)$, we may now use (3.5.34):

$$\langle t, h\rangle = \lim_{\zeta\to 0}\langle t^\zeta, Wh\rangle = \lim_{\zeta\to 0}\Big(\langle t^\zeta, h\rangle - \sum_{|a|\leq\omega}\langle t^\zeta, w_a\rangle\, \partial^a h(0)\Big)\ . \tag{3.5.37}$$

In general, the limit of the individual terms on the right-hand side might not exist. However, if the regularization $\{t^\zeta \,|\, \zeta \in \Omega \setminus \{0\}\}$ is analytic, each term can be expanded in a Laurent series around $\zeta = 0$, and since the overall limit is finite, the principal parts (pp) of these Laurent series must coincide:

$$\big\langle \mathrm{pp}(t^\zeta), h\big\rangle := \mathrm{pp}\big(\langle t^\zeta, h\rangle\big) = \sum_{|a|\leq\omega} \mathrm{pp}\big(\langle t^\zeta, w_a\rangle\big)\, \partial^a h(0)\ , \quad \forall h \in \mathcal{D}(\mathbb{R}^k)\ . \tag{3.5.38}$$

For all h, the l.h.s. is independent of the choice of w_a; hence this must hold also for $\mathrm{pp}\big(\langle t^\zeta, w_a\rangle\big)$. The latter can be understood also as follows: From (3.5.38) we see that $\mathrm{pp}(t^\zeta)$ is a linear combination of derivatives of $\delta(x)$; therefore, all information about w_a that is contained in $\mathrm{pp}\big(\langle t^\zeta, w_a\rangle\big)$ is $\partial^b w_a(0) = \delta^b_a$.

The following proposition, which is a further development of Lemma 4.3 and Corollary 4.4 in [63], treats the difficult case of Problem 3.5.1 – the case $D \in k+\mathbb{N}$ – for which it is a constructive refinement of Proposition 3.2.16.

Proposition 3.5.12 (Minimal subtraction). *Let $t^0 \in \mathcal{D}'(\mathbb{R}^k \setminus \{0\})$ be a distribution which scales almost homogeneously with degree $D \in k + \mathbb{N}$ and power N, and let $\{t^\zeta\}_{\zeta\in\Omega\setminus\{0\}}$ be an analytic regularization of t^0. Then, the following holds true.*

(a) *The principal part $\mathrm{pp}(t^\zeta)$ is a local distribution which scales homogeneously with degree D, explicitly*

$$\mathrm{pp}\big(t^\zeta(x)\big) = \sum_{|a|=D-k} C_a(\zeta)\, \partial^a \delta(x)\ , \quad \textit{where} \quad C_a(\zeta) = (-1)^{|a|}\, \mathrm{pp}\big(\langle t^\zeta, w_a\rangle\big)\ . \tag{3.5.39}$$

(b) *The regular part $\mathrm{rp}(t^\zeta) := t^\zeta - \mathrm{pp}(t^\zeta)$ defines by*

$$\langle t^{\mathrm{MS}}, h\rangle := \lim_{\zeta\to 0} \mathrm{rp}\big(\langle t^\zeta, h\rangle\big)\ , \quad \forall h \in \mathcal{D}(\mathbb{R}^k)\ , \tag{3.5.40}$$

a distinguished extension of t^0 which scales almost homogeneously with degree D. The renormalization prescription (3.5.40) *is called "minimal subtraction".*

(c) *Let R be the order of the pole of t^ζ at the origin (i.e., R is the maximum over $h \in \mathcal{D}(\mathbb{R}^k)$ of the orders of the poles at 0 of the maps* (3.5.35)*). The power of the almost homogeneous scaling of t^{MS} is*

- *N if $N \geq R$,*
- *N or $N+1$ if $N < R$.*

In traditional terminology $(-1)\,\mathrm{pp}(t^\zeta)$ is a "local counter term".

Proof. Ad (a): From (3.5.38) we see that

$$\mathrm{pp}\big(t^\zeta(x)\big) = \sum_{|a|\leq D-k} C_a(\zeta)\,\partial^a\delta(x)\ , \tag{3.5.41}$$

where $C_a(\zeta)$ is given by the asserted expression (3.5.39).[81] That the sum over a runs only over $|a| = D-k$ is shown below in the proof of part (c).

Ad (b): To show that t^{MS} is an extension of t^0 with $\mathrm{sd}(t^{\mathrm{MS}}) = \mathrm{sd}(t^0)$, we compare t^{MS} with the extension t given in (3.5.36), which has scaling degree $\mathrm{sd}(t) = D = \mathrm{sd}(t^0)$, since it scales almost homogeneously with degree D. Using (3.5.37) and (3.5.38), we write t as

$$\begin{aligned}\langle t,h\rangle &= \lim_{\zeta\to 0}\Big(\langle t^\zeta,h\rangle - \sum_{|a|\leq\omega}\Big(\mathrm{pp}\big(\langle t^\zeta,w_a\rangle\big)+\mathrm{rp}\big(\langle t^\zeta,w_a\rangle\big)\Big)\,\partial^a h(0)\Big)\\ &= \langle t^{\mathrm{MS}},h\rangle - \sum_{|a|\leq\omega}\langle t^{\mathrm{MS}},w_a\rangle\,\partial^a h(0)\ ,\quad h\in\mathcal{D}(\mathbb{R}^k)\ ,\quad \omega:=D-k\ . \end{aligned}\tag{3.5.42}$$

Hence, t^{MS} differs from t by a term of the form $t^{\mathrm{MS}}-t=\sum_{|a|\leq\omega} b_a\,\partial^a\delta\ ,\ b_a\in\mathbb{C}$; this yields the claim due to Theorem 3.2.8.

To prove that t^{MS} scales almost homogeneously with degree D, we write

$$t^\zeta(x) = \sum_{r=-R}^{\infty} t_r(x)\,\zeta^r\ ,\quad R\in\mathbb{N},\quad t_{-R}\neq 0\ ; \tag{3.5.43}$$

note that $t^{\mathrm{MS}} = t_0$. Since $D_\zeta = D + D_1\zeta$, we may write

$$(x\partial_x + D_\zeta)^{N+1} = (x\partial_x + D)^{N+1} + \sum_{s=1}^{\min\{R,N+1\}} \zeta^s\,d_s^N\,(x\partial_x + D)^{N+1-s} + \mathcal{O}(\zeta^{R+1})\ ,$$

where $d_s^N := \binom{N+1}{s}(D_1)^s$. Using this and that t^ζ scales almost homogeneously with degree D_ζ and power N, we have

$$\Big((x\partial_x+D)^{N+1}+\sum_{s=1}^{\min\{R,N+1\}}\zeta^s\,d_s^N\,(x\partial_x+D)^{N+1-s}+\mathcal{O}(\zeta^{R+1})\Big)\Big(\sum_{r=-R}^{\infty}t_r(x)\,\zeta^r\Big)=0.$$

[81] Alternatively, this formula for $C_a(\zeta)$ can be obtained directly from (3.5.41) by applying it to w_b and using $\partial^a w_b(0) = \delta^a_b$.

Selecting the terms $\sim \zeta^{-R+p}$, where $0 \leq p \leq R$, we get

$$(x\partial_x + D)^{N+1} t_{-R+p} = - \sum_{s=1}^{\min\{p,N+1\}} d_s^N (x\partial_x + D)^{N+1-s} t_{-R+p-s} . \quad (3.5.44)$$

We are now going to prove

$$(x\partial_x + D)^{N+1+q} t_{-R+q} = 0 , \quad \forall\, 0 \leq q \leq R . \quad (3.5.45)$$

We proceed by induction on q. The case $q = 0$ is obtained by setting $p = 0$ in (3.5.44). Now, we assume that (3.5.45) holds true for $q \in \{0, 1, \ldots, p-1\}$, where $p \leq R$. From (3.5.44) we get

$$(x\partial_x + D)^{N+1+p} t_{-R+p} = - \sum_{s=1}^{\min\{p,N+1\}} d_s^N (x\partial_x + D)^{N+1-s+p} t_{-R+p-s} ,$$

which vanishes by the inductive assumption since $0 \leq p - s \leq p - 1$. This proves (3.5.45).

The particular case $q = R$ of (3.5.45) states that $t_0 = t^{\mathrm{MS}}$ scales almost homogeneously with degree D and power $\leq N + R$.

Ad (c): To get a better bound for this power and to complete the proof of part (a), we use (3.5.41) to write $\mathrm{pp}(t^\zeta)$ as

$$\mathrm{pp}\big(t^\zeta(x)\big) = \sum_{r=-R}^{-1} \zeta^r t_r(x) \quad \text{with} \quad t_r(x) = \sum_{|a| \leq D-k} c_{r,a}\, \partial^a \delta(x) , \quad c_{r,a} \in \mathbb{C} . \quad (3.5.46)$$

From (3.5.45) we know that for $-R \leq r \leq -1$ it holds that

$$\begin{aligned} 0 &= (x\partial_x + D)^{N+1+R+r} \sum_{|a| \leq D-k} c_{r,a}\, \partial^a \delta(x) \\ &= \sum_{|a| \leq D-k} c_{r,a} (-k - |a| + D)^{N+1+R+r}\, \partial^a \delta(x) , \end{aligned}$$

where (3.2.45) is used in the last step. We conclude that $c_{r,a} = 0$ for $|a| < D - k$, i.e., the sum in (3.5.41) is restricted to $|a| = D - k$, as asserted. This implies

$$(x\partial_x + D)\, t_r(x) = 0 \quad \forall\, -R \leq r \leq -1 . \quad (3.5.47)$$

To determine the power of the almost homogeneous scaling of $t^{\mathrm{MS}} = t_0$, we use (3.5.44) for $p = R$:

$$(x\partial_x + D)^{N+1} t^{\mathrm{MS}} = - \sum_{s=1}^{\min\{R,N+1\}} d_s^N (x\partial_x + D)^{N+1-s} t_{-s} . \quad (3.5.48)$$

By means of (3.5.47) we obtain the following results:

- If $R \leq N$ the r.h.s. of (3.5.48) vanishes, since the exponent $N+1-s$ is ≥ 1 for all s.
- If $R > N$ the r.h.s. of (3.5.48) reduces to the term $s = N+1$, which is equal to $-d^N_{N+1}\, t_{-(N+1)}$; using again (3.5.47) we get

$$(x\partial_x + D)^{N+2}\, t^{\mathrm{MS}} = -d^N_{N+1}\,(x\partial_x + D)\, t_{-(N+1)} = 0\ . \qquad \square$$

Remark 3.5.13. Returning to (3.5.42), we see that in the last term on the r.h.s. the sum over a runs actually only over $|a| = \omega$, because both t and t^{MS} scale almost homogeneously with degree D. So we know

$$\lim_{\zeta\to 0} \mathrm{rp}\big(\langle t^\zeta, w_a\rangle\big) = \langle t^{\mathrm{MS}}, w_a\rangle = 0 \quad \forall\, |a| < \omega\ .$$

Since the sum over a in (3.5.41) runs only over $|a| = D - k \equiv \omega$, we also know that

$$\mathrm{pp}\big(\langle t^\zeta, w_a\rangle\big) = 0 \quad \forall\, |a| < \omega\ ; \quad \text{summing up we get} \quad \lim_{\zeta\to 0} \langle t^\zeta, w_a\rangle = 0 \quad \forall\, |a| < \omega\ .$$

Hence, we may restrict the sum over a in (3.5.37) to $|a| = \omega$. This is a peculiarity of analytic regularization: In general, the sum over a in a W-projector yielding an almost homogeneous extension, as in (3.5.36), includes also subtraction terms with $|a| < \omega$; in particular, these terms appear in the improved Epstein–Glaser renormalization treated in Lemma 3.2.19.

Examples. We illustrate analytic regularization and minimal subtraction by some examples and exercises; the first example solves the renormalization problem for a large class of diagrams.

Example 3.5.14 (Massless, primitive divergent diagrams). Cf. [65, Sect. 6.1]. We start with formula (3.3.10) for the restriction of the time-ordered product $T_n^{(m=0)}\big(\otimes_{j=1}^n A_j(x_j)\big)$ to $\mathcal{D}(\check{\mathbb{M}}^n)$. Let $l := n-1$, $y_j := x_j - x_n$ and let $t^0(y_1,\ldots,y_l)$ be the contribution[82] to $\omega_0\big(A_1(x_1) \star_{D^F} \cdots \star_{D^F} A_n(x_n)\big)\big)$ coming from a primitive divergent diagram Γ. That Γ is "divergent" means that it cannot be renormalized by the direct extension, i.e., that $\omega := \omega(\Gamma) := \mathrm{sd}(t^0) - dl \geq 0$. "Primitive divergent" means that, in addition, there is no subdiagram Γ_1 of Γ with less vertices and with $\omega(\Gamma_1) \geq 0$. All divergent 1-loop diagrams are primitive divergent; the setting sun diagram in $d = 4$ dimensions (computed in Exap. 3.5.4) is an example for a primitive divergent 2-loop diagram (see Footn. 79 for the notion "n-loop diagram").

By definition, t^0 is a product of massless Feynman propagators with partial derivatives; for simplicity we assume that the latter are absent:

$$t^0(y_1,\ldots,y_l) = C\,\hbar^S \prod_{s=1}^{S} D^F(x_{i_s} - x_{k_s})\ , \quad i_s < k_s\ \forall s\ , \qquad (3.5.49)$$

[82] We use $\Delta^F_{m=0} = D^F$.

where $C \in \mathbb{R}$ is a combinatorial factor. Because Γ is primitive divergent, t^0 is well defined not only on $\mathcal{D}(\check{\mathbb{R}}^{dl})$, where

$$\check{\mathbb{R}}^{dl} := \{(y_1, \ldots, y_l) \,|\, y_i \neq 0 \wedge y_j \neq y_k \;\; \forall i, \; j < k\} \;, \tag{3.5.50}$$

it even exists in $\mathcal{D}'(\mathbb{R}^{dl} \setminus \{0\})$. From (3.5.49) we see that t^0 scales homogeneously with degree

$$D = S(d-2) \quad \text{and we have} \quad \omega = D - dl \geq 0 \;.$$

Let $t^{\zeta\,0} \in \mathcal{D}'(\mathbb{R}^{dl} \setminus \{0\})$ be the regularization (3.5.31) of t^0 with $Y_j := -(y_j^2 - i\,0)$ and a suitable choice of $j_1, \ldots, j_J$. Since $t^{\zeta\,0}$ scales homogeneously with degree $D_\zeta = D - 2J\zeta \notin dl + \mathbb{N}$ (3.5.33), it has a unique homogeneous extension $t^\zeta \in \mathcal{D}'(\mathbb{R}^{dl})$, which can be computed by formula (3.5.30):

$$\begin{aligned} &t^\zeta(y) \\ &= \frac{1}{\prod_{p=0}^{\omega}(p - \omega + 2J\zeta)} \sum_{r_1 \cdots r_{\omega+1}} \partial_{y_{r_{\omega+1}}} \cdots \partial_{y_{r_1}} \Big(\overline{y_{r_1} \cdots y_{r_{\omega+1}}\, t^0(y) (M^{2J} Y_{j_1} \ldots Y_{j_J})^\zeta} \Big) \end{aligned} \tag{3.5.51}$$

where $\sum_r \partial_{y_r}(y_r \ldots) := \sum_r \partial_\mu^{y_r}(y_r^\mu \ldots)$. As we know from Exap. 3.5.11, $\{t^\zeta\}$ is a regularization of t^0; moreover, we see from (3.5.51) that $\{t^\zeta\}$ is analytic. Hence, Proposition 3.5.12 applies. To compute $t^{\mathrm{MS}} = t_0$ (the latter is defined in (3.5.43)), we use the expansions (3.5.32) and

$$\frac{1}{\prod_{p=0}^{\omega}(p - \omega + 2J\zeta)} = \frac{(-1)^\omega}{\omega!} \Big(\frac{1}{2J\zeta} + \sum_{p=1}^{\omega} \frac{1}{p} + \mathcal{O}(\zeta) \Big) \;,$$

and obtain

$$\begin{aligned} t^{\mathrm{MS}}(y) = \frac{(-1)^\omega}{\omega!} \sum_{r_1 \cdots r_{\omega+1}} \partial_{y_{r_{\omega+1}}} \cdots \partial_{y_{r_1}} \Big[\frac{1}{2J} \Big(\overline{y_{r_1} \cdots y_{r_{\omega+1}}\, t^0(y) \, \log(M^{2J} Y_{j_1} \ldots Y_{j_J})} \Big) \\ + \big(\sum_{p=1}^{\omega} \frac{1}{p}\big) \Big(\overline{y_{r_1} \cdots y_{r_{\omega+1}}\, t^0(y)} \Big) \Big] \,. \end{aligned} \tag{3.5.52}$$

The second term is of the form

$$\frac{(-1)^\omega}{\omega!} \sum_{r_1 \cdots r_{\omega+1}} \partial_{y_{r_{\omega+1}}} \cdots \partial_{y_{r_1}} \Big(\overline{y_{r_1} \cdots y_{r_{\omega+1}}\, t^0(y)} \Big) = \sum_{|a|=\omega} C_a \, \partial^a \delta(y) \tag{3.5.53}$$

for some $C_a \in \mathbb{C}$, for the following reason: For $y \neq 0$ we get

$$\sum_{r_{\omega+1}} \partial_{y_{r_{\omega+1}}} \Big(y_{r_{\omega+1}} \big(y_{r_1} \cdots y_{r_\omega}\, t^0(y) \big) \Big) = 0$$

by using (3.5.27), since $y_{r_1} \ldots y_{r_\omega}\, t^0(y)$ scales homogeneously with degree $D - \omega = dl$. Therefore, the l.h.s. of (3.5.53) is of the form $\sum_a C_a \partial^a \delta(y)$. The sum runs only over $|a| = \omega$, because the l.h.s. of (3.5.53) scales homogeneously with degree $D = dl + \omega$.

Due to the appearance of $\log(M^{2J} Y_{j_1} \ldots Y_{j_J})$ in formula (3.5.52) for t^{MS}, there is a breaking of homogeneous scaling. By straightforward calculation we obtain

$$\rho^D \, t^{\mathrm{MS}}(\rho y) - t^{\mathrm{MS}}(y) = \log \rho \sum_{|a|=\omega} C_a \, \partial^a \delta(y) \;, \tag{3.5.54}$$

where the coefficients C_a are precisely the ones defined in (3.5.53). Part (c) of Proposition 3.5.12 is indeed satisfied: We have $0 = N < R = 1$ and the scaling-power of t^{MS} is $1 = N + 1$.

This example applies also to the leading term (i.e., the term $\sim m^0$) of the Sm-expansion of the T-product $t_\Gamma^{(m)\,0} \in \mathcal{D}'(\mathbb{R}^{dl} \setminus \{0\})$ of a *massive*, primitive divergent diagram Γ, because the leading term is the corresponding massless T-product $t_\Gamma^{(m=0)\,0}$. If, in addition, $\omega(\Gamma) = 0$, the extension of $t_\Gamma^{(m)\,0}$ can be computed as follows: We write the Sm-expansion as $t_\Gamma^{(m)\,0} = t_\Gamma^{(m=0)\,0} + \mathfrak{r}^{(m)\,0}$, then we extend $t_\Gamma^{(m=0)\,0}$ by (3.5.52) and the remainder $\mathfrak{r}^{(m)\,0}$ by the direct extension.

Remark 3.5.15. Reconsidering Prop. 3.2.16 we conclude: For any $t^0 \in \mathcal{D}'(\mathbb{R}^{dl} \setminus \{0\})$ scaling homogeneously with degree $D \in dl + \mathbb{N}$, the universal polynomial $p(\partial)$ (3.2.39) can be computed from t^0 by means of (3.5.53).

Example 3.5.16 (Setting sun with a hat). Cf. [64, Sect. 6]. In $d = 4$ dimensions we compute the massive time-ordered product $t(x_1 - x_3, x_2 - x_3)$ belonging to the diagram

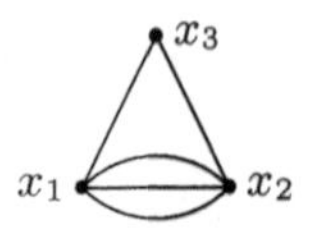

which contains the setting sun diagram t_{ss} as a divergent subdiagram. We will use the result (3.5.14) for $t_{\mathrm{ss}} \equiv t_{\mathrm{ss}}^{m,M}$.

The restriction of t to $\mathcal{D}(\mathbb{R}^8 \setminus \{0\})$ is given by

$$t^0(x,y) = t_{\mathrm{ss}}(x-y)\,\Delta_m^F(x)\,\Delta_m^F(y) \in \mathcal{D}'(\mathbb{R}^8 \setminus \{0\})\ , \quad x := x_1 - x_3\ , \ y := x_2 - x_3\ , \tag{3.5.55}$$

where we omit a combinatorial prefactor and factors $\hbar$. To prove this claim one has to verify that the r.h.s. of (3.5.55) satisfies the causal factorization (3.3.17) on $\mathcal{D}(C_I)$ for all I, cf. Exap. 3.3.7. This verification uses that $t_{\mathrm{ss}}(z) = \Delta_m^+(z)^3$ for $z \notin \overline{V}_-$, as we see from (3.5.10).

We concentrate on the renormalization, which is the extension of t^0 (3.5.55) to $\mathcal{D}'(\mathbb{R}^8)$. The degree D of the Sm-expansion of t^0 is $D = 10$, hence, only the terms $\mathcal{O}(m^0)$ and $\mathcal{O}(m^2)$ cannot be extended by the direct extension; so it suffices to choose the remainder of order $\mathcal{O}(m^4)$. Inserting (3.5.11) and (3.5.14) into (3.5.55) we get:

$$t^0(x,y) = v_0^0(x,y) + m^2\big(v_{2,0}^0(x,y) + v_{2,1}^0(x,y)\,\log\frac{m}{M}\big) + \mathfrak{q}_4^{(m)\,0}(x,y)\,,$$

where we use the letters $(v, \mathfrak{q})$ (instead of $(u, \mathfrak{r})$) to avoid confusion with the distributions appearing in the Sm-expansion of the setting sun diagram. The $v^0_{\dots}$-distributions read:

$$v_0^0(x,y) = u_0(x-y)\,\frac{a^2}{XY}\,,$$
$$v_{2,0}^0(x,y) = u_{2,0}(x-y)\,\frac{a^2}{XY} + u_0(x-y)\,a\Big(\frac{a_0\,\log(M^2Y) + b_0}{X} + \frac{a_0\,\log(M^2X) + b_0}{Y}\Big)\,,$$
$$v_{2,1}^0(x,y) = u_{2,1}(x-y)\,\frac{a^2}{XY} + u_0(x-y)\,2aa_0\Big(\frac{1}{X} + \frac{1}{Y}\Big)\,,$$

where the $u_{\dots}$-distributions are given in (3.5.13) and Y is defined similarly to X (3.5.11).

To compute the almost homogeneous extension of the $v^0_{\dots}$-distributions, we use an analytic regularization of type (3.5.31) which respects the $(x \leftrightarrow y)$-symmetry:

$$v^{\zeta\, 0}(x,y) := v^0(x,y)\,(M^4XY)^\zeta\,, \qquad v = v_0,\, v_{2,0},\, v_{2,1}\,. \tag{3.5.56}$$

The regularized distributions $v^{\zeta\,0}$ scale almost homogeneously with degree $D_\zeta = 8 - 4\zeta$ (for $v^{\zeta\,0}_{2,0}$, $v^{\zeta\,0}_{2,1}$) or $D_\zeta = 10 - 4\zeta$ (for $v_0^{\zeta\,0}$), satisfying $D_\zeta \notin 8 + \mathbb{N}$. Since the power of the almost homogeneous scaling is 1 for u_0 and $u_{2,1}$, and 2 for $u_{2,0}$; we see that this power is $N = 1$ for $v_0^{\zeta\,0}$ and $v^{\zeta\,0}_{2,1}$ and $N = 2$ for $v^{\zeta\,0}_{2,0}$.

In terms of these examples we are now going to explain, how the idea underlying the extension formula (3.5.30) (which applies only to *homogeneously* scaling $t^{\zeta\,0} \in \mathcal{D}'(\mathbb{R}^k \setminus \{0\})$) can be used to extend *almost homogeneously* scaling $t^{\zeta\,0}$. If $D = k$ (i.e., $D = 8$ for the example at hand), the method always works (see [63, Remk. 4.9]): Writing $z := (x,y)$, $\partial_r z_r := \partial_{x^\mu} x^\mu + \partial_{y^\mu} y^\mu$ and $\eta := -4\zeta$, the almost homogeneous scaling of $v^{\zeta\,0}_{2,1}$ can be expressed by

$$0 = (\partial_r z_r + \eta)^2 v^{\zeta\,0}_{2,1}(z) = \eta^2\, v^{\zeta\,0}_{2,1}(z) + (2\eta - 1)\,\partial_r\big(z_r\, v^{\zeta\,0}_{2,1}(z)\big) + \partial_r\partial_s\big(z_r z_s\, v^{\zeta\,0}_{2,1}(z)\big)\,.$$

Using differential renormalization, we obtain the following result for the unique almost homogeneous extension:

$$v^\zeta_{2,1} = \frac{-1}{\eta^2}\left((2\eta - 1)\,\partial_r\overline{(z_r\, v^{\zeta\,0}_{2,1})} + \partial_r\partial_s\overline{(z_r z_s\, v^{\zeta\,0}_{2,1})}\right) \in \mathcal{D}'(\mathbb{R}^8)\,. \tag{3.5.57}$$

Again, the overline denotes the direct extension, which exists since $\mathrm{sd}(z_{r_1}\dots z_{r_l}\, v^{\zeta\,0}) = D_\zeta - l$.

For $v^{\zeta\,0}_{2,0}$ we still have $D = 8$, but the power of the almost homogeneous scaling is $N = 2$. Hence, we have

$$(\partial_r z_r + \eta)^3 v^{\zeta\,0}_{2,0}(z) = 0\,,$$

which yields

$$v^\zeta_{2,0} = \frac{-1}{\eta^3}\left((3\eta^2 - 3\eta + 1)\,\partial_r\overline{(z_r\, v^{\zeta\,0}_{2,0})} + (3\eta - 3)\,\partial_r\partial_s\overline{(z_r z_s\, v^{\zeta\,0}_{2,0})} + \partial_p\partial_r\partial_s\overline{(z_p z_r z_s\, v^{\zeta\,0}_{2,0})}\right). \tag{3.5.58}$$

If $D > k(= 8)$ the method is more involved. For $v_0^{\zeta\,0}$ we have $D - 8 = 2$, hence we need at least $l = 3$ factors z_{r_i} in order that the direct extension $\overline{(z_{r_1}\dots z_{r_l}\, v_0^{\zeta\,0})}$ exists. We proceed as follows: From the almost homogeneous scaling of $v_0^{\zeta\,0}$ with degree $D_\zeta = 10 - 4\zeta$ and power $N = 1$,

$$(\partial_r z_r + 2 + \eta)^2 v_0^{\zeta\,0}(z) = 0\,, \tag{3.5.59}$$

we obtain

$$v_0^{\zeta\,0} = \frac{-1}{(2+\eta)^2}\left((3 + 2\eta)\,\partial_s(z_s\, v_0^{\zeta\,0}) + \partial_r\partial_s(z_r z_s\, v_0^{\zeta\,0})\right).$$

In addition, $(z_s\, v_0^{\zeta\,0})$ scales almost homogeneously with degree $D_\zeta = 9 - 4\zeta$ and the same power $N = 1$:

$$(\partial_r z_r + 1 + \eta)^2\big(z_s\, v_0^{\zeta\,0}(z)\big) = 0\,;$$

this gives

$$z_s\, v_0^{\zeta\,0} = \frac{-1}{(1+\eta)^2}\left((1 + 2\eta)\,\partial_r(z_r z_s\, v_0^{\zeta\,0}) + \partial_p\partial_r(z_p z_r z_s\, v_0^{\zeta\,0})\right).$$

Analogously, the degree of the almost homogeneous scaling of $(z_r z_s\, v_0^{\zeta\,0})$ is $D_\zeta = 8 - 4\zeta$, that is,

$$(\partial_p z_p + \eta)^2 \big(z_r z_s\, v_0^{\zeta\,0}(z)\big) = 0\ ;$$

this yields

$$z_r z_s\, v_0^{\zeta\,0} = \frac{-1}{\eta^2}\Big((2\eta - 1)\,\partial_p(z_p z_r z_s\, v_0^{\zeta\,0}) + \partial_p\partial_q(z_p z_q z_r z_s\, v_0^{\zeta\,0})\Big)\,.$$

Inserting the lower equations into the upper ones and using differential renormalization, we get

$$v_0^\zeta = \frac{1}{\eta^2(1+\eta)^2(2+\eta)^2}\Big((2 + 2\eta - 6\eta^2 - 4\eta^3)\,\partial_p\partial_r\partial_s\overline{(z_p z_r z_s\, v_0^{\zeta\,0})} \\ - (2 + 6\eta + 3\eta^2)\,\partial_p\partial_q\partial_r\partial_s\overline{(z_p z_q z_r z_s\, v_0^{\zeta\,0})}\Big)\,. \qquad (3.5.60)$$

Obviously, the extensions v^ζ scale almost homogeneously with the same degree D_ζ and the same power as the initial $v^{\zeta\,0}$ (in accordance with Proposition 3.2.16); in addition, the maps $\zeta \longmapsto \langle v^\zeta, h\rangle$ are meromorphic in ζ for all $h \in \mathcal{D}(\mathbb{R}^8)$, with a pole at $\zeta = 0$ of order $R = 2$ (for $v_{2,1}^\zeta$ and v_0^ζ) or $R = 3$ (for $v_{2,0}^\zeta$). The latter shows explicitly that this extension method does not work for the unregularized theory (i.e., for $\zeta = 0$).

So $\{v^\zeta\}$ is an analytic regularization of v^0 and Proposition 3.5.12 applies: Minimal subtraction yields an almost homogeneous extension v^{MS} of v^0. Proceeding analogously to Exap. 3.5.14, we expand (in ζ) $(M^4XY)^\zeta$ (see (3.5.32)) and the rational functions of η appearing in (3.5.57), (3.5.58) and (3.5.60), and select the term $\sim \zeta^0$ of the Laurent series for v^ζ. In this way we obtain the following results for the general, almost homogeneous and Lorentz invariant extensions $v = v^{\mathrm{MS}} + \sum_{|a|=\omega} C_a\, \partial^a\delta$, which are $(x \leftrightarrow y)$-invariant:

$$\begin{aligned}
v_{2,1} &= \partial_r\, \overline{\big(z_r\, v_{2,1}^0\, [\tfrac{1}{32}\, \log^2(M^4XY) + \tfrac{1}{2}\, \log(M^4XY)]\big)} \\
&\quad - \partial_r\partial_s\, \overline{\big(z_r z_s\, v_{2,1}^0\, \tfrac{1}{32}\, \log^2(M^4XY)\big)} + C_1\, \delta(x,y) \\
v_{2,0} &= \partial_r\, \overline{\big(z_r\, v_{2,0}^0\, [\tfrac{\log^3(M^4XY)}{384} + \tfrac{3\,\log^2(M^4XY)}{32} + \tfrac{3\,\log(M^4XY)}{4}]\big)} \\
&\quad - \partial_r\partial_s\, \overline{\big(z_r z_s\, v_{2,0}^0\, [\tfrac{3\,\log^3(M^4XY)}{384} + \tfrac{3\,\log^2(M^4XY)}{32}]\big)} \\
&\quad + \partial_p\partial_r\partial_s\, \overline{\big(z_p z_r z_s\, v_{2,0}^0\, \tfrac{\log^3(M^4XY)}{384}\big)} + C_0\, \delta(x,y) \\
v_0 &= \partial_p\partial_r\partial_s\, \overline{\big(z_p z_r z_s\, v_0^0\, [\tfrac{-1}{8} + \tfrac{1}{4}\, \log(M^4XY) + \tfrac{1}{64}\, \log^2(M^4XY)]\big)} \\
&\quad + \partial_q\partial_p\partial_r\partial_s\, \overline{\big(z_q z_p z_r z_s\, v_0^0\, [\tfrac{7}{8} - \tfrac{1}{64}\, \log^2(M^4XY)]\big)} \\
&\quad + C_2\, (\Box_x + \Box_y)\delta(x,y) + C_3\, \partial_\mu^x \partial_y^\mu\, \delta(x,y)\,.
\end{aligned} \qquad (3.5.61)$$

We use here the stronger version of the Sm-expansion axiom: The numbers C_s ($s = 0,1,2,3$) do not depend on m.

We explicitly see that these extensions $v_{\dots}$ scale almost homogeneously with the same degree as the pertinent $v_{\dots}^0$-distributions. Concerning part (c) of Proposition 3.5.12, we have $R > N$ for all three v^ζ. Hence, the power of the almost homogeneous scaling of v^{MS} may be $N+1$; but it may not be larger. However, looking at (3.5.61), we find the estimate:[83]

$$\begin{aligned}[\text{scaling-power of } v^{\mathrm{MS}}] \leq{} & [\text{scaling-power of } v^0 \ (= N)] \\ & + [\text{maximal power of } \log(M^4 XY)] \,, \end{aligned} \tag{3.5.62}$$

according to which there could be terms in (3.5.61) with scaling-power $> N+1$; these terms must cancel out up to terms with scaling-power $\leq N+1$, by identities for the derivatives.

We end up with

$$t(x,y) = v_0(x,y) + m^2\Big(v_{2,0}(x,y) + v_{2,1}(x,y)\,\log\frac{m}{M}\Big) + \mathfrak{q}_4^{(m)}(x,y) \in \mathcal{D}'(\mathbb{R}^8)\,,$$

where $\mathfrak{q}_4^{(m)}$ is the direct extension of $\mathfrak{q}_4^{(m)\,0}$.

Remark 3.5.17 (Sm-expansion axiom versus scaling degree axiom). According to Corollary 3.2.23, the Sm-expansion axiom is a stronger version of the scaling degree renormalization condition (3.2.62): If one replaces in the system of axioms given in Sect. 3.1 [or, in case of T-products, Sect. 3.3], the Sm-expansion, axiom (k) [or axiom (ix), resp.], by the condition (3.2.62) [or (3.3.21)], the set of allowed retarded [or time-odered] products gets truly bigger. To illustrate this we discuss the non-uniqueness of the inductive step $n=2 \to n=3$ for Exap. 3.5.16. Taking into account also the Lorentz invariance and $(x \leftrightarrow y)$-symmetry, the scaling degree condition (3.2.62) leaves the freedom to add a term of the form

$$\Big(f_2\Big(\frac{m}{M}\Big)\,(\square_x + \square_y) + f_3\Big(\frac{m}{M}\Big)\,\partial^x_\mu \partial^\mu_y + m^2\, f_1\Big(\frac{m}{M}\Big)\Big)\,\delta(x,y)\,,$$

where $M > 0$ is a fixed mass scale and f_1, f_2, f_3 are arbitrary functions $f_i\colon \mathbb{R} \longrightarrow \mathbb{C}$ (the values are dimensionless). We have found that the stronger version of the Sm-expansion axiom (axioms (k) and (k′)) restricts these functions to

$$f_2\Big(\frac{m}{M}\Big) = C_2\,, \quad f_3\Big(\frac{m}{M}\Big) = C_3\,, \quad f_1\Big(\frac{m}{M}\Big) = C_0 + C_1\,\log\Big(\frac{m}{M}\Big)\,,$$

with arbitrary constants C_0, C_1, C_2, $C_3 \in \mathbb{C}$. Such a reduction of the freedom of (re)normalization is certainly desirable.

With the weaker version of the Sm-expansion axiom (axiom (k) without (k′)), f_2 and f_3 are restricted in the same way, but $f_1(\frac{m}{M})$ may be an arbitrary polynomial in $\log\frac{m}{M}$ – see (3.2.71)–(3.2.72).

[83] By "scaling-power" we mean the power of the almost homogeneous scaling. As we see from the procedure in the computation of t^{MS}, it holds that [Maximal power of $\log(M^4XY)$] $= R$; hence, the estimate (3.5.62) agrees with the bound for the scaling-power of v^{MS} derived in the proof of part (b) of Proposition 3.5.12.

Exercise 3.5.18 (Massless triangle diagram with subdivergences in $d = 4$ dimensions). The aim is to compute the T-product $t \in \mathcal{D}'(\mathbb{R}^8)$ belonging to the diagram

(3.5.63)

where we omit combinatorial prefactors. The restriction of t to $\mathcal{D}(\mathbb{R}^8 \setminus \{0\})$ is given by

$$t^0(x,y) := t_{\text{fish}}(x)\, t_{\text{fish}}(y)\, D^F(x-y) \;, \quad x := x_1 - x_3 \;, \;\; y := x_2 - x_3 \;, \tag{3.5.64}$$

where $t_{\text{fish}}(x) \in \mathcal{D}'(\mathbb{R}^4)$ is an almost homogeneous extension of $D^F(x)^2 \in \mathcal{D}'(\mathbb{R}^4 \setminus \{0\})$. In Exap. 3.5.4 we have derived

$$t_{\text{fish}}(x) = \frac{1}{(2\pi)^4}\, \square_x \Big(\overline{\frac{\log(M^2X)}{4\,X}} \Big) \;, \qquad X := -(x^2 - i0) \;, \;\; M > 0 \;\; \text{arbitrary.} \tag{3.5.65}$$

(a) Verify that t^0 (3.5.64) fulfills the causal factorization (3.3.17) on $\mathcal{D}(C_{\{x_1,x_3\}})$ (where the notation (3.3.15) is used). Combinatorial factors may be ignored.

(b) Compute the extension $t^{\text{MS}} \in \mathcal{D}'(\mathbb{R}^8)$ (minimal subtraction) of t^0 (3.5.64), by using a Lorentz invariant, analytic regularization of type (3.5.31) which respects the $(x \leftrightarrow y)$-symmetry.

[*Solution*: (a) We write "$\sim$" for "equal up to a combinatorial prefactor". We have to verify that

$$t_{\text{fish}}(x_1 - x_3)\, t_{\text{fish}}(x_2 - x_3)\, D^F(x_1 - x_2) \overset{?}{\sim} \omega_0\Big(\varphi(x_1)\, t_{\text{fish}}(x_1 - x_3)\, \varphi^2(x_3) \star \varphi^3(x_2)\Big)$$

for $\{x_1, x_3\} \cap (x_2 + \overline{V}_-) = \emptyset$. For this configuration we may insert $D^F(x_1 - x_2) = D^+(x_1 - x_2)$ and $t_{\text{fish}}(x_2 - x_3) = D^+(x_3 - x_2)^2$ into the l.h.s. Computing the star product on the r.h.s., we get the same result, namely

$$[\text{r.h.s.}] \sim t_{\text{fish}}(x_1 - x_3)\, D^+(x_1 - x_2)\, D^+(x_3 - x_2)^2 \;.$$

(b) We use the same notations as in Exap. 3.5.16 and, in particular, the same regularization as in (3.5.56). With that, $t^{\zeta\,0}(x,y) := t^0(x,y)\,(M^4XY)^\zeta$ scales also almost homogeneously with degree $D_\zeta = 10 + \eta$, where $\eta := -4\zeta$. Searching the maximal power of $\log\rho$ in $\rho^{D_\zeta}\, t^{\zeta\,0}(\rho x, \rho y)$, one might naively think that each factor t_{fish} contributes a factor $\log\rho$. But, using Exer. 3.5.5 we see that

$$\begin{aligned} &\rho^{D_\zeta}\, t^{\zeta\,0}(\rho x, \rho y) \\ &\quad = (M^4XY)^\zeta\, D^F(x-y)\, \big(t_{\text{fish}}(x) + C_{\text{fish}}\, \delta(x)\, \log\rho\big) \big(t_{\text{fish}}(y) + C_{\text{fish}}\, \delta(y)\, \log\rho\big) \end{aligned}$$

(where $C_{\text{fish}} := \frac{-i}{8\pi^2}$) is a polynomial in $\log\rho$ of degree 1, because this expression is well defined only in $\mathcal{D}'(\mathbb{R}^8 \setminus \{0\})$. So we have

$$(\partial_r z_r + 2 + \eta)^2\, t^{\zeta\,0}(z) = 0 \quad \text{in} \quad \mathcal{D}'(\mathbb{R}^8 \setminus \{0\}) \;,$$

which is the same relation as (3.5.59). Hence, the computation of t^{MS} is precisely the same as that of v_0^{MS} in Exap. 3.5.16; the result for t^{MS} is obtained from the formula for v_0 in (3.5.61) by replacing v_0^0 by t^0 and by setting $C_2 := 0 =: C_3$.]

Exercise 3.5.19. Let $t^0 \in \mathcal{D}'(\mathbb{R}^{dl} \setminus \{0\})$ (where $l \in \mathbb{N}^*$) scale almost homogeneously with degree $D = dl + 2$ and power $N = 2$. In addition, let $t^\zeta \in \mathcal{D}'(\mathbb{R}^{dl})$ be a regularization of t^0 of the type studied in Exaps. 3.5.9 and 3.5.11; that is, t^ζ is the unique almost homogeneous extension of a distribution $t^{\zeta\,0} \in \mathcal{D}'(\mathbb{R}^{dl} \setminus \{0\})$, where the latter is obtained from t^0 by a relation of the form (3.5.31). Compute the minimally subtracted extension $t^{\mathrm{MS}} \in \mathcal{D}'(\mathbb{R}^{dl})$ of t^0 resulting from this regularization.

[*Solution*: We proceed analogously to the computation of the $(m = 0)$-part v_0^{MS} of Exap. 3.5.16 and use the same notations; the only essential difference is that the power of the almost homogeneous scaling of $t^{\zeta\,0}$ is $N = 2$ instead of $N = 1$. So we have

$$(\partial_r z_r + 2 + \eta)^3\, t^{\zeta\,0}(z) = 0\,,$$

which gives

$$t^{\zeta\,0} = \frac{-1}{(2+\eta)^3}\Big((7+9\eta+3\eta^2)\,\partial_s(z_s\, t^{\zeta\,0}) + (3+3\eta)\,\partial_r\partial_s(z_r z_s\, t^{\zeta\,0}) + \partial_p\partial_r\partial_s(z_p z_r z_s\, t^{\zeta\,0})\Big)\,.$$

Analogously, the scaling relations

$$(\partial_r z_r + 1 + \eta)^3 (z_s\, t^{\zeta\,0}(z)) = 0 \quad \text{and} \quad (\partial_p z_p + \eta)^3 (z_r z_s\, t^{\zeta\,0}(z)) = 0$$

yield

$$z_s\, t^{\zeta\,0} = \frac{-1}{(1+\eta)^3}\Big((1+3\eta+3\eta^2)\,\partial_r(z_r z_s\, t^{\zeta\,0}) + 3\eta\,\partial_p\partial_r(z_p z_r z_s\, t^{\zeta\,0}) + \partial_q\partial_p\partial_r(z_q z_p z_r z_s\, t^{\zeta\,0})\Big)$$

and

$$\begin{aligned} z_r z_s\, t^{\zeta\,0} = \frac{-1}{\eta^3}\Big(&(1-3\eta+3\eta^2)\,\partial_p(z_p z_r z_s\, t^{\zeta\,0}) + (-3+3\eta)\,\partial_p\partial_q(z_p z_q z_r z_s\, t^{\zeta\,0}) \\ &+ \partial_q\partial_p\partial_u(z_q z_p z_u z_r z_s\, t^{\zeta\,0})\Big)\,, \end{aligned}$$

respectively. Combining these equations and using differential renormalization, we get

$$\begin{aligned} t^\zeta = \frac{-1}{\eta^3(1+\eta)^3(2+\eta)^3}\Big(&(4+6\eta-9\eta^2-20\eta^3+\mathcal{O}(\eta^4))\,\partial_p\partial_r\partial_s\overline{(z_p z_r z_s\, t^{\zeta\,0})} \\ &+ (-12-42\eta-45\eta^2+20\eta^3+\mathcal{O}(\eta^4))\,\partial_p\partial_q\partial_r\partial_s\overline{(z_p z_q z_r z_s\, t^{\zeta\,0})} \\ &+ (4+18\eta+33\eta^2+24\eta^3+\mathcal{O}(\eta^4))\,\partial_p\partial_q\partial_u\partial_r\partial_s\overline{(z_p z_q z_u z_r z_s\, t^{\zeta\,0})}\Big)\,. \end{aligned}$$

We explicitly see that this regularization $\{t^\zeta\}$ is analytic. For a change, we consider t^ζ as a Laurent series in η (instead of ζ). With that, t^{MS} is the coefficient $\sim \eta^0$. Using the expansion

$$\begin{aligned} \frac{1}{\eta^3(1+\eta)^3(2+\eta)^3}\,(M^4XY)^{-\eta/4} &= \frac{1}{8\eta^3}\Big(1-\eta\big(\tfrac{9}{2}+\tfrac{1}{4}\log(M^4XY)\big) \\ &+ \eta^2\big(12+\tfrac{9}{8}\log(M^4XY)+\tfrac{1}{32}\log^2(M^4XY)\big) \\ &- \eta^3\big(\tfrac{99}{4}+3\log(M^4XY)+\tfrac{9}{64}\log^2(M^4XY)+\tfrac{1}{384}\log^3(M^4XY)\big)+\mathcal{O}(\eta^4)\Big)\,, \end{aligned}$$

we end up with

$$t^{\mathrm{MS}} = \frac{1}{8}\Big(\partial_p\partial_r\partial_s\overline{\big(z_p z_r z_s\, t^0\,[\frac{13}{2} + 3\,\log(M^4XY) + \frac{3}{8}\,\log^2(M^4XY) + \frac{1}{96}\,\log^3(M^4XY)]\big)}$$
$$- \partial_q\partial_p\partial_r\partial_s\overline{\big(z_q z_p z_r z_s\, t^0\,[\frac{31}{2} + \frac{3}{8}\,\log^2(M^4XY) + \frac{1}{32}\,\log^3(M^4XY)]\big)}$$
$$+ \partial_q\partial_p\partial_u\partial_r\partial_s\overline{\big(z_q z_p z_u z_r z_s\, t^0\,[\frac{15}{2} + \frac{1}{96}\,\log^3(M^4XY)]\big)}\Big)\ .$$

The remarks (given after (3.5.61)) about the almost homogeneous scaling of the extensions v (3.5.61) and in particular about the scaling-power, hold also for the extension t^{MS} obtained here.]

Overlapping divergences. We finish Sect. 3.5 by pointing out that the renormalization of diagrams with overlapping divergences is significantly less troublesome in the inductive Epstein–Glaser construction than in traditional renormalization schemes.

Remark 3.5.20. Overlapping divergences caused a lot of problems in the development of renormalization theory.[84] In this remark we explain their role in Epstein–Glaser renormalization. For simplicity we work with T-products (instead of R-products).

Roughly speaking, a Feynman diagram Γ has overlapping divergences, if and only if it contains two divergent subdiagrams[85] $\Gamma_1(\subseteq \Gamma)$ and $\Gamma_2(\subseteq \Gamma)$ such that

$$\Gamma_1 \not\subseteq \Gamma_2 \ \wedge\ \Gamma_2 \not\subseteq \Gamma_1 \ \wedge\ \quad \Gamma_1 \text{ and } \Gamma_2 \text{ share at least one inner line.} \tag{3.5.66}$$

Note that the first two conditions imply that $\Gamma_1 \neq \Gamma$ and $\Gamma_2 \neq \Gamma$. We give some examples:

(a) The setting sun diagram (see (3.2.124)) in $d = 4$ dimensions has overlapping divergences; the overlapping divergent subdiagrams Γ_1 and Γ_2 are both fish diagrams.

(b) The "setting sun with a hat" in $d = 4$ dimensions (computed in Exap. 3.5.16) has several overlapping divergences. To fulfill the criterion (3.5.66) one may choose for Γ_1 and Γ_2 two fish diagrams which are both subdiagrams of the setting sun subdiagram. Alternatively, one may choose

$\Gamma_1 :=$  x_1 x_2 and $\Gamma_2 :=$ x_3 x_1 x_2

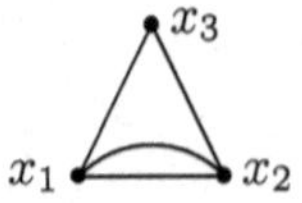

(c) The diagram (3.5.63) in $d = 4$ dimensions, which contains two fish diagrams as divergent subdiagrams, does not have any overlapping divergences, because these fish diagrams do not share an inner line.

[84] For a short overview of the history of overlapping divergences see [174, Sect. 18.2].

[85] The notions "subdiagram" and "divergent diagram" are defined in Remark 3.3.4 and Exap. 3.5.14, respectively.

(d) A prototype of a diagram with overlapping divergences is

$$x_1 \quad x_2 \quad x_3 \quad x_4 \qquad \text{in } d = 6 \text{ dimensions.} \tag{3.5.67}$$

Note that the two triangle subdiagrams, Γ_1 with vertices $\{x_1, x_2, x_3\}$ and Γ_2 with vertices $\{x_2, x_3, x_4\}$, have both singular order $\omega = 0$.

From these examples we see that, in the framework of the inductive Epstein–Glaser construction, the renormalization of a lot of diagrams with overlapping divergences is not significantly more difficult than for other diagrams. For example, this holds for (b) compared with (c) (both have $n = 3$ vertices and $\omega = 2$), or for (a) compared with the fish diagram in $d = 6$ dimensions (both have $n = 2$ vertices and $\omega = 2$, that the number of loops is different is unimportant for the Epstein–Glaser construction).

In the remainder of this remark we study the case in which the divergent and overlapping subdiagrams Γ_1 and Γ_2 have both *less vertices* than the overall diagram Γ for all possible choices of Γ_1 and Γ_2. From the above examples only (d) satisfies this condition. In this case, a simplifying feature of the Epstein–Glaser construction is that *the subdiagrams Γ_1 and Γ_2 are renormalized in preceding inductive steps.* However, frequently the following difficulty appears, which is typical for overlapping divergences of this kind: To write down the distribution $t^0 \in \mathcal{D}'(\mathbb{R}^{d(n-1)} \setminus \{0\})$ which gives the T-product of the overall diagram Γ outside the thin diagonal $x_1 = \cdots = x_n$ (where $x_1, \ldots, x_n$ denotes the vertices of Γ) *one needs a partition of unity*, see (3.3.18).

We illustrate this for the example (d) (cf. [63, Exap. 4.16]): For simplicity we assume that the mass m is $m = 0$. The overall diagram (3.5.67) has $\omega = 2$; we use the notations $x_{kl} := x_k - x_l$ and $\mathbf{x} := (x_{12}, x_{13}, x_{14}) \in \mathbb{R}^{18}$. Let $t_\triangle(x_{12}, x_{13}) \in \mathcal{D}'(\mathbb{R}^{12})$ be the renormalized T-product belonging to the subdiagram Γ_1 (i.e., $t_\triangle(x_{12}, x_{13})$ is an extension of $D^F(x_{12})\, D^F(x_{23})\, D^F(x_{13}) \in \mathcal{D}'(\mathbb{R}^{12} \setminus \{0\})$). With that, the corresponding distribution for Γ_2 is $t_\triangle(x_{42}, x_{43})$. To write down $t^0 \in \mathcal{D}'(\mathbb{R}^{18} \setminus \{0\})$ (belonging to the overall diagram), we consider the expressions

$$t_1(\mathbf{x}) := t_\triangle(x_{12}, x_{13})\, D^F(x_{24})\, D^F(x_{34}) \ ,$$
$$t_2(\mathbf{x}) := t_\triangle(x_{42}, x_{43})\, D^F(x_{12})\, D^F(x_{13}) \ .$$

Restricted[86] to $\mathcal{D}(\check{\mathbb{R}}^{18})$, these distributions $t_1(\mathbf{x})$ and $t_2(\mathbf{x})$ both agree with

$$t_{\text{unren}}(\mathbf{x}) := D^F(x_{12})\, D^F(x_{13})\, D^F(x_{23})\, D^F(x_{24})\, D^F(x_{34}) \ ,$$

[86] See (3.5.50) for the definition of $\check{\mathbb{R}}^{18}$.

that is, both are terms expressing the diagram Γ. They are well defined even in the following spaces:

$$t_1 \in \mathcal{D}'\big(\mathbb{R}^{18} \setminus \{\mathbf{x}\,|\,x_{24} = 0 = x_{34}\}\big) , \quad t_2 \in \mathcal{D}'\big(\mathbb{R}^{18} \setminus \{\mathbf{x}\,|\,x_{12} = 0 = x_{13}\}\big) .$$

But neither t_1 nor t_2 exists in $\mathcal{D}'(\mathbb{R}^{18} \setminus \{0\})$. Hence, we need a partition of unity: For $\mathbf{x} \in \mathbb{R}^{18} \setminus \{0\}$ let

$$1 = f_1(\mathbf{x}) + f_2(\mathbf{x}) \quad \text{with} \quad f_1, f_2 \in C^\infty(\mathbb{R}^{18} \setminus \{0\}) \tag{3.5.68}$$

and

$$\operatorname{supp} f_1 \subseteq \mathbb{R}^{18} \setminus \{\mathbf{x}\,|\,x_{24} = 0 = x_{34}\} , \quad \operatorname{supp} f_2 \subseteq \mathbb{R}^{18} \setminus \{\mathbf{x}\,|\,x_{12} = 0 = x_{13}\} . \tag{3.5.69}$$

Since

$$t_1(\mathbf{x}) = t_{\text{unren}}(\mathbf{x}) = t_2(\mathbf{x}) \quad \text{on} \quad \mathcal{D}(\mathbb{R}^{18} \setminus \{\mathbf{x}\,|\,x_{24} = 0 = x_{34} \vee x_{12} = 0 = x_{13}\}) \tag{3.5.70}$$

we may construct t^0 as follows:[87]

$$t^0(\mathbf{x}) = f_1(\mathbf{x})\,t_1(\mathbf{x}) + f_2(\mathbf{x})\,t_2(\mathbf{x}) \in \mathcal{D}'(\mathbb{R}^{18} \setminus \{0\}) . \tag{3.5.71}$$

Due to (3.5.70), the so-constructed t^0 does not depend on the choice of (f_1, f_2). Therefore, applying analytic regularization and minimal subtraction to this formula for t^0, the resulting t^{MS} is also independent of the choice of (f_1, f_2).

However, to compute t^{MS} we need to know (f_1, f_2) explicitly. An admissible choice for this pair of functions can be obtained as follows: Let χ be a smooth approximation of the Heaviside-function $\theta(x)$ with $\operatorname{supp} \frac{d\chi}{dx} \subseteq [-\varepsilon, \varepsilon]$ for a sufficiently small $\varepsilon > 0$.[88] For $x \in \mathbb{R}^6$ we mean by $|x|$ the Euclidean norm. We set

$$g_2(\mathbf{x}) := \begin{cases} \chi\Big(\frac{|x_{12}|^2 + |x_{13}|^2}{|x_{14}|^2} - a\Big) & \text{if} \quad x_{14} \neq 0 \\ 1 & \text{if} \quad x_{14} = 0 \ \wedge\ \mathbf{x} \neq \mathbf{0} \end{cases} , \tag{3.5.72}$$

with a sufficiently small $a > 0$. Note that $g_2 \in C^\infty(\mathbb{R}^{18} \setminus \{0\})$. Let

$$g_1(\mathbf{x}) \equiv g_1(x_{12}, x_{13}, x_{14}) := g_2(x_{24}, x_{34}, x_{14}) . \tag{3.5.73}$$

---

[87] This is an optimized version of the construction (3.3.18) – optimized in the sense that the partition of unity has a minimal number of elements.

[88] The standard construction of such a function $\chi(x)$ goes as follows: First recall that

$$k\colon \mathbb{R} \longrightarrow \mathbb{R};\ k(x) := \begin{cases} 0 & \text{for} \quad x \leq -\varepsilon \\ e^{-1/(x+\varepsilon)} & \text{for} \quad x > -\varepsilon \end{cases}$$

is smooth. Then

$$\chi(x) := \frac{k(x)}{k(x) + k(-x)}$$

is smooth, equal to 1 for $x \geq \varepsilon$ and equal to 0 for $x \leq -\varepsilon$.

The sets

$$K_j := \{\mathbf{x} \neq \mathbf{0} \,|\, g_j(\mathbf{x}) = 0\} \ , \quad j = 1,2 \ , \tag{3.5.74}$$

are narrow cones around the diagonal $x_{12} = x_{13} = x_{14}$ (for g_1) and the x_{14}-axis (for g_2). Since $K_1 \cap K_2 = \emptyset$, we have $g_1(\mathbf{x}) + g_2(\mathbf{x}) > 0$ for all $\mathbf{x} \neq \mathbf{0}$ and, hence, we may set

$$f_j(\mathbf{x}) := \frac{g_j(\mathbf{x})}{g_1(\mathbf{x}) + g_2(\mathbf{x})} \ , \quad j = 1,2 \ . \tag{3.5.75}$$

Obviously the pair (f_1, f_2) satisfies the required properties (3.5.68) and (3.5.69), and is also $(1 \leftrightarrow 4)$-symmetric and scales homogeneously with degree 0:

$$f_1(\mathbf{x}) = f_2(-x_{24}, -x_{34}, -x_{14}) \ , \quad f_j(\rho\mathbf{x}) = f_j(\mathbf{x}) \quad \forall \rho \neq 0 \ . \tag{3.5.76}$$

It seems that the construction given in this example can be generalized to a systematic construction of a partition of unity for all diagrams which need this tool when we construct $t^0 \in \mathcal{D}'(\mathbb{R}^k \setminus \{0\})$ by Stora's extension method (i.e., the method briefly explained in Sect. 3.3). However, this is not yet worked out.

We recall that the use of a partition of unity can be avoided by working with the distribution splitting method of Epstein and Glaser [66, 148] (see also Exer. 3.5.7); but for diagrams without overlapping divergences of the just treated kind, Stora's extension method is mostly much more efficient for an inductive construction of the T-product.

3.6 The Stückelberg–Petermann renormalization group

References for this section are [27, 55, 57] which use ideas developed in [102].

It is one of the main insights of renormalization theory that the ambiguities of the renormalization process can equivalently be described by redefinitions of the parameters of the given model, and that these redefinitions form a group – the "renormalization group" (RG). This was first observed by E.C.G. Stückelberg and A. Petermann in 1953 [163].

In our framework of causal perturbation theory the corresponding statement is the "Main Theorem of perturbative renormalization". The main difference to traditional formulations of the renormalization group in pQFT is that in causal perturbation theory the parameters of a model are spacetime-dependent coupling constants $\kappa\, g(x)$, that is, they are test functions. Therefore, the renormalization group which governs the change of parameters is more complicated, and it is only in the adiabatic limit $g(x) \to 1$ that the more standard version of the renormalization group will be recovered. Fortunately, an algebraic version of this limit, which circumvents the problem of infrared divergences [24], is sufficient for this purpose [27, 55, 102], as explained in Sects. 3.7–3.8. In this way, one finds an intrinsically local construction of the renormalization group which is suited for theories on curved spacetime and for theories with a bad infrared behavior.

3.6.1 Definition of the Stückelberg–Petermann RG

In terms of the S-matrix $\mathbf{S}(F)$ (Def. 3.3.8) the Main Theorem is essentially the following statement: A change $\mathbf{S} \mapsto \hat{\mathbf{S}}$ of the renormalization prescription can equivalently be described by a renormalization of the interaction $F \mapsto Z(F)$:

$$\hat{\mathbf{S}}(F) = \mathbf{S}\big(Z(F)\big) \quad \forall F \in \mathcal{F}_{\mathrm{loc}} \,. \tag{3.6.1}$$

The definition of the Stückelberg–Petermann RG is such that the set of all maps $Z\colon \mathcal{F}_{\mathrm{loc}} \longrightarrow \mathcal{F}_{\mathrm{loc}}$ appearing in this relation, when $(\mathbf{S}, \hat{\mathbf{S}})$ runs through all admissible pairs of S-matrices, is precisely the Stückelberg–Petermann RG.

Consequently, the definition of the Stückelberg–Petermann RG depends on the set of renormalization conditions for the T-product [or R-product] – see the comment (a) below and Remk. 3.6.6. The following Definition works with the renormalization conditions (v)–(x) [or (f)–(l)] given in Sect. 3.3 [or Sect. 3.1, respectively] *without* the addition (k′) to the Sm-expansion axiom (introduced in Sect. 3.2.6).

Definition 3.6.1. The *Stückelberg–Petermann renormalization group* to the star product $\star_m \equiv \star_{\Delta_m^+}$ is the set $\mathcal{R}$ of all maps $Z \equiv Z_{(m)}\colon \mathcal{F}_{\mathrm{loc}}[\![\hbar,\kappa]\!] \longrightarrow \mathcal{F}_{\mathrm{loc}}[\![\hbar,\kappa]\!]$ satisfying the following properties:[89]

(1) *Analyticity*: Z is analytic in the sense that

$$Z(F) = \sum_{n=0}^{\infty} \frac{1}{n!}\, Z^{(n)}(F^{\otimes n}) \,, \quad \text{where} \quad Z^{(n)} := Z^{(n)}(0) \,:\, \mathcal{F}_{\mathrm{loc}}^{\otimes n} \longrightarrow \mathcal{F}_{\mathrm{loc}}$$

is the nth derivative of $Z(F)$ at $F = 0$ (i.e., $Z^{(n)}$ is linear, symmetrical in all factors and can be computed by $Z^{(n)}(F^{\otimes n}) = \frac{d^n}{d\lambda^n}\big|_{\lambda=0} Z(\lambda F)$; see (A.1.19)–(A.1.22)).

(2) *Lowest orders*: $Z^{(0)} \equiv Z(0) = 0, \;\; Z^{(1)} = \mathrm{Id}$.

(3) *Locality, Translation covariance and Sm-expansion*: For all monomials $A_1, \ldots$ $\ldots, A_n \in \mathcal{P}$ the VEV

$$z^{(n)}_{(m)}(A_1, \ldots, A_n)(x_1 - x_n, \ldots) := \omega_0\Big(Z^{(n)}_{(m)}\big(A_1(x_1) \otimes \cdots \otimes A_n(x_n)\big)\Big) \tag{3.6.2}$$

depends only on the relative coordinates and is of the form

$$z^{(n)}_{(m)}(A_1, \ldots, A_n)(x_1 - x_n, \ldots) = \sum_{l=0}^{\omega} m^l \sum_{p=0}^{P_l} \log^p\Big(\frac{m}{M}\Big) \tag{3.6.3}$$

$$\cdot\, \mathfrak{S}_n \sum_{|a|=\omega-l} C_{l,p,a}(A_1, \ldots, A_n)\, \partial^a \delta(x_1 - x_n, \ldots, x_{n-1} - x_n) \,,$$

[89] For brevity we write $\mathcal{F}_{\mathrm{loc}}$ for $\mathcal{F}_{\mathrm{loc}}[\![\hbar,\kappa]\!]$.

with m-independent numbers $C_{l,p,a}(A_1,\dots,A_n)\in\mathbb{C}$, where $\mathfrak{S}_n$ denotes symmetrization in $(A_1,x_1),\dots,(A_n,x_n)$,

$$\omega\equiv\omega(A_1,\dots,A_n):=\sum_{j=1}^{n}\dim A_j-d(n-1) \tag{3.6.4}$$

and $M>0$ is a fixed mass scale. In addition, for $l=0$ it holds that $P_l=0$.

(4) *Field independence*: $\delta Z/\delta\varphi=0$, which is equivalent to $\delta Z^{(n)}/\delta\varphi=0\ \forall n$, or more explicitly

$$\frac{\delta Z(F)}{\delta\varphi(x)}=\sum_{n=1}^{\infty}\frac{1}{(n-1)!}\,Z^{(n)}\Big(\frac{\delta F}{\delta\varphi(x)}\otimes F^{\otimes(n-1)}\Big)\qquad\forall F\in\mathcal{F}_{\mathrm{loc}}\,. \tag{3.6.5}$$

(5) *Lorentz covariance*: $\beta_\Lambda Z(F)=Z(\beta_\Lambda F)$ for all $\Lambda\in\mathcal{L}_+^\uparrow$, $F\in\mathcal{F}_{\mathrm{loc}}$.

(6) *$*$-structure and Field parity*: $Z(F)^*=Z(F^*)$ and $\alpha\,Z(F)=Z(\alpha F)$, $\forall F\in\mathcal{F}_{\mathrm{loc}}$.

(7) *Field equation*: $Z^{(n)}\big(\varphi(g)\otimes F^{\otimes(n-1)}\big)=0\quad\forall n\geq 2\,,\ g\in\mathcal{D}(\mathbb{M})\,,\ F\in\mathcal{F}_{\mathrm{loc}}$.

(8) *$\hbar$-dependence*: for all monomials $A_1,\dots,A_n$ it holds that

$$z^{(n)}(A_1,\dots,A_n)\sim\hbar^{1-n+\sum_{j=1}^{n}|A_j|/2}\quad\text{if}\quad A_1,\dots,A_n\sim\hbar^0\,, \tag{3.6.6}$$

where we use the convention given in Footn. 33.

As already done in (3.6.2), we will work also with the $\mathcal{F}_{\mathrm{loc}}$-valued distributions $Z^{(n)}(A_1(x_1)\otimes\cdots\otimes A_n(x_n))\in\mathcal{D}'(\mathbb{M}^n,\mathcal{F}_{\mathrm{loc}})$, $A_1,\dots,A_n\in\mathcal{P}$; they are defined analogously to (3.1.3)–(3.1.4) and they satisfy the AWI,

$$\partial_{x_l}Z^{(n)}(\cdots\otimes A(x_l)\otimes\cdots)=Z^{(n)}(\cdots\otimes(\partial A)(x_l)\otimes\cdots)\,,\quad A\in\mathcal{P}\,, \tag{3.6.7}$$

because $Z^{(n)}$ depends on (local) functionals only.

We make some *comments*:

(a) To each renormalization condition for the T-product (or R-product) there is a corresponding defining condition for $\mathcal{R}$. In the definition of $\mathcal{R}$ given here, Translation covariance and Sm-expansion are combined with the basic requirement "Locality"; cf. Remk. 3.6.6.

(b) Assuming that Field independence (and, hence, the Wick expansion (3.6.9) below) hold true, the Field parity property is equivalent to:

$$z^{(n)}(A_1,\dots,A_n)=0\quad\text{if}\quad\sum_{j=1}^{n}|A_j|\quad\text{is odd, }\forall\text{ monomials }A_1,\dots,A_n. \tag{3.6.8}$$

This can be seen by proceeding analogously to the derivation of (3.1.34).

(c) Note that

$$z^{(n)}(A_1,\ldots,A_n)=0 \quad \text{if some } A_j \text{ is a } C\text{-number;}$$

this follows from property (3): If, e.g., A_1 is a C-number, then the VEV (3.6.2) does not depend on x_1, this contradicts (3.6.3).

Considering the VEV of the Field equation property we get

$$z^{(n)}(A_1,\ldots,A_n)=0 \quad \text{if} \quad A_k=\partial^a\varphi \quad \text{for some} \quad k\in\{1,\ldots,n\}\ ,\ a\in\mathbb{N}^d\ .$$

(d) Similarly to the axiom Field independence for the R- or T-product, the Field independence property (4) is equivalent to the validity of the (causal) Wick expansion (3.1.24) for $Z^{(n)}$, $\forall n\geq 2$; that is, for all *monomials* $A_1,\ldots,A_n\in\mathcal{P}$ it holds that

$$\begin{aligned} &Z^{(n)}\big(A_1(x_1),\ldots,A_n(x_n)\big) \\ &\quad=\sum_{\underline{A}_l\subseteq A_l} z^{(n)}(\underline{A}_1,\ldots,\underline{A}_n)(x_1-x_n,\ldots)\,\overline{A}_1(x_1)\cdots\overline{A}_n(x_n)\ . \end{aligned} \tag{3.6.9}$$

Hence, by linearity, $Z^{(n)}$ is completely determined by the vacuum expectation values (3.6.2). In particular, the condition that $z^{(n)}(\ldots)$ depends only on the relative coordinates (part of property (3)) is equivalent to Translation covariance in the form: $\beta_a\circ Z=Z\circ\beta_a$ for all $a\in\mathbb{R}^d$.

We point out that the sum in (3.6.9) is restricted by the conditions

$$\omega(\underline{A}_1,\ldots,\underline{A}_n)\geq 0 \quad\wedge\quad |\underline{A}_l|\geq 2\ \forall 1\leq l\leq n \quad\wedge\quad \sum_{l=1}^n|\underline{A}_l| \text{ is even.} \tag{3.6.10}$$

The first restriction follows from (3.6.3)–(3.6.4), the second and the last one are reformulations of the comments (c) and (b), respectively.

Inserting (3.6.3) into (3.6.9) and taking into account linearity of $Z^{(n)}$, we conclude that

$$Z^{(n)}\big(A_1(x_1),\ldots,A_n(x_n)\big)\Big|_{\mathcal{D}(\mathbb{M}^n\setminus\Delta_n)}=0 \quad \forall A_1,\ldots,A_n\in\mathcal{P}\ , \tag{3.6.11}$$

that is $\operatorname{supp} Z^{(n)}\big(A_1(x_1),\ldots\big)\subseteq\Delta_n$ for the support in the sense of distributions; see (A.1.12) for the notation.

(e) A further consequence of the Field independence (3.6.5) is that $\operatorname{supp} Z(F)\equiv \operatorname{supp}_x\frac{\delta Z(F)}{\delta\varphi(x)}$ is a subset of $\operatorname{supp} F$ (cf. (3.1.21)); taking into account also $Z(F)=F+[\text{higher orders in } F]$, we see that

$$\operatorname{supp} Z(F)=\operatorname{supp} F \quad \forall F\in\mathcal{F}_{\mathrm{loc}}\ . \tag{3.6.12}$$

(f) In the $\hbar$-dependence property, the exponent of $\hbar$ on the r.h.s. of (3.6.6) is an element of $\mathbb{N}^*$; because $z^{(n)}(A_1,\ldots,A_n) = 0$ if $\sum_{j=1}^n |A_j| \notin 2(n+\mathbb{N})$, as we know from the comments (b) and (c). This implies $z^{(n)}(A_1,\ldots,A_n) = \mathcal{O}(\hbar)$ for all monomials $A_1,\ldots,A_n$ which are $\sim \hbar^0$. The latter result can be written as

$$Z^{(n)} = \mathcal{O}(\hbar) \qquad \forall n \geq 2\ , \quad \forall Z \in \mathcal{R}\ , \tag{3.6.13}$$

by taking into account the causal Wick expansion (3.6.9).

(g) For a *massless* model the condition (3.6.3) simplifies to

$$\begin{aligned} &z^{(n)}(A_1,\ldots,A_n)(x_1-x_n,\ldots) \\ &\quad = \mathfrak{S}_n \sum_{|a|=\omega} C_a(A_1,\ldots,A_n)\,\partial^a\delta(x_1-x_n,\ldots,x_{n-1}-x_n)\ ; \end{aligned}$$

hence, $z^{(n)}(A_1,\ldots,A_n)$ *scales homogeneously* with degree $D=\sum_{j=1}^n \dim A_j$ for all monomials $A_1,\ldots,A_n$. Taking into account (3.6.9) and linearity of $Z^{(n)}$, the latter is equivalent to

$$\sigma_\rho \circ Z \circ \sigma_\rho^{-1} = Z \quad \text{for} \quad m = 0\ ; \tag{3.6.14}$$

this is analogous to the equivalence of (3.1.56) and (3.1.57), in particular see (3.1.58)–(3.1.59).

But, proceeding analogously for $m>0$, we conclude from (3.6.3) that $Z^{(n)}_{(m)}$ scales only *almost homogeneously* under $(X,m)\mapsto(\rho X, m/\rho)$ (where $X := (x_1,\ldots,x_n)$):

$$\sigma_\rho \circ Z^{(n)}_{(m/\rho)}\big(\sigma_\rho^{-1}(F)^{\otimes n}\big) \quad \text{is a polynomial in } \log\rho\ ,\ \forall F\in\mathcal{F}_{\mathrm{loc}}\ ,\ n\in\mathbb{N}^*\ , \tag{3.6.15}$$

cf. (3.1.67).

The further comments investigate the structure of the set $\mathcal{R}$.

(h) That $(\mathcal{R},\circ)$ (where $\circ$ denotes the composition of maps) is a *group*, is proved below in Corollary 3.6.4.

(i) For $k\geq 2$ let $\tilde Z^{(k)}$ be a linear and symmetrical map $\tilde Z^{(k)}:\mathcal{F}_{\mathrm{loc}}^{\otimes k}\longrightarrow \mathcal{F}_{\mathrm{loc}}$ and let

$$\tilde Z := \mathrm{Id} + \sum_{k=2}^{\infty}\frac{1}{k!}\,\tilde Z^{(k)}\circ\tau_k\ , \quad \text{where} \quad \tau_k: \begin{cases} \mathcal{F}_{\mathrm{loc}} \longrightarrow \mathcal{F}_{\mathrm{loc}}^{\otimes k} \\ \tau_k(F) := F^{\otimes k}\ . \end{cases} \tag{3.6.16}$$

Note that we do not require that $\tilde Z$ is an element of $\mathcal{R}$. In addition, for $n\in\mathbb{N}^*$ let

$$\tilde Z_n := \mathrm{Id} + \sum_{k=2}^{n}\frac{1}{k!}\,\tilde Z^{(k)}\circ\tau_k \quad \text{be a truncation of } \tilde Z. \tag{3.6.17}$$

From Definition 3.6.1 we conclude that

$$\tilde{Z} \in \mathcal{R} \iff \tilde{Z}_n \in \mathcal{R} \quad \forall n \in \mathbb{N}^* . \tag{3.6.18}$$

(j) A related consequence of Definition 3.6.1 is that $\mathcal{R}$ has the structure of an *affine space* [63]; more precisely we mean the following: Taking into account the defining property (2) and (3.6.13), we write $\mathcal{R}$ as

$$\mathrm{Id} + \hbar\, \mathcal{V}[\![\hbar]\!] := \mathcal{R} . \tag{3.6.19}$$

Then, the statement is that the set $\mathcal{V}[\![\hbar]\!]$ is a *vector space.* This can be verified straightforwardly.

By means of the result (3.6.19), the statement (3.6.18) can equivalently be reformulated as follows:

$$\tilde{Z} \in \mathcal{R} \iff \tilde{Z}^{(k)} \circ \tau_k \in \hbar\, \mathcal{V}[\![\hbar]\!] \quad \forall k \geq 2 .$$

Exercise 3.6.2 (Properties of $Z \in \mathcal{R}$)**.** Let $Z \in \mathcal{R}$ arbitrary. Prove

(a) $Z(F_1 + F_2 + F_3) = Z(F_1 + F_2) - Z(F_2) + Z(F_2 + F_3)$ if $\operatorname{supp} F_1 \cap \operatorname{supp} F_3 = \emptyset$, which expresses Locality of Z, as discussed below in (3.6.33);

(b) and

$$\operatorname{supp}\bigl(Z(F_1 + F_2) - Z(F_1)\bigr) = \operatorname{supp} F_2 \;\Bigl(= \operatorname{supp} Z(F_2)\Bigr) ,$$

which can be understood as a weak substitute for linearity.

[*Solution*: Due to linearity of $Z^{(n)}$, we may assume that $F_j = \int dx\; A_j(x)\, g_j(x)$, where $A_j \in \mathcal{P}$, $g_j \in \mathcal{D}(\mathbb{M})$ and there is no sum over j. Taking into account (3.6.10), we may also assume that $A_j \neq \omega_0(A_j)$; this implies $\operatorname{supp} F_j = \operatorname{supp} g_j$, as we see from (1.3.11).
(a) Let $\operatorname{supp} g_1 \cap \operatorname{supp} g_3 = \emptyset$. Writing

$$Z(F_1 + F_2 + F_3) = (F_1 + F_2) - F_2 + (F_2 + F_3) + \sum_{n=2}^{\infty} \frac{1}{n!} Z^{(n)}\Bigl(\Bigl(\sum_{j=1}^{3} F_j\Bigr)^{\otimes n}\Bigr) \quad \text{with}$$

$$Z^{(n)}\Bigl(\Bigl(\sum_{j=1}^{3} F_j\Bigr)^{\otimes n}\Bigr)$$

$$= \sum_{j_1,\ldots,j_n \in \{1,2,3\}} \int dx_1 \cdots dx_n\; Z^{(n)}\bigl(A_{j_1}(x_1), \ldots, A_{j_n}(x_n)\bigr)\, g_{j_1}(x_1) \cdots g_{j_n}(x_n) ,$$

we conclude from (3.6.11) that the sum over $j_1, \ldots, j_n \in \{1,2,3\}$ can be written as: [Sum over $j_1, \ldots, j_n \in \{1,2\}$]$-$[($j_1 = \ldots = j_n = 2$)-term]+[sum over $j_1, \ldots, j_n \in \{2,3\}$]. This yields

$$Z^{(n)}\bigl((F_1 + F_2 + F_3)^{\otimes n}\bigr) = Z^{(n)}\bigl((F_1 + F_2)^{\otimes n}\bigr) - Z^{(n)}\bigl(F_2^{\otimes n}\bigr) + Z^{(n)}\bigl((F_2 + F_3)^{\otimes n}\bigr) ,$$

from which we get the assertion.

(b) Looking at

$$Z^{(n)}\big((F_1+F_2)^{\otimes n}\big) - Z^{(n)}\big(F_1^{\otimes n}\big)$$
$$= \int dx_1\cdots dx_n \Big[-Z^{(n)}\big(A_1(x_1),\ldots,A_1(x_n)\big)\,g_1(x_1)\cdots g_1(x_n)$$
$$+ \sum_{j_1,\ldots,j_n\in\{1,2\}} Z^{(n)}\big(A_{j_1}(x_1),\ldots,A_{j_n}(x_n)\big)\,g_{j_1}(x_1)\cdots g_{j_n}(x_n)\Big]$$

and taking into account (3.6.11), we see that the expression in $[\cdots]$-brackets fulfills $[\ldots(x_1,\ldots,x_n)] = 0$ if $x_j \notin \operatorname{supp} g_2$ for some $j\in\{1,\ldots,n\}$. Therefore, the integral $\int dx_1\cdots dx_n$ may be restricted to $(\operatorname{supp} g_2)^{\otimes n}$; that is,

$$\operatorname{supp}\Big(Z^{(n)}\big((F_1+F_2)^{\otimes n}\big) - Z^{(n)}\big(F_2^{\otimes n}\big)\Big) \subseteq \operatorname{supp} g_2 = \operatorname{supp} F_2\ .\ ]$$

3.6.2 The Main Theorem of perturbative renormalization

In the framework of causal perturbation theory, a first version of this theorem (including the name) was given by G. Popineau and R. Stora [138], later versions can be found in [88, 102, 136]; we present here the more elaborated version of [55], translated from R-product into T-product – this translation simplifies the central formula (3.6.20), see Exer. 3.6.11.

Theorem 3.6.3 (Main Theorem).

(a) *Given two renormalization prescriptions, i.e., two time-ordered products $T = (T_n)$ and $\widehat{T} = (\widehat{T}_n)$ both fulfilling the axioms, there exists an analytic map $Z\colon \mathcal{F}_{\rm loc} \longrightarrow \mathcal{F}_{\rm loc}$ (i.e., Z satisfies the property* (1) *of Definition* 3.6.1*), which is uniquely determined by*

$$\widehat{T}(e_\otimes^{iF/\hbar}) = T\big(e_\otimes^{iZ(F)/\hbar}\big)\quad \forall F\in\mathcal{F}_{\rm loc}\ ,\qquad \textit{or equivalently}\quad \widehat{\mathbf{S}} = \mathbf{S}\circ Z\ . \tag{3.6.20}$$

This Z is an element of the Stückelberg–Petermann renormalization group $\mathcal{R}$.

(b) *Conversely, given an S-matrix $\mathbf{S}$ fulfilling the axioms for the time-ordered product and an arbitrary $Z\in\mathcal{R}$, the composition $\widehat{\mathbf{S}} := \mathbf{S}\circ Z$ also satisfies these axioms.*

Corollary 3.6.4 (Group properties). *If $Z_1, Z_2\in\mathcal{R}$, then $Z_1\circ Z_2\in\mathcal{R}$. In addition, for any $Z\in\mathcal{R}$, the inverse Z^{-1} is also an element of $\mathcal{R}$. Due to these properties, $(\mathcal{R},\circ)$ is indeed a group.*

Proof of the corollary. Let $Z_1, Z_2\in\mathcal{R}$, and choose an arbitrary S-matrix $\mathbf{S}$. Theorem 3.6.3(b) shows that $\mathbf{S}_1 := \mathbf{S}\circ Z_1$ and $\mathbf{S}_2 := \mathbf{S}_1\circ Z_2$ also satisfy the axioms for time-ordered products. Clearly $\mathbf{S}_2 = \mathbf{S}\circ(Z_1\circ Z_2)$, so it follows from the uniqueness statement of Theorem 3.6.3(a) that $Z_1\circ Z_2\in\mathcal{R}$.

The inverse of Z_1 can be obtained as follows: To the pair $(\mathbf{S}_1,\mathbf{S})$ there exists a unique Z_3 fulfilling $\mathbf{S} := \mathbf{S}_1\circ Z_3$; with that we have $\mathbf{S} = \mathbf{S}\circ(Z_1\circ Z_3)$ and

$\mathbf{S}_1 = \mathbf{S}_1 \circ (Z_3 \circ Z_1)$. Applying the uniqueness statement of Theorem 3.6.3(a) to the pairs $(\mathbf{S}, \mathbf{S})$ and $(\mathbf{S}_1, \mathbf{S}_1)$, we get $Z_1 \circ Z_3 = \mathrm{Id}$ and $Z_3 \circ Z_1 = \mathrm{Id}$, respectively. In addition, Theorem 3.6.3(a) says also that $Z_3 \in \mathcal{R}$. □

We shall now write out $\widehat{T}(e_\otimes^{iF/\hbar}) = T(e_\otimes^{iZ(F)/\hbar})$ (3.6.20) to lowest orders in F, using that Z must be analytic, that is, $Z(F) = Z(0) + \sum_{n\geq 1} \frac{1}{n!} Z^{(n)}(F^{\otimes n})$ where $Z^{(n)}$ is linear and symmetrical.

For $n = 0$ (i.e., $F = 0$) we use $T_0(c) = c\ \forall c \in \mathbb{C}$ and obtain $1 = \widehat{T}(1) = T(e_\otimes^{iZ(0)/\hbar})$, which implies $Z(0) = 0$, as claimed.

For $n = 1$, we get $iF/\hbar = \widehat{T}_1(iF/\hbar) = T_1(iZ^{(1)}(F)/\hbar) = iZ^{(1)}(F)/\hbar$, so that $Z^{(1)}(F) = F$, as claimed.

For $n = 2$, we find that

$$\frac{i^2}{\hbar^2}\widehat{T}_2(F_1 \otimes F_2) = \frac{i^2}{\hbar^2} T_2(F_1 \otimes F_2) + \frac{i}{\hbar} Z^{(2)}(F_1 \otimes F_2). \tag{3.6.21}$$

For $n = 3$, the expansion gives

$$\begin{aligned}\frac{i^3}{\hbar^3}\widehat{T}_3(x_1 \otimes x_2 \otimes x_3) = {}& \frac{i^3}{\hbar^3} T_3(x_1 \otimes x_2 \otimes x_3) + \frac{i^2}{\hbar^2} T_2(Z^{(2)}(x_1 \otimes x_2) \otimes x_3) \\ & + \frac{i^2}{\hbar^2} T_2(Z^{(2)}(x_1 \otimes x_3) \otimes x_2) + \frac{i^2}{\hbar^2} T_2(x_1 \otimes Z^{(2)}(x_2 \otimes x_3)) \\ & + \frac{i}{\hbar} Z^{(3)}(x_1 \otimes x_2 \otimes x_3)\ ,\end{aligned} \tag{3.6.22}$$

where we write x_l for $A_l(x_l)$ to simplify the notation. Here, $Z^{(3)}(x_1 \otimes x_2 \otimes x_3)$ is localized on the thin diagonal $x_1 = x_2 = x_3$, while $T_2(Z^{(2)}(x_1 \otimes x_2) \otimes x_3)$ is localized on the partial diagonal $x_1 = x_2$; and so on.

Proof of Theorem 3.6.3. Ad (b): One verifies straightforwardly that

$$\widehat{T}_n \equiv (-i\hbar)^n\, \widehat{\mathbf{S}}^{(n)}(0) := (-i\hbar)^n\, (\mathbf{S} \circ Z)^{(n)}(0) \tag{3.6.23}$$

fulfills the axioms for the time-ordered product by using the corresponding properties of $\mathbf{S}$ and Z.[90] This is an easy task for the *Initial condition* (axiom (ii)), the *Poincaré covariance* (axiom (vii)),

$$\beta_{\Lambda,a}\widehat{\mathbf{S}}(F) = \beta_{\Lambda,a}\mathbf{S}(Z(F)) = \mathbf{S}(\beta_{\Lambda,a}Z(F)) = \mathbf{S}(Z(\beta_{\Lambda,a}F)) = \widehat{\mathbf{S}}(\beta_{\Lambda,a}F)\ ,$$

the $*$-*structure*,

$$\widehat{\mathbf{S}}(F^*)^{\star-1} = \mathbf{S}(Z(F^*))^{\star-1} = \mathbf{S}(Z(F)^*)^{\star-1} = \mathbf{S}(Z(F))^* = \widehat{\mathbf{S}}(F)^*\ , \tag{3.6.24}$$

and the *Field parity* (axiom (vi)).

[90] We are not aware of any place in the literature, where this verification is presented. To get a better understanding of the Stückelberg–Petermann RG and the Main Theorem, we proceed in detail.

To verify Causality (axiom (iv)) we work with the formulation (3.3.28) of this axiom: Let $\operatorname{supp} H \cap (\operatorname{supp} F + \overline{V}_-) = \emptyset$. From Exer. 3.6.2(a) we then know that $Z(H+F) = Z(H) + Z(F)$. By using also $\operatorname{supp} Z(F) = \operatorname{supp} F$ (3.6.12) and Causality of $\mathbf{S}$, we obtain

$$\begin{aligned}\widehat{\mathbf{S}}(H+F) &= \mathbf{S}\big(Z(H+F)\big) = \mathbf{S}\big(Z(H)+Z(F)\big)\\ &= \mathbf{S}\big(Z(H)\big) \star \mathbf{S}\big(Z(F)\big) = \widehat{\mathbf{S}}(H) \star \widehat{\mathbf{S}}(F)\ .\end{aligned}$$

The verification of *Linearity* (axiom (i)), *Symmetry* (axiom (iii)) and *Field independence* (axiom (v)) is easy for $\widehat{T}_2$ (3.6.21) and $\widehat{T}_3$ (3.6.22); the respective argumentations generalize straightforwardly to $\widehat{T}_n$, $n \in (2+\mathbb{N})$, by using

$$\begin{aligned}\frac{i^n}{\hbar^n}\,\widehat{T}_n(\otimes_{j=1}^n F_j) &:= \frac{i^n}{\hbar^n}\,T_n(\otimes_{j=1}^n F_j)\\ &+ \sum_{\substack{P\in\mathrm{Part}(\{1,\ldots,n\})\\ 1<|P|<n}} \Big(\frac{i}{\hbar}\Big)^{|P|} T_{|P|}\Big(\otimes_{I\in P} Z^{(|I|)}(F_I)\Big) + \frac{i}{\hbar}\,Z^{(n)}(\otimes_{j=1}^n F_j)\ ,\end{aligned} \tag{3.6.25}$$

where $F_I := \otimes_{j\in I} F_j$ and $P \in \mathrm{Part}(\{1,\ldots,n\})$ is a partition of $\{1,\ldots,n\}$ into $|P|$ disjoint subsets. The two terms $|P| = n$ and $|P| = 1$, respectively, are explicitly written out. This formula is the generalization of (3.6.21)–(3.6.22) to arbitrary orders n. Formula (3.6.25) is obtained from (3.6.23) by computing the nth derivative on the r.h.s. by means of the chain rule. In the literature, the chain rule for higher derivatives is sometimes called "Faà di Bruno formula" [68].

To simplify the notations we verify the other axioms only for $\widehat{T}_3$. The *Field equation* (axiom (viii)) is obtained as follows:

$$\begin{aligned}\widehat{T}_3\big(\varphi(g)\otimes F_2\otimes F_3\big) &= T_3\big(\varphi(g)\otimes F_2\otimes F_3\big) - i\hbar\, T_2\big(\varphi(g)\otimes Z^{(2)}(F_2\otimes F_3)\big)\\ &= \varphi(g)\star_{\Delta^F}\Big(T_2(F_2\otimes F_3) - i\hbar\, Z^{(2)}(F_2\otimes F_3)\Big) = \varphi(g)\star_{\Delta^F}\widehat{T}_2(F_2\otimes F_3)\end{aligned}$$

by using the Field equation property (7) of $Z^{(2)}$ and $Z^{(3)}$; with regard to the expression $\varphi(g)\star_{\Delta^F}\cdots$ see the comment to (3.3.20).

$\hbar$-dependence (axiom (x)): We know that the first and the last term on the r.h.s. of (3.6.22) satisfy this condition. Let us study the VEV of, e.g., the second term without the factor i^2; using (3.6.9) it can be written as

$$\hbar^{-2}\sum_{\underline{A}_l\subseteq A_l} \omega_0\Big(T_2\big(z^{(2)}(\underline{A}_1,\underline{A}_2)(x_1-x_2)\,\overline{A}_1(x_1)\overline{A}_2(x_2)\otimes A_3(x_3)\big)\Big)\ , \tag{3.6.26}$$

where $z^{(2)}(\underline{A}_1,\underline{A}_2) \sim \hbar^{-1+(|\underline{A}_1|+|\underline{A}_2|)/2}$ (3.6.6). Using that $\omega_0(T_2(\ldots))$ fulfills the $\hbar$-dependence axiom and that

$$|\underline{A}| + |\overline{A}| = |A| \quad \text{for all monomials } A\ ,$$

which follows from (3.1.25), we see that (3.6.26) is

$$\sim \hbar^{-2}\,\hbar^{-1+(|\underline{A}_1|+|\underline{A}_2|)/2}\,\hbar^{(|\overline{A}_1|+|\overline{A}_2|+|A_3|)/2} = \hbar^{-3+(|A_1|+|A_2|+|A_3|)/2}\;,$$

as claimed.

Sm-expansion (axiom (ix)): Again, investigating the r.h.s. of (3.6.22), the assertion is non-obvious only for the terms containing $Z^{(2)}$. We give the proof for $\omega_0\big(T_2(x_1 \otimes Z^{(2)}(x_2 \otimes x_3))\big)$. Let A_1, A_2, A_3 be monomials. By using that $z^{(2)}(A_2, A_3)$ satisfies the Sm-expansion with degree $\dim A_2 + \dim A_3$, we conclude that $\omega_0\big(A_1(x_1)\star Z^{(2)}(x_2\otimes x_3)\big)$ fulfills the Sm-expansion with degree $\sum_{j=1}^3 \dim A_j$,

$$\omega_0\big(A_1(x_1) \star Z^{(2)}(x_2 \otimes x_3)\big) = \sum_{l=0}^{L} m^l\, u_{1,l}^{(m)}(y_1, y_2) + \mathfrak{r}_{1,L+1}^{(m)}(y_1, y_2) \quad \forall L \in \mathbb{N}\;,$$

where $y_j := x_j - x_3$, by proceeding analogously to (3.2.6)–(3.2.9). This holds also when the order of the factors is reversed: $\omega_0\big(Z^{(2)}(x_2 \otimes x_3) \star A_1(x_1)\big)$ fulfills the Sm-expansion with the same degree; we denote the pertinent $u_l^{(m)}$- and $\mathfrak{r}_{L+1}^{(m)}$-distributions by $u_{2,l}^{(m)}$ and $\mathfrak{r}_{2,L+1}^{(m)}$, respectively. Using (3.3.18) to[91] construct T_2^0, we obtain

$$\begin{aligned}\omega_0\Big(T_2^0\big(A_1(x_1) \star Z^{(2)}(x_2 \otimes x_3)\big)\Big)&\\ = \sum_{l=0}^{L} m^l \Big(f_1(y_1)\, u_{1,l}^{(m)}(y_1, y_2) + f_2(y_1)\, u_{2,l}^{(m)}(y_1, y_2)\Big)&\\ + \Big(f_1(y_1)\, \mathfrak{r}_{1,L+1}^{(m)}(y_1, y_2) + f_2(y_1)\, \mathfrak{r}_{2,L+1}^{(m)}(y_1, y_2)\Big)\;,&\end{aligned} \tag{3.6.27}$$

where $f_1, f_2 \in C^\infty(\mathbb{R}^d \setminus \{0\})$, $f_1(y) + f_2(y) = 1\ \forall y \neq 0$, $\operatorname{supp} f_1 \cap \overline{V}_- = \emptyset$, $\operatorname{supp} f_2 \cap \overline{V}_+ = \emptyset$ and $f_j(\rho y) = f_j(y)\ \forall \rho > 0,\ j = 1, 2$ (cf. (3.5.76)). Due to the latter property of (f_1, f_2) , it is obvious that the scaling properties of $u_{1,l}^{(m)}$ and $u_{2,l}^{(m)}$ go over to $f_1\, u_{1,l}^{(m)} + f_2\, u_{2,l}^{(m)}$, and similarly for the remainders. We conclude that (3.6.27) fulfills the Sm-expansion with degree $\sum_{j=1}^3 \dim A_j$. Finally, $\omega_0\big(T_2(\ldots)\big)$ is an extension of $\omega_0\big(T_2^0(\ldots)\big)$ which maintains the Sm-expansion, cf. Sects. 3.2.3–3.2.4.

Ad (a): The Faà di Bruno formula for $\widehat{\mathbf{S}}^{(n)}(0) = (\mathbf{S} \circ Z)^{(n)}(0)$, given in (3.6.25), is now interpreted differently: $(T_j)_{j\in\mathbb{N}^*}$ and $(\widehat{T}_j)_{j\in\mathbb{N}^*}$ are given and we are searching a Z, i.e., a sequence $(Z^{(j)})_{j\in\mathbb{N}^*}$, solving this equation for all $n \in \mathbb{N}^*$.

---

[91] T_2^0 can be constructed in a simpler way by means of part (b) of Lemma 3.3.2:

$$T_2^0\big(B_1(x_1) \otimes B_2(x_2)\big) = B_1(x_1) \star_{\Delta^F} B_2(x_2)\;, \quad B_1, B_2 \in \mathcal{P}\;.$$

However, in higher orders the use of a partition of unity cannot be avoided, see, e.g., (3.5.71). To indicate the general procedure we work here with a partition of unity.

The solution $(Z^{(j)})_{j\in\mathbb{N}^*}$ can be constructed by *solving* (3.6.25) *inductively*; in addition, this construction shows that the solution is *unique*.

It remains to prove that the so-obtained Z (3.6.16) is an element of $\mathcal{R}$. Obviously, $Z^{(n)}$ is a linear and symmetric map $Z^{(n)}\colon \mathcal{F}_{\mathrm{loc}}^{\otimes n} \longrightarrow \mathcal{F}$, that the range actually fulfills $Z^{(n)}(\mathcal{F}_{\mathrm{loc}}^{\otimes n}) \subseteq \mathcal{F}_{\mathrm{loc}}$ will turn out below. We have already derived the properties $Z(0) = 0$ and $Z^{(1)} = \mathrm{Id}$. To verify that Z fulfills the further defining conditions of an element of $\mathcal{R}$, we show inductively that the pertinent truncation Z_n (3.6.17) fulfills $Z_n \in \mathcal{R}\ \ \forall n \in \{2,3,\ldots\}$. So we assume that $Z_k \in \mathcal{R}\ \ \forall k < n$. Formula (3.6.25) can be written as

$$\Big(\frac{i}{\hbar}\Big)^{1-n} Z^{(n)} = \widehat{T}_n - T_{(n-1),n} \quad \text{where} \quad T_{(n-1)}(e_\otimes^{iF/\hbar}) := (\mathbf{S}\circ Z_{n-1})(F)\ . \quad (3.6.28)$$

For example, $(i/\hbar)^3\, T_{(2),3}(x_1\otimes x_2\otimes x_3)$ is the sum of the first four terms on the r.h.s. of (3.6.22). From part (b) of the theorem we know that $T_{(n-1)} \equiv (T_{(n-1),k})_{k\in\mathbb{N}^*}$ also satisfies the axioms for time-ordered products. In addition, applying (3.6.25) to $T_{(n-1),k} = (\mathbf{S}\circ Z_{n-1})^{(k)}$ for $k \leq n-1$, and using $Z_{n-1}^{(j)} = Z^{(j)}$ for $j \leq n-1$, we conclude that

$$\widehat{T}_k = T_{(n-1),k} \quad \forall 1 \leq k \leq n-1\ . \quad (3.6.29)$$

Summing up, $Z^{(n)}$ is the difference, to nth order, of two T-products ($\hat{T}$ and $T_{(n-1)}$), which agree to lower orders, that is, $Z^{(n)}$ has the form of an indeterminate term in the inductive step $(n-1) \to n$ of the Epstein–Glaser construction. We conclude: For monomials $A_1,\ldots,A_n$ the vacuum expectation value $z^{(n)}(A_1,\ldots,A_n)$ (3.6.2) is of the form (3.2.71)–(3.2.72), that is, $Z^{(n)}$ fulfills the condition "Locality, Translation covariance and Sm-expansion" (3.6.3)–(3.6.4). In addition, $Z^{(n)}$ satisfies the conditions "Field independence" (which implies the causal Wick expansion (3.6.9)), "Lorentz covariance", "Field parity", "Field equation" and "$\hbar$-dependence", because $\widehat{T}_n$ and $T_{(n-1),n}$ fulfill the corresponding conditions. In particular, the relations (3.6.3) and (3.6.9) imply $Z^{(n)}(\mathcal{F}_{\mathrm{loc}}^{\otimes n}) \subseteq \mathcal{F}_{\mathrm{loc}}$.

To verify the "$*$-structure" condition, it is simpler to proceed non-inductively: By reversing the calculation (3.6.24) we get $\mathbf{S}\big(Z(F^*)\big)^{\star-1} = \mathbf{S}\big(Z(F)^*\big)^{\star-1}$, which implies $Z(F^*) = Z(F)^*\ \ \forall F \in \mathcal{F}_{\mathrm{loc}}$.

All these properties of $Z^{(n)}$ can be summarized by $Z^{(n)} \in \hbar\, \mathcal{V}[\![\hbar]\!]$; this implies $Z_n \in \mathcal{R}$. With that, the proof is complete. □

Remark 3.6.5. If the values of the products T_n were *on-shell* fields,[92] that is, $\mathrm{Ran}\, T_n \subseteq \mathcal{F}/\mathcal{J}$ (where $\mathcal{J}$ is the ideal of the free field equation (2.6.2)), then the inductive construction of $(Z^{(n)})_{n\in\mathbb{N}^*}$ becomes much more complicated: Formula (3.6.25) would only determine

$$T_1\big(Z^{(n)}(\otimes_{j=1}^n F_j)\big) = Z^{(n)}(\otimes_{j=1}^n F_j) + \mathcal{J}\ .$$

[92]In contrast to the on-shell T-product T^{on} (introduced in Sect. 3.4), we assume in this remark that the *arguments* of the T-product are still local *off-shell* fields.

Remark 3.6.6 (Larger version of Stückelberg–Petermann group)**.** If one omits some renormalization conditions from the list of axioms for T-products, the Main Theorem still holds true for a *larger version* $\mathcal{R}_0$ *of the Stückelberg–Petermann group*, which is obtained by omitting the pertinent conditions in Definition 3.6.1. A minimal set of axioms, needed to prove the Main Theorem in the above-described way, is given by the basic axioms plus Translation covariance and Field independence. The pertinent defining properties for $\mathcal{R}_0$ are (1), (2), (4) and a substitute for (3) expressing only *Locality and Translation covariance*: for all monomials $A_1,\ldots,A_n \in \mathcal{P}$ the VEV (3.6.2) depends only on the relative coordinates and is of the form

$$z^{(n)}_{(m)}(A_1,\ldots,A_n)(x_1-x_n,\ldots) \tag{3.6.30}$$
$$= \mathfrak{S}_n \sum_{a\in\mathbb{N}^{d(n-1)}} C^a_{(m)}(A_1,\ldots,A_n)\,\partial^a\delta(x_1-x_n,\ldots,x_{n-1}-x_n)\,,$$

with m-dependent coefficients $C^a_{(m)}(A_1,\ldots,A_n)$ (not depending on $(x_1-x_n,\ldots)$).

If one works with time-ordered products fulfilling the axiom *Scaling degree* (3.2.62) (instead of the axiom Sm-expansion), the definition of the pertinent Stückelberg–Petermann group is modified as follows: Property (3) is replaced by the just given formulation of Locality and Translation covariance and by the property

$$\mathrm{sd}\, z^{(n)}_{(m)}(A_1,\ldots,A_n)(x_1-x_n,\ldots) \leq \sum_{j=1}^{n} \dim A_j\,, \quad \forall \text{ monomials } A_1,\ldots,A_n \in \mathcal{P}\,. \tag{3.6.31}$$

The latter condition can equivalently be expressed by the restriction

$$|a| \leq \sum_{j=1}^{n} \dim(A_j) - d(n-1) \tag{3.6.32}$$

of the sum over a in (3.6.30).

Remark 3.6.7 (Alternative formulation of Locality)**.** The formulation of Locality and Translation covariance given in (3.6.3) or (3.6.30), respectively, assumes the validity of the property Field independence, i.e., that $Z^{(n)}(A_1,\ldots,A_n)$ is completely determined by the vacuum expectation values $z^{(n)}(\underline{A}_1,\ldots,\underline{A}_n)$, $\underline{A}_l \subseteq A_l$, see (3.6.9). Alternatively, Locality can be formulated directly in terms of the maps $F \longmapsto Z(F)$, namely by the condition

$$Z(F{+}G{+}H) = Z(F{+}G) - Z(G) + Z(G{+}H) \quad \text{if} \quad \mathrm{supp}\, F \cap \mathrm{supp}\, H = \emptyset\,, \tag{3.6.33}$$

which we proved for $Z \in \mathcal{R}$ (Def. 3.6.1) in Exer. 3.6.2; for details see [27, Sect. 4]. This formulation of Locality does not include Translation covariance and it does not need the validity of the Field independence property. Note the formal similarity of the Locality condition (3.6.33) to the Additivity condition (1.3.21); however, these are conditions on maps of different kind: The domains and the ranges are different.

Exercise 3.6.8 (Equivalent reformulation of the properties "Locality and Translation covariance" and "Scaling degree" of $Z \in \mathcal{R}_0$). Let $\mathcal{R}_0$ be the version of the Stückelberg–Petermann RG (Definition 3.6.1) defined by the properties (1), (2), (4) and the substitute (3.6.30) for (3), and let $Z \in \mathcal{R}_0$.

(a) Show that there exist linear maps $P^n_a : \mathcal{P}^{\otimes n} \longrightarrow \mathcal{P}$ such that

$$Z^{(n)}\big(\otimes_{j=1}^n A_j(x_j)\big) = \mathfrak{S}_n \sum_{a\in\mathbb{N}^{d(n-1)}} \partial^a\delta(x_1 - x_n, \dots, x_{n-1} - x_n)\, P^n_a(\otimes_{j=1}^n A_j)(x_n)\,. \tag{3.6.34}$$

Applying $\omega_0(\,\cdot\,)$ to this equation, we immediately obtain (3.6.30); hence, (3.6.34) is equivalent to (3.6.30) (under the assumption that Z satisfies (1), (2) and (4)).

(b) Prove that, for $Z \in \mathcal{R}_0$, the validity of the Scaling degree property (3.6.31) is equivalent to the following: For all monomials $A_1, \dots, A_n \in \mathcal{P}$ the sum over "a" in (3.6.34) is bounded by

$$|a| \le \sum_{j=1}^n \dim A_j - \dim\big(P^n_a(\otimes_{j=1}^n A_j)\big) - d(n-1)\ , \tag{3.6.35}$$

where $\dim\big(P^n_a(\otimes_{j=1}^n A_j)\big)$ is defined by (3.2.63).

[*Solution*: (a) First let $A_1, \dots, A_n$ be monomials. Inserting (3.6.30) into the causal Wick expansion (3.6.9) and reordering the sums, we get

$$Z^{(n)}\big(\otimes_{j=1}^n A_j(x_j)\big) = \mathfrak{S}_n \sum_{b\in\mathbb{N}^{d(n-1)}} (\partial^b\delta)(x_1 - x_n, \dots, x_{n-1} - x_n)\, Q^n_b(x_1, \dots, x_n)$$

$$\text{where}\quad Q^n_b(x_1, \dots, x_n) := \sum_{\underline{A_j} \subseteq A_j} C^b(\underline{A_1}, \dots, \underline{A_n})\, \overline{A_1}(x_1) \cdots \overline{A_n}(x_n)\,.$$

By iterated use of the Leibniz rule we write

$$\begin{aligned}(\partial^b\delta)(x_1 - x_n, \dots)\, Q^n_b(x_1, \dots, x_n) &= \sum_{a+c=b} c_{a,c}\,\partial^a_{x_1\cdots x_{n-1}}\Big(\delta(x_1 - x_n, \dots)\,\partial^c_{x_1\cdots x_{n-1}} Q^n_b(x_1, \dots, x_n)\Big)\\ &= \sum_{a+c=b} c_{a,c}(\partial^a\delta)(x_1 - x_n, \dots)\,(\partial^c_{x_1\cdots x_{n-1}} Q^n_b)(x_n, \dots, x_n)\ ,\end{aligned}$$

where $c_{a,c}$ is a combinatorial factor containing possibly a sign. Inserting the latter formula into the preceding one, we obtain the assertion for arbitrary monomials $A_1, \dots, A_n$, with

$$P^n_a(\otimes_{j=1}^n A_j)(x_n) := \sum_c c_{a,c}\,(\partial^c_{x_1\cdots x_{n-1}} Q^n_{a+c})(x_n, \dots, x_n)\,.$$

Extending the map P^n_a from monomials to $\mathcal{P}^{\otimes n}$ by linearity, the assertion (3.6.34) holds for all polynomials $A_1, \dots, A_n$, since $Z^{(n)}$ is linear.

(b) We use the same notations as in the solution of part (a), in particular for the indices "a, b, c". According to (3.6.32) the Scaling degree property (3.6.31) is equivalent to

$$|a|+|c| = |a+c| \leq \sum_{j=1}^{n} \dim(\underline{A_j}) - d(n-1) = \sum_{j=1}^{n} (\dim(A_j) - \dim(\overline{A_j})) - d(n-1) \,. \quad (3.6.36)$$

For each monomial

$$M := c_{a,c}\, C^{a+c}(\underline{A_1}, \ldots, \underline{A_n})\, (\partial^{c_1}\overline{A_1}) \cdots (\partial^{c_{n-1}}\overline{A_{n-1}})\, \overline{A_n}$$

(where $c = (c_1, \ldots, c_{n-1})$) contributing to $P_a^n(\otimes_{j=1}^n A_j)$ it holds that

$$\dim(M) = |c| + \sum_{j=1}^{n} \dim(\overline{A_j}) \;;$$

hence the property (3.6.36) can be written as

$$|a| \leq \sum_{j=1}^{n} \dim A_j - \dim(M) - d(n-1) \quad \text{for all } M,$$

which is equivalent to the same condition with $\dim(M)$ replaced by $\dim\big(P_a^n(\otimes_{j=1}^n A_j)\big) := \max \dim(M)$.]

Example 3.6.9 (UV-finite interactions). An interaction $L \in \mathcal{P}$ in d dimensions – for simplicity we assume that L is a monomial – is called

$$\text{"UV-finite" iff} \quad n \dim L < (n-1)d \quad \forall n \geq 2 \,.$$

The reason for this name is that for an UV-finite L the axiom Scaling degree (3.2.62) implies that

$$\operatorname{sd} t(\underline{L}_1, \ldots, \underline{L}_n) \leq n \dim L < (n-1)d \quad \forall \text{ submonomials } \underline{L}_j \subseteq L$$

and, hence, the inductive construction of $\mathbf{S}\big(L(g)\big)$, $g \in \mathcal{D}(\mathbb{M})$, uses solely the direct extension. Consequently, $\mathbf{S}\big(L(g)\big)$ is unique.

For any $Z \in \mathcal{R}$, the properties "Locality and Translation covariance" (3.6.30), "Scaling degree" (3.6.32) and "Field independence" (3.6.9) imply that

$$Z^{(n)}\big(L(g)\big) = 0 \;\; \forall n \geq 2 \,, \quad \text{that is,} \quad Z\big(L(g)\big) = L(g) \quad \text{for all UV-finite interactions } L. \tag{3.6.37}$$

Alternatively, this result can be obtained from part (b) of the Main Theorem and the uniqueness of $\mathbf{S}\big(L(g)\big)$: For any $Z \in \mathcal{R}$ we get $\mathbf{S} \circ Z\big(L(g)\big) = \hat{\mathbf{S}}\big(L(g)\big) = \mathbf{S}\big(L(g)\big)$, which implies $Z\big(L(g)\big) = L(g)$.

Examples for UV-finite interactions are $L = \varphi^n$, $n \in \mathbb{N}^*$ in $d = 2$ dimensions and $L = \varphi^2$ for $d = 3$.

Due to the Main Theorem, the statement that an interaction $\kappa\mathcal{L} \in \mathcal{P}_{\text{bal}}$ is super-renormalizable by power counting (see Proposition 3.2.28), is equivalent to finiteness of the (formal) sum $Z(\kappa\mathcal{L}(g)) = \sum_{n=0}^{\infty} \frac{\kappa^n}{n!} Z^{(n)}(\mathcal{L}(g)^{\otimes n})$, for all $Z \in \mathcal{R}$. The following exercise gives an example.

Exercise 3.6.10 (Renormalization of $\varphi^3_{d=4}$ to all orders)**.** Let $d = 4$ and $m \geq 0$. Show, for all $Z \in \mathcal{R}$, that the Wick expansion of $Z\big(\kappa\,\varphi^3(g)\big)$ reads:

$$Z\big(\kappa\,\varphi^3(g)\big) = \kappa\,\varphi^3(g) + \frac{\kappa^2}{2}\int\Big(\prod_{j=1}^{2} dx_j\; g(x_j)\Big)\Big[z^{(2)}\big(\varphi^3\otimes\varphi^3\big)(x_1-x_2)$$
$$+ 9\,z^{(2)}\big(\varphi^2\otimes\varphi^2\big)(x_1-x_2)\,\varphi(x_1)\varphi(x_2)\Big]$$
$$+ \frac{\kappa^3}{3!}\,9\int\Big(\prod_{j=1}^{3} dx_j\; g(x_j)\Big)\,z^{(3)}\big(\varphi^3\otimes\varphi^3\otimes\varphi^2\big)(x_1-x_3,x_2-x_3)\,\varphi(x_3)$$
$$+ \frac{\kappa^4}{4!}\int\Big(\prod_{j=1}^{4} dx_j\; g(x_j)\Big)\,z^{(4)}\big(\varphi^3\otimes\varphi^3\otimes\varphi^3\otimes\varphi^3\big)(x_1-x_4,\ldots,x_3-x_4)\ , \qquad (3.6.38)$$

where *all* orders are taken into account, i.e., the series ends with the term $\sim\kappa^4$.

[*Solution*: By taking into account the restrictions (3.6.10), all terms contributing to the Wick expansion (3.6.9) of $Z^{(n)}\big((\kappa\,\varphi^3(g))^{\otimes n}\big)$ are written out in (3.6.38), for all orders n. The combinatorial factors in that formula are obtained by a glance at (3.1.23) or (3.1.26).]

Main Theorem in terms of the R-product. The Main Theorem can be translated from T-product into R-product.

Exercise 3.6.11. (a) Let $\widehat{\mathbf{S}} := \mathbf{S}\circ Z$ with $Z \in \mathcal{R}$, and let $\widehat{R}$ and R be the pertinent retarded products, i.e., they are defined by the Bogoliubov formula (3.3.30) in terms of $\widehat{\mathbf{S}}$ and $\mathbf{S}$, respectively. Prove that

$$\widehat{R}\big(e_\otimes^{F/\hbar},G\big) = R\big(e_\otimes^{Z(F)/\hbar},Z'(F)G\big), \quad \text{where} \qquad (3.6.39)$$

$$Z'(F)G := \frac{d}{d\lambda}\Big|_{\lambda=0} Z(F+\lambda G) = \sum_{n=0}^{\infty}\frac{1}{n!}\,Z^{(n+1)}\big(F^{\otimes n}\otimes G\big)\ . \qquad (3.6.40)$$

Using this result, derive the relation

$$\hbar^{-1}\,\widehat{R}_{1,1}(F,G) = \hbar^{-1}\,R_{1,1}(F,G) + Z^{(2)}(F\otimes G)\,. \qquad (3.6.41)$$

(b) Prove that the linear map $Z'(F) : \mathcal{F}_{\mathrm{loc}} \longrightarrow \mathcal{F}_{\mathrm{loc}}$ is invertible and that the inverse is given by

$$Z'(F)^{-1} = (Z^{-1})'\big(Z(F)\big)\ . \qquad (3.6.42)$$

[*Solution*: (a) Combining (3.3.30) and formula (3.6.20) of the Main Theorem, we get

$$\widehat{R}\big(e_\otimes^{F/\hbar},G\big) = \widehat{T}\big(e_\otimes^{iF/\hbar}\big)^{\star-1}\star\frac{\hbar}{i}\frac{d}{d\lambda}\Big|_{\lambda=0}\widehat{T}\big(e_\otimes^{i(F+\lambda G)/\hbar}\big)$$
$$= T\big(e_\otimes^{iZ(F)/\hbar}\big)^{\star-1}\star\frac{\hbar}{i}\frac{d}{d\lambda}\Big|_{\lambda=0} T\big(e_\otimes^{i\,Z(F+\lambda G)/\hbar}\big)$$
$$= T\big(e_\otimes^{iZ(F)/\hbar}\big)^{\star-1}\star T\big(e_\otimes^{i\,Z(F)/\hbar}\otimes Z'(F)G\big) = R\big(e_\otimes^{Z(F)/\hbar},Z'(F)G\big)\ .$$

To derive (3.6.41), we select from (3.6.39) all terms which are of first order in F, by using

$$R(e_\otimes^{Z(F)/\hbar}, Z'(F)G) = Z'(F)G + \hbar^{-1} R_{1,1}(Z(F), Z'(F)G) + \cdots ,$$
$$Z(F) = F + \mathcal{O}(F^2) , \quad Z'(F)G = G + Z^{(2)}(F \otimes G) + \mathcal{O}(F^2) .$$

(b) Invertibility of $Z'(F)$ relies on the invertibility of Z; the inverse $Z'(F)^{-1}$ is obtained by the chain rule:

$$G = \frac{d}{d\lambda}\Big|_{\lambda=0} (Z^{-1} \circ Z)(F + \lambda G) = (Z^{-1})'(Z(F)) \circ Z'(F)\, G ,$$
$$G = \frac{d}{d\lambda}\Big|_{\lambda=0} (Z \circ Z^{-1})(Z(F) + \lambda G) = Z'(F) \circ (Z^{-1})'(Z(F))\, G \quad \forall G \in \mathcal{F}_{\mathrm{loc}} .]$$

Due to (3.6.39), we interpret $F \longmapsto Z(F)$ as the *renormalization of the interaction* and $G \longmapsto Z'(F)G$ as the *field renormalization.*

From the Field equation property of an arbitrary renormalization map $Z \in \mathcal{R}$ we see that

$$Z'(F)(\partial^a \varphi)(g) = (\partial^a \varphi)(g) \quad \forall g \in \mathcal{D}(\mathbb{M}) , \ a \in \mathbb{N}^d ,$$

that is, the field renormalization of a (derivated) basic field is the identity.

By means of Proposition 3.3.14 we conclude: An *equivalent* version of the Main Theorem (Thm. 3.6.3) is obtained by replacing in both parts, (a) and (b), the formula $\widehat{\mathbf{S}} := \mathbf{S} \circ Z$ by (3.6.39).

3.6.3 The Gell-Mann–Low cocycle

M. Gell-Mann and F.E. Low [82] applied the general concept of renormalization group (found by Stückelberg and Petermann [163]) to scaling transformations. The crucial observation is that the *effect of a scaling transformation is precisely a change of the renormalization prescription.* Therefore, by means of the Main Theorem (Thm. 3.6.3), a scaling transformation can equivalently be described by a renormalization of the interaction. In the (algebraic) adiabatic limit, the latter amounts to a change of the physical parameters (Sect. 3.8.1). Proceeding this way, the mass and the coupling constant become scale dependent (Sect. 3.8.2). This "running" of the physical parameters can be understood physically: For example an electron is, according to the laws of QED, always surrounded by a cloud of electron-positron pairs, which are constantly created from the vacuum and immediately annihilated. Since the positrons are attracted and the electrons are repelled by the initial (stable) electron, the charge of the latter is screened. Hence, the observed value of this charge depends on the distance of the observer.

We now start to work this out in our framework, by following essentially [27]; see also [55, 88, 102]. The basic result is that, starting with an S-matrix $\mathbf{S}^m$

satisfying the axioms for time-ordered products, the scaled S-matrix[93]

$$\mathbf{S}^m_\rho := \sigma_\rho \circ \mathbf{S}^{m/\rho} \circ \sigma_\rho^{-1} \tag{3.6.43}$$

fulfills also these axioms (with respect to $\star_m$), that is, it differs from $\mathbf{S}^m$ only by a change of the renormalization prescription. This can be verified straightforwardly. Heuristically, this can be understood as follows: The "unrenormalized" time-ordered product, i.e., the restriction of $T_n^{(m)}\big(A_1(x_1)\otimes\cdots\otimes A_n(x_n)\big)$ to $\mathcal{D}(\check{\mathbb{M}}^n)$, is a multiple $\star_{\Delta^F_m}$-product (see part (b) of Lemma 3.3.2), and the latter is scaling invariant:

$$\begin{aligned}\sigma_\rho\Big(\sigma_\rho^{-1}(A_1(x_1))\star_{\Delta^F_{m/\rho}}\cdots\star_{\Delta^F_{m/\rho}}\sigma_\rho^{-1}(A_n(x_n))\Big)&\\ = A_1(x_1)\star_{\Delta^F_m}\cdots\star_{\Delta^F_m} A_n(x_n) \quad \text{on } \mathcal{D}(\check{\mathbb{M}}^n),&\end{aligned}$$

due to $\rho^{d-2}\,\Delta^F_{m/\rho}(\rho y)=\Delta^F_m(y)$, as one verifies similarly to (3.1.51) (Exer. 3.1.21). Note also that for an arbitrary R-product the difference between

$$R^m_\rho(e^S_\otimes,F) := \sigma_\rho\circ R^{m/\rho}(e_\otimes^{\sigma_\rho^{-1}S},\sigma_\rho^{-1}F) \quad \text{and} \quad R^m(e^S_\otimes,F) \quad \text{vanishes for } \hbar\to 0$$

for all $S,F\in\mathcal{F}_{\mathrm{loc}}$, because of (3.1.55).

Due to the above result, we may apply the Main Theorem: There exists a unique $Z^m_\rho\in\mathcal{R}$ such that

$$\mathbf{S}^m_\rho = \mathbf{S}^m\circ Z^m_\rho\ . \tag{3.6.44}$$

So $(Z^m_\rho-\mathrm{Id})$ describes the breaking of homogeneous scaling under $(X,m)\mapsto(\rho X,\rho^{-1}m)$.

We thus obtain a one-parameter family $\{\,Z^m_\rho\,|\,\rho>0\,\}\subset\mathcal{R}$; but these *do not form a subgroup*: The group law is obstructed by a *cocycle*, the "Gell-Mann–Low cocycle". Concretely, from

$$\begin{aligned}\mathbf{S}^m\circ Z^m_{\rho_1\rho_2} &= \sigma_{\rho_1}\circ\big(\sigma_{\rho_2}\circ\mathbf{S}^{m/\rho_1\rho_2}\circ\sigma_{\rho_2}^{-1}\big)\circ\sigma_{\rho_1}^{-1}\\ &= \sigma_{\rho_1}\circ\mathbf{S}^{m/\rho_1}\circ\big(\sigma_{\rho_1}^{-1}\circ\sigma_{\rho_1}\big)\circ Z^{m/\rho_1}_{\rho_2}\circ\sigma_{\rho_1}^{-1} = \mathbf{S}^m\circ Z^m_{\rho_1}\circ\sigma_{\rho_1}\circ Z^{m/\rho_1}_{\rho_2}\circ\sigma_{\rho_1}^{-1}\end{aligned}$$

and the injectivity of $\mathbf{S}^m$ we conclude that

$$Z^m_{\rho_1\rho_2} = Z^m_{\rho_1}\circ\sigma_{\rho_1}\circ Z^{m/\rho_1}_{\rho_2}\circ\sigma_{\rho_1}^{-1}\ , \tag{3.6.45}$$

which is a cocycle relation. By means of (3.6.14), this simplifies to

$$Z_{\rho_1\rho_2} = Z_{\rho_1}\circ Z_{\rho_2} \quad \text{for} \quad m=0\ , \tag{3.6.46}$$

that is, in the massless case we have a one-parameter group.

[93]Since the mass-dependence of the time-ordered products and of the elements Z of $\mathcal{R}$ plays an important role in the following, we write m as an upper index: $\mathbf{S}^m\equiv\mathbf{S}$ and $Z^m\equiv Z$.

We are now going to show that $Z_\rho^{m\,(n)}$ is a *polynomial in* $\log\rho$ for all $n \geq 2$. This follows from the Sm-expansion axiom in the following way: Due to this axiom, the relation (3.1.67) holds – also for T-products, that is,

$$\sigma_\rho \circ \mathbf{S}^{m/\rho\,(n)}(0) \circ (\sigma_\rho^{-1})^{\otimes n} = (\mathbf{S}^m \circ Z_\rho^m)^{(n)}(0) \tag{3.6.47}$$

is a polynomial in $\log\rho$. Inserting the Faà di Bruno formula (3.6.25) into the r.h.s. and proceeding by induction on n, we get the assertion.

Example 3.6.12. For a massless scalar field φ in $d = 4$ dimensions we obtain

$$\begin{aligned} Z_\rho^{(2)}\big(\varphi^2(x_1)\otimes\varphi^2(x_2)\big) &= \frac{i}{\hbar}\,\sigma_\rho\circ T_2\big(\sigma_\rho^{-1}\varphi^2(x_1)\otimes\sigma_\rho^{-1}\varphi^2(x_2)\big) - T_2\big(\varphi^2(x_1)\otimes\varphi^2(x_2)\big) \\ &= 2i\hbar\left(\rho^4\, t^M_{\text{fish}}(\rho(x_1-x_2)) - t^M_{\text{fish}}(x_1-x_2)\right), \end{aligned} \tag{3.6.48}$$

where $t^M_{\text{fish}} \equiv t_{\text{fish}}$ is given by (3.5.65), and we have used (3.6.21) and that tree diagrams scale homogeneously. In Exer. 3.5.5 we have derived that

$$\rho^4\, t^M_{\text{fish}}(\rho y) - t^M_{\text{fish}}(y) = C_{\text{fish}}\,\delta(y)\,\log\rho\,, \quad C_{\text{fish}} = \frac{-i}{8\,\pi^2}\,. \tag{3.6.49}$$

The r.h.s. of (3.6.49) does not depend on the renormalization mass scale M; that is, the breaking of homogeneous scaling is universal, i.e., the *same* local term for all almost homogeneous extensions. This is an example for the universality of the polynomial $p(\partial)$ introduced in formulas (3.2.39)–(3.2.40) of Proposition 3.2.16. More generally, we conclude from that proposition and the Faà di Bruno formula (3.6.25) that the contribution to Z_ρ is universal for all massless, primitive divergent diagrams[94], in particular for the massless setting sun diagram (see Exer. 3.5.5) and the box diagrams studied in Exer. 3.6.15 below. This statement holds also for the following massive example.

Example 3.6.13. Let φ_1 and φ_2 be two real scalar fields in $d = 4$ dimensions with masses m_1 and m_2, respectively. We aim to compute

$$\begin{aligned} Z_{1\rho}^{(3)\,\mu\nu}(x_1,x_2,x_3) &:= Z_\rho^{(3)}\big(\varphi_1\partial^\mu\varphi_2(x_1)\otimes\varphi_1\partial^\nu\varphi_2(x_2)\otimes\varphi_2^2(x_3)\big) \quad \text{and} \\ Z_{2\rho}^{(3)\,\mu\nu}(x_1,x_2,x_3) &:= Z_\rho^{(3)}\big(\varphi_2\partial^\mu\varphi_1(x_1)\otimes\varphi_2\partial^\nu\varphi_1(x_2)\otimes\varphi_2^2(x_3)\big)\,. \end{aligned}$$

Only triangle diagrams (see (3.2.16)) contribute to $Z_{1\rho}^{(3)}$ and $Z_{2\rho}^{(3)}$, for the following reason: From the proof of the Main Theorem we know that $Z_\rho^{(3)}$ can be written as $Z_\rho^{(3)} = -\hbar^{-2}(\hat{T}_3 - T_{(2),3})$ where $\hat{T}_n^{(m)} := \sigma_\rho\circ T_n^{(m/\rho)}\circ(\sigma_\rho^{-1})^{\otimes n}$ and $\hat{T}_k = T_{(2),k}$ for $k = 1, 2$. The claim follows by using that the inductive step from T_2 to T_3 is unique for tree diagrams and disconnected diagrams, and that all further diagrams are triangles.

With that we obtain

$$Z_{j\rho}^{(3)\,\mu\nu}(x_1,x_2,x_3) = -2\hbar\Big(\rho^8\, t_{j\triangle}^{(m/\rho)\,\mu\nu}(\rho y) - t_{j\triangle}^{(m)\,\mu\nu}(y)\Big)\,, \quad j = 1, 2\,,$$

[94] The notion “primitive divergent diagram” is explained in Exap. 3.5.14.

where $m := (m_1, m_2)$, $y := (y_1, y_2)$, $y_j := x_j - x_3$ and $t_{j\triangle}^{(m)\,\mu\nu}(y)$ is an extension of

$$t_{1\triangle}^{(m)\,\mu\nu\,0}(y) := \partial^\mu \Delta_{m_2}^F(y_1)\, \partial^\nu \Delta_{m_2}^F(y_2)\, \Delta_{m_1}^F(y_1 - y_2) \in \mathcal{D}'(\mathbb{R}^8 \setminus \{0\}) ,$$
$$t_{2\triangle}^{(m)\,\mu\nu\,0}(y) := -\Delta_{m_2}^F(y_1)\, \Delta_{m_2}^F(y_2)\, \partial^\mu \partial^\nu \Delta_{m_1}^F(y_1 - y_2) \in \mathcal{D}'(\mathbb{R}^8 \setminus \{0\}) ,$$

respectively. The latter fulfill

$$\rho^8\, t_{j\triangle}^{(m/\rho)\,\mu\nu\,0}(\rho y) = t_{j\triangle}^{(m)\,\mu\nu\,0}(y) , \quad j = 1, 2 .$$

In addition, $t_{j\triangle}^{(m)\,\mu\nu}$ is constructed from $t_{j\triangle}^{(m)\,\mu\nu\,0}$ by the procedure described in Sect. 3.2.3; hence, it satisfies the Sm-expansion, in particular we may write

$$t_{j\triangle}^{(m)\,\mu\nu}(y) = t_{j\triangle}^{\mu\nu}(y) + \mathfrak{r}_{j\triangle}^{(m)\,\mu\nu}(y) , \quad t_{j\triangle}^{\mu\nu}(y) := t_{j\triangle}^{(m=(0,0))\,\mu\nu}(y) , \quad j = 1, 2 ,$$

and the leading term $t_{j\triangle}^{\mu\nu}(y)$, is an almost homogeneous extension of

$$t_{1\triangle}^{\mu\nu\,0}(y) := \partial^\mu D^F(y_1)\, \partial^\nu D^F(y_2)\, D^F(y_1 - y_2) \in \mathcal{D}'(\mathbb{R}^8 \setminus \{0\}) ,$$
$$t_{2\triangle}^{\mu\nu\,0}(y) := -D^F(y_1)\, D^F(y_2)\, \partial^\mu \partial^\nu D^F(y_1 - y_2) \in \mathcal{D}'(\mathbb{R}^8 \setminus \{0\}) , \tag{3.6.50}$$

respectively, which scales homogeneously: $\rho^8\, t_{j\triangle}^{\mu\nu\,0}(\rho y) = t_{j\triangle}^{\mu\nu\,0}(y)$. The remainders $\mathfrak{r}_{j\triangle}^{(m)\,\mu\nu}$ are obtained from the corresponding unrenormalized terms $\mathfrak{r}_{j\triangle}^{(m)\,\mu\nu\,0} := t_{j\triangle}^{(m)\,\mu\nu\,0} - t_{j\triangle}^{\mu\nu\,0}$ by the direct extension; hence, the scaling relation

$$\rho^8\, \mathfrak{r}_{j\triangle}^{(m/\rho)\,\mu\nu\,0}(\rho y) = \mathfrak{r}_{j\triangle}^{(m)\,\mu\nu\,0}(y) \quad \text{goes over to} \quad \mathfrak{r}_{j\triangle}^{(m)\,\mu\nu} .$$

Therefore, only the leading term of the Sm-expansion contributes to $Z_{j\rho}^{(3)\,\mu\nu}(x_1, x_2, x_3)$:

$$Z_{j\rho}^{(3)}(x_1, x_2, x_3) = -2\hbar\Big(\rho^8\, t_{j\triangle}^{\mu\nu}(\rho y) - t_{j\triangle}^{\mu\nu}(y)\Big) , \quad j = 1, 2 . \tag{3.6.51}$$

We are going to show how the computation of the r.h.s. of (3.6.51) can be traced back to the scaling behaviour of the massless fish diagram (3.6.49) in both cases $j = 1$ and $j = 2$. First note that contraction of $t_{2\triangle}^{\mu\nu\,0}$ with $g_{\mu\nu}$ yields

$$t_{2\triangle\mu}^{\mu\,0}(y) = i\, \delta_{(4)}(y_1 - y_2)\, t_{\text{fish}}^0(y_1) \tag{3.6.52}$$

by using $\square D^F(x) = -i\delta_{(4)}(x)$. Hence, arbitrary almost homogeneous extensions $t_{2\triangle}^{\mu\nu}$ and t_{fish} satisfy the relation

$$t_{2\triangle\mu}^{\mu}(y) - i\, \delta_{(4)}(y_1 - y_2)\, t_{\text{fish}}(y_1) = C\, \delta_{(8)}(y) \quad \text{for some} \quad C \in \mathbb{C} ;$$

obviously, the term on the right side scales homogeneously. We conclude that

$$(y\partial_y + 8)\, t_{2\triangle\mu}^{\mu}(y) = i\, C_{\text{fish}}\, \delta_{(8)}(y) , \quad \text{where} \quad y\partial_y := y_1^\lambda \partial_\lambda^{y_1} + y_2^\lambda \partial_\lambda^{y_2} .$$

Due to Lorentz covariance, the expression $(y\partial_y + 8)\, t_{2\triangle}^{\mu\nu}(y)$ must be $\sim g^{\mu\nu}$; hence we obtain

$$(y\partial_y + 8)\, t_{2\triangle}^{\mu\nu}(y) = i\, \frac{g^{\mu\nu}}{4}\, C_{\text{fish}}\, \delta_{(8)}(y) ,$$

which yields

$$Z^{(3)\,\mu\nu}_{2\rho}(x_1,x_2,x_3) = -\frac{i\hbar}{2}\, C_{\rm fish}\, g^{\mu\nu}\,\delta_{(8)}(y)\,\log\rho\,. \tag{3.6.53}$$

To compute the violation of homogeneous scaling for $t^{\mu\nu}_{1\triangle}$, we introduce

$$\tilde t^{\mu}_{\triangle}(y) := \partial^\mu D^F(y_1)\,D^F(y_2)\,D^F(y_1-y_2) \tag{3.6.54}$$

which exists in $\mathcal{D}'(\mathbb{R}^8)$ by the direct extension (3.2.22) and scales homogeneously:

$$(y\partial_y + 7)\,\tilde t^{\mu}_{\triangle}(y) = 0\,.$$

In $\mathcal{D}'(\mathbb{R}^8\setminus\{0\})$ we find

$$(\partial^\nu_{y_1}+\partial^\nu_{y_2})\tilde t^{\mu\,0}_{\triangle}(y) = -t^{\mu\nu\,0}_{2\triangle}(y_1-y_2,-y_2) + t^{\mu\nu\,0}_{1\triangle}(y)\,.$$

Therefore, arbitrary almost homogeneous extensions fulfill

$$(\partial^\nu_{y_1}+\partial^\nu_{y_2})\,\tilde t^{\mu}_{\triangle}(y) = -t^{\mu\nu}_{2\triangle}(y_1-y_2,-y_2) + t^{\mu\nu}_{1\triangle}(y) + \tilde C\,\delta_{(8)}(y)$$

for some $\tilde C\in\mathbb{C}$. We conclude

$$\begin{aligned} 0 &= (\partial^\nu_{y_1}+\partial^\nu_{y_2})\,(y\partial_y+7)\,\tilde t^{\mu}_{\triangle}(y) = (y\partial_y+8)\,(\partial^\nu_{y_1}+\partial^\nu_{y_2})\,\tilde t^{\mu}_{\triangle}(y)\\ &= -\,(y\partial_y+8)\,t^{\mu\nu}_{2\triangle}(y_1-y_2,-y_2) + (y\partial_y+8)\,t^{\mu\nu}_{1\triangle}(y)\\ &= -\,i\,\frac{g^{\mu\nu}}{4}\,C_{\rm fish}\,\delta_{(8)}(y) + (y\partial_y+8)\,t^{\mu\nu}_{1\triangle}(y)\,. \end{aligned} \tag{3.6.55}$$

We end up with

$$Z^{(3)\,\mu\nu}_{1\rho}(x_1,x_2,x_3) = -\frac{i\hbar}{2}\, C_{\rm fish}\, g^{\mu\nu}\,\delta_{(8)}(y)\,\log\rho\,. \tag{3.6.56}$$

Exercise 3.6.14. Let φ be a real scalar fields in $d=4$ dimensions with mass $m\geq 0$. Compute $Z^{(3)}_\rho\big(\varphi\partial^\mu\varphi(x_1)\otimes\varphi\partial^\nu\varphi(x_2)\otimes\varphi^2(x_3)\big)$.

[*Hint*: Take into account that $\varphi\partial^\mu\varphi = \frac{1}{2}\,\partial^\mu\varphi^2$ and use the AWI (3.1.5).]

[*Solution*: As explained in the preceding example, only triangle diagrams contribute. Hence, we get

$$Z^{(3)}_\rho(\ldots) = \frac{-1}{\hbar^2}\Big(\rho^8\,t^{(m/\rho)}(\varphi\partial^\mu\varphi,\varphi\partial^\nu\varphi,\varphi^2)(\rho y) - t^{(m)}(\varphi\partial^\mu\varphi,\varphi\partial^\nu\varphi,\varphi^2)(y)\Big)\,,$$

where $y := (y_1,y_2) := (x_1-x_3,x_2-x_3)$ and $t^{(m)}(\ldots) := \omega_0\big(T^{(m)}(\cdots)\big)$ (3.2.20). By using the AWI this can be written as

$$Z^{(3)}_\rho(\ldots) = \frac{-1}{4\hbar^2}\,\partial^\mu_{y_1}\partial^\nu_{y_2}\Big(\rho^6\,t^{(m/\rho)}(\rho y) - t^{(m)}(y)\Big)\,,\quad t^{(m)} := t^{(m)}(\varphi^2,\varphi^2,\varphi^2)\,.$$

Since $\mathrm{sd}(t^{(m)})\leq 6<8$ (Corollary 3.2.23), we can obtain $t^{(m)}$ from $t^{(m)\,0}(y)\sim\Delta^F_m(y_1)\,\Delta^F_m(y_2)\,\Delta^F_m(y_1-y_2)$ by the direct extension. Therefore, the homogeneous scaling relation $\rho^6\,t^{(m/\rho)\,0}(\rho y) = t^{(m)\,0}(y)$ holds also for the extension $t^{(m)}$ and, hence, we get $Z^{(3)}_\rho(\ldots) = 0$.]

Exercise 3.6.15. Let $d = 4$. We study the following unrenormalized contributions to a massless time-ordered product:

$$t_{1\square}^{\lambda\nu\,0}(y) := \partial^\lambda\partial^\nu D^F(y_1 - y_2)\,\partial_\mu D^F(y_2 - y_3)\,D^F(y_3)\,\partial^\mu D^F(y_1) \in \mathcal{D}'(\mathbb{R}^{12}\setminus\{0\}) ,$$
$$t_{2\square}^{\lambda\nu\,0}(y) := \partial^\lambda D^F(y_1 - y_2)\,\partial^\nu\partial_\mu D^F(y_2 - y_3)\,D^F(y_3)\,\partial^\mu D^F(y_1) \in \mathcal{D}'(\mathbb{R}^{12}\setminus\{0\}) ,$$

where $y := (y_1, y_2, y_3)$, $y_j := x_j - x_4$. For both terms, the corresponding contraction pattern may be illustrated by a box diagram:

$$\begin{array}{ccc} x_4 & & x_3 \\ \bullet & \text{———} & \bullet \\ | & & | \\ \bullet & \text{———} & \bullet \\ x_1 & & x_2 \end{array} \tag{3.6.57}$$

Let $t_{k\square}^{\lambda\nu} \in \mathcal{D}'(\mathbb{R}^{12})$ $(k = 1, 2)$ be arbitrary almost homogeneous extensions of these distributions, which are also Lorentz covariant. Compute the violation of homogeneous scaling of these extensions – the results and techniques of Exap. 3.6.13 may be used.

[*Solution*: Analogously to (3.6.52), $t_{1\square}^{0}$ is related to $t_{1\triangle}^{0}$ (defined in (3.6.50)) by

$$g_{\lambda\nu}\, t_{1\square}^{\lambda\nu\,0}(y) = -i\delta_{(4)}(y_1 - y_2)\, t_{1\triangle\mu}^{\mu\,0}(y_1, y_1 - y_3)$$

Hence, proceeding similarly to (3.6.52)–(3.6.53) and using the result (3.6.55) for the breaking of homogeneous scaling of $t_{1\triangle}^{0}$, we obtain

$$\rho^{12}\, t_{1\square}^{\lambda\nu}(\rho y) - t_{1\square}^{\lambda\nu}(y) = \frac{1}{4}\, C_{\text{fish}}\, g^{\lambda\nu}\, \delta_{(12)}(y)\, \log\rho \, .$$

To compute the violation of homogeneous scaling of $t_{2\square}$, we proceed as in (3.6.54)–(3.6.55): We take into account that in $\mathcal{D}'(\mathbb{R}^{12}\setminus\{0\})$ it holds that

$$\partial_{y_2}^\nu \left[\partial^\lambda D^F(y_1 - y_2)\,\partial_\mu D^F(y_2 - y_3)\,D^F(y_3)\,\partial^\mu D^F(y_1)\right] = -t_{1\square}^{\lambda\nu\,0}(y) + t_{2\square}^{\lambda\nu\,0}(y) \, .$$

The extension to $\mathcal{D}'(\mathbb{R}^{12})$ of the distribution in $[\cdots]$-brackets can be obtained by the direct extension and, hence, this extension scales homogeneously. We conclude that

$$\rho^{12}\, t_{2\square}^{\lambda\nu}(\rho y) - t_{2\square}^{\lambda\nu}(y) = \rho^{12}\, t_{1\square}^{\lambda\nu}(\rho y) - t_{1\square}^{\lambda\nu}(y) \, .]$$

Example 3.6.16. To give a massive example, in which derivatives of the δ-distribution are involved, let $d = 4$, $m > 0$ and we aim to compute

$$\begin{aligned} Z_\rho^{(2)}\left(\varphi\partial^\mu\partial^\nu\varphi(x_1)\otimes\varphi^2(x_2)\right) &= \frac{i}{\hbar}\left(\sigma_\rho\circ T_2^{(m/\rho)}\circ\sigma_\rho^{-1} - T_2^{(m)}\right)\left(\varphi\partial^\mu\partial^\nu\varphi(x_1)\otimes\varphi^2(x_2)\right) \\ &= 2i\hbar\left(\rho^6\, t_{m/\rho}^{\mu\nu}(\rho y) - t_m^{\mu\nu}(y)\right) , \quad y := x_1 - x_2 , \end{aligned}$$

where $t_m^{\mu\nu}(y) \in \mathcal{D}'(\mathbb{R}^4)$ is an extension of

$$t_m^{\mu\nu\,0}(y) := \Delta_m^F(y)\partial^\mu\partial^\nu\Delta_m^F(y) \in \mathcal{D}'(\mathbb{R}^4\setminus\{0\}) ,$$

fulfilling the Sm-expansion with degree $D = 6$.[95] The violation of homogeneous scaling must be a Lorentz tensor of rank 2 which is a polynomial in derivatives of $\delta(y)$ fulfilling the Sm-expansion with degree $D = 6$. Hence, it is of the form

$$(y\partial_y + 6 - m\partial_m)\, t_m^{\mu\nu}(y) = \Big(C_1\, g^{\mu\nu}\, \Box_y + C_2\, \partial_y^\mu \partial_y^\nu + g^{\mu\nu}\, m^2\, P\big(\log \frac{m}{M}\big)\Big)\, \delta(y)\ , \quad (3.6.58)$$

where $y\partial_y := y_\mu \partial_y^\mu$; moreover, $C_1, C_2 \in \mathbb{C}$ do not depend on m (as required for the leading term of the Sm-expansion), P is a polynomial and M is a fixed mass scale. We renormalize such that the relation

$$g_{\mu\nu}\, t_m^{\mu\nu\, 0}(y) = -m^2\, t_m^0(y)\ , \quad t_m^0(y) := \Delta_m^F(y)^2 \in \mathcal{D}'(\mathbb{R}^4 \setminus \{0\})\ ,$$

is maintained up to a (local) term which scales homogeneously, i.e., which is in the kernel of the operator $(y\partial_y + 6 - m\partial_m)$:

$$g_{\mu\nu}\, t_m^{\mu\nu}(y) + m^2\, t_m(y) = \big(C_3\, \Box + C_4\, m^2\big)\, \delta(y)\ , \quad C_k \in \mathbb{C} \text{ arbitrary.} \quad (3.6.59)$$

This can be reached as follows: For arbitrary extensions $\tilde{t}_m^{\mu\nu}$ and $\tilde{t}_m$ of $t_m^{\mu\nu\, 0}$ and t_m^0, resp., satisfying our axioms, the r.h.s. of (3.6.59) has the following more general form: It is a Lorentz invariant polynomial in derivatives of $\delta(y)$ fulfilling the Sm-expansion with degree $D = 6$, that is, C_4 is replaced by a polynomial $P_1(\log \frac{m}{M})$. Now, the condition (3.6.59) can be fulfilled by a finite renormalization,

$$t_m^{\mu\nu}(y) := \tilde{t}_m^{\mu\nu}(y) + \frac{m^2\, g^{\mu\nu}}{4}\, \Big(C_4 - P_1\big(\log \frac{m}{M}\big)\Big)\, \delta(y) \quad \text{and} \quad t_m := \tilde{t}_m\ ;$$

obviously this change is compatible with our axioms.[96] In addition to (3.6.59), we will use that for the massive fish diagram t_m the violation of homogeneous scaling comes from the leading term in its Sm-expansion, which is the massless fish diagram t_{fish}, i.e.,

$$\rho^4\, t_{m/\rho}(\rho y) - t_m(y) = \rho^4\, t_{\text{fish}}(\rho y) - t_{\text{fish}}(y) = C_{\text{fish}}\, \log \rho\ \delta(y)\ . \quad (3.6.60)$$

This holds for the same reason as (3.6.51).

Combining these results, we obtain

$$\begin{aligned} 0 &= (y\partial_y + 6 - m\partial_m)\Big(g_{\mu\nu}\, t_m^{\mu\nu}(y) + m^2\, t_m(y)\Big) \\ &= \bigg((4C_1 + C_2)\Box + m^2\Big(4\, P\big(\log \frac{m}{M}\big) + C_{\text{fish}}\Big)\bigg)\, \delta(y)\ , \end{aligned}$$

which yields

$$C_1 = -\frac{1}{4}\, C_2\ , \quad P\big(\log \frac{m}{M}\big) = -\frac{1}{4}\, C_{\text{fish}}\ . \quad (3.6.61)$$

[95] The addition (k′) to the Sm-expansion axiom, or more precisely the corresponding condition for time-ordered products, is not used here.

[96] Alternatively, the finite renormalization

$$t_m(y) := \tilde{t}_m(y) + \Big(C_4 - P_1\big(\log \frac{m}{M}\big)\Big)\, \delta(y) \quad \text{and} \quad t_m^{\mu\nu} := \tilde{t}_m^{\mu\nu}$$

yields also (3.6.59); but this t_m violates the Sm-expansion axiom, since $\lim_{m\downarrow 0} t_m$ diverges.

It remains to compute the number C_2. This can be done by setting $m := 0$. We write $t^{\mu\nu}$ for $t^{\mu\nu}_{m=0}$. To extend

$$t^{\mu\nu\,0}(y) := D^F(y)\,\partial^\mu\partial^\nu D^F(y) \in \mathcal{D}'(\mathbb{R}^4\setminus\{0\})\ ,$$

we use $D^F(y) = \frac{-1}{4\pi^2\,(y^2-i0)}$ and formula (3.5.6) to write $t^{\mu\nu\,0}$ as

$$t^{\mu\nu\,0}(y) = \frac{(-g^{\mu\nu}y^2+4y^\mu y^\nu)}{8\pi^4\,(y^2-i0)^4} = \frac{(-g^{\mu\nu}y^2+4y^\mu y^\nu)}{96}\Box_y\Box_y t^0_{\text{fish}}(y)\ .$$

By means of the version (3.5.21) of differential renormalization we obtain

$$t^{\mu\nu}(y) = \frac{(-g^{\mu\nu}y^2+4y^\mu y^\nu)}{96}\,\Box_y\Box_y t_{\text{fish}}(y) \in \mathcal{D}'(\mathbb{R}^4)\ .$$

From this result we conclude that

$$\begin{aligned}(y\partial_y+6)\,t^{\mu\nu}(y) &= \frac{(-g^{\mu\nu}y^2+4y^\mu y^\nu)}{96}\,\Box_y\Box_y\,(y\partial_y+4)\,t_{\text{fish}}(y)\\ &= \frac{C_{\text{fish}}}{96}\,(-g^{\mu\nu}y^2+4y^\mu y^\nu)\,\Box_y\Box_y\delta(y)\\ &= \frac{C_{\text{fish}}}{12}\,(-g^{\mu\nu}\,\Box_y+4\,\partial_y^\mu\partial_y^\nu)\,\delta(y)\ ,\end{aligned} \tag{3.6.62}$$

by using

$$y^\mu y^\nu\,\Box_y\Box_y\delta(y) = (4\,g^{\mu\nu}\,\Box_y+8\,\partial_y^\mu\partial_y^\nu)\,\delta(y)$$

in the last step. On the other hand, taking the limit $m\downarrow 0$ of (3.6.58) and using (3.6.61), we obtain

$$(y\partial_y+6)\,t^{\mu\nu}(y) = \frac{C_2}{4}\,(-g^{\mu\nu}\,\Box_y+4\,\partial_y^\mu\partial_y^\nu)\,\delta(y)\ .$$

Hence, $C_2 = \frac{1}{3}\,C_{\text{fish}}$. We end up with

$$Z_\rho^{(2)}\big(\varphi\partial^\mu\partial^\nu\varphi(x_1)\otimes\varphi^2(x_2)\big) = \frac{i\hbar\,C_{\text{fish}}}{6}\,\Big(-g^{\mu\nu}\,\Box_y+4\,\partial_y^\mu\partial_y^\nu-3\,g^{\mu\nu}\,m^2\Big)\,\delta(y)\,\log\rho\ .$$

Again, we have traced back the task to the breaking of homogeneous scaling of the massless fish diagram $t_{\text{fish}}\equiv t^M_{\text{fish}}$ (3.6.49). If we renormalize the latter with the renormalization mass scale $M := m$, our scaling transformation acts also on M: $M\to M/\rho$. With that, homogeneous scaling is preserved since, e.g., the scaling relation (3.6.62) is replaced by

$$(y\partial_y+6-M\partial_M)t^{\mu\nu\,M}(y) = \frac{(-g^{\mu\nu}y^2+4y^\mu y^\nu)}{96}\,\Box_y\Box_y\,(y\partial_y+4-M\partial_M)t^M_{\text{fish}}(y) = 0\ .$$

That is, $Z_\rho^{(2)}\big(\varphi\partial^\mu\partial^\nu\varphi(x_1)\otimes\varphi^2(x_2)\big)$ vanishes in that case. But, the choice $M := m$ violates the Sm-expansion axiom, because the latter requires that the pertinent massless T-product, $t^{\mu\nu}\equiv t^{\mu\nu\,M}$, does not depend on m. Similarly to (3.5.16), this observation is an example for Remk. 3.2.35.

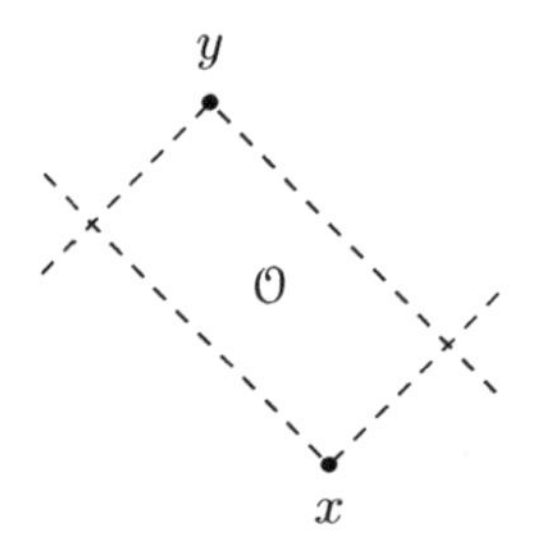

Figure 3.4: A two-dimensional slice of an open "double cone" $\mathcal{O} := (x + V_+) \cap (y + V_-)$.

3.7 The algebraic adiabatic limit

References for this section are [24, 27, 52, 55].

Up to now our renormalization group transformations $Z \in \mathcal{R}$ act on interaction *functionals* $F \in \mathcal{F}_{\text{loc}}$; by Proposition 1.4.3(b) this can be interpreted as an action on explicitly spacetime-dependent interaction Lagrangians $\sum_j \kappa_j g_j(x)\, \mathcal{L}_j(x)$, where $g_j \in \mathcal{D}(\mathbb{M})$, $\mathcal{L}_j \in \mathcal{P}_{\text{bal}}$ and κ_j is a coupling constant. We want to extract from this information the action of the renormalization group on constant Lagrangians $\sum_j \kappa_j \mathcal{L}_j \in \mathcal{P}_{\text{bal}}$ – this is the conventional form of the renormalization group. This requires a kind of adiabatic limit. The usual adiabatic limit $g_j(x) \to 1\ \forall x \in \mathbb{M}$ needs a good infrared behavior (see, e.g., [9, 38, 66, 67, 148]). This restriction can be avoided by using the framework of algebraic quantum field theory (AQFT), also called "local quantum physics" [94]; for a recent and short introduction to AQFT we recommend [76].

A brief outline of algebraic quantum field theory. AQFT aims to understand structural properties of relativistic QFT from first principles, by focussing on two aspects of observables: Their localization in space and time and their algebraic properties.

To explain this more explicitly, we first note that formulating a QFT in terms of fields, a main disadvantage is that different fields may describe the same physics. A way to overcome this redundancy is AQFT: The basic idea is to construct the QFT *locally*. More precisely, the "net of algebras", that is, the map

$$\mathcal{O} \longmapsto \mathcal{A}(\mathcal{O}) , \tag{3.7.1}$$

where $\mathcal{O}$ runs through all open and bounded regions in Minkowski spacetime and $\mathcal{A}(\mathcal{O})$ is the algebra of observables which can be measured within $\mathcal{O}$, "constitutes the intrinsic mathematical description of the theory" (citation from [94]). By definition a net is isotonic, that is, it must hold

$$\mathcal{A}(\mathcal{O}_1) \subseteq \mathcal{A}(\mathcal{O}_2) \quad \text{if} \quad \mathcal{O}_1 \subseteq \mathcal{O}_2 . \tag{3.7.2}$$

In AQFT pertinent *embeddings*, i.e., injective maps

$$\iota_{\mathcal{O}_2,\mathcal{O}_1} : \mathcal{A}_{\mathcal{L}}(\mathcal{O}_1) \hookrightarrow \mathcal{A}_{\mathcal{L}}(\mathcal{O}_2) \ ,$$

must be specified – they contain an essential part of the physical information which is encoded in the net (3.7.1).

R. Haag and D. Kastler formulated axioms for AQFT [93] – these are physically motivated conditions on the net (3.7.1). For brevity we only mention "Locality" (also called "Causality") and "Poincaré covariance":

- *Locality*: If $\mathcal{O}_1$ and $\mathcal{O}_2$ are spacelike separated (i.e., $(x_1-x_2)^2 < 0 \quad \forall (x_1,x_2) \in \mathcal{O}_1 \times \mathcal{O}_2$) and if $(\mathcal{O}_1 \cup \mathcal{O}_2) \subseteq \mathcal{O}$, then

$$[\iota_{\mathcal{O},\mathcal{O}_1}(A_1)\,,\,\iota_{\mathcal{O},\mathcal{O}_2}(A_2)]_\star = 0 \quad \forall (A_1,A_2) \in \mathcal{A}(\mathcal{O}_1) \times \mathcal{A}(\mathcal{O}_2)\ . \tag{3.7.3}$$

- *Poincaré covariance*: For each $L \equiv (\Lambda, a) \in \mathcal{P}_+^\uparrow$ and for each $\mathcal{O}$, there exists an algebra isomorphism

$$\alpha_L^{\mathcal{O}} : \mathcal{A}(\mathcal{O}) \longrightarrow \mathcal{A}(L\mathcal{O}) \tag{3.7.4}$$

with the properties

$$\alpha_{L_2}^{L_1\mathcal{O}} \circ \alpha_{L_1}^{\mathcal{O}} = \alpha_{L_2 L_1}^{\mathcal{O}} \quad \text{and} \tag{3.7.5}$$

$$\alpha_L^{\mathcal{O}_2} \circ \iota_{\mathcal{O}_2,\mathcal{O}_1} = \iota_{L\mathcal{O}_2,L\mathcal{O}_1} \circ \alpha_L^{\mathcal{O}_1}\ . \tag{3.7.6}$$

For our purposes it suffices to define the map (3.7.1) for[97] *open double cones* $\mathcal{O}$; that is, $\mathcal{O}$ is of the form

$$\mathcal{O} = (x + V_+) \cap (y + V_-) \quad \text{for some pair } (x,y) \in \mathbb{M}^2 \text{ fulfilling } y \in (x + V_+); \tag{3.7.7}$$

see Figure 3.4.

An interaction-free example for an AQFT is the net

$$\mathcal{O} \longmapsto \mathcal{A}(\mathcal{O}) := \bigvee\nolimits_\star \mathcal{F}_{\mathrm{loc}}(\mathcal{O})\ , \quad \text{where} \quad \mathcal{F}_{\mathrm{loc}}(\mathcal{O}) := \{\, F \in \mathcal{F}_{\mathrm{loc}} \,\big|\, \operatorname{supp} F \subset \mathcal{O} \,\}\ , \tag{3.7.8}$$

where $\mathcal{O}$ runs through all double cones in Minkowski spacetime and $\bigvee_\star$ means the algebra, under the $\star$-product, generated by members of the indicated set.

Perturbative algebraic quantum field theory. We wish to construct an AQFT, which is generated by our perturbative interacting fields (3.1.2) and which goes over into (3.7.8) when switching off the interaction. For this purpose let $\mathcal{L} \in \mathcal{P}_{\mathrm{bal}}$ be given, with $\mathcal{L} = \mathcal{L}^*$; to simplify the notation, the symbol $\mathcal{L}$ includes the coupling constant κ. We modify (3.7.8) as follows: We replace $F \in \mathcal{F}_{\mathrm{loc}}(\mathcal{O})$ by

[97]The term "double cone" here refers to a bounded region with two conical vertices, rather than its usual meaning of a pair of funnels meeting at a single conical vertex.

the corresponding interacting field $F_{\mathcal{L}(g)}$. The latter fulfills the Locality condition (3.7.3) – as noted in Remk. 3.1.6. This yields the net

$$\mathcal{O} \longmapsto \mathcal{A}_{\mathcal{L}}(\mathcal{O}) := \bigvee_{\star} \{ F_{\mathcal{L}(g)} = R(e_{\otimes}^{\mathcal{L}(g)/\hbar}, F) \,|\, F \in \mathcal{F}_{\rm loc}(\mathcal{O}) \} , \tag{3.7.9}$$

where g is a real-valued test function[98] that is identically 1 on $\mathcal{O}$, precisely:

$$g \in \mathcal{G}(\mathcal{O}) := \{ g \in \mathcal{D}(\mathbb{M}, \mathbb{R}) \,|\, g(x) = 1 \ \forall x \text{ in a neighbourhood of } \overline{\mathcal{O}} \} , \tag{3.7.10}$$

where $\overline{\mathcal{O}}$ is the closure of $\mathcal{O}$. Isotony (3.7.2) of the map (3.7.9) is obvious.

A drawback of this setup is that $F_{\mathcal{L}(g)}$ depends on the restriction of g to $\mathcal{O} + \overline{V}_-$ (by causality); but the algebra $\mathcal{A}_{\mathcal{L}}(\mathcal{O})$ should rather be independent of g. Il'in and Slavnov [108] and, about 20 years later, Brunetti and Fredenhagen [24] showed that this is indeed the case.

Theorem 3.7.1 (Independence from adiabatic switching). *As an abstract algebra, $\mathcal{A}_{\mathcal{L}(g)} := \bigvee_{\star} \{ F_{\mathcal{L}(g)} \,|\, F \in \mathcal{F}_{\rm loc}(\mathcal{O}) \}$ is independent of the choice of $g \in \mathcal{G}(\mathcal{O})$. Concretely, for any $g_1, g_2 \in \mathcal{G}(\mathcal{O})$, there is a unitary[99] element $U_{g_1,g_2} \in \mathcal{F}[\![\kappa]\!]$ such that*

$$U_{g_1,g_2} \star F_{\mathcal{L}(g_1)} \star (U_{g_1,g_2})^{\star -1} = F_{\mathcal{L}(g_2)} , \quad \textit{for all} \quad F \in \mathcal{F}_{\rm loc}(\mathcal{O}) . \tag{3.7.11}$$

Proof. Following [24],[100] we prove the assertion (3.7.11) for the relative S-matrix (3.3.31), which is a more general observable:

$$\mathbf{S}_{\mathcal{L}(g)}(F) := \mathbf{S}\big(\mathcal{L}(g)\big)^{\star -1} \star \mathbf{S}\big(\mathcal{L}(g) + F\big), \quad F \in \mathcal{F}_{\rm loc} . \tag{3.7.12}$$

$F \longmapsto \mathbf{S}_{\mathcal{L}(g)}(F)$ is the generating functional of time-ordered products of the interacting fields $F_{\mathcal{L}(g)}$ (see [66]); in particular we have $F_{\mathcal{L}(g)} = \frac{\hbar}{i} \frac{d}{d\lambda}|_{\lambda=0} \mathbf{S}_{\mathcal{L}(g)}(\lambda F)$, see (3.3.30). The relative S-matrix is introduced already in Exer. 3.3.11 and it will be treated in more detail in (4.2.45) and Remk. 4.2.4.

The proof uses only Causality for time-ordered products in the form of the relation (3.3.27) and unitarity of the S-matrix.

We split $(g_2 - g_1)$ into

$$g_2 - g_1 = a + b \quad \text{with} \quad \operatorname{supp} a \cap (\mathcal{O} + \overline{V}_-) = \emptyset \ \wedge \ \operatorname{supp} b \cap (\mathcal{O} + \overline{V}_+) = \emptyset . \tag{3.7.13}$$

From $\operatorname{supp} \mathcal{L}(a) \cap (\operatorname{supp} F + \overline{V}_-) = \emptyset$ and $\mathcal{L}(g_2) = \mathcal{L}(g_1 + b) + \mathcal{L}(a)$, we conclude

$$\mathbf{S}\big(\mathcal{L}(g_2) + F\big) = \mathbf{S}\big(\mathcal{L}(g_2)\big) \star \mathbf{S}\big(\mathcal{L}(g_1 + b)\big)^{\star -1} \star \mathbf{S}\big(\mathcal{L}(g_1 + b) + F\big) ,$$

and, by $\star$-multiplying with $\mathbf{S}\big(\mathcal{L}(g_2)\big)^{\star -1}$, this gives $\mathbf{S}_{\mathcal{L}(g_2)}(F) = \mathbf{S}_{\mathcal{L}(g_1+b)}(F)$.

[98] Such a g is often taken to be merely a Schwartz function; but test functions are somewhat better suited to an eventual promotion of this theory to curved spaces.

[99] To say that U is *unitary* means that $U^* \star U = U \star U^* = (1, 0, 0, \ldots)$ in $\mathcal{F}[\![\kappa]\!]$.

[100] An alternative proof, relying on the axioms for retarded products, can be found in [52].

Due to $\mathrm{supp}\, F \cap (\mathrm{supp}\, \mathcal{L}(b) + \overline{V}_-) = \emptyset$, we have

$$\mathbf{S}\big(\mathcal{L}(g_1+b)+F\big) = \mathbf{S}\big(\mathcal{L}(g_1)+F\big) \star \mathbf{S}\big(\mathcal{L}(g_1)\big)^{\star -1} \star \mathbf{S}\big(\mathcal{L}(g_1+b)\big)\,.$$

$\star$-multiplying this equation with $\mathbf{S}\big(\mathcal{L}(g_1+b)\big)^{\star-1}$ we obtain

$$\begin{aligned}\mathbf{S}_{\mathcal{L}(g_2)}(F) &= \mathbf{S}_{\mathcal{L}(g_1+b)}(F)\\ &= \Big[\mathbf{S}\big(\mathcal{L}(g_1+b)\big)^{\star-1} \star \mathbf{S}\big(\mathcal{L}(g_1)\big)\Big]\\ &\quad\star \Big[\mathbf{S}\big(\mathcal{L}(g_1)\big)^{\star-1} \star \mathbf{S}\big(\mathcal{L}(g_1)+F\big)\Big] \star \Big[\mathbf{S}\big(\mathcal{L}(g_1)\big)^{\star-1} \star \mathbf{S}\big(\mathcal{L}(g_1+b)\big)\Big]\,.\end{aligned}$$

The expression in the middle $[\ldots]$-bracket is $\mathbf{S}_{\mathcal{L}(g_1)}(F)$. Setting

$$U_{g_1,g_2} := \mathbf{S}\big(\mathcal{L}(g_1+b)\big)^{\star-1} \star \mathbf{S}\big(\mathcal{L}(g_1)\big)\,,$$

the expression in the last $[\ldots]$-bracket is obviously equal to $U_{g_1,g_2}^{\star-1}$; so, we have derived (3.7.11). Due to $\mathbf{S}\big(\mathcal{L}(h)\big)^* = \mathbf{S}\big(\mathcal{L}(h)\big)^{\star-1}$ when $\mathcal{L}=\mathcal{L}^*$, $h=\overline{h}$, we obtain $(U_{g_1,g_2})^* = (U_{g_1,g_2})^{\star-1}$ by using (2.1.10). It is crucial that U_{g_1,g_2} does not depend on F. □

Now for $\mathcal{L} = \kappa L \in \mathcal{P}_{\mathrm{bal}}$, let

$$\mathcal{G}_{\mathcal{L}}(\mathcal{O}) := \{\, G \in \mathcal{D}(\mathbb{M}, \mathcal{P}_{\mathrm{bal}}) \,\big|\, G(x) = \mathcal{L}(x)\ \forall x \text{ in a neighbourhood of } \overline{\mathcal{O}}\,\} \tag{3.7.14}$$

and we will use the notation $\int G := \int_{\mathbb{M}} G(x)\, d^dx$. For example, one could take a G of the form $G(x) = g(x)\,\mathcal{L}(x)$ with $g \in \mathcal{G}(\mathcal{O})$. To define the algebraic adiabatic limit of an interacting field $F_{\int G}$ (where $F \in \mathcal{F}_{\mathrm{loc}}(\mathcal{O})$, $G \in \mathcal{G}_{\mathcal{L}}(\mathcal{O})$) we proceed as follows: To get rid of the algebraically irrelevant choice of a $G \in \mathcal{G}_{\mathcal{L}}(\mathcal{O})$ we admit *all* $G \in \mathcal{G}_{\mathcal{L}}(\mathcal{O})$. This idea can be realized by the following definition.[101]

Definition 3.7.2 (Algebraic adiabatic limit)**.** Let $\mathcal{O}$ be a double cone and $F \in \mathcal{F}_{\mathrm{loc}}(\mathcal{O})$. The *algebraic adiabatic limit with respect to* $\mathcal{O}$ *of the interacting field* $F_{\mathcal{L}(g)}$ ($g \in \mathcal{D}(\mathbb{M})$) is the map

$$F^{\mathcal{O}}_{\mathcal{L}} : \begin{cases} \mathcal{G}_{\mathcal{L}}(\mathcal{O}) \longrightarrow \mathcal{F} \\ G \longmapsto F_{\int G} = R\big(e_\otimes^{\int G/\hbar}, F\big)\,. \end{cases} \tag{3.7.15}$$

The corresponding *algebra of observables localized in* $\mathcal{O}$ is

$$\mathcal{A}_{\mathcal{L}}(\mathcal{O}) := \bigvee\nolimits_{\star} \{\, F^{\mathcal{O}}_{\mathcal{L}} \,|\, F \in \mathcal{F}_{\mathrm{loc}}(\mathcal{O})\,\}\,, \tag{3.7.16}$$

where the operations are defined pointwise:

$$(F^{\mathcal{O}}_{\mathcal{L}} + H^{\mathcal{O}}_{\mathcal{L}})(G) := F^{\mathcal{O}}_{\mathcal{L}}(G) + H^{\mathcal{O}}_{\mathcal{L}}(G)\,, \quad (F^{\mathcal{O}}_{\mathcal{L}} \star H^{\mathcal{O}}_{\mathcal{L}})(G) := F^{\mathcal{O}}_{\mathcal{L}}(G) \star H^{\mathcal{O}}_{\mathcal{L}}(G)$$

for all $G \in \mathcal{G}_{\mathcal{L}}(\mathcal{O})$.

---

[101] Compared with the original paper [24] or with [51, 55], we work here with a somewhat simplified definition of the algebraic adiabatic limit, which is used also in [75].

Given an inclusion of double cones $\mathcal{O}_1 \subseteq \mathcal{O}_2$, we find that $\mathcal{G}_\mathcal{L}(\mathcal{O}_2) \subseteq \mathcal{G}_\mathcal{L}(\mathcal{O}_1)$; with that an embedding can be defined on the generators by restriction of the maps (3.7.15):

$$\iota_{\mathcal{O}_2,\mathcal{O}_1}\colon \begin{cases} \mathcal{A}_\mathcal{L}(\mathcal{O}_1) \hookrightarrow \mathcal{A}_\mathcal{L}(\mathcal{O}_2) \\ F_\mathcal{L}^{\mathcal{O}_1} \longmapsto F_\mathcal{L}^{\mathcal{O}_2} = F_\mathcal{L}^{\mathcal{O}_1}\big|_{\mathcal{G}_\mathcal{L}(\mathcal{O}_2)}\,, \end{cases} \tag{3.7.17}$$

where $F \in \mathcal{F}_{\mathrm{loc}}(\mathcal{O}_1) \subseteq \mathcal{F}_{\mathrm{loc}}(\mathcal{O}_2)$. For $F, H \in \mathcal{F}_{\mathrm{loc}}(\mathcal{O}_1)$ this definition of the embedding gives

$$F_\mathcal{L}^{\mathcal{O}_1} \star H_\mathcal{L}^{\mathcal{O}_1} \equiv \big(F_{\int G_1} \star H_{\int G_1}\big)_{G_1 \in \mathcal{G}_\mathcal{L}(\mathcal{O}_1)} \longmapsto \begin{cases} \iota_{\mathcal{O}_2,\mathcal{O}_1}\big(F_\mathcal{L}^{\mathcal{O}_1} \star H_\mathcal{L}^{\mathcal{O}_1}\big) \\ = \big(F_{\int G_2} \star H_{\int G_2}\big)_{G_2 \in \mathcal{G}_\mathcal{L}(\mathcal{O}_2)}\,. \end{cases}$$

Usually the restriction of a map is not injective. Nevertheless, $\iota_{\mathcal{O}_2,\mathcal{O}_1}$ is injective, as it must be for an embedding, for the following reason: If we know $F_\mathcal{L}^\mathcal{O}(G_0)$ for one $G_0 \in \mathcal{G}_\mathcal{L}(\mathcal{O})$, we know $F_\mathcal{L}^\mathcal{O}(G)$ for all $G \in \mathcal{G}_\mathcal{L}(\mathcal{O})$, namely, $F_\mathcal{L}^\mathcal{O}(G) = U_{G_0,G} \star F_\mathcal{L}^\mathcal{O}(G_0) \star (U_{G_0,G})^{\star -1}$ due to Theorem 3.7.1.

Obviously, the embeddings (3.7.17) satisfy the consistency condition

$$i_{\mathcal{O}_3,\mathcal{O}_2} \circ i_{\mathcal{O}_2,\mathcal{O}_1} = i_{\mathcal{O}_3,\mathcal{O}_1} \quad \text{for} \quad \mathcal{O}_1 \subseteq \mathcal{O}_2 \subseteq \mathcal{O}_3\,.$$

The so-constructed net

$$\mathcal{A}_\mathcal{L}\colon \mathcal{O} \longmapsto \mathcal{A}_\mathcal{L}(\mathcal{O}) \tag{3.7.18}$$

fulfills the Haag–Kastler axioms of algebraic quantum field theory [93], except that there is no suitable norm available on these formal power series. For details we refer to the mentioned references. We only verify Locality and Poincaré covariance: For the former let $\mathcal{O}_1$, $\mathcal{O}_2$ and $\mathcal{O}$ be as in (3.7.3) and let $F_\mathcal{L}^{\mathcal{O}_1} \in \mathcal{A}_\mathcal{L}(\mathcal{O}_1)$, $H_\mathcal{L}^{\mathcal{O}_2} \in \mathcal{A}_\mathcal{L}(\mathcal{O}_2)$; then we get

$$[\iota_{\mathcal{O},\mathcal{O}_1}(F_\mathcal{L}^{\mathcal{O}_1}),\, \iota_{\mathcal{O},\mathcal{O}_2}(H_\mathcal{L}^{\mathcal{O}_2})]_\star(G) = [F_\mathcal{L}^{\mathcal{O}_1}(G), H_\mathcal{L}^{\mathcal{O}_2}(G)]_\star = [F_{\int G}, H_{\int G}]_\star = 0$$

for all $G \in \mathcal{G}_\mathcal{L}(\mathcal{O})$, by using the spacelike commutativity of the interacting fields (Remk. 3.1.6) in the last step.

Turning to Poincaré covariance, it suffices to define the isomorphism $\alpha_L^\mathcal{O}$ ($L \in \mathcal{P}_+^\uparrow$) for the generators $F_\mathcal{L}^\mathcal{O}$ of $\mathcal{A}_\mathcal{L}(\mathcal{O})$. To do this, we first note that the isomorphism[102] $\beta_L \equiv \beta_{\Lambda,a}\colon \mathcal{F} \longrightarrow \mathcal{F}$ (introduced in (3.1.35)) induces an isomorphism $\beta_L^\mathcal{O}\colon \mathcal{F}(\mathcal{O}) \longrightarrow \mathcal{F}(L\mathcal{O})$. Any $G \in \mathcal{G}_\mathcal{L}(\mathcal{O})$ is a finite sum of the form $G(x) = \sum_s g_s(x)\,\mathcal{L}_s(x)$ with $\mathcal{L}_s \in \mathcal{P}_{\mathrm{bal}}$, $\sum_s \mathcal{L}_s = \mathcal{L}$ and $g_s \in \mathcal{G}(\mathcal{O})$. We assume that all $\mathcal{L}_s$ are scalar with respect to Lorentz transformations. Introducing the bijection

$$\beta_{L,\mathcal{L}}^\mathcal{O}\colon \begin{cases} \mathcal{G}_\mathcal{L}(\mathcal{O}) \longrightarrow \mathcal{G}_\mathcal{L}(L\mathcal{O}) \\ \sum_s g_s(x)\,\mathcal{L}_s(x) \longmapsto \sum_s g_s(L^{-1}x)\,\mathcal{L}_s(x)\,, \end{cases}$$

[102] We recall from (3.1.37) that $\beta_L\colon \mathcal{F} \longrightarrow \mathcal{F}$ is an algebra isomorphism with respect to both the classical and the star product.

we obtain

$$\beta_L\Big(\int G\Big) = \int \beta^{\mathcal{O}}_{L,\mathcal{L}}(G) \quad \forall G \in \mathcal{G}_{\mathcal{L}}(\mathcal{O})\ ,\ L \in \mathcal{P}^{\uparrow}_{+}\ .$$

By using Poincaré covariance of the R-products (axiom (h)), we get

$$\begin{aligned}(\beta^{\mathcal{O}}_L F)^{L\mathcal{O}}_{\mathcal{L}}(G) &= (\beta^{\mathcal{O}}_L F)_{\int G} = \beta_L\Big(F_{\int \beta^{L\mathcal{O}}_{L^{-1},\mathcal{L}}(G)}\Big)\\ &= \beta_L \circ F^{\mathcal{O}}_{\mathcal{L}} \circ \beta^{L\mathcal{O}}_{L^{-1},\mathcal{L}}(G) \quad \forall G \in \mathcal{G}_{\mathcal{L}}(L\mathcal{O})\ ,\ F \in \mathcal{F}(\mathcal{O})\ .\end{aligned}$$

Hence, we define

$$\alpha^{\mathcal{O}}_L\ :\ \begin{cases}\mathcal{A}_{\mathcal{L}}(\mathcal{O}) \longrightarrow \mathcal{A}_{\mathcal{L}}(L\mathcal{O})\\ F^{\mathcal{O}}_{\mathcal{L}} \longmapsto \alpha^{\mathcal{O}}_L(F^{\mathcal{O}}_{\mathcal{L}}) := (\beta^{\mathcal{O}}_L F)^{L\mathcal{O}}_{\mathcal{L}} = \beta_L \circ F^{\mathcal{O}}_{\mathcal{L}} \circ \beta^{L\mathcal{O}}_{L^{-1},\mathcal{L}}\ .\end{cases}$$

The property $\alpha^{L_1\mathcal{O}}_{L_2} \circ \alpha^{\mathcal{O}}_{L_1} = \alpha^{\mathcal{O}}_{L_2L_1}$ (3.7.5) relies on $\beta_{L_2} \circ \beta_{L_1} = \beta_{L_2L_1}$ and $\beta^{L_1\mathcal{O}}_{L_1^{-1},\mathcal{L}} \circ \beta^{L_2L_1\mathcal{O}}_{L_2^{-1},\mathcal{L}} = \beta^{L_2L_1\mathcal{O}}_{(L_2L_1)^{-1},\mathcal{L}}$. Finally, the compatibility of $\alpha^{\mathcal{O}}_L$ with the embeddings $\iota_{\mathcal{O}_2,\mathcal{O}_1}$ (3.7.6) can be verified as follows: For $\mathcal{O}_1 \subseteq \mathcal{O}_2$ we have $L\mathcal{O}_1 \subseteq L\mathcal{O}_2$ (hence $\mathcal{G}_{\mathcal{L}}(L\mathcal{O}_2) \subseteq \mathcal{G}_{\mathcal{L}}(L\mathcal{O}_1)$), taking into account also the definition of $\beta^{\mathcal{O}}_{L,\mathcal{L}}$, we get

$$\beta^{L\mathcal{O}_1}_{L^{-1},\mathcal{L}}(G) = \beta^{L\mathcal{O}_2}_{L^{-1},\mathcal{L}}(G) \in \mathcal{G}_{\mathcal{L}}(\mathcal{O}_2) \subseteq \mathcal{G}_{\mathcal{L}}(\mathcal{O}_1) \quad \forall G \in \mathcal{G}_{\mathcal{L}}(L\mathcal{O}_2)\ .$$

Hence, we may write

$$F^{\mathcal{O}_1}_{\mathcal{L}} \circ \beta^{L\mathcal{O}_1}_{L^{-1},\mathcal{L}}(G) = F^{\mathcal{O}_2}_{\mathcal{L}} \circ \beta^{L\mathcal{O}_2}_{L^{-1},\mathcal{L}}(G) \quad \forall G \in \mathcal{G}_{\mathcal{L}}(L\mathcal{O}_2)\ .$$

Applying β_L to this equation, the l.h.s. is equal to

$$\big(\alpha^{\mathcal{O}_1}_L(F^{\mathcal{O}_1}_{\mathcal{L}})\big)(G) = \big((\iota_{L\mathcal{O}_2,L\mathcal{O}_1} \circ \alpha^{\mathcal{O}_1}_L)(F^{\mathcal{O}_1}_{\mathcal{L}})\big)(G)\ ,$$

and the for the r.h.s. we get

$$\big(\alpha^{\mathcal{O}_2}_L(F^{\mathcal{O}_2}_{\mathcal{L}})\big)(G) = \big((\alpha^{\mathcal{O}_2}_L \circ \iota_{\mathcal{O}_2,\mathcal{O}_1})(F^{\mathcal{O}_1}_{\mathcal{L}})\big)(G)\ ,$$

for all $G \in \mathcal{G}_{\mathcal{L}}(L\mathcal{O}_2)$, $F \in \mathcal{F}_{\mathrm{loc}}(\mathcal{O}_1)$. The equality of these two terms yields the assertion (3.7.6). ⊟

Following a standard procedure in AQFT (see, e.g., [51]), the algebra $\mathcal{A}^{\mathrm{loc}}_{\mathcal{L}}$ of "all local observables" is obtained by the inductive limit of the algebras $\mathcal{A}_{\mathcal{L}}(\mathcal{O})$. This limit is *not* the set theoretic union $\cup_{\mathcal{O}} \mathcal{A}_{\mathcal{L}}(\mathcal{O})$; it is defined in terms of the *disjoint union*

$$\bigsqcup_{\mathcal{O}} \mathcal{A}_{\mathcal{L}}(\mathcal{O}) := \bigcup_{\mathcal{O}} \{(A,\mathcal{O}) \,|\, A \in \mathcal{A}_{\mathcal{L}}(\mathcal{O})\}\ ;$$

which is a union of ordered pairs, the union runs through all open double cones $\mathcal{O}$. On this set we consider the equivalence relation "$\sim$" generated by the embeddings: For all $\mathcal{O} \subseteq \mathcal{O}_1$ and all $A \in \mathcal{A}_{\mathcal{L}}(\mathcal{O})$ we define

$$(A,\mathcal{O}) \sim \big(\iota_{\mathcal{O}_1,\mathcal{O}}(A),\mathcal{O}_1\big)\ .$$

The inductive limit is the resulting set of equivalence classes,

$$\mathcal{A}_{\mathcal{L}}^{\mathrm{loc}} := \bigsqcup_{\mathcal{O}} \mathcal{A}_{\mathcal{L}}(\mathcal{O}) \;/\; \sim \,, \tag{3.7.19}$$

which is an algebra. In particular, denoting the equivalence class of $(A, \mathcal{O})$ by $[(A, \mathcal{O})]$, the product is defined by

$$[(A_1, \mathcal{O}_1)] \star [(A_2, \mathcal{O}_2)] := \big[\big(\iota_{\mathcal{O},\mathcal{O}_1}(A_1) \star_{\mathcal{O}} \iota_{\mathcal{O},\mathcal{O}_2}(A_2)\,,\, \mathcal{O}\big)\big]\,,$$

where $\mathcal{O}$ is an open double cone containing $\mathcal{O}_1 \cup \mathcal{O}_2$ and $\star_{\mathcal{O}}$ is the product in $\mathcal{A}_{\mathcal{L}}(\mathcal{O})$. The action of the Poincaré group on the algebra $\mathcal{A}_{\mathcal{L}}^{\mathrm{loc}}$ is defined in an obvious way:

$$\alpha_L \,:\, \begin{cases} \mathcal{A}_{\mathcal{L}}^{\mathrm{loc}} \longrightarrow \mathcal{A}_{\mathcal{L}}^{\mathrm{loc}} \\ [(A, \mathcal{O})] \longmapsto \big[(\alpha_L^{\mathcal{O}}(A), L\mathcal{O})\big] \end{cases} \tag{3.7.20}$$

for all $L \in \mathcal{P}_+^\uparrow$. By means of the weak adiabatic limit one can construct a *vacuum state* on $\mathcal{A}_{\mathcal{L}}^{\mathrm{loc}}$, this is explained in (A.6.20).

Remark 3.7.3. The adiabatic switching of the coupling constant(s), i.e., the replacement of κ by $\kappa g(x)\,,\; g \in \mathcal{D}(\mathbb{M})$, can be interpreted as an IR-regularization of the model. Hence, it might be that the results which one obtains in the adiabatic limit $g(x) \to 1\; \forall x \in \mathbb{M}$ depend on this regularization method. Even if the adiabatic limit exists and is unique in the sense of Epstein and Glaser (see App. A.6), it might be that another IR-regularization method – e.g., by giving the originally massless fields a tiny mass $m > 0$ and performing the limit $m \downarrow 0$ in the end – yields different results. This does certainly not hold for the algebraic structure obtained in the algebraic adiabatic limit, that is, the net (3.7.18), since this structure is independent of the IR-behaviour of the model.

But, for the construction of physically relevant *scattering states* on the algebras $\mathcal{A}_{\mathcal{L}}(\mathcal{O})$, the algebraic adiabatic limit does not suffice, one has to perform the adiabatic limit in the sense of Epstein and Glaser (see App. A.6 and the references [9, 38, 66, 67] and [148, Sects. 3.11-3.12]), which is hard work.

3.8 The renormalization group in the algebraic adiabatic limit

References here are [27, 55, 57, 102].

3.8.1 Renormalization of the interaction

To obtain the traditional form of the RG as a renormalization of the (space and time independent) coupling constants, we investigate the renormalization of the interaction $\kappa L(g) \longmapsto Z\big(\kappa L(g)\big)$ (where Z is an element of the Stückelberg–

Petermann RG $\mathcal{R}$ and $\kappa L \in \mathcal{P}_{\mathrm{bal}}$ is the interaction which is switched on and off by $g \in \mathcal{D}(\mathbb{M})$) in the algebraic adiabatic limit. For this purpose, we first prove an auxiliary lemma.

Lemma 3.8.1 ([55]). *Each $Z^{(n)} \in \mathcal{R}$ admits the expansion*

$$Z^{(n)}\big(A_1(h_1)\otimes\cdots\otimes A_n(h_n)\big) \tag{3.8.1}$$
$$= \sum_{l=0}^{\mathfrak{L}} m^l \sum_{p=0}^{P_l} \log^p\big(\frac{m}{M}\big) \sum_{|a|\leq\alpha_l} \int d^dx\; d_{n,l,p,a}(A_1\otimes\cdots\otimes A_n)(x) \prod_{j=1}^{n} \partial^{a_j} h_j(x)$$

for $A_j \in \mathcal{P}_{\mathrm{bal}}$, $h_j \in \mathcal{D}(\mathbb{M})$; where $a=(a_1,\dots,a_n)\in(\mathbb{N}^d)^n$ and each

$$d_{n,l,p,a}\colon \mathcal{P}_{\mathrm{bal}}^{\otimes n} \longrightarrow \mathcal{P}_{\mathrm{bal}} \quad \textit{is linear and symmetric.}$$

By the latter property we mean

$$d_{n,l,p,a_\pi}(A_{\pi(1)}\otimes\cdots\otimes A_{\pi(n)}) = d_{n,l,p,a}(A_1\otimes\cdots\otimes A_n)\;, \quad a_\pi := (a_{\pi(1)},\dots,a_{\pi(n)})\;,$$

for all $\pi \in S_n$. The upper bounds $\mathfrak{L}$, P_l, $\alpha_l \in \mathbb{N}$ depend on $(A_1,\dots,A_n)$, in particular the sums are finite for a fixed $(A_1,\dots,A_n)$, and for $l=0$ we have $P_l=0$.

In addition, for $A_1,\dots,A_n \in \mathcal{P}_{\mathrm{bal}}\cap\mathcal{P}_{\mathrm{hom}}$ it holds that $d_{n,l,p,a}(A_1\otimes\cdots\otimes A_n) \in \mathcal{P}_{\mathrm{bal}}\cap\mathcal{P}_{\mathrm{hom}}$ and

$$\dim d_{n,l,p,a}(A_1\otimes\cdots\otimes A_n) = \sum_{j=1}^{n} \dim A_j - l - |a| - d(n-1)\;. \tag{3.8.2}$$

Note that $d_{n,l,p,a}(A_1\otimes\cdots\otimes A_n)$ vanishes, if the right-hand side is negative – this shows the dependence of $\mathfrak{L}$ and α_l on $(A_1,\dots,A_n)$.

Proof. Let $A_1,\dots,A_n \in \mathcal{P}_{\mathrm{bal}}$ be given. The idea is to combine the Wick expansion (3.6.9) with the Sm-expansion (3.6.3)–(3.6.4). For this purpose we write $A_j \in \mathcal{P}_{\mathrm{bal}}$ as $A_j = \sum_{k_j} B_{jk_j}$ with monomials B_{jk_j}. Later on we will use linearity of $Z^{(n)}$:

$$Z^{(n)}\big(A_1(h_1)\otimes\cdots\otimes A_n(h_n)\big) = \sum_{k_1,\dots,k_n} Z^{(n)}\big(B_{1k_1}(h_1)\otimes\cdots\otimes B_{nk_n}(h_n)\big)\;. \tag{3.8.3}$$

Following the idea, we obtain

$$Z^{(n)}\big(B_{1k_1}(h_1)\otimes\cdots\otimes B_{nk_n}(h_n)\big)$$
$$= \sum_{\underline{B}_{jk_j}\subseteq B_{jk_j}} \sum_{l=0}^{\omega} m^l \sum_{p=0}^{P_l} \log^p\big(\frac{m}{M}\big) \sum_{|b|=\omega-l} C_{l,p,b}(\underline{B}_{1k_1},\dots,\underline{B}_{nk_n})\, F(h_1,\dots,h_n)\;,$$

where $F(h_1,\dots,h_n) \equiv F^b(\overline{B}_{1k_1},\dots,\overline{B}_{nk_n})(h_1,\dots,h_n) \in \mathcal{F}_{\mathrm{loc}}$ is given by

$$F(h_1,\dots,h_n) := \int dx_1\dots dx_n\; \partial^b\delta(x_1-x_n,\dots)\,\overline{B}_{1k_1}(x_1)\cdots\overline{B}_{nk_n}(x_n)$$
$$\cdot\, h_1(x_1)\cdots h_n(x_n)\;,$$

and $\omega := \sum_j \dim \underline{B}_{jk_j} - d(n-1)$; in addition, for $l = 0$ we have $P_l = 0$. By integrating out the δ-distribution and with integration by parts, $F(h_1, \dots, h_n)$ can uniquely be written as

$$F(h_1,\dots,h_n) = \sum_{a=(a_1,\dots,a_n)} \int d^dx\; B_a(x) \prod_{j=1}^{n} \partial^{a_j} h_j(x) =: \sum_a B_a\Big(\prod_j \partial^{a_j} h_j\Big)$$

with $B_a \in \mathcal{P}_{\mathrm{bal}} \cap \mathcal{P}_{\mathrm{hom}}$ for all a – this is an example for part (b) of Proposition 1.4.3. For clarity we will write $B_a^b(\overline{B}_{1k_1}, \dots, \overline{B}_{nk_n}) := B_a$. Reordering the sums and setting

$$\begin{aligned} &d_{n,l,p,a}(A_1 \otimes \cdots \otimes A_n) \\ &\quad := \sum_{k_1,\dots,k_n} \sum_{\underline{B}_{jk_j} \subseteq B_{jk_j}} \sum_{|b|=\omega-l} C_{l,p,b}(\underline{B}_{1k_1}, \dots, \underline{B}_{nk_n}) B_a^b(\overline{B}_{1k_1}, \dots, \overline{B}_{nk_n}) \end{aligned}$$

we obtain (3.8.1). Linearity and symmetry of the maps $d_{n,l,p,a}$ follow from the corresponding properties of $Z^{(n)}$.

Finally let $A_1, \dots, A_n \in \mathcal{P}_{\mathrm{bal}} \cap \mathcal{P}_{\mathrm{hom}}$. Then it holds that $\dim B_{jk_j} = \dim A_j$ for all k_j. By direct computation we get on the one hand

$$\begin{aligned} \sigma_\rho^{-1} F(h_1,\dots,h_n) &= \rho^{\sum_j \dim \overline{B}_{jk_j} - d + |b|} F(h_{1\rho},\dots,h_{n\rho}) \\ &= \rho^{\sum_j \dim \overline{B}_{jk_j} - d + |b|} \sum_a B_a\Big(\prod_j \partial^{a_j} h_{j\rho}\Big) \end{aligned}$$

and on the other hand

$$\sigma_\rho^{-1} F(h_1,\dots,h_n) = \sigma_\rho^{-1} \sum_a B_a\Big(\prod_j \partial^{a_j} h_j\Big) = \sum_a \rho^{\dim B_a - d + |a|} B_a\Big(\prod_j \partial^{a_j} h_{j\rho}\Big)$$

where $h_{j\rho}(x) := h_j(x/\rho)$. We conclude that

$$\sum_j \dim \overline{B}_{jk_j} + |b| = \dim B_a^b(\overline{B}_{1k_1}, \dots, \overline{B}_{nk_n}) + |a| \;, \quad \forall a \;.$$

Using $\dim \overline{B}_{jk_j} = \dim A_j - \dim \underline{B}_{jk_j}$ and $\sum_j \dim \underline{B}_{jk_j} = \omega + d(n-1) = |b| + l + d(n-1)$, we end up with

$$\dim B_a^b(\overline{B}_{1k_1}, \dots, \overline{B}_{nk_n}) = \sum_{j=1}^{n} \dim A_j - l - |a| - d(n-1) \;,$$

which yields (3.8.2). □

To study the renormalization of the interaction in the algebraic adiabatic limit, let $G = g\,L \in \mathcal{G}_L(\mathcal{O})$, see (3.7.14), where L does not include the coupling constant κ. Then the term $Z^{(n)}\big(L(g)^{\otimes n}\big)$ can be expanded as

$$Z^{(n)}\big(L(g)^{\otimes n}\big)=\sum_{l\geq 0} m^l \sum_{p=0}^{P_l^n} \log^p\big(\frac{m}{M}\big)\int dx\,\Big\{d_{n,l,p,0}(L^{\otimes n})(x)\,\big(g(x)\big)^n + \sum_{|a|\geq 1} d_{n,l,p,a}(L^{\otimes n})(x)\prod_{j=1}^{n}\partial^{a_j}g(x)\Big\}\,,$$

where we indicate by the notation P_l^n that $P_l^n\equiv P_l$ may depend on n. If x lies in $\overline{\mathcal{O}}$, then $\big(g(x)\big)^n=1$ and the derivatives $\partial^{a_j}g(x)$ vanish for $|a_j|\geq 1$, so only the first term in braces counts. We therefore find that

$$Z\big(\kappa L(g)\big)=\kappa L(g)+\sum_{n=2}^{\infty}\frac{\kappa^n}{n!}\,Z^{(n)}\big(L(g)^{\otimes n}\big)=\int dx\,\Big\{\kappa L(x)g(x)$$

$$+\sum_{n\geq 2,\,l}\frac{\kappa^n m^l}{n!}\sum_{p=0}^{P_l^n}\log^p\big(\frac{m}{M}\big)\,d_{n,l,p,0}(L^{\otimes n})(x)\,\big(g(x)\big)^n+\sum_{n\geq 2}\kappa^n\sum_r A_{n,r}(x)\,f_{n,r}(x)\Big\}$$

for some $A_{n,r}\in\mathcal{P}_{\mathrm{bal}}$ and some test functions $f_{n,r}$ which vanish on $\overline{\mathcal{O}}$. For $x\in\overline{\mathcal{O}}$ the term in braces on the right-hand side has the constant value

$$\mathfrak{z}(\kappa L):=\kappa L+\sum_{n\geq 2,\,l}\frac{\kappa^n m^l}{n!}\sum_{p=0}^{P_l^n}\log^p\big(\frac{m}{M}\big)\;d_{n,l,p,0}(L^{\otimes n})\in\kappa\,\mathcal{P}_{\mathrm{bal}}[\![\kappa]\!]\,,\qquad(3.8.4)$$

that is, it belongs to $\mathcal{G}_{\mathfrak{z}(\kappa L)}(\mathcal{O})$.

A general $G\in\mathcal{G}_L(\mathcal{O})$ is a finite sum $G(x)=\sum_s g_s(x)\,L_s(x)$ with $g_s\in\mathcal{G}(\mathcal{O})$, $L_s\in\mathcal{P}_{\mathrm{bal}}$ and $\sum_s L_s=L$. Generalizing the above procedure to such a G, we obtain

$$Z\Big(\kappa\int G\Big)=\kappa\int\widehat{G}\,,\quad\text{with a unique}\quad\kappa\widehat{G}\in\mathcal{G}_{\mathfrak{z}(\kappa L)}(\mathcal{O})\,,$$

where $\mathfrak{z}(\kappa L)$ is given by the same formula (3.8.4). (The uniqueness is due to Proposition 1.4.3(b).) In this way, we have defined a map

$$\mathfrak{z}:\begin{cases}\kappa\,\mathcal{P}_{\mathrm{bal}}[\![\kappa]\!]\longrightarrow\kappa\,\mathcal{P}_{\mathrm{bal}}[\![\kappa]\!]\\ \kappa L\longmapsto\mathfrak{z}(\kappa L)=\kappa L+O(\kappa^2)\,,\end{cases}\qquad(3.8.5)$$

which is the *renormalization of the interaction in the algebraic adiabatic limit.* The latter statement will be explained more in detail later on by means of Theorem 3.8.6 (given in Sect. 3.8.3). The map $\mathfrak{z}$ (3.8.5) is *universal*, in the sense that it does not actually depend on the region $\mathcal{O}$; note also that $\mathfrak{z}$ is uniquely determined by Z, that is, by a pair $(T,\widehat{T})$ or $(R,\widehat{R})$.

The correspondence $\gamma\colon \mathcal{R} \longrightarrow \gamma(\mathcal{R})\,;\ Z \longmapsto \mathfrak{z}$ is actually a *homomorphism*. In other words, it satisfies

$$\gamma(Z_2 \circ Z_1) = \gamma(Z_2) \circ \gamma(Z_1)\ , \tag{3.8.6}$$

for the following reason:[103] for $G \in \mathcal{G}_{\kappa L}(\mathcal{O})$ we have $Z_1(\int G) = \int G_1$ with $G_1 \in \mathcal{G}_{\gamma(Z_1)(\kappa L)}(\mathcal{O})$, and $Z_2(\int G_1) = \int G_2$ with $G_2 \in \mathcal{G}_{\gamma(Z_2)(\gamma(Z_1)(\kappa L))}(\mathcal{O})$. Since $\int G_2 = Z_2 \circ Z_1(\int G)$, we also have $G_2 \in \mathcal{G}_{\gamma(Z_2 \circ Z_1)(\kappa L)}(\mathcal{O})$. Hence the following diagram is commutative:

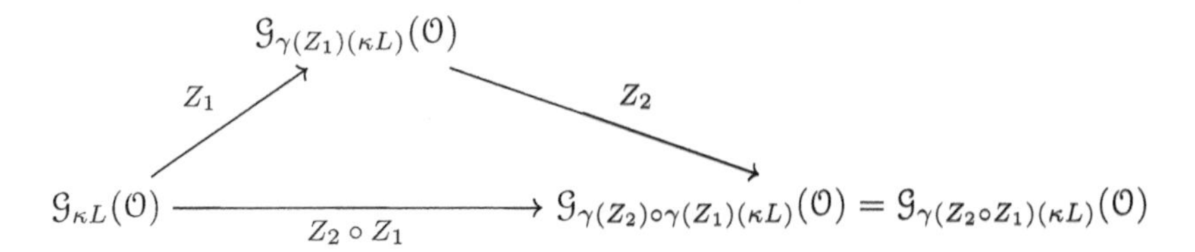

The property (3.8.6) implies that

$$\begin{aligned} &\gamma(Z_2) \circ \gamma(Z_1) \in \gamma(\mathcal{R})\ , \quad \mathrm{Id}_{\kappa\, \mathcal{P}_{\mathrm{bal}}[\![\kappa]\!]} = \gamma(\mathrm{Id}_{\mathcal{F}_{\mathrm{loc}}}) \in \gamma(\mathcal{R}) \\ &\text{and} \quad \gamma(Z)^{-1} = \gamma(Z^{-1}) \in \gamma(\mathcal{R}) \end{aligned}$$

for all $Z_1, Z_2, Z \in \mathcal{R}$. That is, $\gamma(\mathcal{R})$ is also a group: This is the *renormalization group in the adiabatic limit*. We point out that we obtain this group without really performing the adiabatic limit $g(x) \to 1$ (for all x); the *algebraic* adiabatic limit suffices for this purpose.

By performing the latter limit, we arrive at the conventional formulation of the renormalization group as a *renormalization of the coupling constants* – the maps $\mathfrak{z} \in \gamma(\mathcal{R})$ can be interpreted in this way, see Exap. 3.8.2 and Sect. 3.8.2.

Let $L \in \mathcal{P}_{\mathrm{bal}} \cap \mathcal{P}_{\mathrm{hom}}$. From (3.8.2) and (3.8.4) we immediately find:

$$\dim L \le d \quad \Longrightarrow \quad \dim \mathfrak{z}(\kappa L) \le d\ , \tag{3.8.7}$$

where d is the number of spacetime dimensions and by $\dim \mathfrak{z}(\kappa L)$ we mean the maximum of the mass dimensions of all field monomials contributing to the formal power series $\mathfrak{z}(\kappa L)$, analogously to (3.2.63). Hence, a renormalizable (by power counting) interaction remains renormalizable under a renormalization group transformation. For a *massless* model we obtain a stronger result:[104]

$$L \in \mathcal{P}_{\mathrm{bal}} \cap \mathcal{P}_d \Longrightarrow \mathfrak{z}(\kappa L) \in \mathcal{P}_{\mathrm{bal}} \cap \mathcal{P}_d \quad \text{for} \quad m = 0\ . \tag{3.8.8}$$

Example 3.8.2 (Renormalization of $\varphi^4_{d=4}$). An important example is the renormalization of $L = \varphi^4$ in $d = 4$ dimensions. Since $Z \in \mathcal{R}$ fulfills the "Field parity" condition, $\mathfrak{z}(\kappa\varphi^4)$ contains solely terms which are even in φ. Taking into account also the properties

[103] For simplicity we hide κ in G.

[104] We recall that $\mathcal{P}_j$ ($j \in \mathbb{N}$) denotes the vector space spanned by all monomials $A \in \mathcal{P}$ with $\dim A = j$.

"Lorentz covariance" and "$*$-structure" of Z, and using (3.6.13), (3.8.5), $\dim \mathfrak{z}(\kappa\varphi^4) \leq 4$ and Exap. 1.4.2, we conclude

$$\mathfrak{z}(\kappa\varphi^4) = \kappa\varphi^4\,(1+\kappa\hbar\, a)\varphi^4 + \kappa^2\hbar\, b\big((\partial\varphi)^2 - \varphi\Box\varphi\big) + \kappa^2\hbar\; c\Big(\log\frac{m}{M}\Big)\, m^2\varphi^2 + \kappa^2\hbar\; e\Big(\log\frac{m}{M}\Big)\, m^4 \qquad (3.8.9)$$

with $c\big(\log\frac{m}{M}\big) := \sum_{p\geq 0} c_p \log^p\frac{m}{M}$ and $e\big(\log\frac{m}{M}\big) := \sum_{p\geq 0} e_p \log^p\frac{m}{M}$, for certain coefficients $a, b, c_p, e_p \in \mathbb{R}[\![\kappa,\hbar]\!]$. Again, $M > 0$ is some mass scale. The m^4-term, which is a C-number, comes from vacuum diagrams, and is physically irrelevant, since $R_{n,1}(\cdots\otimes c\otimes\cdots) = 0$ for $c\in\mathbb{C}$, see Remk. 3.1.4. One can interpret a as "coupling constant renormalization", b as "wave function renormalization" and $c\big(\log\frac{m}{M}\big)$ as "mass renormalization" – see Sect. 3.8.2.

For the *massless* φ_4^4-model, we know that $\mathfrak{z}(\kappa\varphi^4) \in \mathcal{P}_{\mathrm{bal}}\cap\mathcal{P}_4$, hence, $\mathfrak{z}(\kappa\varphi^4)$ is a linear combination of the fields φ^4 and $\big((\partial\varphi)^2 - \varphi\Box\varphi\big)$. Alternatively, this result can be obtained by performing the limit $m\downarrow 0$ of formula (3.8.9).

To give an explicit computation we study, as in Sect. 3.6.3, the scaling as renormalization transformation and assume $m > 0$. We use the stronger version of the Sm-expansion axiom. From (3.6.44) and (3.6.21) we know that

$$\begin{aligned} &Z_\rho^{m\,(2)}\big(\varphi^4(x_1)\otimes\varphi^4(x_2)\big) \\ &= \frac{i}{\hbar}\Big(\sigma_\rho\circ T_2^{m/\rho}\big(\sigma_\rho^{-1}\varphi^4(x_1),\,\sigma_\rho^{-1}\varphi^4(x_2)\big) - T_2^m\big(\varphi^4(x_1),\,\varphi^4(x_2)\big)\Big) \\ &= \frac{36i}{\hbar}\Big\{\rho^4\, t^{m/\rho}(\varphi^2,\varphi^2)(\rho y) - t^m(\varphi^2,\varphi^2)(y)\Big\}\,\varphi^2(x_1)\varphi^2(x_2) \\ &\quad + \frac{16i}{\hbar}\Big\{\rho^6\, t^{m/\rho}(\varphi^3,\varphi^3)(\rho y) - t^m(\varphi^3,\varphi^3)(y)\Big\}\,\varphi(x_1)\varphi(x_2) \\ &\quad + \frac{i}{\hbar}\Big\{\rho^8\, t^{m/\rho}(\varphi^4,\varphi^4)(\rho y) - t^m(\varphi^4,\varphi^4)(y)\Big\}, \qquad y := x_1 - x_2\,, \end{aligned} \qquad (3.8.10)$$

where we have taken into account that the corresponding tree diagrams scale homogeneously and, hence, only the following diagrams contribute:

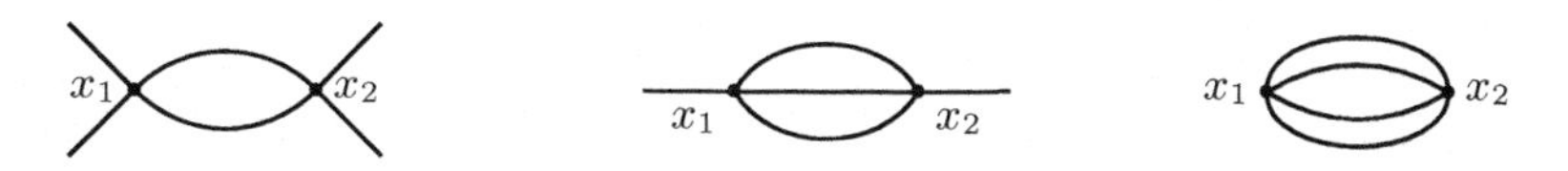

Inserting the results (3.6.60), (3.5.15) and the prefactors appearing in $t^m(\varphi^2,\varphi^2) = 2\hbar^2\, t_{\mathrm{fish}} + \mathcal{O}(m^2)$ and $t^m(\varphi^3,\varphi^3) = 6\hbar^3\, t^m_{\mathrm{ss}}$, we obtain

$$\begin{aligned} &Z_\rho^{m\,(2)}\big(\varphi^4(x_1)\otimes\varphi^4(x_2)\big) = 72\, i\,\hbar\, C_{\mathrm{fish}}\,\log\rho\;\delta(y)\,\varphi^2(x_1)\varphi^2(x_2) \qquad (3.8.11) \\ &+ 96\, i\,\hbar^2\,\Big(K_2\,\log\rho\;(\Box\delta)(y) \\ &\qquad + m^2\Big[K_1\,\log^2\rho + (K_3 - C_1)\,\log\rho + K_4\,\log\frac{m}{M}\;\log\rho\Big]\,\delta(y)\Big)\,\varphi(x_1)\varphi(x_2) \\ &+ \hbar^3\,\Big(m^4\, P_0\Big(\log\frac{m}{M},\log\rho\Big)\,\delta(y) + m^2\, P_2\Big(\log\frac{m}{M},\log\rho\Big)\,\Box\delta(y) + p_4\,\log\rho\;\Box\Box\delta(y)\Big)\;, \end{aligned}$$

where $p_4\in\mathbb{C}$ and P_0, P_2 are polynomials in the two variables $\log\frac{m}{M}$ and $\log\rho$. The form of the $\hbar^3$-term follows from the Lorentz invariance, the defining property (3) of the Stückelberg–Petermann group (Definition 3.6.1) and the polynomial dependence of $Z_\rho^{m\,(2)}$ on $\log\rho$ (see (3.6.47)). The $p_4\,\Box\Box\delta$-term is the breaking of homogeneous scaling of $(D^F)^4|_{\text{renormalized}}$, which is $\sim \log\rho$ according to Proposition 3.2.16. Next we insert the values for C_{fish} (3.6.49) and $K_1,\dots,K_4$, the latter are computed in Exer. 3.5.5.

Turning to the algebraic adiabatic limit, we compute $d_{n=2,l,p,a=0}(\varphi^4\otimes\varphi^4)$ for $l=0,2,4$. This is obvious for the terms without derivatives of $\delta(y)$. The terms $m^2\,P_2(\dots)\,\Box\delta$ and $p_4\,\Box\Box\delta$ do not contribute, since we are interested in the $(a=0)$-terms only. With regard to the $K_2\,\Box\delta\,\varphi\varphi$-term we take into account that $d_{n,l,p,a}(\varphi^4\otimes\varphi^4)\in\mathcal{P}_{\text{bal}}$ and write

$$\begin{aligned}&-\int dx_1dx_2\ (\Box\delta)(y)\,\varphi(x_1)\varphi(x_2)\,h_1(x_1)h_2(x_2) \qquad (3.8.12)\\ &\quad=\int dx\,\Big(\frac{1}{2}\big((\partial^\mu\varphi)\partial_\mu\varphi-\varphi\Box\varphi\big)(x)\,h_1(x)h_2(x)+\sum_{|a_1|+|a_2|>0}\cdots\,\partial^{a_1}h_1(x)\partial^{a_2}h_2(x)\Big)\,.\end{aligned}$$

Inserting these results into (3.8.4), we get the renormalization of the interaction:[105]

$$\begin{aligned}\mathfrak{z}_\rho\big(\kappa\varphi^4/4!\big)&=\zeta(\rho)\,\kappa\varphi^4/4!+\xi(\rho)\,\big((\partial\varphi)^2-\varphi\Box\varphi\big)+\chi(\rho)\,m^2\varphi^2\\ &\quad+\hbar\Big\{\Big(\frac{\kappa\hbar}{4!}\Big)^2\frac{1}{2}\,P_0\Big(\log\frac{m}{M},\log\rho\Big)+\mathcal{O}\big((\kappa\hbar)^3\big)\Big\}\,m^4\,,\end{aligned}\qquad (3.8.13)$$

where

$$\begin{aligned}\zeta(\rho)&=1+\frac{\kappa\hbar\cdot 3}{4\,(2\pi)^2}\,\log\rho+\mathcal{O}\big((\kappa\hbar)^2\big)\,,\quad \xi(\rho)=\frac{(\kappa\hbar)^2}{2^7\cdot 3\,(2\pi)^4}\,\log\rho+\mathcal{O}\big((\kappa\hbar)^3\big)\,,\\ \chi(\rho)&=\frac{(\kappa\hbar)^2}{2^4\cdot(2\pi)^4}\Big(\frac{-\log^2\rho}{2}+(\tilde{C}_1-\log 2+\gamma)\,\log\rho+\log\frac{m}{M}\,\log\rho\Big)+\mathcal{O}\big((\kappa\hbar)^3\big)\,.\end{aligned}$$

Here, $\tilde{C}_1$ is related to the renormalization constant C_1 (3.5.13) by $C_1=:\frac{3i}{2^6\pi^4}\,\tilde{C}_1$. As indicated, $\zeta(\rho)$, $\xi(\rho)$, $\chi(\rho)$ and the term in braces are formal power series in the variable $\kappa\hbar$; this can be seen as follows: By using (3.6.6) and (3.6.9) we get

$$\kappa^n\,Z^{(n)}\big(\varphi^4(g)^{\otimes n}\big)=(\kappa\hbar)^n\,\Big(\hbar^{-1}\,\mathcal{O}(\varphi^4)+\hbar^0\,\mathcal{O}(\varphi^2)+\hbar\,\mathcal{O}(\varphi^0)\Big)\,,\qquad (3.8.14)$$

where we mean by $\mathcal{O}(\varphi^s)$ a linear combination of monomials $A\in\mathcal{P}$ with $|A|=s$. The functions $\rho\longmapsto\zeta(\rho)$, $\rho\longmapsto\xi(\rho)$ and $\rho\longmapsto\chi(\rho)$ are called the "running coupling parameters", they make up the "renormalization group flow".

For the same reasons as for the m^4-term in $\mathfrak{z}_\rho\big(\kappa\varphi^4/4!\big)$, $\chi(\rho)$ is a polynomial in the two variables $\log\frac{m}{M}$ and $\log\rho$, to each order in $\kappa\hbar$. In the next section we will see that the mass renormalization is of a much simpler form, if $\chi(\rho)$ does not depend on m, i.e., does not contain any powers of $\log\frac{m}{M}$. As we see, this assumption is not fulfilled when we work with the stronger version of the Sm-expansion axiom. But, at least to order $(\kappa\hbar)^2$, it can be satisfied if we require only the weaker version of this axiom. Namely, in this case, the non-uniqueness of $t_{\text{ss}}^{m,M}$ (3.5.14) reads

$$C\,\Box_y\delta(y)+m^2\,Q\Big(\log\frac{m}{M}\Big)\,\delta(y)\quad\text{where}\quad Q(x)=\sum_{j=0}^{J}C_jx^j\quad\text{for some } J<\infty\,,$$

[105] Following usual conventions, we replace κ by $\kappa/4!$.

as explained directly before Exer. 3.5.5. Hence, in (3.8.11) the term in [...]-brackets is replaced by

$$\Big[K_1 \log^2\rho + K_3 \log\rho + K_4 \log\frac{m}{M}\ \log\rho + Q\big(\log\frac{m}{M} - \log\rho\big) - Q\big(\log\frac{m}{M}\big)\Big] . \quad (3.8.15)$$

The condition that in this expression the terms $\sim \log^p \frac{m}{M}$ (with $p \geq 1$) cancel, restricts the coefficients C_j as follows: C_0 and C_1 are unrestricted, $C_2 = K_4/2$ and $C_j = 0\ \forall j \geq 3$. Inserting these values and $K_1 = -K_4/2$ (Exer. 3.5.5) into (3.8.15) we get

$$\Big[\dots\Big] = (K_3 - C_1)\log\rho ,$$

which yields

$$\chi(\rho) = (\kappa\hbar)^2\,\hat{C}_1 \log\rho , \quad \text{where} \quad \hat{C}_1 := \frac{i(K_3 - C_1)}{12} = \frac{\tilde{C}_1 - \log 2 + \gamma}{2^4\cdot(2\pi)^4} , \quad (3.8.16)$$

with the same $\tilde{C}_1$ as in (3.8.13). Note that the terms $\sim \log^2\rho$ cancel.

Exercise 3.8.3 (Renormalization of φ_4^3, continued). Let $d = 4$, $m > 0$ and $Z \in \mathcal{R}$. Taking into account all orders in κ, show that $\mathfrak{z}(\kappa\varphi^3) \equiv \gamma(Z)(\kappa\varphi^3)$ is of the form

$$\mathfrak{z}(\kappa\varphi^3) = \kappa\varphi^3 + \kappa^2\Big(\hbar^2 m^2 P\big(\log\frac{m}{M}\big) + \hbar C_1\varphi^2\Big) + \kappa^3\hbar^2 C_2\varphi + \kappa^4\hbar^3 C_3 \quad (3.8.17)$$

for some numbers C_1, C_2, $C_3 \in \mathbb{R}$ and some polynomial P with real coefficients, by using Exer. 3.6.10.

[*Solution*: From (3.6.3)–(3.6.4), (3.6.6) and $Z(F)^* = Z(F^*)$ we know that

$$\begin{aligned}
z^{(2)}(\varphi^3\otimes\varphi^3)(x_1 - x_2) &= \hbar^2\Big(a_0\,\Box\delta(x_1 - x_2) + m^2 Q\big(\log\frac{m}{M}\big)\,\delta(x_1 - x_2)\Big) ,\\
z^{(2)}(\varphi^2\otimes\varphi^2)(x_1 - x_2) &= \hbar\, a_1\,\delta(x_1 - x_2) ,\\
z^{(3)}(\varphi^3\otimes\varphi^3\otimes\varphi^2)(x_1 - x_3, x_2 - x_3) &= \hbar^2 a_2\,\delta(x_1 - x_3, x_2 - x_3) ,\\
z^{(4)}(\varphi^3\otimes\varphi^3\otimes\varphi^3\otimes\varphi^3)(x_1 - x_4, \dots, x_3 - x_4) &= \hbar^3 a_3\,\delta(x_1 - x_4, \dots, x_3 - x_4)
\end{aligned}$$

for some a_0, a_1, a_2, $a_3 \in \mathbb{R}$ and some real polynomial Q. We insert these results into (3.6.38), integrate out the δ-distributions and perform the algebraic adiabatic limit. The term $\sim a_0 \int dx\ (g\Box g)(x)$ does not contribute to this limit. In this way we obtain (3.8.17), with $Q/2 = P$, $\frac{9}{2}\,a_1 = C_1$, $\frac{3}{2}\,a_2 = C_2$, $\frac{1}{24}\,a_3 = C_3$.]

3.8.2 Wave function, mass and coupling constant renormalization

In this subsection, we follow essentially [27, Sect. 7, App. C], see also [65].

In general the renormalized interaction $\mathfrak{z}(\kappa L)$ (3.8.5) contains novel interaction terms. For a renormalizable (by power counting) interaction it is possible to absorb $\big(\mathfrak{z}(\kappa L) - \kappa L\big)$ by redefinitions of the field ("wave function renormalization"), the mass ("mass renormalization") and the coupling constant ("coupling constant renormalization"), that is, after these redefinitions[106] the total Lagrangian $\big(L_0 - \mathfrak{z}_\rho(\kappa L)\big)$ has the same form as the initial one $\big(L_0 - \kappa L\big)$.

We work out these concepts in terms of two models: Massive $\varphi^4_{d=4}$ and massless $\varphi^3_{d=6}$ with the scaling as renormalization transformation. The interesting property of the latter model is that the "β-function", which describes the change of the renormalized (or "running") coupling constant under scaling, is < 0 to leading order – this exhibits asymptotic freedom.[107]

The massive $\varphi^4_{d=4}$-model. Ignoring the physically irrelevant C-number term $\sim m^4$ of $\mathfrak{z}_\rho\big(\kappa\varphi^4/4!\big)$, formula (3.8.13) suggests that the renormalization group flow has 3 parameters: $\zeta(\rho)$ for the φ^4-component, $\xi(\rho)$ for the $\big((\partial\varphi)^2 - \varphi\Box\varphi\big) \simeq 2(\partial\varphi)^2$-component and $\chi(\rho)$ for the $m^2\varphi^2$-component. However, the model has only two physically relevant parameters: The renormalized coupling constant κ_ρ and the renormalized (or "running") mass m_ρ; this can be seen by performing the wave function, mass and coupling constant renormalization. In detail: Introducing a new field φ_ρ, which is of the form

$$\varphi_\rho(x) := f(\rho)\,\varphi(x) \tag{3.8.18}$$

(where $f : (0,\infty) \longrightarrow \mathbb{R}$ is a suitable function), and the running mass m_ρ and the running coupling constant κ_ρ, we can achieve that $\big(L_0 - \mathfrak{z}_\rho(\kappa\varphi^4/4!)\big)$ has the same form as $\big(L_0 - \kappa\varphi^4/4!\big)$, explicitly:

$$\frac{1}{2}\big((\partial\varphi)^2 - m^2\varphi^2\big) - \mathfrak{z}_\rho\big(\kappa\varphi^4/4!\big) \simeq \frac{1}{2}\big((\partial\varphi_\rho)^2 - m_\rho^2\varphi_\rho^2\big) - \frac{\kappa_\rho\varphi_\rho^4}{4!}\,. \tag{3.8.19}$$

In particular we absorb the novel bilinear interaction terms $\xi(\rho)\,\big((\partial\varphi)^2 - \varphi\Box\varphi\big) + \chi(\rho)\,m^2\varphi^2$ in the free Lagrangian. The symbol "$\simeq$" means, that we ignore the C-number term $\hbar\,\{\ldots\}\,m^4$ appearing in $\mathfrak{z}_\rho\big(\kappa\varphi^4/4!\big)$ and that we identify $\big((\partial\varphi)^2 - \varphi\Box\varphi\big)$ with $2\,(\partial\varphi)^2$. The latter is justified by the following: If we do not require that $\mathfrak{z}_\rho$ takes values in $\mathcal{P}_{\mathrm{bal}}$, we may replace the r.h.s. of (3.8.12) by

$$\int dx\,\Big((\partial\varphi)^2(x)\,h_1(x)h_2(x) + \sum_{|a_1|+|a_2|>0} \cdots\,\partial^{a_1}h_1(x)\partial^{a_2}h_2(x)\Big)\,.$$

The condition (3.8.19) is an equation for polynomials in $\varphi, \partial\varphi$; equating the coefficients the implicit definition (3.8.19) of the running quantities turns into the following explicit formulas:

---

[106] We recall that L_0 denotes the free Lagrangian.

[107] Asymptotic freedom is the property that the interaction between particles becomes weaker as the distance between the particles decreases; asymptotically they behave as free particles. Asymptotic freedom is a feature of quantum chromodynamics (QCD), the QFT of the strong interaction between quarks and gluons.

- For the wave function renormalization

$$(\partial\varphi)^2 \left(\frac{1}{2} - 2\,\xi(\rho)\right) = \frac{(\partial\varphi_\rho)^2}{2} \quad \Rightarrow \quad \varphi_\rho(x) = \varphi(x)\,\sqrt{1 - 4\,\xi(\rho)}\;; \tag{3.8.20}$$

- for the mass renormalization

$$m^2\varphi^2\left(\frac{1}{2} + \chi(\rho)\right) = \frac{m_\rho^2}{2}\,\varphi_\rho^2 = \frac{m_\rho^2}{2}\,(1 - 4\,\xi(\rho))\,\varphi^2 \;\Rightarrow\; m_\rho = m\sqrt{\frac{1 + 2\,\chi(\rho)}{1 - 4\,\xi(\rho)}}\;; \tag{3.8.21}$$

- for the coupling constant renormalization

$$\zeta(\rho)\,\kappa\varphi^4 = \kappa_\rho\varphi_\rho^4 = \kappa_\rho\left(1 - 4\,\xi(\rho)\right)^2\varphi^4 \quad \Rightarrow \quad \kappa_\rho = \kappa\,\frac{\zeta(\rho)}{\left(1 - 4\,\xi(\rho)\right)^2}\,. \tag{3.8.22}$$

Now we explicitly see: If $\chi(\rho)$ does not depend on m, the function $m \longmapsto m_\rho(m)$ is *linear*, otherwise this function is much more complicated. Note that $\kappa \longmapsto \kappa(\kappa_\rho)$ is under no circumstances linear, because $\zeta(\rho)$ and $\xi(\rho)$ are nontrivial formal power series in $\kappa\hbar$.

Inserting the lowest-order expressions for $\xi(\rho)$, $\zeta(\rho)$ (3.8.13) and $\chi(\rho)$ (3.8.16), we obtain

$$\begin{aligned}
\varphi_\rho(x) &= \varphi(x)\Big(1 - 2\xi(\rho) + \mathcal{O}\big((\kappa\hbar)^4\big)\Big) = \varphi(x)\Big(1 - \frac{(\kappa\hbar)^2}{2^6\cdot 3\,(2\pi)^4}\,\log\rho + \mathcal{O}\big((\kappa\hbar)^3\big)\Big),\\
m_\rho &= m\,\Big(1 + \chi(\rho) + 2\xi(\rho) + \mathcal{O}\big((\kappa\hbar)^4\big)\Big)\\
&= m\,\left(1 + \frac{(\kappa\hbar)^2}{2^4\cdot(2\pi)^4}\left(\tilde{C}_1 - \log 2 + \gamma + \frac{1}{12}\right)\log\rho + \mathcal{O}\big((\kappa\hbar)^3\big)\right)\,,\\
\kappa_\rho &= \kappa\,\zeta(\rho)\Big(1 + 8\,\xi(\rho) + \mathcal{O}\big((\kappa\hbar)^4\big)\Big) = \kappa\Big(1 + \frac{\kappa\hbar\cdot 3}{4\,(2\pi)^2}\,\log\rho + \mathcal{O}\big((\kappa\hbar)^2\big)\Big)\,.
\end{aligned}$$

Note that the term $\sim \kappa\,(\kappa\hbar)^2$ of κ_ρ is the sum of two contributions: The $(\kappa\hbar)^2$-term of $\zeta(\rho)$, which we have not computed, and the leading term of $8\,\xi(\rho)$.

The β-function is defined as

$$\beta(\kappa) := \rho\,\frac{d}{d\rho}\Big|_{\rho=1}\,\kappa_\rho\,. \tag{3.8.23}$$

To lowest order we get

$$\beta(\kappa) = \kappa\Big(\frac{\kappa\hbar\cdot 3}{4\,(2\pi)^2} + \mathcal{O}\big((\kappa\hbar)^2\big)\Big)\;;$$

some comments on this result are given below after (3.8.30).

Remark 3.8.4 ("Perturbative agreement"). By the renormalization of the wave function and mass, we change the splitting of the total Lagrangian $\big(L_0 - \mathfrak{z}_\rho(\kappa L)\big)$ into a free and interacting part, i.e., we change the starting point for the perturbative expansion. To justify this, one has to show that the two perturbative QFTs given by the splittings $\big(L_0 - \mathfrak{z}_\rho(\kappa L)\big)$ and $\big(L_{0,\rho} - \kappa_\rho L_\rho\big)$ (the latter is a shorthand notation for the r.h.s. of (3.8.19)), respectively, have the same physical content. This statement can be viewed as an application of the "Principle of Perturbative Agreement" of Hollands and Wald, which is used in [103] as an additional renormalization condition.

A proof that the "old" perturbative QFT (given by $\big(L_0 + \mathfrak{z}_\rho(L)\big)$) and the "new" one (given by $\big(L_{0,\rho} - \kappa_\rho L_\rho\big)$) are physically equivalent, is beyond the scope of this book. Using the framework of perturbative algebraic QFT (introduced in Sect. 3.7), the following conjecture has been formulated and verified for a few examples [26]: Given a renormalization prescription (i.e., a T-product) for the old perturbative QFT, there exists a renormalization prescription for the new perturbative QFT, such that the pertinent nets of local observables in the algebraic adiabatic limit are equivalent. The corresponding isomorphisms can be chosen such that local fields are identified with local fields modulo the free field equation. For recent progresses see [172] and [37].

Obviously, a necessary condition in order that the requirement $L_0 - \mathfrak{z}_\rho(\kappa L) = L_{0,\rho} - \kappa_\rho L_\rho$ can be fulfilled is that $\dim \mathfrak{z}_\rho(\kappa L) = \dim L$; for a renormalizable (by power counting) interaction (i.e., $\dim L = d$) the latter condition is satisfied, due to (3.8.7).

The massless $\varphi^3_{d=6}$-model. For simplicity we assume $m = 0$. In $d = 6$ dimensions Definition 3.1.18 yields

$$\dim \varphi = 2 \quad \text{and it holds that} \quad D^F(y) = \frac{1}{4\pi^3\,(y^2 - i0)^2}\,, \tag{3.8.24}$$

the latter formula can be found, e.g., in [27, App. A].

Renormalization of the interaction: To obtain the form of $\mathfrak{z}_\rho(\kappa\varphi^3)$ to lowest orders we study $Z_\rho^{(2)}\big(\varphi^3(x_1) \otimes \varphi^3(x_2)\big)$ and $Z_\rho^{(3)}\big(\varphi^3(x_1) \otimes \varphi^3(x_2) \otimes \varphi^3(x_3)\big)$. The contributing diagrams are

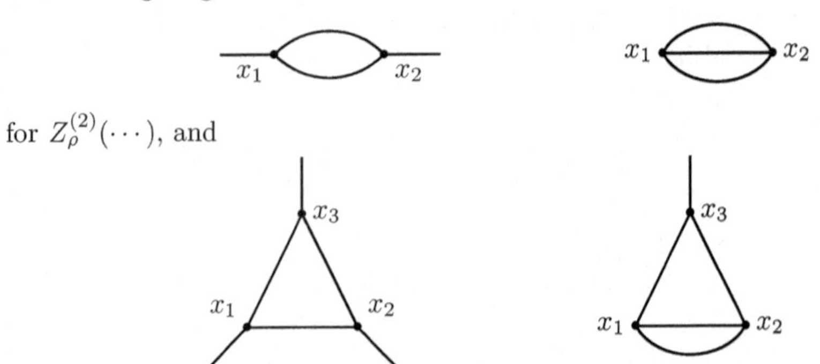

for $Z_\rho^{(2)}(\cdots)$, and

plus [two times the last diagram with cyclic permutations of (x_1, x_2, x_3)] for $Z_\rho^{(3)}(\cdots)$, respectively. So we get

$$\begin{aligned} Z_\rho^{(2)}\big(\varphi^3(x_1)\otimes\varphi^3(x_2)\big) &= \frac{i}{\hbar}\Big[9\Big(\rho^8\,t(\varphi^2,\varphi^2)(\rho y)-t(\varphi^2,\varphi^2)(y)\Big)\,\varphi(x_1)\varphi(x_2) \\ &\qquad + \Big(\rho^{12}\,t(\varphi^3,\varphi^3)(\rho y)-t(\varphi^3,\varphi^3)(y)\Big)\Big] \\ &= i\hbar\, a\,\log\rho\,(\Box\delta)(y)\,\varphi(x_1)\varphi(x_2)+i\hbar^2\, b\,\log\rho\,(\Box\Box\Box\delta)(y)\,, \quad y:=x_1-x_2\,, \end{aligned} \tag{3.8.25}$$

and

$$\begin{aligned} &Z_\rho^{(3)}\big(\varphi^3(x_1)\otimes\varphi^3(x_2)\otimes\varphi^3(x_3)\big) \\ &= \frac{i^2}{\hbar^2}\Big[27\Big(\rho^{12}\,t(\varphi^2,\varphi^2,\varphi^2)(\rho y_1,\rho y_2))-t(\varphi^2,\varphi^2,\varphi^2)(y_1,y_2)\Big) \\ &\qquad\qquad\qquad\qquad \cdot\varphi(x_1)\varphi(x_2)\varphi(x_3) \\ &\quad +3\Big(\rho^{16}\,t(\varphi^3,\varphi^3,\varphi^2)(\rho y_1,\rho y_2)-t(\varphi^3,\varphi^3,\varphi^2)(y_1,y_2)\Big)\,\varphi(x_3) \\ &\qquad\qquad\qquad\qquad + \mathrm{cycl}(x_1,x_2,x_3)\Big] \\ &= -\hbar c\,\log\rho\,\delta(y_1,y_2)\,\varphi(x_1)\varphi(x_2)\varphi(x_3) \\ &\quad -\hbar^2\Big[\sum_{r,s,u,v=1}^{2}\big(e_1^{rsuv}\log\rho+e_2^{rsuv}\log^2\rho\big)\,\big(\partial_\mu^{y_r}\partial^\mu_{y_s}\partial_\nu^{y_u}\partial^\nu_{y_v}\delta(y_1,y_2)\big)\,\varphi(x_3) \\ &\qquad\qquad + \mathrm{cycl}(x_1,x_2,x_3)\Big]\,, \qquad y_1:=x_1-x_3\,,\quad y_2:=x_2-x_3\,, \end{aligned} \tag{3.8.26}$$

where "$\mathrm{cycl}(x_1,x_2,x_3)$" stands for two terms which are obtained from the term $\sim\varphi(x_3)$ by cyclic permutations of (x_1,x_2,x_3). The numbers $a,b,c,e_1^{\cdots},e_2^{\cdots}\in\mathbb{C}$ can be computed; this is partially done below. We have taken into account Proposition 3.2.16 and that the unrenormalized distributions $t^0(\varphi^2,\varphi^2)$, $t^0(\varphi^3,\varphi^3)\in\mathcal{D}'(\mathbb{R}^6\setminus\{0\})$, $t^0(\varphi^2,\varphi^2,\varphi^2)\in\mathcal{D}'(\mathbb{R}^{12}\setminus\{0\})$ scale homogeneously and that

$$t^0(\varphi^3,\varphi^3,\varphi^2)(y_1,y_2)\sim t(\varphi^2,\varphi^2)(y_1-y_2)\,D^F(y_1)\,D^F(y_2)\in\mathcal{D}'(\mathbb{R}^{12}\setminus\{0\})$$

scales almost homogeneously with power $N=1$.

To perform the algebraic adiabatic limit we transform the term $\sim(\Box\delta)\,\varphi\varphi$ of $Z_\rho^{(2)}(\cdots)$ by means of the relation (3.8.12). Obviously the $(\Box\Box\Box\delta)$-term of $Z_\rho^{(2)}(\cdots)$ does not contribute to $\mathfrak{z}_\rho(\kappa\varphi^3)$. Also the terms $\sim e_j^{\cdots}\,(\partial\partial\partial\partial\delta)\,\varphi$ of $Z_\rho^{(3)}(\cdots)$ do not contribute, due to

$$\begin{aligned} &\int dx_1dx_2dx_3\;\big(\partial_\mu^{y_1}\partial^\mu_{y_2}\partial_\nu^{y_1}\partial^\nu_{y_2}\delta(y_1,y_2)\big)\,\varphi(x_3)\,h(x_1)h(x_2)h(x_3) \\ &\qquad = \int dx\;\varphi(x)\,(\partial_\mu\partial_\nu h)(x)\,(\partial^\mu\partial^\nu h)(x)\,h(x) \end{aligned}$$

and similar identities. Summing up, we obtain[108]

$$\mathfrak{z}_\rho\big(\kappa\varphi^3/3!\big) = \zeta(\rho)\,\kappa\varphi^3/3! + \xi(\rho)\,\big((\partial\varphi)^2 - \varphi\Box\varphi\big)\,, \quad \text{where} \tag{3.8.27}$$
$$\zeta(\rho) = \Big\{1 + \kappa^2\hbar\,\tilde{c}\,\log\rho + \mathcal{O}\big((\kappa^2\hbar)^2\big)\Big\}\,, \quad \tilde{c} = -\frac{c}{2^3\cdot 3^3}$$
$$\xi(\rho) = \Big\{\kappa^2\hbar\,\tilde{a}\,\log\rho + \mathcal{O}\big((\kappa^2\hbar)^2\big)\Big\}\,, \quad \tilde{a} = -\frac{ia}{2^4\cdot 3^2}\,.$$

The numbers $\tilde{a}, \tilde{c} \in \mathbb{C}$ are computed below. Also to higher orders, $\mathfrak{z}_\rho\big(\kappa\varphi^3/3!\big)$ is a linear combination of $\kappa\varphi^3/3!$ and $\big((\partial\varphi)^2 - \varphi\Box\varphi\big)$, because $\mathfrak{z}_\rho\big(\kappa\varphi^3/3!\big) \in \mathcal{P}_{\text{bal}} \cap \mathcal{P}_6$ (as noted after (3.8.7)) and since $\{\kappa\varphi^3/3!\,,\ \big((\partial\varphi)^2 - \varphi\Box\varphi\big)\}$ is a basis of $\mathcal{P}_{\text{bal}} \cap \mathcal{P}_6$, as one verifies easily. The terms in braces are formal power series in the variable $\kappa^2\hbar$; this can be seen analogously to (3.8.14).

Wave function and coupling constant renormalization, and the β-function: The massless φ^3_6-model has only one physically relevant parameter: The running coupling constant κ_ρ, because the $\xi(\rho)\,\big((\partial\varphi)^2 - \varphi\Box\varphi\big)$-term can be absorbed in the free part of the Lagrangian by the wave function renormalization. Proceeding analogously to (3.8.18)–(3.8.19), we implicitly define φ_ρ and κ_ρ by requiring

$$\frac{(\partial\varphi)^2}{2} - \mathfrak{z}_\rho\big(\kappa\varphi^3/3!\big) \simeq \frac{(\partial\varphi_\rho)^2}{2} - \frac{\kappa_\rho\varphi_\rho^3}{3!}\,. \tag{3.8.28}$$

This yields for the wave function renormalization precisely the same formula as for the φ^4_4-model, namely (3.8.20); and for the coupling constant renormalization we get

$$\zeta(\rho)\,\kappa\varphi^3 = \kappa_\rho\varphi_\rho^3 = \kappa_\rho\,\big(1 - 4\,\xi(\rho)\big)^{\frac{3}{2}}\,\varphi^3 \quad \Rightarrow \quad \kappa_\rho(x) = \kappa\,\frac{\zeta(\rho)}{\big(1 - 4\,\xi(\rho)\big)^{\frac{3}{2}}}\,. \tag{3.8.29}$$

To lowest orders we obtain

$$\varphi_\rho(x) = \varphi(x)\,\Big(1 - 2\xi(\rho) + \mathcal{O}\big((\kappa^2\hbar)^2\big)\Big) = \varphi(x)\,\Big(1 - \frac{\kappa^2\hbar}{3\cdot 2^8\,\pi^3}\,\log\rho + \mathcal{O}\big((\kappa^2\hbar)^2\big)\Big)\,,$$
$$\kappa_\rho = \kappa\,\zeta(\rho)\Big(1 + 6\,\xi(\rho) + \mathcal{O}\big((\kappa^2\hbar)^2\big)\Big) = \kappa\Big(1 - \frac{\kappa^2\hbar\cdot 3}{32\cdot(2\pi)^3}\,\log\rho + \mathcal{O}\big((\kappa^2\hbar)^2\big)\Big)\,,$$

where we have inserted (3.8.27) and the values for $\tilde{a}$ (3.8.34) and $\tilde{c}$ (3.8.38) computed below. For the β-function we get

$$\beta(\kappa) := \rho\,\frac{d}{d\rho}\Big|_{\rho=1}\,\kappa_\rho = \kappa\Big(-\frac{\kappa^2\hbar\cdot 3}{32\cdot(2\pi)^3} + \mathcal{O}\big((\kappa^2\hbar)^2\big)\Big)\,. \tag{3.8.30}$$

In contrast to the φ^4_4-model, the wave function renormalization contributes to the leading term of the β-function. The result (3.8.30) agrees with formula (3.4.64)

[108] We replace κ by $\kappa/3!$, in order that we can better compare with the literature.

in [127], although the renormalization group flow generated by scaling is defined there by a different method, namely by conventional dimensional regularization[109] with minimal subtraction, which is a momentum space method.

Usually, the sign of the β-function is positive, as, e.g., for the φ_4^4-model; this means that the strength of the interaction increases if we study the model at smaller distances. However, for the φ_6^3-model $\beta(\kappa)$ is negative to leading order – this is an indication for asymptotic freedom.

Computation of $\tilde{a}$: In $\mathcal{D}'(\mathbb{M}_6 \setminus \{0\})$ we have

$$t^0(\varphi^2,\varphi^2)(y) = 2\hbar^2 \left(D^F(y)\right)^2 = \frac{\hbar^2}{8\pi^6}\, t_1^0(y) \quad \text{with} \quad t_1^0(y) := \frac{1}{(y^2 - i0)^4} \ . \tag{3.8.31}$$

Let $t_1 \in \mathcal{D}'(\mathbb{M}_6)$ be an almost homogeneous extension of t_1^0. Due to Proposition 3.2.16 and the Lorentz invariance, the breaking of homogeneous scaling of t_1 is of the form

$$\rho\partial_\rho \left(\rho^8\, t_1(\rho y)\right) = a_1\, \square_y \delta(y) \ . \tag{3.8.32}$$

The computation of the constant $a_1 \in \mathbb{C}$ can be traced back to the breaking of homogeneous scaling of a simpler extension problem, namely the extension of

$$t^0(y) := \frac{1}{(y^2 - i0)^3} \in \mathcal{D}'(\mathbb{M}_6 \setminus \{0\}) \quad \text{to} \quad t \in \mathcal{D}'(\mathbb{M}_6) \ .$$

Applying differential renormalization to (3.5.9), we get an explicit formula for t:

$$t(y) = -\square_y \overline{\left(\frac{\log(-M^2(y^2 - i0))}{8\,(y^2 - i0)^2}\right)} \ ,$$

from which we derive

$$\rho\partial_\rho \left(\rho^6\, t(\rho y)\right) = -\square_y \overline{\left(\frac{1}{4(y^2 - i0)^2}\right)} = -\pi^3\, \square_y D^F(y) = i\pi^3\, \delta(y) \ . \tag{3.8.33}$$

To obtain the constant a_1 appearing in (3.8.32) from this result, we multiply (3.8.32) with y^2: On the l.h.s. we use that

$$y^2\, t_1(y) = t(y) + a_2\, \delta(y)$$

for some $a_2 \in \mathbb{C}$, since both $y^2\, t_1(y)$ and $t(y)$ are almost homogeneous extensions of $t^0(y)$. Hence, the l.h.s. gives

$$y^2\, \rho\partial_\rho \left(\rho^8\, t_1(\rho y)\right) = \rho\partial_\rho \left(\rho^6\, t(\rho y)\right) = i\pi^3\, \delta(y) \ .$$

The r.h.s. turns into

$$a_1\, y^2\, \square_y \delta(y) = a_1\, (\square_y y^2)\, \delta(y) = a_1\, 12\, \delta(y) \ .$$

[109] Here, we do not mean dimensional regularization in the Epstein–Glaser framework, which proceeds in position space, as worked out in [63].

Equating the last two equations we get $a_1 = \frac{i\pi^3}{12}$. Taking into account all prefactors, we end up with

$$\tilde{a} = \frac{1}{3 \cdot 2^9\, \pi^3} \ . \tag{3.8.34}$$

Computation of $\tilde{c}$: This task is technically more demanding. In $\mathcal{D}'(\mathbb{R}^{12} \setminus \{0\})$ it holds that

$$t^0(\varphi^2, \varphi^2, \varphi^2)(y_1, y_2) = 8\hbar^3\, t^0_\triangle(y_1, y_2)$$
$$\text{where} \quad t^0_\triangle(y_1, y_2) := D^F(y_1)\, D^F(y_2)\, D^F(y_1 - y_2) \ .$$

Due to Proposition 3.2.16, there exists an almost homogeneous extension $t_\triangle \in \mathcal{D}'(\mathbb{R}^{12})$ of $t^0_\triangle$ satisfying

$$\rho\partial_\rho \left(\rho^{12} t_\triangle(\rho y_1, \rho y_2)\right) = c_1\, \delta(y_1, y_2) \ ,$$

with a number $c_1 \in \mathbb{C}$. To compute this number we write

$$\begin{aligned}\rho\partial_\rho \left(\rho^{12} t_\triangle(\rho y_1, \rho y_2)\right) &= (y_1^\mu \partial_\mu^{y_1} + y_2^\mu \partial_\mu^{y_2} + 12) t_\triangle(y_1, y_2) \\ &= \partial_\mu^{y_1} \left(y_1^\mu t_\triangle(y_1, y_2)\right) + \partial_\mu^{y_2} \left(y_2^\mu t_\triangle(y_1, y_2)\right) \\ &= \partial_\mu^{y_1} \overline{\left(y_1^\mu t^0_\triangle(y_1, y_2)\right)} + \partial_\mu^{y_2} \overline{\left(y_2^\mu t^0_\triangle(y_1, y_2)\right)} \ .\end{aligned}$$

In the last step we take into account that the singular order of $y_j^\mu t_\triangle(y_1, y_2)$ is $\omega = \mathrm{sd}\big(y_j^\mu t_\triangle(y_1, y_2)\big) - 12 = -1$; therefore, $y_j^\mu t_\triangle(y_1, y_2)$ is equal to the (unique) direct extension of $y_j^\mu t^0_\triangle(y_1, y_2)$.

Next we rewrite $t^0_\triangle(y_1, y_2)$ by using the explicit form of the Feynman propagator D^F (3.8.24) and the method of Feynman parameters (proved below in Exer. 3.8.5):

$$t^0_\triangle(y_1, y_2) = \frac{5!}{(4\pi^3)^3} \int_{z_1, \ldots, z_6 \geq 0} dz_1 \cdots dz_6$$
$$\cdot \frac{\delta(1 - z_1 - \cdots - z_6)}{\left((y_1^2 - i0)^2(z_1 + z_2) + (y_2^2 - i0)^2(z_3 + z_4) + ((y_1 - y_2)^2 - i0)(z_5 + z_6)\right)^6} \ .$$

Since the distribution to be integrated depends only on

$$\alpha := z_1 + z_2 \ , \quad \beta := z_3 + z_4 \ , \quad \gamma := z_5 + z_6 \ ,$$

we may use that

$$\int_0^\infty dz_1 \int_0^\infty dz_2\ f(z_1 + z_2) = \int_0^\infty d\alpha\ f(\alpha) \int_0^\alpha dz_2 = \int_0^\infty d\alpha\ \alpha\, f(\alpha) \ .$$

This yields

$$t^0_\triangle(y_1, y_2) = \frac{5!}{(4\pi^3)^3} \int_{\alpha, \beta, \gamma \geq 0} d\alpha\, d\beta\, d\gamma\ \alpha\beta\gamma\, \frac{\delta(1 - \alpha - \beta - \gamma)}{(z^T G z - i0)^6} \ ,$$

where we use matrix notation and the definitions $z^T := (y_1, y_2)$ and

$$G := \begin{pmatrix} (\alpha+\gamma)\,\mathbf{g} & -\gamma\,\mathbf{g} \\ -\gamma\,\mathbf{g} & (\beta+\gamma)\,\mathbf{g} \end{pmatrix} \quad \text{with} \quad \mathbf{g} := \mathrm{diag}(1,-1,-1,-1,-1,-1) \in \mathbb{R}^{6\times 6} \; .$$

Summing up we may write

$$c_1\,\delta(y_1,y_2) = \frac{5!}{(4\pi^3)^3} \int_{\alpha,\beta,\gamma\geq 0} d\alpha\, d\beta\, d\gamma\; \delta(1-\alpha-\beta-\gamma)\,\alpha\beta\gamma\,\partial_{z^j}\overline{\Big(\frac{z^j}{(z^TGz-i0)^6}\Big)}\;,$$

where $\partial_z z \equiv \partial_{z^j} z^j := \partial^{y_1}_\mu y_1^\mu + \partial^{y_2}_\mu y_2^\mu$.

To compute $\partial_z\,\overline{\big(z\,(z^TGz-i0)^{-6}\big)}$ we use the following formula for the fundamental solution of the Laplacian on the pseudo-Riemannian space $\mathbb{R}^{2,10}$:

$$\square^{2,10}_x\,\overline{\Big(\frac{1}{(x^2-i0)^5}\Big)} = -\frac{20\,\pi^6}{5!}\,\delta_{(12)}(x) \quad \text{where} \tag{3.8.35}$$

$$x^2 \equiv x_\mu x^\mu := \sum_{k=1}^{2}(x^k)^2 - \sum_{k=3}^{12}(x^k)^2\;, \quad \square^{2,10}_x \equiv \partial^x_\mu\partial^\mu_x := \sum_{k=1}^{2}\partial^2_{x^k} - \sum_{k=3}^{12}\partial^2_{x^k}\;.$$

Similarly to the identities (3.5.18) in $\mathbb{M}_4 \equiv \mathbb{R}^{1,3}$ and (3.8.33) in $\mathbb{M}_6 \equiv \mathbb{R}^{1,5}$, this formula is of type $\square_x\,\overline{\big((x^2-i0)^{1-\frac{d}{2}}\big)} \sim \delta(x)$ in $\mathbb{R}^{d-s,s}$, where $d>2$ is even; for a proof we refer to [27, Lemma C.2]. Due to

$$\partial^x_\mu\,\overline{\Big(\frac{1}{(x^2-i0)^5}\Big)} = \overline{\Big(\partial^x_\mu\frac{1}{(x^2-i0)^5}\Big)} = -10\,\overline{\Big(\frac{x_\mu}{(x^2-i0)^6}\Big)}\;,$$

we may write

$$\square^{2,10}_x\overline{\Big(\frac{1}{(x^2-i0)^5}\Big)} = -10\,\partial^\mu_x\,\overline{\Big(\frac{x_\mu}{(x^2-i0)^6}\Big)}\;. \tag{3.8.36}$$

Now, $G\in\mathbb{R}^{12\times 12}$ is invertible and symmetric and one verifies that 2 of the eigenvalues of G are >0 and the other 10 are <0. Therefore, G is diagonalizable by an orthogonal matrix $R\in O(12)$: $R^T G\,R = \mathrm{diag}(d_1,\ldots,d_{12})$; in addition, one can choose R such that $d_1, d_2 > 0$ and $d_3,\ldots,d_{12}<0$. Defining

$$D := \mathrm{diag}(d_1^{-\frac12}, d_2^{-\frac12}, |d_3|^{-\frac12},\ldots,|d_{12}|^{-\frac12}) \quad \text{and} \quad L := (RD)^{-1}\in GL(12,\mathbb{R})\;,$$

we obtain

$$G = L^T\eta\,L \quad \text{with} \quad \eta = \mathrm{diag}(1,1,-1,-1,\ldots,-1)\in\mathbb{R}^{12\times 12}\;.$$

Hence, setting

$$x := Lz\;, \quad \text{we get} \quad z^TGz = x^T\,\eta\,x \equiv x_\mu x^\mu \equiv x^2\;.$$

We conclude that

$$\partial_{z^j}\overline{\Big(\frac{z^j}{(z^TGz-i0)^6}\Big)}=\partial_{x^\mu}\overline{\Big(\frac{x^\mu}{(x^2-i0)^6}\Big)}=\frac{2\,\pi^6}{5!}\,\delta_{(12)}(x)$$
$$=\frac{2\,\pi^6}{5!\,\sqrt{\det G}}\,\delta_{(12)}(z)\;, \tag{3.8.37}$$

where $\det G$ means the determinant of the matrix G. In the last step we use that

$$\delta_{(12)}(Lz)=\frac{1}{|\det L|}\,\delta_{(12)}(z)\quad\text{and}\quad\det G=\det L^T\cdot\det\eta\cdot\det L=(\det L)^2>0\;.$$

To compute $\det G$, we transform G into an upper triangle matrix, by adding $\frac{\gamma}{\alpha+\gamma}\cdot$ [line k] to the line $(6+k)$ for all $1\le k\le 6$. From the resulting triangle matrix, $\det G$ is obtained as the product of the coefficients on the diagonal:

$$\det G=(\alpha+\gamma)^6\Big(\beta+\gamma+\frac{\gamma}{\alpha+\gamma}(-\gamma)\Big)^6=(\alpha\beta+\alpha\gamma+\gamma\beta)^6\;.$$

Summing up we get

$$c_1=\frac{1}{2^5\,\pi^3}\,I\quad\text{with}\quad I:=\int_{\alpha,\beta,\gamma\ge 0}d\alpha\,d\beta\,d\gamma\;\delta(1-\alpha-\beta-\gamma)\,\frac{\alpha\beta\gamma}{(\alpha\beta+\alpha\gamma+\gamma\beta)^3}.$$

To compute

$$I=\int_{\alpha,\beta\ge 0}d\alpha\,d\beta\;\theta(1-\alpha-\beta)\,\frac{\alpha\beta\gamma}{(\alpha\beta+\alpha\gamma+\gamma\beta)^3}\Big|_{\gamma=1-\alpha-\beta}$$

we substitute $\alpha=:\lambda\kappa$, $\beta=:(1-\lambda)\kappa$ with $\lambda,\kappa\in[0,1]$; the upper bounds $\lambda\le 1$ and $\kappa\le 1$ result from $\beta\ge 0$ and $\gamma=1-\kappa\ge 0$, respectively. By using

$$d\alpha\,d\beta=\Big|\det\Big(\frac{\partial(\alpha,\beta)}{\partial(\lambda,\kappa)}\Big)\Big|\;d\lambda\,d\kappa=\kappa\;d\lambda\,d\kappa\;,$$

we get

$$I=\int_0^1 d\lambda\int_0^1 d\kappa\,\frac{\kappa^3\lambda(1-\lambda)(1-\kappa)}{\big(\lambda\kappa^2(1-\lambda)+\lambda\kappa(1-\kappa)+\kappa(1-\lambda)(1-\kappa)\big)^3}$$
$$=\int_0^1 d\lambda\;a\int_0^1 d\kappa\,\frac{(1-\kappa)}{\big(1+(a-1)\kappa\big)^3}\;,\quad\text{where}\quad a\equiv a(\lambda):=\lambda(1-\lambda)\;.$$

By using partial fraction expansion,

$$\frac{(1-\kappa)}{\big(1+(a-1)\kappa\big)^3}=\frac{1}{a-1}\Big(\frac{a}{\big(1+(a-1)\kappa\big)^3}-\frac{1}{\big(1+(a-1)\kappa\big)^2}\Big)$$

the κ-integration can easily be done:

$$\int_0^1 d\kappa \, \frac{(1-\kappa)}{\big(1+(a-1)\kappa\big)^3} = \frac{1}{2a} \, .$$

Hence, $I = \frac{1}{2}$, which gives $c_1 = \frac{1}{2^6 \, \pi^3}$. Inserting all prefactors, we finally get

$$\tilde{c} = -c_1 = \frac{-1}{2^6 \, \pi^3} . \tag{3.8.38}$$

Exercise 3.8.5 (Feynman parameters)**.** Prove the integral formula

$$\frac{1}{b_1 b_2 \cdots b_n} = (n-1)! \int_0^\infty dz_1 \int_0^\infty dz_2 \cdots \int_0^\infty dz_n \, \frac{\delta(1-z_1-z_2-\cdots-z_n)}{(b_1 z_1 + b_2 z_2 + \cdots + b_n z_n)^n} \tag{3.8.39}$$

for $b_1, \ldots, b_n \in \mathbb{C} \setminus \{0\}$, by proceeding by induction on n.

[*Solution*: Let us assume that the assertion (3.8.39) holds true for $n \le N-1$ for some $N-1 \in \mathbb{N}^*$. The validity for $n = N-1$ can be written as

$$\frac{1}{b_1 \cdots b_{N-2} b_{N-1}} = (N-2)! \int_0^\infty dz_1 \cdots \int_0^\infty dz_{N-2} \, \theta(z) \, \frac{1}{(b + b_{N-1} z)^{N-1}} \, , \quad \text{where}$$
$$b := b_1 z_1 + \cdots + b_{N-2} z_{N-2} \, , \qquad z := 1 - z_1 - \cdots - z_{N-2} \, , \tag{3.8.40}$$

after removing the δ-distribution by the z_{N-1}-integration. To prove the assertion for $n = N$ we proceed analogously: Performing the integration $\int_0^\infty dz_N \, \delta(z - z_{N-1} - z_N) \cdots$, the r.h.s. of (3.8.39) is equal to

$$(N-1)! \int_0^\infty dz_1 \cdots \int_0^\infty dz_{N-2} \, \theta(z) \int_0^z dz_{N-1} \, \frac{1}{\big(b + b_{N-1} z_{N-1} + b_N (z - z_{N-1})\big)^N} \, .$$

The z_{N-1}-integration can be computed by means of the substitution $y := b + b_{N-1} z_{N-1} + b_N (z - z_{N-1})$:

$$= \frac{(N-1)!}{b_{N-1} - b_N} \int_0^\infty dz_1 \cdots \int_0^\infty dz_{N-2} \, \theta(z) \int_{b + b_N z}^{b + b_{N-1} z} dy \, \frac{1}{y^N}$$
$$= \frac{(N-2)!}{b_{N-1} - b_N} \int_0^\infty dz_1 \cdots \int_0^\infty dz_{N-2} \, \theta(z) \Big(\frac{1}{(b + b_N z)^{N-1}} - \frac{1}{(b + b_{N-1} z)^{N-1}} \Big) \, .$$

To compute the remaining integrals we use the inductive assumption (3.8.40); this yields

$$= \frac{1}{b_{N-1} - b_N} \Big(\frac{1}{b_1 \cdots b_{N-2} b_N} - \frac{1}{b_1 \cdots b_{N-2} b_{N-1}} \Big) = \frac{1}{b_1 \cdots b_{N-2} b_{N-1} b_N} \, .]$$

3.8.3 Field renormalization

In Sect. 3.8.1 we have seen that, in the algebraic adiabatic limit, the renormalization of the interaction $\kappa L(g) \longmapsto Z\big(\kappa L(g)\big)$ can be interpreted as a renormalization $\mathfrak{z}$ (3.8.5) of (constant) field polynomials. This holds true also for the renormalization of the field $A(h) \longmapsto Z'\big(\kappa L(g)\big) A(h)$ (3.6.40), where $A \in \mathcal{P}_{\mathrm{bal}}$, $h \in \mathcal{D}(\mathbb{M})$.

To derive this result, let again $\mathcal{O}$ be an open double cone, $h \in \mathcal{D}(\mathcal{O})$ and take $g\,L \in \mathcal{G}_L(\mathcal{O})$ (3.7.14). Using (3.6.40) and Lemma 3.8.1, we compute:

$$Z'\big(\kappa L(g)\big)A(h) = \frac{d}{d\lambda}\bigg|_{\lambda=0} Z\big(\kappa L(g) + \lambda A(h)\big) = \sum_{n=0}^{\infty} \frac{\kappa^n}{n!}\, Z^{(n+1)}\big(L(g)^{\otimes n} \otimes A(h)\big)$$

$$= \sum_{n,l,p,(b,a)} \frac{\kappa^n m^l}{n!} \log^p\Big(\frac{m}{M}\Big) \int dx\, d_{n+1,l,p,(b,a)}(L^{\otimes n} \otimes A)(x)\, \partial^a h(x) \prod_{j=1}^{n} \partial^{b_j} g(x)\,,$$

where $b \in (\mathbb{N}^d)^n$, $a \in \mathbb{N}^d$. We know that $\mathrm{supp}\big(\partial^a h \prod_j \partial^{b_j} g\big) \subseteq \mathrm{supp}(\partial^a h) \subset \mathcal{O}$; but on $\mathcal{O}$, every nontrivial derivative $\partial^{b_j} g$ vanishes, so that $\partial^a h \prod_j \partial^{b_j} g = 0$ if $|b| > 0$. Therefore,

$$Z'\big(\kappa L(g)\big)A(h) = \sum_{n,l,p,a} \frac{\kappa^n m^l}{n!} \log^p\Big(\frac{m}{M}\Big) \int dx\, d_{n+1,l,p,(0,a)}(L^{\otimes n} \otimes A)(x)\, \partial^a h(x)\,.$$

Integrating by parts to obtain a field polynomial integrated with $h(x)$ (rather than with $\partial^a h(x)$), this becomes

$$Z'\big(\kappa L(g)\big)A(h) = \int dx\, \big(\mathfrak{z}^{(1)}(\kappa L)A\big)(x)\, h(x)\,, \quad \text{where}$$

$$\mathfrak{z}^{(1)}(\kappa L)A := \sum_{n,l,p,a} \frac{\kappa^n m^l}{n!} \log^p\Big(\frac{m}{M}\Big)\, (-1)^{|a|}\, \partial^a d_{n+1,l,p,(0,a)}(L^{\otimes n} \otimes A) \in \mathcal{P}[\![\kappa]\!]\,. \tag{3.8.41}$$

(Notice that the differentiated polynomials on the right-hand side need no longer be balanced.)

We have thereby arrived at a *linear* map

$$\mathfrak{z}^{(1)}(\kappa L)\colon \begin{cases} \mathcal{P}_{\mathrm{bal}} \longrightarrow \mathcal{P}[\![\kappa]\!] \\ A \longmapsto \mathfrak{z}^{(1)}(\kappa L)A = A + O(\kappa)\,, \end{cases} \tag{3.8.42}$$

which is *universal*, i.e., independent of $\mathcal{O}$, and is uniquely determined by Z or $(R, \widehat{R})$.

Let $L,\, A \in \mathcal{P}_{\mathrm{bal}} \cap \mathcal{P}_{\mathrm{hom}}$. From (3.8.2) and (3.8.41) we conclude

$$\dim L \le d \quad \Longrightarrow \quad \dim \mathfrak{z}^{(1)}(\kappa L)A = \dim A\,, \tag{3.8.43}$$

where d is the number of spacetime dimensions and $\dim \mathfrak{z}^{(1)}(\kappa L)A$ is understood similarly to $\dim \mathfrak{z}(\kappa L)$ in (3.8.7). In a *massless* model solely the terms with $l = 0$ and $p = 0$ contribute to the expansion (3.8.1); therefore we obtain:

$$L \in \mathcal{P}_d \cap \mathcal{P}_{\mathrm{bal}} \,\wedge\, A \in \mathcal{P}_j \cap \mathcal{P}_{\mathrm{bal}} \quad \Longrightarrow \quad \mathfrak{z}^{(1)}(\kappa L)A \in \mathcal{P}_j\,, \tag{3.8.44}$$

cf. (3.8.8).

The map $\mathfrak{z}^{(1)}(\kappa L)$ (3.8.42) may be called the *field renormalization in the algebraic adiabatic limit*, due to Theorem 3.8.6, to which we turn now. The central formula (3.6.20) of the Main Theorem, which can equivalently be written in terms of interacting fields as

$$\widehat{A(h)}_{\kappa L(g)} \equiv \widehat{R}\big(e_\otimes^{\kappa L(g)/\hbar}, A(h)\big) = \Big(Z'\big(\kappa L(g)\big)A(h)\Big)_{Z(\kappa L(g))}$$

(see Exer. 3.6.11), goes over, in the algebraic adiabatic limit, to the following theorem (cf. [27, Thm. 6.5], which is a refinement of results derived in [55, 102]).

Theorem 3.8.6 (Algebraic renormalization group equation). *Let $\widehat{\mathcal{A}}_{\kappa L}(\mathcal{O})$ and $\mathcal{A}_{\kappa L}(\mathcal{O})$ denote the algebra of observables localized in the double cone $\mathcal{O}$ (3.7.16) which are obtained by using the renormalization prescriptions (i.e., retarded products) $\widehat{R}$ and R, respectively. In addition, let $Z \in \mathcal{R}$ be the unique renormalization group transformation relating $\widehat{R}$ and R by formula (3.6.39); and let $\mathfrak{z} = \gamma(Z)$ (3.8.5) and $\mathfrak{z}^{(1)}(\kappa L)$ (3.8.42) be the pertinent renormalization of the interaction and field, respectively, in the algebraic adiabatic limit. This Z induces an isomorphism $\beta_Z = (\beta_Z^{\mathcal{O}})_{\mathcal{O}}$ from the net $\widehat{\mathcal{A}}_{\kappa L}$ to the net $\mathcal{A}_{\mathfrak{z}(\kappa L)}$, that is, the maps*

$$\beta_Z^{\mathcal{O}} : \widehat{\mathcal{A}}_{\kappa L}(\mathcal{O}) \longrightarrow \mathcal{A}_{\mathfrak{z}(\kappa L)}(\mathcal{O}) \,, \tag{3.8.45}$$

are algebra-isomorphisms which satisfy

$$\iota_{\mathcal{O}_2,\mathcal{O}_1} \circ \beta_Z^{\mathcal{O}_1} = \beta_Z^{\mathcal{O}_2} \circ \hat{\iota}_{\mathcal{O}_2,\mathcal{O}_1} \quad \text{for all} \quad \mathcal{O}_1 \subseteq \mathcal{O}_2 \,, \tag{3.8.46}$$

where $\hat{\iota}_{\mathcal{O}_2,\mathcal{O}_1}$ and $\iota_{\mathcal{O}_2,\mathcal{O}_1}$ are the embeddings (3.7.17) belonging to the nets $\widehat{\mathcal{A}}_{\kappa L}$ and $\mathcal{A}_{\mathfrak{z}(\kappa L)}$, respectively. On the generators (i.e., the interacting fields in the algebraic adiabatic limit, see (3.7.15)–(3.7.16)) the isomorphisms $\beta_Z^{\mathcal{O}}$ are given by the relation

$$\beta_Z^{\mathcal{O}}\Big(\widehat{A(h)}_{\kappa L}^{\mathcal{O}}\Big) = \Big(\mathfrak{z}^{(1)}(\kappa L)A\Big)(h)_{\mathfrak{z}(\kappa L)}^{\mathcal{O}} \tag{3.8.47}$$

for all double cones $\mathcal{O}$, all $A, L \in \mathcal{P}_{\mathrm{bal}}$, and all $h \in \mathcal{D}(\mathcal{O})$.

In order that the nets $\widehat{\mathcal{A}}_{\kappa L}$ and $\mathcal{A}_{\mathfrak{z}(\kappa L)}$ describe the same physics, it does not suffice that the algebras $\widehat{\mathcal{A}}_{\kappa L}(\mathcal{O})$ and $\mathcal{A}_{\mathfrak{z}(\kappa L)}(\mathcal{O})$ are isomorphic for all $\mathcal{O}$, the compatibility with the embeddings, i.e., the relation (3.8.46), must also be fulfilled for a suitable choice of the isomorphisms $\beta_Z^{\mathcal{O}}$.

Note also that for a nontrivial $Z \in \mathcal{R}$ it may happen that $\gamma(Z)(\kappa L) = \kappa L$ for some $L \in \mathcal{P}_{\mathrm{bal}}$, and even then the pertinent net-isomorphism β_Z is in general nontrivial, see [55, Thm. 5.1].

Proof. Every $Z \in \mathcal{R}$ induces a map

$$f_Z^{\mathcal{O}} : \mathcal{G}_{\kappa L}(\mathcal{O}) \longrightarrow \mathcal{G}_{\gamma(Z)(\kappa L)}(\mathcal{O}) \quad \text{given uniquely by} \quad Z\Big(\int G\Big) = \int f_Z^{\mathcal{O}}(G)$$

for all $G \in \mathcal{G}_{\kappa L}(\mathcal{O})$, which is bijective, since it has a unique inverse:

$$(f_Z^{\mathcal{O}})^{-1} = f_{Z^{-1}}^{\mathcal{O}} : \mathcal{G}_{\gamma(Z)(\kappa L)}(\mathcal{O}) \longrightarrow \mathcal{G}_{\gamma(Z^{-1})\circ\gamma(Z)(\kappa L)}(\mathcal{O}) = \mathcal{G}_{\kappa L}(\mathcal{O}) \; .$$

The main point of the proof is that we have shown that any $\widehat{R}(e_\otimes^{\int G}, F)$, with $G \in \mathcal{G}_{\kappa L}(\mathcal{O})$ and $F \in \mathcal{F}_{\mathrm{loc}}(\mathcal{O})$, can be written as

$$\widehat{R}\Big(e_\otimes^{\int G}, F\Big) = R\Big(e_\otimes^{\int f_Z^{\mathcal{O}}(G)}, \mathfrak{z}^{(1)}(F)\Big) \; , \tag{3.8.48}$$

where we omit $\hbar$ for better readability. Here and in the following, the notation $\mathfrak{z}^{(1)}(F)$ is defined as follows: Using that any $F \in \mathcal{F}_{\mathrm{loc}}(\mathcal{O})$ can uniquely be written as $F = \sum_j A_j(h_j)$ with $A_j \in \mathcal{P}_{\mathrm{bal}}$, $h_j \in \mathcal{D}(\mathcal{O})$ (Proposition 1.4.3(b)), we set $\mathfrak{z}^{(1)}(F) := \sum_j \big(\mathfrak{z}^{(1)}(\kappa L) A_j\big)(h_j)$.

Next we recall that the interacting field $F_{\kappa L}^{\mathcal{O}}$ in the algebraic adiabatic limit is the map given in (3.7.15), and by $\widehat{F}_{\kappa L}^{\mathcal{O}} \big(\in \widehat{\mathcal{A}}_{\kappa L}(\mathcal{O})\big)$ we mean the same map with R replaced by $\widehat{R}$. To construct a map $\beta_Z^{\mathcal{O}}$ (3.8.45) satisfying the relation (3.8.47), we introduce the map

$$\begin{aligned} &\tilde\beta_Z^{\mathcal{O}} : \{\, \widehat{F}_{\kappa L}^{\mathcal{O}} \,|\, F \in \mathcal{F}_{\mathrm{loc}}(\mathcal{O}) \,\} \longrightarrow \{\, F_{\mathfrak{z}(\kappa L)}^{\mathcal{O}} \,|\, F \in \mathcal{F}_{\mathrm{loc}}(\mathcal{O}) \,\} \; ; \\ &\tilde\beta_Z^{\mathcal{O}}\Big(\big(G \mapsto \widehat{R}(e_\otimes^{\int G}, F)\big)_{G\in\mathcal{G}_{\kappa L}(\mathcal{O})}\Big) := \Big(f_Z^{\mathcal{O}}(G) \mapsto R(e_\otimes^{\int f_Z^{\mathcal{O}}(G)}, \mathfrak{z}^{(1)}(F))\Big)_{G\in\mathcal{G}_{\kappa L}(\mathcal{O})} \end{aligned} \tag{3.8.49}$$

for all $F \in \mathcal{F}_{\mathrm{loc}}(\mathcal{O})$, where we use that $f_Z^{\mathcal{O}}\big(\mathcal{G}_{\kappa L}(\mathcal{O})\big) = \mathcal{G}_{\mathfrak{z}(\kappa L)}(\mathcal{O})$. In the argument of $\tilde\beta_Z^{\mathcal{O}}$ we may replace $\widehat{R}(e_\otimes^{\int G}, F)$ by $R(e_\otimes^{\int f_Z^{\mathcal{O}}(G)}, \mathfrak{z}^{(1)}(F))$, due to (3.8.48). Moreover, from $\mathfrak{z}^{(1)}(F) = Z'(\int G)F$ (3.8.41) and the invertibility of $Z'(\int G)$ (Exer. 3.6.11(b)) we know that

$$\{\, \mathfrak{z}^{(1)}(F) \,|\, F \in \mathcal{F}_{\mathrm{loc}}(\mathcal{O}) \,\} = \mathcal{F}_{\mathrm{loc}}(\mathcal{O}) \; .$$

Combining these facts, we may write (3.8.49) also as

$$\tilde\beta_Z^{\mathcal{O}}\Big(\big(G \mapsto R(e_\otimes^{\int f_Z^{\mathcal{O}}(G)}, F)\big)_{G\in\mathcal{G}_{\kappa L}(\mathcal{O})}\Big) = \Big(f_Z^{\mathcal{O}}(G) \mapsto R(e_\otimes^{\int f_Z^{\mathcal{O}}(G)}, F)\Big)_{G\in\mathcal{G}_{\kappa L}(\mathcal{O})}$$

for all $F \in \mathcal{F}_{\mathrm{loc}}(\mathcal{O})$. Taking also into account that $f_Z^{\mathcal{O}}$ is bijective, we conclude that $\tilde\beta_Z^{\mathcal{O}}$ is a bijection from the generating set of $\widehat{\mathcal{A}}_{\kappa L}(\mathcal{O})$ to the generating set of $\mathcal{A}_{\mathfrak{z}(\kappa L)}(\mathcal{O})$; the inverse is

$$(\tilde\beta_Z^{\mathcal{O},\kappa L})^{-1} = \tilde\beta_{Z^{-1}}^{\mathcal{O},\mathfrak{z}(\kappa L)} \; , \quad \text{where we write} \quad \tilde\beta_Z^{\mathcal{O},\kappa L} := \tilde\beta_Z^{\mathcal{O}}$$

to indicate the (κL)-dependence of $\tilde\beta_Z^{\mathcal{O}}$. It follows that $\tilde\beta_Z^{\mathcal{O}}$ can be extended to an algebra-isomorphism $\beta_Z^{\mathcal{O}} : \widehat{\mathcal{A}}_{\kappa L}(\mathcal{O}) \longrightarrow \mathcal{A}_{\mathfrak{z}(\kappa L)}(\mathcal{O})$.

It suffices to verify the compatibility with the embeddings (3.8.46) for the generators, i.e., for the maps $(\tilde{\beta}_Z^{\mathcal{O}})$. Let $\mathcal{O}_1 \subseteq \mathcal{O}_2$. Comparing

$$\begin{aligned}
&\tilde{\beta}_Z^{\mathcal{O}_2} \circ \hat{\iota}_{\mathcal{O}_2,\mathcal{O}_1}\Big(\mathcal{G}_{\kappa L}(\mathcal{O}_1) \ni G_1 \mapsto \widehat{R}\Big(e_\otimes^{\int G_1}, F\Big)\Big) \\
&\quad = \tilde{\beta}_Z^{\mathcal{O}_2}\Big(\mathcal{G}_{\kappa L}(\mathcal{O}_2) \ni G_2 \mapsto \widehat{R}\Big(e_\otimes^{\int G_2}, F\Big)\Big) \\
&\quad = \Big(\mathcal{G}_{\mathfrak{z}(\kappa L)}(\mathcal{O}_2) \ni f_Z^{\mathcal{O}_2}(G_2) \mapsto R\Big(e_\otimes^{\int f_Z^{\mathcal{O}_2}(G_2)}, \mathfrak{z}^{(1)}(F)\Big)\Big)
\end{aligned}$$

with

$$\begin{aligned}
&\iota_{\mathcal{O}_2,\mathcal{O}_1} \circ \tilde{\beta}_{\mathcal{O}_1}\Big(\mathcal{G}_{\kappa L}(\mathcal{O}_1) \ni G_1 \mapsto \widehat{R}\Big(e_\otimes^{\int G_1}, F\Big)\Big) \\
&\quad = \iota_{\mathcal{O}_2,\mathcal{O}_1}\Big(\mathcal{G}_{\mathfrak{z}(\kappa L)}(\mathcal{O}_1) \ni f_Z^{\mathcal{O}_1}(G_1) \mapsto R\Big(e_\otimes^{\int f_Z^{\mathcal{O}_1}(G_1)}, \mathfrak{z}^{(1)}(F)\Big)\Big) \\
&\quad = \Big(\mathcal{G}_{\mathfrak{z}(\kappa L)}(\mathcal{O}_2) \ni f_Z^{\mathcal{O}_1}(G_2) \mapsto R\Big(e_\otimes^{\int f_Z^{\mathcal{O}_1}(G_2)}, \mathfrak{z}^{(1)}(F)\Big)\Big) \quad \text{where } G_2 \in \mathcal{G}_{\kappa L}(\mathcal{O}_2),
\end{aligned}$$

we see that two resulting maps are equal, by using that $f_Z^{\mathcal{O}_2}(G_2) = f_Z^{\mathcal{O}_1}(G_2)$ for $G_2 \in \mathcal{G}_{\kappa L}(\mathcal{O}_2)$. The latter follows immediately from the definition of $f_Z^{\mathcal{O}}$. □

Example 3.8.7 (Field renormalization for a scaling transformation)**.** As in Exap. 3.8.2, we study, for a *massless* theory in $d = 4$ dimensions, the scaling (with $\rho > 0$) as renormalization transformation. To compute the renormalization of the field $A = \varphi^4$ for the interaction $\kappa L = \kappa\varphi^4$ in the algebraic adiabatic limit, let g and h be as in the derivation of (3.8.41). We obtain

$$\begin{aligned}
Z'_\rho(\kappa\,\varphi^4(g))\,\varphi^4(h) &= \varphi^4(h) + \kappa \int dx\; h(x) \int dx_1\; g(x_1)\, Z_\rho^{(2)}\left(\varphi^4(x_1)\otimes\varphi^4(x)\right) + \mathcal{O}(\kappa^2) \\
&= \int dx\; h(x)\left[\varphi^4(x) + \kappa\Big(\frac{36\,\hbar}{(2\pi)^2}\log\rho\;\varphi^4(x) - \frac{6\,\hbar^2}{(2\pi)^4}\log\rho\;\varphi\square\varphi(x)\Big) + \mathcal{O}(\kappa^2)\right],
\end{aligned} \tag{3.8.50}$$

where we have inserted the limit $m \downarrow 0$ of (3.8.11), performed the x_1-integration and used that $g(x) = 1 =$ constant (and, hence, $\partial^\mu g(x) = 0$) for $x \in \operatorname{supp} h$. In particular the $(p_4\square\square\delta)$-term of (3.8.11) does not contribute. We end up with

$$\mathfrak{z}_\rho^{(1)}(\kappa\varphi^4)\,\varphi^4 = \Big\{1+\kappa\hbar\,\frac{36}{(2\pi)^2}\log\rho+\mathcal{O}\big((\kappa\hbar)^2\big)\Big\}\,\varphi^4+\hbar\Big\{-\kappa\hbar\,\frac{6}{(2\pi)^4}\log\rho+\mathcal{O}\big((\kappa\hbar)^2\big)\Big\}\,\varphi\square\varphi\,.$$

The terms in braces are formal power series in $\kappa\hbar$; this can be seen analogously to (3.8.14).

Exercise 3.8.8 (Field renormalization for a scaling transformation, continued)**.** Continuing the preceding example, compute

$$\mathfrak{z}_\rho^{(1)}(\kappa\varphi^4)\,\varphi^2 \quad\text{and}\quad \mathfrak{z}_\rho^{(1)}(\kappa\varphi^4)\,\varphi^3\,,$$

up to first order in κ, where still $m = 0$ and $d = 4$.

[*Solution*: In the intermediate result of (3.8.50) we have to replace $Z_\rho^{(2)}\big(\varphi^4(x_1)\otimes\varphi^4(x)\big)$ by $Z_\rho^{(2)}\big(\varphi^4(x_1)\otimes\varphi^2(x)\big)$ or $Z_\rho^{(2)}\big(\varphi^4(x_1)\otimes\varphi^3(x)\big)$, respectively. Only one diagram contributes to $Z_\rho^{(2)}(\varphi^4,\varphi^2)$, namely

and the contributions to $Z_\rho^{(2)}(\varphi^4,\varphi^3)$ come from the diagrams

Hence, we obtain

$$\begin{aligned} Z_\rho^{(2)}\big(\varphi^4(x_1)\otimes\varphi^2(x)\big) &= \frac{6}{\hbar}\left[\rho^4 r(\varphi^2,\varphi^2)(\rho(x_1-x)) - r(\varphi^2,\varphi^2)(x_1-x)\right]\varphi^2(x_1) \\ &= \frac{6\,\hbar}{(2\pi)^2}\,\log\rho\;\,\delta(x_1-x)\,\varphi^2(x_1)\,, \end{aligned}$$

and

$$\begin{aligned} Z_\rho^{(2)}\big(\varphi^4(x_1)\otimes\varphi^3(x)\big) &= \frac{18}{\hbar}\left[\rho^4 r(\varphi^2,\varphi^2)(\rho(x_1-x)) - r(\varphi^2,\varphi^2)(x_1-x)\right]\varphi(x)\varphi^2(x_1) \\ &\quad + \frac{4}{\hbar}\left[\rho^6 r(\varphi^3,\varphi^3)(\rho(x_1-x)) - r(\varphi^3,\varphi^3)(x_1-x)\right]\varphi(x_1) \\ &= \frac{18\,\hbar}{(2\pi)^2}\,\log\rho\;\,\delta(x_1-x)\,\varphi^3(x_1) - \frac{3\,\hbar^2}{2(2\pi)^4}\,\log\rho\;\,\Box\delta(x_1-x)\,\varphi(x_1)\,, \end{aligned}$$

respectively, where – for a change – we compute $Z^{(2)}$ from $R_{1,1}$ (see (3.6.41)) and use the results (3.2.119), (3.2.122) and (3.2.123). Continuing similarly to the preceding example, we get

$$\begin{aligned} \mathfrak{z}_\rho^{(1)}(\kappa\varphi^4)\,\varphi^2 &= \left\{1+\kappa\hbar\,\frac{6}{(2\pi)^2}\,\log\rho + \mathcal{O}\big((\kappa\hbar)^2\big)\right\}\varphi^2 \\ \mathfrak{z}_\rho^{(1)}(\kappa\varphi^4)\,\varphi^3 &= \left\{1+\kappa\hbar\,\frac{18}{(2\pi)^2}\,\log\rho + \mathcal{O}\big((\kappa\hbar)^2\big)\right\}\varphi^3 \\ &\quad + \hbar\left\{\kappa\hbar\,\frac{-3}{2(2\pi)^4}\,\log\rho + \mathcal{O}\big((\kappa\hbar)^2\big)\right\}\Box\varphi\,. \end{aligned}$$

Again, the terms in braces are formal power series in $\kappa\hbar$.]

3.9 A version of Wilson's renormalization group and of the flow equation

In this section we essentially follow [27, Sect. 5.2] and [61].

3.9.1 Outline of the procedure

The renormalization group (RG) in the sense of Wilson [171] relies usually on a functional integral approach; it describes the dependence of the theory on a cutoff Λ, which one introduces to avoid UV-divergences.[110] A major example for an UV-regularization is a discretization of Minkowski spacetime, that is, UV-finiteness is achieved by replacing the Minkowski space by a lattice. IR-divergences are avoided by a finite extension of the lattice. Setting $\Lambda := \{\text{lattice constant}\}^{-1}$, the "continuum limit" $\Lambda \to \infty$ of the suitably renormalized functional integral (i.e., Λ-dependent counter terms are added) yields an approach for the model in Minkowski spacetime, or in a finite subvolume of the Minkowski space if the extension of the lattice is kept finite.

Another UV-regularization method is to modify the (Feynman) propagator by cutting off the momenta above a scale Λ. It is rather obvious how to do this for a massive model on a Euclidean space (in place of the Minkowski space), see Exap. 3.9.2 and, e.g., [147]; however, in Minkowski spacetime such a momentum cutoff is more tricky [112] – see Exap. 3.9.3.

Regularized Feynman propagator yields regularized S-matrix. We imitate Wilson's RG in the framework of causal perturbation theory by picking up the idea of regularizing the Feynman propagator: Let $(p_\Lambda)_{\Lambda>0}$ be a family of *smooth functions*, $p_\Lambda \in C^\infty(\mathbb{R}^d)$, which approximates Δ_m^F, as roughly introduced in Remk. 3.3.3. More precisely we assume[111]

$$\lim_{\Lambda\to\infty} p_\Lambda = \Delta_m^F \quad \text{in the Hörmander topology [27, 107]}, \tag{3.9.1}$$

and for $\Lambda \to 0$ it is required that

$$p_0 := \lim_{\Lambda\downarrow 0} p_\Lambda = 0 \quad \text{in the Hörmander topology.} \tag{3.9.2}$$

[110] In the conventional literature one mostly needs also an IR-cutoff, to ensure IR-finiteness. In causal perturbation theory the test function g, which switches the interaction, acts as an IR-cutoff. Since, in this section, we do not study the adiabatic limit $g(x) \to 1$, IR-finiteness is guaranteed. So we concentrate on the UV-behaviour.

[111] Necessary conditions for $\lim_{\Lambda\to\infty} p_\Lambda = p$ in the Hörmander topology are:

- Convergence in the weak topology of $\mathcal{D}'(\mathbb{R}^d)$ (A.1.15) (or $\mathcal{S}'(\mathbb{R}^d)$, resp.), and
- convergence in $\mathcal{E}(K)$ for each compact set $K \subset \mathbb{R}^d$ which does not intersect the singular support of p (see (A.3.1) for the definition of the latter). For the non-expert reader: convergence in $\mathcal{E}(K)$ means that $(p_\Lambda - p)$ and all its derivatives converge to 0 *uniformly* on K.

Motivated by $\Delta^F_m(-z) = \Delta^F_m(z)$, we additionally assume that

$$p_\Lambda(-z) = p_\Lambda(z) \qquad \forall z \in \mathbb{R}^d \,. \tag{3.9.3}$$

Why do we choose the Hörmander topology? Convergence in $\mathcal{D}'(\mathbb{R}^d)$ (or $\mathcal{S}'(\mathbb{R}^d)$, resp.) is too weak for the proof of Theorem 3.9.4 below. The latter proof uses the following:[112] if $t_{1,\Lambda} \in \mathcal{D}'(\mathbb{R}^k)$, $t_{2,\Lambda} \in \mathcal{D}'(\mathbb{R}^l)$ and $\lim_{\Lambda\to\infty} t_{j,\Lambda} = t_j$ for $j = 1, 2$, then $\lim_{\Lambda\to\infty} t_{1,\Lambda} \cdot t_{2,\Lambda} = t_1 \cdot t_2$, where the appearing products of distributions are a mixture of the tensor product and the pointwise product and we assume that these products exist. Convergence in the Hörmander topology suffices for this relation.

We introduce the *regularized time-ordered product of* n*th order*:

$$T_{\Lambda,n}\colon \begin{cases} \mathcal{F}^{\otimes n} \longrightarrow \mathcal{F} \\ T_{\Lambda,n}(F_1 \otimes \cdots \otimes F_n) := F_1 \star_\Lambda \cdots \star_\Lambda F_n \,, \end{cases} \qquad \text{where} \quad \star_\Lambda := \star_{p_\Lambda} \tag{3.9.4}$$

is defined in Exer. 2.3.1. Due to (3.9.1)–(3.9.2), $\star_\Lambda$ interpolates between the pointwise (or "classical") product (1.2.5) ($\Lambda = 0$) and the unrenormalized time-ordered product $\star_{\Delta^F}$ (3.3.10) ($\Lambda = \infty$). Apart from the much larger domain, $T_{\Lambda,n}$ is obtained from the unrenormalized T-product (3.3.10) by replacing Δ^F_m by p_Λ. In contrast to the latter product and to the product $\mathcal{T}_n$ (2.3.10), $T_{\Lambda,n}$ is well defined on the whole space $\mathcal{F}^{\otimes n}$, because $p_\Lambda \in C^\infty(\mathbb{R}^d)$. We also point out that the definition (3.9.4) uses associativity of $\star_\Lambda$, which follows from the equivalence to the classical product, shown in Exer. 2.3.1 (formula (2.3.3)).

The generating functional of the regularized time-ordered product is the *regularized* S*-matrix*:

$$\mathbf{S}_\Lambda\colon \begin{cases} \mathcal{F} \longrightarrow \mathcal{F} \\ \mathbf{S}_\Lambda(F) := 1 + \sum_{n=1}^\infty \frac{i^n}{\hbar^n\, n!}\, T_{\Lambda,n}(F^{\otimes n}) =: e_{\star_\Lambda}^{iF/\hbar} \,. \end{cases} \tag{3.9.5}$$

The expression $e_{\star_\Lambda}^{iF/\hbar}$ is a suggestive shorthand notation for the series. In contrast to the exact S-matrix $\mathbf{S}$ (Definition 3.3.8), the domain of $\mathbf{S}_\Lambda$ is $\mathcal{F}$ (and not only $\mathcal{F}_{\mathrm{loc}}$) and $\mathbf{S}_\Lambda$ is invertible with respect to "$\circ$" (composition of maps), as shown below in Sect. 3.9.2. Since $\star_\Lambda$ is associative and commutative, the factorization

$$\mathbf{S}_\Lambda(F + G) = \mathbf{S}_\Lambda(F) \star_\Lambda \mathbf{S}_\Lambda(G) \tag{3.9.6}$$

holds true for all $F, G \in \mathcal{F}$.

For $F \in \mathcal{F}_{\mathrm{loc}}$ the limit $\Lambda \to \infty$ of $\mathbf{S}_\Lambda(F)$ does not exist in general – convergence in an appropriate sense can be achieved by adding suitable, Λ-dependent, local counter terms to the interaction F; essentially this is the message of Theorem 3.9.4.

---

[112] I thank Urs Schreiber for pointing this out [151, Chap. 16].

Effective potential and flow equation. Very heuristically, the idea of an *effective potential* V_Λ *at scale* Λ[113] can be explained as follows: We interpret Λ as an energy scale and p_Λ as propagator with an UV-cutoff at energy Λ. Now we split the "quantum effects" (more precisely, the degrees of freedom to be integrated out in a functional integral approach) into the ones above Λ and the ones below Λ. The former are taken into account by replacing the original interaction V by the effective potential V_Λ, and the latter are considered in the application of $\mathbf{S}_\Lambda$ (i.e., the S-matrix with an energy cutoff at Λ) to V_Λ.

So we define V_Λ as a function of $V = \kappa L(g)$ (where $g \in \mathcal{D}(\mathbb{M})$ and $L \in \mathcal{P}_{\mathrm{bal}}$ is the interaction Lagrangian) by the condition that the cutoff theory $\mathbf{S}_\Lambda$ with interaction V_Λ agrees with the exact theory $\mathbf{S}$ with interaction V, that is:

Definition 3.9.1. [27] The *effective potential* $V_\Lambda \in \mathcal{F}$ *at scale* $\Lambda \geq 0$ is defined by

$$\mathbf{S}_\Lambda(V_\Lambda) := \mathbf{S}(V) \ , \quad \text{or explicitly} \quad V_\Lambda \equiv V_\Lambda(V) := \mathbf{S}_\Lambda^{-1} \circ \mathbf{S}\,(V) \ . \tag{3.9.7}$$

We point out that in general V_Λ is a *non-local* interaction.

The problem with this Definition is that usually $\mathbf{S}$ is unknown. Therefore, one computes V_Λ by solving *Polchinski's flow equation* [112, 137, 147], which is a differential equation for V_Λ and can be derived from Definition 3.9.1 (see Sect. 3.9.3); it has the form

$$\frac{d}{d\Lambda} V_\Lambda = F_\Lambda(V_\Lambda \otimes V_\Lambda) \ , \tag{3.9.8}$$

where the map $F_\Lambda : \mathcal{F}^{\otimes 2} \longrightarrow \mathcal{F}$ is linear and explicitly known. Integration of the flow equation yields V_Λ, the integration constants are determined by appropriate boundary values. The non-uniqueness of the latter corresponds to the non-uniqueness of $\mathbf{S}$ in Definition 3.9.1.

From V_Λ the S-matrix may be obtained by the limit $\lim_{\Lambda\downarrow 0} e^{iV_\Lambda(V)/\hbar} = \mathbf{S}(V)$, with respect to the pointwise convergence in $\mathcal{F}$ (1.2.3). Heuristically, this can be seen as follows: Inserting $p_0 = 0$ into (3.9.5) we get $\mathbf{S}_0(F) = e^{iF/\hbar}$ (the exponential series with respect to the classical product), hence

$$\mathbf{S}_0^{-1}(1+F) = \tfrac{\hbar}{i}\,\log(1+F) := \tfrac{\hbar}{i}\big(F - \tfrac{1}{2}\,F\cdot F + \tfrac{1}{3}\,F\cdot F\cdot F - \cdots\big)$$

which gives $V_0(V) = \frac{\hbar}{i}\,\log\circ\mathbf{S}(V)$.

Definition 3.9.1 implies the relation

$$V_\Lambda = W_{\Lambda,\Lambda_0}(V_{\Lambda_0}) \quad \text{for} \quad W_{\Lambda,\Lambda_0} := \mathbf{S}_\Lambda^{-1} \circ \mathbf{S}_{\Lambda_0} \ , \tag{3.9.9}$$

[113] In Sect. 3.9 we denote the interaction by the letter "V" instead of "S", to avoid confusion with the letter $\mathbf{S}$ denoting the S-matrix. Following the naming in [27] and [147] we use the name "effective potential" in the framework of causal perturbation theory and the name "effective action" for the corresponding quantity in the functional integral approach (the latter is sketched in Sect. 3.9.5).

that is, W_{Λ,Λ_0} is the "flow of the effective potential from Λ_0 to Λ". In the functional integral approach the flow is usually defined for $\Lambda \leq \Lambda_0$ only; hence, it is non-invertible. In our framework this restriction does not appear: W_{Λ,Λ_0} is well defined for all $\Lambda, \Lambda_0 \in [0,\infty)$ and

$$W_{\Lambda,\Lambda_0}^{-1} = \mathbf{S}_{\Lambda_0}^{-1} \circ \mathbf{S}_\Lambda = W_{\Lambda_0,\Lambda}$$

exists. The set of flow operators,

$$\mathcal{W} := \{W_{\Lambda,\Lambda_0} \mid \Lambda, \Lambda_0 \in [0,\infty)\} \;, \tag{3.9.10}$$

is a *version of Wilson's RG.* The "product"

$$W_{\Lambda,\Lambda_1} \circ W_{\Lambda_1,\Lambda_0} = W_{\Lambda,\Lambda_0}$$

is again a flow operator. However, $(\mathcal{W},\circ)$ is in general not a group in the mathematical sense, since $W_{\Lambda_1,\Lambda_2} \circ W_{\Lambda_3,\Lambda_4}$ does not lie in $\mathcal{W}$ for $\Lambda_2 \neq \Lambda_3$. In the functional integral approach the situation is even worse, due to the non-invertibility of the flow; nevertheless one can interpret Wilson's RG as a "semigroup", see, e.g., [147, Chap. 3].

3.9.2 The regularized S-matrix

Proof of invertibility. First we prove the above-claimed *invertibility of* $\mathbf{S}_\Lambda$ (3.9.5) *with respect to the composition of maps*:

Proof. As shown in Exer. 2.3.1 the product $\star_\Lambda$ is equivalent to the classical product:

$$F \star_\Lambda G = \tau_\Lambda (\tau_\Lambda^{-1} F \cdot \tau_\Lambda^{-1} G) \quad \text{where} \quad \tau_\Lambda := e^{\Gamma_{p_\Lambda}} \,, \quad \tau_\Lambda^{-1} := e^{-\Gamma_{p_\Lambda}} \tag{3.9.11}$$

and the notations (2.3.1)–(2.3.2) are used. With that $\mathbf{S}_\Lambda$ can be written as

$$\mathbf{S}_\Lambda = \tau_\Lambda \circ \exp \circ \tfrac{i}{\hbar}\, \tau_\Lambda^{-1} \;, \quad \text{where} \quad \exp F := 1 + F + F\cdot F/2! + \cdots \;, \tag{3.9.12}$$

from which it is obvious that $\mathbf{S}_\Lambda$ is invertible:

$$\mathbf{S}_\Lambda^{-1} = \tfrac{\hbar}{i}\, \tau_\Lambda \circ \log \circ \tau_\Lambda^{-1} \;. \tag{3.9.13}$$

□

For $\Lambda \downarrow 0$ we have $\lim_{\Lambda\downarrow 0} p_\Lambda = p_0 = 0$, which gives $\lim_{\Lambda\downarrow 0} \tau_\Lambda = \tau_0 = \mathrm{Id}$. Hence we obtain

$$\lim_{\Lambda\downarrow 0} \mathbf{S}_\Lambda = \mathbf{S}_0 = \exp(\tfrac{i}{\hbar}\,\cdot) \;, \qquad \lim_{\Lambda\downarrow 0} \mathbf{S}_\Lambda^{-1} = \mathbf{S}_0^{-1} = \tfrac{\hbar}{i} \log \;. \tag{3.9.14}$$

Here and in the following, convergence of a family (M_Λ) of maps $M_\Lambda : \mathcal{F} \longrightarrow \mathcal{F}$ is meant in the pointwise sense, that is, $M_\Lambda(F)(h)$ converges in $\mathbb{C}$ for all $F \in \mathcal{F}$ and for all $h \in \mathcal{C}$.

Explicit examples for UV-regularizations. We are going to give two examples for UV-regularizations of the (Feynman) propagator. Since we will use Fourier transformation, we work with distribution spaces $\mathcal{S}'(\dots)$ in place of $\mathcal{D}'(\dots)$.

Example 3.9.2 (Scalar Euclidean QFT with mass $m > 0$). The Epstein–Glaser renormalization can be formulated also for scalar quantum field theories on spaces of Euclidean signature; this is worked out in [113]. For non-coinciding points the "Euclidean time-ordered product" $E_n^{(m)}$ is the n-fold product $\star_p$ with propagator

$$p(x) := \frac{1}{(2\pi)^d} \int d^dk \, \frac{e^{-ikx}}{k^2+m^2} \in \mathcal{S}'(\mathbb{R}^d) \, , \quad kx := k_0x_0 + \vec{k}\vec{x} \, , \; k^2 := k_0^2 + \vec{k}^2 \, . \quad (3.9.15)$$

More precisely, let $\mathbb{E}$ be $\mathbb{R}^d$ with the Euclidean metric and let $\check{\mathbb{E}}^n := \{(x_1,\dots,x_n) \in \mathbb{E}^n \mid x_l \neq x_j \ \forall l < j\}$. Then

$$E_n^{(m)}\big(A_1(x_1)\otimes\cdots\otimes A_n(x_n)\big) = A_1(x_1) \star_p \cdots \star_p A_n(x_n) \quad \text{on } \mathcal{S}(\check{\mathbb{E}}^n) \qquad (3.9.16)$$

for all $A_1,\dots,A_n \in \mathcal{P}$, analogously to Lemma 3.3.2(b). Similarly to Minkowski spacetime, the expression on the r.h.s. of (3.9.16) is not defined on $\mathcal{S}(\mathbb{E}^n)$; for example, for $d = 4$ the powers $p(x)^j$, $j \geq 2$, exist only in $\mathcal{S}'(\mathbb{R}^d \setminus \{0\})$ but not in $\mathcal{S}'(\mathbb{R}^d)$. Again, renormalization is the extension to coinciding points. In the Epstein–Glaser construction of the sequence $(E_n^{(m)})_{n\in\mathbb{N}^*}$ (more precisely, proceeding by induction on n and requiring Translation covariance (3.1.40) and Field independence, i.e., the causal Wick expansion (3.1.24)), renormalization amounts to the extension of distributions $e^0 \in \mathcal{S}'(\mathbb{R}^{d(n-1)} \setminus \{0\})$ to $e \in \mathcal{S}'(\mathbb{R}^{d(n-1)})$.

However, replacing the propagator $p \in \mathcal{S}'(\mathbb{R}^d)$ by a smooth function $p_\Lambda \in C^\infty(\mathbb{R}^d)$, the r.h.s. of (3.9.16) is well defined on $\mathcal{S}(\mathbb{E}^n)$. Since for any distribution $f \in \mathcal{S}'(\mathbb{R}^d)$, compactness of the support of its Fourier transform $\hat{f}$ implies smoothness of f,[114] we can obtain an UV-regularized propagator p_Λ from p by cutting off the momenta above a scale $\Lambda > 0$:

$$\hat{p}_\Lambda(k) := \frac{1}{(2\pi)^{d/2}\,(k^2+m^2)}\, K\Big(\frac{k^2}{\Lambda^2}\Big) \in \mathcal{S}(\mathbb{R}^d) \, , \quad k^2 := k_0^2 + \vec{k}^2 \, , \qquad (3.9.17)$$

where K is a smooth version of a step function: Following [147] we require that $K \in \mathcal{C}^\infty(\mathbb{R}_0^+, [0,1])$ with

$$K(x) = \begin{cases} 0 & \text{if} \quad x \geq 4 \\ 1 & \text{if} \quad x \leq 1 \end{cases}$$

and $K'(x) < 0 \ \forall x \in (1,4)$.

We are going to verify that this p_Λ satisfies the above-made assumptions. Since $\hat{p}_\Lambda(k) \in \mathcal{S}(\mathbb{R}^d)$ we have $p_\Lambda(x) \in \mathcal{S}(\mathbb{R}^d)$,[115] and $\hat{p}_\Lambda(-k) = \hat{p}_\Lambda(k)$ implies $p_\Lambda(-x) = p_\Lambda(x)$. We easily see that

$$\lim_{\Lambda\to\infty} \hat{p}_\Lambda(k) = \hat{p}(k) \, , \qquad \lim_{\Lambda\to 0} \hat{p}_\Lambda(k) = 0 \qquad (3.9.18)$$

---

[114] Actually, "rapid decay in all directions" of $\hat{f}$ suffices for smoothness of f, see App. A.3.

[115] In the examples treated here, the regularized propagator, $p_\Lambda(x)$ (3.9.17) or $p_{\varepsilon,\Lambda}(x)$ (3.9.21), is not only smooth, as required, it has the additional nice property that it decays rapidly, which is advantageous for some technical purposes.

with respect to the weak topology of $\mathcal{S}'(\mathbb{R}^d)$. Due to continuity of the inverse Fourier transformation from $\mathcal{S}'(\mathbb{R}^d)$ to $\mathcal{S}'(\mathbb{R}^d)$, these convergence statements hold also in x-space with respect to the weak topology. One can show that they are valid also with respect to the Hörmander topology in x-space.

To give a more explicit impression with what kind of objects we are dealing, we derive x-space formulas for the Euclidean propagator $p_d \equiv p$, by computing the inverse Fourier transformation (3.9.15). Due to rotation symmetry, $p_d(x)$ depends only on $|x| := \sqrt{x_0^2 + \vec{x}^{\,2}}$; hence, we may assume $x = (|x|, \vec{0})$. We first compute the k_0-integral, that is, the inner integral in

$$p_d(x) = \frac{|S^{d-2}|}{(2\pi)^d} \int_0^\infty d|\vec{k}|\; |\vec{k}|^{d-2}\, 2 \int_0^\infty dk_0\; \frac{\cos(k_0|x|)}{k_0^2 + |\vec{k}|^2 + m^2} \in \mathcal{S}'(\mathbb{R}^d)\,,$$

by using $\int_0^\infty dk_0\, \frac{\cos(ak_0)}{k_0^2+1} = \frac{\pi}{2}\, e^{-|a|}$. Rewriting the $|\vec{k}|$-integral by means of the substitution $q := \sqrt{|\vec{k}|^2 + m^2}$, we get

$$p_d(x) = \frac{2\,\pi^{\frac{d+1}{2}}}{(2\pi)^d\, \Gamma(\frac{d-1}{2})} \int_m^\infty dq\, (q^2 - m^2)^{\frac{d-3}{2}}\, e^{-q\,|x|} \in \mathcal{S}'(\mathbb{R}^d)\,,$$

for $d \geq 2$. In low dimensions the remaining integral can easily be computed: For $d = 2, 3$ we obtain

$$p_2(x) = (2\pi)^{-1}\, K_0(m\,|x|)$$

(where K_0 is a modified Bessel function of second kind) and

$$p_3(x) = \frac{1}{4\pi}\, \frac{e^{-m\,|x|}}{|x|}\,,$$

respectively. In $d = 4$-dimensions the computation of this integral is more difficult, it results

$$p_4(x) = \frac{1}{4\pi^2\, x^2} + m^2\, f(-m^2x^2)\, \log\Big(\frac{m^2x^2}{4}\Big) + m^2\, f_1(-m^2x^2)\,,$$

where $x^2 := x_0^2 + \vec{x}^{\,2}$ and f, f_1 are the analytic functions defined in (2.2.4) and (2.2.5), respectively. For a general $d \geq 2$ the result reads

$$p_d(x) = (2\pi)^{-d/2} \Big(\frac{m}{|x|}\Big)^{d/2-1} K_{d/2-1}(m|x|)\,.$$

These formulas agree with the analytic continuation of the corresponding formulas for the Wightman two-point function in the Minkowski space; for example, $p_4(x)$ is obtained from the formula for $\Delta_m^{+\,(d=4)}(x)$ (2.2.3) by replacing x_0 by ix_0.

Obviously, the propagators p_2, p_3 and p_4 have an integrable singularity at $x = 0$ and p_2, p_3 decay exponentially for $|x| \to \infty$.

Example 3.9.3 (ε-regularized relativistic theory with $m > 0$). This example is taken from [112], for simplicity we assume $d = 4$.

ε-regularization of the relativistic theory means that the Minkowski metric is replaced by

$$i\eta_\varepsilon := i\, \mathrm{diag}(\varepsilon - i, \varepsilon + i, \varepsilon + i, \varepsilon + i) \quad \text{with} \quad \varepsilon > 0\,.$$

With that we have

$$k\eta_\varepsilon k := k_0^2\, (\varepsilon - i) + \vec{k}^{\,2}\, (\varepsilon + i)\,.$$

Note that $k\eta_\varepsilon k \neq 0$ for $k \neq 0$ and that

$$\mathrm{Re}\,(k\eta_\varepsilon k + (\varepsilon + i)m^2) = \varepsilon\,(k_0^2 + \vec{k}^{\,2} + m^2) \geq \varepsilon\, m^2 \quad \forall k \in \mathbb{R}^4\ . \tag{3.9.19}$$

The Feynman propagator p_ε of the ε-regularized relativistic theory is defined in momentum space by the formula

$$\hat{p}_\varepsilon(k) := \frac{1}{(2\pi)^2\,\big(k\eta_\varepsilon k + (\varepsilon + i)m^2\big)}\ . \tag{3.9.20}$$

For $\varepsilon \downarrow 0$ the family $(\hat{p}_\varepsilon)_{\varepsilon>0}$ of analytic functions converges to the Minkowski space Feynman propagator,

$$\lim_{\varepsilon\downarrow 0} \hat{p}_\varepsilon(k) = \frac{i}{(2\pi)^2\,(k^2 - m^2 + i0)} \equiv \hat{\Delta}^F_m(k)\ ,$$

with respect to the weak topology of $\mathcal{S}'(\mathbb{R}^d)$ and, hence, this holds also in x-space: $\lim_{\varepsilon\downarrow 0} p_\varepsilon(x) = \Delta^F_m(x)$ in $\mathcal{S}'(\mathbb{R}^d)$. Whether this holds even with respect to the Hörmander topology is a more difficult question, which is not yet answered – as far as we know.

Since $\hat{p}_\varepsilon(k)$ does not rapidly decay for $\|k\| \to \infty$, $p_\varepsilon(x)$ is non-smooth. Therefore, we introduce, for fixed $\varepsilon > 0$, an UV-cutoff Λ for the propagator p_ε (3.9.20), by an exponential damping in momentum space:

$$\hat{p}_{\varepsilon,\Lambda}(k) := e^{-\Lambda^{-2}\left(k\eta_\varepsilon k + (\varepsilon+i)m^2\right)}\,\hat{p}_\varepsilon(k) = \frac{-i}{(2\pi)^2}\int_{\Lambda^{-2}}^{\infty} d\alpha\ e^{-\alpha\left(k\eta_\varepsilon k + (\varepsilon+i)m^2\right)}\ , \tag{3.9.21}$$

where we take into account (3.9.19). We see that $p_{\varepsilon,\Lambda}(-x) = p_{\varepsilon,\Lambda}(x)$ and that $\hat{p}_{\varepsilon,\Lambda}(k) \in \mathcal{S}(\mathbb{R}^d)$, which implies $p_{\varepsilon,\Lambda}(x) \in \mathcal{S}(\mathbb{R}^d)$. Note that we need both $\varepsilon > 0$ and $\Lambda < \infty$ in order that $p_{\varepsilon,\Lambda}(x)$ is smooth, as required for a regularization of the Feynman propagator.

We are now going to study the behaviour of $p_{\varepsilon,\Lambda}(x)$ for $\Lambda \to \infty$ and $\Lambda \to 0$, respectively, where $\varepsilon > 0$ is fixed. We claim

$$\lim_{\Lambda\to\infty} p_{\varepsilon,\Lambda}(x) = p_\varepsilon(x) \quad \text{and} \quad \lim_{\Lambda\to 0} p_{\varepsilon,\Lambda}(x) = 0$$

with respect to the weak topology of $\mathcal{S}'(\mathbb{R}^d)$ and also with respect to the Hörmander topology. Very roughly this can be seen as follows: Using (3.9.19) we get

$$|\hat{p}_{\varepsilon,\Lambda}(k)| = e^{-\Lambda^{-2}\,\varepsilon\,(k_0^2+\vec{k}^{\,2})}\ e^{-\Lambda^{-2}\,\varepsilon\, m^2}\ |\hat{p}_\varepsilon(k)|\ . \tag{3.9.22}$$

The first factor on the r.h.s. is a Euclidean momentum cutoff, similarly to the function $K\Big(\frac{k_0^2+\vec{k}^{\,2}}{\Lambda^2}\Big)$ in the regularized Euclidean propagator (3.9.17); therefore, the convergence behaviour for $\Lambda \to \infty$ is essentially the same as in the Euclidean case treated above. For $\Lambda \to 0$ the behaviour of $p_{\varepsilon,\Lambda}(x)$ is dominated by the prefactor $e^{-\Lambda^{-2}\,\varepsilon\, m^2}$, which is the second factor on the r.h.s. of (3.9.22) and remains unchanged under inverse Fourier transformation.

Which axioms hold also for $\mathbf{S}_\Lambda$? Although $T_{\Lambda,n}$ and T_n have different domains, it makes sense to investigate whether the axioms defining $(T_n)_{n\in\mathbb{N}^*}$ (Sect. 3.3.2) are preserved under the regularization, i.e., are valid also for $(T_{\Lambda,n})_{n\in\mathbb{N}^*}$.

Obviously, $T_{\Lambda,n} : \mathcal{F}^{\otimes n} \longrightarrow \mathcal{F}$ satisfies the basic axioms Linearity, Initial condition and Symmetry; the latter relies on (3.9.3). But Causality does certainly not hold with respect to the usual star product $\star_{\Delta^+}$, because $p_\Lambda(x) \neq$

$\theta(x^0)\,\Delta^+(x) + \theta(-x^0)\,\Delta^+(-x)$. Nevertheless, $T_{\Lambda,n}$ respects Causality in the following sense: Using associativity of $\star_\Lambda$ we get[116]

$$T_{\Lambda,n}\big(\otimes_{r=1}^n A_r(x_r)\big) = T_{\Lambda,k}\big(\otimes_{j=1}^k A_j(x_j)\big) \star_\Lambda T_{\Lambda,n-k}\big(\otimes_{l=k+1}^n A_l(x_l)\big) \quad \text{on } \mathcal{D}(\mathbb{M}^n).$$

The point is that for $\{x_1,\dots,x_k\} \cap \big(\{x_{k+1},\dots,x_n\} + \overline{V}_-\big) = \emptyset$ the product $\star_\Lambda$ of the two T_Λ-products on the r.h.s. can be interpreted as an approximation of $\star_{\Delta^+}$, because the arguments of the pertinent propagators p_Λ fulfill $(x_j - x_l) \notin \overline{V}_-$ (where $1 \le j \le k$, $k+1 \le l \le n$) and on $\mathcal{D}(\mathbb{R}^d \backslash \overline{V}_-)$ it holds that $\lim_{\Lambda\to\infty} p_\Lambda = \Delta^F = \Delta^+$, as we know from (2.3.9).

Turning to the renormalization conditions, we first note that Field independence (and, hence, the causal Wick expansion) is satisfied: For $T_{\Lambda,2}$ this is an immediate consequence of the definition of $\star_\Lambda$ (2.1.5); and due to associativity, more precisely $T_{\Lambda,n}(F_1 \otimes \cdots \otimes F_n) = T_{\Lambda,n-1}(F_1 \otimes \cdots \otimes F_{n-1}) \star_\Lambda F_n$, the $(n=2)$-reasoning can be used to verify the statement for $(T_{\Lambda,n})_{n\ge 3}$ by induction on n.

It is easy to see that $T_{\Lambda,n}$ fulfills also the renormalization conditions Translation covariance, Field parity (the equivalent formulation (3.1.34) holds obviously true), $\hbar$-dependence and the Off-shell field equation in integrated form, that is, formula (3.3.20) with p_Λ in place of Δ^F_m. Concerning the latter note that generically it holds that $(\Box + m^2)p_\Lambda(x) \ne -i\delta(x)$, since p_Λ is smooth; therefore, $T_{\Lambda,n}$ violates the Off-shell field equation in *differential* form, by which we mean

$$\begin{aligned}(\Box_x + m^2)\,T_n(\varphi(x) \otimes F_1 \otimes \cdots) &= (\Box_x + m^2)\varphi(x)\; T_{n-1}(F_1 \otimes \cdots) \\ &\quad - i\hbar\,\frac{\delta}{\delta\varphi(x)} T_{n-1}(F_1 \otimes \cdots)\,.\end{aligned}$$

Generically an UV-cutoff breaks homogeneous scaling and the Lorentz invariance of the (Feynman) propagator; e.g., the preceding Exaps. 3.9.2 and 3.9.3 fulfill

$$\rho^{d-2}\, p_{\rho^{-1}m,\Lambda}(\rho x) = p_{m,\rho\Lambda}(x) \ne p_{m,\Lambda}(x)\;; \tag{3.9.23}$$

and for $\varepsilon > 0$ the propagators p_ε (3.9.20) and $p_{\varepsilon,\Lambda}$ (3.9.21) are not Lorentz invariant. Therefore, $T_{\Lambda,n}$ usually fails to satisfy the renormalization conditions Sm-expansion and Lorentz covariance. But the axiom Scaling degree (i.e., the condition (3.2.62) with $t(A_1,\dots,A_n)$ in place of $r(A_1,\dots,A_n)$), which is weaker than the Sm-expansion axiom, is trivially satisfied by $T_{\Lambda,n}$. Namely, because

$$t_\Lambda(A_1,\dots,A_n)(x_1 - x_n,\dots) = \sum_J C_J \prod_{j\in J} \partial^{a_{J,j}} p_\Lambda(x_{J,j,1} - x_{J,j,2})$$

(where $C_J \in \mathbb{C}$ and $x_{J,j,1} \ne x_{J,j,2}$) is a smooth function, we have $\mathrm{sd}\, t_\Lambda(A_1,\dots,A_n) \le 0 \le \sum_k \dim A_k$, see Exer. 3.2.7(b)(i). Finally, a discussion of the validity of the $*$-structure axiom would be more involved – we omit it here.

---

[116] As we know from Exer. 3.3.10, this formula can be understood as an equivalent reformulation of the factorization (3.9.6) of $\mathbf{S}_\Lambda$.

As a simple application, we prove that the effective potential V_Λ (3.9.7) fulfills Field parity and Translation covariance: These symmetries for $\mathbf{S}_\Lambda$,

$$\alpha \circ \mathbf{S}_\Lambda = \mathbf{S}_\Lambda \circ \alpha \;, \qquad \beta_a \circ \mathbf{S}_\Lambda = \mathbf{S}_\Lambda \circ \beta_a \quad \forall a \in \mathbb{R}^d \;,$$

imply

$$\mathbf{S}_\Lambda^{-1} \circ \alpha = \alpha \circ \mathbf{S}_\Lambda^{-1} \;, \qquad \mathbf{S}_\Lambda^{-1} \circ \beta_a = \beta_a \circ \mathbf{S}_\Lambda^{-1} \;,$$

respectively. Since $\mathbf{S}$ satisfies also these symmetries, we obtain

$$\gamma \circ V_\Lambda(V) = \mathbf{S}_\Lambda^{-1} \circ \gamma \circ \mathbf{S}(V) = V_\Lambda(\gamma V) \;, \quad \text{for } \gamma = \alpha \text{ and } \gamma = \beta_a \;, \forall V \in \mathcal{F}_{\mathrm{loc}} \;.$$

Consistent scheme for addition of local counter terms such that cutoff can be removed. Widespread, traditional renormalization methods in pQFT work with a regularization of the S-matrix to avoid UV-divergences (cf. Remk. 3.3.3), e.g., dimensional regularization or Pauli–Villars regularization. It is a main and deep result of perturbative renormalization, that convergence of the limit removing the cutoff can consistently be achieved, by adding suitable local counter terms to the interaction.

Causal perturbation theory does neither involve any regularization nor divergent counter terms. Nevertheless, the following theorem states that any solution $\mathbf{S}$ of our axioms (given in Sect. 3.3.2) can be obtained by starting with a regularized S-matrix $\mathbf{S}_\Lambda$ of type (3.9.5), adding suitable, local counter terms to the interaction and finally performing the limit $\Lambda \to \infty$ which removes the cutoff. The addition of local counter terms is consistent, that is, it yields the wanted convergence of the limit $\Lambda \to \infty$ simultaneously for all orders n of $T_{\Lambda,n}$ and for all arguments $F_1 \otimes \cdots \otimes F_n \in \mathcal{F}_{\mathrm{loc}}^{\otimes n}$. The theorem states also that the addition of local counter terms can be described by an element Z_Λ of the Stückelberg–Petermann RG. More precisely, we mean here the larger version $\mathcal{R}_0$ of this group (introduced in Remk. 3.6.6) belonging to the renormalization conditions valid for $\mathbf{S}_\Lambda$ (since we want to maintain these properties in the addition of local counter terms), that is, Field independence, Translation covariance, Field parity, $\hbar$-dependence, and Scaling degree; for simplicity we omit the Off-shell field equation.

Theorem 3.9.4 (Addition of local counter terms). *Given*

- *an arbitrary solution* $\mathbf{S}$ *of the axioms for* T*-products – here we mean the complete set of axioms or a subset which contains the axioms defining* $\mathcal{R}_0$ –
- *and an approximation* $(p_\Lambda)_{\Lambda>0}$ *of the Feynman propagator fulfilling the above-assumed properties (see Sect.* 3.9.1*),*

there exists a family $Z_\Lambda \in \mathcal{R}_0$ *such that*

$$\lim_{\Lambda\to\infty} \mathbf{S}_\Lambda \circ Z_\Lambda = \mathbf{S} \;. \tag{3.9.24}$$

Convergence of this limit is understood in the following sense: $\lim_{\Lambda\to\infty} \mathbf{S}_\Lambda \circ Z_\Lambda(0) = \mathbf{S}(0) = 1$ *and, for all* $k \in \mathbb{N}^*$ *and for all monomials* $A_1, \ldots, A_k \in \mathcal{P}$*, the* *distributions*

$$\left(\tfrac{i}{\hbar}\right)^k t_\Lambda(A_1,\dots,A_k)(x_1-x_k,\dots) := \omega_0\Big((\mathbf{S}_\Lambda\circ Z_\Lambda)^{(k)}\big(\otimes_{i=1}^k A_i(x_i)\big)\Big) \in \mathcal{D}'(\mathbb{R}^{d(k-1)}) \tag{3.9.25}$$

converge for $\Lambda\to\infty$ *to*

$$\left(\tfrac{i}{\hbar}\right)^k t(A_1,\dots,A_k)(x_1-x_k,\dots) := \omega_0\Big(\mathbf{S}^{(k)}\big(\otimes_{i=1}^k A_i(x_i)\big)\Big) \in \mathcal{D}'(\mathbb{R}^{d(k-1)}) \tag{3.9.26}$$

with respect to the Hörmander topology.

Convergence in this sense implies convergence in the pointwise sense:

$$\lim_{\Lambda\to\infty} \mathbf{S}_\Lambda\circ Z_\Lambda(V)(h) = \mathbf{S}(V)(h) \quad \forall V\in\mathcal{F}_{\mathrm{loc}}\ ,\ h\in\mathcal{C}\ . \tag{3.9.27}$$

To see this we may use the causal Wick expansion (3.1.24) for $\mathbf{S}^{(k)}$ and also for $(\mathbf{S}_\Lambda\circ Z_\Lambda)^{(k)}$; the latter is allowed since $\mathbf{S}_\Lambda$ and Z_Λ satisfy Field independence and, hence, this holds true also for $(\mathbf{S}_\Lambda\circ Z_\Lambda)^{(k)}$ for all $k\geq 1$. We may assume that $V=\kappa\int dx\ A(x)\,g(x)$, with[117] $A\equiv A(\varphi)\in\mathcal{P}$ a monomial and $g\in\mathcal{D}(\mathbb{M})$. So we get

$$(\mathbf{S}_\Lambda\circ Z_\Lambda)^{(k)}(V^{\otimes k})(h) = \left(\tfrac{i\kappa}{\hbar}\right)^k \sum_{\underline{A}_j\subseteq A}\int dx_k\ \Big\langle t_\Lambda(\underline{A}_1,\dots,\underline{A}_k)(\mathbf{y}), f(\mathbf{y},x_k)\Big\rangle_{\mathbf{y}}\ ,$$

where $\mathbf{y}:=(y_j)_{j=1}^{k-1}\in\mathbb{R}^{d(k-1)}$, $y_j:=x_j-x_k$, and

$$f(\mathbf{y},x_k) := \overline{A}_1(h)(y_1+x_k)\cdots\overline{A}_k(h)(x_k)\,g(y_1+x_k)\cdots g(x_k)\in\mathcal{D}(\mathbb{R}^{dk})\ ;$$

and for $\mathbf{S}^{(k)}(V^{\otimes k})(h)$ we get an analogous formula, the only difference is that $t_\Lambda(\underline{A}_1,\dots,\underline{A}_k)$ is replaced by $t(\underline{A}_1,\dots,\underline{A}_k)$. Since convergence of the limit $\lim_{\Lambda\to\infty} t_\Lambda(\underline{A}_1,\dots,\underline{A}_k) = t(\underline{A}_1,\dots,\underline{A}_k)$ in the Hörmander topology implies convergence of the same limit in the weak topology of $\mathcal{D}'(\mathbb{R}^{d(k-1)})$ (A.1.15) (as already mentioned in Footn. 111), we obtain the assertion (3.9.27). ⊟

If not any solution $\mathbf{S}$ (of the considered axioms) is given, we nevertheless know that such a solution exists (as explained in Sect. 3.3.2) and, in this case, the theorem says that *given an approximation* $(p_\Lambda)_{\Lambda>0}$ *of the Feynman propagator, there exists a consistent addition of local counter terms to the interaction (which can be described by a family* $Z_\Lambda\in\mathcal{R}_0$*) such that we obtain a solution of our axioms when removing the cutoff; and any solution of our axioms can be obtained this way.*

To explain the scheme of cancellation of divergences we write out $\mathbf{S}_\Lambda\circ Z_\Lambda(F)$ to lowest orders (cf. (3.6.21)–(3.6.22)):

$$\mathbf{S}_\Lambda\circ Z_\Lambda(F) = \frac{i}{\hbar}F + \frac{1}{2!}\Big(\frac{i^2}{\hbar^2}T_{\Lambda,2}(F^{\otimes 2}) + \frac{i}{\hbar}Z_\Lambda^{(2)}(F^{\otimes 2})\Big)$$
$$+\frac{1}{3!}\Big(\frac{i^3}{\hbar^3}T_{\Lambda,3}(F^{\otimes 3}) + 3\,\frac{i^2}{\hbar^2}T_{\Lambda,2}\big(Z_\Lambda^{(2)}(F^{\otimes 2})\otimes F\big) + \frac{i}{\hbar}Z_\Lambda^{(3)}(F^{\otimes 3})\Big) + \cdots\ .$$

[117] Note that $A(\varphi)(h)=A(h)$, $h\in\mathcal{C}$, differs from the notation $A(g)=\int dx\ A(x)\,g(x)$, $g\in\mathcal{D}(\mathbb{M})$, frequently used in this book.

So, ignoring prefactors i and $\hbar^{-1}$, $Z_\Lambda^{(2)}(F^{\otimes 2})$ removes the divergences of $T_{\Lambda,2}(F^{\otimes 2})$ localized on Δ_2; $3\,T_{\Lambda,2}(Z_\Lambda^{(2)}(F^{\otimes 2})\otimes F)$ removes the "subdivergences" of $T_{\Lambda,3}(F^{\otimes 3})$, that is, the divergences of $T_{\Lambda,3}(F^{\otimes 3})$ localized on partial diagonals, which come from divergent subdiagrams; and $Z_\Lambda^{(3)}(F^{\otimes 3})$ removes the "overall divergences" of $T_{\Lambda,3}(F^{\otimes 3})$ localized on Δ_3.

This theorem was first formulated in [27] and proved in [63, App. A]. The proof proceeds by an inductive construction of the sequence $(Z_\Lambda^{(n)})$, which is analogous to the proof of part (a) of the Main Theorem – it also relies on the Faà di Bruno formula (3.6.25) (in a somewhat modified form). But, the difficult step is to show that the so-constructed Z_Λ lies in $\mathcal{R}_0$. On the level of the C-number distributions (which depend on the relative coordinates) this can be done by using the W-projection (3.2.29):

$$W_\omega\colon \begin{cases} \mathcal{D}(\mathbb{R}^{dn}) \longrightarrow \mathcal{D}_\omega(\mathbb{R}^{dn}) \\ W_\omega h(x) := h(x) - \sum_{|a|\le\omega} w_a(x)\,(-1)^{|a|}\,\langle \partial^a\delta, h\rangle \end{cases} \tag{3.9.28}$$

for $\omega \ge 0$. In [63, App. A] the C-number distributions are introduced by an expansion of the (regularized) S-matrix as a sum over (Feynman) diagrams; here we proceed differently: The C-number distributions are the coefficients in the causal Wick expansion (3.1.24).

Proof[118]. We are going to construct the sequence $(Z_\Lambda^{(n)})_{n\in\mathbb{N}}$, where $Z_\Lambda^{(n)} \in \hbar\,\mathcal{V}[\![\hbar]\!]$ (3.6.19) for $n \ge 2$, by induction on n. More precisely, we construct the truncations $Z_{\Lambda,n} \in \mathcal{R}$ of Z_Λ (3.6.17) by induction on n such that

$$\lim_{\Lambda\to\infty} (\mathbf{S}_\Lambda \circ Z_{\Lambda,n})^{(k)} = \mathbf{S}^{(k)} \quad \forall k \le n \tag{3.9.29}$$

with convergence in the above-explained sense.

Due to the requirement $Z_\Lambda \in \mathcal{R}$, we set $Z_\Lambda^{(0)} := 0$ and $Z_\Lambda^{(1)} := \mathrm{Id}$; as discussed for the Main Theorem (see the explanations above (3.6.21)), these values satisfy (3.9.29) for $n = 0$ and $n = 1$.

Let (3.9.29) hold true for a certain $n \ge 1$. We are searching a $Z_\Lambda^{(n+1)} \in \hbar\,\mathcal{V}[\![\hbar]\!]$ such that

$$\lim_{\Lambda\to\infty} (\mathbf{S}_\Lambda \circ Z_{\Lambda,n+1})^{(n+1)} = \mathbf{S}^{(n+1)} . \tag{3.9.30}$$

Note that $\lim_{\Lambda\to\infty}(\mathbf{S}_\Lambda \circ Z_{\Lambda,n+1})^{(k)} = \mathbf{S}^{(k)}\ \forall k \le n$ is satisfied for any $Z_\Lambda^{(n+1)}$, because $(\mathbf{S}_\Lambda \circ Z_{\Lambda,n+1})^{(k)} = (\mathbf{S}_\Lambda \circ Z_{\Lambda,n})^{(k)}$ for $k \le n$.

We construct $Z_\Lambda^{(n+1)}$ by means of the causal Wick expansion (3.6.9); due to that, the condition Field independence on $Z_{\Lambda,n+1}$ (3.6.5) will be satisfied. Let

[118] Essential parts of the version of proof given here are due to Urs Schreiber [151, Chap. 16].

$\mathbf{y} := (y_1, \ldots, y_n)$, $y_j := x_j - x_{n+1}$; we introduce the notation (3.9.25) also for $(\mathbf{S}_\Lambda \circ Z_{\Lambda,p})$:

$$\left(\tfrac{i}{\hbar}\right)^k t_{\Lambda,p}(A_1, \ldots, A_k)(x_1 - x_k, \ldots) := \omega_0\Big((\mathbf{S}_\Lambda \circ Z_{\Lambda,p})^{(k)}\big(\otimes_{i=1}^k A_i(x_i)\big)\Big)$$

for all $k \geq 2$, $p \geq 1$. Using also the notation (3.9.26), the explicit formulation of the requirement (3.9.30) on $Z_{\Lambda,n+1}$ reads

$$\lim_{\Lambda\to\infty} t_{\Lambda,n+1}(A_1, \ldots, A_{n+1})(\mathbf{y}) = t(A_1, \ldots, A_{n+1})(\mathbf{y})$$

in the Hörmander topology, for all monomials $A_1, \ldots, A_{n+1} \in \mathcal{P}$. Looking at the Faà di Bruno formula (3.6.25) for $(\mathbf{S}_\Lambda \circ Z_{\Lambda,n+1})$, we see that

$$t_{\Lambda,n+1}(A_1, \ldots, A_{n+1})(\mathbf{y}) = t_{\Lambda,n}(A_1, \ldots, A_{n+1})(\mathbf{y}) + \left(\tfrac{\hbar}{i}\right)^n z_\Lambda^{(n+1)}(A_1, \ldots, A_{n+1})(\mathbf{y})$$

where the notation (3.6.2) is used. So for all monomials $A_1, \ldots, A_{n+1} \in \mathcal{P}$ we have to find a $k_\Lambda(A_1, \ldots, A_{n+1}) \in \mathcal{D}'(\mathbb{R}^{dn})$ such that, setting

$$\begin{aligned}\left(\tfrac{\hbar}{i}\right)^n z_\Lambda^{(n+1)}(A_1, \ldots, A_{n+1})(\mathbf{y}) &:= t(A_1, \ldots, A_{n+1})(\mathbf{y}) \\ &\quad - t_{\Lambda,n}(A_1, \ldots, A_{n+1})(\mathbf{y}) + k_\Lambda(A_1, \ldots, A_{n+1})(\mathbf{y}) \,, \end{aligned} \tag{3.9.31}$$

the following statements hold true:

(i) $z_\Lambda^{(n+1)}$ is of the form

$$z_\Lambda^{(n+1)}(A_1, \ldots, A_k)(\mathbf{y}) = \mathfrak{S}_{n+1} \sum_{|a|\leq\omega} C^a(A_1, \ldots, A_{n+1})\, \partial^a \delta(\mathbf{y})$$

for some coefficients $C^a(A_1, \ldots, A_{n+1}) \in \mathbb{C}$, where $\omega := \sum_{j=1}^{n+1} \dim A_j - dn$ and $\mathfrak{S}_{n+1}$ denotes symmetrization in $(A_1, x_1), \ldots, (A_{n+1}, x_{n+1})$; and

(ii)
$$\lim_{\Lambda\to\infty} k_\Lambda(A_1, \ldots, A_{n+1})(\mathbf{y}) = 0$$

with respect to the Hörmander topology.

According to Lemma 3.2.15, there exists, for each $(n+1)$-tupel $(A_1, \ldots, A_{n+1})$ of field monomials, a W-projector $W_\omega(A_1, \ldots, A_{n+1})$ (3.9.28) (where again $\omega := \sum_{j=1}^{n+1} \dim A_j - dn$) such that

$$t(A_1, \ldots, A_{n+1}) = \mathfrak{S}_{n+1}\, t^0(A_1, \ldots, A_{n+1}) \circ W_\omega(A_1, \ldots, A_{n+1}) \,;$$

more precisely, by $t^0(\cdots)(\mathbf{y})$ we mean the direct extension of $\omega_0\big(T^0_{n+1}(\cdots)\big)(\mathbf{y})$ to $\mathcal{D}_\omega(\mathbb{R}^{dn})$, see (3.2.27), and $T^0_{n+1} := T_{n+1}|_{\mathcal{D}(\mathbb{M}^{n+1}\setminus\Delta_{n+1})}$. We claim that

$$\begin{aligned} &k_\Lambda(A_1, \ldots, A_{n+1}) \\ &\quad := -t(A_1, \ldots, A_{n+1}) + \mathfrak{S}_{n+1}\, t_{\Lambda,n}(A_1, \ldots, A_{n+1}) \circ W_\omega(A_1, \ldots, A_{n+1}) \\ &\quad = \mathfrak{S}_{n+1}\left(-t^0(A_1, \ldots, A_{n+1}) + t_{\Lambda,n}(A_1, \ldots, A_{n+1})\right) \circ W_\omega(A_1, \ldots, A_{n+1}) \end{aligned} \tag{3.9.32}$$

solves (i) and (ii).

Verification of (i): Inserting (3.9.32) into (3.9.31) and using that $(\mathbf{S}_\Lambda \circ Z_{\Lambda,n})^{(n+1)}$ is symmetric and, hence,

$$t_{\Lambda,n}(A_1,\ldots,A_{n+1}) = \mathfrak{S}_{n+1}\Big(t_{\Lambda,n}(A_1,\ldots,A_{n+1}) \circ W_\omega(A_1,\ldots,A_{n+1})\Big) + \mathfrak{S}_{n+1}\Big(t_{\Lambda,n}(A_1,\ldots,A_{n+1}) \circ (\mathrm{Id} - W_\omega(A_1,\ldots,A_{n+1}))\Big)\,,$$

we get

$$\Big(\frac{\hbar}{i}\Big)^n z_\Lambda^{(n+1)}(A_1,\ldots,A_k) = -\mathfrak{S}_{n+1}\Big(t_{\Lambda,n}(A_1,\ldots,A_{n+1}) \circ \big(\mathrm{Id} - W_\omega(A_1,\ldots,A_{n+1})\big)\Big)\,. \tag{3.9.33}$$

Taking into account (3.9.28), we explicitly see that (i) is indeed satisfied.

Summing up, the soconstructed $Z_{\Lambda,n+1}$ fulfills the following defining conditions on an element of $\mathcal{R}_0$: The Analyticity (1), the Lowest orders property (2) and the Field independence (4) (all three required in Definition 3.6.1) and, in addition, the Locality and Translation covariance (3.6.30) and the Scaling degree property (3.6.32).

Also the remaining defining conditions of $\mathcal{R}_0$ – Field parity and $\hbar$-dependence – hold true for $Z_{\Lambda,n+1}$. Namely, since $Z_{\Lambda,n}$ has these properties by induction and S_Λ satisfies the corresponding axioms, we conclude that

$$t_{\Lambda,n}(A_1,\ldots,A_{n+1}) = 0 \quad \text{if} \quad \sum_{j=1}^{n+1} |A_j| \quad \text{is odd, and}$$

$$t_{\Lambda,n}(A_1,\ldots,A_{n+1}) \sim \hbar^{\sum_{j=1}^{n+1}|A_j|/2}$$

for all monomials $A_1,\ldots,A_{n+1} \sim \hbar^0$. From this and (3.9.33) we immediately see that $z_\Lambda^{(n+1)}$ fulfills the properties Field parity (3.6.8) and $\hbar$-dependence (3.6.6).

Verification of (ii): We will prove that

$$\lim_{\Lambda\to\infty} t_{\Lambda,n}(A_1,\ldots,A_{n+1}) \circ W_\omega(A_1,\ldots,A_{n+1}) = t^0(A_1,\ldots,A_{n+1}) \circ W_\omega(A_1,\ldots,A_{n+1})$$

in the Hörmander topology. Because for both $t_{\Lambda,n}(A_1,\ldots,A_{n+1})$ and $t^0(A_1,\ldots$ $\ldots,A_{n+1})$ the extension from $\mathcal{D}'(\mathbb{R}^{dn}\setminus\{0\})$ to $\mathcal{D}'_\omega(\mathbb{R}^{dn})$ is uniquely given by the direct extension (see (3.2.26)), it suffices to show that

$$\lim_{\Lambda\to\infty} t_{\Lambda,n}(A_1,\ldots,A_{n+1})(\mathbf{y}) = t^0(A_1,\ldots,A_{n+1})(\mathbf{y}) \quad \text{on} \quad \mathcal{D}(\mathbb{R}^{dn}\setminus\{0\}) \tag{3.9.34}$$

in the Hörmander topology. For this purpose we note that the open cover $\cup_I C_I = \mathbb{M}^{n+1}\setminus\Delta_{n+1}$ (3.3.16) can also be written as an open cover in the relative coordinates, that is, a cover of $\mathbb{R}^{dn} \setminus \{0\}$:

$$\bigcup_{\substack{I\subset\{1,\dots,n+1\}\\ I\neq\emptyset,\, I^c\neq\emptyset}} C_I^0 = \mathbb{R}^{dn} \setminus \{0\} \quad \text{where}$$

$$C_I^0 := \{\mathbf{y} \in \mathbb{R}^{dn} \,|\, y_i - y_j \notin \overline{V}_- \;\; \forall i \in I,\, j \in I^c\} \quad \text{with} \quad y_{n+1} := 0 \,.$$

So we have to verify (3.9.34) on each $\mathcal{D}(C_I^0)$. To simplify the notations, we do this for the case $I = \{1,\dots,k\}$ with $1 \le k \le n$.

To compute $t^0(A_1,\dots,A_{n+1})|_{\mathcal{D}(C_I^0)}$, we take into account that $T^0_{n+1}|_{\mathcal{D}(C_I)}$ is given by (3.3.17). Inserting the causal Wick expansion (3.1.24) for $T_{|I|}$ and $T_{|I^c|}$ into that formula and computing the star product, we obtain the following formula on $\mathcal{D}(C_I^0)$:

$$t^0(A_1,\dots,A_{n+1})(\mathbf{y}) = \sum n^R_{(i_r),(j_r),(a_r)} \hbar^R\, t(\underline{A}_1,\dots,\underline{A}_k)(y_1 - y_k,\dots)$$
$$\cdot \Big(\prod_{r=1}^{R} \partial^{a_r} \Delta^F(y_{i_r} - y_{j_r})\Big)\, t(\underline{A}_{k+1},\dots,\underline{A}_{n+1})(y_{k+1},\dots)\,, \tag{3.9.35}$$

where $i_r \in I$, $j_r \in I^c$ for all r and $n^R_{(i_r),(j_r),(a_r)}$ is a combinatorial factor. Here, we have used that $\Delta^+(y) = \Delta^F(y)$ for $y \notin \overline{V}_-$ (2.3.9).

To get an analogous formula for $t_{\Lambda,n}(A_1,\dots,A_{n+1})$ on $\mathcal{D}(C_I^0)$, we use the formulation (3.6.33) of locality of $Z_{\Lambda,n}$ and the factorization (3.9.6) of $\mathbf{S}_\Lambda$ to obtain

$$\begin{aligned}(\mathbf{S}_\Lambda \circ Z_{\Lambda,n})(F+G) &= \mathbf{S}_\Lambda\big(Z_{\Lambda,n}(F) + Z_{\Lambda,n}(G)\big)\\ &= (\mathbf{S}_\Lambda \circ Z_{\Lambda,n})(F) \star_\Lambda (\mathbf{S}_\Lambda \circ Z_{\Lambda,n})(G) \quad \text{if} \quad \operatorname{supp} F \cap \operatorname{supp} G = \emptyset\,.\end{aligned}$$

Proceeding similarly to the second part of the solution of Exer. 3.3.10, we conclude that

$$(\mathbf{S}_\Lambda\circ Z_{\Lambda,n})^{(n+1)}\big(\otimes_{i=1}^{n+1} A_i(x_i)\big) =$$
$$(\mathbf{S}_\Lambda \circ Z_{\Lambda,n})^{(k)}\big(\otimes_{i=1}^{k} A_i(x_i)\big) \star_\Lambda (\mathbf{S}_\Lambda \circ Z_{\Lambda,n})^{(n+1-k)}\big(\otimes_{j=k+1}^{n+1} A_j(x_j)\big) \tag{3.9.36}$$

on $\mathcal{D}(C_I)$. Since $\mathbf{S}_\Lambda$ and $Z_{\Lambda,n}$ satisfy Field independence, this holds true also for $\mathbf{S}_\Lambda \circ Z_{\Lambda,n}$ and, hence, $(\mathbf{S}_\Lambda \circ Z_{\Lambda,n})^{(l)}$ fulfills the causal Wick expansion (3.1.24) for all $l \in \mathbb{N}^*$. Therefore, taking the VEV of (3.9.36) we obtain

$$t_{\Lambda,n}(A_1,\dots,A_{n+1})(\mathbf{y}) = \sum n^R_{(i_r),(j_r),(a_r)} \hbar^R\, t_{\Lambda,n}(\underline{A}_1,\dots,\underline{A}_k)(y_1 - y_k,\dots)$$
$$\cdot \Big(\prod_{r=1}^{R} \partial^{a_r} p_\Lambda(y_{i_r} - y_{j_r})\Big)\, t_{\Lambda,n}(\underline{A}_{k+1},\dots,\underline{A}_{n+1})(y_{k+1},\dots)\,, \tag{3.9.37}$$

on $\mathcal{D}(C_I^0)$, with the same notations as in (3.9.35). From the inductive assumption (3.9.29), we know that

$$\lim_{\Lambda\to\infty} t_{\Lambda,n}(A_1,\dots,A_k) = t(A_1,\dots,A_k) \quad \forall k \le n$$

and for all monomials $A_1, \ldots A_k \in \mathcal{P}$, with respect to the Hörmander topology. Using additionally that in this topology the multiplication of distributions and partial derivatives commute with taking limits (as mentioned directly below (3.9.3)), we obtain

$$\lim_{\Lambda\to\infty} t_{\Lambda,n}(A_1,\ldots,A_{n+1})(\mathbf{y}) = \sum n^R_{\ldots}\,\hbar^R \Big(\lim_{\Lambda\to\infty} t_{\Lambda,n}(\underline{A}_1,\ldots,\underline{A}_k)(y_1-y_k,\ldots)\Big)$$
$$\cdot\Big(\prod_{r=1}^R \partial^{a_r}\big(\lim_{\Lambda\to\infty} p_\Lambda(y_{i_r}-y_{j_r})\big)\Big)\Big(\lim_{\Lambda\to\infty} t_{\Lambda,n}(\underline{A}_{k+1},\ldots,\underline{A}_{n+1})(y_{k+1},\ldots)\Big)$$
$$= t^0(A_1,\ldots,A_{n+1})(\mathbf{y})$$

on $\mathcal{D}(C^0_I)$, where in the last step formula (3.9.35) is taken into account. □

Example 3.9.5 (UV-finite interactions – continued). For an UV-finite interaction L (introduced in Exap. 3.6.9) the limit

$$\lim_{\Lambda\to\infty} \mathbf{S}_\Lambda\big(L(g)\big) = \mathbf{S}\big(L(g)\big) \qquad g \in \mathcal{D}(\mathbb{M})\ , \tag{3.9.38}$$

exists without adding any counter terms, it can be performed in a naive way by replacing p_Λ by Δ^F_m. This statement follows directly from the definition of an UV-finite interaction. Alternatively, it is also contained in the theorem given here: The reasons yielding the relation $Z\big(L(g)\big) = L(g)$ (3.6.37) hold not only for all $Z \in \mathcal{R}$, they hold even for all Z lying in the larger version $\mathcal{R}_0$ of the Stückelberg–Petermann group appearing here; hence, formula (3.9.24) reduces to (3.9.38) when applied to an UV-finite $L(g)$.

As an application of Theorem 3.9.4 we are going to explain, how one can construct the Gell-Mann–Low cocycle $Z^m_\rho \in \mathcal{R}$ (3.6.44) from the counter terms $Z^m_\Lambda \in \mathcal{R}_0$ [27, Sect. 5.2]. Proceeding analogously to Exer. 3.1.21(a), we derive from the scaling behaviour of $p_{m,\Lambda}$ (3.9.23) the relation

$$\sigma_\rho(F \star_{\rho^{-1}m,\Lambda} G) = (\sigma_\rho F) \star_{m,\rho\Lambda} (\sigma_\rho G) \quad \text{where} \quad \star_{m,\Lambda} := \star_{p_{m,\Lambda}}\ ,$$

which implies

$$\sigma_\rho \circ \mathbf{S}^{m/\rho}_\Lambda \circ \sigma_\rho^{-1} = \mathbf{S}^m_{\rho\Lambda}\ . \tag{3.9.39}$$

Using that a scaling transformation commutes with the limit $\Lambda \to \infty$ in (3.9.24), we get

$$\mathbf{S}^m \circ Z^m_\rho = \sigma_\rho \circ \mathbf{S}^{m/\rho} \circ \sigma_\rho^{-1} = \lim_{\Lambda\to\infty} \sigma_\rho \circ \mathbf{S}^{m/\rho}_\Lambda \circ \sigma_\rho^{-1} \circ \sigma_\rho \circ Z^{m/\rho}_\Lambda \circ \sigma_\rho^{-1}$$
$$= \lim_{\Lambda_1=\Lambda\to\infty} \mathbf{S}^m_{\rho\Lambda_1} \circ Z^m_{\rho\Lambda_1} \circ (Z^m_{\rho\Lambda})^{-1} \circ \sigma_\rho \circ Z^{m/\rho}_\Lambda \circ \sigma_\rho^{-1}\ .$$

Assuming that the convergence behaviour is sufficiently well, we replace the "diagonal limit" $\lim_{\Lambda_1=\Lambda\to\infty}\cdots$ by $\lim_{\Lambda\to\infty}\lim_{\Lambda_1\to\infty}\cdots$. Taking into account additionally that $\lim_{\Lambda_1\to\infty} \mathbf{S}^m_{\rho\Lambda_1} \circ Z^m_{\rho\Lambda_1} = \mathbf{S}^m$ and that $\mathbf{S}^m$ is injective, we conclude that

$$Z^m_\rho = \lim_{\Lambda\to\infty} (Z^m_{\rho\Lambda})^{-1} \circ \sigma_\rho \circ Z^{m/\rho}_\Lambda \circ \sigma_\rho^{-1}\ . \tag{3.9.40}$$

In this way the Gell-Mann–Low cocycle Z^m_ρ can be obtained from the counter terms Z^m_Λ.

3.9.3 Effective potential and flow equation

Comments on the definition of the effective potential. To explain more in detail the definition of the effective potential V_Λ (Definition 3.9.1), we recall that $\mathbf{S}_\Lambda$ is explicitly known and invertible and that $\mathbf{S}$ exists (although it is usually unknown). With that the effective potential V_Λ at scale Λ is well defined by (3.9.7). We also recall that in general $V_\Lambda \notin \mathcal{F}_{\mathrm{loc}}$. Similarly to $\mathbf{S}(V)$, V_Λ can be viewed as a formal power series in $\hbar$, or in κ (since $V \sim \kappa$), or in both. For the lowest terms of the expansion in κ we obtain

$$V_\Lambda(V) := \mathbf{S}_\Lambda^{-1} \circ \mathbf{S}\,(V) = V + \mathcal{O}(\kappa^2)\ , \tag{3.9.41}$$

by using the axiom (ii) Initial condition and (3.9.13).

The limit $\Lambda \downarrow 0$ of V_Λ is unproblematic, from (3.9.14) we see that

$$\lim_{\Lambda\downarrow 0} V_\Lambda = \tfrac{\hbar}{i}\, \log \circ \mathbf{S}(V) = \mathbf{S}_0^{-1} \circ \mathbf{S}\,(V) \equiv V_0(V) \equiv V_0 \tag{3.9.42}$$

in the pointwise sense (1.2.3). This can equivalently be written as

$$e^{iV_0(V)/\hbar} = \lim_{\Lambda\downarrow 0} e^{iV_\Lambda(V)/\hbar} = \mathbf{S}(V)\ . \tag{3.9.43}$$

That is, $e^{iV_\Lambda/\hbar}$ *is an approximation for the S-matrix* $\mathbf{S}(V)$ *for small* $\Lambda > 0$.

We point out that the construction of V_Λ requires renormalization, since V_Λ is defined in terms of the renormalized S-matrix. In the *functional integral approach* [112, 147], the corresponding renormalization is performed by introducing an UV-cutoff Λ_0. The so regularized effective potential is a 2-parametric family $\big(G_{\Lambda,\Lambda_0}(V)\big)_{\Lambda\leq\Lambda_0}$. The existence of the limit $\Lambda_0 \to \infty$ of $G_{\Lambda,\Lambda_0}(V)$ (which removes the UV-cutoff Λ_0) requires the addition of suitable, Λ_0-dependent, local counter terms to V. Roughly, the so renormalized, Λ-dependent family

$$\Big(\lim_{\Lambda_0\to\infty} \tfrac{i}{\hbar}\, G_{\Lambda,\Lambda_0}(V + \Lambda_0\text{-dependent, local counter terms})\Big)_{\Lambda>0}$$

corresponds to our effective potential V_Λ, for details see Sect. 3.9.5.

Remark 3.9.6. In general, the limit $\lim_{\Lambda\to\infty} V_\Lambda(V)$ does not exist; here and in the following also this limit is understood in the pointwise sense (1.2.3). On a heuristic level, convergence can be achieved by adding suitable, Λ-dependent, local counter terms. By means of Theorem 3.9.4 this can be described as follows: Let $Z_\Lambda \in \mathcal{R}_0$ be an addition of local counter terms yielding that S-matrix $\mathbf{S}$ which is used in the definition of $V_\Lambda(V)$, that is, $\lim_{\Lambda\to\infty} \mathbf{S}_\Lambda \circ Z_\Lambda = \mathbf{S}$ in the pointwise sense (3.9.27). With that we get

$$\begin{aligned}\lim_{\Lambda\to\infty} V_\Lambda\big(Z_\Lambda^{-1}(V)\big) &= \lim_{\Lambda\to\infty} \mathbf{S}_\Lambda^{-1} \circ \mathbf{S} \circ Z_\Lambda^{-1}(V)\\ &= \lim_{\Lambda\to\infty}\lim_{\Lambda_1\to\infty} \mathbf{S}_\Lambda^{-1} \circ \mathbf{S}_{\Lambda_1} \circ Z_{\Lambda_1} \circ Z_\Lambda^{-1}(V)\ .\end{aligned} \tag{3.9.44}$$

Proceeding heuristically, we replace $\lim_{\Lambda\to\infty}\lim_{\Lambda_1\to\infty}\cdots$ by the "diagonal limit" $\lim_{\Lambda=\Lambda_1\to\infty}\cdots$; this yields

$$\lim_{\Lambda\to\infty} V_\Lambda\big(Z_\Lambda^{-1}(V)\big) \overset{?}{=} V \in \mathcal{F}_{\mathrm{loc}} \tag{3.9.45}$$

in the pointwise sense (1.2.3). We point out: If this relation holds true, the non-local terms appearing in $V_\Lambda\big(Z_\Lambda^{-1}(V)\big)$ cancel in the limit $\Lambda\to\infty$.

To second order in κ, the relation (3.9.45) holds true for all interactions $V\sim\kappa$, as we show below by explicit computation of $(V_\Lambda\circ Z_\Lambda^{-1})^{(2)}(V^{\otimes 2})$. But, to third order in κ the limit $\lim_{\Lambda\to\infty}(V_\Lambda\circ Z_\Lambda^{-1})^{(3)}(V^{\otimes 3})$ does in general not exist, see Exer. 3.9.8(b).

Flow equation. Adapting Wilson's concept of renormalization group and effective Lagrangian to *renormalization of a perturbative field theory*, Polchinski developed the flow equation [137]. It is the basis of a powerful method for statistical mechanics and QFT on Euclidean space, which is very useful for conceptual issues, e.g. an investigation of "perturbative renormalizability" of a given interaction;[119] for an introductory overview see [147]. The method was extended to perturbative QFT on Minkowski spacetime by Keller, Kopper and Schophaus [112]. The flow equation (also called "Renormalization Group equation") is a differential equation for the effective potential as a function of the UV-cutoff Λ.

In our framework, the flow equation is obtained by computing the derivative $\frac{d}{d\Lambda}$ of the effective potential V_Λ by using Definition 3.9.1; comparing with the traditional literature (see, e.g., [147]) it corresponds to the "flow equation in Wick ordered form".

Theorem 3.9.7 (Flow equation, [27, 61]). *The effective potential V_Λ solves the differential equation*

$$\frac{d}{d\Lambda}V_\Lambda = -\frac{i}{2}\int dx\,dy\,\frac{d\,p_\Lambda(x-y)}{d\Lambda}\,\frac{\delta V_\Lambda}{\delta\varphi(x)}\star_\Lambda\frac{\delta V_\Lambda}{\delta\varphi(y)} \tag{3.9.46}$$

$$= -\frac{i}{2\hbar}\,\frac{d}{d\lambda}\Big|_{\lambda=\Lambda}(V_\Lambda\star_\lambda V_\Lambda)\,. \tag{3.9.47}$$

Proof. [120] From $\mathbf{S}_\Lambda(V)=e_{\star_\Lambda}^{iV/\hbar}$ we see that

$$\frac{d}{d\lambda}\mathbf{S}_\Lambda(V_\lambda)=\frac{i}{\hbar}\,\frac{dV_\lambda}{d\lambda}\star_\Lambda\mathbf{S}_\Lambda(V_\lambda)\,.$$

With that, Definition 3.9.1 yields

$$0=\frac{d}{d\Lambda}\mathbf{S}_\Lambda(V_\Lambda)=\frac{d}{d\lambda}\Big|_{\lambda=\Lambda}\mathbf{S}_\lambda(V_\Lambda)+\frac{i}{\hbar}\,\frac{dV_\Lambda}{d\Lambda}\star_\Lambda\mathbf{S}_\Lambda(V_\Lambda)\,. \tag{3.9.48}$$

[119]The notion "perturbative renormalizability" is explained directly below (3.9.74).
[120]The proofs given in [27] and [61] differ somewhat, we follow the latter version.

Due to $\mathbf{S}_\Lambda(F) = 1+\mathcal{O}(F)$ the inverse regularized S-matrix $\mathbf{S}_\Lambda(F)^{\star_\Lambda -1}$ exists – here we mean the inverse with respect to $\star_\Lambda$. Hence we may write formula (3.9.48) as

$$\frac{dV_\Lambda}{d\Lambda} = -\frac{\hbar}{i}\,\frac{d}{d\lambda}\Big|_{\lambda=\Lambda} \mathbf{S}_\lambda(V_\Lambda)\star_\Lambda \mathbf{S}_\Lambda(V_\Lambda)^{\star_\Lambda -1}\,. \tag{3.9.49}$$

Using formula (2.1.7) for $\star_\Lambda$, that is,

$$F\star_\Lambda G = \mathcal{M}\circ e^{\hbar D_\Lambda}(F\otimes G) \quad \text{where} \quad D_\Lambda := \int dx\,dy\;p_\Lambda(x-y)\,\frac{\delta}{\delta\varphi(x)}\otimes\frac{\delta}{\delta\varphi(y)}\,,$$

we obtain

$$\begin{aligned}\frac{d}{d\Lambda}\frac{F\star_\Lambda F}{2} &= \mathcal{M}\circ e^{\hbar D_\Lambda}\Big(\frac{\hbar}{2}\int dx\,dy\;\frac{d\,p_\Lambda(x-y)}{d\Lambda}\Big(\frac{\delta F}{\delta\varphi(x)}\otimes\frac{\delta F}{\delta\varphi(y)}\Big)\Big)\\ &= \frac{\hbar}{2}\int dx\,dy\;\frac{d\,p_\Lambda(x-y)}{d\Lambda}\,\frac{\delta F}{\delta\varphi(x)}\star_\Lambda\frac{\delta F}{\delta\varphi(y)}\,,\end{aligned} \tag{3.9.50}$$

and for n factors

$$\frac{d}{d\Lambda}\frac{T_{\Lambda,n}(F^{\otimes n})}{n!} = \frac{\hbar}{2\,(n-2)!}\int dx\,dy\,\frac{d\,p_\Lambda(x-y)}{d\Lambda}\,\frac{\delta F}{\delta\varphi(x)}\star_\Lambda\frac{\delta F}{\delta\varphi(y)}\star_\Lambda F\star_\Lambda\cdots\star_\Lambda F \tag{3.9.51}$$

(2 factors $\frac{\delta F}{\delta\varphi}$ and $(n-2)$ factors F). Multiplying this equation with $n!$, the combinatorial factor appearing on the r.h.s. is $\binom{n}{2}$, which is the number of possibilities to choose the two F's which are derivated from the set of n F's.[121] Summing over n we get

$$\frac{d}{d\Lambda}\mathbf{S}_\Lambda(F) = -\frac{1}{2\hbar}\int dx\,dy\,\frac{d\,p_\Lambda(x-y)}{d\Lambda}\,\frac{\delta F}{\delta\varphi(x)}\star_\Lambda\frac{\delta F}{\delta\varphi(y)}\star_\Lambda\mathbf{S}_\Lambda(F)\,. \tag{3.9.53}$$

Inserting this into (3.9.49) it results (3.9.46), from which we obtain (3.9.47) by using (3.9.50). □

Construction of V_Λ by solving the flow equation inductively. Usually $\mathbf{S}$ is unknown, only $V = \kappa L(g)$ (with $L\in\mathcal{P}_{\mathrm{bal}}$, $g\in\mathcal{D}(\mathbb{M})$) and $(p_\Lambda)_{\Lambda>0}$ are given, and from that V_Λ is computed by solving the flow equation. In perturbation theory this amounts to an inductive construction of V_Λ as a formal power series in κ:

$$V_\Lambda(V) = V + \sum_{n=2}^{\infty}\frac{1}{n!}\,V_\Lambda^{(n)}(V^{\otimes n})\,,\quad \text{where}\quad V_\Lambda^{(n)}(V^{\otimes n}) := V_\Lambda^{(n)}(0)(V^{\otimes n}) \sim \kappa^n$$

---

[121] For $n=3$ formula (3.9.51) can be obtained from (2.4.5):

$$\begin{aligned}\frac{d}{d\Lambda}\,F\star_\Lambda F\star_\Lambda F &= \mathcal{M}\circ(\mathrm{id}\otimes\mathcal{M})\circ e^{\hbar(D_\Lambda^{12}+D_\Lambda^{13}+D_\Lambda^{23})}\circ\hbar\Big(\frac{d\,D_\Lambda^{12}}{d\Lambda}+\frac{d\,D_\Lambda^{13}}{d\Lambda}+\frac{d\,D_\Lambda^{23}}{d\Lambda}\Big)(F\otimes F\otimes F)\\ &= 3\hbar\int dx\,dy\,\frac{d\,p_\Lambda(x-y)}{d\Lambda}\,\frac{\delta F}{\delta\varphi(x)}\star_\Lambda\frac{\delta F}{\delta\varphi(y)}\star_\Lambda F\,.\end{aligned} \tag{3.9.52}$$

and the values $V_\Lambda^{(0)} = 0$ and $V_\Lambda^{(1)}(V) = V$ (known from (3.9.41)) are taken into account. With that the perturbative version of the flow equation reads

$$\frac{1}{n!}\frac{d}{d\Lambda}V_\Lambda^{(n)} = -\sum_{k=1}^{n-1}\frac{i}{2\hbar\, k!\,(n-k)!}\,\frac{d}{d\lambda}\Big|_{\lambda=\Lambda}(V_\Lambda^{(k)} \star_\lambda V_\Lambda^{(n-k)})\,. \tag{3.9.54}$$

Proceeding inductively, we start with $V_\Lambda^{(1)} = V$ and assume that $V_\Lambda^{(k)}$ is known for all $k < n$. Then, the r.h.s. is known and, hence, an integration yields $V_\Lambda^{(n)}$. A major problem is the determination of the integration constants by suitable boundary values. Usually the value (3.9.42) at $\Lambda = 0$ does not help, because it contains the unknown $\mathbf{S}$. Comparing with the definition of $V_\Lambda\big(\kappa L(g)\big)$ (Definition 3.9.1), the *freedom in the choice of boundary values corresponds precisely to the non-uniqueness of* $\mathbf{S}\big(\kappa L(g)\big)$ *coming from the extension of* $T_n\big(L(x_1)\otimes\cdots\otimes L(x_n)\big)$ *to coinciding points.*

To *second order* the flow equation (3.9.54) can easily be solved, because it has the simple form

$$\frac{d}{d\Lambda}V_\Lambda^{(2)}(V^{\otimes 2}) = -\frac{i}{\hbar}\,\frac{d}{d\Lambda}\;V \star_\Lambda V\,.$$

Integration yields

$$\begin{aligned} V_\Lambda^{(2)}(V^{\otimes 2}) &= V_{\Lambda_0}^{(2)}(V^{\otimes 2}) + \int_{\Lambda_0}^{\Lambda} d\lambda\;\frac{d}{d\lambda}V_\lambda^{(2)}(V^{\otimes 2}) \\ &= V_{\Lambda_0}^{(2)}(V^{\otimes 2}) + \frac{i}{\hbar}\Big(V \star_{\Lambda_0} V - V \star_\Lambda V\Big)\,. \end{aligned} \tag{3.9.55}$$

Assuming that we know $\mathbf{S}^{(2)}(V^{\otimes 2}) = \frac{i^2}{\hbar^2}\,T_2(V^{\otimes 2})$ – an assumption which is usually not satisfied in practice – we can determine the integration constant $V_{\Lambda_0}^{(2)}(V^{\otimes 2})$ by choosing $\Lambda_0 = 0$: Writing

$$\begin{aligned} V_0^{(2)}(V^{\otimes 2}) &= \frac{\hbar}{i}\,(\log\circ\mathbf{S})^{(2)}(V^{\otimes 2}) = \frac{\hbar}{i}\,\Big(\log\big(\mathrm{Id} + \mathcal{T}(V)\big)\Big)^{(2)} \\ &= \frac{\hbar}{i}\,\Big(\mathcal{T}(V) - \frac{\mathcal{T}(V)\cdot\mathcal{T}(V)}{2} + \cdots\Big)^{(2)}\,, \end{aligned}$$

where

$$\mathcal{T}(V) := \frac{i}{\hbar}\,V + \frac{i^2}{2!\,\hbar^2}\,T_2(V^{\otimes 2}) + \cdots\,,$$

we get

$$V_0^{(2)}(V^{\otimes 2}) = \frac{i}{\hbar}\,\Big(T_2(V^{\otimes 2}) - V\cdot V\Big)\,. \tag{3.9.56}$$

Inserting this value into (3.9.55) and using $V \star_{\Lambda_0} V|_{\Lambda_0=0} = V\cdot V$, we end up with

$$V_\Lambda^{(2)}(V^{\otimes 2}) = \frac{i}{\hbar}\Big(T_2(V^{\otimes 2}) - V \star_\Lambda V\Big)\,. \tag{3.9.57}$$

With that, the relation (3.9.45) describing the behaviour of V_Λ for $\Lambda \to \infty$, can be explicitly verified to second order in κ, that is,

$$0 \overset{?}{=} \lim_{\Lambda\to\infty} (V_\Lambda \circ Z_\Lambda^{-1})^{(2)}(V^{\otimes 2}) = \lim_{\Lambda\to\infty} \left(V_\Lambda^{(2)}(V^{\otimes 2}) - Z_\Lambda^{(2)}(V^{\otimes 2})\right) . \tag{3.9.58}$$

Here we use that $Z_\Lambda(V) = V + \frac{1}{2!} Z_\Lambda^{(2)}(V^{\otimes 2}) + \mathcal{O}(\kappa^3)$ implies $Z_\Lambda^{-1}(V) = V - \frac{1}{2!} Z_\Lambda^{(2)}(V^{\otimes 2}) + \mathcal{O}(\kappa^3)$. In addition, $Z_\Lambda^{(2)}(V^{\otimes 2})$ is determined by $\mathbf{S}^{(2)} = \lim_{\Lambda\to\infty}(\mathbf{S}_\Lambda \circ Z_\Lambda)^{(2)}$, that is,

$$\frac{i^2}{\hbar^2} T_2(V^{\otimes 2}) = \frac{i^2}{\hbar^2} V \star_\Lambda V + \frac{i}{\hbar} Z_\Lambda^{(2)}(V^{\otimes 2}) + \mathfrak{o}(\Lambda) , \tag{3.9.59}$$

where $\mathfrak{o}(\Lambda)$ stands for terms vanishing for $\Lambda \to \infty$. Inserting the so-obtained expression for $Z_\Lambda^{(2)}(V^{\otimes 2})$ and the formula for $V_\Lambda^{(2)}(V^{\otimes 2})$ (3.9.57) into (3.9.58), we see that the latter relation holds indeed true.

Exercise 3.9.8.

(a) Compute $V_\Lambda^{(3)}(V^{\otimes 3})$ by solving the flow equation inductively, that is, the second-order result (3.9.57) may be used. Determine the integration constant by the boundary value at $\Lambda = 0$, assuming that $T_2(V^{\otimes 2})$ and $T_3(V^{\otimes 3})$ are known.

(b) Investigate the limit $\lim_{\Lambda\to\infty}(V_\Lambda \circ Z_\Lambda^{-1})^{(3)}(V^{\otimes 3})$ by using the result of part (a).

[*Solution*: (a) Since $p_\Lambda(-x) = p_\Lambda(x)$ the product $\star_\Lambda$ is commutative. With that the flow equation to third order, i.e., (3.9.54) with $n = 3$, can be written as

$$\begin{aligned} \frac{d}{d\Lambda} V_\Lambda^{(3)}(V^{\otimes 3}) &= -\frac{3i}{\hbar} \frac{d}{d\lambda}\Big|_{\lambda=\Lambda} V \star_\lambda V_\Lambda^{(2)}(V^{\otimes 2}) \\ &= \frac{3}{\hbar^2}\left(\frac{d}{d\Lambda} V \star_\Lambda T_2(V^{\otimes 2}) - \frac{d}{d\lambda}\Big|_{\lambda=\Lambda} V \star_\lambda (V \star_\Lambda V)\right) , \end{aligned}$$

where (3.9.57) is inserted. Now, we use the formula

$$\frac{d}{d\lambda}\Big|_{\lambda=\Lambda} V \star_\lambda (V \star_\Lambda V) = \frac{1}{3}\frac{d}{d\Lambda} V \star_\Lambda V \star_\Lambda V ,$$

which follows from (3.9.52). With that the flow equation can easily be integrated:

$$\begin{aligned} V_\Lambda^{(3)}(V^{\otimes 3}) = V_{\Lambda_0}^{(3)}(V^{\otimes 3}) + \frac{1}{\hbar^2}\Big(&- 3\,V \star_{\Lambda_0} T_2(V^{\otimes 2}) + V \star_{\Lambda_0} V \star_{\Lambda_0} V \\ &+ 3\,V \star_\Lambda T_2(V^{\otimes 2}) - V \star_\Lambda V \star_\Lambda V\Big) . \end{aligned}$$

To determine the integration constant $V_{\Lambda_0}^{(3)}(V^{\otimes 3})$ we choose $\Lambda_0 = 0$ and insert the value

$$V_0^{(3)}(V^{\otimes 3}) = \frac{\hbar}{i}(\log \circ \mathbf{S})^{(3)}(V^{\otimes 3}) = \frac{-1}{\hbar^2}\Big(T_3(V^{\otimes 3}) - 3\,V \cdot T_2(V^{\otimes 2}) + 2\,V \cdot V \cdot V\Big) ,$$

which is obtained by expanding similarly to the derivation of (3.9.56). Using also $F \star_{\Lambda_0} G|_{\Lambda_0=0} = F \cdot G$, we get the final result

$$V_\Lambda^{(3)}(V^{\otimes 3}) = \frac{1}{\hbar^2}\Big(-T_3(V^{\otimes 3}) - V \cdot V \cdot V + 3\,V \star_\Lambda T_2(V^{\otimes 2}) - V \star_\Lambda V \star_\Lambda V\Big) . \tag{3.9.60}$$

(b) Applying the chain rule (Faà di Bruno formula, cf. (3.6.25)) to $(V_\Lambda \circ Z_\Lambda^{-1})^{(3)}(V^{\otimes 3})$ we obtain

$$(V_\Lambda \circ Z_\Lambda^{-1})^{(3)}(V^{\otimes 3}) = V_\Lambda^{(3)}(V^{\otimes 3}) + 3\, V_\Lambda^{(2)}\Big((Z_\Lambda^{-1})^{(2)}(V^{\otimes 2}) \otimes V\Big) + (Z_\Lambda^{-1})^{(3)}(V^{\otimes 3}) \,. \quad (3.9.61)$$

Below we derive that

$$(Z_\Lambda^{-1})^{(3)}(V^{\otimes 3}) = \frac{1}{\hbar^2}\Big(T_3(V^{\otimes 3}) - V \star_\Lambda V \star_\Lambda V\Big) - \frac{3i}{\hbar}\, T_2\Big((Z_\Lambda^{-1})^{(2)}(V^{\otimes 2}) \otimes V\Big) + \mathfrak{o}(\Lambda)\,. \quad (3.9.62)$$

Inserting (3.9.57), (3.9.60) and (3.9.62) into (3.9.61) some terms cancel out and it remains

$$(V_\Lambda \circ Z_\Lambda^{-1})^{(3)}(V^{\otimes 3}) = \frac{1}{\hbar^2}\Big(-V \cdot V \cdot V + 3\, V \star_\Lambda T_2(V^{\otimes 2}) - 2\, V \star_\Lambda V \star_\Lambda V\Big) - \frac{3i}{\hbar}\,(Z_\Lambda^{-1})^{(2)}(V^{\otimes 2}) \star_\Lambda V + \mathfrak{o}(\Lambda)\,.$$

Taking into account that we know from (3.9.59) that

$$(Z_\Lambda^{-1})^{(2)}(V^{\otimes 2}) = -Z_\Lambda^{(2)}(V^{\otimes 2}) = \frac{i}{\hbar}\Big(-T_2(V^{\otimes 2}) + V \star_\Lambda V\Big) + \mathfrak{o}(\Lambda) \in \mathcal{F}_{\mathrm{loc}}\,, \quad (3.9.63)$$

a further pair of terms cancels out and we end up with

$$(V_\Lambda \circ Z_\Lambda^{-1})^{(3)}(V^{\otimes 3}) = \frac{1}{\hbar^2}\Big(-V \cdot V \cdot V + V \star_\Lambda V \star_\Lambda V\Big) + \mathfrak{o}(\Lambda)\,.$$

In general $V \star_\Lambda V \star_\Lambda V$ diverges for $\Lambda \to \infty$, hence, this holds true also for $(V_\Lambda \circ Z_\Lambda^{-1})^{(3)}(V^{\otimes 3})$.

It remains to derive (3.9.62):[122] applying the chain rule to $0 = (Z_\Lambda \circ Z_\Lambda^{-1})^{(3)}(V^{\otimes 3})$ and $\mathbf{S}^{(3)} = (\mathbf{S}_\Lambda \circ Z_\Lambda)^{(3)} + \mathfrak{o}(\Lambda)$ we get

$$(Z_\Lambda^{-1})^{(3)}(V^{\otimes 3}) = -Z_\Lambda^{(3)}(V^{\otimes 3}) - 3\, Z_\Lambda^{(2)}\Big((Z_\Lambda^{-1})^{(2)}(V^{\otimes 2}) \otimes V\Big) \quad (3.9.64)$$

and

$$\frac{i^3}{\hbar^3}\, T_3(V^{\otimes 3}) = \frac{i^3}{\hbar^3}\, V \star_\Lambda V \star_\Lambda V + \frac{3i^2}{\hbar^2}\, Z_\Lambda^{(2)}(V^{\otimes 2}) \star_\Lambda V + \frac{i}{\hbar}\, Z_\Lambda^{(3)}(V^{\otimes 3}) + \mathfrak{o}(\Lambda)\,,$$

respectively. Taking into account (3.9.63), the latter equation yields

$$Z_\Lambda^{(3)}(V^{\otimes 3}) = \frac{1}{\hbar^2}\Big(-T_3(V^{\otimes 3}) - 2\, V \star_\Lambda V \star_\Lambda V + 3\, T_2(V^{\otimes 2}) \star_\Lambda V\Big) + \mathfrak{o}(\Lambda)\,. \quad (3.9.65)$$

Writing the second term on the r.h.s. of (3.9.64) as

$$\begin{aligned} Z_\Lambda^{(2)}&\Big((Z_\Lambda^{-1})^{(2)}(V^{\otimes 2}) \otimes V\Big) \\ &= \frac{i}{\hbar}\Big(T_2\Big((Z_\Lambda^{-1})^{(2)}(V^{\otimes 2}) \otimes V\Big) - (Z_\Lambda^{-1})^{(2)}(V^{\otimes 2}) \star_\Lambda V\Big) + \mathfrak{o}(\Lambda) \\ &= \frac{i}{\hbar}\, T_2\Big((Z_\Lambda^{-1})^{(2)}(V^{\otimes 2}) \otimes V\Big) - \frac{1}{\hbar^2}\Big(T_2(V^{\otimes 2}) - V \star_\Lambda V\Big) \star_\Lambda V + \mathfrak{o}(\Lambda) \end{aligned}$$

by using twice (3.9.63), and inserting this equation and (3.9.65) into (3.9.64), we indeed obtain (3.9.62).]

[122] On a heuristic level, the relation (3.9.62) can be obtained much faster: From $\mathbf{S} = \mathbf{S}_\Lambda \circ Z_\Lambda + \mathfrak{o}(\Lambda)$ we get $\mathbf{S}_\Lambda^{(3)} = (\mathbf{S} \circ Z_\Lambda^{-1})^{(3)} + (\mathfrak{o}(\Lambda) \circ Z_\Lambda^{-1})^{(3)}$. Neglecting the second term on the r.h.s., the chain rule yields (3.9.62). But, it is unclear whether $\mathfrak{o}(\Lambda) \circ Z_\Lambda^{-1} \overset{?}{=} \mathfrak{o}(\Lambda)$, because in general Z_Λ^{-1} diverges for $\Lambda \to \infty$.

3.9.4 A version of Wilson's renormalization "group"

In this section we study more in detail the imitation $\mathcal{W}$ of Wilson's RG introduced in (3.9.9)–(3.9.10).

Connection between the RGs of Stückelberg–Petermann and Wilson. On a heuristic level we can establish a connection between the renormalization groups of Stückelberg–Petermann and Wilson.

Conjecture 3.9.9. *The restriction of the flow operators* $(W_{\Lambda,\Lambda_0})_{\Lambda,\Lambda_0>0}$ (3.9.9) *to* $\mathcal{F}_{\rm loc}$ *can be approximated by the* 2*-parametric subfamily* $(Z_\Lambda\circ Z_{\Lambda_0}^{-1})_{\Lambda,\Lambda_0>0}$ *of the Stückelberg–Petermann renormalization group* $\mathcal{R}_0$ *for* Λ,Λ_0 *big enough, more precisely*

$$\lim_{\Lambda\to\infty}\Big(W_{\Lambda,\lambda\Lambda}\Big|_{\mathcal{F}_{\rm loc}}-Z_\Lambda\circ Z_{\lambda\Lambda}^{-1}\Big)=0\quad \textit{for all fixed } \lambda\in(0,\infty)$$

in the pointwise sense (see (3.9.27)*).*

In particular this shows that the Stückelberg–Petermann group $\mathcal{R}_0$ is much larger than the version $\mathcal{W}$ (3.9.10) of Wilson's RG.

Justification. Let $V_\Lambda(V)$ be defined in terms of a given S-matrix $\mathbf{S}$ by (3.9.7) and let $Z_\Lambda\in\mathcal{R}_0$ be an addition of local counter terms yielding this $\mathbf{S}$ according to Theorem 3.9.4. We know that

$$0=\lim_{\Lambda\to\infty}\Big(\mathbf{S}_{\lambda\Lambda}\Big|_{\mathcal{F}_{\rm loc}}\circ Z_{\lambda\Lambda}-\mathbf{S}_\Lambda\Big|_{\mathcal{F}_{\rm loc}}\circ Z_\Lambda\Big)\,.$$

We conjecture that this implies

$$\begin{aligned}0&=\lim_{\Lambda\to\infty}\mathbf{S}_\Lambda^{-1}\circ\Big(\mathbf{S}_{\lambda\Lambda}\Big|_{\mathcal{F}_{\rm loc}}\circ Z_{\lambda\Lambda}-\mathbf{S}_\Lambda\Big|_{\mathcal{F}_{\rm loc}}\circ Z_\Lambda\Big)\circ Z_{\lambda\Lambda}^{-1}\\&=\lim_{\Lambda\to\infty}\Big(W_{\Lambda,\lambda\Lambda}\Big|_{\mathcal{F}_{\rm loc}}-Z_\Lambda\circ Z_{\lambda\Lambda}^{-1}\Big)\,;\end{aligned}$$

however, the problem is that both $\mathbf{S}_\Lambda^{-1}$ and $Z_{\lambda\Lambda}^{-1}$ diverge for $\Lambda\to\infty$. So mathematically the conjecture is still on shaky ground, but physically it is just a version of what is generally believed. ⊟

Regularized S-matrix with UV- and IR-cutoff and its relation to the flow operator. In the functional integral approach one usually regularizes the (Feynman) propagator by cutting off both the large and the small values of the pertinent momentum, more precisely, of $\|k\|\equiv\sqrt{k_0^2+\vec{k}^2}$. Following [112, 147] this can be done as follows: The propagator

$$p_{\Lambda,\Lambda_0}:=p_{\Lambda_0}-p_\Lambda\quad(0<\Lambda\le\Lambda_0<\infty)\tag{3.9.66}$$

has an UV-cutoff Λ_0 and an IR-cutoff Λ; the double limit $\Lambda_0\to\infty$ and $\Lambda\to 0$ of p_{Λ,Λ_0} converges to Δ_m^F in the Hörmander topology.

Analogously to (3.9.5) we introduce the pertinent regularized S-matrix

$$\mathbf{S}_{\Lambda,\Lambda_0}: \begin{cases} \mathcal{F} \longrightarrow \mathcal{F} \\ \mathbf{S}_{\Lambda,\Lambda_0}(F) := e^{iF/\hbar}_{\star_{\Lambda,\Lambda_0}} \equiv 1 + \frac{i}{\hbar}\, F + \frac{i^2}{2!\,\hbar^2}\, F \star_{\Lambda,\Lambda_0} F + \cdots \,, \end{cases} \tag{3.9.67}$$

where $\star_{\Lambda,\Lambda_0} := \star_{p_{\Lambda,\Lambda_0}}$; similarly to (3.9.12) it satisfies

$$\mathbf{S}_{\Lambda,\Lambda_0} = \tau_{\Lambda,\Lambda_0} \circ \exp \circ \tfrac{i}{\hbar}\, \tau_{\Lambda,\Lambda_0}^{-1} \,, \qquad \tau_{\Lambda,\Lambda_0} := e^{\Gamma_{p_{\Lambda,\Lambda_0}}} \,. \tag{3.9.68}$$

To find a relation between $\mathbf{S}_{\Lambda,\Lambda_0}$ and the flow operator W_{Λ,Λ_0} we first study the special case $\Lambda = 0$. In this case we have $p_{0,\Lambda_0} = p_{\Lambda_0}$, hence $S_{0,\Lambda_0} = S_{\Lambda_0}$. Taking into account also $\mathbf{S}_0^{-1} = \frac{\hbar}{i}\log$ (3.9.14), we get

$$W_{0,\Lambda_0} \equiv \mathbf{S}_0^{-1} \circ \mathbf{S}_{\Lambda_0} = \tfrac{\hbar}{i}\, \log \circ \mathbf{S}_{0,\Lambda_0} \,. \tag{3.9.69}$$

For the flow from Λ_0 to an arbitrary $\Lambda \in [0,\infty)$ we assert:

$$W_{\Lambda,\Lambda_0} \equiv \mathbf{S}_\Lambda^{-1} \circ \mathbf{S}_{\Lambda_0} = \tfrac{\hbar}{i}\, \tau_\Lambda \circ \log \circ \mathbf{S}_{\Lambda,\Lambda_0} \circ \tau_\Lambda^{-1} \,, \tag{3.9.70}$$

which is a generalization of (3.9.69) due to $\tau_0 = e^{\Gamma_{p_0}} = \mathrm{Id}$ (see (2.3.2)). We point out that the maps $F \longmapsto W_{\Lambda,\Lambda_0}(F)$ and $F \longmapsto \frac{\hbar}{i}\log \circ \mathbf{S}_{\Lambda,\Lambda_0}(F)$ (where $F \in \mathcal{F}$) agree up to a similarity transformation by τ_Λ .

Exercise 3.9.10. Prove the assertion (3.9.70).

[*Solution*: We first note $\Gamma_{p_{\Lambda,\Lambda_0}} = \Gamma_{p_{\Lambda_0}} - \Gamma_{p_\Lambda}$ and, hence,

$$\tau_{\Lambda,\Lambda_0} = \exp(\Gamma_{p_{\Lambda_0}} - \Gamma_{p_\Lambda}) = \tau_\Lambda^{-1} \circ \tau_{\Lambda_0} = \tau_{\Lambda_0} \circ \tau_\Lambda^{-1} \,.$$

Due to (3.9.68) and (3.9.12) it holds that

$$\mathbf{S}_{\Lambda,\Lambda_0} = \tau_{\Lambda,\Lambda_0} \circ \exp \circ \tfrac{i}{\hbar}\, \tau_{\Lambda,\Lambda_0}^{-1} = \tau_\Lambda^{-1} \circ \mathbf{S}_{\Lambda_0} \circ \tau_\Lambda \,. \tag{3.9.71}$$

With that and with (3.9.13) we get

$$\mathbf{S}_\Lambda^{-1} \circ \mathbf{S}_{\Lambda_0} \circ \tau_\Lambda = \tfrac{\hbar}{i}\, \tau_\Lambda \circ \log \circ \tau_\Lambda^{-1} \circ \mathbf{S}_{\Lambda_0} \circ \tau_\Lambda = \tfrac{\hbar}{i}\, \tau_\Lambda \circ \log \circ \mathbf{S}_{\Lambda,\Lambda_0} \,,$$

which is equivalent to the assertion (3.9.70).]

3.9.5 Comparison with the functional integral approach

In this section we give a kind of translation of quantities introduced in our formalism into the functional integral approach. Our formalism corresponds to a functional integral in Minkowski spacetime, which is mathematically much more problematic than a Euclidean functional integral. To avoid these mathematical difficulties, we compare our formalism with a Euclidean functional integral. To introduce the latter, we study, for simplicity, the model of one real, massive, scalar field on the Euclidean space $\mathbb{R}^d$. We assume that the reader is familiar with the basics of the functional integral approach, as introduced, e.g., in [147]. To a far extent, this section is based on that reference, also concerning the notations.

We point out: *Similarly to the off-shell field formalism developed in this book, the functional integral approach does not use the Fock space; it works with field configurations being not restricted by any field equation.*

Let $\mathcal{C}_E$ be the configuration space of a Euclidean real scalar field $\varphi(x)$ on $\mathbb{R}^d$. More precisely, we require $\mathcal{C}_E \subset C^\infty(\mathbb{R}^d, \mathbb{R})$ and, in order that the action[123]

$$S(h) := \int d^dx \,\Big(\tfrac{1}{2}\,(\nabla h(x))^2 + \tfrac{m^2}{2}\,h(x)^2 + \kappa\, h(x)^j\Big)$$
$$= \int d^dx \,\Big(h(x)\,\tfrac{1}{2}(-\triangle + m^2)h(x) + \kappa\, h(x)^j\Big)\ , \quad j \in \mathbb{N}^*\ ,$$

is well defined for all $h \in \mathcal{C}_E$, we also require that the configurations $h \in \mathcal{C}_E$ decay sufficiently fast for $|x| \to \infty$. In contrast to causal perturbation theory, the interaction is not adiabatically switched off, i.e., κ is not multiplied with $g(x)$ (where $g \in \mathcal{D}(\mathbb{M})$). We assume $m > 0$. Similarly to (1.1.2), the basic field $\varphi(x)$ is the evaluation functional on $\mathcal{C}_E$ and the definitions (1.1.3) and (1.1.4) are used also in the framework at hand.

In the functional integral approach, the *effective action* $G_{\Lambda,\Lambda_0}(h)$, $h \in \mathcal{C}_E$, is defined as follows: Let $d\mu_{p_{\Lambda,\Lambda_0}}(\mathfrak{h})$ be the *Gaussian* measure given by the covariance (i.e., propagator) p_{Λ,Λ_0}, which is a smooth approximation of the Euclidean propagator (3.9.15). To be explicit, let p_{Λ,Λ_0} be given in terms of p_Λ by (3.9.66), with p_Λ defined in (3.9.17). We assume that the measure is normalized: $\int d\mu_{p_{\Lambda,\Lambda_0}}(\mathfrak{h}) = 1$. In addition let $\kappa\, V^{(\Lambda_0)}$ be the interaction;[124] we assume that $V^{(\Lambda_0)}(h)$ is a polynomial in h. Further important explanations about $V^{(\Lambda_0)}$ are given below. With that, $G_{\Lambda,\Lambda_0}(h)$, is defined by the functional integral[125]

$$e^{G_{\Lambda,\Lambda_0}(h)} := \int d\mu_{p_{\Lambda,\Lambda_0}}(\mathfrak{h})\; e^{-\kappa V^{(\Lambda_0)}(\mathfrak{h}+h)}\ , \quad h \in \mathcal{C}_E\ . \tag{3.9.72}$$

Since p_{Λ,Λ_0} contains an UV-cutoff Λ_0 and an IR-cutoff Λ, we may heuristically say that the "degrees of freedom in the region $\Lambda^2 < k^2 \le \Lambda_0^2$ of momentum space are integrated out". In order that the functional integral (3.9.72) is well defined, we assume that $\kappa \ge 0$ and that $V^{(\Lambda_0)}(\varphi)$ is bounded below. This assumption may be dropped, if κ is a formal parameter. Then, the functional integral (3.9.72) is understood in the sense of formal perturbation theory, that is, as a formal power series in κ:

$$e^{G_{\Lambda,\Lambda_0}(h)} = 1 + \sum_{n=1}^{\infty} \frac{(-\kappa)^n}{n!} \int d\mu_{p_{\Lambda,\Lambda_0}}(\mathfrak{h})\ \big(V^{(\Lambda_0)}(\mathfrak{h}+h)\big)^n\ . \tag{3.9.73}$$

We take this point of view.

---

[123]By "$\triangle$" we mean the Laplace operator: $\triangle := \nabla \cdot \nabla$.

[124]In this section we change the notation a bit: The factor κ (for the coupling constant), which is contained in V in the preceding sections, is written explicitly.

[125]The condition for the configurations that the action must be finite is used only for the "background field" h and not for the configurations $\mathfrak{h}$ over which we integrate.

Graphically $e^{G_{\Lambda,\Lambda_0}(h)}$ is the sum of all Feynman diagrams with vertices $\kappa V^{(\Lambda_0)}$, internal lines symbolizing p_{Λ,Λ_0} and external lines symbolizing the field configuration h. By using the linked cluster theorem (treated in formula (4.4.2) and Exer. 4.4.2), we conclude that $G_{\Lambda,\Lambda_0}(h)$ is the sum of all *connected* Feynman diagrams of this kind.

The interaction $V^{(\Lambda_0)}$ is usually local and depends on Λ_0, since it is normally ordered with respect to p_{Λ,Λ_0} (or p_{0,Λ_0}) and because it contains Λ_0-dependent local counter terms as explained in (3.9.74)–(3.9.75) below. Normal ordering may be defined by induction on the number of fields: $\Omega_p(1) := 1$, $\Omega_p\big(\varphi(x)\big) := \varphi(x)$ and

$$\begin{aligned}\Omega_p\Big(\prod_{k=1}^{n+1}\varphi(x_k)\Big) :=&\,\Omega_p\Big(\prod_{k=1}^{n}\varphi(x_k)\Big)\cdot\varphi(x_{n+1})\\ &-\sum_{j=1}^{n}p(x_j-x_{n+1})\cdot\Omega_p\Big(\prod_{k=1,\,k\neq j}^{n}\varphi(x_k)\Big)\,.\end{aligned}$$

So for the φ^j-model the unrenormalized interaction (i.e., without counter terms) reads

$$\Omega_p(V)(h)=\int d^dx\;\Omega_p\big(\varphi(x)^j\big)(h)=\int d^dx\;\Omega_p\big(h(x)^j\big)\quad\forall h\in\mathcal{C}_E\,.$$

Computing $\frac{\partial}{\partial\Lambda}$ of the functional integral (3.9.72) one derives the flow equation. Let $G^{(n)}_{\Lambda,\Lambda_0}$ be that term of G_{Λ,Λ_0} which is of nth order in κ. The flow equation expresses $\frac{\partial G^{(n)}_{\Lambda,\Lambda_0}}{\partial\Lambda}$ in terms of lower-order terms $G^{(k)}_{\Lambda,\Lambda_0}$, $k<n$, similarly to (3.9.54). Hence, also here, the flow equation can be solved by induction on n:

$$G^{(n)}_{\Lambda,\Lambda_0}(h)=G^{(n)}_{\Lambda_0,\Lambda_0}(h)-\int_{\Lambda}^{\Lambda_0}d\Lambda'\;\frac{\partial G^{(n)}_{\Lambda',\Lambda_0}}{\partial\Lambda'}(h)\quad\forall h\in\mathcal{C}_E\,,$$

where $\frac{\partial G^{(n)}_{\Lambda',\Lambda_0}}{\partial\Lambda'}$ is expressed in terms of inductively known terms by the flow equation. There appears the crucial question, how to choose the boundary value G_{Λ_0,Λ_0}; that is $V^{(\Lambda_0)}$, since

$$G_{\Lambda_0,\Lambda_0}=-\kappa V^{(\Lambda_0)}\,,$$

as we see from $p_{\Lambda_0,\Lambda_0}=0$ and (3.9.72). Choosing for $V^{(\Lambda_0)}$ the unrenormalized, normally ordered interaction $\Omega_p(V)$, the limit $\lim_{\Lambda_0\to\infty}G_{\Lambda,\Lambda_0}$ (understood in the pointwise sense analogously to (1.2.3)) does not exist in general, due to the usual UV-divergences. Analogously to Theorem 3.9.4, one can achieve that this limit exists by adding suitable Λ_0-dependent local counter terms to $\Omega_p(V)$:

$$-\kappa V^{(\Lambda_0)}=-\kappa\,\Omega_p(V)+\kappa^2\cdot[\Lambda_0\text{-dependent local counter terms}]\,.\tag{3.9.74}$$

An interaction is called "perturbatively renormalizable" if a *finite* number of counter terms suffices for convergence of this limit, where each counter term may be a formal power series in κ. For example, the φ^4-interaction in $d = 4$ dimensions is perturbatively renormalizable in this sense, since only three counter terms are required:

$$G_{\Lambda_0,\Lambda_0} = \kappa\,\Omega_p(\varphi^4)\Big(-1 + \sum_{r\geq 1} c_{\Lambda_0,r}\kappa^r\Big) \tag{3.9.75}$$
$$+ m^2\,\Omega_p(\varphi^2)\sum_{r\geq 2} a_{\Lambda_0,r}\kappa^r + \Omega_p\big((\partial\varphi)^2\big)\sum_{r\geq 2} b_{\Lambda_0,r}\kappa^r\ ,$$

where $a_{\Lambda_0,r}$, $b_{\Lambda_0,r}$, $c_{\Lambda_0,r}$ are numbers generally diverging for $\Lambda_0 \to \infty$, cf. (3.8.9).

We are now going to compare with our formalism.

- **τ_{Λ,Λ_0} and $\mathbf{S}_{\Lambda,\Lambda_0}$ as functional integrals:** Let τ_{Λ,Λ_0} be defined as in (3.9.68). For $F \in \mathcal{F}$

$$(\tau_{\Lambda,\Lambda_0}F)(h) \quad \text{corresponds to} \quad \int d\mu_{p_{\Lambda,\Lambda_0}}(\mathfrak{h})\ F(\mathfrak{h}+h)\ , \quad h \in \mathcal{C}_E\ , \tag{3.9.76}$$

since both expressions are the sum over all possible contractions of φ in $F(\varphi)$ with propagator p_{Λ,Λ_0} and finally each uncontracted φ is replaced by h, see the explanation after Exer. 2.3.1.

Moreover, let $\mathbf{S}_{\Lambda,\Lambda_0}$ be the regularized S-matrix with propagator p_{Λ,Λ_0}. Then, for $V \in \mathcal{F}$,

$$\mathbf{S}_{\Lambda,\Lambda_0}(\kappa V)(h) \quad \text{corresponds to} \quad \int d\mu_{p_{\Lambda,\Lambda_0}}(\mathfrak{h})\, e^{-\kappa\,\Omega_{p_{\Lambda,\Lambda_0}}(V)(\mathfrak{h}+h)}, \tag{3.9.77}$$

since for both expressions the term $\sim \kappa^n$ is the sum over all contractions (with propagator p_{Λ,Λ_0}) between n vertices, each vertex given by V and, again, in the end each uncontracted φ is replaced by h. We point out that in the functional integral self-contractions of a vertex (i.e., tadpoles) drop out due to the normal ordering of V with respect to p_{Λ,Λ_0}. Note also that, in contrast to (3.9.72), the interaction V in the functional integral (3.9.77) is *localized* (since $V \in \mathcal{F}$); this adiabatic switching off does not cause any problems.[126]

Even for $F, V \in \mathcal{F}_{\text{loc}}$ all expressions in (3.9.76) and (3.9.77) are well defined (that is, renormalization is not needed at this level), since $p_{\Lambda,\Lambda_0} \in C^\infty(\mathbb{R}^d)$.

---

[126] Normal ordering of $V = \sum_k \langle f_k, \varphi^{\otimes k}\rangle \in \mathcal{F}$ is defined in the natural way:

$$\Omega_p(V) := \sum_k \int dx_1 \cdots dx_k\ f_k(x_1,\ldots,x_k)\,\Omega_p\big((\varphi(x_1)\cdots\varphi(x_k)\big)\ .$$

- **Effective potential and effective action:** Our "effective potential" is related to the "effective action" (3.9.72) of the functional integral approach, to wit,

$$\tfrac{i}{\hbar}\,V_\Lambda(\kappa V) := \tfrac{i}{\hbar}\,\mathbf{S}_\Lambda^{-1}\circ \mathbf{S}(\kappa V) \text{ corresponds roughly to } G_{\Lambda,\infty} := \lim_{\Lambda_0\to\infty} G_{\Lambda,\Lambda_0}\ , \tag{3.9.78}$$

where again the limit is understood in the pointwise sense. The factor "i" appearing in this correspondence is due to the fact that V_Λ is the effective potential of a relativistic QFT, whereas G_{Λ,Λ_0} is the effective action of a Euclidean QFT. For $\Lambda = 0$ the expressions (3.9.78) agree (apart from the different frameworks): Namely in our formalism we have the value

$$e^{iV_0(\kappa V)(h)/\hbar} = \mathbf{S}(\kappa V)(h)\ , \tag{3.9.79}$$

as we know from (3.9.43); this is a main justification to interpret $V_\Lambda(\kappa V)$ as effective potential. On the other side, the functional integral

$$\lim_{\Lambda_0\to\infty} e^{G_{0,\Lambda_0}(h)} = \lim_{\Lambda_0\to\infty} \int d\mu_{p_{0,\Lambda_0}}(\mathfrak{h})\, e^{-\kappa V^{(\Lambda_0)}(\mathfrak{h}+h)} \tag{3.9.80}$$

gives also the renormalized S-matrix to the interaction κV, evaluated at the field configuration $h \in \mathcal{C}_E$. Here we assume that the pointwise limit $\Lambda \downarrow 0$ exists and may be computed by setting $\Lambda = 0$. Since we assume $m > 0$, this assumption is usually satisfied, in particular this holds true for the φ^4-interaction in $d = 4$ dimensions.

- **UV-finite models:** To investigate the connection between $V_\Lambda(V)$ and $G_{\Lambda,\infty}$ for $\Lambda > 0$, i.e., to clarify the statement (3.9.78), we make the simplifying assumption that the interaction V is UV-finite, as introduced in Exap. 3.6.9. For such an interaction, $V^{(\Lambda_0)}$ (3.9.74) depends on Λ_0 only by normal ordering. If the latter is done with respect to p_{Λ,Λ_0}, that is, $V^{(\Lambda_0)} = \Omega_{p_{\Lambda,\Lambda_0}}(V)$, we see from (3.9.72) and (3.9.77) that

$$S_{\Lambda,\Lambda_0}(\kappa V)(h) \quad \text{corresponds to} \quad e^{G_{\Lambda,\Lambda_0}(h)}\ . \tag{3.9.81}$$

From (3.9.38) we know that

$$\mathbf{S}(\kappa V) = \lim_{\Lambda_0\to\infty} \mathbf{S}_{\Lambda_0}(\kappa V)\ . \tag{3.9.82}$$

Taking into account these relations and (3.9.71), we see that $G_{\Lambda,\infty}$ corresponds to the pointwise limit

$$\begin{aligned}\lim_{\Lambda_0\to\infty} \log\circ \mathbf{S}_{\Lambda,\Lambda_0}(\kappa V) &= \lim_{\Lambda_0\to\infty} \log\circ\tau_\Lambda^{-1}\circ \mathbf{S}_{\Lambda_0}\circ\tau_\Lambda(\kappa V)\\ &= \log\circ\tau_\Lambda^{-1}\circ\mathbf{S}\circ\tau_\Lambda(\kappa V)\ .\end{aligned} \tag{3.9.83}$$

In the last step we use that under the map τ_Λ UV-finiteness of κV is maintained. The result (3.9.83) is equal to

$$\begin{aligned}\tau_\Lambda^{-1}\circ\tau_\Lambda\circ\log\circ\tau_\Lambda^{-1}\circ\mathbf{S}\circ\tau_\Lambda(\kappa V) &= \tfrac{i}{\hbar}\,\tau_\Lambda^{-1}\circ\mathbf{S}_\Lambda^{-1}\circ\mathbf{S}\circ\tau_\Lambda(\kappa V)\\ &= \tfrac{i}{\hbar}\,\tau_\Lambda^{-1}\circ V_\Lambda\circ\tau_\Lambda(\kappa V)\ ,\end{aligned} \tag{3.9.84}$$

by applying (3.9.13). Summing up, the map

$$\kappa V \longmapsto G_{\Lambda,\infty} \quad \text{corresponds to the map} \quad \kappa V \longmapsto \log \circ \tau_\Lambda^{-1} \circ \mathbf{S} \circ \tau_\Lambda(\kappa V)$$

(3.9.83), which agrees with the map

$$\kappa V \longmapsto \tfrac{i}{\hbar}\, V_\Lambda(\kappa V) \quad \text{up to a similarity transformation by } \tau_\Lambda$$

(3.9.84); such a transformation is a matter of convention. For $\Lambda = 0$, we have $\tau_{\Lambda=0} = \mathrm{Id}$; so, in this case, we are back to the correspondence of the functional integral (3.9.80) with our term (3.9.79), which holds true for *any* interaction κV.

The similarity between

$$\kappa V \longmapsto \tfrac{i}{\hbar}\, V_\Lambda(\kappa V) \quad \text{and} \quad \kappa V \longmapsto \lim_{\Lambda_0 \to \infty} \log \circ \mathbf{S}_{\Lambda,\Lambda_0}(\kappa V)$$

can be understood as the similarity between $\frac{i}{\hbar}\, W_{\Lambda,\Lambda_0}$ and $\log \circ \mathbf{S}_{\Lambda,\Lambda_0}$ pointed out in (3.9.70). Namely, applying the latter relation to an UV-finite interaction κV and performing the limit $\Lambda_0 \to \infty$ we get:

$$\lim_{\Lambda_0 \to \infty} \tfrac{i}{\hbar}\, W_{\Lambda,\Lambda_0}(\kappa V) = \tfrac{i}{\hbar}\, V_\Lambda(\kappa V) = \lim_{\Lambda_0 \to \infty} \tau_\Lambda \circ \log \circ \mathbf{S}_{\Lambda,\Lambda_0} \circ \tau_\Lambda^{-1}(\kappa V)\ .$$

Chapter 4

Symmetries – the Master Ward Identity

For this chapter the references are [16, 17, 53, 54]. However, at first sight, [53] differs quite a lot from the presentation given here.[127]

A fundamental problem in the quantization of a classical field theory is that, in general, not all classical symmetries can be maintained: Due to the distributional character of quantum fields the arguments valid for classical field theory cannot be applied. In the inductive construction of the retarded or time-ordered product (given in Sect. 3.2 or 3.3, resp.) the decisive step, in which classical symmetries can be violated, is the extension to the thin diagonal, see Sects. 3.2.2 and 3.2.7. Therefore, symmetries are renormalization conditions, as mentioned in Sect. 3.1.4.

In Sect. 3.2.7 we have shown that symmetries with respect to a group G can be maintained in the extension to the thin diagonal, if all finite-dimensional representations of G are completely reducible. However, physically important symmetries are not of this kind, e.g., scaling symmetries, the conservation of certain currents and gauge (or BRST [7, 8, 164]) invariance.

Therefore, in this chapter, we return to the maintenance of symmetries in the process of quantization, from a general point of view: The Master Ward Identity (MWI) is a universal formulation of symmetries – it is the straightforward generalization to QFT of the most general classical identity for local fields which can be obtained from the off-shell field equation and the fact that classical fields may be multiplied pointwise. It is a highly nontrivial renormalization condition, which cannot always be fulfilled due to the well-known anomalies appearing in QFT.

In conventional renormalization methods (e.g., BPHZ renormalization or dimensional regularization) the question whether certain Ward identities can be preserved in the process of renormalization, is usually analyzed by means of the

[127]It is impossible to formulate the Master Ward Identity if the arguments of the R- (or T-) products are pure on-shell fields. In [53] the arguments of the T-products are on-shell fields, however, generalized by an additional, symbolical derivative ("external derivative") with respect to which the free field equations do not hold.

M. Dütsch, *From Classical Field Theory to Perturbative Quantum Field Theory*, Progress in Mathematical Physics 74, https://doi.org/10.1007/978-3-030-04738-2_4

Quantum Action Principle (QAP), which was derived by Lowenstein [125] and Lam [119], and proved in several renormalization schemes, e.g., in dimensional regularization [15]. The QAP characterizes the possible violations of Ward identities; it can be used to derive necessary and sufficient conditions for the existence of a renormalization maintaining the Ward identities. The solvability of these conditions frequently amounts to cohomological questions involving the infinitesimal symmetry operators which appear in the considered Ward identities. This program is called "algebraic renormalization", for an overview and a bunch of applications we refer to the book [135]. We will prove a version of the QAP in terms of the "anomalous MWI" (Sect. 4.3) and we will mimic the procedure of algebraic renormalization for the MWI in the framework of causal perturbation theory (Sects. 4.4–4.5).

4.1 Derivation of the Master Ward Identity in classical field theory

The defining property (ii) of a retarded wave operator (Definition 1.6.2) is the off-shell field equation. We start with its perturbative formulation (1.7.7) or (1.9.11):

$$R_{\rm cl}\Big(e_\otimes^S, \frac{\delta(S_0+S)}{\delta\varphi(x)}\Big) = \frac{\delta S_0}{\delta\varphi(x)}\,. \tag{4.1.1}$$

Let $Q \in \mathcal{P}$. We multiply (4.1.1) with $R_{\rm cl}(e_\otimes^S, Q(x))$ and use the perturbative formulation (1.7.6) of the crucial factorization property (1.6.5) of classical fields:

$$\begin{aligned} R_{\rm cl}\Big(e_\otimes^S, Q(x)\cdot\frac{\delta(S_0+S)}{\delta\varphi(x)}\Big) &= R_{\rm cl}\big(e_\otimes^S, Q(x)\big)\cdot R_{\rm cl}\Big(e_\otimes^S, \frac{\delta(S_0+S)}{\delta\varphi(x)}\Big) \\ &= R_{\rm cl}\big(e_\otimes^S, Q(x)\big)\cdot\frac{\delta S_0}{\delta\varphi(x)}\,, \end{aligned} \tag{4.1.2}$$

where " $\cdot$ " means the classical product. This is the "Master Ward Identity" (MWI). A crucial point is that the right-hand side vanishes "modulo the free field equation", i.e., when restricted to $\mathcal{C}_{S_0}$.

To write the MWI in the form given in [16, 17, 54], we introduce the functional

$$A = \int dx\, \sum_{i=1}^K h_i(x)\, Q_i(x)\, \frac{\delta S_0}{\delta\varphi(x)} \quad \text{with} \quad K<\infty\,, \quad Q_i \in \mathcal{P}\,, \tag{4.1.3}$$

and arbitrary test functions $h_i \in \mathcal{D}(\mathbb{M})$; and a corresponding derivation

$$\delta_{\vec{h}\cdot\vec{Q}} := \int dx\, \sum_{i=1}^K h_i(x)\, Q_i(x)\, \frac{\delta}{\delta\varphi(x)}\,. \tag{4.1.4}$$

Obviously we may write $A = \delta_{\vec{h}\cdot\vec{Q}} S_0$. With that we obtain from (4.1.2) the *MWI for the symmetry given by $\vec{Q}$ and the interaction S*:

$$R_{\mathrm{cl}}\big(e_\otimes^S, (A + \delta_{\vec{h}\cdot\vec{Q}} S)\big) = \int dx \sum_{i=1}^K h_i(x)\, R_{\mathrm{cl}}\big(e_\otimes^S, Q_i(x)\big) \cdot \frac{\delta S_0}{\delta\varphi(x)} \in \mathfrak{J}\,, \tag{4.1.5}$$

where $\mathfrak{J} \equiv \mathfrak{J}^{(m)}$ (2.6.2) is the ideal of $(\mathfrak{F}, \cdot)$ generated by the free field equation.

By construction, the particular case $K = 1$, $Q = 1$ must give the off-shell field equation (4.1.1) smeared out with $h(x)$; this holds indeed true since $R_{\mathrm{cl}}\big(e_\otimes^S, 1\big) = 1$.

Remark 4.1.1. Obviously the functional A given in (4.1.3) is an element of $\mathfrak{J} \cap \mathfrak{F}_{\mathrm{loc}}$. Vice versa, we conclude from (1.3.8) and (2.6.2) that every $A \in \mathfrak{J} \cap \mathfrak{F}_{\mathrm{loc}}$ can be written in the form (4.1.3). But, it may happen that the Q_i are not uniquely determined by A. E.g., for a complex scalar field (see Exap. 1.3.2) we have $S_0 = \int dx\, \big(\partial^\mu\phi^*(x)\partial_\mu\phi(x) - m^2\,\phi^*(x)\phi(x)\big)$; with that, $A := \int dx\, h(x)\, \frac{\delta S_0}{\delta\phi(x)}\, (\Box + m^2)\phi(x)$ can also be written as $A = \int dx\, h(x)\, \frac{\delta S_0}{\delta\phi^*(x)}\, (\Box + m^2)\phi^*(x)$. However, in [54] a *unique* prescription is given to write every $A \in \mathfrak{J} \cap \mathfrak{F}_{\mathrm{loc}}$ in the form (4.1.3). With this prescription one may write δ_A instead of $\delta_{\vec{h}\cdot\vec{Q}}$, as it is done in [16, 17, 54]. Every $A \in \mathfrak{J} \cap \mathfrak{F}_{\mathrm{loc}}$ gives rise to a corresponding case of the MWI.

4.2 The Master Ward Identity as a universal renormalization condition

4.2.1 Formulation of the MWI

We transfer the classical MWI (4.1.5) to perturbative QFT by replacing each $R_{\mathrm{cl}}\big(e_\otimes^S, F\big)$ by $R\big(e_\otimes^{S/\hbar}, F\big)$:

$$R\big(e_\otimes^{S/\hbar}, (A + \delta_{\vec{h}\cdot\vec{Q}} S)\big) = \int dx \sum_{i=1}^K h_i(x)\, R\big(e_\otimes^{S/\hbar}, Q_i(x)\big) \cdot \frac{\delta S_0}{\delta\varphi(x)} \in \mathfrak{J}\,. \tag{4.2.1}$$

This is the *quantum MWI* to the interaction S and for the symmetry given by A or $\vec{h}\cdot\vec{Q}$. The crucial difference to the classical MWI (4.1.5) is, that the factorization (1.6.5) does not hold in QFT; therefore, the quantum MWI (4.2.1) does not follow from the off-shell field equation. Instead, as we will see, it is a renormalization condition, which cannot always be satisfied. Its applications reach so far that one may call it a universal formulation of symmetries [53, 54].

To simplify the notation, we shall almost always consider the case $K = 1$; the generalization to $1 \le K < \infty$ is obvious.

The MWI can easily be translated into the time-ordered product: We insert Bogoliubov's formula (3.3.30)

$$R(e_\otimes^{S/\hbar}, F) = \overline{T}(e_\otimes^{-iS/\hbar}) \star T(e_\otimes^{iS/\hbar} \otimes F)$$

(where again $\overline{T}(e_\otimes^{-iG}) := T(e_\otimes^{iG})^{\star -1}$) into (4.2.1) and use the identity

$$(F \star G) \cdot \frac{\delta S_0}{\delta \varphi} = F \star \big(G \cdot \frac{\delta S_0}{\delta \varphi}\big) \qquad \forall F, G \in \mathcal{F} \ . \tag{4.2.2}$$

The latter relies on the fact that on the r.h.s. a contraction of F with $\frac{\delta S_0}{\delta \varphi}$ vanishes, since the propagator Δ_m^+ of the star product (2.1.5) is a solution of the free field equation, see (2.1.11). Proceeding this way, we obtain

$$\begin{aligned} &\overline{T}\big(e_\otimes^{-iS/\hbar}\big) \star T\big(e_\otimes^{iS/\hbar} \otimes (A + \delta_{hQ} S)\big) \\ &\qquad = \overline{T}\big(e_\otimes^{-iS/\hbar}\big) \star \bigg(\int dx\, h(x)\, T\big(e_\otimes^{iS/\hbar} \otimes Q(x)\big) \cdot \frac{\delta S_0}{\delta \varphi(x)} \bigg) . \end{aligned}$$

Now, $\star$-multiplication with $T\big(e_\otimes^{iS/\hbar}\big)$ from the left yields the MWI for the time-ordered product:

$$T\big(e_\otimes^{iS/\hbar} \otimes (A + \delta_{hQ} S)\big) = \int dx\, h(x)\, T\big(e_\otimes^{iS/\hbar} \otimes Q(x)\big) \cdot \frac{\delta S_0}{\delta \varphi(x)} . \tag{4.2.3}$$

For later purpose we write this identity to nth order and with non-diagonal entries: Inserting $S := \sum_{j=1}^n \lambda_j F_j$ (where $F_j \in \mathcal{F}_{\mathrm{loc}}$) into (4.2.3) and applying the derivative $\frac{\partial^n}{\partial\lambda_1 \cdots \partial\lambda_n}\big|_{\lambda_1 = \cdots = \lambda_n = 0}$ we obtain

$$\begin{aligned} &T_{n+1}(F_1 \otimes \cdots \otimes F_n \otimes A) + \frac{\hbar}{i} \sum_{l=1}^n T_n(F_1 \otimes \cdots \otimes \delta_{hQ} F_l \otimes \cdots \otimes F_n) \\ &\qquad = \int dx\ h(x)\, T_{n+1}\big(F_1 \otimes \cdots \otimes F_n \otimes Q(x)\big) \cdot \frac{\delta S_0}{\delta \varphi(x)} . \end{aligned} \tag{4.2.4}$$

4.2.2 Verification that the MWI is a renormalization condition

To show that the MWI is indeed a renormalization condition, we work with the time-ordered product. This is somewhat simpler than the corresponding argumentation in terms of the retarded product.

We recall from Sect. 3.3: The time-ordered product $T \equiv (T_n)_{n \in \mathbb{N}}$ can be constructed directly from its axioms by induction on n. In the inductive step $(n-1) \to n$ the basic axioms determine $T_n\big(\otimes_{i=1}^n A_i(x_i)\big)$ uniquely on $\mathcal{D}(\mathbb{M}^n \setminus \Delta_n)$, all further axioms restrict only the renormalization, that is, the extension to $\mathcal{D}(\mathbb{M}^n)$ (which is non-unique) – therefore they are called "renormalization conditions".

Hence, to verify that an axiom (A) is a renormalization condition, one usually has to show that the validity of (A) for all T-products of lower orders, i.e., $(T_k)_{k<n}$, implies the validity of (A) for $T_n|_{\mathcal{D}(\mathbb{M}^n\setminus\Delta_n)}$.

For the MWI this procedure has to be modified a bit, since (4.2.4) contains T-products of different orders: Our method is induction on the order in the *interaction* (which is given in (4.2.4) by the fields $(F_j)_{1\leq j\leq n}$). More in detail we proceed as follows:

- In a first step we construct the T-product $T\equiv(T_n)_{n\in\mathbb{N}^*}$ by induction on n, such that all renormalization conditions except the MWI are fulfilled.
- In a second step we modify the so-obtained T-product such that it satisfies also the MWI, that is, we perform a finite renormalization $\mathbf{S}\longmapsto\mathbf{S}\circ Z$ for a suitable Z lying in the Stückelberg–Petermann RG. Here we proceed by induction on the order in the interaction, except for Sect. 4.5 where we use induction on the order in $\hbar$.

Now, let $S=\int dx\; g(x)\,L(x)$, $g\in\mathcal{D}(\mathbb{M})$, $L\in\mathcal{P}$ be the interaction. With that, the MWI (4.2.4) to nth order in g can be written as

$$\begin{aligned}&\Big(\frac{i}{\hbar}\Big)^n\,T_{n+1}\Big(\bigotimes_{j=1}^n L(x_j)\otimes Q(y)\cdot\frac{\delta S_0}{\delta\varphi(y)}\Big)\\&+\sum_{l=1}^n\Big(\frac{i}{\hbar}\Big)^{n-1}\,T_n\Big(\bigotimes_{j=1,\,j\neq l}^n L(x_j)\otimes Q(y)\cdot\frac{\delta L(x_l)}{\delta\varphi(y)}\Big)\\&\qquad\qquad=\Big(\frac{i}{\hbar}\Big)^n\,T_{n+1}\Big(\bigotimes_{j=1}^n L(x_j)\otimes Q(y)\Big)\cdot\frac{\delta S_0}{\delta\varphi(y)}\end{aligned}\tag{4.2.5}$$

From (1.3.19) we know that

$$\frac{\delta L(x)}{\delta\varphi(y)}=\sum_a B_a(y)\,\partial^a\delta(y-x)\quad\text{for some}\quad B_a\in\mathcal{P}\,.\tag{4.2.6}$$

Due to this, $T_n\big(\cdots\otimes Q(y)\cdot\frac{\delta L(x)}{\delta\varphi(y)}\big)$ is well defined, it is equal to $\sum_a T_n\big(\cdots\otimes(QB_a)(y)\big)\cdot\partial^a\delta(y-x)$.

Proceeding by induction on the order in g, we assume that the T-products to lower orders are renormalized such that the MWI holds to all lower orders in g. We are going to show that this implies that (4.2.5) is fulfilled on $\mathcal{D}(\mathbb{M}^{n+1}\setminus\Delta_{n+1})$. Namely, for $(x_1,\dots,x_n,y)\notin\Delta_{n+1}$ there exists a $K\subset\{1,\dots,n\}$ with $K^{\mathrm{c}}:=(\{1,\dots,n\}\setminus K)\neq\emptyset$ such that either $(\{x_k\,|\,k\in K^{\mathrm{c}}\}+\overline{V}_+)\cap(\{x_j\,|\,j\in K\}\cup\{y\})=\emptyset$ or $(\{x_k\,|\,k\in K^{\mathrm{c}}\}+\overline{V}_-)\cap(\{x_j\,|\,j\in K\}\cup\{y\})=\emptyset$. We treat the first case, the second case is completely analogous. Using causal factorization of the time-ordered product (3.3.5) and $\frac{\delta L(x)}{\delta\varphi(y)}=0$ for $x\neq y$, the l.h.s. of (4.2.5) can be written as

$$T\Big(\bigotimes_{j\in K^c} L(x_j)\Big)\star\Big[\Big(\frac{i}{\hbar}\Big)^n T\Big(\bigotimes_{k\in K} L(x_k)\otimes Q(y)\cdot\frac{\delta S_0}{\delta\varphi(y)}\Big) + \Big(\frac{i}{\hbar}\Big)^{n-1}\sum_{l\in K} T\Big(\bigotimes_{k\in K,\,k\neq l} L(x_k)\otimes Q(y)\cdot\frac{\delta L(x_l)}{\delta\varphi(y)}\Big)\Big], \tag{4.2.7}$$

which is equal to

$$\Big(\frac{i}{\hbar}\Big)^n T\Big(\bigotimes_{j\in K^c} L(x_j)\Big)\star\Big[T\Big(\bigotimes_{k\in K} L(x_k)\otimes Q(y)\Big)\cdot\frac{\delta S_0}{\delta\varphi(y)}\Big] \tag{4.2.8}$$

due to the validity of the MWI (4.2.5) to order $|K|(<n)$. By means of (4.2.2) and causal factorization, the last expression agrees with the right-hand side of (4.2.5). This proves that the MWI is indeed a renormalization condition.

The crucial and difficult question is, whether there exists a renormalization (i.e., an extension to the thin diagonal) of these time-ordered products, which is compatible with all other renormalization conditions and which fulfills (4.2.5). Before studying this question, we give some applications of the MWI.

4.2.3 A few applications of the MWI

The physically most important application of the MWI is BRST symmetry [7, 8], see [53, 54]. We treat here a few simpler examples.

Many applications of the MWI can be understood by means of Noether's Theorem for classical field theory: If the total action (S_0+S) of a model is invariant under a certain transformation of the basic fields φ_j, there exists a corresponding conserved Noether current j^μ. The conservation of this current relies on the validity of the field equations, more precisely, there exist field polynomials $Q_j\in\mathcal{P}$ such that

$$\partial_\mu j^\mu(x)=\sum_j Q_j(x)\,\frac{\delta(S_0+S)}{\delta\varphi_j(x)}\ . \tag{4.2.9}$$

The classical current conservation $\partial_\mu j^\mu_S(x)=0$ (where Definition 1.6.1 is used) can equivalently be written in the perturbative framework:

$$\partial^x_\mu R_{\rm cl}\big(e_\otimes^S, j^\mu(x)\big)\big|_{\mathcal{C}_0}=0\ . \tag{4.2.10}$$

Due to (1.7.5), the identity (4.2.10) is the restriction to $\mathcal{C}_0$ of the MWI

$$\partial^x_\mu R_{\rm cl}\big(e_\otimes^S, j^\mu(x)\big)=R_{\rm cl}\Big(e_\otimes^S,\sum_j Q_j(x)\,\frac{\delta(S_0+S)}{\delta\varphi_j(x)}\Big) = \sum_j R_{\rm cl}\big(e_\otimes^S, Q_j(x)\big)\cdot\frac{\delta S_0}{\delta\varphi_j(x)}\ . \tag{4.2.11}$$

Example 4.2.1 (Scaling invariance expressed by conservation of the dilatation current). The key to express scaling invariance in terms of the MWI is the sketched connection (4.2.9)–(4.2.11) of Noether's Theorem to the MWI. Since scaling invariance cannot be maintained under quantization, we investigate clFT only.

We will work with the infinitesimal version $s : \mathcal{F} \longrightarrow \mathcal{F}$ of the scaling transformation σ_ρ^{-1} (Def. 3.1.20):

$$s\,F = \frac{\partial}{\partial\rho}\Big|_{\rho=1} \sigma_\rho^{-1} F\ , \quad \forall F \in \mathcal{F}\ .$$

Due to (3.1.49), s is a derivation; this holds also for the application to field polynomials:

$$s(B_1 B_2)(x) = \big(s\,B_1(x)\big) \cdot B_2(x) + B_1(x) \cdot s\,B_2(x)\ , \quad \forall B_1, B_2 \in \mathcal{P}\ . \tag{4.2.12}$$

For a homogeneous field polynomial $B(x)$, the relation (3.1.50) yields

$$s\,B(x) = (\dim B + x^\nu \partial_\nu^x) B(x)\ , \quad \forall B \in \mathcal{P}_{\mathrm{hom}}\ . \tag{4.2.13}$$

From this formula and $\dim \partial^\mu B = \dim B + 1$ for $B \in \mathcal{P}_{\mathrm{hom}}$ we conclude that

$$s(\partial^\mu B(x)) = \partial_x^\mu (s\,B(x))\ , \quad \forall B \in \mathcal{P}_{\mathrm{hom}}\ ; \tag{4.2.14}$$

actually, by linearity of s, this relation holds even for all $B \in \mathcal{P}$.

Let $\mathcal{O}$ be a double cone and $g \in \mathcal{G}(\mathcal{O})$ (see (3.7.10)). We formally[128] write the total action as

$$S_0 + S = \int d^d x\ \mathcal{L}_{\mathrm{tot}}(x)\ , \quad \mathcal{L}_{\mathrm{tot}}(x) = L_0(x) - \kappa\, g(x)\, L_{\mathrm{int}}(x)\ ,$$

where $g \in \mathcal{D}(\mathbb{M})$, and we assume that L_0 and L_{int} contain zeroth and first derivatives of φ only.

We first study the *massless case* $m = 0$. The crucial assumption is that $L_0, L_{\mathrm{int}} \in \mathcal{P}_d$ (see Def. 3.1.18), where d is the dimension of the Minkowski space. Setting $g_\rho(x) = g(\rho x)$ and writing S_g for S, we formally obtain scaling invariance of the total action (up to scaling of the switching function g)

$$\sigma_\rho^{-1}(S_0 + S_{g_\rho}) = \int d^d x\ \rho^d \left(L_0(\rho x) - \kappa\, g(\rho x)\, L_{\mathrm{int}}(\rho x)\right) = S_0 + S_g\ .$$

A rigorous derivation and formulation of scaling invariance can be given by

$$s\,\mathcal{L}_{\mathrm{tot}}(x) = (d + x^\nu \partial_\nu^x)\mathcal{L}_{\mathrm{tot}}(x) = \partial_\nu I^\nu(x) \quad \text{with} \quad I^\nu(x) := x^\nu\, \mathcal{L}_{\mathrm{tot}}(x)\ , \quad \forall x \in \mathcal{O}\ . \tag{4.2.15}$$

To derive the corresponding Noether current – the "dilatation current", we compute $s\,\mathcal{L}_{\mathrm{tot}}(x)$ in a different way by using the properties (4.2.12) and (4.2.14) of s:

$$\begin{aligned} s\,\mathcal{L}_{\mathrm{tot}}(x) &= \frac{\partial \mathcal{L}_{\mathrm{tot}}}{\partial \varphi}(x)\, s\varphi(x) + \frac{\partial \mathcal{L}_{\mathrm{tot}}}{\partial(\partial_\mu \varphi)}(x)\, s\big(\partial_\mu \varphi(x)\big) \\ &= \partial_\mu^x \Big(\frac{\partial \mathcal{L}_{\mathrm{tot}}}{\partial(\partial_\mu \varphi)}(x)\, s\varphi(x)\Big) + \Big[\frac{\partial \mathcal{L}_{\mathrm{tot}}}{\partial \varphi}(x) - \partial_\mu \frac{\partial \mathcal{L}_{\mathrm{tot}}}{\partial(\partial_\mu \varphi)}(x)\Big]\, s\varphi(x)\ . \end{aligned} \tag{4.2.16}$$

---

[128] "Formally" means here that we ignore the IR-problem of the integration over x, discussed in Sect. 1.5.

Since the term in $[\cdots]$-brackets is equal to $\frac{\delta(S_0+S)}{\delta\varphi(x)}$, we indeed obtain the relation (4.2.9):

$$\partial_\mu j^\mu(x) = \frac{\delta(S_0+S)}{\delta\varphi(x)}\, Q(x) \quad \forall x \in \mathcal{O}\,, \quad \text{where} \quad Q(x) := s\varphi(x) \tag{4.2.17}$$

is the infinitesimal transformation of the basic field, and with the dilatation current

$$j^\mu(x) := I^\mu(x) - \frac{\partial \mathcal{L}_{\mathrm{tot}}}{\partial(\partial_\mu\varphi)}(x)\, s\varphi(x)\,. \tag{4.2.18}$$

The derivation δ_{hQ} (introduced in (4.1.4)) agrees with s in the following sense: If we choose $h(x) = 1\ \forall x \in \overline{\mathcal{O}}$ we have

$$\delta_{hQ} F = \int dx\; s\varphi(x)\, \frac{\delta F}{\delta\varphi(x)} = s\,F \qquad \forall F \in \mathcal{F}(\mathcal{O})\,. \tag{4.2.19}$$

As explained in (4.2.10)–(4.2.11), the conservation of the dilatation current for $x \in \mathcal{O}$ can be understood as the restriction to $\mathcal{C}_0$ of the MWI:

$$\int dx\; h(x)\, \partial^x_\mu R_{\mathrm{cl}}\big(e^S_\otimes, j^\mu(x)\big) = \int dx\; h(x) R_{\mathrm{cl}}\big(e^S_\otimes, s\varphi(x)\big)\cdot \frac{\delta S_0}{\delta\varphi(x)}\,, \quad \forall h \in \mathcal{D}(\mathcal{O})\,;$$

the restriction to $h \in \mathcal{D}(\mathcal{O})$ is necessary since scaling invariance (4.2.15) holds true only for $x \in \mathcal{O}$. The procedure (4.2.15)–(4.2.18) is a standard method to derive Noether's Theorem in clFT, it is of general applicability (see Exaps. 4.2.2 and 4.2.3), in particular it works for BRST symmetry (see, e.g., [56, Sect. 3]).

Turning to the *massive case* $m \geq 0$, we admit that L_0 and L_{int} are of the form

$$B^m := \sum_{l=0}^{d} m^{d-l}\, B_l \quad \text{with} \quad B_l \in \mathcal{P}_l\,.$$

Due to

$$\sigma_\rho^{-1} B^{\rho m}(x) = \rho^d\, B^m(\rho x)\,, \tag{4.2.20}$$

the total action $(S_0^m + S_g^m)$ is then invariant under the scaling $(x, m) \mapsto (\rho x, \rho^{-1} m)$, up to scaling of g. Formally, with $g_\rho(x) := g(\rho x)$,

$$\sigma_\rho^{-1}\big(S_0^{\rho m} + S_{g_\rho}^{\rho m}\big) = \int d^dx\, \Big(\sigma_\rho^{-1} L_0^{\rho m}(x) - \kappa\, g(\rho x)\, \sigma_\rho^{-1} L_{\mathrm{int}}^{\rho m}(x)\Big) = S_0^m + S_g^m\,. \tag{4.2.21}$$

From (4.2.20) we get

$$(m\partial_m + s) B^m(x) = \frac{\partial}{\partial\rho}\Big|_{\rho=1} \sigma_\rho^{-1} B^{\rho m}(x) = (d + x^\nu \partial^x_\nu) B^m(x)\,.$$

Hence, the rigorous derivation and formulation of scaling invariance of the total action given in (4.2.15), can be generalized to the massive case by

$$(m\partial_m + s)\, \mathcal{L}^m_{\mathrm{tot}}(x) = (d + x^\nu \partial^x_\nu) \mathcal{L}^m_{\mathrm{tot}}(x) = \partial^\nu I^m_\nu(x) \quad \text{with} \quad I^m_\nu(x) := x_\nu\, \mathcal{L}^m_{\mathrm{tot}}(x) \tag{4.2.22}$$

for all $x \in \mathcal{O}$.

Formula (4.2.16) holds true also for the computation of $s\,\mathcal{L}^m_{\mathrm{tot}}(x)$; hence, for $x \in \mathcal{O}$, we get

$$\partial^\nu I^m_\nu(x) = m\partial_m\, \mathcal{L}^m_{\mathrm{tot}}(x) + \partial^x_\mu \Big(\frac{\partial \mathcal{L}^m_{\mathrm{tot}}}{\partial(\partial_\mu\varphi)}(x)\, s\varphi(x)\Big) + \frac{\delta(S_0^m + S^m)}{\delta\varphi(x)}\, s\varphi(x)\,.$$

By using the definition of the dilatation current (4.2.18), we find the relation

$$\partial^\mu j^m_\mu(x) - m\partial_m \mathcal{L}^m_{\rm tot}(x) = \frac{\delta(S_0+S)}{\delta\varphi(x)}\,Q(x) \quad \forall x\in\mathcal{O}\,, \quad \text{with} \quad Q(x):=s\varphi(x)\,. \quad (4.2.23)$$

Therefore, conservation of the dilatation current is broken in a well-controlled way; which can be expressed by the MWI

$$\partial^\mu_x R_{\rm cl}\big(e^S_\otimes, j^m_\mu(x)\big) - R_{\rm cl}\big(e^S_\otimes, m\partial_m \mathcal{L}^m_{\rm tot}(x)\big) = R_{\rm cl}\big(e^S_\otimes, s\varphi(x)\big)\cdot\frac{\delta S_0}{\delta\varphi(x)} \quad \forall x\in\mathcal{O}\,. \quad (4.2.24)$$

A prime example is $L_{\rm int}(x)=\varphi^4(x)$ in $d=4$ dimensions. By using the expression (1.5.1) for L_0 we get the explicit results

$$m\partial_m \mathcal{L}^m_{\rm tot}(x) = -m^2\varphi^2(x) \quad \text{and, for } x\in\mathcal{O},$$
$$j^m_\mu(x) = x_\mu\Big(\tfrac{1}{2}\big((\partial\varphi)^2(x) - m^2\,\varphi^2(x)\big) - \kappa\varphi^4(x)\Big) - \partial_\mu\varphi(x)\,(1+x\partial_x)\varphi(x)$$

(with the shorthand notation $ab:=a^\mu b_\mu$); the formula for the dilatation current applies also to the massless case $m=0$.

Example 4.2.2 (Scalar model with global $O(N)$-symmetry). [16] We consider a multiplet of N real scalar fields $\varphi(x)\equiv\big(\varphi_i(x)\big)_{i=1}^N : C^\infty(\mathbb{M},\mathbb{R}^N)\longrightarrow\mathbb{R}^N$, transforming under the defining (or "fundamental") representation of $O(N)$ – the group of orthogonal $N\times N$ matrices:

$$\varphi_i(x)\longrightarrow M_{ij}\varphi_j(x)\,, \quad M\in O(N).^{129} \quad (4.2.25)$$

Since M does not depend on x, one calls this a "global" transformation. Let $\{X^a\,|\,a=1,\ldots,\frac{1}{2}N(N-1)\}$ be a basis of the Lie algebra $\mathfrak{o}(N)$ of $O(N)$ and f^{abc} the corresponding structure constants,

$$[X^a,X^b]=f^{abc}X^c \qquad \forall a,b,c\,. \quad (4.2.26)$$

The main defining property of $\mathfrak{o}(N)$ is the skew-symmetry

$$X^a_{ij}=-X^a_{ji} \qquad \forall i,j\in\{1,\ldots,N\}\,,\ \forall a\,. \quad (4.2.27)$$

The relevant space $\mathcal{F}$ of fields is the set of all functionals

$$F=\sum_n\langle f_n^{i_1\ldots i_n},\varphi_{i_1}\otimes\cdots\otimes\varphi_{i_n}\rangle : C^\infty(\mathbb{M},\mathbb{R}^N)\longrightarrow\mathbb{R}\,,$$

where $f_n^{i_1\ldots i_n}(x_1,\ldots,x_n)$ is symmetric under permutations of $(i_1,x_1),\ldots,(i_n,x_{i_n})$ and satisfies the further properties listed in Definition 1.2.1; and $\mathcal{F}_{\rm loc}$ is defined analogously to Definition 1.3.4, the only difference is that $\mathcal{P}$ is now the space of polynomials in $\partial^a\varphi_i$ where $a\in\mathbb{N}^d$ and $i\in\{1,\ldots,N\}$. The field transformation (4.2.25) induces the following

[129] Repeated indices are summed over.

infinitesimal transformation on $\mathcal{F}$:

$$\begin{aligned} s^a &:= s(X^a) : \mathcal{F} \longrightarrow \mathcal{F} \ ; \\ sF &:= \frac{d}{d\lambda}\Big|_{\lambda=0} \sum_n \langle f_n^{i_1\dots i_n}, (e^{\lambda X^a})_{i_1 j_1}\varphi_{j_1} \otimes \cdots \otimes (e^{\lambda X^a})_{i_n j_n}\varphi_{j_n}\rangle \\ &= \sum_n \sum_{k=1}^n \langle f_n^{i_1\dots i_n}, \varphi_{i_1} \otimes \cdots \otimes X^a_{i_k j}\varphi_j \otimes \cdots \otimes \varphi_{i_n}\rangle \\ &= \sum_n n\, \langle f_n^{i_1\dots i_n}, X^a_{i_1 j\varphi_j} \otimes \varphi_{i_2} \otimes \cdots \otimes \varphi_{i_n}\rangle \end{aligned}$$

for all $a = 1, \dots, \frac{1}{2}N(N-1)$; in particular note the formula $(s^a\varphi)_i(x) = X^a_{ij}\varphi_j(x)$.

We study the free action

$$S_0 = \int dx\, L_0(x) \ , \quad L_0 := \tfrac{1}{2}\left(\partial^\mu\varphi_i(x)\partial_\mu\varphi_i(x) - m^2\varphi_i(x)\varphi_i(x)\right), \tag{4.2.28}$$

which is invariant under the transformation (4.2.25). For each s^a there is a corresponding Noether current j^a_μ, which can be computed by formula (4.2.18): Since $s^a L_0(x) = 0\ \forall x \in \mathbb{M}$ we have $I^\mu = 0$ in that formula and, hence,

$$j^a_\mu(x) := -\frac{\partial L_0}{\partial(\partial^\mu\varphi_i)}(x)\,(s^a\varphi)_i(x) = -\partial_\mu\varphi_i(x)\,X^a_{ij}\,\varphi_j(x)\,, \quad a = 1,\dots,\frac{1}{2}N(N-1)\ . \tag{4.2.29}$$

Formula (4.2.17) applies also:

$$\partial^\mu j^a_\mu(x) = \frac{\delta S_0}{\delta\varphi_i(x)}(s^a\varphi_i)(x) = \frac{\delta S_0}{\delta\varphi_i(x)}\,X^a_{ij}\,\varphi_j(x)$$

Hence, the functionals

$$A^a := \int dx\ h(x)\,\partial^\mu j^a_\mu(x) = \int dx\ h(x)\,Q^a_i(x)\,\frac{\delta S_0}{\delta\varphi_i(x)} \quad \text{with} \quad Q^a_i(x) := X^a_{ij}\,\varphi_j(x) \tag{4.2.30}$$

are elements of $\mathcal{J} \cap \mathcal{F}_{\mathrm{loc}}$ for all $h \in \mathcal{D}(\mathbb{M})$, where the ideal $\mathcal{J}$ of $(\mathcal{F}, \cdot)$ is the space of functionals $\sum_n \langle(\Box_1 + m^2) f_n^{i_1\dots i_n}, \varphi_{i_1} \otimes \cdots \otimes \varphi_{i_n}\rangle$ with $f_n^{i_1\dots i_n}$ as above – similarly to (2.6.2). The corresponding derivations (4.1.4) read

$$\delta_{hQ^a} = \int dx\ h(x)\,X^a_{ij}\,\varphi_j(x)\,\frac{\delta}{\delta\varphi_i(x)}\,. \tag{4.2.31}$$

The relation (4.2.19) holds true also here: $\delta_{hQ^a}F = s^a F$, if $h(x) = 1\ \forall x \in \operatorname{supp} F$.

Now let $S \in \mathcal{F}_{\mathrm{loc}}$ be an interaction of the form

$$S = -\kappa \int dx\, g(x)\, L_{\mathrm{int}}(x)\ , \quad L_{\mathrm{int}}(x) := \Big(\sum_i \varphi_i(x)\varphi_i(x)\Big)^n \quad \text{with} \quad g \in \mathcal{D}(\mathbb{M})\,,\ n \in \mathbb{N}^*\,.$$

By using (4.2.27) we verify that

$$\delta_{hQ^a} L_{\mathrm{int}}(x) = 0\ , \quad \forall x \in \mathbb{M}\ ,\ \forall a\ ,\ \forall h \in \mathcal{D}(\mathbb{M})\ ;$$

hence $\delta_{hQ^a} S = 0\ \forall h$, $\forall a$ and $s^a L_{\mathrm{int}}(x) = 0\ \forall x$, $\forall a$.

Applying the derivation of the Noether current (4.2.15)–(4.2.18) to the interacting model $\mathcal{L}_{\mathrm{tot}}(x) = L_0(x) - \kappa\, g(x)\, L_{\mathrm{int}}(x)$ studied here, we find the simplifications that $s^a \mathcal{L}_{\mathrm{tot}}(x) = 0\ \forall x \in \mathbb{M}$ and $\frac{\partial \mathcal{L}_{\mathrm{tot}}}{\partial(\partial^\mu \varphi_i)} = \frac{\partial L_0}{\partial(\partial^\mu \varphi_i)}$. Therefore, the Noether currents $(j^a_\mu)_S(x)$ of the interacting model are given by the same formula (4.2.29) as for the free model, and in clFT they are conserved modulo the field equations for all $x \in \mathbb{M}$, that is, also in the region where $\partial g(x) \neq 0$. This result can very easily be obtained in an alternative way, by arguing with the classical MWI: Namely, using (4.2.30) and $\delta_{hQ^a} S = 0$ the latter says

$$\int dx\ h(x)\, \partial^\mu_x R_{\mathrm{cl}}\big(e^S_\otimes, j^a_\mu(x)\big) = R_{\mathrm{cl}}\big(e^S_\otimes, A^a\big) = \int dx\ h(x)\, R_{\mathrm{cl}}\big(e^S_\otimes, Q^a_i(x)\big) \cdot \frac{\delta S_0}{\delta \varphi_i(x)}$$

for all $h \in \mathcal{D}(\mathbb{M})$ and all $a = 1, \ldots, \frac{1}{2}N(N-1)$.

In Sect. 4.5.2 we will study the question whether the interacting QFT-model is also $O(N)$-invariant – more precisely, whether the interacting currents $(j^a_\mu)_S(x) = R\big(e^{S/\hbar}_\otimes, j^a_\mu(x)\big)$ are conserved when restricted to $\mathcal{C}_0$. This holds indeed true, if the MWI

$$\int dx\ h(x)\, \partial^\mu_x R\big(e^{S/\hbar}_\otimes, j^a_\mu(x)\big) = -\int dx\ h(x)\, R\big(e^{S/\hbar}_\otimes, X^a_{ij}\varphi_j(x)\big) \cdot (\square + m^2)\varphi_i(x) \quad (4.2.32)$$

is satisfied for all $h \in \mathcal{D}(\mathbb{M})$ and all a. When referring to this MWI, we speak of *one* (Master) Ward Identity, although these are actually $\frac{1}{2}N(N-1)$ identities – one for each a.

Example 4.2.3 (Non-Abelian matter currents). [53] [130] This application starts also with a conserved current of the free theory, but the corresponding interacting current is not conserved, even in clFT, due to $\delta_{hQ} S \neq 0$. We work in $d = 4$ dimensions. To describe Dirac spinors, we use the formalism introduced in Sect. 5.1.1; in particular, complex conjugation is denoted by z^c instead of $\overline{z}$, according to Footn. 144.

We consider N Dirac spinor fields $(\psi_\alpha)_{\alpha=1}^N$ and the currents[131]

$$j^\mu_a := \overline{\psi}_\alpha \wedge \gamma^\mu \frac{(\lambda_a)_{\alpha\beta}}{2}\, \psi_\beta \qquad (4.2.33)$$

(we use matrix notation for the spinor structure), where $-i\lambda_a/2 \in \mathbb{C}^{N\times N}$, $a = 1, \ldots, N^2 - 1$, is a basis of the Lie algebra $su(N)$. That is, the λ-matrices satisfy the relations

$$[\lambda_a, \lambda_b] = 2i f_{abc} \lambda_c\ , \quad \lambda_a^{cT} = \lambda_a \quad \text{and} \quad \mathrm{tr}\, \lambda_a = 0 \quad \forall a, b\ , \qquad (4.2.34)$$

where (f_{abc}) are the structure constants belonging to this basis; they are totally antisymmetric. The conventions are such that we may choose the generalized Gell-Mann matrices for the (λ_a). Under an $SU(n)$-transformation, the Dirac fields transform with the fundamental representation and the gauge fields $A \equiv (A^\mu_a)$ (defined in Exer. 1.5.2, see also Sect. 5.1.3) with the adjoint representation; we denote these representations by ρ_1 and ρ_2, respectively. That is, the infinitesimal transformations read:

$$(s_b \psi)_\alpha \equiv \rho_1(\lambda_b)\, \psi_\alpha = -\frac{i}{2}\, (\lambda_b)_{\alpha\beta}\, \psi_\beta\ , \quad (s_b \overline{\psi})_\alpha \equiv \rho_1(\lambda_b)\, \overline{\psi}_\alpha = \frac{i}{2}\, \overline{\psi}_\beta\, (\lambda_b)_{\beta\alpha}\ ,$$
$$(s_b A)^\mu_a \equiv \rho_2(\lambda_b)\, A^\mu_a = -f_{bac}\, A^\mu_c\ ,$$

[130] To simplify the formulas we set $\hbar = 1$.
[131] Again, repeated indices are summed over.

from which we obtain

$$(s_b j)^\mu_a = f_{abc}\, j^\mu_c \;, \tag{4.2.35}$$

by using that s_b is a derivation and (4.2.34).

We study an $SU(N)$-invariant action:

$$S_0 := \int dy\, \Big(L_0(\psi,\overline{\psi})(y) + L_0(A,u,\tilde{u})(y)\Big)\,, \quad L_0(\psi,\overline{\psi})(y) := \overline{\psi}_\alpha(y)\wedge(i\partial\!\!\!/ - m)\psi_\alpha(y)\;,$$

$$S := -\,\kappa \int dy\, g(y)\, \Big(j^\mu_a(y) A_{a\mu}(y) + L_1(A,u,\tilde{u})(y)\Big)\;, \tag{4.2.36}$$

with $\partial\!\!\!/ := \gamma_\mu \partial^\mu$, and where $L_0(A,u,\tilde{u})$ and $L_1(A,u,\tilde{u})$ are suitable polynomials in the gauge fields and the anticommuting Faddeev–Popov ghost fields u_a, $\tilde{u}_a$ (introduced in Sect. 5.1.2); the precise form of L_0 and L_1 is irrelevant in the following, we only require $s_b\, L_0(A,u,\tilde{u}) = 0$ and $s_b\, L_1(A,u,\tilde{u}) = 0$ for all b. QED (for $N=1$) and QCD (for $N=3$) fit in this framework: The quark fields ψ_α are in the fundamental representation of $SU(3)$.

By means of[132]

$$\frac{\delta S_0}{\delta \overline{\psi}_\alpha(x)} = (i\partial\!\!\!/ - m)\psi_\alpha(x) \quad \text{and} \quad \frac{\delta_r S_0}{\delta \psi_\alpha(x)} = -\big(i\partial^\mu \overline{\psi}_\alpha(x)\gamma_\mu + m\overline{\psi}_\alpha(x)\big) \tag{4.2.37}$$

and $(\lambda_a^c)_{\alpha\beta} = (\lambda_a)_{\beta\alpha}$ we obtain

$$\partial_\mu j^\mu_a(x) = \frac{\delta_r S_0}{\delta \psi_\alpha(x)} \wedge Q_{a\,\alpha}(x) + \overline{Q}_{a\,\alpha}(x) \wedge \frac{\delta S_0}{\delta \overline{\psi}_\alpha(x)}\;, \tag{4.2.38}$$

where

$$Q_{a\,\alpha}(x) := \frac{i}{2}\,(\lambda_a)_{\alpha\beta}\,\psi_\beta(x) = -(s_a\psi)_\alpha \quad \text{and} \quad \overline{Q}_{a\,\alpha} := Q^\dagger_{a\,\alpha}\,\gamma_0 = -(s_a\overline{\psi})_\alpha \tag{4.2.39}$$

with $Q^\dagger$ being the adjoint matrix in spinor space of Q. That is, the currents j^μ_a are conserved modulo the free field equations.

We point out that j^μ_a (4.2.33) is the Noether current of the free model w.r.t. the infinitesimal field transformation $(-s_a^{\psi,\overline{\psi}})$ generated by $s_a^{\psi,\overline{\psi}}\varphi := s_a\varphi$ for $\varphi = \psi,\overline{\psi}$ and $s_a^{\psi,\overline{\psi}}\phi := 0$ for $\phi = A^\nu_b,\, u_b,\, \tilde{u}_b$. Namely, by using $-s_a^{\psi,\overline{\psi}}\big(L_0(\psi,\overline{\psi}) + L_0(A,u,\tilde{u})\big) = 0$, the expression for j^μ_a (4.2.33) can be obtained from the formula for the Noether current (4.2.18).

In view of the MWI, we introduce the functionals[133]

$$\mathcal{A}_a := \int dx\; h(x)\, \partial_\mu j^\mu_a(x)$$

and the corresponding derivations

$$\delta_{hQ_a} F = \int dx\; h(x)\,\Big(\frac{\delta_r F}{\delta \psi_\alpha(x)} \wedge Q_{a\,\alpha}(x) + \overline{Q}_{a\,\alpha}(x) \wedge \frac{\delta F}{\delta \overline{\psi}_\alpha(x)}\Big)\,. \tag{4.2.40}$$

From (4.2.39) we conclude: If $h(x) = 1\; \forall x \in \operatorname{supp} F$ we have $\delta_{hQ_a} F = -s_a^{\psi,\overline{\psi}} F$, in accordance with (4.2.19).

---

[132] $\frac{\delta_r}{\delta\psi}$ is the functional derivative acting from the right-hand side, defined in (5.1.33).

[133] To avoid confusion with the gauge field we write $\mathcal{A}$ instead of A.

The interaction S (4.2.36) is $SU(N)$-invariant, $s_a S = 0\ \forall a$, but it is not $s_a^{\psi,\overline{\psi}}$-invariant: Proceeding analogously to the derivation of (4.2.35) we obtain

$$\delta_{hQ_a} S = \frac{i\kappa}{4} \int dx\ h(x)\, g(x)\, A_{b\mu}(x) \overline{\psi}(x) \wedge \gamma^\mu \left(-\lambda_b \lambda_a + \lambda_a \lambda_b\right) \psi(x)$$
$$= -\kappa\, f_{abc} \int dx\ h(x)\, g(x)\, A_{b\mu}(x)\, j_c^\mu(x)\ . \tag{4.2.41}$$

Note that the last expression is equal to $-\kappa \int dx\ h(x)\, g(x)\, (s_a A)_{c\mu}(x)\, j_c^\mu(x)$, as it must be due to $s_a(A_{c\mu} j_c^\mu) = 0$.

By using the result (4.2.41), the MWI for Q_a reads

$$\int dx\ h(x)\, \partial_\mu^x R\big(e_\otimes^S, j_a^\mu(x)\big) \equiv R\big(e_\otimes^S, \mathcal{A}_a\big) \tag{4.2.42}$$
$$= ka\, f_{abc} \int dx\ h(x)\, g(x)\, R\big(e_\otimes^S, (A_{b\mu} j_c^\mu)(x)\big)$$
$$+ \int dx\ h(x) \Big(\frac{\delta_r S_0}{\delta \psi_\alpha(x)} \wedge R\big(e_\otimes^S, Q_{a\,\alpha}(x)\big) + R\big(e_\otimes^S, \overline{Q}_{a\,\alpha}(x)\big) \wedge \frac{\delta S_0}{\delta \overline{\psi}_\alpha(x)}\Big)\ .$$

For $N = 1$ the first term on the right-hand side is absent and the MWI (4.2.42) yields the well-known current conservation of QED, $\partial_\mu^x j_S^\mu(x)|_{\mathfrak{C}_0} = 0$, which expresses gauge invariance of the model – see Sect. 5.2. But for $N \geq 2$, the interacting current is not conserved:

$$\partial_\mu^x\, j^\mu_{a\,S}(x)\big|_{\mathfrak{C}_0} = \kappa\, g(x)\, f_{abc}\, (A_{b\mu} j_c^\mu)_S(x)\big|_{\mathfrak{C}_0}\ . \tag{4.2.43}$$

An interesting expression (discussed, e.g., in formulas (11-88)–(11-89) of [109]) is

$$\partial_\mu^x\, \mathcal{T}\big(j^\mu_{a\,S}(x) \otimes j^\nu_{b\,S}(y)\big) - \mathcal{T}\big(\partial_\mu^x\, j^\mu_{a\,S}(x) \otimes j^\nu_{b\,S}(y)\big)\ , \tag{4.2.44}$$

where $\mathcal{T}\big(B_{1\,S}(x) \otimes B_{2\,S}(y)\big)$ denotes the time-ordered product of the interacting fields $B_{1\,S}(x)$ and $B_{2\,S}(y)$. In causal perturbation theory, the time-ordered product of the interacting fields $F_{1\,S}, \ldots, F_{l\,S}$ is defined by

$$T_S\big(F_1 \otimes \ldots \otimes F_l\big) := \overline{T}\big(e_\otimes^{-iS}\big) \star T\big(e_\otimes^{iS} \otimes F_1 \otimes \ldots \otimes F_l\big)$$
$$= \frac{\partial^l}{i^n\, \partial\lambda_1 \ldots \partial\lambda_l}\bigg|_{\lambda_1 = \cdots = \lambda_l = 0} \mathbf{S}_S\Big(\sum_{j=1}^l \lambda_j F_j\Big)\ , \tag{4.2.45}$$

where $F_1, \ldots, F_l, S \in \mathcal{F}_{\text{loc}}$ and $\mathbf{S}_S(F)$ is the relative S-matrix studied in Exers. 3.3.11 and 3.3.16 and in (3.7.12). Note that T_S is symmetric:

$$T_S\big(F_{\pi 1} \otimes \cdots \otimes F_{\pi l}\big) = T_S\big(F_1 \otimes \cdots \otimes F_l\big) \quad \text{for all} \quad \pi \in S_l.$$

Remark 4.2.4. In Exer. 3.3.11 we have proved that $\mathbf{S}_S(\cdot)$ satisfies the causality relation (3.3.27). This implies that T_S fulfills causal factorization, i.e., (3.3.5) with T_S in place of T, by means of the same argumentation as for: [(3.3.27) for $\mathbf{S}(\cdot)$] $\Rightarrow$ [(3.3.5) for T]. Proceeding analogously to part (a) of Lemma 3.3.2, we conclude that

$$T_S\big(B_1(x_1) \otimes \cdots \otimes B_l(x_l)\big) = B_{1\,S}(x_1) \star B_{2\,S}(x_2) \star \cdots \star B_{l\,S}(x_l)$$

if $x_j \notin (\{x_{j+1}, \ldots, x_l\} + \overline{V}_-)$ for all $1 \leq j \leq l-1$, where $B_1, \ldots, B_l \in \mathcal{P}$. This relation justifies the name "time-ordered product of interacting fields" for T_S (4.2.45).

Exercise 4.2.5. Let $\mathcal{A} = \int dx\; h(x)\, Q(x)\, \frac{\delta S_0}{\delta\varphi(x)}$ and δ_{hQ} the pertinent derivation (4.1.4). Derive the relation

$$T_S(\mathcal{A}\otimes F) = -T_S(\delta_{hQ}S\otimes F) + i\,(\delta_{hQ}F)_S + \int dx\; h(x)\, T_S\big(Q(x)\otimes F\big)\cdot \frac{\delta S_0}{\delta\varphi(x)} \tag{4.2.46}$$

from the MWI.

[*Solution*: We start with the MWI in the form (4.2.3), replace S by $S+\lambda F$ and compute the derivative $\frac{d}{i\,d\lambda}|_{\lambda=0}$; this yields

$$\begin{aligned} T\big(e_\otimes^{iS}\otimes F\otimes \mathcal{A}\big) = &- T\big(e_\otimes^{iS}\otimes F\otimes \delta_{hQ}S\big) + i\,T\big(e_\otimes^{iS}\otimes \delta_{hQ}F\big)\\ &+ \int dx\; h(x)\, T\big(e_\otimes^{iS}\otimes F\otimes Q(x)\big)\cdot \frac{\delta S_0}{\delta\varphi(x)}\,. \end{aligned}$$

By $\star$-multiplying this equation from the l.h.s. with $\overline{T}\big(e_\otimes^{-iS}\big)$ and by using (4.2.2) we get the assertion (4.2.46).]

To study the term (4.2.44) in the framework of causal perturbation theory, we first remark that, due the AWI, it holds that

$$\partial_\mu^x T_S\big(j_a^\mu(x)\otimes j_b^\nu(y)\big) = T_S\big((\partial_\mu j_a^\mu)(x)\otimes j_b^\nu(y)\big)\,.$$

Taking into account (4.2.43) and that the book [109] uses an on-shell formalism, the term corresponding to (4.2.44) is

$$\partial_\mu^x T_S\big(j_a^\mu(x)\otimes j_b^\nu(y)\big)\big|_{\mathcal{C}_{S_0}} - \kappa\, g(x)\, f_{acd}\, T_S\big((A_{c\mu}j_d^\mu)(x)\otimes j_b^\nu(y)\big)\big|_{\mathcal{C}_{S_0}}\,. \tag{4.2.47}$$

To compute this term by means of the MWI (4.2.46), we need $\delta_{hQ_a} j_b^\nu$: By a modification of (4.2.41) we get

$$\delta_{hQ_a} j_b^\nu(y) = f_{abc}\int dx\; h(x)\,\delta(x-y)\, j_c^\nu(y)\,.$$

With that we obtain

$$\begin{aligned} \int dx\; h(x)\,\partial_\mu^x T_S\big(j_a^\mu(x)\otimes j_b^\nu(y)\big) &\equiv T_S\big(\mathcal{A}_a\otimes j_b^\nu(y)\big) \qquad (4.2.48)\\ &\overset{(4.2.46)}{=} \int dx\; h(x)\Big(\kappa\, g(x)\, f_{acd}\, T_S\big((A_{c\mu}j_d^\mu)(x)\otimes j_b^\nu(y)\big) + i\,\delta(x-y)\, f_{abc}\, j_c^\nu{}_S(y)\\ &\quad + \frac{\delta_r S_0}{\delta\psi_\alpha(x)}\wedge T_S\big(Q_{a\,\alpha}(x)\otimes j_b^\nu(y)\big) + T_S\big(\overline{Q}_{a\,\alpha}(x)\otimes j_b^\nu(y)\big)\wedge \frac{\delta S_0}{\delta\overline{\psi}_\alpha(x)}\Big)\,. \end{aligned}$$

Restricting to $\mathcal{C}_{S_0}$, this yields

$$(4.2.47) = i\,\delta(x-y)\, f_{abc}\, j_c^\nu{}_S(y)\big|_{\mathcal{C}_{S_0}}\,, \tag{4.2.49}$$

This result agrees with what is postulated in [109]. It holds true, if the relevant cases of the MWI (i.e., (4.2.42) and (4.2.48)) are fulfilled by suitable renormalization of the time-ordered products.

Exercise 4.2.6 (Ward identity for a complex scalar field). Let ϕ be a complex scalar field as introduced in Exap. 1.3.2. The free action

$$S_0 := \int dx\ \big(\partial_\mu \phi^*(x)\, \partial^\mu \phi(x) - m^2 \phi^*(x)\, \phi(x)\big)\,.$$

is invariant under the global $U(1)$-transformation

$$\phi(x) \longmapsto \phi_\alpha(x) := e^{i\alpha}\phi(x)\ , \quad \phi^*(x) \longmapsto \phi^*_\alpha(x) := e^{-i\alpha}\phi^*(x)\ .$$

The pertinent Noether current can be computed by means of (4.2.18): Up to an irrelevant sign we obtain

$$j^\mu(x) := i\big(\phi(x)\, \partial^\mu \phi^*(x) - \phi^*(x)\, \partial^\mu \phi(x)\big)\,;$$

note that j^μ is real: $j^\mu(x)^* = j^\mu(x)$.

Let $P_1, \ldots, P_n \in \mathcal{P}$ be polynomials in ϕ and ϕ^* only, for simplicity we assume that they do not contain any derivatives of ϕ, ϕ^*. Let θ be the charge number operator:

$$i\theta\, P_j(\phi, \phi^*) := \frac{d}{d\alpha}\Big|_{\alpha=0} P_j(\phi_\alpha, \phi^*_\alpha)\ , \quad \text{that is,} \quad \theta = \phi\, \frac{\partial}{\partial \phi} - \phi^*\, \frac{\partial}{\partial \phi^*}\,,$$

or more explicitly: $\theta\big(\phi^r (\phi^*)^s\big) = (r-s)\, \phi^r (\phi^*)^s$. Show that the MWI belonging to the symmetry $A = \int dx\ h(x)\, \partial_\mu j^\mu(x)$ ($h \in \mathcal{D}(\mathbb{M})$ arbitrary) can be written as

$$\begin{aligned}
&\partial^x_\mu T\big(P_1(x_1) \otimes \cdots \otimes P_n(x_n) \otimes j^\mu(x)\big)\\
&\qquad - \hbar \sum_{l=1}^n \delta(x - x_l)\, T\big(P_1(x_1) \otimes \cdots \otimes (\theta P_l)(x_l) \otimes \cdots \otimes P_n(x_n)\big)\\
&\qquad = i\, T\big(P_1(x_1) \otimes \cdots \otimes P_n(x_n) \otimes \phi(x)\big) \cdot (\square + m^2)\phi^*(x)\\
&\qquad - i\, T\big(P_1(x_1) \otimes \cdots \otimes P_n(x_n) \otimes \phi^*(x)\big) \cdot (\square + m^2)\phi(x)\ .
\end{aligned}$$

We point out: The r.h.s. vanishes when we restrict the functionals to $\mathcal{C}_{S_0}$.

[*Solution*: By using $\frac{\delta S_0}{\delta \phi(x)} = -(\square + m^2)\phi^*(x)$ and $\frac{\delta S_0}{\delta \phi^*(x)} = -(\square + m^2)\phi(x)$ we get

$$\partial_\mu j^\mu(x) = Q(x)\, \frac{\delta S_0}{\delta \phi(x)} + Q^*(x)\, \frac{\delta S_0}{\delta \phi^*(x)} \quad \text{where} \quad Q(x) := -i\, \phi(x)\ .$$

We apply the MWI in the form (4.2.4), with F_j replaced by $P_j(x_j)$ and A replaced by $\partial_\mu j^\mu(x)$ (i.e., we omit $\int dx\ h(x)$):

$$\begin{aligned}
&T\big(P_1(x_1) \otimes \cdots \otimes P_n(x_n) \otimes \partial_\mu j^\mu(x)\big)\\
&\quad + \frac{\hbar}{i} \sum_{l=1}^n T_n\Big(P_1(x_1) \otimes \cdots \otimes \Big(Q(x)\, \frac{\delta P_l(x_l)}{\delta \phi(x)} + Q^*(x)\, \frac{\delta P_l(x_l)}{\delta \phi^*(x)}\Big) \otimes \cdots \otimes P_n(x_n)\Big)\\
&= T\big(P_1(x_1) \otimes \cdots \otimes P_n(x_n) \otimes Q(x)\big) \cdot \frac{\delta S_0}{\delta \phi(x)}\\
&\quad + T\big(P_1(x_1) \otimes \cdots \otimes P_n(x_n) \otimes Q^*(x)\big) \cdot \frac{\delta S_0}{\delta \phi^*(x)}\ .
\end{aligned}$$

The assertion follows by using the AWI and

$$Q(x)\, \frac{\delta P_l(x_l)}{\delta \phi(x)} + Q^*(x)\, \frac{\delta P_l(x_l)}{\delta \phi^*(x)} = -i\, \delta(x - x_l)\ (\theta P_l)(x_l)\ .]$$

Remark 4.2.7. It is a peculiarity of the examples treated in this section that the symmetry Q is of zeroth order in the coupling constant κ and, hence, $\delta_{hQ} \sim \kappa^0$. However, in general Q may be a formal power series in κ: $Q = \sum_{n=0}^{\infty} \kappa^n Q_n$; with this we have

$$\delta_{hQ} = \sum_{n=0}^{\infty} \kappa^n \, \delta_{hQ_n} \, , \quad \text{where} \quad \delta_{hQ_n} := \int dy \; h(y) \, Q_n(y) \, \frac{\delta}{\delta\varphi(y)} \; .$$

An important application of the MWI for which higher orders of κ appear in δ_{hQ} is BRST symmetry [54, Sect. 5]. There, δ_{hQ} is a localized version of the usual BRST transformation s, which is, e.g., for QED (see Sect. 5.5.2) and for Yang–Mills fields of the form $s = s_0 + \kappa s_1$.

4.3 Anomalous Master Ward Identity (Quantum Action Principle)

In Sects. 4.3–4.5 we follow [16, 17].

Let be given an interaction $S \in \mathcal{F}_{\mathrm{loc}}$ and a symmetry, i.e., a $Q \in \mathcal{P}$. To fulfill the pertinent MWI one usually proceeds as follows: One starts with a renormalization of the retarded or time-ordered product, which fulfills all other renormalization conditions. Then the MWI for this Q is in general violated. The theorem given in this section describes the structure of the violating term (also called "anomalous term"). By using this information, one then seeks a finite renormalization which maintains all other renormalization conditions – i.e., a map $Z \in \mathcal{R}$ in the language of the Main Theorem (Thm. 3.6.3)[134] – and which removes the violating term.

Theorem 4.3.1 (Anomalous Master Ward Identity or "Quantum Action Principle").

Part (a): *Let $(R_{n,1})_{n\in\mathbb{N}}$ be a retarded product satisfying the basic axioms, Translation covariance and Field independence. Then there exists a unique sequence of* linear *maps $\Delta \equiv (\Delta^n)_{n\in\mathbb{N}}$ (called "anomaly map"),*

$$\begin{aligned} &\Delta^n : \mathcal{P}^{\otimes(n+1)} \longrightarrow \mathcal{D}'(\mathbb{M}^{n+1}, \mathcal{F}_{\mathrm{loc}}) \; ; \\ &\qquad \otimes_{j=1}^n L_j \otimes Q \mapsto \Delta^n\big(\otimes_{j=1}^n L_j(x_j); Q(y)\big) \equiv \Delta^n\big((\otimes_{j=1}^n L_j) \otimes Q\big)(x_1, \ldots, x_n, y) \end{aligned} \tag{4.3.1}$$

(where $\mathcal{D}'(\mathbb{M}^{n+1}, \mathcal{F}_{\mathrm{loc}})$ is the space of $\mathcal{F}_{\mathrm{loc}}$-valued distributions on $\mathcal{D}(\mathbb{M}^{n+1})$), which are invariant with respect to permutations of the first n factors,

$$\Delta^n\big(\otimes_{j=1}^n L_{\pi j}(x_{\pi j}); Q(y)\big) = \Delta^n\big(\otimes_{j=1}^n L_j(x_j); Q(y)\big) \; , \quad \forall \pi \in S_n \; , \tag{4.3.2}$$

[134] If one works with a reduced list of renormalization conditions for the retarded product, $\mathcal{R}$ is replaced by a larger version, as explained in Remk. 3.6.6.

and which fulfill the "anomalous MWI":

$$R\Big(e_\otimes^{S/\hbar}, A+\delta_{hQ}S+\Delta(e_\otimes^S;hQ)\Big)=\int dy\, h(y)R\Big(e_\otimes^{S/\hbar},Q(y)\Big)\cdot\frac{\delta S_0}{\delta\varphi(y)}. \qquad (4.3.3)$$

Here we assume that the interaction S is of the form $S=\kappa\int dx\; g(x)\,L(x)$ with some $L\in\mathcal{P}$, $g\in\mathcal{D}(\mathbb{M})$; in addition, $A:=\int dx\; h(x)\,Q(x)\,\frac{\delta S_0}{\delta\varphi(x)}$ and δ_{hQ} is the corresponding derivation (4.1.4)*. We also use the obvious notations*

$$\Delta(e_\otimes^S;hQ):=\sum_{n=0}^{\infty}\frac{1}{n!}\int dy\; h(y)\,\Delta^n\big(S^{\otimes n};Q(y)\big)\in\mathcal{F}_{\mathrm{loc}}[\![\kappa]\!]\,,$$

$$\Delta^n\big(S^{\otimes n};Q(y)\big):=\kappa^n\int dx_1\dots dx_n\; g(x_1)\dots g(x_n)\,\Delta^n\big(\otimes_{j=1}^n L(x_j);Q(y)\big)\,.$$

The so-defined maps Δ^n have the following properties:

(i) $\Delta^0=0$;

(ii) **Locality and Translation covariance:** *There exist* linear *maps $P_a^n:\mathcal{P}^{\otimes(n+1)}\to\mathcal{P}$ which are uniquely determined by*

$$\Delta^n\big(\otimes_{j=1}^n L_j(x_j);Q(y)\big)=\sum_{a\in(\mathbb{N}^d)^n}\partial^a\delta(x_1-y,\dots,x_n-y)\,P_a^n(\otimes_{j=1}^n L_j;Q)(y)\,, \qquad (4.3.4)$$

where the sum over "a" is finite. The permutation invariance (4.3.2) *translates into*

$$P_{\pi a}^n(\otimes_{j=1}^n L_{\pi j};Q)=P_a^n(\otimes_{j=1}^n L_j;Q)\ ,\ \ \forall\pi\in S_n\ ,\ \ \pi a:=(a_{\pi 1},\dots,a_{\pi n})\ .$$

(iii) $\Delta^n\big(\otimes_{j=1}^n L_j(x_j);Q(y)\big)=\mathcal{O}(\hbar)\quad\forall n>0$ *if* $L_j\sim\hbar^0,\ Q\sim\hbar^0$.

(iv) **Field independence:**

$$\frac{\delta\Delta^n\Big(\bigotimes_{j=1}^n L_j(x_j);Q(y)\Big)}{\delta\varphi(z)}=\sum_{s=1}^n\Delta^n\Big(\bigotimes_{j(\neq s)}L_j(x_j)\otimes\frac{\delta L_s(x_s)}{\delta\varphi(z)};Q(y)\Big)+\Delta^n\Big(\bigotimes_j L_j(x_j);\frac{\delta Q(y)}{\delta\varphi(z)}\Big)\,. \qquad (4.3.5)$$

Part (b): *If the retarded product satisfies further renormalization conditions, all the maps $(\Delta^n)_{n\in\mathbb{N}}$ have corresponding properties:*

- *The Scaling degree property* (3.2.62) *– which follows from axiom* (k) *or, alternatively, can be used as a substitute for* (k) *– implies that on the right-hand side of* (4.3.4) *the sum $\sum_a$ is restricted by*

$$|a|+\dim\big(P_a^n(\otimes_{j=1}^n L_j;Q)\big)\le\sum_{j=1}^n\dim(L_j)+\dim(Q)+\frac{d+2}{2}-d\,n\,, \qquad (4.3.6)$$

where the mass dimension of a polynomial *is defined by* (3.2.63). *For a renormalizable interaction (that is,* $\dim(L) \leq d$*) this implies*

$$|a| + \dim\big(P^n_a(L^{\otimes n}; Q)\big) \leq \dim(Q) + \frac{d+2}{2}\,.$$

- *Lorentz covariance (axiom* (h)*) yields*

$$\beta_L\,\Delta\big(e_\otimes^S; hQ\big) = \Delta\big(e_\otimes^{\beta_L S}; h\,\beta_L Q\big)\,, \quad \forall L \in L_+^\uparrow\,;$$

- *the* $*$*-structure (axiom* (g)*) gives*

$$\Delta(e_\otimes^S; hQ)^* = \Delta(e_\otimes^{S^*}; \overline{h}\,Q^*)\,;$$

- *and from the Off-shell field equation (axiom* (j)*) it follows that*

$$\Delta(e_\otimes^S; h\,1) = 0\,, \quad \forall S \in \mathcal{F}_{\mathrm{loc}}\,,\; h \in \mathcal{D}(\mathbb{M})\,. \tag{4.3.7}$$

A further useful property of the anomaly map Δ *coming from the Off-shell field equation is worked out in Exer.* 4.3.3.

Note that the anomalous MWI (4.3.3) differs from the MWI (4.2.1) only by the local term $\Delta(e_\otimes^S; hQ)$, which clearly depends on the chosen renormalization for the retarded product. Therefore, the equality $\Delta(e_\otimes^S; hQ) = 0$ is a sufficient condition for the validity of the MWI for Q and S; it is also necessary due to the uniqueness of the maps Δ^n.

The main message of this theorem is that the *anomalous term is a local interacting field*, namely $R\big(e_\otimes^{S/\hbar}, \Delta(e_\otimes^S; hQ)\big)$ is the *interacting field* to the interaction S and corresponding to the local functional $\Delta(e_\otimes^S; hQ)$.

This theorem plays an important role in the proof of BRST symmetry for quantum Yang–Mills fields in curved spacetime given by S. Hollands [105], and in the analysis of the Batalin–Vilkovisky formalism in the framework of causal perturbation theory given by K. Fredenhagen and K. Rejzner [74].

The name "Quantum Action Principle (QAP)" refers to the famous Quantum Action Principle of Lowenstein [125] and Lam [119]; Theorem 4.3.1 is the corresponding statement in the framework of causal perturbation theory.

The anomalous MWI (4.3.3) can equivalently be written in terms of time-ordered products: Proceeding as in Sect. 4.2 we obtain

$$T\Big(e_\otimes^{iS/\hbar} \otimes \big(A + \delta_{hQ}S + \Delta(e_\otimes^S; hQ)\big)\Big) = \int dy\, h(y)\, T\big(e_\otimes^{iS/\hbar} \otimes Q(y)\big) \cdot \frac{\delta S_0}{\delta\varphi(y)}\,. \tag{4.3.8}$$

Before turning to the proof of the theorem, we mention a few consequences:

- From (4.3.4) we immediately see that $\Delta(e_\otimes^S; hQ)$ is translation invariant: $\beta_a\,\Delta(e_\otimes^S; hQ) = \Delta(e_\otimes^{\beta_a S}; h\,\beta_a Q)\,,\ \forall a \in \mathbb{R}^d\,.$

- The Field independence (4.3.5) implies that $\Delta(e_\otimes^S; hQ)$ satisfies the causal Wick expansion (3.1.23).
- By using that the values of $|a|$ appearing on the r.h.s. of (4.3.4) are bounded by the r.h.s. of (4.3.6), we conclude that

$$\begin{aligned}&\mathrm{sd}\,\omega_0\Big(\Delta^n\big(\otimes_{j=1}^n L_j(x_j); Q(y)\big)\Big)\\ &\qquad\leq \sum_{j=1}^n \dim(L_j) + \dim(Q) + \frac{d+2}{2}\,.\end{aligned} \tag{4.3.9}$$

In addition, we point out the similarity of properties (4.3.4) and (4.3.6) of Δ to formulas (3.6.34) and (3.6.35) expressing the properties "Locality and Translation covariance" and "Scaling degree" of an element Z of the Stückelberg–Petermann group. At first glance one might think that formula (4.3.6) has an additional term: $\frac{d+2}{2}$. However, since in the MWI there are terms in which Q is accompanied by $\frac{\delta S_0}{\delta\varphi}$, we have to consider $\dim(Q\,\frac{\delta S_0}{\delta\varphi}) = \dim(Q) + \frac{d+2}{2}$ (instead of only $\dim(Q)$) when comparing with (3.6.35).

Sketch of the proof. Ad (a): The procedure is analogous to the proof of part (a) of the Main Theorem (Thm. 3.6.3): Writing the implicit definition (4.3.8) of the sequence $(\Delta^l)_{l\in\mathbb{N}}$ to an arbitrary, fixed order in S (i.e., in g or in κ), we get a recursion relation which yields a unique inductive construction of (Δ^l). In detail: Omitting $\int dy\; h(y)$ in the anomalous MWI (4.3.8), we obtain[135]

$$\begin{aligned}\Delta^n\big(S^{\otimes n}; Q(y)\big) &= T\Big((iS/\hbar)^{\otimes n}\otimes Q(y)\Big)\cdot\frac{\delta S_0}{\delta\varphi(y)}\\ &\quad - T\Big((iS/\hbar)^{\otimes n}\otimes Q(y)\cdot\frac{\delta S_0}{\delta\varphi(y)}\Big)\\ &\quad - n\,T\Big((iS/\hbar)^{\otimes n-1}\otimes Q(y)\cdot\frac{\delta S}{\delta\varphi(y)}\Big)\\ &\quad - \sum_{l=0}^{n-1}\binom{n}{l}\,T\Big((iS/\hbar)^{\otimes n-l}\otimes \Delta^l\big(S^{\otimes l}; Q(y)\big)\Big)\,,\end{aligned} \tag{4.3.10}$$

where the $(l = n)$-summand of the sum appearing in the last term is written on the l.h.s. of the equation and $T_1\big(\Delta^n(\cdot)\big) = \Delta^n(\cdot)$ is used.

Taking into account linearity and permutation invariance (4.3.2), we extend this relation (4.3.10) to non-diagonal entries and write it in terms of the distribu-

[135] Alternatively, the proof can be given in terms of retarded products, by starting with (4.3.3), as it is done in [16]. This simplifies the proof of the $*$-structure property of $\Delta(e_\otimes^S; hQ)$, but the proof of part (a) is somewhat more involved.

tional kernels – analogously to (4.2.5):

$$\Delta^n\big(\otimes_{j=1}^n L_j(x_j);Q(y)\big)=\Big(\frac{i}{\hbar}\Big)^n T\Big(\bigotimes_{j=1}^n L_j(x_j)\otimes Q(y)\Big)\cdot\frac{\delta S_0}{\delta\varphi(y)} \tag{4.3.11}$$
$$-\Big(\frac{i}{\hbar}\Big)^n T\Big(\bigotimes_{j=1}^n L_j(x_j)\otimes Q(y)\cdot\frac{\delta S_0}{\delta\varphi(y)}\Big)$$
$$-\sum_{l=1}^n\Big(\frac{i}{\hbar}\Big)^{n-1} T\Big(\bigotimes_{j=1\,(j\neq l)}^n L_j(x_j)\otimes Q(y)\cdot\frac{\delta L_l(x_l)}{\delta\varphi(y)}\Big)$$
$$-\sum_{I\subset\{1,\dots,n\}\,,\,I^c\neq\emptyset}\Big(\frac{i}{\hbar}\Big)^{|I^c|} T\Big(\bigotimes_{k\in I^c} L_k(x_k)\otimes\Delta^{|I|}(\otimes_{j\in I}L_j(x_j);Q(y))\Big)\,.$$

We immediately get $\Delta^0=0$. In order that this relation gives a unique inductive construction of the sequence $(\Delta^l)_{l\in\mathbb{N}}$, the sum of distributions on the right-hand side of (4.3.11) must take values in $\mathcal{F}_{\mathrm{loc}}$. Assuming that this holds true, the so-obtained maps $\Delta^n\colon\mathcal{P}^{\otimes(n+1)}\longrightarrow\mathcal{D}'(\mathbb{M},\mathcal{F}_{\mathrm{loc}})$ are obviously linear and permutation invariant (4.3.2).

The main task is to prove that $\Delta^n(\otimes_{j=1}^n L_j;Q)$, defined inductively by (4.3.11), satisfies "Locality and Translation covariance" (4.3.4); the latter implies that $\Delta^n(\otimes_{j=1}^n L_j;Q)$ takes values in $\mathcal{F}_{\mathrm{loc}}$. The first and main step of the proof of (4.3.4) is to verify that[136]

$$\operatorname{supp}\Delta^n(\otimes_{j=1}^n L_j;Q)\subseteq\Delta_{n+1}\,. \tag{4.3.12}$$

To prove this we proceed analogously to (4.2.7)–(4.2.8): We show inductively that $\Delta^n\big(\otimes_{j=1}^n L_j(x_j);Q(y)\big)=0$ if there exists a $K\subset\{1,\dots,n\}$ with $K^c\neq\emptyset$ and $(\{x_k\,|\,k\in K^c\}+\overline{V}_+)\cap(\{x_j\,|\,j\in K\}\cup\{y\})=\emptyset$. (Again, the case "$y$ lies in the later set" is completely analogous.) If Δ^l vanishes to lower orders $l<n$, this verification is given by (4.2.7)–(4.2.8). To include the lower orders $\Delta^{|I|}$-terms, we use that, for the considered configuration, the range of the sum in the last term of (4.3.11) is reduced to $I\subseteq K$, since $\operatorname{supp}\Delta^{|I|}(\dots)\subseteq\Delta_{|I|+1}$ by the inductive assumption. Due to that, the causal factorization of this last term reads

$$T\Big(\bigotimes_{j\in K^c} L_j(x_j)\Big)\star\sum_{I\subseteq K}\Big(\frac{i}{\hbar}\Big)^{|K^c|+|K\setminus I|} T\Big(\bigotimes_{i\in K\setminus I} L_i(x_i)\otimes\Delta^{|I|}\big(\otimes_{s\in I}L_s(x_s);Q(y)\big)\Big)\,. \tag{4.3.13}$$

Now, in (4.2.7) and (4.2.8) we throughout replace $L(x_j)$ by $L_j(x_j)$. With that and by using causal factorization, the sum of the second, third and fourth term

[136] Mind the difference: Δ^n with an upper index n denotes the nth anomaly map (4.3.1); but by Δ_k with a lower index k we mean the thin diagonal in $\mathbb{M}^k$ (A.1.12). Note also that we mean the support in the sense of distributions, that is, the support of $\Delta^n(\cdots)$ itself and not of $\delta\Delta^n(\cdots)/\delta\varphi$ – Definition 1.3.3 does not apply here.

on the r.h.s. of (4.3.11) is equal to minus (4.2.7) minus (4.3.13). By means of the anomalous MWI (4.3.11) to order $|K|$, the latter sum is equal to minus (4.2.8). Finally this last expression is equal to minus the first term on the r.h.s. of (4.3.11), as explained after (4.2.8). Therefore, the r.h.s. of (4.3.11) vanishes on $\mathcal{D}(\mathbb{M}^{n+1} \setminus \Delta_{n+1})$ and, hence, this holds also for $\Delta^n\big(\otimes_{j=1}^n L_j(x_j); Q(y)\big)$.

To show that (4.3.12) implies (4.3.4) we proceed as follows: $\Delta^n(\otimes_{j=1}^n L_j; Q)$ is, according to its inductive definition (4.3.11), a distribution on $\mathcal{D}(\mathbb{M}^{n+1})$ which takes values in $\mathcal{F}$. Hence, according to Definition 1.2.1, it is of the form

$$\Delta^n\big(\otimes_{j=1}^n L_j(x_j); Q(y)\big) = \sum_k \int dz_1 \cdots dz_k \cdot f_k^n(\otimes_{j=1}^n L_j \otimes Q)(x_1, \ldots, x_n, y, z_1, \ldots, z_k)\, \varphi(z_1) \cdots \varphi(z_k) \ ,$$

where $f_k^n(\otimes_{j=1}^n L_j \otimes Q)(x_1, \ldots, x_n, y, z_1, \ldots, z_k) \in \mathcal{D}'(\mathbb{M}^{n+k+1})$ depends linearly on

$(\otimes_{j=1}^n L_j \otimes Q)$, is invariant under permutations of the pairs $(L_1, x_1), \ldots, (L_n, x_n)$ and is symmetric in $z_1, \ldots, z_k$. Due to (4.3.12) the support of $f_k^n(\otimes_{j=1}^n L_j \otimes Q)$ is contained in $\Delta_{n+1} \times \mathbb{M}^k$; but, to obtain the assertion (4.3.4), we have to show that $f_k^n(\otimes_{j=1}^n L_j \otimes Q)$ is a linear combination of derivatives of $\delta(x_1 - y, \ldots, x_n - y, z_1 - y, \ldots, z_k - y)$. For this purpose we take into account that the T-products on the r.h.s. of (4.3.11) are translation covariant and satisfy Field independence and that, by induction, these properties hold also for $\Delta^{|I|}$ if $|I| < n$. We conclude that $f_k^n(\otimes_{j=1}^n L_j \otimes Q)$ depends only on the relative coordinates and that

$$\frac{\delta^r \Delta^n\big(\otimes_{j=1}^n L_j(x_j); Q(y)\big)}{\delta\varphi(u_1) \cdots \delta\varphi(u_r)} = 0 \quad \text{if} \quad \{u_1, \ldots, u_r\} \not\subseteq \{x_1, \ldots, x_n, y\} \ , \quad \forall r \in \mathbb{N}^* \ ,$$

respectively, where we also use that $\frac{\delta B(x)}{\delta\varphi(u)} = 0$ if $u \neq x$ for any $B \in \mathcal{P}$. It follows that

$$\operatorname{supp} f_k^n(\otimes_{j=1}^n L_j \otimes Q)(x_1 - y, \ldots, z_k - y) \subseteq \{0\} \ ;$$

hence, $f_k^n(\otimes_{j=1}^n L_j \otimes Q)$ is indeed a linear combination of derivatives of $\delta(x_1 - y, \ldots, z_k - y)$ – see (A.1.17).

Uniqueness of $P_a^n(\otimes_{j=1}^n L_j; Q)(y)$ is due to the following: If some family $(B_a)_{a \in (\mathbb{N}^d)^n}$ of field polynomials $B_a \in \mathcal{P}$ fulfills

$$0 \stackrel{!}{=} \int dx_1 \cdots dx_n dy \sum_a \partial^a \delta(x_1 - y, \ldots, x_n - y)\, B_a(y)\, h(x_1, \ldots, x_n, y)$$

$$\equiv \sum_a (-1)^{|a|} \int dy\ B_a(y)\, \partial^a_{x_1, \ldots, x_n} h(x_1, \ldots, x_n, y)\big|_{x_1 = \ldots = x_n = y} \quad \forall h \in \mathcal{D}(\mathbb{M}^{n+1}) \ ,$$

then it must hold $B_a = 0\ \forall a$.

The important property (iii) of Δ^n follows immediately from the validity of the classical MWI and (3.1.71).

Field independence of Δ^n (property (iv)) can be proved inductively by means of formula (4.3.11) and Field independence of the T-products; note that the two terms with the derivative acting on $\frac{\delta S_0}{\delta\varphi(y)}$ cancel out, since $\frac{\delta^2 S_0}{\delta\varphi(z)\delta\varphi(y)} = -(\square + m^2)\,\delta(z-y)$ is a C-number (i.e., does not contain φ) and, hence, it does not matter whether $\frac{\delta^2 S_0}{\delta\varphi\,\delta\varphi}$ stands inside or outside of $T(\cdots)$.

Ad (b): The statement about the Off-shell field equation is trivial, because this field equation is precisely the MWI for $Q=1$. Lorentz covariance and $*$-structure of Δ^n follow from the inductive construction (4.3.11) (see Footn. 135). The proof of the mass dimension bound (4.3.6) is nontrivial, we refer to [16, 17]. As a substitute we prove a corresponding statement for the *massless* theory – this is done in Proposition 4.3.2 below. □

In Theorem 4.3.1 we describe the scaling behaviour of $\Delta^n(\otimes_{j=1}^n L_j;Q)$ only by the relation (4.3.6), which yields the bound (4.3.9) on the scaling degree of $\omega_0(\Delta^n(\ldots))$. The reason for this is that, for $m>0$, we are faced with m-dependent inhomogeneous polynomials: $S_0^{(m)} \equiv S_0 \notin \mathcal{P}_{\mathrm{hom}}$, $\frac{\delta S_0^{(m)}}{\delta\varphi} \notin \mathcal{P}_{\mathrm{hom}}$ and, hence, also $A^{(m)} \equiv A$ lies in general not in $\mathcal{P}_{\mathrm{hom}}$. A relation expressing the scaling behaviour of $\Delta^n(\otimes_{j=1}^n L_j;Q)$ can most probably be derived, if the mass m of the underlying star product, which is also the mass appearing in $S_0^{(m)}$, is also scaled, and if the retarded (or time-ordered) product fulfills the Sm-expansion axiom; but such a relation would be rather involved (cf. [16, Sect. 5.2]).

However, for $m=0$ the values of the anomaly map, i.e., $\Delta^n(\otimes_{j=1}^n L_j;Q)$, scale homogeneously, if suitable assumptions are made. This is essentially the message of the following proposition.

Proposition 4.3.2 (Homogeneous scaling of the anomalous term for $m=0$)**.** *Let $m=0$ and let $L_j, Q \in \mathcal{P}_{\mathrm{hom}}$ $(1\le j\le n)$ be given. In addition, we assume that the T-product (or R-product) fulfills (besides the axioms needed for part* (a) *of Thm.* 4.3.1) *the Sm-expansion axiom; for $m=0$ the latter is simply the condition of almost homogeneous scaling:*

$$(\rho\partial_\rho)^N\,\sigma_\rho\circ T_n(\sigma_\rho^{-1}F_1\otimes\cdots\otimes\sigma_\rho^{-1}F_n)=0 \tag{4.3.14}$$

for a sufficiently large $N\in\mathbb{N}^*$ (which depends on $F_1,\ldots,F_n\in\mathcal{F}_{\mathrm{loc}}$), as explained in Remk. 3.1.23 *and* (3.1.67)*.*

Then, the field polynomials $P_a^n(\otimes_{j=1}^n L_j;Q)$ appearing in (4.3.4) *are elements of $\mathcal{P}_{\mathrm{hom}}$ and the bound* (4.3.6) *on $|a|+\dim\big(P_a^n(\otimes_{j=1}^n L_j;Q)\big)$ may be replaced by the equality*

$$|a|+\dim\big(P_a^n(\otimes_{j=1}^n L_j;Q)\big)=\sum_{j=1}^n \dim(L_j)+\dim(Q)+\frac{d+2}{2}-d\,n\,. \tag{4.3.15}$$

This implies that $\Delta^n(\otimes_{j=1}^n L_j; Q)$ scales homogeneously:

$$\rho^{\sum_{j=1}^n \dim(L_j)+\dim(Q)+\frac{d+2}{2}}\,\sigma_\rho\circ\Delta^n\big(\otimes_{j=1}^n L_j(\rho x_j);Q(\rho y)\big)=\Delta^n\big(\otimes_{j=1}^n L_j(x_j);Q(y)\big) \tag{4.3.16}$$

Proof. The idea of proof is that, in the inductive construction (4.3.11) of Δ^n, almost homogeneous scaling of the T-products implies almost homogeneous scaling of Δ^n. Then, taking into account the structure (4.3.4) of Δ^n, it follows that the scaling of Δ^n is even homogeneous.

To work this out in detail, we first note the relations

$$\begin{aligned}&\rho^{\dim L}\,L(\rho x)=\sigma_\rho^{-1}\,L(x)\quad\text{in particular}\quad \rho^{\frac{d+2}{2}}\,\frac{\delta S_0}{\delta\varphi(\rho y)}=\sigma_\rho^{-1}\,\frac{\delta S_0}{\delta\varphi(y)}\,,\\ &\rho^{\dim(L)+\frac{d+2}{2}}\,\frac{\delta L(\rho x)}{\delta\varphi(\rho y)}=\sigma_\rho^{-1}\,\frac{\delta L(x)}{\delta\varphi(y)}\quad\text{for}\quad L\in\mathcal{P}_{\mathrm{hom}}\,,\end{aligned} \tag{4.3.17}$$

where we take into account that $\frac{\delta S_0}{\delta\varphi(y)}=-\Box\varphi(y)$. The last identity can be obtained from (3.1.53).

Proceeding by induction on n, we use that Δ^n is given by (4.3.11); this yields:

$$\begin{aligned}&\rho^{\sum_{j=1}^n \dim(L_j)+\dim(Q)+\frac{d+2}{2}}\,\sigma_\rho\circ\Delta^n\big(\otimes_{j=1}^n L_j(\rho x_j);Q(\rho y)\big)\\ &\quad=\Big(\frac{i}{\hbar}\Big)^n\,\sigma_\rho\circ T\Big(\bigotimes_j\rho^{\dim(L_j)}\,L_j(\rho x_j)\otimes\rho^{\dim(Q)}\,Q(\rho y)\Big)\cdot\rho^{\frac{d+2}{2}}\,\sigma_\rho\,\frac{\delta S_0}{\delta\varphi(\rho y)}\\ &\quad\quad-\Big(\frac{i}{\hbar}\Big)^n\,\sigma_\rho\circ T\Big(\bigotimes_j\rho^{\dim(L_j)}\,L_j(\rho x_j)\otimes\rho^{\dim(Q)}\,Q(\rho y)\cdot\rho^{\frac{d+2}{2}}\,\frac{\delta S_0}{\delta\varphi(\rho y)}\Big)\\ &\quad\quad-\sum_{l=1}^n\Big(\frac{i}{\hbar}\Big)^{n-1}\,\sigma_\rho\circ T\Big(\bigotimes_{j(\neq l)}\rho^{\dim(L_j)}\,L_j(\rho x_j)\\ &\qquad\qquad\qquad\otimes\rho^{\dim(Q)}\,Q(\rho y)\cdot\rho^{\dim(L_l)+\frac{d+2}{2}}\,\frac{\delta L_l(\rho x_l)}{\delta\varphi(\rho y)}\Big)\\ &\quad\quad-\sum_{I\subseteq\{1,\dots,n\}\,,\,I^c\neq\emptyset}\Big(\frac{i}{\hbar}\Big)^{|I^c|}\,\sigma_\rho\circ T\Big(\bigotimes_{k\in I^c}\rho^{\dim(L_k)}\,L_k(\rho x_k)\\ &\qquad\qquad\qquad\otimes\rho^{\sum_{j\in I}\dim(L_j)+\dim(Q)+\frac{d+2}{2}}\,\Delta^{|I|}\big(\otimes_{j\in I}L_j(\rho x_j);Q(\rho y)\big)\Big)\,.\end{aligned}$$

The terms on the r.h.s. can be written such that their ρ-dependence is of the form $\sigma_\rho\circ T_r\circ(\sigma_\rho^{-1})^{\otimes r}$, by using formulas (4.3.17) and, in the last term, the inductive assumption (i.e., (4.3.16) with $|I|$ in place of n). Then, almost homogeneous scaling of the T-product (4.3.14) implies that

$$(\rho\partial_\rho)^N\Big(\rho^{\sum_{j=1}^n \dim(L_j)+\dim(Q)+\frac{d+2}{2}}\,\sigma_\rho\circ\Delta^n\big(\otimes_{j=1}^n L_j(\rho x_j);Q(\rho y)\big)\Big)=0 \tag{4.3.18}$$

for a sufficiently large $N\in\mathbb{N}^*$.

Finally we insert formula (4.3.4) into (4.3.18). Since we do not know yet that $P^n_a := P^n_a(\otimes_{j=1}^n L_j;Q)$ is an element of $\mathcal{P}_{\mathrm{hom}}$, we write

$$P^n_a = \sum_k P^n_{a,k} \quad \text{with} \quad P^n_{a,k} \in \mathcal{P}_{\mathrm{hom}}\ , \quad \text{hence } \sigma_\rho P^n_a(\rho y) = \sum_k \rho^{-\dim(P^n_{a,k})}\, P^n_{a,k}(y)\ .$$

With this we obtain

$$(\rho\partial_\rho)^N\Big(\rho^{\sum_{j=1}^n \dim(L_j)+\dim(Q)+\frac{d+2}{2}-|a|-dn-\dim(P^n_{a,k})}$$
$$\cdot \sum_{a,k}\partial^a\delta(x_1-y,\dots,x_n-y)\,P^n_{a,k}(y)\Big) = 0\ . \qquad (4.3.19)$$

We conclude that the exponent of ρ vanishes; hence, $\dim(P^n_{a,k})$ does not depend on k, that is, $P^n_a \in \mathcal{P}_{\mathrm{hom}}$. In addition, the vanishing of this exponent yields the restriction (4.3.15) on the sum over a, and it also implies that (4.3.19) holds true already for $N = 1$. Consequently, the latter holds also for (4.3.18) – this is the assertion (4.3.16). □

Exercise 4.3.3 (Property of the anomaly map Δ coming from the Off-shell field equation). Let a symmetry $Q \in \mathcal{P}$ be given. A widespread method to find a finite renormalization which removes the anomalous terms $\Delta^n\big(\otimes_j F_j; Q(y)\big)$ is to follow the inductive Epstein–Glaser construction of the T-products, that is, to proceed by induction on n. So we assume that $T_1,\dots,T_n$ are constructed and that

$$\Delta^k\big(\otimes_{j=1}^{n-1}F_j;Q(y)\big) = 0 \qquad \forall F_j \in \mathcal{F}_{\mathrm{loc}}\ , \quad \forall k<n\ .$$

Considering the MWI to order n, given in (4.2.4), for the case that at least one of the factors F_j is equal to $\varphi(x_j)$, the T-products of order $(n+1)$ appearing in this MWI are *uniquely* determined by induction and by the Off-shell field equation axiom. Prove that this MWI holds true, i.e., that

$$\Delta^n\Big(\varphi(x)\otimes\bigotimes_{j=1}^{n-1}F_j;Q(y)\Big) = 0 \qquad \forall F_j \in \mathcal{F}_{\mathrm{loc}}\ .$$

[*Solution*: The MWI we want to prove reads

$$0 \overset{?}{=} T_{n+1}\Big(\varphi(x)\otimes\bigotimes_{j=1}^{n-1}F_j\otimes Q(y)\Big)\cdot\frac{\delta S_0}{\delta\varphi(y)} - T_{n+1}\Big(\varphi(x)\otimes\bigotimes_{j=1}^{n-1}F_j\otimes Q(y)\,\frac{\delta S_0}{\delta\varphi(y)}\Big)$$
$$+\,i\hbar\Big[\sum_{l=1}^{n-1}T_n\Big(\varphi(x)\otimes\bigotimes_{j=1\,(j\neq l)}^{n-1}F_j\otimes Q(y)\,\frac{\delta F_l}{\delta\varphi(y)}\Big) + \delta(x-y)\,T_n\Big(\bigotimes_{j=1}^{n-1}F_j\otimes Q(y)\Big)\Big]\ .$$

To show that the r.h.s. vanishes, we insert the Off-shell field equation (3.3.20) and, for the first term, we use the relation

$$\frac{\delta S_0}{\delta\varphi(y)}\cdot\frac{\delta}{\delta\varphi(z)}T\big(\cdots\otimes Q(y)\big) = \frac{\delta}{\delta\varphi(z)}\Big\{\frac{\delta S_0}{\delta\varphi(y)}\cdot T\big(\cdots\otimes Q(y)\big)\Big\}$$
$$+\,(\square_z+m^2)\delta(z-y)\,T\big(\cdots\otimes Q(y)\big)\ .$$

Proceeding in this way, the assertion reads

$$0 \stackrel{?}{=} \Big[\varphi(x) + \hbar \int dz\ \Delta^F(x-z)\,\frac{\delta}{\delta\varphi(z)}\Big]\Big\{T_n\Big(\bigotimes_{j=1}^{n-1} F_j \otimes Q(y)\Big)\cdot \frac{\delta S_0}{\delta\varphi(y)}$$

$$- T_n\Big(\bigotimes_{j=1}^{n-1} F_j \otimes Q(y)\,\frac{\delta S_0}{\delta\varphi(y)}\Big) + i\hbar\sum_{l=1}^{n-1} T_{n-1}\Big(\bigotimes_{j=1\,(j\neq l)}^{n-1} F_j \otimes Q(y)\,\frac{\delta F_l}{\delta\varphi(y)}\Big)\Big\}$$

$$+ \hbar\Big\{\int dz\ \Delta^F(x-z)\,(\Box_z + m^2)\delta(z-y) + i\,\delta(x-y)\Big\}\,T_n\Big(\bigotimes_{j=1}^{n-1} F_j \otimes Q(y)\Big)\,.$$

Both $\{\cdots\}$-brackets vanish; for the first one this is due to the MWI to order $(n-1)$ and for the second one this is due to $(\Box + m^2)\Delta^F = -i\delta$.]

4.4 Reduction of the MWI to the "quantum part"

Since the MWI holds true in classical field theory, the quantum MWI is automatically satisfied for tree diagrams; working in terms of the R-product, this statement follows from Claim 3.1.28 (in Sect. 3.1.7), however, it holds also for the MWI in terms of the T-product. Therefore, there is some redundancy if one formulates the MWI in terms of the whole time-ordered product. In this section we show that there is an equivalent reformulation of the (anomalous) MWI in terms of the "quantum part" of the T-product. Mostly this simplifies the search for a finite renormalization $Z \in \mathcal{R}$ which removes the anomalous term from the anomalous MWI. And it is the so-obtained formulation of the anomalous MWI which corresponds to the formulation of the QAP mostly used in algebraic renormalization, see, e.g., [135].

4.4.1 Proper vertices

Proper vertices[137] are an old and standard tool in pQFT. Usually they are described in terms of the "vertex functional" $\Gamma(h)$ (where $h \in \mathcal{S}(\mathbb{M},\mathbb{R})$ is a field configuration), which is defined as the Legendre transform $j \to h$ of the generating functional $Z(j)$ of the connected Green's functions (where j is the "classical source" of φ). This procedure is due to G. Jona-Lasinio [110], it can be found in many textbooks, e.g., [135], [109, Sect. 6-2-2].

In the framework of causal perturbation theory, proper vertices (or the "proper interaction") have been introduced by proceeding more elementary [16, 17]. In terms of the R-product the basic idea is the following: An interacting quantum field can be rewritten as a *classical* field with a *non-local interaction* ("proper

[137] In the literature proper vertices (or the "proper interaction") are sometimes called "effective vertices" (or "effective interaction" resp.). However, differently to what we are doing here, the notion "effective field theory" usually means an *approximation* to the perturbation series. For this reason we omit the word "effective" and use the terminology of [109, Sect. 6-2-2].

interaction") and a non-local field vertex ("proper field vertex"), which agree to lowest order in $\hbar$ with the original local interaction and the original local field vertex, respectively. Hence, the proper interaction and the proper field vertex can be understood as the "quantum parts" of the interacting quantum field. In terms of the diagrammatic interpretation of the unrenormalized R-product, this rewriting can be justified as follows: Since R_{cl} is the sum of all (connected) tree diagrams (Claim 3.1.28), we interpret each diagram as tree diagram with non-local vertices ("proper vertices") given by the 1-particle-irreducible (1PI) subdiagrams.[138]

This structural decomposition of Feynman diagrams into [tree diagrams] "times" [1PI-diagrams] can just as well be done for the unrenormalized T-product, and it is this latter form of proper vertices which is well known in the traditional literature. Since the T-product is totally symmetric, it is simpler to introduce proper vertices in terms of the T-product than in terms of the R-product and, hence, we work with the former. (For a study of proper vertices belonging to the R-product we refer to [16, App.].)

A main motivation to introduce proper vertices is that the renormalization of an arbitrary diagram reduces to the renormalization of its 1PI-subdiagrams, because tree diagrams can be renormalized by the direct extension (3.2.22). Note that also some 1PI diagrams can be renormalized by the direct extension – see Exap. 3.2.10.

We are now going to formulate this diagrammatic intuition about proper vertices in terms of algebraic definitions for the renormalized T-product. First we eliminate all disconnected diagrams. This elimination is not necessary, proper vertices can also be defined in terms of the complete T-product – the two definitions agree, see [16]. However, working with the connected part T^c of a T-product (Definition 4.4.1), we can interpret the connected tree diagrams as the "classical limit" of T^c (formula (4.4.5) below); hence, all quantum effects are described by diagrams containing at least one 1PI-subdiagram. In addition, proceeding this way, it is much easier to verify that the proper interaction is of the form [original local interaction] + [quantum corrections] (see Exer. 4.4.7 below).

The connected part of a T-product. The following Definition does not use any diagrammatic interpretation of the T-product.

Definition 4.4.1. The *connected part* T^c of a time-ordered product T can be defined recursively by

$$T_n^c(\otimes_{j=1}^n F_j) := T_n(\otimes_{j=1}^n F_j) - \sum_{|P|\geq 2} \prod_{J\in P} T^c_{|J|}(\otimes_{j\in J} F_j)\,, \quad n \in \mathbb{N}^*\,, \tag{4.4.1}$$

[138]The defining criterion for a 1-particle-irreducible Feynman diagram is that it cannot be split into two disconnected parts by cutting only one inner line. Feynman diagrams, which do not satisfy this criterion are called "1-particle-reducible". Trivial examples for 1-particle-reducible diagrams are disconnected diagrams or connected tree diagrams. Every 1PI diagram is a loop diagram; but, there are connected loop diagrams which are not 1PI, see Footn. 34.

where P is a partition of $\{1,\dots,n\}$ into $|P|$ subsets (which are labeled by J) and $\prod$ means the classical product. For $n=0$ we set $T_0^c(c) := 0\ \forall c\in\mathbb{C}$.

An immediate consequence of this definition is that $T_n^c : \mathcal{F}_{\mathrm{loc}}^{\otimes n} \longrightarrow \mathcal{F}$ satisfies the axioms (i) (Linearity), (ii) (Initial condition, explicitly: $T_1^c(F)=F$) and (iii) (Symmetry) for a T-product. Due to linearity, we may use the notations (A.1.7)–(A.1.8) also for T^c.

One can prove that T and T^c are related by the *"linked cluster theorem"*:

$$T(e_\otimes^{iF}) = \exp_\bullet\big(T^c(e_\otimes^{iF})\big)\,, \tag{4.4.2}$$

where "$\exp_\bullet$" denotes the exponential function w.r.t. the classical product.

Exercise 4.4.2.

(a) Express $T_n^c(F^{\otimes n})$ in terms of $\{T_l(F^{\otimes l})\,|\,1\le l\le n\}$ for $n=2,3$.

(b) Verify the linked cluster theorem (4.4.2) to third order in F, by using the results of part (a).

[*Solution*: (a) $T_2^c(F^{\otimes 2}) = T_2(F^{\otimes 2}) - F\cdot F$ and

$$T_3^c(F^{\otimes 3}) = T_3(F^{\otimes 3}) - 3\,T_2(F^{\otimes 2})\cdot F + 2\,F\cdot F\cdot F\,.$$

(b) Denoting by $[\dots]_3$ the selection of the term of third order in F, we get

$$\begin{aligned}[\exp_\bullet(T^c(e_\otimes^{iF}))]_3 &= [T^c(e_\otimes^{iF})]_3 + \frac{1}{2}\,[T^c(e_\otimes^{iF})\cdot T^c(e_\otimes^{iF})]_3 + \frac{1}{6}\,[T^c(e_\otimes^{iF})\cdot T^c(e_\otimes^{iF})\cdot T^c(e_\otimes^{iF})]_3\\ &= \frac{i^3}{6}\,\Big(T_3^c(F^{\otimes 3}) + 3\,T_2^c(F^{\otimes 2})\cdot F + F\cdot F\cdot F\Big) = \frac{i^3}{6}\,T_3(F^{\otimes 3})\,.\end{aligned}$$

In the last equality-sign we use the results of part (a).]

Definition 4.4.1 agrees with the diagrammatic understanding of "connected" in the following sense: restricted to $\mathcal{D}(\check{\mathbb{M}}^n)$ (defined in Lemma 3.3.2), $T_n^c\big(A_1(x_1)\otimes\cdots\otimes A_n(x_n)\big)$ is precisely the contribution of all connected diagrams to the unrenormalized T-product $A_1(x_1)\star_{\Delta_m^F}\cdots\star_{\Delta_m^F} A_n(x_n)$ (3.3.10). This statement can easily be verified to lowest orders by using the results of Exer. 4.4.2(a).

Remark 4.4.3 (Main Theorem in terms of T^c)**.** Using the linked cluster theorem (4.4.2), formula (3.6.20) of the Main Theorem translates into the equivalent relation

$$\widehat{T}^c\big(e_\otimes^{iF/\hbar}\big) = T^c\big(e_\otimes^{iZ(F)/\hbar}\big)\,,\quad \forall F\in\mathcal{F}_{\mathrm{loc}}\,. \tag{4.4.3}$$

The time-ordered connected tree product. Proceeding analogously to Sect. 3.1.6 and the proof of Claim 3.1.28, we are now going to explain that, for $A_1,\dots,A_n\sim\hbar^0$, it holds that

$$T_n^c\big(A_1(x_1)\otimes\cdots\otimes A_n(x_n)\big) = \mathcal{O}(\hbar^{n-1})\,. \tag{4.4.4}$$

In terms of diagrams one can argue as follows: For the unrenormalized T-product (i.e., restricted to $\mathcal{D}(\check{\mathbb{M}}^n)$) each diagram contributing to the l.h.s. has at least

$(n-1)$ inner lines and each inner line has precisely one factor $\hbar$,[139] and there are no further $\hbar$. Hence, the unrenormalized time-ordered product fulfills (4.4.4). Due to the axiom (x) for the T-product, this relation is maintained in the process of renormalization.

For the unrenormalized T_n^c we additionally see that the term of lowest order in $\hbar$ of (4.4.4), i.e., the limit

$$\hbar^{(n-1)} \lim_{\hbar\to 0} \hbar^{-(n-1)}\, T_n^c\big(A_1(x_1)\otimes\cdots\otimes A_n(x_n)\big)\big|_{\mathcal{D}(\check{\mathbb{M}}^n)}\ ,$$

selects precisely the contribution of all connected tree diagrams, because the connected tree diagrams are precisely the connected diagrams with $(n-1)$ inner lines. This is the motivation to define the "*connected tree part*" $T^c_{\mathrm{tree},n}$ of the renormalized T-product T_n by

$$T^c_{\mathrm{tree},n} := \hbar^{(n-1)} \lim_{\hbar\to 0} \hbar^{-(n-1)}\, T_n^c\ :\ \mathcal{F}_{\mathrm{loc}}^{\otimes n} \longrightarrow \mathcal{F}\ . \tag{4.4.5}$$

Hence, T^c_{tree} is the "classical part" of T^c, and connected loop diagrams are of higher orders in $\hbar$. Since the contributions of tree diagrams to $A_1(x_1)\star_{\Delta_m^F}\cdots\star_{\Delta_m^F} A_n(x_n)$ are given by pure tensor products of distributions (pointwise products of distributions do not appear) and, hence, are well defined on $\mathcal{D}(\mathbb{M}^n)$, we may write

$$T^c_{\mathrm{tree},n}\big(A_1(x_1)\otimes\cdots\otimes A_n(x_n)\big) = A_1(x_1)\star_{\Delta_m^F}\cdots\star_{\Delta_m^F} A_n(x_n)\big|_{\text{connected tree diagrams}} \tag{4.4.6}$$

in $\mathcal{D}'(\mathbb{M}^n)$, that is, $T^c_{\mathrm{tree},n}$ may be computed by the Feynman rules.

We point out that the definition (4.4.5) works only for *local* entries. But, since in general proper vertices are non-local, we need the "*time-ordered connected tree product*", which is the generalization of the connected tree part (4.4.5) to *non-local* entries and is also denoted by $T^c_{\mathrm{tree},n}$. The definition of $T^c_{\mathrm{tree},n} : \mathcal{F}^{\otimes n} \longrightarrow \mathcal{F}$ can be given by means of formula (4.4.6):

$$T^c_{\mathrm{tree},n}(\otimes_{j=1y}^n F_j) := F_1 \star_{\Delta_m^F}\cdots\star_{\Delta_m^F} F_n\big|_{\text{connected tree diagrams}} \qquad \forall F_1,\ldots,F_n\in\mathcal{F}\ . \tag{4.4.7}$$

To give a more explicit, analytic definition we proceed recursively: For an arbitrary connected tree diagrams of order $(n+1)$ with (possibly non-local) vertices $F_1,\ldots,F_n,F_{n+1}$, the additional vertex F_{n+1} is connected to the other vertices as shown in Figure 4.1. This figure has to be understood with the specifications $k\in\{1,\ldots,n\}$, $I_j\neq\emptyset\ \forall j$ and $I_1\sqcup\cdots\sqcup I_k=\{1,\ldots,n\}$, where $\sqcup$ means the disjoint union; moreover we use the notation $F_I := \otimes_{j\in I}F_j$. This diagrammatic consideration yields the following definition:

Definition 4.4.4. The *time-ordered connected tree product* $T^c_{\mathrm{tree}} = (T^c_{\mathrm{tree},n})_{n\in\mathbb{N}}$, where $T^c_{\mathrm{tree},n} : \mathcal{F}^{\otimes n}\longrightarrow\mathcal{F}$, is defined recursively by $T^c_{\mathrm{tree},0}(c) := 0\ \forall c\in\mathbb{C}$,

[139] The latter is due to the factor $\hbar$ accompanying each propagator H_m in the definition of the star product (Definition 2.1.1, (2.1.5)).

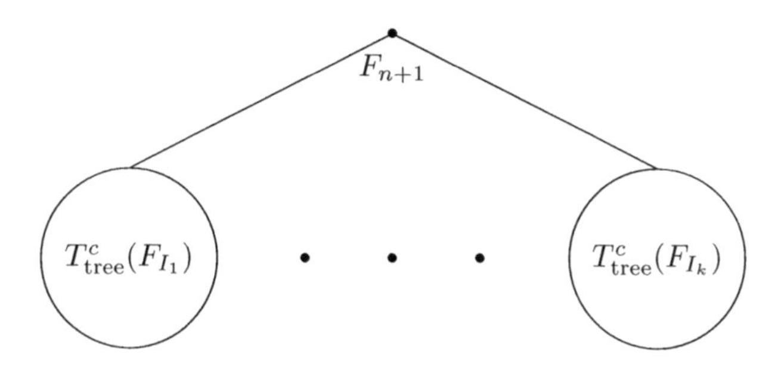

Figure 4.1: Structure of an arbitrary connected tree diagram

$T^c_{\text{tree},1}(F) := F$ and

$$T^c_{\text{tree},n+1}(\otimes_{j=1}^{n+1} F_j) := \sum_{k=1}^{n} \int dx_1 \ldots dx_k\, dy_1 \ldots dy_k\, \frac{\delta^k F_{n+1}}{\delta\varphi(x_1)\ldots\delta\varphi(x_k)}$$
$$\cdot \prod_{j=1}^{k} \Delta_m^F(x_j - y_j)\, \frac{\hbar^k}{k!} \sum_{I_1\sqcup\cdots\sqcup I_k=\{1,\ldots,n\}} \frac{\delta T^c_{\text{tree},|I_1|}(F_{I_1})}{\delta\varphi(y_1)} \cdot \ldots \cdot \frac{\delta T^c_{\text{tree},|I_k|}(F_{I_k})}{\delta\varphi(y_k)}\,, \tag{4.4.8}$$

with Δ_m^F being the Feynman propagator (2.3.8). Note that in the sum over $I_1, \ldots$ $\ldots, I_k$ the order of $I_1, \ldots, I_k$ is distinguished and, hence, there is a factor $\frac{1}{k!}$.

From (4.4.8) we inductively see that $T^c_{\text{tree},n+1} \sim \hbar^n$ – in agreement with (4.4.5). For local entries, $F_1, \ldots, F_n \in \mathcal{F}_{\text{loc}}$, the two definitions (4.4.5) and (4.4.8) of $T^c_{\text{tree},n}(F_1 \otimes \cdots \otimes F_n)$ indeed agree, since both definitions are equivalent to (4.4.7).

From the inductive definition (4.4.8) we see that $T^c_{\text{tree},n} : \mathcal{F}^{\otimes n} \longrightarrow \mathcal{F}$ is *linear*; in addition, $T^c_{\text{tree},n}$ is also totally symmetric – this follows from (4.4.7), since $F_1 \star_{\Delta_m^F} \cdots \star_{\Delta_m^F} F_n$ is invariant under permutations of $F_1, \ldots, F_n$. Due to linearity we may use the notations (A.1.7)–(A.1.8) also for T^c_{tree}. Hence, for example, we may write

$$T^c_{\text{tree}}(e_\otimes^F \otimes G) = \frac{d}{d\lambda}\Big|_{\lambda=0} T^c_{\text{tree}}(e_\otimes^{F+\lambda G})$$
$$= G + \sum_{n=1}^{\infty} \frac{1}{n!} T^c_{\text{tree},n+1}(F^{\otimes n} \otimes G)\,, \quad \forall F, G \in \mathcal{F}\,. \tag{4.4.9}$$

Exercise 4.4.5 (Properties of T^c_{tree}).

(a) Verify to third order that T^c_{tree} is non-associative, that is,

$$T^c_{\text{tree}}\big(F_1 \otimes T^c_{\text{tree}}(F_2 \otimes F_3)\big) \neq T^c_{\text{tree}}\big(T^c_{\text{tree}}(F_1 \otimes F_2) \otimes F_3\big)\,, \quad F_1, F_2, F_3 \in \mathcal{F}\,,$$

and that both sides of this equation differ from $T^c_{\text{tree},3}(F_1 \otimes F_2 \otimes F_3)$.

(b) An additional hint that T^c_{tree} is the "classical part" of T^c is the following result: Similarly to the factorization of the classical retarded product (1.7.6), T^c_{tree} fulfills

$$T^c_{\text{tree}}\Big(e_\otimes^{iS/\hbar}\otimes FG\Big)=T^c_{\text{tree}}\Big(e_\otimes^{iS/\hbar}\otimes F\Big)\cdot T^c_{\text{tree}}\Big(e_\otimes^{iS/\hbar}\otimes G\Big)\,. \tag{4.4.10}$$

Prove this relation.

[*Solution*: (a) Compared with $T^c_{\text{tree},3}(F_1\otimes F_2\otimes F_3)$ the contributions of the diagrams

F_1 — F_2, F_1 — F_3 and F_1 — F_3, F_2 — F_3

are missing in $T^c_{\text{tree}}(F_1\otimes T^c_{\text{tree}}(F_2\otimes F_3))$ and $T^c_{\text{tree}}(T^c_{\text{tree}}(F_1\otimes F_2)\otimes F_3)$, respectively.
(b) To prove (4.4.10) to nth order in S, we start with

$$\frac{1}{n!}T^c_{\text{tree}}\Big(S^{\otimes n}\otimes FG\Big)=\sum_{k=1}^n\int dx_1\dots dy_k\,\frac{\delta^k(FG)}{\delta\varphi(x_1)\dots\delta\varphi(x_k)}\prod_{j=1}^k\Delta^F_m(x_j-y_j)$$
$$\cdot\sum_{n_1+\dots+n_k=n,\,n_j>0}\frac{\hbar^k}{k!n_1!\dots n_k!}\,\frac{\delta T^c_{\text{tree}}(S^{\otimes n_1})}{\delta\varphi(y_1)}\cdots\frac{\delta T^c_{\text{tree}}(S^{\otimes n_k})}{\delta\varphi(y_k)}\,.$$

Using the Leibniz rule

$$\frac{\delta^k(FG)}{\delta\varphi(x_1)\cdots\delta\varphi(x_k)}=\mathfrak{S}_k\sum_{r=0}^k\binom{k}{r}\frac{\delta^r F}{\delta\varphi(x_1)\cdots\delta\varphi(x_r)}\cdot\frac{\delta^{k-r}G}{\delta\varphi(x_{r+1})\cdots\delta\varphi(x_k)}\,,$$

where $\mathfrak{S}_k$ denotes symmetrization in $x_1,\dots,x_k$, this is equal to

$$\frac{1}{n!}\Big(T^c_{\text{tree}}(S^{\otimes n}\otimes F)\,G+F\,T^c_{\text{tree}}(S^{\otimes n}\otimes G)\Big)$$
$$+\sum_{k=2}^n\int dx_1\dots dy_k\sum_{r=1}^{k-1}\frac{\delta^r F}{\delta\varphi(x_1)\dots\delta\varphi(x_r)}\,\frac{\delta^{k-r}G}{\delta\varphi(x_{r+1})\dots\delta\varphi(x_k)}\prod_{j=1}^k\Delta^F_m(x_j-y_j)$$
$$\cdot\sum_{l=r}^{n-k+r}\sum_{n_1+\dots+n_r=l}\frac{\hbar^r}{r!\,n_1!\dots n_r!}\,\frac{\delta T^c_{\text{tree}}(S^{\otimes n_1})}{\delta\varphi(y_1)}\cdots\frac{\delta T^c_{\text{tree}}(S^{\otimes n_r})}{\delta\varphi(y_r)}$$
$$\cdot\sum_{n_{r+1}+\dots+n_k=n-l}\frac{\hbar^{k-r}}{(k-r)!\,n_{r+1}!\dots n_k!}\,\frac{\delta T^c_{\text{tree}}(S^{\otimes n_{r+1}})}{\delta\varphi(y_{r+1})}\cdots\frac{\delta T^c_{\text{tree}}(S^{\otimes n_k})}{\delta\varphi(y_k)}\,.$$

Reordering the sums we see that this expression agrees with

$$\frac{1}{n!}\Big(T^c_{\text{tree}}(S^{\otimes n}\otimes F)\,G+F\,T^c_{\text{tree}}(S^{\otimes n}\otimes G)\Big)$$
$$+\sum_{l=1}^{n-1}\frac{1}{l!\,(n-l)!}\Big(T^c_{\text{tree}}(S^{\otimes l}\otimes F)\cdot T^c_{\text{tree}}(S^{\otimes(n-l)}\otimes G)\Big)\,.\,]$$

Proper vertices. We describe proper vertices in terms of the "vertex functional" Γ, which is defined by the following proposition.

Proposition 4.4.6 ([16]). *There exists a* totally symmetric *and* linear *map*[140]

$$\Gamma\colon \begin{cases} \mathcal{T}(\mathcal{F}_{\mathrm{loc}}) \longrightarrow \mathcal{F} \\ e_\otimes^S \longmapsto \Gamma(e_\otimes^S) \equiv \sum_{n=1}^\infty \frac{1}{n!}\,\Gamma(S^{\otimes n}) \,, \end{cases} \tag{4.4.11}$$

which is uniquely determined by $\Gamma(1) := 0$ *and*

$$T^c\big(e_\otimes^{iS/\hbar}\big) = T^c_{\mathrm{tree}}\Big(e_\otimes^{i\Gamma(e_\otimes^S)/\hbar}\Big)\,. \tag{4.4.12}$$

Before giving the proof we comment on this definition of Γ.

- Given Γ, T^c is uniquely determined by (4.4.12) and, due to the linked cluster theorem (4.4.2), this holds also for T.
- The basic relation
$$\Gamma(S) = S \tag{4.4.13}$$
is obtained from (4.4.12) by selecting the terms of first order in S and by using $\Gamma(e_\otimes^{\lambda S}) = \lambda\,\Gamma(S) + \mathcal{O}(\lambda^2)$.
- The *unrenormalized vertex functional, i.e.,* $\Gamma\big(\otimes_{j=1}^n L(x_j)\big)\big|_{\mathcal{D}(\check{\mathbb{M}}^n)}$ *with* $L \in \mathcal{P}$, *can be interpreted as "proper vertex of order n", that is, as the sum of all 1PI-diagrams contributing to* $\big(\frac{i}{\hbar}\big)^{n-1}\,T_n^c\big(\otimes_{j=1}^n L(x_j)\big)\big|_{\mathcal{D}(\check{\mathbb{M}}^n)}$. To prove this statement, we proceed by induction on n and write (4.4.12) to nth order in $S = \int dx\; L(x)\,g(x)$ (with $g \in \mathcal{D}(\mathbb{M})$) by using the Faà di Bruno formula, see (3.6.25):
$$\begin{aligned}\Big(\frac{i}{\hbar}\Big)^{n-1} T_n^c\big(\otimes_{j=1}^n L(x_j)\big)\big|_{\mathcal{D}(\check{\mathbb{M}}^n)} &= \Gamma\big(\otimes_{j=1}^n L(x_j)\big)\big|_{\mathcal{D}(\check{\mathbb{M}}^n)} \\ &\quad + \sum_{|P|\geq 2}\Big(\frac{i}{\hbar}\Big)^{|P|-1} T^c_{\mathrm{tree},|P|}\Big(\bigotimes_{J\in P}\Gamma\big(\otimes_{j\in J}L(x_j)\big)\big|_{\mathcal{D}(\check{\mathbb{M}}^{|J|})}\Big)\,,\end{aligned} \tag{4.4.14}$$
where P has the same meaning as in (4.4.1), cf. (4.4.16). On the right-hand side, the first term can be interpreted as the $(|P| = 1)$-term. Now, $T^c_{\mathrm{tree},|P|}\big(\bigotimes_{J\in P}\Gamma(\otimes_{j\in J}L(x_j))\big)$ is the sum of all connected tree diagrams with $|P|$ vertices $\big(\Gamma(\otimes_{j\in J}L(x_j))\big)_{J\in P}$, which may be non-local. If $x_i \neq x_j\ \forall i<j$, the latter are proper vertices of order $|J|(<n)$; this is known by induction. Therefore, the term given by the sum over $|P| \geq 2$ can be interpreted as the sum of all 1-particle-reducible diagrams (see Footn. 138) contributing to $\big(\frac{i}{\hbar}\big)^{n-1}\,T_n^c\big(\otimes_j L(x_j)\big)\big|_{\mathcal{D}(\check{\mathbb{M}}^n)}$. We conclude that the remaining term on the r.h.s., i.e. $\Gamma\big(\otimes_{j=1}^n L(x_j)\big)\big|_{\mathcal{D}(\check{\mathbb{M}}^n)}$, is the contribution of all 1PI-diagrams.

[140] We recall from (A.1.10) that Γ is uniquely determined on the whole domain $\mathcal{T}(\mathcal{F}_{\mathrm{loc}})$ by its values on the subset $\{\, e_\otimes^S \mid S \in \mathcal{F}_{\mathrm{loc}} \,\}$, since Γ is required to be totally symmetric and linear.

- For later purpose we mention: Since T^c, T^c_{tree} and Γ are linear and totally symmetric, the defining relation (4.4.12) implies

$$T^c\big(e_\otimes^{iS/\hbar}\otimes F\big)=\frac{\hbar}{i}\frac{d}{d\lambda}\Big|_{\lambda=0}T^c\big(e_\otimes^{i(S+\lambda F)/\hbar}\big)=\frac{\hbar}{i}\frac{d}{d\lambda}\Big|_{\lambda=0}T^c_{\mathrm{tree}}\Big(e_\otimes^{i\Gamma(e_\otimes^{(S+\lambda F)})/\hbar}\Big)$$
$$=T^c_{\mathrm{tree}}\Big(e_\otimes^{i\Gamma(e_\otimes^S)/\hbar}\otimes\Gamma(e_\otimes^S\otimes F)\Big)\,. \tag{4.4.15}$$

- Relation to the Main Theorem (Thm. 3.6.3): Writing any element Z of the Stückelberg–Petermann renormalization group $\mathcal{R}$ as $D(e_\otimes^S):=Z(S)$, with a linear and symmetric map $D:\mathcal{T}(\mathcal{F}_{\mathrm{loc}})\longrightarrow\mathcal{F}_{\mathrm{loc}}$ the Main Theorem formula, $\widehat{T}^c(e_\otimes^{iS/\hbar})=T^c(e_\otimes^{iD(e_\otimes^S)/\hbar})$ (4.4.3), is of a similar form as the definition of Γ (4.4.12); a main difference is that the values of Γ are in general *non-local.*

Proof. We prove the proposition by constructing $\Gamma(\otimes_{j=1}^n F_j)$ by induction on n. We start with $\Gamma(1)=0$ and $\Gamma(S)=S$. The inductive step is uniquely given by the defining relation (4.4.12): By using total symmetry and linearity of T^c, T^c_{tree} and Γ, this relation yields

$$\Gamma(\otimes_{j=1}^n F_j)=\Big(\frac{i}{\hbar}\Big)^{n-1}T^c_n(\otimes_{j=1}^n F_j)-\sum_{|P|\geq 2}\Big(\frac{i}{\hbar}\Big)^{|P|-1}T^c_{\mathrm{tree},|P|}\Big(\bigotimes_{J\in P}\Gamma(\otimes_{j\in J}F_j)\Big) \tag{4.4.16}$$

similarly to (4.4.14); again P has the same meaning as in (4.4.1). Obviously, the so-constructed Γ is totally symmetric and linear. So the procedure is analogous to the inductive construction of the map $Z\equiv(Z^{(n)})$ in the proof of part (a) of the Main Theorem. □

Exercise 4.4.7. Prove the relations

$$\Gamma(e_\otimes^S)=S+\mathcal{O}(\hbar)\,,\quad \Gamma(e_\otimes^S\otimes F)=F+\mathcal{O}(\hbar)\quad\text{if}\quad F,S\sim\hbar^0\,, \tag{4.4.17}$$

by proceeding inductively and using (4.4.16) and $(T^c_n-T^c_{\mathrm{tree},n})=\mathcal{O}(\hbar^n)$.

[*Solution*: To prove $\Gamma(S^{\otimes n})=\mathcal{O}(\hbar)\ \forall n\geq 2$ by induction on n, we rewrite the recursion relation (4.4.16) in the form

$$\Gamma(S^{\otimes n})=\Big(\frac{i}{\hbar}\Big)^{n-1}(T^c_n-T^c_{\mathrm{tree},n})(S^{\otimes n})$$
$$-\sum_{k=2}^{n-1}\sum_{\substack{l_1+\dots+l_k=n\\ l_j\geq 1}}\frac{i^{k-1}\,n!}{\hbar^{k-1}\,k!\,l_1!\dots l_k!}T^c_{\mathrm{tree},k}\Big(\Gamma(S^{\otimes l_1})\otimes\dots\otimes\Gamma(S^{\otimes l_k})\Big)\,.$$

Since at least one l_j is ≥ 2 we have $\Gamma(S^{\otimes l_1})\otimes\dots\otimes\Gamma(S^{\otimes l_k})=\mathcal{O}(\hbar)$ and since $T^c_{\mathrm{tree},k}\sim\hbar^{k-1}$ we see that the sum in the second line is $\mathcal{O}(\hbar)$. Moreover, from (4.4.4) and (4.4.5) we know that $(T^c_n-T^c_{\mathrm{tree},n})=\mathcal{O}(\hbar^n)$. Summing up, we get $\Gamma(S^{\otimes n})=\mathcal{O}(\hbar)$.
By using that, we conclude $\Gamma\big(e_\otimes^S\otimes F\big)=\frac{d}{d\lambda}\big|_{\lambda=0}\Gamma\big(e_\otimes^{(S+\lambda F)}\big)=F+\mathcal{O}(\hbar)$.]

The relation $\Gamma(e_\otimes^S) = S + \mathcal{O}(\hbar)$ may be read as $\Gamma(e_\otimes^S) = S+$ [quantum corrections]. Motivated by this, $\Gamma\big(e_\otimes^S\big)$ is also called the "proper interaction (corresponding to the classical interaction S)".

The validity of renormalization conditions for the time-ordered product $T \equiv T^{(m)}$ is equivalent to corresponding properties of Γ (see [16, Lemma 6]). As an example we discuss almost homogeneous scaling, which is a consequence of the Sm-expansion axiom (cf. (3.1.67)): If

$$\sigma_\rho \circ T_n^{(\rho^{-1}m)}\big((\sigma_\rho^{-1}F)^{\otimes n}\big) = T_n^{(m)}(F^{\otimes n}) + \mathcal{O}(\log\rho) \;\; \text{is a polynomial in } \log\rho \tag{4.4.18}$$

for all $F \in \mathcal{F}_{\text{loc}}$, $n \in \mathbb{N}$, we see from the definition (4.4.1) that $T^c \equiv T^{c\,(m)}$ scales also almost homogeneously, i.e., satisfies also (4.4.18). Moreover, $T_{\text{tree}}^{c\,(m)}$ scales even homogeneously,

$$\sigma_\rho \circ T_{\text{tree}}^{c\,(\rho^{-1}m)}\big(e_\otimes^{\sigma_\rho^{-1}F}\big) = T_{\text{tree}}^{c\,(m)}\big(e_\otimes^F\big) \qquad \forall F \in \mathcal{F}\,;$$

this follows inductively from the definition (4.4.8) since $\rho^{d-2}\,\Delta_{m/\rho}^F(\rho x) = \Delta_m^F(x)$. Using these results, we inductively conclude from (4.4.16) that $\Gamma \equiv \Gamma^{(m)}$ scales also almost homogeneously:

$$\sigma_\rho \circ \Gamma^{(\rho^{-1}m)}\big((\sigma_\rho^{-1}S)^{\otimes n}\big) = \Gamma^{(m)}(S^{\otimes n}) + \mathcal{O}(\log\rho) \;\; \text{is a polynomial in } \log\rho \tag{4.4.19}$$

for all $S \in \mathcal{F}_{\text{loc}}$, $n \in \mathbb{N}^*$. The argumentation can easily be reversed, to show that (4.4.19) is also sufficient for (4.4.18).

If the Sm-expansion axiom is replaced by the (weaker) axiom Scaling degree (3.2.62), the corresponding property of Γ reads:

$$\text{sd}\Big(\omega_0\big(\Gamma(A_1,\ldots,A_n)\big)(x_1-x_n,\ldots)\Big) \leq \sum_{j=1}^n \dim A_j\,, \quad \forall A_1,\ldots,A_n \in \mathcal{P}_{\text{hom}}\,. \tag{4.4.20}$$

This can also be proven inductively, by using the recursion relation (4.4.16). In addition we take into account that $\omega_0\big(T_{\text{tree}}^c(\cdots)\big)$ is a tensor product of distributions t_j and that $\text{sd}(\otimes_j t_j) = \sum_j \text{sd}(t_j)$ (Exer. 3.2.7(c)(v)).

Exercise 4.4.8 (Off-shell field equation in terms of Γ). Prove the following statement: Assume that a time-ordered product fulfills the Off-shell field equation (axiom (viii) in Sect. 3.3). Then, the corresponding vertex functional Γ satisfies the relation

$$\Gamma\big(e_\otimes^S \otimes \varphi(h)\big) = \varphi(h) \quad \forall S \in \mathcal{F}_{\text{loc}},\; h \in \mathcal{D}(\mathbb{M})\,. \tag{4.4.21}$$

This is the Off-shell field equation in terms of proper vertices.

[*Solution*: We prove

$$\Gamma\big(\varphi(h)\otimes F_1 \otimes\cdots\otimes F_n\big) = 0 \quad \forall n \geq 1$$

by induction on n. By using the inductive assumption, the recursion relation (4.4.16) takes the form

$$\Gamma\big(\varphi(h)\otimes F_1\otimes\cdots\otimes F_n\big)=\Big(\frac{i}{\hbar}\Big)^n T^c\big(\varphi(h)\otimes F_1\otimes\cdots\otimes F_n\big)\\ -\sum_P\Big(\frac{i}{\hbar}\Big)^{|P|} T^c_{\text{tree}}\Big(\varphi(h)\otimes\bigotimes_{J\in P}\Gamma(F_J)\Big) \tag{4.4.22}$$

where $F_J:=\otimes_{j\in J}F_j$ and P runs through all partitions of $\{1,\dots,n\}$; in particular there is a $(|P|=1)$-contribution. Next we take into account that T^c and T^c_{tree} fulfill the Off-shell field equation, that is,

$$T^c(\varphi(h)\otimes F_1\otimes\cdots\otimes F_n)=\hbar\int dx\,dy\;h(x)\,\Delta^F(x-y)\,\frac{\delta}{\delta\varphi(y)}T^c(F_1\otimes\cdots\otimes F_n)$$

and the same equation for T^c_{tree}. In the latter case $F_1,\dots,F_n$ may be non-local. With that, the r.h.s. of (4.4.22) is equal to

$$i\int dx\,dy\,h(x)\,\Delta^F(x-y)\,\frac{\delta}{\delta\varphi(y)}\bigg(\Big(\frac{i}{\hbar}\Big)^{n-1} T^c(F_1\otimes\cdots\otimes F_n)\\ -\sum_{|P|\geq 1}\Big(\frac{i}{\hbar}\Big)^{|P|-1} T^c_{\text{tree}}\Big(\bigotimes_{J\in P}\Gamma(F_J)\Big)\bigg)=0\,.$$

In the last step, we use again the relation (4.4.16).]

In the proper vertex formalism a finite renormalization $\mathbf{S}\mapsto\widehat{\mathbf{S}}=\mathbf{S}\circ Z$ of the T-product (where $Z\in\mathcal{R}$) is equivalent to a finite renormalization $\Gamma\mapsto\widehat{\Gamma}$ of the corresponding vertex functional. The following exercise gives the relation between $\widehat{\Gamma}$ and Γ:

Exercise 4.4.9 (Main Theorem in terms of Γ).

(a) Prove inductively the following auxiliary relation:

$$\mathcal{T}\Big(e_\otimes^{\sum_{n=1}^\infty F_n\kappa^n}\Big)=\mathcal{T}\Big(e_\otimes^{\sum_{n=1}^\infty G_n\kappa^n}\Big)\implies F_n=G_n\;\;\forall n\,,$$

where $\mathcal{T}$ stands for T, T^c or T^c_{tree}. A crucial point is that the formal power series $\sum_{n=1}^\infty F_n\kappa^n$ and $\sum_{n=1}^\infty G_n\kappa^n$ vanish to zeroth order in κ, as it holds for $\Gamma(e_\otimes^S)$ since $S\sim\kappa$.

(b) Show that formula (3.6.20) of the Main Theorem is equivalent to

$$\widehat{\Gamma}(e_\otimes^S)=\Gamma(e_\otimes^{Z(S)})\qquad\forall S\in\mathcal{F}_{\text{loc}}\,. \tag{4.4.23}$$

[*Solution*: (a) By using that

$$\mathcal{T}\Big(e_\otimes^{\sum_{n\geq 1}F_n\kappa^n}\Big)=\sum_{n\geq 1}F_n\kappa^n+\sum_{k=2}^\infty\frac{1}{k!}\sum_{n_1,\dots,n_k\geq 1}\kappa^{n_1+\cdots+n_k}\,\mathcal{T}_k(F_{n_1}\otimes\cdots\otimes F_{n_k})\,,$$

the assumption to nth order in κ gives

$$F_n - G_n = \sum_{k=2}^{n} \frac{1}{k!} \sum_{\substack{n_1+\cdots+n_k=n \\ n_j \geq 1}} \mathcal{T}_k(G_{n_1} \otimes \cdots \otimes G_{n_k} - F_{n_1} \otimes \cdots \otimes F_{n_k}) \,.$$

For $n = 1$, the r.h.s. vanishes. Proceeding by induction, we assume $F_l = G_l \ \forall l < n$. This implies $F_n - G_n = 0$ since $n_j < n \ \forall j$.
(b) Due to Remk. 4.4.3 it suffices to prove (4.4.3) $\iff$ (4.4.23). The assertion "$\Longrightarrow$" is obtained from

$$T^c_{\text{tree}}\Big(e_\otimes^{i\widehat{\Gamma}(e_\otimes^S)/\hbar}\Big) = \widehat{T}^c\big(e_\otimes^{iS/\hbar}\big) = T^c\big(e_\otimes^{iZ(S)/\hbar}\big) = T^c_{\text{tree}}\Big(e_\otimes^{i\Gamma(e_\otimes^{Z(S)})/\hbar}\Big) \tag{4.4.24}$$

(where (4.4.12) is used) by applying part (a). To verify the claim "$\Longleftarrow$" we simply rearrange (4.4.24).]

Given Γ and $\widehat{\Gamma}$, formula (4.4.23) can be used for a unique inductive construction of $(Z^{(n)})_{n\in\mathbb{N}^*}$: Due to the similarity of (4.4.23) to the Main Theorem formula $\widehat{T}(e_\otimes^{iF/\hbar}) = T\big(e_\otimes^{iZ(F)/\hbar}\big)$, this construction is analogous to (3.6.25):

$$Z^{(n)}(\otimes_{j=1}^n F_j) = \widehat{\Gamma}(\otimes_{j=1}^n F_j) - \Gamma(\otimes_{j=1}^n F_j) - \sum_{\substack{P\in\text{Part}(\{1,\ldots,n\}) \\ 1<|P|<n}} \Gamma\Big(\otimes_{I\in P} Z^{(|I|)}(F_I)\Big) \,. \tag{4.4.25}$$

In particular, $Z = (Z^{(n)})_{n\in\mathbb{N}^*}$ is uniquely determined by the pair $(\Gamma, \widehat{\Gamma})$.

Analogously to the conventions for R- and T-products we sometimes write the expression $\int dx\; g(x)\,\Gamma\big(B(x)\otimes F_2\ldots\big)$ for $\Gamma\big(B(g)\otimes F_2\ldots\big)$, where $B\in\mathcal{P}$, $g\in\mathcal{D}(\mathbb{M})$, see (3.1.3)–(3.1.4). Since Γ depends only on *functionals*, it fulfills the AWI:

$$\partial_x^\mu \Gamma\big(B(x)\otimes F_2\ldots\big) = \Gamma\big((\partial^\mu B)(x)\otimes F_2\ldots\big) \,.$$

4.4.2 The Master Ward Identity in terms of proper vertices

Since, as we will show, the MWI holds for connected tree diagrams with non-local vertices, the anomalous MWI can equivalently be reformulated in terms of proper vertices. We do this in several steps.

Proof of the MWI for connected tree diagrams. Since T^c_{tree} is the classical part of the time-ordered product, we can proceed analogously to the derivation of the *classical* MWI (in Sect. 4.1). An essential new feature is that *non-local functionals*, $S\in\mathcal{F}$ and $A = \int dx\; h(x)Q(x)\frac{\delta S_0}{\delta\varphi(x)} \in \mathcal{J}$ (with $h\in\mathcal{D}(\mathbb{M})$), may appear as entries of T^c_{tree}; in case of A this means that $Q(h)\equiv\int dx\; h(x)Q(x)\in\mathcal{F}$ needs not to lie in $\mathcal{F}_{\text{loc}}$. As above, let $\delta_{hQ} := \int dx\, h(x)Q(x)\,\frac{\delta}{\delta\varphi(x)}$.

Exercise 4.4.10.

(a) Prove the "classical field equation in terms of T^c_{tree}":

$$T^c_{\text{tree}}\Big(e_\otimes^{iS/\hbar}\otimes\frac{\delta(S_0+S)}{\delta\varphi(x)}\Big) = \frac{\delta S_0}{\delta\varphi(x)}\,,\qquad S\in\mathcal{F}\,. \tag{4.4.26}$$

(b) Prove the MWI for connected tree diagrams:

$$T^{\mathrm{c}}_{\mathrm{tree}}\Big(e_\otimes^{iS/\hbar}\otimes(A+\delta_{hQ}S)\Big)=\int dx\;h(x)\,T^{\mathrm{c}}_{\mathrm{tree}}\Big(e_\otimes^{iS/\hbar}\otimes Q(x)\Big)\cdot\frac{\delta S_0}{\delta\varphi(x)}\,. \tag{4.4.27}$$

[*Hint*: Use (4.4.10).]

[*Solution*: (a) To zeroth order in S, the assertion (4.4.26) follows from (4.4.9). To nth order we have to show

$$-\frac{i}{\hbar}\,T^{\mathrm{c}}_{\mathrm{tree},n+1}\Big(S^{\otimes n}\otimes\frac{\delta S_0}{\delta\varphi(x)}\Big)=n\,T^{\mathrm{c}}_{\mathrm{tree},n}\Big(S^{\otimes(n-1)}\otimes\frac{\delta S}{\delta\varphi(x)}\Big)\,,\quad\forall n\geq 1\,.$$

The r.h.s. is equal to $\frac{\delta}{\delta\varphi(x)}\,T^{\mathrm{c}}_{\mathrm{tree},n}(S^{\otimes n})$, because $T^{\mathrm{c}}_{\mathrm{tree},n}$ satisfies Field independence; the latter follows inductively from (4.4.8). To compute the l.h.s. we use again (4.4.8) and $\frac{\delta^2 S_0}{\delta\varphi(x_1)\delta\varphi(x)}=-(\Box+m^2)\delta(x_1-x)$; with that the l.h.s. is equal to

$$i\int dx_1dy_1\;\big((\Box+m^2)\delta\big)(x_1-x)\,\Delta^F_m(x_1-y_1)\,\frac{\delta\,T^{\mathrm{c}}_{\mathrm{tree},n}(S^{\otimes n})}{\delta\varphi(y_1)}=\frac{\delta\,T^{\mathrm{c}}_{\mathrm{tree},n}(S^{\otimes n})}{\delta\varphi(x)}\,;$$

in the last step $(\Box+m^2)\Delta^F_m(x)=-i\,\delta(x)$ is applied.
(b) Due to (4.4.10) we may proceed analogously to (4.1.2):

$$\begin{aligned}T^{\mathrm{c}}_{\mathrm{tree}}\Big(e_\otimes^{iS/\hbar}\otimes Q(x)\cdot\frac{\delta(S_0+S)}{\delta\varphi(x)}\Big)&=T^{\mathrm{c}}_{\mathrm{tree}}\big(e_\otimes^{iS/\hbar}\otimes Q(x)\big)\cdot T^{\mathrm{c}}_{\mathrm{tree}}\Big(e_\otimes^{iS/\hbar}\otimes\frac{\delta(S_0+S)}{\delta\varphi(x)}\Big)\\&=T^{\mathrm{c}}_{\mathrm{tree}}\big(e_\otimes^{iS/\hbar}\otimes Q(x)\big)\cdot\frac{\delta S_0}{\delta\varphi(x)}\,.\end{aligned}$$

Applying the integration $\int dx\,h(x)\dots$ to this equation, we get the assertion.]

Translation of the anomalous MWI from T into T^{c}

Exercise 4.4.11. Let $S\in\mathcal{F}_{\mathrm{loc}}$ and $Q\in\mathcal{P}$, hence $A:=\int dx\,h(x)Q(x)\frac{\delta S_0}{\delta\varphi(x)}\in\mathcal{J}\cap\mathcal{F}_{\mathrm{loc}}$. Derive the anomalous MWI for T^{c}:

$$T^{\mathrm{c}}\Big(e_\otimes^{iS/\hbar}\otimes\big(A+\delta_{hQ}S+\Delta(e_\otimes^S;hQ)\big)\Big)=\int dx\,h(x)\,T^{\mathrm{c}}\big(e_\otimes^{iS/\hbar}\otimes Q(x)\big)\cdot\frac{\delta S_0}{\delta\varphi(x)}\,. \tag{4.4.28}$$

[*Solution*: Using the linked cluster theorem (4.4.2) we get

$$\begin{aligned}T\big(e_\otimes^{iF}\otimes G\big)&=-i\,\frac{d}{d\lambda}\Big|_{\lambda=0}T\big(e_\otimes^{i(F+\lambda G)}\big)=-i\,\frac{d}{d\lambda}\Big|_{\lambda=0}\exp_\bullet\big(T^{\mathrm{c}}(e_\otimes^{i(F+\lambda G)})\big)\\&=T\big(e_\otimes^{iF}\big)\cdot T^{\mathrm{c}}\big(e_\otimes^{iF}\otimes G\big)\,.\end{aligned} \tag{4.4.29}$$

Inserting this relation into the anomalous MWI (4.3.8), we get the assertion (4.4.28) multiplied with $T\big(e_\otimes^{iS/\hbar}\big)$ by the classical product. Since $S\sim\kappa$ we have $T\big(e_\otimes^{iS/\hbar}\big)=1+\sum_{n=1}^\infty F_n\kappa^n$ for some $F_n\in\mathcal{F}$. Therefore, the factor $T\big(e_\otimes^{iS/\hbar}\big)$ is invertible with respect to the classical product (see (A.1.3)) and, hence, it can be removed by multiplication with $T\big(e_\otimes^{iS/\hbar}\big)^{-1}$.]

Remark 4.4.12 (Time-ordered tree product). By means of the linked cluster theorem we may define the *time-ordered tree product* $T_{\text{tree}} : \mathcal{T}(\mathcal{F}) \longrightarrow \mathcal{F}$:

$$1 + \sum_{n=1}^{\infty} \frac{i^n}{n!} T_{\text{tree},n}(F^{\otimes n}) \equiv T_{\text{tree}}(e_{\otimes}^{iF}) := \exp_{\bullet}\big(T^c_{\text{tree}}(e_{\otimes}^{iF})\big) \quad \forall F \in \mathcal{F}\,, \tag{4.4.30}$$

note that the entries may be non-local. Analogously to (4.4.6) it holds that

$$T_{\text{tree},n}\big(F_1 \otimes \cdots \otimes F_n\big) = F_1 \star_{\Delta_m^F} \cdots \star_{\Delta_m^F} F_n\big|_{\text{tree diagrams}}\,.$$

By reversing the argumentation of the preceding exercise we find: The validity of the MWI for T^c_{tree}, which we have proved above, implies the validity of the MWI for T_{tree}. In particular we use here that, due to the definition of T_{tree} (4.4.30), the relation (4.4.29) holds also for $(T_{\text{tree}}, T^c_{\text{tree}})$ in place of (T, T^c).

Translation of the MWI from T^c into Γ. Inserting (4.4.15) into both sides of the anomalous MWI (4.4.28), we obtain

$$\begin{aligned}\int dx\; h(x)\, T^c_{\text{tree}}\Big(e_{\otimes}^{i\Gamma(e_{\otimes}^S)/\hbar} \otimes \Gamma\Big(e_{\otimes}^S \otimes \Big[Q(x)\,\frac{\delta(S_0+S)}{\delta\varphi(x)} + \Delta\big(e_{\otimes}^S; Q(x)\big)\Big]\Big)\Big)\\ = \int dx\; h(x)\, T^c_{\text{tree}}\Big(e_{\otimes}^{i\Gamma(e_{\otimes}^S)/\hbar} \otimes \Gamma\big(e_{\otimes}^S \otimes Q(x)\big)\Big) \cdot \frac{\delta S_0}{\delta\varphi(x)}\,.\end{aligned} \tag{4.4.31}$$

Now we use the MWI for T^c_{tree} (4.4.27): We insert for S and $Q(x)$ the non-local functionals $\Gamma(e_{\otimes}^S)$ and $\Gamma(e_{\otimes}^S \otimes Q(x))$, respectively. Proceeding this way, we obtain that the r.h.s. of (4.4.31) is equal to

$$\int dx\; h(x)\, T^c_{\text{tree}}\Big(e_{\otimes}^{i\Gamma(e_{\otimes}^S)/\hbar} \otimes \Gamma\big(e_{\otimes}^S \otimes Q(x)\big) \cdot \frac{\delta\big(S_0 + \Gamma(e_{\otimes}^S)\big)}{\delta\varphi(x)}\Big)\,. \tag{4.4.32}$$

By using an auxiliary result (formulated and proved below in (4.4.34)), we conclude:

Proposition 4.4.13 (Anomalous MWI in terms of Γ, [16, 17]). *The anomalous MWI in Thm.* 4.3.1 (*given in* (4.3.3) *or* (4.3.8), *respectively*) *can equivalently be expressed in terms of proper vertices:*

$$\Gamma\Big(e_{\otimes}^S \otimes \Big[Q(x)\,\frac{\delta(S_0+S)}{\delta\varphi(x)} + \Delta\big(e_{\otimes}^S; Q(x)\big)\Big]\Big) = \Gamma\big(e_{\otimes}^S \otimes Q(x)\big) \cdot \frac{\delta\big(S_0 + \Gamma(e_{\otimes}^S)\big)}{\delta\varphi(x)}\,. \tag{4.4.33}$$

Also the original version of the Quantum Action Principle of Lowenstein [125] and Lam [119] is formulated in terms of proper vertices.

In the last step of the derivation of (4.4.33) we use the following auxiliary result: If there is some formal power series $\sum_{n=1}^{\infty} F_n \kappa^n$ (note that $F_0 = 0$) such that

$$T^c_{\text{tree}}\Big(e_{\otimes}^{\sum_{n=1}^{\infty} F_n \kappa^n} \otimes \sum_{n=0}^{\infty} G_n \kappa^n\Big) = 0\,, \quad \text{then} \quad G_n = 0 \;\; \forall n\,. \tag{4.4.34}$$

Proof. We proceed analogously to the inductive proof in Exer. 4.4.9(a): To zeroth order in κ the assumption is simply the relation $G_0 = 0$. To nth order it reads

$$G_n + \sum_{k=1}^{n} \frac{1}{k!} \sum_{n_1+\cdots+n_k+l=n} T^c_{\mathrm{tree},k+1}(F_{n_1} \otimes \cdots \otimes F_{n_k} \otimes G_l) = 0 \ ,$$

where $l \geq 0$ and $n_j \geq 1 \ \forall j$. Due to the latter we have $l < n$. Inductively we know that $G_l = 0 \ \forall l < n$; therefore, we obtain $G_n = 0$. □

4.5 Removing violations of the Master Ward Identity

We continue to follow [16].

A crucial and difficult problem in perturbative renormalization reads: How can we fulfill the MWI for a given interaction S and a given symmetry (i.e., a given $Q \in \mathcal{P}$) to all orders?

We recall from Sect. 4.3 that a widespread strategy is to start with an arbitrary renormalization prescription fulfilling all other renormalization conditions; with that the considered MWI is in general broken, that is, the MWI contains an anomalous term $R\Big(e_\otimes^{S/\hbar}, \Delta\big(e_\otimes^S; Q(x)\big)\Big)$. One then searches for a finite renormalization which removes the anomalous term. There are symmetries Q and interactions S for which this cannot be completely done. The non-removable $R\Big(e_\otimes^{S/\hbar}, \Delta\big(e_\otimes^S; Q(x)\big)\Big)$ are the famous anomalies of perturbative QFT, e.g., the scaling anomaly (cf. Sect. 3.2.7) or the axial anomaly in axial QED (see formula (5.2.30)).

4.5.1 Proceeding analogously to algebraic renormalization

The formal equivalence of the anomalous MWI in terms of proper vertices (given in Proposition 4.4.13) and the QAP makes it possible to apply basic techniques of *algebraic renormalization* (see, e.g., [135]) within the framework of causal perturbation theory. Algebraic renormalization proceeds *by induction on the power of* $\hbar$. The question whether one can find a finite renormalization which removes the possible anomaly (i.e., the local interacting field violating the Ward identity) then amounts – by means of the QAP – to purely algebraic problems, often cohomological properties of the underlying symmetry group (see, e.g., [5, 96] for detailed studies of the BRST cohomology).

General procedure. We will apply this strategy to the MWI in terms of proper vertices (which is given by (4.4.33) with $\Delta\big(e_\otimes^S; Q(x)\big) := 0$). A main difference between our MWI-formalism and algebraic renormalization is that we work with *compactly supported* interactions

$$S = \kappa\, L(g) \equiv \kappa \int dx\; g(x)\, L(x) \ , \quad g \in \mathcal{D}(\mathbb{M}) \ , \ L \in \mathcal{P} \ , \tag{4.5.1}$$

and *localized* symmetry transformations δ_{hQ}. To bring our formalism into line with algebraic renormalization with respect to the latter point, we will replace $h \in \mathcal{D}(\mathbb{M})$ by the constant function $h(x) = 1 \; \forall x \in \mathbb{M}$ at a certain stage – see (4.5.15).

We will work with the axiom Scaling degree instead of the Sm-expansion. Given $Q, L \in \mathcal{P}$ – the switching function g in (4.5.1) is arbitrary, we are going to investigate whether possible anomalies of the corresponding MWI can be removed by finite renormalizations of Γ (4.4.23). Proceeding by induction on the power of $\hbar$, we assume that for a given vertex functional Γ_k (which satisfies the renormalization conditions Field independence, Poincaré covariance, $*$-structure, Field parity, Off-shell field Equation, $\hbar$-dependence and Scaling degree) the MWI is violated only by terms of order $\hbar^l$ with $l \geq k$. This means that in the anomalous MWI for Γ_k,

$$\Gamma_k\Big(e_\otimes^S \otimes Q(x)\,\frac{\delta(S_0+S)}{\delta\varphi(x)}\Big) + \Gamma_k\Big(e_\otimes^S \otimes \Delta_k\big(e_\otimes^S; Q(x)\big)\Big) = \\ = \Gamma_k\big(e_\otimes^S \otimes Q(x)\big) \cdot \frac{\delta\big(S_0 + \Gamma_k(e_\otimes^S)\big)}{\delta\varphi(x)} \,, \qquad (4.5.2)$$

the violating term can be written as

$$\Gamma_k\Big(e_\otimes^S \otimes \Delta_k\big(e_\otimes^S; Q(x)\big)\Big) = \Delta_k\big(e_\otimes^S; Q(x)\big) + \mathcal{O}(\hbar^{k+1}) \,, \quad \Delta_k\big(e_\otimes^S; Q(x)\big) = \mathcal{O}(\hbar^k) \,, \qquad (4.5.3)$$

where (4.4.17) is used. Note that, due to statement (iii) of Thm. 4.3.1(a), any vertex functional Γ fulfills this assumption (4.5.2)–(4.5.3) for $k = 1$ and, hence, can be used as Γ_1.

To fulfill the MWI to kth order in $\hbar$ and to maintain the other renormalization conditions, we have to find an element Z_k of the Stückelberg–Petermann renormalization group $\mathcal{R}_0$ (we mean the version of this group belonging to the above-mentioned renormalization conditions, see Remk. 3.6.6) such that

$$\Gamma_{k+1}\Big(e_\otimes^S \otimes Q(x)\,\frac{\delta(S_0+S)}{\delta\varphi(x)}\Big) + \mathcal{O}(\hbar^{k+1}) \stackrel{!}{=} \Gamma_{k+1}\big(e_\otimes^S \otimes Q(x)\big) \cdot \frac{\delta\big(S_0 + \Gamma_{k+1}(e_\otimes^S)\big)}{\delta\varphi(x)} \,, \qquad (4.5.4)$$

with Γ_{k+1} given by

$$\Gamma_{k+1}(e_\otimes^F) := \Gamma_k\Big(e_\otimes^{Z_k(F)}\Big) \quad \forall F \in \mathcal{F}_{\mathrm{loc}} \,, \qquad (4.5.5)$$

where (4.4.23) is used. In order that the MWI is maintained to lower orders in $\hbar$, we additionally require that $Z_k \in \mathcal{R}_0$ is of the form

$$Z_k = \mathrm{Id} + Z_k^{>} \,, \quad Z_k^{>}(F) = \mathcal{O}(\hbar^k) \,, \quad \forall F \in \mathcal{F}_{\mathrm{loc}} \,, \; F \sim \hbar^0 \,. \qquad (4.5.6)$$

We recall that writing

$$Z_k^{>}(F) = \sum_{n=2}^{\infty} \frac{1}{n!}\, Z_k^{(n)}(F^{\otimes n})$$

the derivatives $Z_k^{(n)}$ of Z_k at $F = 0$ are linear and symmetrical. The relation (4.5.6) implies

$$e_\otimes^{Z_k(F)} = e_\otimes^F + e_\otimes^F \otimes_{\text{sym}} Z_k^>(F) + \mathcal{O}(\hbar^{k+1}) \tag{4.5.7}$$

(where $\otimes_{\text{sym}}$ denotes the symmetrized tensor product), from which we obtain

$$\Gamma_{k+1}(e_\otimes^S) = \Gamma_k(e_\otimes^S) + \Gamma_k\big(e_\otimes^S \otimes Z_k^>(S)\big) + \mathcal{O}(\hbar^{k+1}) = \Gamma_k(e_\otimes^S) + Z_k^>(S) + \mathcal{O}(\hbar^{k+1}) \tag{4.5.8}$$

by using again (4.4.17). The identity (4.5.8) yields

$$\Gamma_{k+1}(e_\otimes^S \otimes F) = \frac{d}{d\lambda}\Big|_{\lambda=0} \Gamma_{k+1}(e_\otimes^{S+\lambda F}) = \Gamma_k(e_\otimes^S \otimes F) + Z_k^{>\prime}(S)F + \mathcal{O}(\hbar^{k+1}) \tag{4.5.9}$$

where we define

$$Z_k^{>\prime}(S)F := \frac{d}{d\lambda}\Big|_{\lambda=0} Z_k^>(S+\lambda F) = \sum_{n=1}^\infty \frac{1}{n!}\, Z_k^{(n+1)}(S^{\otimes n} \otimes F) = \mathcal{O}(\hbar^k)\,, \tag{4.5.10}$$

analogously to (3.6.40). Now we insert the relations (4.5.8)–(4.5.9) into our requirement (4.5.4) to express everywhere Γ_{k+1} by Γ_k and $Z_k^>$, $Z_k^{>\prime}$:

$$\begin{aligned}
-\Gamma_k\Big(e_\otimes^S \otimes Q(x)\,\frac{\delta(S_0+S)}{\delta\varphi(x)}\Big) &+ \Gamma_k\big(e_\otimes^S \otimes Q(x)\big)\cdot \frac{\delta\big(S_0+\Gamma_k(e_\otimes^S)\big)}{\delta\varphi(x)} \\
&\stackrel{!}{=} Z_k^{>\prime}(S)\big[Q(x)\,\frac{\delta(S_0+S)}{\delta\varphi(x)}\big] - \Gamma_k\big(e_\otimes^S \otimes Q(x)\big)\cdot \frac{\delta Z_k^>(S)}{\delta\varphi(x)} \\
&\quad - \big[Z_k^{>\prime}(S)Q(x)\big]\cdot \frac{\delta\big(S_0+\Gamma_k(e_\otimes^S)\big)}{\delta\varphi(x)} + \mathcal{O}(\hbar^{k+1})\,,
\end{aligned}$$

where we have taken into account that $\big[Z_k^{>\prime}(S)Q(x)\big]\cdot \frac{\delta Z_k^>(S)}{\delta\varphi(x)} = \mathcal{O}(\hbar^{2k})$. Due to the inductive assumption (4.5.2)–(4.5.3), the l.h.s. is equal to $\Delta_k\big(e_\otimes^S; Q(x)\big) + \mathcal{O}(\hbar^{k+1})$; and on the r.h.s. we use the relations

$$\Gamma_k\big(e_\otimes^S \otimes Q(x)\big)\cdot \frac{\delta Z_k^>(S)}{\delta\varphi(x)} = Q(x)\cdot \frac{\delta Z_k^>(S)}{\delta\varphi(x)} + \mathcal{O}(\hbar^{k+1})$$

$$\text{and}\quad \frac{\delta\big(S_0+\Gamma_k(e_\otimes^S)\big)}{\delta\varphi(x)} = \frac{\delta(S_0+S)}{\delta\varphi(x)} + \mathcal{O}(\hbar)\,.$$

So we end up with the condition

$$\begin{aligned}
\Delta_k\big(e_\otimes^S; Q(x)\big) &\stackrel{!}{=} Z_k^{>\prime}(S)\big[Q(x)\,\frac{\delta(S_0+S)}{\delta\varphi(x)}\big] - Q(x)\cdot \frac{\delta Z_k^>(S)}{\delta\varphi(x)} \\
&\quad - \big[Z_k^{>\prime}(S)Q(x)\big]\cdot \frac{\delta(S_0+S)}{\delta\varphi(x)} + \mathcal{O}(\hbar^{k+1})
\end{aligned} \tag{4.5.11}$$

on $Z_k^>$. The term we want to eliminate, $\Delta_k\big(e_\otimes^S; Q(x)\big) = \mathcal{O}(\hbar^k)$, is given by the inductively known Γ_k – this is the way we interpret (4.5.2)–(4.5.3). If we succeed to find a map $Z_k^>$ solving (4.5.11) and fulfilling the above-mentioned properties, then the pertinent finite renormalization $Z_k \in \mathcal{R}_0$ removes the "anomaly" $\Delta_k\big(e_\otimes^S; Q(x)\big)$ to order $\hbar^k$. However, since $Z_k^>$ appears in (4.5.11) several times and in different forms, it seems almost impossible to discuss the existence of solutions in general.

Simplifying assumptions. To simplify the search for a $Z_k^>$ solving (4.5.11), we make the following assumptions:

- *Symmetry* $Q \in \mathcal{P}$: We assume that

$$Q(x)\frac{\delta S_0}{\delta\varphi(x)} = \partial_\mu j^\mu(x) \quad \text{for some } j^\mu \in \mathcal{P} \text{ which is } \textit{bilinear} \text{ in the basic fields.} \tag{4.5.12}$$

  Consequently, *Q is linear in the basic fields*, that is, $Q(x) = \sum_a C_a \partial^a\varphi(x)$ (with $C_a \in \mathbb{C}$) for the model of one real scalar field. Usually, this assumption is satisfied in the following way: $Q(x)$ is the infinitesimal version of a *linear* transformation of a basic field under which S_0 is invariant and j^μ is the corresponding conserved Noether current , see (4.2.17) – Exaps. 4.2.1, 4.2.2 and 4.2.3 and Exer. 4.2.6 are of this type.
- *Interaction* $S \in \mathcal{F}_{\mathrm{loc}}$: We assume

$$Q(x)\,\frac{\delta S}{\delta\varphi(x)} = 0 \qquad \forall x\;. \tag{4.5.13}$$

  In contrast to Exap. 4.2.2, this assumption is not fulfilled in Exaps. 4.2.1 and 4.2.3.

With these assumptions the condition (4.5.11) on $Z_k^>$ simplifies to

$$\Delta_k(e_\otimes^S; Q(x)) = \partial_\mu^x\big(Z_k^{>\prime}(S)j^\mu(x)\big) - Q(x)\frac{\delta Z_k^>(S)}{\delta\varphi(x)} + \mathcal{O}(\hbar^{k+1})\,. \tag{4.5.14}$$

Here we use

$$Z_k^{>\prime}(S)Q(x) = 0 \quad \text{and} \quad Z_k^{>\prime}(S)\big(\partial_\mu^x j^\mu(x)\big) = \partial_\mu^x\big(Z_k^{>\prime}(S)j^\mu(x)\big)\;.$$

From (4.5.10) we see that these two relations follow from the defining properties Field equation (Property (7) in Definition 3.6.1) and AWI (formula (3.6.7)), respectively, of the elements of the relevant Sückelberg–Petermann group $\mathcal{R}_0$.

To bring the condition (4.5.14) more into line with algebraic renormalization, we integrate it over x. The integrals

$$\int dx\,\Delta_k\big(e_\otimes^S; Q(x)\big) \quad \text{and} \quad \delta_Q\, Z_k^>(S) := \int dx\; Q(x)\frac{\delta Z_k^>(S)}{\delta\varphi(x)}$$

are finite, since

$$\text{for } x \notin \operatorname{supp} S \quad \text{it holds that} \quad \Delta_k\big(e_\otimes^S; Q(x)\big) = 0 \quad \text{and} \quad \frac{\delta Z_k^>(S)}{\delta\varphi(x)} = 0\;;$$

these two equations are due to the Locality of Δ_k (4.3.4) and the Field independence of Z_k (Property (4) in Definition 3.6.1). From (4.5.10) and the Locality of Z_k (3.6.34) we see that

$$Z_k^{>\prime}(S)j^\mu(x) = 0 \quad \text{for} \quad x \notin \operatorname{supp} S \ ;$$

hence, $\int dx\, \partial_\mu^x \big(Z_k^{>\prime}(S)j^\mu(x)\big) = 0$. Summing up, we obtain

$$\int dx\ \Delta_k\big(e_\otimes^S; Q(x)\big) = -\delta_Q\, Z_k^>(S) + \mathcal{O}(\hbar^{k+1}) \quad \text{where} \quad S = \kappa\, L(g)\ , \tag{4.5.15}$$

which is a necessary condition for $Z_k^>(S)$.

To first order in S (i.e., in κ) $Z_k^>(S)$ vanishes. Therefore, the condition (4.5.15) can only be satisfied if[141]

$$\int dx\ \Delta_k^1\big(S; Q(x)\big) = \mathcal{O}(\hbar^{k+1})\ .$$

One can even fulfill the stronger condition $\Delta_k^1\big(S; Q(x)\big) = \mathcal{O}(\hbar^{k+1})$: We first perform a finite renormalization of the T-product to second order, which maintains the relevant renormalization conditions and removes the term $\sim \hbar^k$ of $\Delta_k^1\big(S; Q(x)\big)$, we refer to [16, Sect. 5.4.1]. (The explicit proof given there does not need the assumptions (4.5.12)–(4.5.13), it shows that the MWI can generally be fulfilled to first order in S – for all Q and for all S.)

Such a finite renormalization of T_2 modifies also the T-product to higher orders and $\Delta_k^n(S^{\otimes n}; Q(x))$ for $n \geq 2$. However, since this finite renormalization is the addition of local terms of order $\mathcal{O}(\hbar^k)$, the relation $\Delta_k\big(e_\otimes^S; Q(x)\big) = \mathcal{O}(\hbar^k)$ is preserved.

A main problem is that (4.5.15) is not sufficient for (4.5.14). To discuss this, let us assume that we have found a $\frac{\delta Z_k^>(\kappa\, L(g))}{\delta\varphi} \sim \hbar^k$ solving (4.5.15) for the particular pair (L, Q) given from the outset and for all $g \in \mathcal{D}(\mathbb{M})$, and which is compatible with the requirement $Z_k \in \mathcal{R}_0$. We point out that the map $Z_k^>$ is not completely determined by that; uniquely fixed are only $\delta_Q\, Z_k^>\big(\kappa\, L(\cdot)\big)$ and, due to $Z_k \in \mathcal{R}_0$, some structural properties of $\frac{\delta Z_k^>(\kappa\, L(\cdot))}{\delta\varphi}$, notably "Locality" (3.6.34) and "Scaling degree" (3.6.35). But, in particular

$$Z_k^{>\prime}(S)\, j^\mu(x) = \sum_{n=1}^\infty \frac{1}{n!}\, Z_k^{(n+1)}\big(S^{\otimes n} \otimes j^\mu(x)\big)\ , \quad \text{with} \quad S = \kappa\, L(g)\ ,$$

may still be indeterminate. Starting with such a $\frac{\delta Z_k^>(\kappa\, L(\cdot))}{\delta\varphi}$, how can we construct a solution Z_k of (4.5.14) which has also the other required properties? To answer this question we use the following lemma – results of this kind are frequently used to solve problems of this type, cf. Sect. 5.2.2.

[141] We recall $\Delta_k\big(e_\otimes^S; Q(x)\big) = \sum_{n=1}^\infty \frac{1}{n!}\, \Delta_k^n(S^{\otimes n}; Q(x))$ from Thm. 4.3.1(a).

Lemma 4.5.1. *Let*

$$f(y,x_1,\dots,x_n)=\sum_{a\in(\mathbb{N}^d)^n}\partial^a\delta(x_1-y,\dots,x_n-y)\,P_a(y)\quad where\quad P_a\in\mathcal{P}\tag{4.5.16}$$

and let

$$\int dy\;f(y,x_1,\dots,x_n)=0\;.\tag{4.5.17}$$

Then there exist field polynomials $U^\mu_a\in\mathcal{P}$ such that

$$f(y,x_1,\dots,x_n)=\partial^y_\mu\Big[\sum_{a\in(\mathbb{N}^d)^n}\partial^a\delta(x_1-y,\dots,x_n-y)\,U^\mu_a(y)\Big]\;.\tag{4.5.18}$$

It is remarkable that the $\mathcal{F}_{\mathrm{loc}}$-valued distribution in the $[\cdots]$-brackets is again supported on the thin diagonal, since $(\partial^a)\delta$-distributions can be generated by computing the divergence $\partial^y_\mu t^\mu(y,\dots)$ of a non-local $t^\mu(y,\dots)$. To give an example for that, let $P\in\mathcal{P}$ arbitrary and

$$\begin{aligned}f(y,x)&:=\partial^\nu\delta(x-y)\,P(y)-\delta(x-y)\,\partial^\nu P(y)=-\partial^\nu_y\big[\delta(x-y)\,P(y)\big]\\&=(\partial^\nu\delta)(x-y)\,P(x)\;.\end{aligned}\tag{4.5.19}$$

Although $f(x,y)$ satisfies the assumptions of the lemma, it can also be written as

$$f(x,y)=\partial^y_\mu t^\mu(x,y)\quad\text{with}\quad t^\mu(x,y):=(\partial^\mu\partial^\nu D^{\mathrm{ret}})(x-y)\,P(x)\;.$$

In [16, Lemma 14] Lemma 4.5.1 is used but not proved, we close this gap here.

Proof. Let $(B_j)_{j\in\mathbb{N}}$ be a basis of $\mathcal{P}_{\mathrm{bal}}$. Due to Proposition 1.4.3(a) any $P_a\in\mathcal{P}$ can uniquely be written as

$$P_a(\varphi)=\sum_{j,b}c_{ajb}\;\partial^bB_j(\varphi)\quad\text{with coefficients}\quad c_{ajb}\in\mathbb{C}\;,$$

where $b\in\mathbb{N}^d$. By using this result, we write

$$f(\varphi)(y,x_1,\dots,x_n)=\sum_{j,b}\partial^bB_j(\varphi)(y)\;\;p_{jb}(\partial_{x_1},\dots,\partial_{x_n})\delta(x_1-y,\dots,x_n-y)\,,\tag{4.5.20}$$

where $p_{jb}(\partial_{x_1},\dots,\partial_{x_n})$ is a polynomial in the dn variables $(\partial^\mu_{x_k})^{\mu=0,\dots,d-1}_{k=1,\dots,n}$. In order that the Fourier transformation is well defined, we apply the functional $f(y,x_1,\dots)$

to an h which is an element of $\mathcal{S}(\mathbb{M},\mathbb{R})$. We obtain

$$\begin{aligned}\widehat{f(h)}(k,p_1,\ldots,p_n) &= \frac{1}{(2\pi)^{d(n+1)/2}}\int dy\, dx_1\cdots dx_n\; e^{i(ky+p_1x_1+\cdots+p_nx_n)}\\ &\quad\cdot\sum_{j,b}\partial^b B_j(h)(y)\;\, p_{jb}(\partial_{x_1},\ldots,\partial_{x_n})\delta(x_1-y,\ldots,x_n-y)\\ &= \frac{1}{(2\pi)^{dn/2}}\sum_{j,b} p_{jb}(\,-ip_1,\ldots,-ip_n)\,\big(-i(k+q)\big)^b\,\widehat{B_j(h)}(k+q)\big|_{q:=p_1+\cdots+p_n}\,.\end{aligned} \tag{4.5.21}$$

Working with the independent variables $(k,q,p_1,\ldots,p_{n-1})$ (instead of $(k,p_1,\ldots,p_n)$) we write the latter expression as

$$\frac{1}{(2\pi)^{dn/2}}\sum_j \tilde{P}_j(k,q,p_1,\ldots,p_{n-1})\,\widehat{B_j(h)}(k+q)\;,$$

where $\tilde{P}_j$ is a polynomial in the indicated $d(n+1)$ variables. The property (4.5.17) translates into

$$\sum_j \tilde{P}_j(0,q,p_1,\ldots,p_{n-1})\,\widehat{B_j(h)}(q)=0\quad\forall q,p_1,\ldots,p_{n-1}\;,\;\forall h\in\mathcal{S}(\mathbb{M},\mathbb{R})\;,$$

which implies

$$\tilde{P}_j(0,q,p_1,\ldots,p_{n-1})=0\quad\forall q,p_1,\ldots,p_{n-1}\;,\;\forall j\;,$$

since the balanced fields $(B_j(\varphi))_{j\in\mathbb{N}}$ are linearly independent. Expanding with respect to k, we conclude that there exist polynomials $R^\mu_j(k,q,p_1,\ldots,p_{n-1})$ such that

$$\tilde{P}_j(k,q,p_1,\ldots,p_{n-1})=-ik_\mu\,R^\mu_j(k,q,p_1,\ldots,p_{n-1})\;.$$

Interpreting $R^\mu_j(k,q,p_1,\ldots,p_{n-1})$ as a polynomial in the independent variables $(k+q,p_1,\ldots,p_{n-1},q-p_1-\cdots-p_{n-1})$ we may write

$$R^\mu_j(k,q,p_1,\ldots,p_{n-1})=\sum_b\big(-i(k+q)\big)^b\,\tilde{R}^\mu_{jb}(-ip_1,\ldots,-ip_n)\big|_{p_n:=q-p_1-\cdots-p_{n-1}}\;,$$

where the sum is finite and $\tilde{R}^\mu_{jb}$ is a polynomial in the indicated variables. Summing up we have

$$\begin{aligned}&\widehat{f(h)}(k,p_1,\ldots,p_n)\\ &\quad=\frac{-ik_\mu}{(2\pi)^{dn/2}}\sum_{j,b}\tilde{R}^\mu_{jb}(-ip_1,\ldots,-ip_n)\,\big(-i(k+q)\big)^b\,\widehat{B_j(h)}(k+q)\big|_{q:=p_1+\cdots+p_n}\;;\end{aligned}$$

which implies

$$f(\varphi)(y,x_1,\dots,x_n)=\partial^y_\mu\Big(\sum_{j,b}\partial^b B_j(\varphi)(y)\ \tilde{R}^\mu_{jb}(\partial_{x_1},\dots,\partial_{x_n})\delta(x_1-y,\dots,x_n-y)\Big)\,,$$

as we see by comparing with (4.5.21). Similarly to the equality of (4.5.20) and (4.5.16), the last expression can be rewritten in the form (4.5.18). □

Returning to the question given directly above Lemma 4.5.1 we note that (4.5.15) can be written as

$$\int dy\left(\Delta^n_k\big(\otimes^n_{j=1}L(x_j);Q(y)\big)+Q(y)\frac{\delta Z^{(n)}_k\big(\otimes^n_{j=1}L(x_j)\big)}{\delta\varphi(y)}\right)=\mathcal{O}(\hbar^{k+1})\quad\forall n\geq 2\,.$$

From (4.3.4) we know that $\Delta^n_k\big(\otimes^n_{j=1}L(x_j);Q(y)\big)$ is of the form (4.5.16) and, due to the Locality of $\frac{\delta Z^{(n)}_k(S^{\otimes n})}{\delta\varphi}$ (3.6.34), the term $Q(y)\frac{\delta Z^{(n)}_k\left(\otimes^n_{j=1}L(x_j)\right)}{\delta\varphi(y)}$ can also be transformed into this form by using identities of type (4.5.19). From the lemma we conclude that there exists

$$K^{n\,\mu}_{L,Q}(x_1,\dots,x_n,y):=\mathfrak{S}_n\sum_a\partial^a\delta(x_1-y,\dots,x_n-y)\,U^\mu_a(y)\in\mathcal{D}'(\mathbb{M}^{n+1},\mathcal{F}_{\mathrm{loc}})\tag{4.5.22}$$

for $0\leq\mu\leq d-1$ (where $\mathfrak{S}_n$ denotes symmetrization in $x_1,\dots,x_n$), such that[142]

$$\int dy\ h(y)\Big(\Delta^n_k\big(S^{\otimes n};Q(y)\big)+Q(y)\frac{\delta Z^{(n)}_k(S^{\otimes n})}{\delta\varphi(y)}\Big)=-K^{n\,\mu}_{L,Q}(g^{\otimes n}\otimes\partial_\mu h)+\mathcal{O}(\hbar^{k+1})\tag{4.5.23}$$

for all $n\geq 2$, $g,h\in\mathcal{D}(\mathbb{M})$. Defining

$$Z^{(n+1)}_k\big(S^{\otimes n}\otimes j^\mu(y)\big):=\int dx_1\cdots dx_n\ g(x_1)\cdots g(x_n)\,K^{n\,\mu}_{L,Q}(x_1,\dots,x_n,y)\,,\tag{4.5.24}$$

for all $n\geq 2$, $g\in\mathcal{D}(\mathbb{M})$, the condition (4.5.14) is satisfied. This definition respects (4.5.6), since $\Delta^n_k\big(S^{\otimes n};Q\big)$ and $\frac{\delta Z^{(n)}_k(S^{\otimes n})}{\delta\varphi}$ are of order $\hbar^k$ and, hence, $K^{n\,\mu}_{L,Q}(x_1,\dots,x_n,y)=\mathcal{O}(\hbar^k)$. Note that we require the relation (4.5.23) only for the particular pair (L,Q) given from the outset, and also the definition (4.5.24) is made only for this particular pair.

The crucial question is whether this definition is compatible with the requirement that Z_k is an element of the relevant Stückelberg–Petermann group $\mathcal{R}_0$. From (4.5.22) we know that (4.5.24) satisfies the "Locality" property (3.6.34). The properties "Poincaré covariance", "$*$-structure", "Field parity" and "$\hbar$-dependence"

[142] As usual we write $K^n(g_1\otimes\cdots\otimes g_{n+1}):=\int\Big(\prod_{j=1}^{n+1}dx_j\,g_j(x_j)\Big)\ K^n(x_1,\dots,x_{n+1})$.

can straightforwardly be verified by using corresponding properties of both $\Delta_k\big(e_\otimes^S; Q(y)\big)$ and $\frac{\delta Z_k^{(n)}(S^{\otimes n})}{\delta\varphi}$.

The problem with the definition (4.5.24) comes from the condition that $Z_k^{(n)}$ must satisfy the Field independence, that is, the causal Wick expansion (3.6.9). To discuss this, we split the polynomials L and j into monomials: $L = \sum_r L_r$ and $j = \sum_s j_s$. By the causal Wick expansion, $Z_k^{(n)}\big(\otimes_{l=1}^n L(x_l)\big)$ and $Z_k^{(n)}\big(\otimes_{l=1}^{n-1} L(x_l)\otimes j(x)\big)$ are uniquely determined in terms of the VEVs

$$z_k^{(n)}\big(\underline{L}_{r_1},\ldots,\underline{L}_{r_n}\big) \quad \text{and} \quad z_k^{(n)}\big(\underline{L}_{r_1},\ldots,\underline{L}_{r_{n-1}}, j_s\big)\;, \quad \underline{L}_{r_k}\subseteq L_{r_k}\;, \tag{4.5.25}$$

respectively. To fulfill the Field equation property, we set $z_k^{(n)}(\ldots,\partial^a\varphi,\ldots) := 0$ for all $a\in\mathbb{N}^d$. With that, true submonomials of j_s need not to be considered (since we assume that j_s is bilinear in φ), and only $\underline{L}_{r_k}$ which are at least bilinear in φ are relevant. If $j_s \subseteq L_r$ for some s and r, the definition (4.5.24) is plagued with the following problems:

- Since $z_k^{(n)}\big(\underline{L}_{r_1},\ldots,\underline{L}_{r_{n-1}}, j_s\big)$ appears in the causal Wick expansion of both, $Z_k^{(n)}(\otimes_{l=1}^{n-1} L_{r_l}\otimes L_r)$ and $Z_k^{(n)}(\otimes_{l=1}^{n-1} L_{r_l}\otimes j_s)$, these two cannot be defined independently.
- Among the $z_k^{(n)}\big(\underline{L}_{r_1},\ldots,\underline{L}_{r_{n-1}}, j_s\big)$ determining $Z_k^{(n)}\big(\otimes_{l=1}^{n-1} L(x_l)\otimes j_s(x)\big)$ there appear distributions $z_k^{(n)}\big(\underline{L}_{r_1},\ldots,\underline{L}_{r_t}, j_s,\ldots,j_s\big)$ with $0\le t\le n-2$. The latter must be symmetrical in the $(n-t)$ entries j_s. However, in general there is no hint that $K_{L,Q}^{n\,\mu}$ has a corresponding symmetry.

For the relevant MWI of QED (called "QED-MWI", see Sect. 5.2), L and j are monomials and it holds that $j\subset L$, that is, the just mentioned problems appear. Although the proof of the QED-MWI given in Sect. 5.2.2 uses a different method – it proceeds by induction on the order in S – essentially the same problems reappear; they are solved by a case by case study.

For the scalar $O(N)$-model, introduced in Exap. 4.2.2, we have $j_s \not\subseteq L = \big(\sum_i(\varphi_i)^2\big)^n$ $\forall s$; hence, the method applies. This holds also for Exer. 4.2.6, if we choose $P_j := L$ (for all $1\le j\le n$) with $\theta L = 0$ in order to fulfill the assumption (4.5.13).

If these problems do not appear or can be solved, one finally verifies that the VEVs $z_k^{(n)}\big(\underline{L}_{r_1},\ldots,\underline{L}_{r_{n-1}}, j_s\big)$ (4.5.25) satisfy the property "Scaling degree" (3.6.31). This can be done straightforwardly by means of the definition of $Z_k^{(n)}(L^{\otimes(n-1)}\otimes j)$ (4.5.23)–(4.5.24), using also the causal Wick expansion of $\Delta_k^n(S^{\otimes n}; Q)$ and $\frac{\delta Z_k^{(n)}(S^{\otimes n})}{\delta\varphi}$ and the upper bounds (4.3.9) and (3.6.31), respectively, for the scaling degree of the coefficients appearing in these two expansions.

4.5.2 Proof of the relevant MWI for the scalar $O(N)$-model

We follow [16, Sect. 5.4.4].

We return to the scalar $O(N)$-model treated in Exap. 4.2.2. Since the group $O(N)$ is compact, the MWI (4.2.32) can be maintained in the process of renormalization (i.e., the extension to the thin diagonal) by integration over the group $O(N)$, as explained in (3.2.91). To illustrate the developed formalism we proceed alternatively.

Let

$$S = \kappa\, L(g) \ , \quad L := \Big(\sum_i (\varphi_i)^2\Big)^2 \quad \text{and } g \in \mathcal{D}(\mathbb{M}) \text{ arbitrary,}$$

the $\frac{1}{2}N(N-1)$ symmetries under investigation are given by $Q^a(x) = X^a\varphi(x)$ (4.2.30). We recall that the assumptions (4.5.12)–(4.5.13) are satisfied:

$$Q^a(x)\,\frac{\delta S_0}{\delta\varphi(x)} = \partial^\mu j^a_\mu(x) \quad \text{and} \quad Q^a(x)\,\frac{\delta S}{\delta\varphi(x)} = 0 \quad \forall 1 \le a \le \tfrac{1}{2}N(N-1)\ .$$

Following the method of proof given in the preceding section, the anomalous MWI for Γ_k (4.5.2), which determines $\Delta_k\big(e_\otimes^S; X^a\varphi(x)\big) = \mathcal{O}(\hbar^k)$, can be written as

$$\begin{aligned}\Delta_k\big(e_\otimes^S; X^a\varphi(x)\big) :=\ &-\partial_x^\mu\Big(\Gamma_k\big(e_\otimes^S \otimes j^a_\mu(x)\big) - j^a_\mu(x)\Big)\\ &+ X^a_{ij}\varphi_j(x)\cdot\frac{\delta\Gamma_k(e_\otimes^S)}{\delta\varphi_i(x)} + \mathcal{O}(\hbar^{k+1})\ ,\end{aligned} \tag{4.5.26}$$

by using

$$\Gamma_k\big(e_\otimes^S \otimes \partial^\mu j^a_\mu(x)\big) = \partial_x^\mu\,\Gamma_k\big(e_\otimes^S \otimes j^a_\mu(x)\big) \quad \text{and} \quad \Gamma_k\big(e_\otimes^S \otimes Q^a(x)\big) = Q^a(x)\ ,$$

which follow from the AWI for Γ_k and from the Field equation for Γ_k (4.4.21), respectively. By a finite renormalization of the T-product to second order, we reach that to first order in κ

$$\Delta_k^1\big(S; X^a\varphi(x)\big) = \mathcal{O}(\hbar^{k+1})\ .$$

Note that, for a fixed "a"; the last entry of Δ_k has N components: $Q^a = X^a\varphi := (X^a_{ij}\varphi_j)_{i=1,\dots,N} \in \mathcal{P}^N$, but $\Delta_k\big(e_\otimes^S; X^a\varphi(x)\big) \in \mathcal{F}_{\mathrm{loc}}[\![\kappa]\!]$ has only one component.

To find a finite renormalization Z_k which removes $\Delta_k\big(e_\otimes^S; X^a\varphi(x)\big)$, we only need to find an admissible solution $Z_k^>(S)$ of (4.5.15), that is,

$$\Delta_k^a(e_\otimes^S) = -\delta^a Z_k^>(S) + \mathcal{O}(\hbar^{k+1}) \quad \forall 1 \le a \le \tfrac{1}{2}N(N-1)\ ,\ g \in \mathcal{D}(\mathbb{M})\ , \tag{4.5.27}$$

where

$$\Delta_k^a(e_\otimes^S) := \int dx\ \Delta_k\big(e_\otimes^S; X^a\varphi(x)\big) \quad \text{and} \quad \delta^a := \int dx\ X^a_{ij}\,\varphi_j(x)\,\frac{\delta}{\delta\varphi_i(x)}\ .$$

Having found such a $Z_k^{>}(S)$, the definition of $Z_k^{(n+1)}(S^{\otimes n} \otimes j^\mu(x))$ (4.5.24) is unproblematic, as explained above; the resulting $Z_k^{>}$ solves (4.5.14).

To solve (4.5.27) we temporarily restrict the functionals (1.2.1) to the space $\mathcal{D}(\mathbb{M}, \mathbb{R})$. This permits us to integrate (4.5.26) over $x \in \mathbb{M}$ without meeting any infrared divergences, ending up with the equation

$$\delta^a \Gamma_k(e_\otimes^S) = \Delta_k^a(e_\otimes^S) + \mathcal{O}(\hbar^{k+1}) . \tag{4.5.28}$$

Furthermore, using (4.2.26) we obtain the identity

$$[\delta^a, \delta^b] = -f^{abc}\delta^c , \tag{4.5.29}$$

which we insert into $[\delta^a, \delta^b]\,\Gamma_k(e_\otimes^S)$. This yields the *consistency condition*

$$\delta^a \Delta_k^b(e_\otimes^S) - \delta^b \Delta_k^a(e_\otimes^S) = -f^{abc}\, \Delta_k^c(e_\otimes^S) + \mathcal{O}(\hbar^{k+1}) . \tag{4.5.30}$$

From $S = \kappa\, L(g)$, $g \in \mathcal{D}(\mathbb{M})$ and the Locality of Δ_k (4.3.4), we conclude that $\Delta_k^a(e_\otimes^S) \in \mathcal{F}_{\mathrm{loc}}[\![\kappa]\!]$. Taking into account Exer. 1.3.5, we see that also $\delta^b \Delta_k^a(e_\otimes^S) \in \mathcal{F}_{\mathrm{loc}}[\![\kappa]\!]$. That is, each term in (4.5.30) is of the form (1.3.8). Therefore, the consistency condition (4.5.30) holds true on the entire configuration space $\mathcal{C} \equiv C^\infty(\mathbb{M}, \mathbb{R})$, i.e., the restriction of the functionals to $\mathcal{D}(\mathbb{M}, \mathbb{R})$ can be omitted.

Local fields of the form

$$\Delta_k^a = \delta^a F_k \quad \text{for some} \quad F_k \in \mathcal{F}_{\mathrm{loc}} \ , \quad F_k = \mathcal{O}(\hbar^k) \ , \tag{4.5.31}$$

obviously solve the consistency condition (4.5.30).

As roughly explained below, *every local field solving* (4.5.30) *is of the form* (4.5.31). In particular this holds for the integrated anomaly $\Delta_k^a(e_\otimes^S)$ (up to terms of order $\mathcal{O}(\hbar^{k+1})$), since it is given by (4.5.26), hence satisfies (4.5.28) and, therefore, solves (4.5.30). So we conclude that

$$\Delta_k^a(e_\otimes^S) = \delta^a \sum_{n=2}^{\infty} \frac{1}{n!}\, \hat{\Delta}_k^n(S^{\otimes n}) + \mathcal{O}(\hbar^{k+1}) \tag{4.5.32}$$

for some sequence

$$\hat{\Delta}_k^n(S^{\otimes n}) = \kappa^n \int \prod_{j=1}^{n} \big(dx_j\, g(x_j)\big)\ \hat{\Delta}_k^n\big(\otimes_{j=1}^n L(x_j)\big) \in \mathcal{F}_{\mathrm{loc}} \ , \tag{4.5.33}$$

where $g \in \mathcal{D}(\mathbb{M})$ is arbitrary. In particular we use here the following: Since $g^{\otimes n} \longmapsto \Delta_k^{a\,n}(S^{\otimes n})$ is linear and continuous (i.e., a $\mathcal{F}_{\mathrm{loc}}$-valued distribution), we can choose $\hat{\Delta}_k^n(S^{\otimes n})$ such that this holds also for $g^{\otimes n} \longmapsto \hat{\Delta}_k^n(S^{\otimes n})$. In addition, one verifies that the properties of $\Delta_k^a(e_\otimes^S)$ (see Thm. 4.3.1) imply corresponding properties of

$\hat{\Delta}_k^n(S^{\otimes n})$; in particular, from (4.3.4) and (4.3.6) we conclude[143] that

$$\hat{\Delta}_k^n\big(\otimes_{j=1}^n L(x_j)\big) = \mathfrak{S}_n \sum_{b\in(\mathbb{N}^d)^{n-1}} \partial^b \delta(x_1 - x_n, \ldots, x_{n-1} - x_n)\, P_{k\,b}^n(x_n) \quad (4.5.35)$$

for some $P_{k\,b}^n \in \mathcal{P}$ (where $\mathfrak{S}_n$ denotes symmetrization in $x_1, \ldots, x_n$), with the sum bounded by

$$|b| \le n \dim(L) - \dim(P_{k\,b}^n) - d(n-1)\ . \quad (4.5.36)$$

Then, the definition

$$Z_k^{>}(S) := -\sum_{n=2}^{\infty} \frac{1}{n!}\, \hat{\Delta}_k^n(S^{\otimes n}) \quad (4.5.37)$$

yields a solution of (4.5.27) which fulfills all requirements, in particular the properties "Locality and Translation covariance" (3.6.34) and "Scaling degree" (3.6.35). We remark that $Z_k^{>}(S)$ is not uniquely determined.

Finally, the above-given claim about the solutions of the consistency condition (4.5.30) relies on deep results of Lie algebra cohomology (see [135, Sect. 3.5.2] and references cited there): Local fields solving the consistency condition (4.5.30) are called "cocycles" and trivial solutions of the form (4.5.31) are called "coboundaries". Now, $\mathfrak{o}(N)$ is semisimple and the first cohomology of a semisimple Lie algebra which takes values in a finite-dimensional representation is trivial (this is "Whitehead's Lemma"), that is, every cocycle is a coboundary. Since one can show that there exists a finite dimensional subspace V of $\mathcal{F}_{\mathrm{loc}}$ containing $\Delta_k^a(e_\otimes^S)$ and

[143] In detail, the conclusion goes as follows: Inserting (4.3.4) and (4.5.33) into (4.5.32), we may omit $\int \prod_{j=1}^n (dx_j\, g(x_j))$ (since $g \in \mathcal{D}(\mathbb{M})$ is arbitrary) and obtain

$$\begin{aligned}\int dy\; X_{ij}^a\, \varphi_j(y)\, &\frac{\delta \hat{\Delta}_k^n\big(\otimes_{j=1}^n L(x_j)\big)}{\delta \varphi_i(y)} \\ &= \mathfrak{S}_n \int dy \sum\nolimits_{b,c} \partial^b \delta(x_1 - x_n, \ldots, x_{n-1} - x_n)\, \partial^c \delta(x_n - y)\, \tilde{P}_{k\,bc}^{an}(y) \\ &= \mathfrak{S}_n \sum\nolimits_b \partial^b \delta(x_1 - x_n, \ldots, x_{n-1} - x_n) \sum\nolimits_c \partial^c \tilde{P}_{k\,bc}^{an}(x_n)\ .\end{aligned}$$

From this equation we conclude that $\hat{\Delta}_k^n\big(\otimes_{j=1}^n L(x_j)\big)$ is of the form (4.5.35), because, given $\sum_c \partial^c \tilde{P}_{k\,bc}^{an}$, it is an easy task to find a $P_{k\,b}^n \in \mathcal{P}$ solving

$$\int dy\; X_{ij}^a\, \varphi_j(y)\, \frac{\delta P_{k\,b}^n(x)}{\delta \varphi_i(y)} = \sum\nolimits_c \partial^c \tilde{P}_{k\,bc}^{an}(x) \qquad \forall a,\, b\ . \quad (4.5.34)$$

The Scaling degree property of $\Delta_k(e_\otimes^S; Q^a)$ (4.3.6) says that

$$|b| + |c| + \dim(\tilde{P}_{k\,bc}^{an}) \le n \dim(L) + \dim(Q^a) + \tfrac{d+2}{2} - dn \qquad \forall a,\, b,\, c\ .$$

From (4.5.34) we see that

$$\dim(P_{k\,b}^n) \le \max_{c,a}\big(|c| + \dim(\tilde{P}_{k\,bc}^{an})\big) \quad \text{and, since} \quad \dim(Q^a) = \tfrac{d-2}{2}\ ,$$

we conclude that the range of the sum over b in (4.5.35) is bounded by (4.5.36).

satisfying $\delta^b V \subseteq V$ for all $1 \leq a, b \leq \frac{1}{2}N(N-1)$, Whitehead's Lemma can be applied to the representation $\mathfrak{o}(N) \ni X^a \longmapsto (-\delta^a : V \to V)$.

A further illustration of the MWI formalism is the relevant MWI for QED, formulated and proved in Sect. 5.2.

Chapter 5

Quantum Electrodynamics

QED is the prime example for the application of QFT to elementary particle physics. From a pragmatic point of view the construction of QED is solved since the works of Tomonaga, Schwinger, Feynman and Dyson in the middle of the last century: The predictions (e.g., on the magnetic moment of the electron), obtained by computing the lowest orders of the perturbation series, are in perfect agreement with experiment. However, fundamental theoretical questions remained: First of all, there is the question of convergence of the perturbation series, which is still unsolved and which we do not touch in this book. Whereas the problem of ultraviolet divergences can be reduced to a freedom of finite renormalizations by the Epstein–Glaser method, the infrared problem is only partially solved, cf. [157]. One aspect is that charged particles cannot be eigenstates of the mass operator (they have to be "infraparticles" [30, 152]). Another aspect of the infrared problem are the divergences which appear in the adiabatic limit $g(x) \to 1$ of the S-matrix, cf. Sect. 3.7 and App. A.6. In QED these divergences are logarithmic and cancel in the cross section. (This is shown at least to lowest orders of the perturbation series [148].) Moreover, Blanchard–Seneor [9] and Duch [38] proved that the adiabatic limit defining Green and Wightman functions exists and is unique for QED, see App. A.6.

A perturbative construction of QED by means of the Epstein–Glaser renormalization is given in [148]. A main merit of this book is that the lowest orders of the S-matrix of QED are explicitly calculated – this demonstrates that the Epstein–Glaser method is not only well suited to clarify conceptual issues, but also that it is useful for practical computations. Our construction of QED differs from the procedure in that book surely by the application of deformation quantization; hence, Fock space will be used only at a late stage, for the representation of the algebra of interacting fields. But, a physically more important difference is that our construction is *local*, it relies on the algebraic adiabatic limit (see Sect. 3.7). We do not perform the adiabatic limit in the sense of Epstein and Glaser (see App. A.6), as done in [148], and hence we do not meet any IR-divergences. For

M. Dütsch, *From Classical Field Theory to Perturbative Quantum Field Theory*,
Progress in Mathematical Physics 74, https://doi.org/10.1007/978-3-030-04738-2_5

this reason, our procedure can be applied to massless non-Abelian gauge theories, in particular to quantum chromodynamics (QCD).

More in detail, for an arbitrary finite region $\mathcal{O}$, for simplicity we take a double cone (3.7.7), we construct the algebra $\mathcal{A}_{\mathcal{L}}(\mathcal{O})$ of observables localized in $\mathcal{O}$ by proceeding as follows: First we perform the algebraic adiabatic limit for the interacting fields localized in $\mathcal{O}$. Then, we eliminate the non-observable fields; a crucial input for this step is that the observables should be BRST-invariant. To construct physical states, we proceed as in (2.5.4): We represent the algebra $\mathcal{A}_{\mathcal{L}}(\mathcal{O})$ in a pre Hilbert space. The main problem in the construction of such a representation is that the *field* algebra (which contains non-observable fields) cannot be nontrivially represented in a pre Hilbert space, only a representation in a space with an *indefinite inner product* is possible ("positivity problem"). However, as we will demonstrate, the algebra of observables, which is a kind of sub-algebra of the field algebra, possesses a representation in a pre Hilbert space.

We restrict our study of QED to $d = 4$ dimensions; essentially we follow [48]. We put the focus more on conceptual issues than on computational techniques.

5.1 Deformation quantization for fermionic and gauge fields

The main aim of this section is to generalize deformation quantization of the free real scalar field, developed in Chaps. 1 and 2, to the basic fields appearing in a BRST formulation of QED: These are the Dirac spinors, the photon field and the Faddeev–Popov ghost fields. We will also explain the modifications appearing in the axioms for the R- and T-product and in their inductive constructions.

5.1.1 Dirac spinors

To some extent, the stuff presented here is a mathematically more elementary version of [142]. We assume that the reader is familiar with a description of Dirac spinors in the framework of relativistic quantum mechanics.

A Dirac spinor $\psi(x)$ is a complex field with 4 components, which is subject to *Fermi statistics*. In detail, the formalism introduced for a real scalar field is modified as follows.

Single spinor field. A single spinor is the evaluation functional

$$\psi(x)\colon \begin{cases} C^\infty(\mathbb{M},\mathbb{C}^4) \longrightarrow \mathbb{C}^4 \\ h \equiv (h_k)_{k=1,\dots,4} \longmapsto \psi(x)(h) = h(x)\ . \end{cases} \tag{5.1.1}$$

Under the Lorentz group $\mathcal{L}_+^\uparrow$ (or more precisely $SL(2,\mathbb{C})$), $\psi(x)$ transforms according to the representation $D^{(\frac{1}{2},0)} \oplus D^{(0,\frac{1}{2})}$; details are given in (5.1.23)–(5.1.24) below.

Let $\gamma^\mu \in \mathbb{C}^{4\times 4}$, $\mu = 0, 1, 2, 3$, be the γ-matrices, which satisfy the anticommutation relation

$$[\gamma^\mu, \gamma^\nu]_+ := \gamma^\mu\gamma^\nu + \gamma^\nu\gamma^\mu = 2\, g^{\mu\nu}\, 1_{4\times 4} \,. \tag{5.1.2}$$

We will exclusively work with representations of the γ-matrices which are unitarily equivalent to the Chiral-representation: $\gamma^\mu = U\,\gamma^\mu_{\text{Chiral}}\,U^{-1}$ for some $U \in U(4)$; for example the Dirac- or the Majorana-representation, see, e.g., [109, App. A2]. These representations have the nice property that

$$\gamma^0\,\gamma^\mu\,\gamma^0 = \gamma^{\mu\,\dagger}\,, \quad \text{or equivalently} \quad \gamma^{0\,\dagger} = \gamma^0\,, \quad \gamma^{k\,\dagger} = -\gamma^k \quad \text{for} \quad k = 1, 2, 3\,, \tag{5.1.3}$$

where $\dagger$ means the adjoint matrix in spinor space, i.e., in $\mathbb{C}^{4\times 4}$. For a Lorentz covariant vector $a \equiv (a^\mu)_{\mu=0,1,2,3}$ we use the standard notation $\not{a} := a^\mu\gamma_\mu$. In addition,[144] let $\overline{\psi}(x) := \psi^\dagger(x)\,\gamma^0$. Explicitly, $\overline{\psi}(x)$ is the functional

$$\overline{\psi}(x)\colon \begin{cases} C^\infty(\mathbb{M}, \mathbb{C}^4) \longrightarrow \mathbb{C}^{1\times 4} \\ h \longmapsto \overline{\psi}(x)(h) = h^\dagger(x)\,\gamma^0 = \left(\sum_{k=1}^4 h_k(x)^c\,(\gamma^0)_{kj}\right)_{j=1,\dots,4}\,, \end{cases}$$

which is anti-linear, due to complex conjugation. In the following $\overline{\psi}(x)$ and $\psi(x)$ are treated as *independent* fields. This is analogous to the independence of $\phi(x)$ and $\phi^*(x)$ for a complex scalar field – see Exap. 1.3.2.

Classical product. According to Fermi statistics, the classical product of Dirac spinors must be *anticommutative.* Therefore, our bosonic definition (1.2.5) cannot be transferred to fermionic fields. So we are looking for an alternative definition of the bosonic classical product which indicates how to proceed for fermionic fields: The bosonic classical product can be defined as symmetrized tensor product. The n-fold product of a real scalar field φ reads

$$\begin{aligned} &\varphi(x_1) \otimes_{\text{sym}} \cdots \otimes_{\text{sym}} \varphi(x_n) := \mathfrak{S}_n\, \varphi(x_1) \otimes \cdots \otimes \varphi(x_n)\colon C^\infty(\mathbb{M}, \mathbb{R})^{\times n} \longrightarrow \mathbb{R}\,; \\ &(h^1, \dots, h^n) \longmapsto \frac{1}{n!} \sum_{\pi\in S_n} h^{\pi 1}(x_1) \cdot \;\cdots\; \cdot h^{\pi n}(x_n)\,, \end{aligned} \tag{5.1.4}$$

where $\mathfrak{S}_n$ denotes symmetrization of the n-fold tensor product. The symmetrized tensor product is certainly commutative; it is also associative, in particular the relation

$$\begin{aligned} &\Big(\varphi(x_1)\otimes_{\text{sym}} \cdots \otimes_{\text{sym}} \varphi(x_r)\Big) \otimes_{\text{sym}} \Big(\varphi(x_{r+1}) \otimes_{\text{sym}} \cdots \otimes_{\text{sym}} \varphi(x_{r+s})\Big) \\ &\qquad = \varphi(x_1) \otimes_{\text{sym}} \cdots \otimes_{\text{sym}} \varphi(x_{r+s}) \end{aligned} \tag{5.1.5}$$

[144] In the presence of spinor fields we denote complex conjugation by an upper index "c", i.e., z^c instead of $\overline{z}$; for example, $(\overline{\psi})_k(x) = \sum_l \psi_l(x)^c\,(\gamma^0)_{lk}$.

holds true for all $r, s \in \mathbb{N}^*$. Definition 1.2.1 can be modified accordingly: A bosonic field is then a functional of the form

$$F(\varphi): \begin{cases} \bigoplus_{n=1}^{\infty} C^\infty(\mathbb{M}, \mathbb{R})^{\times n} \longrightarrow \mathbb{C} \\ F(\varphi) := \sum_{n=0}^{N} \langle f_n(x_1, \ldots, x_n), \varphi(x_1) \otimes_{\text{sym}} \cdots \otimes_{\text{sym}} \varphi(x_n) \rangle_{(x_1,\ldots,x_n)} \end{cases},$$

where $N < \infty$ and f_n is as in Definition 1.2.1. Motivated by (5.1.5), the classical product $F \cdot G \equiv F \otimes_{\text{sym}} G$ is defined by

$$\langle f_r, \varphi^{\otimes_{\text{sym}} r} \rangle \otimes_{\text{sym}} \langle g_s, \varphi^{\otimes_{\text{sym}} s} \rangle := \langle f_r \otimes g_s, \varphi^{\otimes_{\text{sym}}(r+s)} \rangle \, : \, C^\infty(\mathbb{M}, \mathbb{R})^{\times (r+s)} \longrightarrow \mathbb{C} \tag{5.1.6}$$

and by bilinearity.

Turning to Dirac spinor fields, the symmetrized tensor product of basic fields is replaced by the antisymmetric wedge product of basic fields (which is essentially the antisymmetrized tensor product). The wedge product is also *associative.* In detail, the classical product of n basic fields is defined to be a map

$$\overline{\psi}(x_1) \wedge \cdots \wedge \overline{\psi}(x_l) \wedge \psi(x_{l+1}) \wedge \cdots \wedge \psi(x_n) \, : \, C^\infty(\mathbb{M}, \mathbb{C}^4)^{\times n} \longrightarrow \mathbb{C}^{(4^n)} \tag{5.1.7}$$

(for all $0 \le l \le n$) which is given by

$$\begin{aligned} &\overline{\psi}_{k_1}(x_1) \wedge \cdots \wedge \overline{\psi}_{k_l}(x_l) \wedge \psi_{k_{l+1}}(x_{l+1}) \wedge \cdots \wedge \psi_{k_n}(x_n) \\ &\quad := n! \, \overline{\psi}_{k_1}(x_1) \otimes_{\text{as}} \cdots \otimes_{\text{as}} \overline{\psi}_{k_l}(x_l) \otimes_{\text{as}} \psi_{k_{l+1}}(x_{l+1}) \otimes_{\text{as}} \cdots \otimes_{\text{as}} \psi_{k_n}(x_n) \, , \end{aligned} \tag{5.1.8}$$

where $\otimes_{\text{as}}$ denotes the antisymmetrized tensor product, that is,

$$a_1 \otimes_{\text{as}} \cdots \otimes_{\text{as}} a_n := \frac{1}{n!} \sum_{\pi \in S_n} \operatorname{sgn} \pi \;\, a_{\pi 1} \otimes \cdots \otimes a_{\pi n} \tag{5.1.9}$$

with $\operatorname{sgn} \pi$ being the signature of the permutation π. Explicitly the mapping in (5.1.8) reads

$$\begin{aligned} &\overline{\psi}_{k_1}(x_1) \wedge \cdots \wedge \psi_{k_n}(x_n)(h^1, \ldots, h^n) \\ &\qquad := \sum_{\pi \in S_n} \operatorname{sgn} \pi \sum_{r_1, \ldots} h^{\pi 1}_{r_1}(x_1)^c \, (\gamma^0)_{r_1 k_1} \cdot \cdots \cdot h^{\pi n}_{k_n}(x_n) \, . \end{aligned} \tag{5.1.10}$$

Remark 5.1.1. An alternative formulation of the classical product of spinor fields, which makes explicit that $\psi(x)$ and $\overline{\psi}(x)$ are independent fields, is as follows: Let

$$\mathcal{C}_1 := C^\infty(\mathbb{M}, \mathbb{C}^4) \oplus C^\infty(\mathbb{M}, \mathbb{C}^4) \tag{5.1.11}$$

and for

$$H = (h_1, h_2) \in \mathcal{C}_1 \quad \text{set} \quad \psi(x)(H) := h_1(x) \, , \quad \overline{\psi}(x)(H) := h_2(x) \, .$$

Instead of (5.1.7)–(5.1.8), we interpret $\overline{\psi}(x_1) \wedge \cdots \wedge \psi(x_n)$ as a functional on $\mathcal{C}_1^{\otimes_{\mathrm{as}} n} \equiv \mathcal{C}_1 \otimes_{\mathrm{as}} \cdots \otimes_{\mathrm{as}} \mathcal{C}_1$:

$$\begin{aligned} &\overline{\psi}(x_1) \wedge \cdots \wedge \overline{\psi}(x_l) \wedge \psi(x_{l+1}) \wedge \cdots \wedge \psi(x_n) \,:\, \mathcal{C}_1^{\otimes_{\mathrm{as}} n} \longrightarrow \mathbb{C}^{(4^n)}\,; \\ &\overline{\psi}_{k_1}(x_1) \wedge \cdots \wedge \psi_{k_n}(x_n)(H^1, \ldots, H^n) \\ &\qquad := \sum_{\pi \in S_n} \operatorname{sgn} \pi \; h^{\pi 1}_{2,k_1}(x_1) \cdot \; \cdots \; \cdot h^{\pi n}_{1,k_n}(x_n) \in \mathbb{C}\,, \end{aligned} \tag{5.1.12}$$

where $(h_1^j, h_2^j) := H^j$. On the one hand the doubling in (5.1.11) makes the formalism more complicated, on the other hand the map (5.1.12) has the advantage of being *linear*; the latter is also due to a modification of the domain of $\overline{\psi}(x_1) \wedge \cdots \wedge \psi(x_n)$: We choose $\mathcal{C}_1^{\otimes_{\mathrm{as}} n}$ instead of $\mathcal{C}_1^{\times n}$. We illustrate linearity by an example:

$$\begin{aligned} &\overline{\psi}(x_1) \wedge \psi(x_2)\big((\alpha H^1 + H^3) \otimes_{\mathrm{as}} H^2\big) \\ &\qquad = \tfrac{1}{2}\big((\alpha h_2^1 + h_2^3)(x_1)\, h_1^2(x_2) - h_2^2(x_1)\,(\alpha h_1^1 + h_1^3)(x_2)\big) \\ &\qquad = \alpha\, \overline{\psi}(x_1) \wedge \psi(x_2)\big(H^1 \otimes_{\mathrm{as}} H^2\big) + \overline{\psi}(x_1) \wedge \psi(x_2)\big(H^3 \otimes_{\mathrm{as}} H^2\big)\,, \qquad \alpha \in \mathbb{C}\,. \end{aligned}$$

Functional derivative. Since $\overline{\psi}(x)$ and $\psi(x)$ are independent fields, the functional derivative of monomials in $\overline{\psi}(x)$ and $\psi(x)$ is defined as follows:

$$\begin{aligned} &\frac{\delta}{\delta \overline{\psi}_r(y)}\, \overline{\psi}_{k_1}(x_1) \wedge \cdots \wedge \overline{\psi}_{k_l}(x_l) \wedge \psi_{k_{l+1}}(x_{l+1}) \wedge \cdots \wedge \psi_{k_n}(x_n) := \sum_{j=1}^{l} (-1)^{j-1} \\ &\cdot \delta_{k_j,r}\delta(x_j - y)\, \overline{\psi}_{k_1}(x_1) \wedge \cdots \widehat{\overline{\psi}_{k_j}(x_j)} \cdots \wedge \overline{\psi}_{k_l}(x_l) \wedge \psi_{k_{l+1}}(x_{l+1}) \wedge \cdots \wedge \psi_{k_n}(x_n)\,, \end{aligned} \tag{5.1.13}$$

$$\begin{aligned} &\frac{\delta}{\delta \psi_s(y)}\, \overline{\psi}_{k_1}(x_1) \wedge \cdots \wedge \overline{\psi}_{k_l}(x_l) \wedge \psi_{k_{l+1}}(x_{l+1}) \wedge \cdots \wedge \psi_{k_n}(x_n) := \sum_{j=l+1}^{n} (-1)^{j-1} \\ &\cdot \delta_{k_j,s}\delta(x_j - y)\, \overline{\psi}_{k_1}(x_1) \wedge \cdots \wedge \overline{\psi}_{k_l}(x_l) \wedge \psi_{k_{l+1}}(x_{l+1}) \cdots \widehat{\psi_{k_j}(x_j)} \cdots \wedge \psi_{k_n}(x_n)\,, \end{aligned} \tag{5.1.14}$$

see Footn. 27 for the notation and cf. the corresponding definition (1.3.7) for a complex scalar field. The sign $(-1)^{j-1}$ in (5.1.13)–(5.1.14) is motivated as follows: The derivative w.r.t. the factor at the jth position is performed after anticommuting this factor to the first position. For this reason, (5.1.13)–(5.1.14) is also called the functional derivative acting from the left-hand side; cf. (5.1.33). Note the change of sign in

$$\frac{\delta^2}{\delta\psi_r(y_1)\,\delta\psi_s(y_2)} := \frac{\delta}{\delta\psi_r(y_1)}\,\frac{\delta}{\delta\psi_s(y_2)} = -\frac{\delta}{\delta\psi_s(y_2)}\,\frac{\delta}{\delta\psi_r(y_1)} =: -\frac{\delta^2}{\delta\psi_s(y_2)\,\delta\psi_r(y_1)}\,,$$

and similarly for $\delta^2/\delta\psi\,\delta\overline{\psi}$ and $\delta^2/\delta\overline{\psi}\,\delta\overline{\psi}$.

Spinor field space. The *classical configuration space* is an infinite direct sum:

$$\mathcal{C}_{\rm spinor} := \bigoplus_{n=1}^{\infty} C^\infty(\mathbb{M},\mathbb{C}^4)^{\times n}\ . \tag{5.1.15}$$

Definition 5.1.2. The *spinor field space* $\mathcal{F}_{\rm spinor}$ is the set of functionals

$$F\colon \mathcal{C}_{\rm spinor} \longrightarrow \mathbb{C}$$

of the form[145]

$$\begin{aligned} F = f_0 \oplus \bigoplus_{n=1}^{N}\sum_{l=0}^{n} \int dx_1\dots dx_n \sum_{k_1,\dots,k_n} f_{n,l}^{k_1,\dots,k_n}(x_1,\dots,x_n)\\ \cdot\overline{\psi}_{k_1}(x_1)\wedge\dots\wedge\overline{\psi}_{k_l}(x_l)\wedge\psi_{k_{l+1}}(x_{l+1})\wedge\dots\wedge\psi_{k_n}(x_n)\\ =: \bigoplus_{n=0}^{N}\sum_{l=0}^{n}\langle f_{n,l}^{k_1,\dots,k_n},\overline{\psi}_{k_1}\wedge\dots\wedge\overline{\psi}_{k_l}\wedge\psi_{k_{l+1}}\wedge\dots\wedge\psi_{k_n}\rangle \end{aligned} \tag{5.1.16}$$

with $N<\infty$ and $f_0\in\mathbb{C}$. For $n\geq 1$, each $f_{n,l}^{k_1,\dots,k_n}$ is a $\mathbb{C}$-valued distribution with compact support, $f_{n,l}^{k_1,\dots,k_n}(x_1,\dots,x_n)$ is totally antisymmetric under permutations of $(k_1,x_1),\dots,(k_l,x_l)$ and permutations of $(k_{l+1},x_{l+1}),\dots,(k_n,x_n)$, and it fulfills the wave front set condition (1.2.2). We will write $\mathcal{F}'_{\rm spinor}(\mathbb{M}^n)$ for the space of distributions of this kind.

The application of an $F\in\mathcal{F}_{\rm spinor}$ to an $h\in\mathcal{C}_{\rm spinor}$ is defined in the standard way: For

$$F = f_0\oplus\bigoplus_{n=1}^{N} F_n\in\mathcal{F}_{\rm spinor}\ ,\quad F_n := \sum_{l=0}^{n} F_{n,l}\ ,\quad h=\bigoplus_{n=1}^{\infty} h_n\in\mathcal{C}_{\rm spinor}$$

we set

$$F(h) := f_0 + \sum_{n=1}^{N} F_n(h_n) = f_0 + \sum_{n=1}^{N}\sum_{l=0}^{n} F_{n,l}(h_n)\ ,$$

where $F_{n,l}$ is the summand with indices (n,l) in (5.1.16). And, for $h_n = (h^1,\dots$ $\dots,h^n)$, $F_{n,l}(h_n)$ is obtained by applying the distribution $f_{n,l}^{k_1,\dots,k_n}(x_1,\dots,x_n)$ to the "test function" (5.1.10) and by summing over $k_1,\dots,k_n$.

Functional derivatives of $F = f_0\oplus\bigoplus_{n=1}^{N}F_n\in\mathcal{F}_{\rm spinor}$ are defined by (5.1.13)–(5.1.14) and

$$\begin{aligned} &\frac{\delta^{p+q}F}{\delta\overline{\psi}(y_1)\cdots\delta\overline{\psi}(y_p)\delta\psi(z_1)\cdots\delta\psi(z_q)}\\ &:= \bigoplus_{n=p+q}^{N}\sum_{l=p}^{n-q}\Big\langle f_{n,l}^{k_1,\dots,k_n}, \frac{\delta^{p+q}(\overline{\psi}_{k_1}\wedge\dots\wedge\overline{\psi}_{k_l}\wedge\psi_{k_{l+1}}\wedge\dots\wedge\psi_{k_n})}{\delta\overline{\psi}(y_1)\cdots\delta\overline{\psi}(y_p)\delta\psi(z_1)\cdots\delta\psi(z_q)}\Big\rangle\ . \end{aligned} \tag{5.1.17}$$

[145] To distinguish the direct sum from other sums, we write $v\oplus w$ for an element of $V\oplus W$ (where $v\in V,\ w\in W$), instead of the usual notation $v+w$.

For example, a functional derivative with respect to ψ may be written as

$$\begin{aligned}
&\Big\langle \frac{\delta}{\delta\psi_s(x)} \Big\langle f_{n,l}^{k_1,\dots,k_n}(x_1,\dots,x_n), \\
&\qquad\qquad \overline{\psi}_{k_1}(x_1)\wedge\cdots\wedge\psi_{k_n}(x_n)\Big\rangle_{(x_1,\dots,x_n)}, g_s(x)\Big\rangle_x (h^1,\dots,h^{n-1}) \\
&= (n-l)(-1)^{n-1}\sum_{\pi\in S_{n-1}} \operatorname{sgn}\pi\,\Big\langle f_{n,l}^{k_1,\dots,k_n}(x_1,\dots,x_n), \\
&\qquad (h^{\pi 1}(x_1)^c\gamma^0)_{k_1}\cdot\ldots\cdot h_{k_{n-1}}^{\pi(n-1)}(x_{n-1})\cdot g_{k_n}(x_n)\Big\rangle_{(x_1,\dots,x_n)}
\end{aligned} \tag{5.1.18}$$

for $g\in C^\infty(\mathbb{M},\mathbb{C}^4)$; in the application of formula (5.1.14) we have replaced the sum over j by $(n-l)$ times the $(j=n)$-term, by using the antisymmetry of the distributions $f_{n,l}$. It is only the case $l=0$ in which formula (5.1.18) takes the much simpler form

$$\Big\langle \frac{\delta F_{n,0}}{\delta\psi_s(x)}, g_s(x)\Big\rangle_x (h^1,\dots,h^{n-1}) = F_{n,0}(g,h^1,\dots,h^{n-1})\ ;$$

and analogously only for $l=n$ we may write

$$\Big\langle \frac{\delta F_{n,n}}{\delta\overline{\psi}_s(x)}, g_s(x)\Big\rangle_x (h^1,\dots,h^{n-1}) = F_{n,n}(g,h^1,\dots,h^{n-1})\ .$$

For $F,G\in\mathcal{F}_{\text{spinor}}$, the *classical product* $F\wedge G\in\mathcal{F}_{\text{spinor}}$ is defined by

$$\begin{aligned}
&\big\langle f(x_1,\dots,x_r),\overline{\psi}(x_1)\wedge\cdots\wedge\psi(x_r)\big\rangle_x \wedge \big\langle g(y_1,\dots,y_s),\overline{\psi}(y_1)\wedge\cdots\wedge\psi(y_s)\big\rangle_y \\
&\quad := \big\langle f(x_1,\dots,x_r)\,g(y_1,\dots,y_s),\overline{\psi}(x_1)\wedge\cdots\wedge\psi(x_r)\wedge\overline{\psi}(y_1)\wedge\cdots\wedge\psi(y_s)\big\rangle_{x,y}
\end{aligned} \tag{5.1.19}$$

and by bilinearity, in analogy to (5.1.6).

An *even-odd grading* of $\mathcal{F}_{\text{spinor}}$ can be introduced, expressing the splitting into Bose- and Fermi-fields:

$$\begin{aligned}
&\mathcal{F}_{\text{spinor}} = \mathcal{F}^+_{\text{spinor}}\oplus\mathcal{F}^-_{\text{spinor}} \quad \text{with} && (5.1.20)\\
&\mathcal{F}^+_{\text{spinor}} := \mathbb{C}\oplus\Big[\{\,F_n\,\big|\,n\in\mathbb{N}^* \text{ is even}\,\}\Big]\ ,\quad \mathcal{F}^-_{\text{spinor}} := \Big[\{\,F_n\,\big|\,n\in\mathbb{N} \text{ is odd}\,\}\Big]\ ,
\end{aligned}$$

where $F_n\in\mathcal{F}_{\text{spinor}}$ is defined as above, that is, it is a sum of n-fold $\wedge$-products of ψ, $\overline{\psi}$ (each integrated out with some $f_{n,l}$), and $[-]$ denotes the linear span. Note that

$$\begin{cases} F_1\wedge F_2\,,\ G_1\wedge G_2\in\mathcal{F}^+_{\text{spinor}} \\ F\wedge G\,,\ G\wedge F\in\mathcal{F}^-_{\text{spinor}} \end{cases} \qquad \forall F,F_j\in\mathcal{F}^-_{\text{spinor}}\,,\ \ G,G_j\in\mathcal{F}^+_{\text{spinor}}\ . \tag{5.1.21}$$

With that the defining properties of an even-odd grading (also called "$\mathbb{Z}_2$-grading") are satisfied. Actually, the relations (5.1.21) hold also with respect to the star-product, introduced below in (5.1.34).

The space of *local spinor fields* $\mathcal{F}_{\text{spinor,loc}}$ is the subspace of $\mathcal{F}_{\text{spinor}}$ given by the functionals F of the form

$$f_0 \oplus \bigoplus_{n=1}^{N} \sum_{l=0}^{n} \int dx\; g_{n,l,a_1,\dots,a_n}^{k_1,\dots,k_n}(x) \cdot \partial^{a_1}\overline{\psi}_{k_1}(x) \wedge \dots \wedge \partial^{a_l}\overline{\psi}_{k_l}(x) \wedge \partial^{a_{l+1}}\psi_{k_{l+1}}(x) \wedge \dots \wedge \partial^{a_n}\psi_{k_n}(x) \;, \tag{5.1.22}$$

where repeated indices are summed over and $f_0 \in \mathbb{C}$, $g_{n,l,a_1,\dots,a_n}^{k_1,\dots,k_n} \in \mathcal{D}(\mathbb{M})$ and $N < \infty$. In addition, $g_{n,l,a_1,\dots,a_n}^{k_1,\dots,k_n}$ is totally antisymmetric under permutations of $(k_1, a_1), \dots, (k_l, a_l)$ and permutations of $(k_{l+1}, a_{l+1}), \dots, (k_n, a_n)$.

The (linear) action of the proper, orthochronous *Poincaré group* $\mathcal{P}_+^\uparrow$ on $\mathcal{F}_{\text{spinor}}$ – more precisely, of $SL(2,\mathbb{C}) \times \mathbb{R}^d$ – is defined analogously to (3.1.35):

$$F \longmapsto \beta_{A,a} F := f_0 \oplus \bigoplus_{n=1}^{N} \sum_{l=0}^{n} \int dx_1 \dots dx_n \sum_{k_1,\dots,k_n} f_{n,l}^{k_1,\dots,k_n}(x_1,\dots,x_n) \cdot \big(\overline{\psi}(\Lambda(A)x_1 + a)S(A)\big)_{k_1} \wedge \dots \wedge \big(S(A^{-1})\psi(\Lambda(A)x_{l+1} + a)\big)_{k_{l+1}} \wedge \cdots \tag{5.1.23}$$

for F given by (5.1.16). Here we use the representation

$$D^{(\frac{1}{2},0)} \oplus D^{(0,\frac{1}{2})} : \begin{cases} SL(2,\mathbb{C}) \longrightarrow \mathbb{C}^{4\times 4} \\ A \longmapsto S(A) := \begin{pmatrix} A & 0 \\ 0 & (A^{-1})^\dagger \end{pmatrix} \end{cases} \tag{5.1.24}$$

and $A \longmapsto \Lambda(A)$ is the usual group homomorphism from $SL(2,\mathbb{C})$ onto $\mathcal{L}_+^\uparrow$, see, e.g., [148, 162].

$*$-operation. The $*$-operation is defined as follows: For a single spinor it is given by taking the adjoint in spinor space,

$$\psi(x)^* := \psi^\dagger(x) = \overline{\psi}(x)\gamma^0 \quad \text{and} \quad \overline{\psi}(x)^* := (\psi^\dagger(x)\gamma^0)^\dagger = \gamma^0\psi(x) \;, \tag{5.1.25}$$

and for $F = \sum_{n=0}^{N}\sum_{l=0}^{n} \langle f_{n,l}^{k_1,\dots,k_n}, \overline{\psi}_{k_1} \wedge \dots \wedge \overline{\psi}_{k_l} \wedge \psi_{k_{l+1}} \wedge \dots \wedge \psi_{k_n}\rangle$ we define

$$F^* := \sum_{n=0}^{N}\sum_{l=0}^{n} \Big\langle \overline{f_{n,l}^{k_1,\dots,k_n}(x_1,\dots,x_n)}, (\overline{\psi}(x_n)\gamma^0)_{k_n} \wedge \dots \wedge (\overline{\psi}(x_{l+1})\gamma^0)_{k_{l+1}} \wedge (\gamma^0\psi)_{k_l}(x_l) \wedge \dots \wedge (\gamma^0\psi)_{k_1}(x_1)\Big\rangle_{x_1,\dots,x_n} .$$

Since the order of the factors $\psi, \overline{\psi}$ is reversed, we conclude that

$$(F \wedge G)^* = G^* \wedge F^* \;. \tag{5.1.26}$$

For example, the electromagnetic current of the free theory,

$$j^\mu(x) := \overline{\psi}(x) \wedge \gamma^\mu \psi(x) := \sum_{j,l=1}^{4} (\gamma^\mu)_{jl}\, \overline{\psi}_j(x) \wedge \psi_l(x) \,, \tag{5.1.27}$$

fulfills $j^\mu(x)^* = j^\mu(x)$, due to $\gamma^{\mu\,\dagger}\gamma^0 = \gamma^0\gamma^\mu$ (5.1.3). Note that $j^\mu(x)^c = -j^\mu(x)$; hence, if we would define the $*$-operation only by complex conjugation (as we do for the complex scalar field, see Exap. 1.3.2), we would have $\omega(j^\mu) \in i\mathbb{R}$ for any state ω (Definition 2.5.1), instead of the expected relation $\omega(j^\mu) \in \mathbb{R}$.

Free field equations. The *free action* for Dirac spinors reads

$$S_0^{\mathrm{spinor}} \equiv S_0(\psi, \overline{\psi}) := \int dy\, \overline{\psi}(y) \wedge (i\partial\!\!\!/_y - m\, 1_{4\times 4})\psi(y) \,, \tag{5.1.28}$$

for brevity we will omit $1_{4\times 4}$. The pertinent free field equations are the *Dirac equation* and its adjoint:

$$0 = \frac{\delta S_0}{\delta \overline{\psi}(x)} = (i\partial\!\!\!/ - m)\psi(x) \quad \text{and} \quad 0 = \frac{\delta S_0}{\delta \psi(x)} = i\partial^\mu \overline{\psi}(x)\gamma_\mu + m\overline{\psi}(x) \,. \tag{5.1.29}$$

The subspace $\mathcal{C}_{\mathrm{spinor},S_0}$ of $\mathcal{C}_{\mathrm{spinor}}$ of solutions of the Dirac equation is defined by

$$\mathcal{C}_{\mathrm{spinor},S_0} := \bigoplus_{n=1}^{\infty} C^\infty_{S_0}(\mathbb{M}, \mathbb{C}^4)^{\times n} \,, \quad \text{where}$$
$$C^\infty_{S_0}(\mathbb{M}, \mathbb{C}^4) := \{\, h \in C^\infty(\mathbb{M}, \mathbb{C}^4) \,|\, (i\partial\!\!\!/_x - m)h(x) = 0 \,\} \,.$$

Note that any $h \in C^\infty_{S_0}(\mathbb{M}, \mathbb{C}^4)$ satisfies also the adjoint Dirac equation.

The ideal $\mathcal{J}^{(m)}_{\mathrm{spinor}}$ of $(\mathcal{F}_{\mathrm{spinor}}, \wedge)$ generated by the Dirac equation is the set of functionals of the form

$$\sum_{n=1}^{N}\sum_{l=0}^{n} \Big\langle f_{n,l}(x_1, \ldots, x_n), \\ \big(i\partial^\mu_{x_1}\overline{\psi}(x_1)\gamma_\mu + m\overline{\psi}(x_1)\big) \wedge \overline{\psi}(x_2) \wedge \cdots \wedge \psi(x_{l+1}) \wedge \cdots \wedge \psi(x_n)\Big\rangle_x$$

or

$$\sum_{n=1}^{N}\sum_{l=0}^{n} \Big\langle f_{n,l}(x_1, \ldots, x_n), \overline{\psi}(x_1) \wedge \cdots \wedge \overline{\psi}(x_l) \wedge (i\partial\!\!\!/_{x_{l+1}} - m)\psi(x_{l+1}) \wedge \psi(x_{l+2}) \wedge \cdots \Big\rangle_x \,,$$

where $f_{n,l}$ is as above. Similarly to scalar fields, it holds that

$$F\big|_{\mathcal{C}_{\mathrm{spinor},S_0}} = G\big|_{\mathcal{C}_{\mathrm{spinor},S_0}} \iff (F - G) \in \mathcal{J}^{(m)}_{\mathrm{spinor}}$$

for $F, G \in \mathcal{F}_{\mathrm{spinor}}$.

Star product. The star product has two different propagators, the Wightman two-point functions for spinor fields, which are defined by

$$\begin{aligned}
S^+(x) \equiv S_m^+(x) &:= (i\partial\!\!\!/_x + m)\Delta_m^+(x) \\
&= \frac{1}{(2\pi)^3}\int d^4p\,(p\!\!\!/ + m)\,\theta(p^0)\,\delta(p^2 - m^2)\,e^{-ipx}\;, \\
S^-(x) \equiv S_m^-(x) &:= -\,(i\partial\!\!\!/_x + m)\Delta_m^+(-x) \\
&= \frac{1}{(2\pi)^3}\int d^4p\,(p\!\!\!/ - m)\,\theta(p^0)\,\delta(p^2 - m^2)\,e^{ipx}\;.
\end{aligned} \tag{5.1.30}$$

Note that $S^\pm$ is a matrix in spinor space: $S^\pm \in \mathcal{D}'(\mathbb{M}, \mathbb{C}^{4\times 4})$. We point out that $S^-(x) \neq -S^+(-x)$. Analogously to $\Delta_m^+(x)$, the propagators S_m^+, S_m^- fulfill the free field equations, which are the Dirac equation and its adjoint:

$$(i\partial\!\!\!/_x - m)S_m^\pm(x) = 0 \quad \text{and} \quad i\partial_x^\mu S_m^\pm(x)\gamma_\mu - m\,S_m^\pm(x) = 0\;; \tag{5.1.31}$$

this follows from

$$(i\partial\!\!\!/_x - m)(i\partial\!\!\!/_x + m) = -(\Box + m^2)\;. \tag{5.1.32}$$

To define the star product we introduce the *functional derivative acting from the right-hand side*:

$$\frac{\delta_r}{\delta\psi(y)}\,\overline{\psi}(x_1)\wedge\cdots\wedge\psi(x_n) := (-1)^{n-1}\,\frac{\delta}{\delta\psi(y)}\,\overline{\psi}(x_1)\wedge\cdots\wedge\psi(x_n) \tag{5.1.33}$$

and similarly for $\delta_r/\delta\overline{\psi}(y)$. The purpose of the factor $(-1)^{n-1}$ is that the term $\sim\delta(x_{n-l} - y)$, i.e., the term coming from the functional derivative acting on the factor at the $(l+1)$th position *from the right-hand side*, has the sign $(-1)^l$ – in accordance with the prescription that the derivative w.r.t. this factor is performed after anticommuting it to the position furthest to the right. For the higher derivatives we write

$$\frac{\delta_r^k}{\delta\psi(y_k)\cdots\delta\psi(y_1)} := \frac{\delta_r}{\delta\psi(y_k)}\cdots\frac{\delta_r}{\delta\psi(y_1)}\;,$$

for example,

$$\begin{aligned}
&\frac{\delta_r^2}{\delta\psi(y_2)\,\delta\overline{\psi}(y_1)}\,\overline{\psi}(x_1)\wedge\cdots\wedge\psi(x_n) \\
&\qquad = (-1)^{(n-1)+(n-2)}\,\frac{\delta^2}{\delta\psi(y_2)\,\delta\overline{\psi}(y_1)}\,\overline{\psi}(x_1)\wedge\cdots\wedge\psi(x_n)\;.
\end{aligned}$$

Definition 5.1.3 (Star product for Dirac spinors). The star product as a map $\star : \mathcal{F}_{\text{spinor}}[\![\hbar]\!] \times \mathcal{F}_{\text{spinor}}[\![\hbar]\!] \longrightarrow \mathcal{F}_{\text{spinor}}[\![\hbar]\!]$ is defined by the formula

$$F \star G := \sum_{n,k=0}^{\infty} \frac{\hbar^{n+k}}{n!k!} \int dx_1 \cdots dx_{n+k}\, dy_1 \cdots dy_{n+k} \tag{5.1.34}$$

$$\cdot \frac{\delta_r^{n+k} F}{\delta\psi_{t_1}(x_1)\cdots(n)\,\delta\overline{\psi}_{u_1}(x_{n+1})\cdots(k)} \wedge \prod_{j=1}^{n} S^+_{t_j s_j}(x_j - y_j)$$

$$\cdot \prod_{l=1}^{k} S^-_{v_l u_l}(y_{n+l} - x_{n+l}) \frac{\delta^{n+k} G}{\delta\overline{\psi}_{s_1}(y_1)\cdots(n)\,\delta\psi_{v_1}(y_{n+1})\cdots(k)} ,$$

where $\delta^n/\delta\psi_{t_1}(x_1)\cdots(n) := \delta^n/\delta\psi_{t_1}(x_1)\cdots\delta\psi_{t_n}(x_n)$. Note that the argument of S^- is $(y_{\dots} - x_{\dots})$ (and not $(x_{\dots} - y_{\dots})$). For brevity we shall write $\mathcal{F}_{\text{spinor}}$ for $\mathcal{F}_{\text{spinor}}[\![\hbar]\!]$, as we do for scalar fields.

Example 5.1.4. For j^μ being the electromagnetic current of the free theory (5.1.27) we obtain

$$\begin{aligned} j^\mu(x) \star j^\nu(y) &= j^\mu(x) \wedge j^\nu(y) + \hbar\,\overline{\psi}(x) \wedge \gamma^\mu\, S^+(x-y)\,\gamma^\nu\, \psi(y) \\ &\quad - \hbar\,\overline{\psi}(y) \wedge \gamma^\nu\, S^-(y-x)\,\gamma^\mu\, \psi(x) + \hbar^2\, \mathrm{tr}\big(\gamma^\mu\, S^+(x-y)\,\gamma^\nu\, S^-(y-x)\big) , \end{aligned}$$

where we use matrix notation for the spinors and $\mathrm{tr}(\cdot)$ denotes the trace in $\mathbb{C}^{4\times 4}$.

By using the wave front set properties of $F, G \in \mathcal{F}_{\text{spinor}}$ and

$$\mathrm{WF}(S^\pm) = \mathrm{WF}\big((i\partial\!\!\!/_x + m)\Delta^+_m(\pm x)\big) \subseteq \mathrm{WF}(\Delta^+)$$

(see App. A.3, property (c)), one verifies that $F \star G$ exists and that $F \star G$ lies again in $\mathcal{F}_{\text{spinor}}$; the argumentation is essentially the same as for the real scalar field (Sect. 2.4).

Let us check whether our requirements on the star product, given in Sect. 2.1, are satisfied also by the star product for spinor fields (Definition 5.1.3) – of course, the formulation of some requirements needs to be adapted. *Bilinearity* and

$$F \star G = F \wedge G + \mathcal{O}(\hbar) \quad \text{for} \quad F, G \sim \hbar^0$$

are obvious. *Associativity* can be proved by essentially the same method as for the real scalar field (Sect. 2.4); the relation of the "graded" commutator to the Poisson bracket is explained below, see (5.1.36)–(5.1.40). Due to (5.1.31), the star product is compatible with the free field equation, that is,

$$(i\partial\!\!\!/_x - m)\big(\psi(x) \star F\big)\big|_{\mathcal{C}_{\text{spinor},S_0}} = 0 , \quad \big(F \star \overline{\psi}(x)\big)\big|_{\mathcal{C}_{\text{spinor},S_0}} (i\overleftarrow{\partial}^\mu_x \gamma_\mu + m) = 0$$

for all $F \in \mathcal{F}_{\text{spinor}}$, and analogously for $(\overline{\psi} \star F)$ and $(F \star \psi)$. Since $S^\pm$ is Poincaré covariant, this holds also for the star product:

$$(\beta_{A,a} F) \star (\beta_{A,a} G) = \beta_{A,a}(F \star G) \quad \forall F, G \in \mathcal{F}_{\text{spinor}} , \quad (A, a) \in SL(2, \mathbb{C}) \times \mathbb{R}^d ,$$

where (5.1.23) is used. Finally, for scalar fields, the compatibility of the star product with the $*$-operation is an immediate consequence of $\Delta^+(x)^c = \Delta^+(-x)$ (2.1.10), but for spinor fields the corresponding proof requires some work:

Lemma 5.1.5. *For $F, G \in \mathfrak{F}_{\text{spinor}}$ it holds that*

$$(F \star G)^* = G^* \star F^* \; . \tag{5.1.35}$$

Proof. One verifies straightforwardly the equations

$$\begin{aligned} \Big(\frac{\delta G}{\delta \psi_v(y)}\Big)^* &= \frac{\delta_r G^*}{\delta \psi_v^\dagger(y)} = (\gamma^0)_{vv_1}\, \frac{\delta_r G^*}{\delta \overline{\psi}_{v_1}(y)} \; , \\ \Big(\frac{\delta G}{\delta \overline{\psi}_s(y)}\Big)^* &= \frac{\delta_r G^*}{\delta \overline{\psi}_s^\dagger(y)} = \frac{\delta_r G^*}{\delta \psi_{s_1}(y)}\,(\gamma^0)_{s_1 s} \; , \\ \overline{S^+(x)_{ts}} &= \big(S^+(x)^\dagger\big)_{st} = \big(\gamma^0\, S^+(-x)\,\gamma^0\big)_{st} \end{aligned}$$

and the same relations with $\delta/\delta\psi \leftrightarrow \delta_r/\delta\psi$, $\delta/\delta\overline{\psi} \leftrightarrow \delta_r/\delta\overline{\psi}$ and with S^- in place of S^+, respectively. Using these identities and $(F \wedge G)^* = G^* \wedge F^*$ (5.1.26) and the definition of the star product (5.1.34), the assertion follows: For example we get

$$\begin{aligned} &\Big(\frac{\delta_r}{\delta\psi_t(x_1)}\frac{\delta_r F}{\delta\overline{\psi}_u(x_2)} \wedge S^+_{ts}(x_1-y_1)\, S^-_{vu}(y_2-x_2)\, \frac{\delta}{\delta\overline{\psi}_s(y_1)}\frac{\delta G}{\delta\psi_v(y_2)}\Big)^* \\ &\quad = \frac{\delta_r}{\delta\psi_{s_1}(y_1)}\frac{\delta_r G^*}{\delta\overline{\psi}_{v_1}(y_2)} \wedge (\gamma^0)_{s_1 s}\,(\gamma^0)_{vv_1}\, \big(\gamma^0\, S^+(y_1-x_1)\,\gamma^0\big)_{st} \\ &\qquad \cdot \big(\gamma^0\, S^-(x_2-y_2)\,\gamma^0\big)_{uv}\,(\gamma^0)_{u_1 u}\,(\gamma^0)_{tt_1}\, \frac{\delta}{\delta\overline{\psi}_{t_1}(x_1)}\frac{\delta F^*}{\delta\psi_{u_1}(x_2)} \\ &\quad = \frac{\delta_r}{\delta\psi_{s_1}(y_1)}\frac{\delta_r G^*}{\delta\overline{\psi}_{v_1}(y_2)} \wedge S^+_{s_1 t_1}(y_1-x_1)\, S^-_{u_1 v_1}(x_2-y_2)\, \frac{\delta}{\delta\overline{\psi}_{t_1}(x_1)}\frac{\delta F^*}{\delta\psi_{u_1}(x_2)} \; . \qquad \square \end{aligned}$$

Graded commutator and Poisson bracket. By means of the Bose–Fermi grading (5.1.20), we define a graded commutator:[146] for $(F, G) \in \mathfrak{F}^p_{\text{spinor}} \times \mathfrak{F}^q_{\text{spinor}}$ we set

$$[F, G]_\star := \begin{cases} F \star G - G \star F & \text{if} \quad (p,q) = (+,+),(+,-),(-,+) \\ F \star G + G \star F & \text{if} \quad (p,q) = (-,-) \end{cases} \tag{5.1.36}$$

and extend this definition to $F, G \in \mathfrak{F}_{\text{spinor}}$ by bilinearity of the star product.

According to the principles of deformation quantization (2.0.1), the formula for the *Poisson bracket* $\{F, G\}$, for $F, G \in \mathfrak{F}_{\text{spinor}}$, can be obtained from $[F, G]_\star$ by selecting all terms with precisely one contraction:

$$\{F, G\} = \lim_{\hbar\to 0} \frac{1}{i\hbar}\,[F, G]_\star \; , \quad \text{if} \quad F, G \sim \hbar^0 \; . \tag{5.1.37}$$

[146] Here and in the following, "graded" means "modified by signs which are due to Fermi statistics".

We point out that the Poisson bracket may be symmetric: $\{F,G\}=\{G,F\}$ for $(F,G)\in\mathcal{F}^-_{\text{spinor}}\times\mathcal{F}^-_{\text{spinor}}$. In particular, we obtain

$$\begin{aligned}\{\psi_k(x),\overline{\psi}_j(y)\}=\{\overline{\psi}_j(y),\psi_k(x)\}&=\frac{1}{i}\left(S^+_{kj}(x-y)+S^-_{kj}(x-y)\right)\\&=(i\partial\!\!\!/_x+m)_{kj}\Delta(x-y)=:S_{kj}(x-y)\,.\end{aligned}\tag{5.1.38}$$

The distribution $S\equiv S_m$ appearing on the r.h.s., is the "(anti)commutator function" for spinor fields, which is the propagator for the Poisson bracket. From the properties of $\Delta_m(x)$ and (5.1.32) we conclude that

$$\operatorname{supp}S_m\subseteq(\overline{V}_+\cup\overline{V}_-)\,,\quad(i\partial\!\!\!/_x-m)S_m(x)=0=i\partial^\mu_x S_m(x)\gamma_\mu-m\,S_m(x)\tag{5.1.39}$$

and

$$S_m(x)=\frac{-i}{(2\pi)^3}\int d^4p\;(p\!\!\!/+m)\,\operatorname{sgn}p^0\,\delta(p^2-m^2)\,e^{-ipx}\neq-S_m(-x)\,.$$

From (5.1.37) we get the following explicit formula for the Poisson bracket:[147]

$$\{F,G\}=\int dx\,dy\left(\frac{\delta_r F}{\delta\psi_t(x)}\wedge S_{ts}(x-y)\,\frac{\delta G}{\delta\overline{\psi}_s(y)}\pm(F\leftrightarrow G)\right)\tag{5.1.40}$$

for $(F,G)\in\mathcal{F}^p_{\text{spinor}}\times\mathcal{F}^q_{\text{spinor}}$, where we have "+" for $(p,q)=(-,-)$ and "−" for $(p,q)=(+,+),(+,-),(-,+)$. In the derivation of this formula, we take into account that $\delta F/\delta\psi=\delta_r F/\delta\psi$ for $F\in\mathcal{F}^-_{\text{spinor}}$, and $\delta F/\delta\psi=-\delta_r F/\delta\psi$ for $F\in\mathcal{F}^+_{\text{spinor}}$, and similarly for $\delta/\delta\overline{\psi}$.

Let us verify that formula (5.1.40) satisfies the usual properties of a Poisson bracket, that is, the properties listed in part (ii) of Proposition 1.8.5 adapted to the Bose–Fermi grading: *Bilinearity* and *graded skew-symmetry* are obvious. For the *graded Leibniz rule* and the *graded Jacobi identity* we can argue as for the real scalar field: The former follows from the graded Leibniz rule for $\delta(G\wedge H)/\delta\psi$ and $\delta(G\wedge H)/\delta\overline{\psi}$, respectively, and the latter is the graded Jacobi identity for the graded commutator to lowest non-vanishing order in $\hbar$, see (2.4.7)–(2.4.8).

Exercise 5.1.6 (Spacelike commutativity). Prove that Lemma 2.4.1 holds true also for the graded commutator (5.1.36) of spinor fields.

[*Solution*: Let $(F,G)\in\mathcal{F}^p_{\text{spinor}}\times\mathcal{F}^q_{\text{spinor}}$ with $p,q\in\{+,-\}$. By inserting the definition of the star product (Definition 5.1.3) into $[F,G]_\star$ (5.1.36), using

$$\frac{\delta_r^{n+k}G}{\delta\psi\cdots\delta\overline{\psi}\cdots}\wedge\frac{\delta^{n+k}F}{\delta\overline{\psi}\cdots\delta\psi\cdots}=\pm\frac{\delta_r^{n+k}F}{\delta\psi\cdots\delta\overline{\psi}\cdots}\wedge\frac{\delta^{n+k}G}{\delta\overline{\psi}\cdots\delta\psi\cdots}$$

[147] If one starts with clFT and then applies deformation quantization (as we do in Chaps. 1 and 2 for the real scalar field), formula (5.1.40) is the *definition* of the Poisson bracket.

and handling the signs with care for all possible values of (p,q), we obtain:

$$[F,G]_\star = \ldots \int dx_1 \cdots dy_1 \cdots \frac{\delta_r^{n+k} F}{\delta\psi_{t_1}(x_1)\cdots\delta\overline{\psi}_{u_1}(x_{n+1})\cdots} \wedge \frac{\delta^{n+k} G}{\delta\overline{\psi}_{s_1}(y_1)\cdots\delta\psi_{v_1}(y_{n+1})\cdots}$$
$$\cdot \Big(\prod_{j=1}^{n} S^+_{t_j s_j}(x_j - y_j) \prod_{l=1}^{k} S^-_{v_l u_l}(y_{n+l} - x_{n+l}) - (-1)^{n+k}\left(S^+ \leftrightarrow S^-\right)\Big) .$$

The term in the last line vanishes due to $S^+(z) = -S^-(z)$ for $z^2 < 0$, which we know from (5.1.38)–(5.1.39).]

Exercise 5.1.7 (Charge belonging to the free electromagnetic current). In this exercise all functionals are restricted to $\mathcal{C}_{\mathrm{spinor},S_0}$, we write $F_0 := F|_{\mathcal{C}_{\mathrm{spinor},S_0}}$. We introduce the charge

$$Q_0^\psi := \int d\vec{x}\; j^0(c,\vec{x})_0 \;, \quad \text{where} \quad j^\mu(x)_0 := \overline{\psi}_0(x) \wedge \gamma^\mu \psi_0(x) \tag{5.1.41}$$

and $c \in \mathbb{R}$ is fixed; we will study the graded commutator

$$[Q_0^\psi \,,\, F_0]_\star := \int d\vec{x}\; [j^0(c,\vec{x})_0 \,,\, F_0]_\star \;, \qquad F \in \mathcal{F} . \tag{5.1.42}$$

The existence of integrals of the kind (5.1.41) is explained after (5.5.25) for the free Kugo–Ojima charge.

This exercise investigates only the commutator (5.1.42): Explain why the region of integration is bounded and why this commutator does not depend on the choice of $c \in \mathbb{R}$.

In addition, derive the commutation relation

$$\begin{aligned}&[Q_0^\psi \,, \overline{\psi}_0(x_1) \wedge \cdots \wedge \overline{\psi}_0(x_s) \wedge \psi_0(y_1) \wedge \cdots \wedge \psi_0(y_r)]_\star \\ &\quad = \hbar\,(s-r)\,\overline{\psi}_0(x_1) \wedge \cdots \wedge \overline{\psi}_0(x_s) \wedge \psi_0(y_1) \wedge \cdots \wedge \psi_0(y_r) .\end{aligned} \tag{5.1.43}$$

Hence, for on-shell fields the operator $F_0 \longmapsto \frac{-1}{\hbar}\,[Q_0^\psi, F_0]_\star$ may be interpreted as charge number operator, as introduced later in (5.1.99).

[*Solution*: Due to Exer. 5.1.6 the region of integration in (5.1.42) is a subset of

$$\{\, \vec{x} \,|\, (c,\vec{x}) \in \operatorname{supp} F + (\overline{V}_+ \cup \overline{V}_-)\,\} ,$$

which is bounded for all $c \in \mathbb{R}$, since $\operatorname{supp} F$ is bounded. Using $\partial_x^\mu\,[j_\mu(x)_0 \,,\, F_0]_\star = 0$ and Gauss' integral theorem (A.1.13), we find that $[Q_0^\psi \,,\, F_0]_\star$ (5.1.42) does not depend on c. Hence, considering

$$[Q_0^\psi \,,\, \overline{\psi}_0(y)]_\star = i\hbar \int_{x^0=c} d\vec{x}\; \overline{\psi}_0(x)\gamma^0\, S(x-y) ,$$

we may choose $c := y^0$. By using (A.2.14) we get

$$\gamma^0\, S(0,\vec{z}) = (i\partial_0 - i\gamma^0\gamma_k\partial_k + \gamma^0 m)\,\Delta(0,\vec{z}) = -i\,\delta(\vec{z}) ,$$

which yields $[Q_0^\psi \,,\, \overline{\psi}_0(y)]_\star = \hbar\,\overline{\psi}_0(y)$. Analogously we derive $[Q_0^\psi \,,\, \psi_0(y)]_\star = -\hbar\,\psi_0(y)$. The assertion follows from the derivation property of the commutator, that is,

$$[G \,,\, F_1 \wedge \cdots \wedge F_n]_\star = \sum_{j=1}^{n} F_1 \wedge \cdots \wedge [G \,,\, F_j]_\star \wedge \cdots \wedge F_n$$

for any $G \in \mathcal{F}^+_{\mathrm{spinor}}$.]

Retarded product. The axioms for the retarded product, given in Sect. 3.1, are modified as follows – we only explain the changes. We start with the basic axioms:

- *Domain and range*: Let $\mathcal{F}^{\pm}_{\text{spinor,loc}} := \mathcal{F}_{\text{spinor,loc}} \cap \mathcal{F}^{\pm}_{\text{spinor}}$. The retarded product $R_{n,1}$ is a collection of 2^{n+1} *linear* maps

$$R^{(p_1,\ldots,p_{n+1})}_{n,1} : \mathcal{F}^{p_1}_{\text{spinor,loc}} \otimes \cdots \otimes \mathcal{F}^{p_{n+1}}_{\text{spinor,loc}} \to \mathcal{F}^{p}_{\text{spinor}} \text{ where } p := p_1 \cdot \cdots \cdot p_{n+1} \tag{5.1.44}$$

and $p_1, \ldots, p_{n+1} \in \{+,-\} \equiv \{1,-1\}$; the upper indices $(p_1, \ldots, p_{n+1})$ are mostly omitted. Since the sum of a Bose and a Fermi field is ill defined – only the *direct sum* (5.1.20) makes sense – one has to distinguish these 2^{n+1} maps.

- *Graded symmetry*: Axiom (b) is now a graded symmetry,

$$R^{(p_{\pi 1},\ldots,p_{\pi n},p_{n+1})}_{n,1}(F_{\pi 1} \otimes \cdots \otimes F_{\pi n}, F) = \text{sgn}_g \pi \, R^{(p_1,\ldots,p_n,p_{n+1})}_{n,1}(F_1 \otimes \cdots \otimes F_n, F) \tag{5.1.45}$$

for all $\pi \in S_n$. By $\text{sgn}_g \pi$ we mean the "graded signature" of π, which depends on π and on the signs $(p_1, \ldots, p_n)$ and is a generalization of the signature of π. To define it, write π as a product of transpositions, $\pi = t_1 \circ \cdots \circ t_k$. If t_j is a transposition of two Fermi fields we set $\text{sgn}_g t_j := -1$ and in all other cases (i.e., for transpositions of Bose with Bose fields and Bose with Fermi fields) we set $\text{sgn}_g t_j = 1$. Then we set $\text{sgn}_g \pi := \text{sgn}_g t_1 \cdot \cdots \cdot \text{sgn}_g t_k$. It is well known that $\text{sgn}_g \pi$ is *uniquely* determined by this prescription, although the representation of π as a product of transpositions and even the number k of transpositions are non-unique. In particular, if $p_1 = \ldots = p_n = -1$, we have $\text{sgn}_g \pi = \text{sgn}\, \pi$.

We point out that interacting fields[148]

$$G_F \equiv R\big(e_\otimes^{F/\hbar}, G\big) \equiv \sum_{n=0}^{\infty} \frac{1}{n!\,\hbar^n} R_{n,1}\big(F^{\otimes n}, G\big)$$

are useful *only if F is a Bose field,* since for $F \in \mathcal{F}^{-}_{\text{spinor,loc}}$ and $G \in \mathcal{F}^{\pm}_{\text{spinor,loc}}$ we obtain

$$R\big(e_\otimes^{F/\hbar}, G\big) = G \oplus \frac{1}{\hbar} R_{1,1}(F, G) , \tag{5.1.46}$$

where "$\oplus$" stands for the direct sum (5.1.20) of Bose and Fermi fields. Nevertheless, a generating functional for the retarded or time-ordered product of spinor fields can be introduced by means of the η-trick explained below, see Remk. 5.1.10.

- Turning to the *GLZ relation* (axiom (e)), let us first study the case in which the interaction is a Bose field. Then, the commutator is replaced by the

[148] Since the letter S is used for the spinor propagators, we denote the interaction by F.

graded commutator (5.1.36); more precisely, for $F \in \mathcal{F}^{+}_{\text{spinor,loc}}$ and $(G,H) \in \mathcal{F}^{p}_{\text{spinor,loc}} \times \mathcal{F}^{q}_{\text{spinor,loc}}$, the GLZ relation reads

$$\frac{1}{i}\big[R(e_\otimes^{F/\hbar},G),R(e_\otimes^{F/\hbar},H)\big]_\star = R(e_\otimes^{F/\hbar}\otimes G,H) \mp R(e_\otimes^{F/\hbar}\otimes H,G)\ , \quad (5.1.47)$$

where $[\cdot,\cdot]_\star$ is the graded commutator. If $(p,q)=(-,-)$, the l.h.s. is symmetric under $G \leftrightarrow H$ and, hence, there is a "+" for the second term on the r.h.s.. For all other values of (p,q), the l.h.s. is antisymmetric and we have the "−" sign on the r.h.s..

To determine the signs appearing in the GLZ relation in the general case, i.e., for $R(F_1\otimes\cdots\otimes F_{n-1},F_n)$ with $F_j \in \mathcal{F}^{p_j}_{\text{spinor,loc}}$, $p_j \in \{+,-\}$ arbitrary, we use a formal trick ("η-trick"), which is often used to compute signs coming from Fermi statistics, see, e.g., [109, Sect. 4-2-2]. Let $(\eta_j)_{j=1}^J$, where $J \leq \infty$, be the generators of a Grassmann algebra, that is, they are anticommuting variables:

$$\eta_j\eta_k = -\eta_k\eta_j \quad \forall 1 \leq j \leq k \leq J \leq \infty\ .$$

For the abstract definition of a Grassmann algebra we refer to [96, Sect. 6.2.1]. Frequently the product of Grassmann variables is defined more explicitly as the wedge product of differential forms (see, e.g., [173, Sect. 9], [148, App. D]); following usual conventions we write $\eta_j\eta_k$ instead of $\eta_j\wedge\eta_k$. The fields

$$\tilde{F}_j := \begin{cases} 1\otimes F_j & \text{if} \quad p_j = + \\ \eta_j\otimes F_j & \text{if} \quad p_j = -\ , \end{cases} \quad (5.1.48)$$

obey Bose statistics. Defining

$$\begin{aligned} &(a_1\otimes F_1)\star(a_2\otimes F_2) := (a_1a_2)\otimes(F_1\star F_2) \quad \text{where} \quad a_j\in\{1,\eta_j\}\ , \\ &R_{n-1,1}\big((a_1\otimes F_1),\ldots;(a_n\otimes F_n)\big) := (a_1\ldots a_n)\otimes R_{n-1,1}(F_1,\ldots;F_n)\ , \end{aligned} \quad (5.1.49)$$

we obtain relations as, e.g.,

$$\begin{aligned} &R(\eta_1\otimes F_1, 1\otimes F_2;\eta_4\otimes F_4)\star R(\eta_3\otimes F_3;\eta_5\otimes F_5) \\ &\quad = (\eta_1\eta_4\eta_3\eta_5)\otimes\big(R(F_1,F_2;F_4)\star R(F_3,F_5)\big) \\ &\quad = -(\eta_1\eta_3\eta_4\eta_5)\otimes\big(R(F_1,F_2;F_4)\star R(F_3;F_5)\big)\ . \end{aligned}$$

The retarded product $R(\tilde{F}_1,\ldots,\tilde{F}_{n-1};\tilde{F}_n)$ fulfills the GLZ relation for Bose fields, as it is given in (3.1.11b)–(3.1.12). From that the signs of the GLZ relation for $R(F_1,\ldots,F_{n-1};F_n)$ can be derived by bringing the η's in the same order in all terms.

With this trick, the graded symmetry of the retarded product of a mixture of Bose and Fermi fields, given in (5.1.45), can be traced back to the symmetry of the retarded product of Bose fields, and similarly for the time-ordered product (see (5.1.59) below). We illustrate the procedure by an example.

Example 5.1.8. For $F_j \in \mathcal{F}^-_{\text{spinor,loc}}$ $(j = 1, 2, 3, 4)$ we get the following signs in the GLZ relation:

$$R(F_1,F_2,F_3;F_4) + R(F_1,F_2,F_4;F_3) = -i\big([R(F_1,F_2,F_3);R(F_4)]_+ \\ - [R(F_1;F_3),R(F_2;F_4)] + [R(F_2;F_3),R(F_1;F_4)] + [R(F_3),R(F_1,F_2;F_4)]_+\big) ,$$

where $[\cdot,\cdot]$ denotes the commutator and $[\cdot,\cdot]_+$ the anticommutator. For example, the second term on the r.h.s. is obtained as follows:

$$\begin{aligned}[R(\tilde{F}_1,\tilde{F}_3),R(\tilde{F}_2,\tilde{F}_4)] &= (\eta_1\eta_3\eta_2\eta_4)\otimes\big(R(F_1,F_3)\star R(F_2,F_4)\big) \\ &\quad - (\eta_2\eta_4\eta_1\eta_3)\otimes\big(R(F_2,F_4)\star R(F_1,F_3)\big) \\ &= -(\eta_1\eta_2\eta_3\eta_4)\otimes[R(F_1,F_3),R(F_2,F_4)] .\end{aligned}$$

The renormalization conditions are modified as follows:

- The *Field independence* (axiom (f)) and the $*$-*structure* (axiom (g)) have an additional sign, which is due to Fermi statistics: The former axiom reads

$$\frac{\delta}{\delta\psi(x)}R_{n-1,1}(F_1\otimes\cdots\otimes F_n) = \sum_{l=1}^{n}\text{sgn}_g\pi_l\; R_{n-1,1}\Big(F_1\otimes\cdots\otimes\frac{\delta F_l}{\delta\psi(x)}\otimes\cdots\otimes F_n\Big)$$

and similarly for $\delta/\delta\overline{\psi}$, where π_l is the permutation

$$\pi_l\,:\,(\psi,F_1,\ldots,F_n)\longmapsto(F_1,\ldots,F_{l-1},\psi,F_l,\ldots,F_n) .$$

Similarly to scalar field theory, Field independence is equivalent to the *causal Wick expansion* (3.1.24)

The additional sign for the $*$-structure axiom (3.1.29) can be motivated from the formula for the unrenormalized retarded product (3.1.69) (where $[\cdot,\cdot]_\star$ means now the graded commutator) and from the following relations which hold for $(F,G)\in\mathcal{F}^p_{\text{spinor}}\times\mathcal{F}^q_{\text{spinor}}$:

$$[F,G]_\star\in\mathcal{F}^{p\cdot q}_{\text{spinor}}\quad\text{and}\quad\Big(\frac{1}{i}\,[F,G]_\star\Big)^* = \pm\frac{1}{i}\,[F^*,G^*]_\star ,$$

where the minus-sign holds for $(p,q)=(-,-)$ and the plus-sign for all other cases; the latter relation follows immediately from Lemma 5.1.5. Computing in this way the $*$-conjugation of formula (3.1.69), we obtain

$$R_{n-1,1}(F_1\otimes\cdots\otimes F_n)^* = (-1)^{\lfloor f/2\rfloor}\,R_{n-1,1}(F_1^*\otimes\cdots\otimes F_n^*) , \qquad (5.1.50)$$

for the unrenormalized R-product, where f is the number of Fermi fields in $F_1,\ldots,F_n$ and $\lfloor\cdot\rfloor$ denotes the integer part. The $*$-structure axiom requires that this formula holds also for the renormalized R-product.

In the following exercise, we derive the sign in (5.1.50) in an alternative way, by using the η-trick (5.1.48) with the additional definitions:

$$\big((a_{j_1}\cdots a_{j_k})\otimes F\big)^* := (a_{j_1}\cdots a_{j_k})^*\otimes F^*\ ,\qquad (a_{j_1}\cdots a_{j_k})^* := a_{j_k}\cdots a_{j_1}$$

for $a_j\in\{1,\eta_j\}$ and $F\in\mathcal{F}_{\mathrm{spinor}}$. In particular, it follows $a_j^*=a_j$.

Exercise 5.1.9. Derive formula (5.1.50) from the bosonic version of the $*$-structure axiom, by using the η-trick.

[*Solution*: From

$$\begin{aligned}(a_n\cdots a_1)\otimes R_{n-1,1}(F_1\otimes\cdots\otimes F_n)^* &= \big((a_1\cdots a_n)\otimes R_{n-1,1}(F_1\otimes\cdots\otimes F_n)\big)^*\\ &= R_{n-1,1}(\tilde F_1\otimes\cdots\otimes\tilde F_n)^* = R_{n-1,1}(\tilde F_1^*\otimes\cdots\otimes\tilde F_n^*)\\ &= (a_1\cdots a_n)\otimes R_{n-1,1}(F_1^*\otimes\cdots\otimes F_n^*)\end{aligned}$$

we see that the sign we want to compute is the graded signature of the permutation $\pi:(a_1,\ldots,a_n)\longmapsto(a_n,\ldots,a_1)$. We may omit the a_j which are equal to 1, that is, we may replace π by $\tilde\pi:(\eta_{j_1},\ldots,\eta_{j_f})\longmapsto(\eta_{j_f},\ldots,\eta_{j_1})$ where $j_k<j_{k+1}$ for all k. With that, the graded signature of π is equal to the signature of $\tilde\pi$, and for the latter we get

$$\operatorname{sgn}\tilde\pi = (-1)^{f(f-1)/2} = \begin{cases}(-1)^{f\cdot f/2-f/2}=(-1)^{f/2} & \text{if } f \text{ is even}\\ (-1)^{(f-1)\cdot(f-1)/2+(f-1)/2}=(-1)^{(f-1)/2} & \text{if } f \text{ is odd,}\end{cases}$$

which agrees with the sign in (5.1.50).]

- Let $F\in\mathcal{F}^+_{\mathrm{spinor,loc}}$. There are two *Off-shell field equations* corresponding to $\frac{\delta_r(S_0+F)}{\delta\overline{\psi}(x)}$ and $\frac{\delta_r(S_0+F)}{\delta\psi(x)}$, respectively – in view of the generalization to $F\in\mathcal{F}^-_{\mathrm{spinor,loc}}$, given in (5.1.57) below, it is better to work with the functional derivative acting from the right-hand side:

$$\begin{aligned}(i\partial\!\!\!/-m)\psi_F(x) &= (i\partial\!\!\!/-m)\psi(x)+\Big(\frac{\delta_r F}{\delta\overline{\psi}(x)}\Big)_F\ ,\\ i\partial^\mu\overline{\psi}_F(x)\gamma_\mu+m\overline{\psi}_F(x) &= i\partial^\mu\overline{\psi}(x)\gamma_\mu+m\overline{\psi}(x)+\Big(\frac{\delta_r F}{\delta\psi(x)}\Big)_F\ .\end{aligned}\qquad(5.1.51)$$

- The *Sm-expansion axiom* can be required also for Dirac spinors, details are given below.
- The substitute for the Field parity axiom (3.1.30) (or (3.1.34)) is *Charge number conservation*, which we introduce in (5.1.101); and in Step 1 of Sect. 5.2.2 we explain how this renormalization condition can be fulfilled and in which sense it plays the same role as the Field parity axiom (Remk. 5.2.4).

Similarly to the doublet (S^+, S^-) of propagators for the star product, there are two different kinds of propagators for the *classical* retarded product:[149] by using the GLZ relation and Causality we get

$$\begin{aligned} R_{\mathrm{cl}\,1,1}\big(\overline{\psi}_j(y),\psi_k(x)\big) &= \frac{1}{\hbar}\,R_{1,1}\big(\overline{\psi}_j(y),\psi_k(x)\big) = \frac{1}{i\hbar}\,[\overline{\psi}_j(y),\psi_k(x)]_\star\,\theta(x^0-y^0) \\ &= S_{kj}(x-y)\,\theta(x^0-y^0) =: S^{\mathrm{ret}}_{kj}(x-y)\ , \\ R_{\mathrm{cl}\,1,1}\big(\psi_k(x),\overline{\psi}_j(y)\big) &= \frac{1}{\hbar}\,R_{1,1}\big(\psi_k(x),\overline{\psi}_j(y)\big) = \frac{1}{i\hbar}\,[\psi_k(x),\overline{\psi}_j(y)]_\star\,\theta(y^0-x^0) \\ &= S_{kj}(x-y)\,\theta(y^0-x^0) =: -S^{\mathrm{av}}_{kj}(x-y)\ ; \end{aligned} \tag{5.1.52}$$

S^{ret} and S^{av} are called the "retarded and advanced propagator for spinor fields", respectively. For scalar fields we have not introduced the advanced propagator, since it is given by $\Delta^{\mathrm{av}}_m(x) := \Delta^{\mathrm{ret}}_m(-x)$, but for spinor fields it holds that $S_m(-x) \neq -S_m(x)$ and, hence, $S^{\mathrm{av}}_m(x) \neq S^{\mathrm{ret}}_m(-x)$. By using (5.1.38) and $\big(\partial^0_z\theta(z^0)\big)\,\Delta_m(z) = \delta(z^0)\,\Delta_m(z) = 0$ (see (A.2.14)), we obtain

$$\begin{aligned} S^{\mathrm{ret}}_m(z) &= \theta(z^0)\,(i\partial\!\!\!/_z + m)\,\Delta_m(z) = (i\partial\!\!\!/_z + m)\,\big(\theta(z^0)\,\Delta_m(z)\big) \\ &= (i\partial\!\!\!/_z + m)\,\Delta^{\mathrm{ret}}_m(z) = \frac{1}{(2\pi)^4}\int d^4p\,\frac{(p\!\!\!/ + m)}{p^2 - m^2 + ip^0 0}\,e^{-ipz}\ , \end{aligned} \tag{5.1.53}$$

and by proceeding analogously we get

$$S^{\mathrm{av}}_m(z) = (i\partial\!\!\!/_z + m)\,\Delta^{\mathrm{ret}}_m(-z) = \frac{1}{(2\pi)^4}\int d^4p\,\frac{(p\!\!\!/ + m)}{p^2 - m^2 - ip^0 0}\,e^{-ipz}\ . \tag{5.1.54}$$

From the defining properties of Δ^{ret}_m (1.8.1) and (5.1.32) we conclude

$$(i\partial\!\!\!/_z - m)S^{\mathrm{ret/av}}_m(z) = \delta(z) = i\partial^\mu_z S^{\mathrm{ret/av}}_m(z)\gamma_\mu - m\,S^{\mathrm{ret/av}}_m(z) \tag{5.1.55}$$

and

$$\mathrm{supp}\,S^{\mathrm{ret}}_m \subseteq \overline{V}_+\ ,\quad \mathrm{supp}\,S^{\mathrm{av}}_m \subseteq \overline{V}_-\ . \tag{5.1.56}$$

The relation

$$S_m(z) = S^{\mathrm{ret}}_m(z) - S^{\mathrm{av}}_m(z)$$

can be interpreted as the splitting of an anticommutator (support $\subseteq (\overline{V}_+ \cup \overline{V}_-)$) into its retarded part (support $\subseteq \overline{V}_+$) and its advanced part (support $\subseteq \overline{V}_-$).

[149] Due to the *h-dependence axiom* (axiom (l)), the statement (3.1.71) about the classical limit holds also in the presence of spinor fields, in particular we use here that $\lim_{\hbar\to 0}\frac{1}{\hbar}\,R_{1,1}(S,F) = R_{\mathrm{cl}\,1,1}(S,F)$ if $S, F \sim \hbar^0$.

As an illustration for $S^{\rm ret}$ and $S^{\rm av}$, we given the Off-shell field equation axiom in integrated form and generalized to $F_j \in \mathcal{F}^{p_j}_{\rm spinor,loc}$, $p_j \in \{+,-\}$ arbitrary:

$$\begin{aligned} R_{n,1}\big(F_1\otimes\cdots\otimes F_n,\psi(x)\big) &= \hbar\int dy\; S^{\rm ret}(x-y) \\ &\quad\cdot\sum_{l=1}^n \operatorname{sgn}_g\pi_l\; R_{n-1,1}\Big(F_1\otimes\cdots\widehat{F_l}\cdots\otimes F_n,\frac{\delta_r F_l}{\delta\overline{\psi}(y)}\Big)\;, \\ R_{n,1}\big(F_1\otimes\cdots\otimes F_n,\overline{\psi}(x)\big) &= -\hbar\int dy \\ &\quad\cdot\sum_{l=1}^n \operatorname{sgn}_g\pi_l\; R_{n-1,1}\Big(F_1\otimes\cdots\widehat{F_l}\cdots\otimes F_n,\frac{\delta_r F_l}{\delta\psi(y)}\Big)\,S^{\rm av}(y-x)\;, \end{aligned} \tag{5.1.57}$$

where π_l is the permutation $(1,\ldots,n)\longmapsto(1,\ldots\widehat{l}\ldots,n,l)$. For $F\in\mathcal{F}^+_{\rm spinor,loc}$, we see that these formulas are indeed the integrated form of the field equations (5.1.51) for the interacting fields $\psi_F(x) = R\big(e_\otimes^{F/\hbar},\psi(x)\big)$ and $\overline{\psi}_F(x) = R\big(e_\otimes^{F/\hbar},\overline{\psi}(x)\big)$, respectively, by using (5.1.55).

Turning to the *Sm-expansion axiom* (axiom (k)), we first define the mass dimension of Dirac spinors in $d=4$ dimensions:

$$\dim\partial^a\psi := \dim\partial^a\overline{\psi} := \frac{3}{2}+|a|\;, \tag{5.1.58}$$

in agreement with the procedure given in Remk. 3.1.19.

From Lemma 3.1.26(3) we conclude that $S^{\rm ret}(z) = (i\partial\!\!\!/_z+m)\,\Delta^{\rm ret}_m(z)$ fulfills the Sm-expansion with degree $2+1=3$. Since $\dim\overline{\psi}+\dim\psi$ is also equal to 3, the VEV $r(\overline{\psi},\psi)(z) = \hbar\,S^{\rm ret}(z)$ fulfills the Sm-expansion axiom.

The formulation of this axiom for (real) scalar fields (given in Sect. 3.1.5) can verbatim be used for spinor fields – this holds not only for the example $r(\overline{\psi},\psi)$, it is generally true. Also for spinor fields, the Sm-expansion axiom is required for all $r(A_1,\ldots,A_n)$, where $A_1,\ldots,A_n$ are arbitrary field monomials. Again this axiom plays the role of a renormalization condition.

When using the stronger version of the Sm-expansion axiom (axiom (k) with the addition (k′)), a significant difference to the case of scalar fields is the following: For scalar fields in $d=4$ dimensions solely *even* powers of m appear in the Sm-expansion of $r(A_1,\ldots,A_n)$, as explained in Remk. 3.2.26. But for Dirac spinors in $d=4$ dimensions also *odd* powers of m appear – already in the Sm-expansion of $S_m^{\rm ret/av}$ all natural numbers appear as powers of m, cf. Exer. 3.1.24.

Time-ordered product. The axioms for the time-ordered product are modified accordingly. We give the changes only for the basic axioms and the field equation: T_n is a collection of 2^n *linear* maps

$$T_n^{(p_1,\ldots,p_n)}\;:\;\mathcal{F}^{p_1}_{\rm spinor,loc}\otimes\cdots\otimes\mathcal{F}^{p_n}_{\rm spinor,loc}\longrightarrow\mathcal{F}^{p}_{\rm spinor}\quad\text{where}\quad p:=p_1\cdot\;\cdots\;\cdot p_n \tag{5.1.59}$$

and $p_1, \ldots, p_n \in \{+, -\}$, which is *graded symmetric* in *all* its arguments. In particular, for

$$F \in \mathcal{F}^-_{\text{spinor,loc}} \quad \text{we get} \quad \mathbf{S}(F) \equiv T\big(e_\otimes^{iF/\hbar}\big) = 1 \oplus \frac{i}{\hbar} F \,. \tag{5.1.60}$$

Again, fermionic signs can be managed by the η-trick, with the definition:

$$T_n\big((a_1 \otimes F_1) \otimes \cdots \otimes (a_n \otimes F_n)\big) := (a_1 \ldots a_n) \otimes T_n(F_1 \otimes \cdots \otimes F_n) \tag{5.1.61}$$

for $a_j \in \{1, \eta_j\}$ and $F_j \in \mathcal{F}^\pm_{\text{spinor,loc}}$.

Due to Fermi statistics, the *Causality axiom* has an additional sign:

$$\begin{aligned} &T_n\big(A_1(x_1), \ldots, A_n(x_n)\big) \\ &= \operatorname{sgn}_g \pi \; T_k\big(A_{\pi 1}(x_{\pi 1}), \ldots, A_{\pi k}(x_{\pi k})\big) \star T_{n-k}\big(A_{\pi(k+1)}(x_{\pi(k+1)}), \ldots, A_{\pi n}(x_{\pi n})\big) \end{aligned} \tag{5.1.62}$$

whenever $\{x_{\pi 1}, \ldots, x_{\pi k}\} \cap \big(\{x_{\pi(k+1)}, \ldots, x_{\pi n}\} + \overline{V}_-\big) = \emptyset$ for a permutation $\pi \in S_n$.

For the propagator of the time-ordered product, i.e., the Feynman propagator of spinor fields, we obtain

$$\begin{aligned} -\omega_0\Big(T_2\big(\overline{\psi}_k(y) \otimes \psi_j(x)\big)\Big) &= \omega_0\Big(T_2\big(\psi_j(x) \otimes \overline{\psi}_k(y)\big)\Big) \\ &= \omega_0\Big(\psi_j(x) \star \overline{\psi}_k(y)\Big)\, \theta(x^0 - y^0) - \omega_0\Big(\overline{\psi}_k(y) \star \psi_j(x)\Big)\, \theta(y^0 - x^0) \\ &= \hbar\, \big(S^+_{jk}(x-y)\, \theta(x^0 - y^0) - S^-_{jk}(x-y)\, \theta(y^0 - x^0)\big) \\ &=: \hbar\, S^F_{jk}(x-y) \,. \end{aligned} \tag{5.1.63}$$

By using the definition of $S^\pm_m$ (5.1.30), relations for the scalar Feynman propagator Δ^F_m (see App. A.2) and

$$\begin{aligned} \big(\partial^0_z \theta(z^0)\big)\, \Delta^+_m(z) + \big(\partial^0_z \theta(-z^0)\big)\, \Delta^+_m(-z) &= \delta(z^0)\, \big(\Delta^+_m(z) - \Delta^+_m(-z)\big) \\ &= i\, \delta(z^0)\, \Delta_m(z) = 0 \end{aligned}$$

(where (A.2.14) is used), we find

$$\begin{aligned} S^F_m(z) &= \theta(z^0)\, (i\partial\!\!\!/_z + m)\, \Delta^+_m(z) + \theta(-z^0)\, (i\partial\!\!\!/_z + m)\, \Delta^+_m(-z) \\ &= (i\partial\!\!\!/_z + m)\, \big(\theta(z^0)\, \Delta^+_m(z) + \theta(-z^0)\, \Delta^+_m(-z)\big) \\ &= (i\partial\!\!\!/_z + m)\, \Delta^F_m(z) \\ &= \frac{i}{(2\pi)^4} \int d^4p\, \frac{(p\!\!\!/ + m)}{p^2 - m^2 + i0}\, e^{-ipz} \,. \end{aligned} \tag{5.1.64}$$

The relations (5.1.32) and $(\Box + m^2)\Delta^F_m = -i\delta$ yield

$$(i\partial\!\!\!/_z - m) S^F_m(z) = i\, \delta(z) = i\partial^\mu_z S^F_m(z) \gamma_\mu - m\, S^F_m(z) \,. \tag{5.1.65}$$

Finally, using $i\,S_m = (S_m^+ + S_m^-)$ and $S_m^{\mathrm{ret/av}}(z) = \pm\theta(\pm z^0)\,S_m(z)$, we may write

$$\begin{aligned} S_m^F(z) &= \theta(z^0)\,\big(S_m^+(z) + S_m^-(z)\big) - \big(\theta(-z^0) + \theta(z^0)\big)\,S_m^-(z) \\ &= i\,S_m^{\mathrm{ret}}(z) - S_m^-(z) \end{aligned} \tag{5.1.66}$$

and

$$\begin{aligned} S_m^F(z) &= \big(\theta(z^0) + \theta(-z^0)\big)\,S_m^+(z) - \theta(-z^0)\,\big(S_m^+(z) + S_m^-(z)\big) \\ &= S_m^+(z) + i\,S_m^{\mathrm{av}}(z)\,. \end{aligned} \tag{5.1.67}$$

We point out that $S_m^F(-z) \neq S_m^F(z)$.

The *Off-shell field equation axiom* reads

$$\begin{aligned} T_{n+1}\big(\psi(x)\otimes F_1\otimes\cdots\otimes F_n\big) &= \psi(x)\wedge T_n(F_1\otimes\cdots\otimes F_n) \\ &\quad + \hbar\int dy\; S^F(x-y)\frac{\delta}{\delta\overline{\psi}(y)}T_n(F_1\otimes\cdots\otimes F_n)\;, \\ T_{n+1}\big(F_1\otimes\cdots\otimes F_n\otimes\overline{\psi}(x)\big) &= T_n(F_1\otimes\cdots\otimes F_n)\wedge\overline{\psi}(x) \\ &\quad + \hbar\int dy\;\frac{\delta_r}{\delta\psi(y)}T_n(F_1\otimes\cdots\otimes F_n)\,S^F(y-x)\;, \end{aligned} \tag{5.1.68}$$

for $F_1,\ldots,F_n \in \mathcal{F}^{\pm}_{\mathrm{spinor,loc}}$.

Remark 5.1.10 (Generating functional for the R- and T-product)**.** The triviality of $R\big(e_\otimes^{F/\hbar},G\big)$ and $\mathbf{S}(F)$ for $F\in\mathcal{F}^-_{\mathrm{spinor,loc}}$, see (5.1.46) and (5.1.60), respectively, can be circumvented by using the η-trick: Let $\tilde{F}_j$ be defined as in (5.1.48) and let

$$\tilde{F} := \sum_{j=1}^{J}\lambda_j\tilde{F}_j = \sum_{j=1}^{J}\lambda_j(a_j\otimes F_j)\;,\quad a_j\in\{1,\eta_j\}\;,\;\;\lambda_j\in\mathbb{R}\;,\;\;J<\infty\;.$$

Since $\tilde{F}$ is bosonic, the definition of the S-matrix (Definition 3.3.8),

$$\mathbf{S}(\tilde{F}) := 1 + \sum_{n=1}^{\infty}\frac{i^n}{n!\,\hbar^n}\,T_n(\tilde{F}^{\otimes n}) \equiv T\big(e_\otimes^{i\tilde{F}/\hbar}\big)\;, \tag{5.1.69}$$

yields a generating functional for the T-products $T_n(F_{j_1}\otimes\cdots\otimes F_{j_n})$:[150]

$$(a_{j_1}\cdots a_{j_n})\otimes T_n(F_{j_1}\otimes\cdots\otimes F_{j_n}) = \frac{\hbar^n}{i^n}\,\frac{\partial^n}{\partial\lambda_{j_1}\cdots\partial\lambda_{j_n}}\Big|_{\lambda_1=\ldots=\lambda_J=0}\mathbf{S}(\tilde{F}) \tag{5.1.70}$$

[150] A version of this procedure is introduced in [148, App. D]: Local fermionic field polynomials $A_j(x)$ are integrated out with *Grassmann-valued* test functions $g_j(x)$, that is, the values of the functions g_j are anticommuting variables. With that,

$$\sum_j A_j(g_j) := \sum_j\int dx\;A_j(x)\,g_j(x)\quad\text{is bosonic;}$$

hence, $\mathbf{S}\big(\sum_j A_j(g_j)\big)$ is the generating functional of the T-products $T_n\big(A_{j_1}(x_1),\ldots,A_{j_n}(x_n)\big)$.

by using (A.1.10). Similarly, $R(e_\otimes^{\tilde{F}/\hbar}, G)$ is a generating functional for the R-products $R_{n,1}(F_{j_1} \otimes \cdots \otimes F_{j_n}, G)$; cf. (3.1.8).

Note: If all F_j are fermionic, i.e., $a_j = \eta_j$ for all $1 \leq j \leq J$, then the sum in (5.1.69) is finite, it stops at $n = J$; both $\eta_{j_1} \cdots \eta_{j_n}$ and $T_n(F_{j_1} \otimes \cdots \otimes F_{j_n})$ vanish for $n > J$.

Antichronological product. As for Bose fields, the $\overline{T}$-product is uniquely determined by the T-product. By using the η-trick (5.1.48), the bosonic definition (3.3.34) applies:

$$\overline{T}\big(e_\otimes^{-i\tilde{F}/\hbar}\big) := T\big(e_\otimes^{i\tilde{F}/\hbar}\big)^{\star-1} \; ; \tag{5.1.71}$$

and from this the $\overline{T}$-product of spinor fields is obtained by requiring (5.1.61) also for $\overline{T}_n$. Since T- and $\overline{T}$-products of fermionic fields vanish if at least two factors are equal, we rewrite this definition more explicitly, by using (3.3.37) for non-diagonal entries.

Definition 5.1.11 (Antichronological product $\overline{T}$). Let $F_j \in \mathcal{F}^{\pm}_{\text{spinor,loc}}$ and let $\tilde{F}_j$ be as in (5.1.48). For $p_1, \ldots, p_n \in \{+,-\}$ we define

$$\overline{T}_n^{(p_1,\ldots,p_n)} : \mathcal{F}^{p_1}_{\text{spinor,loc}} \otimes \cdots \otimes \mathcal{F}^{p_n}_{\text{spinor,loc}} \longrightarrow \mathcal{F}^{p}_{\text{spinor}} \quad (\text{where } p := p_1 \cdot \cdots \cdot p_n)$$

by requiring the relation (5.1.61) also for $\overline{T}_n$ and by

$$\overline{T}_n(\tilde{F}_1 \otimes \cdots \otimes \tilde{F}_n) := \sum_{r=1}^{n} (-1)^{n-r} \sum_{\substack{I_1 \sqcup \cdots \sqcup I_r = \{1,\ldots,n\} \\ |I_j| \geq 1}} T_{|I_1|}(\tilde{F}_{I_1}) \star \cdots \star T_{|I_r|}(\tilde{F}_{I_r}) \, , \tag{5.1.72}$$

where "$\sqcup$" means the disjoint union and $\tilde{F}_I := \otimes_{j \in I} \tilde{F}_j$.

In particular, formula (5.1.72) says that $\overline{T}_n^{(p_1,\ldots,p_n)}$ is *linear* and that the collection $\overline{T}_n \equiv (\overline{T}_n^{(p_1,\ldots,p_n)})_{p_1,\ldots,p_n \in \{+,-\}}$ is *graded symmetric*.

The reasoning given in (3.3.36) applies also here: The definition (5.1.71) implies that causal factorization of $T_n(\tilde{F}_1 \otimes \cdots \otimes \tilde{F}_n)$ (3.3.5) is equivalent to anticausal factorization of $\overline{T}_n(\tilde{F}_1 \otimes \cdots \otimes \tilde{F}_n)$ (3.3.35). It is instructive to check this explicitly to lowest orders:

Exercise 5.1.12.

(a) Verify to orders $n = 2$ and $n = 3$ that $\overline{T}_n\big(\tilde{F}_1 \otimes \cdots \otimes \tilde{F}_n\big)$, defined in (5.1.72), fulfills the *anticausal factorization* (3.3.35).

(b) Compute $\omega_0\Big(\overline{T}_2\big(\psi_j(x) \otimes \overline{\psi}_k(y)\big)\Big)$ and express the result in terms of Δ^{AF}, which is defined in (3.3.38).

[*Solution*: (a) For $n = 2$: Let $\operatorname{supp} F_1 \cap (\operatorname{supp} F_2 + \overline{V}_-) = \emptyset$. By using $T_1(\tilde{F}) = \tilde{F} = \overline{T}_1(\tilde{F})$ and the causal factorization of T_2, we obtain

$$\overline{T}_2(\tilde{F}_1 \otimes \tilde{F}_2) = -T_2(\tilde{F}_1 \otimes \tilde{F}_2) + \tilde{F}_1 \star \tilde{F}_2 + \tilde{F}_2 \star \tilde{F}_1 = \overline{T}_1(\tilde{F}_2) \star \overline{T}_1(\tilde{F}_1) \, .$$

For $n = 3$: Let $\operatorname{supp} F_1 \cap \big((\operatorname{supp} F_2 \cup \operatorname{supp} F_3) + \overline{V}_-\big) = \emptyset$. By applying the causal factorization of T_2 and T_3, we get

$$\begin{aligned}\overline{T}_3(\tilde{F}_1 \otimes \tilde{F}_2 \otimes \tilde{F}_3) &= T_3(\tilde{F}_1 \otimes \tilde{F}_2 \otimes \tilde{F}_3) \\ &\quad - \tilde{F}_1 \star T_2(\tilde{F}_2 \otimes \tilde{F}_3) - T_2(\tilde{F}_2 \otimes \tilde{F}_3) \star \tilde{F}_1 - \tilde{F}_2 \star T_2(\tilde{F}_1 \otimes \tilde{F}_3) \\ &\quad - T_2(\tilde{F}_1 \otimes \tilde{F}_3) \star \tilde{F}_2 - \tilde{F}_3 \star T_2(\tilde{F}_1 \otimes \tilde{F}_2) - T_2(\tilde{F}_1 \otimes \tilde{F}_2) \star \tilde{F}_3 \\ &\quad + \tilde{F}_1 \star \tilde{F}_2 \star \tilde{F}_3 + \tilde{F}_1 \star \tilde{F}_3 \star \tilde{F}_2 + \tilde{F}_2 \star \tilde{F}_1 \star \tilde{F}_3 \\ &\quad + \tilde{F}_2 \star \tilde{F}_3 \star \tilde{F}_1 + \tilde{F}_3 \star \tilde{F}_1 \star \tilde{F}_2 + \tilde{F}_3 \star \tilde{F}_2 \star \tilde{F}_1 \\ &= \big(-T_2(\tilde{F}_2 \otimes \tilde{F}_3) + \tilde{F}_2 \star \tilde{F}_3 + \tilde{F}_3 \star \tilde{F}_2\big) \star \tilde{F}_1 \\ &= \overline{T}_2(\tilde{F}_2 \otimes \tilde{F}_3) \star \overline{T}_1(\tilde{F}_1)\ .\end{aligned}$$

(b) By using Definitions 5.1.11 and 5.1.3 and identities for the spinor propagators given in this section, we obtain

$$\begin{aligned}\omega_0\Big(\overline{T}_2\big(\psi_j(x) \otimes \overline{\psi}_k(y)\big)\Big) &= \omega_0\Big(-T_2\big(\psi_j(x) \otimes \overline{\psi}_k(y)\big) + \psi_j(x) \star \overline{\psi}_k(y) - \overline{\psi}_k(y) \star \psi_j(x)\Big) \\ &= \hbar\big(-S^F_{jk}(x-y) + S^+_{jk}(x-y) - S^-_{jk}(x-y)\big) \\ &= \hbar\,(i\partial\!\!\!/_x + m)_{jk}\,\big(-\Delta^F(x-y) + \Delta^+(x-y) + \Delta^+(y-x)\big) \\ &= \hbar\,(i\partial\!\!\!/_x + m)_{jk}\,\Delta^{AF}(x-y) =: \hbar\, S^{AF}_{jk}(x-y)\ .\end{aligned}$$

In the last step we have taken into account the definitions (2.3.8) and (3.3.38) of Δ^F and Δ^{AF}, respectively, and $\theta(x^0 - y^0) + \theta(y^0 - x^0) = 1$. The resulting propagator S^{AF} is called the "anti-Feynman propagator for spinor fields".]

Remark 5.1.13 (Bogoliubov formula for spinor fields)**.** One verifies straightforwardly that the axioms for the R- and T-product of spinor fields (given in this section) imply that the pertinent $R_{n-1,1}(\tilde{F}_1 \otimes \cdots, \tilde{F}_n)$ (5.1.49) and $T_n(\tilde{F}_1 \otimes \cdots \otimes \tilde{F}_n)$ (5.1.61) satisfy the axioms for the R- and T-product of bosonic fields (given in Sect. 3.1 and 3.3.2, respectively) with suitable modifications concerning the domain and range of the maps $R_{n-1,1}$ and T_n. Therefore, the R- and T-products of the $\tilde{F}$-fields fulfill Proposition 3.3.14, that is, they are connected by the Bogoliubov formula:

$$R_{n,1}(\tilde{F}_1 \otimes \cdots \otimes \tilde{F}_n, \tilde{G}) = i^n \sum_{I \subseteq \{1,\ldots,n\}} (-1)^{|I|}\, \overline{T}_{|I|}(\tilde{F}_I) \star T_{|I^c|+1}(\tilde{F}_{I^c} \otimes \tilde{G})\ , \quad (5.1.73)$$

where $\tilde{F}_J := \otimes_{j \in J} \tilde{F}_j$; note that in the sum an $(I = \emptyset)$-term appears. From (5.1.73) we can read off the Bogoliubov formula for R- and T-products of spinor fields including all signs coming from Fermi statistics.

Some techniques to compute T- or R-products containing spinor fields are explained in Sect. 5.1.7.

5.1.2 Faddeev–Popov ghosts: A pair of anticommuting scalar fields

Generally, "ghost fields" are unphysical or unobservable fields, which may be bosonic or fermionic; they are a mathematical tool for a consistent quantization of spin 1 or spin 2 fields. An example for a bosonic ghost field is the "Stückelberg field", which is a real scalar field and is usually introduced for the quantization of a *massive* spin 1 field, see, e.g., [58, Sect. 3]. The Faddeev–Popov ghost fields are a pair[151] $(u, \tilde{u})$ of complex scalar fields, which obey Fermi statistics. They were first introduced by Feynman [72], to remove an inconsistency appearing in naive one-loop calculations for Yang–Mills theories. Faddeev and Popov [69] introduced these ghost fields more systematically: They used them to solve the problem that in the functional integral quantization of Yang–Mills theories one has to integrate over *classes of gauge equivalent fields* instead of integrating over all gauge fields. A few years later, Becchi, Rouet and Stora [7, 8] and Tyutin [164] developed a rigorous quantization method for (massless and massive) spin 1 fields – the famous "BRST quantization" – for which the Faddeev–Popov ghosts are a crucial ingredient, as explained in Sect. 5.4.1.

In QED the anticommuting Faddeev–Popov ghost fields do not couple (i.e., they do not appear in the interaction Lagrangian) and could therefore be eliminated. However, in view of the generalization to non-Abelian gauge theories, we aim to formulate the gauge structure by means of the BRST formalism, for this purpose we need these ghost fields.

In detail, we introduce a pair $U(x) := \begin{pmatrix} u(x) \\ \tilde{u}(x) \end{pmatrix}$, of complex scalar fields,[152]

$$U(x)\colon \begin{cases} C^\infty(\mathbb{M}, \mathbb{C}^2) \longrightarrow \mathbb{C}^2 \\ H \equiv \begin{pmatrix} h \\ \tilde{h} \end{pmatrix} \longmapsto U(x)(H) = H(x) = \begin{pmatrix} h(x) \\ \tilde{h}(x) \end{pmatrix} \end{cases},$$

$$\beta_{\Lambda,a} u(x) = u(\Lambda x + a)\ , \quad \beta_{\Lambda,a} \tilde{u}(x) = \tilde{u}(\Lambda x + a) \qquad \forall (\Lambda, a) \in \mathcal{P}_+^\uparrow\ ;$$

$U_1 = u$ is called *"ghost field"* and $U_2 = \tilde{u}$ *"anti-ghost field"*. We point out: Although u and $\tilde{u}$ are spin zero fields obeying Fermi statistics, they do not violate the famous Spin-Statistics Theorem (see, e.g., [162]), because they do not satisfy all assumptions of this theorem, for details we refer to [117] and [149, Sect. 1.2].

Similarly to spinor fields, the classical product of ghost fields is the antisymmetric $\wedge$-product. The *configuration space* reads

$$\mathcal{C}_{\text{ghost}} := \bigoplus_{n=1}^{\infty} C^\infty(\mathbb{M}, \mathbb{C}^2)^{\times n}\ ; \tag{5.1.74}$$

[151] For Yang–Mills theories (introduced in Exer. 1.5.2) there is, for each component A_a of the gauge field, a pair $(u_a, \tilde{u}_a)$ of ghost fields.

[152] That we write the pair $(u, \tilde{u})$ as a two-dimensional vector is only for notational convenience, it does not have any physical significance.

the *ghost field space* $\mathcal{F}_{\text{ghost}}$ is the set of functionals $F : \mathcal{C}_{\text{ghost}} \longrightarrow \mathbb{C}$ of the form

$$F = f_0 \oplus \bigoplus_{s=1}^{S} \int dz_1 \ldots dz_s \sum_{j_i=1,2} f_s^{j_1,\ldots,j_s}(z_1,\ldots,z_s)\, U_{j_1}(z_1) \wedge \cdots \wedge U_{j_s}(z_s)$$

$$=: \bigoplus_{s=0}^{S} \sum_{j_1,\ldots,j_s} \left\langle f_s^{j_1,\ldots,j_s}, U_{j_1} \wedge \cdots \wedge U_{j_s} \right\rangle , \quad S < \infty , \tag{5.1.75}$$

where $f_0 \in \mathbb{C}$ and $f_s^{j_1,\ldots,j_s} \in \mathcal{F}'_{\text{ghost}}(\mathbb{M}^s)$. The latter space is defined analogously to the distribution space $\mathcal{F}'_{\text{spinor}}(\mathbb{M}^n)$ in (5.1.16); in particular $f_s^{j_1,\ldots,j_s}(z_1,\ldots,z_s)$ is totally antisymmetric under permutations of the pairs $(z_1, j_1), \ldots, (z_s, j_s)$.

The functional derivative with respect to $u(x)$ or $\tilde{u}(x)$ of a ghost field $F \in \mathcal{F}^{\text{ghost}}$ (acting from the left- or right-hand side) is defined similarly to spinor fields, see (5.1.14), (5.1.17) and (5.1.33).

In contrast to the model of a bosonic complex scalar field (see Exap. 1.3.2), the $*$-*operation* is not defined by complex conjugation, and $\tilde{u}$ is not the $*$-conjugate of u: A frequently used convention is

$$u^*(x) := u(x) \quad \text{and} \quad \tilde{u}^*(x) := -\tilde{u}(x)$$

and generally we define

$$F^* := \bigoplus_{s=0}^{S} \sum_{j_1,\ldots,j_s} \left\langle \overline{f_s^{j_1,\ldots,j_s}}, U_{j_1}^* \wedge \cdots \wedge U_{j_s}^* \right\rangle , \quad U^*(x)(H) = \begin{pmatrix} h(x) \\ -\tilde{h}(x) \end{pmatrix} . \tag{5.1.76}$$

Similarly to the photon field, the ghost fields are massless, that is, the *free action* reads

$$S_0^{\text{ghost}} \equiv S_0(u, \tilde{u}) := \int dx\; \partial^\mu \tilde{u}(x) \wedge \partial_\mu u(x) , \tag{5.1.77}$$

which yields the *free field equations*

$$\Box u(x) = 0 , \qquad \Box \tilde{u}(x) = 0$$

and motivates the values

$$\dim u := 1 , \quad \dim \tilde{u} := 1$$

for the mass dimension (according to Remk. 3.1.19).

The only non-vanishing contractions of the star product are

$$\omega_0\big(u(x) \star \tilde{u}(y)\big) := \hbar\, D^+(x-y) , \quad \omega_0\big(\tilde{u}(x) \star u(y)\big) := -\hbar\, D^+(x-y) , \tag{5.1.78}$$

which yields the graded commutation relation

$$\begin{aligned} [u(x), \tilde{u}(y)]_\star \equiv& u(x) \star \tilde{u}(y) + \tilde{u}(y) \star u(x) \\ =& \hbar \left(D^+(x-y) - D^+(y-x) \right) = i\hbar\, D(x-y) . \end{aligned} \tag{5.1.79}$$

Generally, the *star product* is defined analogously to (5.1.34), that is,

$$F \star G := \sum_{n,k=0}^{\infty} \frac{\hbar^{n+k}(-1)^k}{n!k!} \int dx_1 \cdots dx_{n+k}\, dy_1 \cdots dy_{n+k} \tag{5.1.80}$$

$$\cdot \frac{\delta_r^{n+k} F}{\delta u(x_1) \cdots (n)\, \delta\tilde{u}(x_{n+1}) \cdots (k)} \wedge \prod_{j=1}^{n+k} D^+(x_j - y_j)\, \frac{\delta^{n+k} G}{\delta\tilde{u}(y_1) \cdots (n)\, \delta u(y_{n+1}) \cdots (k)}$$

for $F, G \in \mathcal{F}_{\text{ghost}}$.

In our treatment of QED solely time-ordered (or retarded) products $T_n(F_1 \otimes \cdots \otimes F_n)$ appear for which at most one factor F_k contains the ghost fields u or $\tilde{u}$. Due to linearity of T, we may assume that this sole factor is of the form

$$F_k = \int dx\; g(x)\, Q(u, \tilde{u})(x) \otimes P(\psi, \overline{\psi}, A)(x)\ ,$$

where $g \in \mathcal{D}(\mathbb{M})$ and Q and P are polynomials (with respect to the $\wedge$-product) in partial derivatives of the fields written in their arguments, cf. (5.1.22) and (5.1.96). For such T-products we may require

$$T_n(\cdots \otimes F_k \otimes \cdots) = \int dx\; g(x)\, Q(u, \tilde{u})(x) \otimes T_n(\cdots \otimes P(\psi, \overline{\psi}, A)(x) \otimes \cdots)\ ; \tag{5.1.81}$$

that is, the *ghost fields behave as classical (or "external") fields.* To see how the requirement (5.1.81) can be fulfilled, we follow the inductive construction of the T-product: In the inductive step outside the thin diagonal (see (3.3.18)) the condition (5.1.81) cannot be violated, because there is no contraction partner for $Q(u, \tilde{u})(x)$; hence, (5.1.81) is a renormalization condition. To extend $Q(u, \tilde{u})(x) \otimes T_n^0(\cdots \otimes P(\psi, \overline{\psi}, A)(x) \otimes \cdots)$ to the thin diagonal, we simply extend $T_n^0(\cdots \cdots \otimes P(\psi, \overline{\psi}, A) \otimes \cdots)$ such that the latter extension satisfies all other renormalization conditions – this yields (5.1.81).

5.1.3 The photon field

The photon field is a *gauge field.* Gauge theories are field theories in which different field configurations describe the same observable; such configurations are connected by a "gauge transformation". In the classical theory the Cauchy problem[153] is well posed for the observables, but in general not for the non-observable, "gauge dependent" basic fields $A^\mu(x)$, due to the existence of time-dependent gauge transformations.

[153] For the free photon field the Cauchy problem for $A^\mu(x)$ is the initial value problem given by the free field equation (5.1.84) below (which is a partial differential equation of second order) and the initial values $A^\mu(t_0, \vec{x}) = f^\mu(\vec{x})$ and $\partial^0 A^\mu(t_0, \vec{x}) = f_1^\mu(\vec{x})$, where the functions $f^\mu(\vec{x})$ and $f_1^\mu(\vec{x})$ are given and t_0 is some fixed time.

Attempts to quantize the observables directly have not yet been completely satisfactory and, hence, we do not follow that way. In perturbation theory the non-uniqueness of the solutions of the Cauchy problem for $A^\mu(x)$ shows up already for the free gauge fields. We will solve this problem in the usual way: We will modify the classical, free action by adding a *gauge fixing term* such that the dynamics of the classical observables of the free theory is not changed and the Cauchy problem for $A^\mu(x)$ becomes unique.

In the perturbative interacting theory, the observables should be independent on how the gauge fixing for the free theory is done ("gauge independence"). In the framework of causal perturbation theory this has been shown for the physical S-matrix (i.e., the S-matrix as an operator $S_{\text{phys}} : \mathcal{H} \longrightarrow \mathcal{H}$, where $\mathcal{H}$ is the pre Hilbert space of physical states, introduced in Sect. 5.4.2) in reference [1], under the assumption that the IR-behaviour is unproblematic.

Gauge fixing. In our formalism, the *photon field* $\big(A^\mu(x)\big)_{\mu=0,1,2,3}$ is the evaluation functional

$$A(x)\colon \begin{cases} \mathcal{C}_{\text{photon}} := C^\infty(\mathbb{M},\mathbb{R}^4) \longrightarrow \mathbb{R}^4 \\ h \equiv (h^\mu)_{\mu=0,1,2,3} \longmapsto A(x)(h) = h(x) \ , \end{cases} \tag{5.1.82}$$

cf. Exer. 1.5.2, which is a Lorentz vector:

$$\beta_{\Lambda,a} A^\mu(x) := (\Lambda^{-1})^\mu{}_\nu\, A^\nu(\Lambda x + a) \qquad \forall (\Lambda, a) \in \mathcal{P}_+^\uparrow \ .$$

Using $F^{\mu\nu} := \partial^\mu A^\nu - \partial^\nu A^\mu$, the *free action*

$$\begin{aligned} S_0^{\text{photon}} \equiv S_0(A) :=& -\frac{1}{4}\int dx\; F^{\mu\nu}(x) F_{\mu\nu}(x) \\ =& \ \frac{1}{2}\int dx\, A_\mu(x)\, D^{\mu\nu} A_\nu(x) \ , \quad D^{\mu\nu} := g^{\mu\nu}\Box - \partial^\mu\partial^\nu \ , \end{aligned} \tag{5.1.83}$$

yields the free field equation

$$0 = \frac{\delta S_0}{\delta A_\mu(x)} = D^{\mu\nu} A_\nu(x) \ . \tag{5.1.84}$$

Here, we understand S_0 as "generalized Lagrangian" in the sense of Remk. 1.5.1.

Taking into account $D^{\mu\nu}\partial_\nu = 0$, we can now explicitly see that the solutions of the Cauchy problem are non-unique: If $h \equiv (h^\nu) \in \mathcal{C}_{\text{photon}}$ is a solution, then this holds also for $\big(h^\nu(x) + \partial^\nu H(x)\big)$ for an arbitrary $H \in C^\infty(\mathbb{M},\mathbb{R})$ fulfilling $\partial^\nu H(t_0,\vec{x}) = 0$ and $\partial^0\partial^\nu H(t_0,\vec{x}) = 0$. These conditions do not determine $H(x)$ in the complement of an arbitrary small neighbourhood of the set $\{x = (t_0,\vec{x}) \,|\, \vec{x} \in \mathbb{R}^3\}$. Note that $h^\nu \longmapsto (h^\nu + \partial^\nu H)$ is a time-dependent gauge transformation of a field configuration.

The usual way out is to modify the free action by adding a Lorentz invariant *gauge fixing term*: $D^{\mu\nu}$ is replaced by $D^{\mu\nu}_\lambda := D^{\mu\nu} + \lambda\partial^\mu\partial^\nu$ where $\lambda \in \mathbb{R}\setminus\{0\}$ is an arbitrary constant. That is, the gauge fixing term reads

$$S_0^{\text{gf}} = \frac{-\lambda}{2} \int dx\, \big(\partial_\mu A^\mu(x)\big)^2 \,. \tag{5.1.85}$$

For the Euler–Lagrange equation belonging to $(S_0^{\text{photon}} + S_0^{\text{gf}})$, i.e., the *free field equation*, we obtain

$$0 = D_\lambda^{\mu\nu} A_\nu(x) = \Box A^\mu - (1-\lambda)\partial^\mu \partial_\nu A^\nu \,. \tag{5.1.86}$$

The mass dimension of the photon field is defined to be

$$\dim A^\mu := 1 \,, \quad \mu = 0,1,2,3 \,,$$

for any value of λ, in agreement with Remk. 3.1.19.

For simplicity let us choose $\lambda = 1$, which is referred to as "Feynman gauge". The generalization to arbitrary values of λ can be found, e.g., in [1]. With $\lambda = 1$, the free field equation simplifies to

$$\Box A^\mu(x) = 0 \,, \quad \mu = 0,1,2,3 \,. \tag{5.1.87}$$

That is, the free photon field can be interpreted as a quadruplet of *independent* real, massless scalar fields – as long as we ignore the behaviour under Lorentz transformations. For this reason, and since also the star product is diagonal in the components of the photon field (i.e., contractions of different components do not appear – see below), the generalization of our formalism from one real scalar field (given in Chaps. 1 to 3) to the photon field in Feynman gauge is obvious to a far extent – we only mention some basic points.

Photon field space and star product. The *photon field space* $\mathcal{F}_{\text{photon}}$ is the set of functionals $F\colon \mathcal{C}_{\text{photon}} \longrightarrow \mathbb{C}$ of the form

$$\begin{aligned} F &= \sum_{r=0}^{R} \int dy_1 \cdots dy_r \; f_r^{\mu_1,\dots,\mu_r}(y_1,\dots,y_r)\, A_{\mu_1}(y_1)\cdots A_{\mu_r}(y_r) \\ &=: \sum_{r=0}^{R} \big\langle f_r^{\mu_1,\dots,\mu_r}, A_{\mu_1}\cdots A_{\mu_r} \big\rangle \end{aligned} \tag{5.1.88}$$

where $R < \infty$, $f_0 \in \mathbb{C}$ and $f_r^{\mu_1,\dots,\mu_r} \in \mathcal{F}'_{\text{photon}}(\mathbb{M}^r)$. The space $\mathcal{F}'_{\text{photon}}(\mathbb{M}^r)$ is defined analogously to $\mathcal{F}'(\mathbb{M}^n)$ in Definition 1.2.1; the only difference is that "symmetry" of $f_r^{\mu_1,\dots,\mu_r}(y_1,\dots,y_r)$ is meant with respect to permutations of the pairs $(y_1,\mu_1),\dots,(y_r,\mu_r)$.

The $*$-*operation* is introduced similarly to (1.2.7):

$$F^* := \sum_{r=0}^{R} \big\langle \overline{f_r^{\mu_1,\dots,\mu_r}}, A_{\mu_1}\cdots A_{\mu_r} \big\rangle \,, \quad \text{in particular} \quad A^\mu(x)^* = A^\mu(x) \,.$$

Finally, the propagator of the *star product* is defined to be

$$\omega_0\big(A^\mu(x) \star A^\nu(y)\big) := -\hbar g^{\mu\nu}\, D^+(x-y)\ ; \tag{5.1.89}$$

and generally we define

$$F \star G := \sum_{n=0}^{\infty} \frac{\hbar^n\,(-1)^n}{n!} \int dx_1 \cdots dx_n\, dy_1 \cdots dy_n \tag{5.1.90}$$
$$\cdot \frac{\delta^n F}{\delta A_{\mu_1}(x_1) \cdots \delta A_{\mu_n}(x_n)} \prod_{l=1}^{n} D^+(x_l - y_l)\, \frac{\delta^n G}{\delta A^{\mu_1}(y_1) \cdots \delta A^{\mu_n}(y_n)}$$

for $F, G \in \mathcal{F}_{\text{photon}}$.

5.1.4 Definition of the retarded and time-ordered product in QED

In the BRST formulation of a gauge theory (or vector boson theory), each gauge (or vector boson) field $A(x) \equiv (A^\mu(x))_{\mu=0,\ldots,3}$ is accompanied by one pair $\big(u(x), \tilde{u}(x)\big)$ of Faddeev–Popov ghosts.[154] So for QED we have three different kinds of basic fields: A pair $\big(\psi(x), \overline{\psi}(x)\big)$ of Dirac spinors, the photon field and a pair of Faddeev–Popov ghosts.

Field space of QED. The field space of QED is the tensor product

$$\mathcal{F}_{\text{QED}} := \mathcal{F}_{\text{photon}} \otimes \mathcal{F}_{\text{spinor}} \otimes \mathcal{F}_{\text{ghost}}\ . \tag{5.1.91}$$

A general $F \in \mathcal{F}_{\text{QED}}$ is a functional of the form

$$F = \sum_{p=1}^{P} F_{p,A} \otimes F_{p,\psi} \otimes F_{p,u}\ : \tag{5.1.92}$$

$$\begin{cases} \mathcal{C}_{\text{QED}} := \mathcal{C}_{\text{photon}} \times \mathcal{C}_{\text{spinor}} \times \mathcal{C}_{\text{ghost}} \longrightarrow \mathbb{C} \\ F(h_A, H_\psi, H_u) := \sum_{p=1}^{P} F_{p,A}(h_A) \cdot F_{p,\psi}(H_\psi) \cdot F_{p,u}(H_u)\ , \end{cases}$$

with $F_{p,A} \in \mathcal{F}_{\text{photon}}$, $F_{p,\psi} \in \mathcal{F}_{\text{spinor}}$, $F_{p,u} \in \mathcal{F}_{\text{ghost}}$ and $P < \infty$. More explicitly, an arbitrary $F \in \mathcal{F}_{\text{QED}}$ can be written as

$$F = \sum_{r=0}^{R} \bigoplus_{n=0}^{N} \sum_{l=0}^{n} \bigoplus_{s=0}^{S} \int dy_1 \cdots dy_r\, dx_1 \cdots dx_n\, dz_1 \cdots dz_s \tag{5.1.93}$$
$$\cdot f^{\mu_1,\ldots,k_1,\ldots,j_1,\ldots}_{r,n,l,s}(y_1,\ldots,y_r,x_1,\ldots,x_n,z_1,\ldots,z_s) \cdot A_{\mu_1}(y_1) \cdots (r)$$
$$\otimes \overline{\psi}_{k_1}(x_1) \wedge \cdots (l) \wedge \psi_{k_{l+1}}(x_{l+1}) \wedge \cdots (n) \otimes U_{j_1}(z_1) \wedge \cdots (s)$$
$$=: \sum_r \bigoplus_n \sum_l \bigoplus_s \big\langle f^{\mu_1,\ldots,k_1,\ldots,j_1,\ldots}_{r,n,l,s}, A_{\mu_1} \cdots \otimes \overline{\psi}_{k_1} \wedge \cdots \wedge \psi_{k_{l+1}} \wedge \cdots \otimes U_{j_1} \wedge \cdots \big\rangle$$

[154]This statement includes the case in which A^μ is massive. Since the mass term $\frac{1}{2} m^2 A^\mu A_\mu$ of the free Lagrangian is not gauge invariant, we avoid the notion "massive gauge field" and call A^μ a "massive vector boson field", cf. Remk. 5.4.1.

where repeated indices are summed over, $R, N, S < \infty$, $f_{0,0,0,0} \in \mathbb{C}$ and

$$f_{r,n,l,s}^{\mu_1,\dots,k_1,\dots,j_1,\dots} \in \mathcal{F}'_{\text{photon}}(\mathbb{M}^r) \otimes \mathcal{F}'_{\text{spinor}}(\mathbb{M}^n) \otimes \mathcal{F}'_{\text{ghost}}(\mathbb{M}^s) \,. \tag{5.1.94}$$

The shorthand notations (r), (l), (n) and (s) for $A_{\mu_r}(y_r)$, $\overline{\psi}_{k_l}(x_l)$, $\psi_{k_n}(x_n)$ and $U_{j_s}(z_s)$, respectively, are used. To verify the statement (5.1.93), we take into account that any $f_{r,n,l,s}^{\mu_1,\dots,k_1,\dots,j_1,\dots}$ of the kind (5.1.94) can be written as

$$f_{r,n,l,s}^{\mu_1,\dots,k_1,\dots,j_1,\dots}(y,x,z) = \sum_{p=1}^{P} g_{p,r}^{\mu_1,\dots}(y)\, d_{p,n,l}^{k_1,\dots}(x)\, e_{p,s}^{j_1,\dots}(z) \,, \tag{5.1.95}$$

with $g_{p,r}^{\mu_1,\dots} \in \mathcal{F}'_{\text{photon}}(\mathbb{M}^r)$, $d_{p,n,l}^{k_1,\dots} \in \mathcal{F}'_{\text{spinor}}(\mathbb{M}^n)$ and $e_{p,s}^{j_1,\dots} \in \mathcal{F}'_{\text{ghost}}(\mathbb{M}^s)$ and $P < \infty$. Inserting (5.1.95) into (5.1.93), we see that (5.1.93) is indeed of the form (5.1.92).

The subspace $\mathcal{F}_{\text{QED}}^{\text{loc}}(\subset \mathcal{F}_{\text{QED}})$ of *local* QED fields is the set of functionals of the form

$$\sum_{r=0}^{R} \bigoplus_{n=0}^{N} \sum_{l=0}^{n} \bigoplus_{s=0}^{S} \int dx\; g_{r,n,l,s,a,b,c}^{\mu_1,\dots,k_1,\dots,j_1,\dots}(x)\, \partial^{a_1} A_{\mu_1}(x) \cdots (r) \tag{5.1.96}$$
$$\otimes\, \partial^{b_1}\overline{\psi}_{k_1}(x) \wedge \cdots (l) \wedge \partial^{b_{l+1}}\psi_{k_{l+1}}(x) \wedge \cdots (n) \otimes \partial^{c_1} U_{j_1}(x) \wedge \cdots (s) \,,$$

where $a \equiv (a_1, \dots, a_r)$, $b \equiv (b_1, \dots, b_n)$, $c \equiv (c_1, \dots, c_s)$, $a_i, b_i, c_i \in \mathbb{N}^4$ and $g_{r,n,l,s,a,b,c}^{\mu_1,\dots,k_1,\dots,j_1,\dots} \in \mathcal{D}(\mathbb{M})$; this is a generalization of (5.1.22).

The classical and the star product are defined as maps $\mathcal{F}_{\text{QED}} \times \mathcal{F}_{\text{QED}} \longrightarrow \mathcal{F}_{\text{QED}}$ by

$$(F_A \otimes F_\psi \otimes F_u) \cdot (G_A \otimes G_\psi \otimes G_u) := (F_A \cdot G_A) \otimes (F_\psi \wedge G_\psi) \otimes (F_u \wedge G_u) \,,$$
$$(F_A \otimes F_\psi \otimes F_u) \star (G_A \otimes G_\psi \otimes G_u) := (F_A \star G_A) \otimes (F_\psi \star G_\psi) \otimes (F_u \star G_u) \,,$$

where $F_A, G_A \in \mathcal{F}_{\text{photon}}$, $F_\psi, G_\psi \in \mathcal{F}_{\text{spinor}}$, $F_u, G_u \in \mathcal{F}_{\text{ghost}}$, and by bilinearity. So the appearance of the $\wedge$-product is suppressed in the notation for the classical product: We simply write $FG \equiv F \cdot G$ for $F, G \in \mathcal{F}_{\text{QED}}$. Note also that $\psi(x)$ and $u(y)$ commute with respect to the star product – although both are Fermi fields:

$$(1 \otimes \psi(x) \otimes 1) \star (1 \otimes 1 \otimes u(y)) = 1 \otimes \psi(x) \otimes u(y) = (1 \otimes 1 \otimes u(y)) \star (1 \otimes \psi(x) \otimes 1) \,.$$

To define the support of a QED field $F \in \mathcal{F}_{\text{QED}}$ we generalize Def. 1.3.3:

$$\operatorname{supp} F := \bigcup_{\phi = A^\mu, \overline{\psi}, \psi, \tilde{u}, u} \operatorname{supp} \frac{\delta F}{\delta \phi(\cdot)} \,. \tag{5.1.97}$$

Retarded and time-ordered product. The R- and T-product to order $n-1$ or n, respectively, are maps

$$R_{n-1,1},\, T_n \,:\, (\mathcal{F}^{\mathrm{loc}}_{\mathrm{QED}})^{\otimes n} \longrightarrow \mathcal{F}_{\mathrm{QED}}$$

which satisfy the axioms given in Sect. 3.1 or 3.3, respectively; the modifications due the presence of spinor fields are explained in Sect. 5.1.1 – it is to a far extent obvious how to generalize them from a model containing solely spinor fields to QED. The Fermi statistics of the ghost fields is irrelevant for the R- and T-products of interest, since in QED the arguments of these products contain ghost fields in at most one factor, see (5.1.81).

In addition to the system of axioms explained in Sect. 5.1.1, we require the following renormalization conditions for QED:

- The relevant MWI for QED (5.2.7) ("QED-MWI") – formulated and proved in Sect. 5.2,
- the nearly obvious condition (5.1.81) that ghost fields behave as classical fields.
- The Field parity axiom (3.1.30) is replaced by corresponding requirements for each kind of fields; namely, Field parity with respect to the photon field, Charge number conservation – a condition which deals with [the number of ψ] minus [the number of $\overline{\psi}$] – and an analogous condition for the ghost fields. The latter is irrelevant for QED, since ghost fields do not couple, see (5.1.81).

  Field parity with respect to the photon field is the requirement

$$\alpha_A \circ T_n = T_n \circ \alpha_A^{\otimes n}\ , \quad \text{where} \quad \alpha_A\, F(A^\mu,\psi,\overline{\psi},u,\tilde{u}) := F(-A^\mu,\psi,\overline{\psi},u,\tilde{u}) \tag{5.1.98}$$

  for all $F \equiv F(A^\mu,\psi,\overline{\psi},u,\tilde{u}) \in \mathcal{F}_{\mathrm{QED}}$. Analogously to (3.1.34), an equivalent formulation of this renormalization condition can be given in terms of the VEV of the T-product.

  To formulate *Charge number conservation* we introduce the *charge number operator*:

$$\theta\colon \begin{cases} \mathcal{F}_{\mathrm{QED}} \longrightarrow \mathcal{F}_{\mathrm{QED}} \\ \theta F := \int dx\, \left(\frac{\delta_r F}{\delta\psi(x)} \wedge \psi(x) - \overline{\psi}(x) \wedge \frac{\delta F}{\delta\overline{\psi}(x)}\right)\,, \end{cases} \tag{5.1.99}$$

  which we already introduced – in modified form – for the complex scalar field in Exer. 4.2.6. Note that the integral in (5.1.99) exists, since the $\mathcal{F}_{\mathrm{QED}}$-valued distributions $\frac{\delta_r F}{\delta\psi(x)}$ and $\frac{\delta F}{\delta\overline{\psi}(x)}$ have compact support. A functional of the form

$$\sum_{r=0}^{R}\bigoplus_{s=0}^{S} \langle f^{\mu_1,\dots,k_1,\dots,j_1,\dots}_{r,n,l,s}, A_{\mu_1}\cdots(r)\otimes\overline{\psi}_{k_1}\wedge\cdots(l)\wedge\psi_{k_{l+1}}\wedge\cdots(n)\otimes U_{j_1}\wedge\cdots(s)\rangle$$

is an eigenvector of θ with eigenvalue

$$\text{charge number} = (n-l) - l = [\text{number of } \psi] - [\text{number of } \overline{\psi}] \;;$$

cf. the relation (1.3.3). One easily verifies that θ satisfies the Leibniz rule with respect to the star product:

$$\theta(F \star G) = (\theta F) \star G + F \star (\theta G) \qquad \forall F, G \in \mathcal{F}_{\mathrm{QED}} \,. \tag{5.1.100}$$

We require *Charge number conservation*: Since for non-coinciding points the time-ordered product can be expressed as a multiple $\star$-product (Lemma 3.3.2(a)) and due to (5.1.100), this axiom reads

$$\theta\, T_n(F_1 \otimes \cdots \otimes F_n) = \sum_{j=1}^{n} T_n(F_1 \otimes \cdots \otimes \theta F_j \otimes \cdots \otimes F_n) \,. \tag{5.1.101}$$

In Step 1 of the proof of the QED-MWI (Sect. 5.2.2) we show by means of the causal Wick expansion that it suffices to require this condition for the vacuum expectation value of the T-products (see (5.2.18)), and explain how this simplified version of (5.1.101) can be satisfied.

- Besides charge number conservation, which concerns only the spinor fields, and Field parity with respect to the photon field, one traditionally requires two further discrete symmetries for QED which act nontrivially on both the spinor and the photon field(s): *Charge conjugation invariance* and *PCT-symmetry.* We treat them in the next two sections.

The interaction of QED reads

$$\mathcal{L}(g) := e\, L(g) := e \int dx\; g(x)\, L(x) \;, \quad L(x) := A^\mu(x) \otimes j_\mu(x) \otimes 1 = L(x)^* \;, \tag{5.1.102}$$

where e is the elementary charge, j_μ is the current (5.1.27) and $g \in \mathcal{D}(\mathbb{M})$. Comparing with (1.5.4), we see that a sign is absorbed in our definition of e, that is, $-\kappa = e > 0$, which is the electric charge of a positron. We point out that L is renormalizable by power counting: $\dim L = 4$.

To shorten the notation we will omit the tensor products between the different kinds of fields and we freely (anti-)commute the factors; e.g., instead of (5.1.102) we write

$$L = j_\mu A^\mu = \overline{\psi} A\!\!\!/ \psi$$

for the interaction Lagrangian of QED.

Ghost number operator. Finally, in view of the BRST formalism, we introduce the *ghost number operator* θ_u, which is defined analogously to the charge number operator (5.1.99):

$$\theta_u : \begin{cases} \mathcal{F}_{\mathrm{QED}} \longrightarrow \mathcal{F}_{\mathrm{QED}} \\ \theta_u F := \int dx\, \left(\tilde{u}(x) \wedge \frac{\delta F}{\delta \tilde{u}(x)} - u(x) \wedge \frac{\delta F}{\delta u(x)} \right) . \end{cases} \tag{5.1.103}$$

Every $F \in \mathfrak{F}_{\rm QED}$ which is a *monomial in* $u, \tilde{u}$, is an eigenvector of θ_u; the eigenvalue is called the "ghost number of F" and is denoted by

$$\delta(F) := [\text{number of } \tilde{u} \text{ in } F] - [\text{number of } u \text{ in } F] \ . \tag{5.1.104}$$

If F and G are eigenvectors of θ_u, then $F \wedge G$ and $F \star G$ are also eigenvectors of θ_u with eigenvalues

$$\delta(F \wedge G) = \delta(F) + \delta(G) \ , \qquad \delta(F \star G) = \delta(F) + \delta(G) \ , \tag{5.1.105}$$

as one verifies easily. From $u^* = u$ and $\tilde{u}^* = -\tilde{u}$ we see that

$$\delta(F^*) = \delta(F) \ . \tag{5.1.106}$$

We introduce an even-odd grading of $\mathfrak{F}_{\rm QED}$ with respect to the ghost number:

$$\mathfrak{F}_{\rm QED} = \mathfrak{F}_{\rm QED}^{\rm even} \oplus \mathfrak{F}_{\rm QED}^{\rm odd} \quad \text{where} \quad (-1)^{\delta(F)} = \begin{cases} +1 & \forall F \in \mathfrak{F}_{\rm QED}^{\rm even} \\ -1 & \forall F \in \mathfrak{F}_{\rm QED}^{\rm odd} \ . \end{cases} \tag{5.1.107}$$

Note that the Bose–Fermi grading of $\mathfrak{F}_{\rm spinor}$, defined in (5.1.20), can be understood analogously: It is the splitting of $\mathfrak{F}_{\rm spinor}$ into the two subspace spanned by the eigenvectors of θ whose eigenvalues are even or odd, respectively.

5.1.5 Charge conjugation invariance

Charge conjugation is a transformation that mutually exchanges all particles with their corresponding antiparticles; hence, it changes the sign of all charges: In QED this is only the electric charge, but generally these are all charges relevant to any interaction of the model.

To define charge conjugation for Dirac spinors we need the charge conjugation matrix $C \in \mathbb{C}^{4\times 4}$, which is invertible and satisfies the relation

$$C^{-1}\gamma^\mu C = -\gamma^{\mu T} \ . \tag{5.1.108}$$

For simplicity we work here in the Dirac-, Majorana- or Chiral-representation of the γ-matrices, see, e.g., [109, App. A2]. These representations are unitarily equivalent to each other, fulfill the relation (5.1.3) and the charge conjugation matrix C has the nice properties[155]

$$C^{\rm c} = C = -C^T \ , \quad C^{-1} = -C \ . \tag{5.1.109}$$

In view of the definition of charge conjugation as a map $\beta_C : \mathfrak{F}_{\rm QED} \longrightarrow \mathfrak{F}_{\rm QED}$ we first introduce the charge conjugated spinors (cf., e.g., [148, Sect. 4.4])

$$\beta_C\psi(x) := C\,\overline{\psi}(x)^T = C\,\gamma^{0\,T}\,\psi^{\rm c}(x) \ , \quad \beta_C\overline{\psi}(x) := \psi(x)^T\,C \ . \tag{5.1.110}$$

[155] We recall that the upper index "c" denotes complex conjugation.

Proceeding straightforwardly, we obtain

$$\overline{\beta_C\psi(x)} \equiv \big(\beta_C\psi(x)\big)^\dagger\gamma^0 = \psi(x)^T\,\gamma^{0\,T}\,C^+\gamma^0 = \beta_C\overline{\psi}(x)\ ,$$
$$\beta_C^2\psi(x) = \beta_C\big(C\,\overline{\psi}(x)^T\big) = C\big(\beta_C\overline{\psi}(x)\big)^T = \psi(x) \quad \text{and}$$
$$\beta_C^2\overline{\psi}(x) = \beta_C\big(\psi(x)^T\,C\big) = \big(\beta_C\psi(x)\big)^T C = \overline{\psi}(x)\ .$$

In terms of the Dirac equation in relativistic quantum mechanics, charge conjugation can be understood as follows (cf., e.g., [148, Sect. 1.5]): If $(h_1, h) \in \mathfrak{C}_{\mathrm{photon}} \times \mathfrak{C}_{\mathrm{spinor},1}$ (where $\mathfrak{C}_{\mathrm{spinor},1} := C^\infty(\mathbb{M}, \mathbb{C}^4)$) is a solution of the Dirac equation in the presence of the electromagnetic field,

$$\Big((i\partial\!\!\!/ + eA\!\!\!/ - m)\psi(x)\Big)(h_1, h) = 0\ , \tag{5.1.111}$$

then (h_1, h) solves also the same equation for $\beta_C\psi$ in place of ψ and with e replaced by $(-e)$:

$$\Big((i\partial\!\!\!/ - eA\!\!\!/ - m)\beta_C\psi(x)\Big)(h_1, h) = 0\ .$$

This result is obtained straightforwardly by manipulating (5.1.111) as follows: First we apply complex conjugation, then we multiply with $C\gamma^{0\,T}$ form the left and use $C\,\gamma^{0\,T}\,\gamma^{\mu\,c} = -\gamma^\mu\,C\,\gamma^{0\,T}$.

In order that the Dirac equation (5.1.111) is charge conjugation invariant, i.e., the validity of this equation for (A^μ, ψ) is equivalent to the validity of the *same* equation for $(\beta_C A^\mu, \beta_C\psi)$, we define

$$\beta_C A^\mu(x) := -A^\mu(x)\ . \tag{5.1.112}$$

Turning to the Faddeev–Popov ghost fields, the restriction of β_C to a single ghost field is defined to be the identity map: $\beta_C u(x) := u(x)\ ,\ \beta_C\tilde{u}(x) := \tilde{u}(x)\ .$

With these preparations, we define $\beta_C : \mathcal{F}_{\mathrm{QED}} \longrightarrow \mathcal{F}_{\mathrm{QED}}$ by

$$\begin{aligned}\beta_C F := \sum_{r=0}^{M}\bigoplus_{n=0}^{N}\sum_{l=0}^{n}\bigoplus_{s=0}^{S}\big\langle f_{r,n,l,s}^{\mu_1,\dots,k_1,\dots,j_1,\dots}, \beta_C A_{\mu_1}\cdots\beta_C A_{\mu_r} \\ \otimes \beta_C\overline{\psi}_{k_1}\wedge\cdots\wedge\beta_C\overline{\psi}_{k_l}\wedge\beta_C\psi_{k_{l+1}}\wedge\cdots\wedge\beta_C\psi_{k_n}\otimes U_{j_1}\wedge\cdots\wedge U_{j_s}\big\rangle\end{aligned} \tag{5.1.113}$$

for F given by (5.1.93). Obviously, β_C *is linear* and has the properties

$$\beta_C^2 = \mathrm{Id} \quad \text{and} \quad \beta_C(FG) = (\beta_C F)(\beta_C G)\ ;$$

the latter relation holds also with respect to the star product:

$$\beta_C(F \star G) = (\beta_C F) \star (\beta_C G)\ , \tag{5.1.114}$$

as we will show in Exer. 5.1.14. From the definition (5.1.113) we see that

$$\frac{\delta(\beta_C F)}{\delta\big(\beta_C\varphi(x)\big)} = \beta_C\Big(\frac{\delta F}{\delta\varphi(x)}\Big) \quad \text{for} \quad \varphi = \psi,\ \overline{\psi},\ A\ , \tag{5.1.115}$$

from which we conclude

$$\frac{\delta(\beta_C F)}{\delta A(x)} = -\beta_C\Big(\frac{\delta F}{\delta A(x)}\Big)\,,$$

$$\frac{\delta(\beta_C F)}{\delta \psi_l(x)} = \int dy\; \frac{\delta(\beta_C F)}{\delta(\beta_C\overline{\psi})_j(y)}\,\frac{\delta(\beta_C\overline{\psi})_j(y)}{\delta\psi_l(x)} = C_{lj}\,\beta_C\Big(\frac{\delta F}{\delta\overline{\psi}_j(x)}\Big)\,,$$

$$\frac{\delta(\beta_C F)}{\delta \overline{\psi}_l(x)} = \int dy\; \frac{\delta(\beta_C F)}{\delta(\beta_C\psi)_j(y)}\,\frac{\delta(\beta_C\psi)_j(y)}{\delta\overline{\psi}_l(x)} = \beta_C\Big(\frac{\delta F}{\delta\psi_j(x)}\Big)\,C_{jl}\,. \tag{5.1.116}$$

Exercise 5.1.14.

(a) Prove that the interaction Lagrangian of QED, $L = j_\mu A^\mu$, is charge conjugation invariant: $\beta_C L = L$.

(b) Verify the relation (5.1.114).

[*Solution*: (a) For the current $j^\mu = \overline{\psi}\wedge\gamma^\mu\psi$ we obtain

$$\beta_C j^\mu = \psi^T C\wedge\gamma^\mu C\overline{\psi}^T = -\overline{\psi}C^T\gamma^{\mu T}\wedge C^T\psi = \overline{\psi}\wedge C\gamma^{\mu T}C^{-1}\psi = -j^\mu\,.$$

Taking into account (5.1.112), this implies the assertion.
(b) Proceeding by induction on the number of contractions, we have to show

$$\sum_{r,s}\beta_C\Big(\frac{\delta_r F}{\delta\varphi_r(x)}\Big)\,H_{rs}(x-y)\,\beta_C\Big(\frac{\delta G}{\delta\varphi_s(y)}\Big) \overset{?}{=} \sum_{r,s}\frac{\delta_r(\beta_C F)}{\delta\varphi_r(x)}\,H_{rs}(x-y)\,\frac{\delta(\beta_C G)}{\delta\varphi_s(y)}\,, \tag{5.1.117}$$

where $\varphi_1 := \psi\,,\ \varphi_2 := \overline{\psi}\,,\ \varphi_3 := A$ and

$$H_{rs}(x-y) := \omega_0\big(\varphi_r(x)\star\varphi_s(y)\big)\,,$$

that is, $H_{12}(z) = S^+(z)\,,\ H_{21}(z) = S^-(-z)^T\,,\ H_{33}^{\mu\nu}(z) = g^{\mu\nu}\,D^+(z)$ and all other H_{rs} vanish.
Let us verify (5.1.117): For a photon contraction, i.e., $(r,s)=(3,3)$, the relation (5.1.117) is obvious, due to (5.1.116). The validity of (5.1.117) for the other terms is more involved: The term $(r,s)=(2,1)$ of the l.h.s. is equal to the $(1,2)$-term of the r.h.s.; this can be seen as follows. Using (5.1.116), this assertion is equivalent to

$$\beta_C\Big(\frac{\delta_r F}{\delta\overline{\psi}_k(x)}\Big)\,S^-_{lk}(y-x)\,\beta_C\Big(\frac{\delta G}{\delta\psi_l(y)}\Big) \overset{?}{=} \beta_C\Big(\frac{\delta_r F}{\delta\overline{\psi}_k(x)}\Big)\,C_{jk}\,S^+_{jh}(x-y)\,C_{lh}\,\beta_C\Big(\frac{\delta G}{\delta\psi_l(y)}\Big)\,.$$

This relation holds indeed true, since

$$C^T\,S^+(-z)\,C^T = C\,(-i\partial\!\!\!/_z + m)\,C\,\Delta^+(-z) = -(i\partial\!\!\!/_z^T + m)\Delta^+(-z) = S^-(z)^T\,.$$

Analogously, one verifies that the $(1,2)$-term of the l.h.s. is equal to the $(2,1)$-term of the r.h.s..]

Motivated by (5.1.114), we require the following additional axiom – for simplicity we work with the T-product:

(xi) **Charge conjugation invariance:** The axiom reads

$$\beta_C \circ T_n = T_n \circ \beta_C^{\otimes n} \qquad \forall n \in \mathbb{N}^* \ . \tag{5.1.118}$$

Remark 5.1.15. By using the Field parity axiom with respect to the photon field (5.1.98), we can equivalently require that $\beta_C \circ \alpha_A$ commutes with T_n. Mathematically, $\beta_C \circ \alpha_A$ is a simpler transformation, since it concerns only the spinor fields. But for physics β_C is more significant: It can be interpreted as charge conjugation and it is a symmetry of QED, it leaves (the free and the interacting part of) the Lagrangian of QED invariant.

How to fulfill this axiom? Following the inductive Epstein–Glaser construction of the T-product, we proceed by induction on n. Assuming that all axioms are satisfied for all T_k with $k < n$, we conclude from (3.3.18) and (5.1.114) that $T_n^0(\otimes_{j=1}^n B_j(x_j)) \in \mathcal{D}'(\mathbb{M}^n \setminus \Delta_n, \mathcal{F}_{\mathrm{QED}})$ fulfills Charge conjugation invariance. Hence, the new axiom is indeed a renormalization condition.

Let $T_n(\dots) \in \mathcal{D}'(\mathbb{M}^n, \mathcal{F}_{\mathrm{QED}})$ be an extension of $T_n^0(\dots)$ which satisfies all other axioms. The symmetrized expression

$$T_n^{\mathrm{sym}} := \tfrac{1}{2}\big(T_n + \beta_C \circ T_n \circ \beta_C^{\otimes n}\big) \tag{5.1.119}$$

is an extension of T_n^0 (since $\beta_C \circ T_n \circ \beta_C^{\otimes n}$ is an extension of $\beta_C \circ T_n^0 \circ \beta_C^{\otimes n} = T_n^0$) and obviously it fulfills Charge conjugation invariance.

It remains to show that the other renormalization conditions are preserved under the symmetrization (5.1.119). Since $\frac{\delta}{\delta\psi} \circ \beta_C \neq \beta_C \circ \frac{\delta}{\delta\psi}$, this is non-obvious for the Field independence. However, by using (5.1.116) we get

$$\begin{aligned}
\frac{\delta}{\delta\psi_l(x)}\,\beta_C\,T_n(\beta_C F_1 \otimes \cdots \otimes \beta_C F_n) &= C_{lj}\,\beta_C\,\frac{\delta}{\delta\overline{\psi}_j(x)}\,T_n(\beta_C F_1 \otimes \cdots \otimes \beta_C F_n)\\
&= \sum_{k=1}^n \beta_C\,T_n(\beta_C F_1 \otimes \cdots \otimes C_{lj}\,\frac{\delta(\beta_C F_k)}{\delta\overline{\psi}_j(x)} \otimes \cdots \otimes \beta_C F_n)\\
&= \sum_{k=1}^n \beta_C\,T_n(\beta_C F_1 \otimes \cdots \otimes \beta_C\Big(\frac{\delta F_k}{\delta\psi_l(x)}\Big) \otimes \cdots \otimes \beta_C F_n)\ ,
\end{aligned} \tag{5.1.120}$$

and analogously for $\frac{\delta}{\delta\overline{\psi}}$ and $\frac{\delta}{\delta A}$. We conclude that T_n^{sym} fulfills the Field independence.

Remark 5.1.16 (Furry's theorem)**.** Let $L = j^\mu A_\mu$ be the interaction Lagrangian of QED. A well-known consequence of Charge conjugation invariance and $\beta_C L = L$ is Furry's Theorem:

$$t(j^{\mu_1}, \dots, j^{\mu_r}, L, \dots, L) = 0 \quad \text{if } r \text{ is odd.} \tag{5.1.121}$$

It follows from the fact that in the causal Wick expansion of $T_n\big(L(x_1)\otimes\cdots\otimes L(x_n)\big)$ the summand

$$t(j^{\mu_1}, \dots, j^{\mu_r}, L, \dots, L)(x_1 - x_n, \dots, x_{n-1} - x_n)\, A_{\mu_1}(x_1) \cdots A_{\mu_r}(x_r)$$

must be charge conjugation invariant individually:

$$t(\dots)\,A_{\mu_1}\cdots A_{\mu_r} = \beta_C\left(t(\dots)\,A_{\mu_1}\cdots A_{\mu_r}\right) = (-1)^r\,t(\dots)\,A_{\mu_1}\cdots A_{\mu_r}\ .$$

Exercise 5.1.17. Let

$$\gamma^5 := i\gamma^0\gamma^1\gamma^2\gamma^3\ ,\quad j_a^\mu(x) := \overline{\psi}(x)\wedge\gamma^\mu\gamma^5\psi(x)\ ,\quad L_a(x) := j_a^\mu(x)\,A_\mu(x)\ . \tag{5.1.122}$$

j_a is called the "axial current" and "axial QED" is the model with interaction L_a. Is there a restriction of

$$t(j_a^{\mu_1},\dots,j_a^{\mu_r},L_a,\dots,L_a)\ ,\quad \text{where the argument } L_a \text{ appears } (n-r) \text{ times,}$$

coming from Charge conjugation invariance?

[*Solution*: Proceeding analogously to Exer. 5.1.14(a), we obtain

$$\beta_C j_a^\mu = j_a^\mu \quad \text{by using} \quad C\,\gamma^{5\,T}\gamma^{\mu\,T}\,C^{-1} = \gamma^\mu\gamma^5\ ,$$

which follows from the relations (5.1.2) and (5.1.108)–(5.1.109). Hence, $\beta_C L_a = -L_a$. Using $\omega_0\circ\beta_C = \omega_0$, we get

$$\begin{aligned} t(j_a^{\mu_1},\dots,j_a^{\mu_r},L_a,\dots,L_a)(x_1-x_n,\dots) &= \omega_0\Big(\beta_C\,T_n\big(j_a^{\mu_1}(x_1),\dots,L_a(x_n)\big)\Big) \\ &= \omega_0\Big(T_n\big(\beta_C j_a^{\mu_1}(x_1),\dots,\beta_C L_a(x_n)\big)\Big) \\ &= (-1)^{n-r}\,t(j_a^{\mu_1},\dots,j_a^{\mu_r},L_a,\dots,L_a)(x_1-x_n,\dots)\ . \end{aligned}$$

Hence, $t(j_a^{\mu_1},\dots,j_a^{\mu_r},L_a,\dots,L_a)$ vanishes if $(n-r)$ is odd. Actually this is precisely the same condition as it results from Field parity with respect to the photon field (5.1.98), since $\omega_0\circ\alpha_A = \omega_0$ and $\alpha_A j_a^\mu = j_a^\mu$, $\alpha_A L_a = -L_a$.]

5.1.6 The PCT-Theorem

This section is based on [148, Sect. 4.4] and [62].

In the abbreviation "PCT",[156] the letter "C" stands for charge conjugation (introduced in the preceding section), "P" for the parity transformation (or space reflection) $(t,\vec{x})\longmapsto(t,-\vec{x})$, and "$T$" for the time reversal $(t,\vec{x})\longmapsto(-t,\vec{x})$. In QED one can renormalize the T-product such that it is symmetric with respect to each of these transformations individually; for a proof of this statement in the framework of causal perturbation theory we refer to [148, Sect. 4.4]. But in weak interactions neither C, nor P, nor T is a symmetry; this holds only for the composition PCT. PCT-symmetry has been proved under very general assumptions; in particular, it is a consequence of the Wightman axioms [162]. For this reason we will give a (to a far extent) model independent proof of PCT-symmetry.

[156] In the literature "CPT" in place of "PCT" is widespread; we prefer the latter abbreviation to avoid confusion with "Causal Perturbation Theory".

***P*- and *T*-transformation.** However, to motivate the (weak) assumptions we will need for this proof, we first define the P- and T-transformation for QED. Again, to simplify the formulas, we proceed in the Dirac-, Majorana- or Chiral-representation of the γ-matrices. Let $\gamma^5 \in \mathbb{C}^{4\times 4}$ be the matrix introduced in (5.1.122) and C the charge conjugation matrix (5.1.108). One verifies straightforwardly the identities

$$[\gamma^5,\gamma^\mu]_+ = 0\ , \quad \gamma^5\gamma^5 = 1_{4\times 4}\ , \quad {\gamma^5}^\dagger = \gamma^5, \quad \gamma^5{\gamma^\mu}^T\gamma^5 = -{\gamma^\mu}^T, \quad C^{-1}\gamma^5 C = {\gamma^5}^T \tag{5.1.123}$$

as a consequence of the relations (5.1.2)–(5.1.3) and (5.1.108); to obtain the second last identity we also use ${\gamma^5}^T = \pm\gamma^5$, where the sign depends on the representation. In this section the relations (5.1.2), (5.1.3) and (5.1.123) for the γ-matrices and (5.1.108)–(5.1.109) for the charge conjugation matrix C will be used without referring to them.

The following Definition can be motivated in terms of the 1-particle theory, i.e., the Dirac equation in relativistic quantum mechanics, see, e.g., [148, Sect. 1.5].

Definition 5.1.18 (P- and T-transformation for QED). The P- and T-transformation are maps $\beta_P, \beta_T : \mathcal{F}_{\mathrm{QED}} \longrightarrow \mathcal{F}_{\mathrm{QED}}$ which are defined by their action on the basic fields

$$\begin{aligned}
\beta_P A^\mu(x) &:= g^{\mu\mu}\, A^\mu(x_P) \quad \text{(without summation over } \mu), \quad \beta_P U(x) := U(x_P)\ , \\
\beta_P \psi(x) &:= i\gamma^0\psi(x_P)\ , \quad \beta_P\overline{\psi}(x) := -i\overline{\psi}(x_P)\gamma^0 \text{ where} \quad x_P := (x^0, -\vec{x})\ ; \\
\beta_T A^\mu(x) &:= g^{\mu\mu}\, A^\mu(x_T) \quad \text{(without summation over } \mu), \quad \beta_T U(x) := U(x_T)\ , \\
\beta_T \psi(x) &:= \gamma^5 C\psi(x_T)\ , \quad \beta_T\overline{\psi}(x) := \overline{\psi}(x_T)C^{-1}\gamma^5 \text{ where} \quad x_T := (-x^0, \vec{x})\ ;
\end{aligned} \tag{5.1.124}$$

and by

$$\begin{aligned}
\beta_P F := \sum_{r=0}^{M}\bigoplus_{n=0}^{N}\sum_{l=0}^{n}\bigoplus_{s=0}^{S}\big\langle f_{r,n,l,s}^{\mu_1,\dots,k_1,\dots,j_1,\dots}(y_1,\dots,x_1,\dots,z_1,\dots)\,,\, \beta_P A_{\mu_1}(y_1)\cdots \\
\otimes\, \beta_P\overline{\psi}_{k_1}(x_1)\wedge\cdots\wedge\beta_P\psi_{k_{l+1}}(x_{l+1})\wedge\cdots\otimes\beta_P U_{j_1}(z_1)\wedge\cdots\big\rangle
\end{aligned} \tag{5.1.125}$$

and

$$\begin{aligned}
\beta_T F := \sum_{r=0}^{M}\bigoplus_{n=0}^{N}\sum_{l=0}^{n}\bigoplus_{s=0}^{S}\big\langle f_{r,n,l,s}^{\mu_1,\dots,k_1,\dots,j_1,\dots}(y_1,\dots,x_1,\dots,z_1,\dots)^c\,,\, \beta_T A_{\mu_1}(y_1)\cdots \\
\otimes\, \beta_T\overline{\psi}_{k_1}(x_1)\wedge\cdots\wedge\beta_T\psi_{k_{l+1}}(x_{l+1})\wedge\cdots\otimes\beta_T U_{j_1}(z_1)\wedge\cdots\big\rangle\ .
\end{aligned} \tag{5.1.126}$$

Obviously, β_P *is linear*, but β_T *is antilinear*,[157] due to the complex conjugation of $f_{\cdots}^{\cdots}$ in (5.1.126). An immediate consequence of this Definition are the relations

$$\beta_P(FG) = (\beta_P F)(\beta_P G) \quad \text{and} \quad \beta_T(FG) = (\beta_T F)(\beta_T G)$$

[157] "Antilinearity" means $\beta_T(\lambda F + G) = \lambda^c\,\beta_T(F) + \beta_T(G) \quad \forall F, G \in \mathcal{F}_{\mathrm{QED}}\,,\ \lambda \in \mathbb{C}$.

for the classical product. The definitions are such that

$$\beta_P \psi^\dagger(x) = \big(\beta_P \psi(x)\big)^\dagger \quad \text{and} \quad \beta_T \psi^\dagger(x) = \big(\beta_T \psi(x)\big)^\dagger \ ; \tag{5.1.127}$$

namely, by using (5.1.124) and (anti-)linearity of $\beta_{P,T}$, we obtain

$$\begin{aligned} \beta_P \psi^\dagger(x) &= \beta_P\big(\overline{\psi}(x)\big)\gamma^0 = -i\overline{\psi}(x_P) = \big(i\gamma^0\psi(x_P)\big)^\dagger = \big(\beta_P\psi(x)\big)^\dagger \ , \\ \beta_T \psi^\dagger(x) &= \beta_T\big(\overline{\psi}(x)\big)\gamma^{0c} = \overline{\psi}(x_T)C^{-1}\gamma^5\gamma^{0T} = \overline{\psi}(x_T)\gamma^0 C^{-1}\gamma^5 \\ &= \psi^\dagger(x_T)C^\dagger\gamma^{5\dagger} = \big(\beta_T\psi(x)\big)^\dagger \ . \end{aligned}$$

The P- and T-transformation satisfy the following basic relations.

Lemma 5.1.19 (Algebraic relations of β_P and β_T)**.** *With θ being the charge number operator, defined in* (5.1.99)*, and β_C the charge conjugation* (5.1.113)*, the following relations hold true:*

$$\text{(a)}\ \beta_P \circ \beta_P = (-1)^\theta \ , \quad \beta_T \circ \beta_T = (-1)^\theta \ , \tag{5.1.128}$$

$$\text{(b)}\ \beta_C \circ \beta_P = \beta_P \circ \beta_C \ , \quad \beta_C \circ \beta_T = \beta_T \circ \beta_C \ , \quad \beta_T \circ \beta_P = (-1)^\theta \circ \beta_P \circ \beta_T \ , \tag{5.1.129}$$

$$\text{(c)}\ \beta_P(F \star G) = (\beta_P F) \star (\beta_P G) \ , \quad \beta_T(F \star G) = (\beta_T F) \star (\beta_T G) \tag{5.1.130}$$

for all $F, G \in \mathcal{F}_{\text{QED}}$.

Exercise 5.1.20. Prove this lemma, by proceeding in part (c) analogously to the proof of the corresponding relation for β_C (5.1.114).

[*Solution:* (a) and (b): These identities need to be verified only on the basic fields, as we see from (5.1.125)–(5.1.126). For $A^\mu(y)$ and $U(z)$ these verifications are trivial; so it remains to prove the assertions[158]

$$\begin{aligned} &\beta_P^2\psi(x) \overset{?}{=} -\psi(x) \ , \quad \beta_T^2\psi(x) \overset{?}{=} -\psi(x) \ , \quad \beta_C\beta_P\psi(x) \overset{?}{=} \beta_P\beta_C\psi(x) \ , \\ &\beta_C\beta_T\psi(x) \overset{?}{=} \beta_T\beta_C\psi(x) \ , \quad \beta_T\beta_P\psi(x) \overset{?}{=} -\beta_P\beta_T\psi(x) \ , \end{aligned}$$

and the same assertions for $\overline{\psi}$ in place of ψ. This can be done by straightforward calculations, for example:

$$\begin{aligned} \beta_T^2\psi(x) &= \beta_T\big(\gamma^5 C\psi(x_T)\big) = \gamma^{5c}C^c\gamma^5 C\psi(x) = \gamma^{5^T}C\gamma^5 C\psi(x) = -\psi(x) \ , \\ \beta_T\beta_C\psi(x) &= \beta_T\big(C\overline{\psi}(x)^T\big) = C^c\big(\overline{\psi}(x_T)^T C^{-1}\gamma_5\big)^T = C\gamma^{5^T}C\overline{\psi}(x_T)^T \\ &= \gamma^5 C C\overline{\psi}(x_T)^T = \beta_C\big(\gamma^5 C\psi(x_T)\big) = \beta_C\beta_T\psi(x) \ , \\ \beta_T\beta_P\psi(x) &= \beta_T\big(i\gamma^0\psi(x_P)\big) = -i\gamma^{0c}\gamma^5 C\psi(-x) = -i\gamma^5 C\gamma^0\psi(-x) \\ &= -\beta_P\big(\gamma^5 C\psi(x_T)\big) = -\beta_P\beta_T\psi(x) \ . \end{aligned}$$

---

[158] For brevity, we omit "∘".

The $\overline{\psi}$-assertions can be traced back to the corresponding relations for ψ by using (5.1.127), for example,

$$\beta_T^2\overline{\psi}(x) = \beta_T\big((\beta_T\psi^\dagger(x))\gamma^{0c}\big) = \beta_T\big((\beta_T\psi(x))^\dagger\big)\gamma^0 = \big(\beta_T^2\psi(x)\big)^\dagger\gamma^0 = -\overline{\psi}(x)\ .$$

Here we use that $\beta_T\psi(x)$ is of the form $\beta_T\psi(x) = M_T\psi(x_T)$ for some $M_T \in \mathbb{C}^{4\times 4}$ and, hence, the relation (5.1.127) implies $\beta_T\big((\beta_T\psi(x))^\dagger\big) = \big(\beta_T^2\psi(x)\big)^\dagger$ due to the antilinearity of β_T.

(c) We proceed by induction on the number of contractions. With that, the proof of the assertion for β_T amounts to the verification of

$$\int dx\,dy\;\beta_T\Big(\frac{\delta_r F}{\delta\psi_t(x)}\Big)\,\big(S^+_{ts}(x-y)\big)^c\,\beta_T\Big(\frac{\delta G}{\delta\overline{\psi}_s(y)}\Big) \tag{5.1.131}$$
$$\stackrel{?}{=} \int dx\,dy\;\frac{\delta_r(\beta_T F)}{\delta\psi_j(x_T)}\,S^+_{jl}(x_T-y_T)\,\frac{\delta(\beta_T G)}{\delta\overline{\psi}_l(y_T)}\ ,$$

and analogous identities for the $\overline{\psi}$-ψ and the A-A contraction. Note that on the right-hand side we have renamed the integration variables: $(x,y)\to(x_T,y_T)$. Obviously, the relation (5.1.115) holds also with β_T (or β_P) in place of β_C and with that we find

$$\frac{\delta_r(\beta_T F)}{\delta\psi_j(x_T)} = \int dz\;\frac{\delta_r(\beta_T F)}{\delta(\beta_T\psi)_t(z)}\;\frac{\delta(\beta_T\psi)_t(z)}{\delta\psi_j(x_T)} = \beta_T\Big(\frac{\delta_r F}{\delta\psi_t(x)}\Big)(\gamma^5 C)_{tj}\ ,$$
$$\frac{\delta(\beta_T G)}{\delta\overline{\psi}_l(y_T)} = \int dz\;\frac{\delta(\beta_T G)}{\delta(\beta_T\overline{\psi})_s(z)}\;\frac{\delta(\beta_T\overline{\psi})_s(z)}{\delta\overline{\psi}_l(y_T)} = (C^{-1}\gamma^5)_{ls}\,\beta_T\Big(\frac{\delta G}{\delta\overline{\psi}_s(y)}\Big)\ ,$$

analogously to (5.1.116). Due to these equations, the claim (5.1.131) is equivalent to

$$S^+(z)^c \stackrel{?}{=} \gamma^5 C\,S^+(z_T)\,C^{-1}\gamma^5\ . \tag{5.1.132}$$

We are going to show that the latter identity follows from

$$S^+(z) = (i\partial\!\!\!/_z + m)\Delta^+(z) \quad\text{and}\quad \Delta^+(z) = \Delta^+(z_P)\ ,\quad \Delta^+(z)^c = \Delta^+(-z)\ ,$$

by treating the terms in $(i\partial\!\!\!/_z + m)$ individually. For the terms $\sim m$ this is obvious: $\Delta^+(z)^c = \Delta^+(-z) = \Delta^+(z_T)$. For the terms $\sim \partial_{z_0}$ we get

$$-i\gamma_0^c\,\partial_{z_0}\Delta^+(-z) \stackrel{?}{=} i\gamma^5 C\gamma_0 C^{-1}\gamma^5\,(-\partial_{z_0})\Delta^+(z_T)\ .$$

This equation holds indeed true, since $\gamma^5 C\gamma_0 C^{-1}\gamma^5 = -\gamma^5\gamma_0^T\gamma^5 = \gamma_0^T = \gamma_0^c$. Finally, for the terms $\sim \partial_{z_k}$ we have to show

$$-i\gamma_k^c\,\partial_{z_k}\Delta^+(-z) \stackrel{?}{=} i\gamma^5 C\gamma_k C^{-1}\gamma^5\,\partial_{z_k}\Delta^+(z_T)\ .$$

This identity is verified as follows: $\gamma^5 C\gamma_k C^{-1}\gamma^5 = -\gamma^5\gamma_k^T\gamma^5 = \gamma_k^T = -\gamma_k^c$.

For a $\overline{\psi}$-ψ contraction (in place of (5.1.131)) the proof is analogous, the main difference is that in (5.1.132) $S^+(z)$ is replaced by $S^-(z) = -(i\partial\!\!\!/_z + m)\Delta^+(-z)$; for an A-A contraction the verification is clearly simpler.

The proof of the corresponding assertion for β_P (5.1.130) can be given by walking along the same lines; the claim (5.1.132) is replaced by

$$S^\pm(z) \stackrel{?}{=} \gamma^0\,S^\pm(x_P)\,\gamma^0\ ,$$

which can be verified by proceeding analogously to above.]

Exercise 5.1.21 (Group generated by β_P, β_T and β_C). Compute the group table of the group generated by

(a) β_P and β_T;

(b) β_P, β_T and β_C .

We denote these groups by G_8 and G_{16}, respectively.

[*Solution*: (a) Using the results of the parts (a) and (b) of Lemma 5.1.19 and noting that $(-1)^\theta$ commutes with β_P and β_T and that

$$\beta_P^{-1} = \beta_P^3 = (-1)^\theta\, \beta_P \;, \quad \beta_T^{-1} = \beta_T^3 = (-1)^\theta\, \beta_T \;,$$

the table can easily be completed:

G_8	Id	β_P	β_P^{-1}	β_T	β_T^{-1}	$(-1)^\theta$	$\beta_P\beta_T$	$\beta_T\beta_P$
Id	Id	β_P	β_P^{-1}	β_T	β_T^{-1}	$(-1)^\theta$	$\beta_P\beta_T$	$\beta_T\beta_P$
β_P	β_P	$(-1)^\theta$	Id	$\beta_P\beta_T$	$\beta_T\beta_P$	β_P^{-1}	β_T^{-1}	β_T
β_P^{-1}	β_P^{-1}	Id	$(-1)^\theta$	$\beta_T\beta_P$	$\beta_P\beta_T$	β_P	β_T	β_T^{-1}
β_T	β_T	$\beta_T\beta_P$	$\beta_P\beta_T$	$(-1)^\theta$	Id	β_T^{-1}	β_P	β_P^{-1}
β_T^{-1}	β_T^{-1}	$\beta_P\beta_T$	$\beta_T\beta_P$	Id	$(-1)^\theta$	β_T	β_P^{-1}	β_P
$(-1)^\theta$	$(-1)^\theta$	β_P^{-1}	β_P	β_T^{-1}	β_T	Id	$\beta_T\beta_P$	$\beta_P\beta_T$
$\beta_P\beta_T$	$\beta_P\beta_T$	β_T	β_T^{-1}	β_P^{-1}	β_P	$\beta_T\beta_P$	$(-1)^\theta$	Id
$\beta_T\beta_P$	$\beta_T\beta_P$	β_T^{-1}	β_T	β_P	β_P^{-1}	$\beta_P\beta_T$	Id	$(-1)^\theta$

Looking at this table, a mathematician realizes that G_8 is isomorphic to the *quaternion group*

$$Q_8 := \{1, (-1), i, (-i), j, (-j), k, (-k)\} \;;$$

the identifications may be chosen as follows: $\mathrm{Id} \simeq 1$, $(-1)^\theta \simeq (-1)$, $\beta_P \simeq i$, $\beta_P^{-1} \simeq (-i)$, $\beta_T \simeq j$, $\beta_T^{-1} \simeq (-j)$, $\beta_P\beta_T \simeq k$, $\beta_T\beta_P \simeq (-k)$.
(b) Since β_C commutes with β_P and β_T, it commutes with all elements of G_8. Taking into account additionally that $\beta_C^2 = \mathrm{Id}$, we see that $C_2 := (\{\mathrm{Id}, \beta_C\}, \circ)$ is an invariant subgroup of G_{16}. Consequently, $G_{16} = C_2 \times G_8$, where "$\times$" means the direct product of groups. Denoting the elements of G_8 by $\{g_a \,|\, a = 1, \ldots, 8\}$, the group table can be written as

G_{16}	g_b	$\beta_C g_b$
g_a	$g_a g_b$	$\beta_C g_a g_b$
$\beta_C g_a$	$\beta_C g_a g_b$	$g_a g_b$

where $a, b \in \{1, \ldots, 8\}$ arbitrary.]

The *PCT-transformation*

$$\beta := \beta_P \circ \beta_C \circ \beta_T \,:\, \mathfrak{F}_{\mathrm{QED}} \longrightarrow \mathfrak{F}_{\mathrm{QED}} \quad \text{is } \textit{antilinear}, \tag{5.1.133}$$

because β_P, β_C are linear and β_T is antilinear. We claim that the QED-interaction $L(g) = \int dx\; j^\mu(x) A_\mu(x)\, g(x)$ is PCT-invariant in the sense that

$$\beta\, L(x) = L(-x)\;, \quad \text{or} \quad \beta\, L(g) = L(\tilde{g}) \quad \text{with} \quad \tilde{g}(x) := g(-x)^c\;. \tag{5.1.134}$$

Exercise 5.1.22. Verify the claim (5.1.134).

[*Solution*: By straightforward calculation we obtain

$$\beta\, A^\mu(x) = -A^\mu(-x)\;, \quad \beta\, \psi(x) = i\gamma^5\, \psi^\dagger(-x)^T\;, \quad \beta\, \overline{\psi}(x) = i\, \psi(-x)^T\, \gamma^{0T}\gamma^5\;, \tag{5.1.135}$$

which yields

$$\beta\, j^\mu(x) = -\psi(-x)^T\gamma^{0T}\gamma^5\gamma^{\mu c}\gamma^5\psi^\dagger(-x)^T = \psi(-x)^T\gamma^{\mu T}\gamma^{0T}\psi^\dagger(-x)^T = -j^\mu(-x)$$

by using $\gamma^{\mu c} = \gamma^{\mu\dagger T} = (\gamma^0\gamma^\mu\gamma^0)^T$. So we indeed get $\beta\, L(x) = L(-x)$ and $\beta\, L(g) = (\beta L)(g^c) = L(\tilde{g})$.]

***PCT*-symmetry.** In this paragraph we study a *general model*, with basic fields $\phi_j(x)$ $(j = 1, \ldots J)$, fulfilling the following assumptions:

Assumptions 5.1.23.

(i) *The PCT-transformation is a map $\beta \equiv \beta_P \circ \beta_C \circ \beta_T : \mathcal{F} \longrightarrow \mathcal{F}$ of the form*

$$\beta \sum_n \langle f_n^{j_1,\ldots,j_n}, \phi_{j_1} \otimes \cdots \otimes \phi_{j_n} \rangle := \sum_n \langle f_n^{j_1,\ldots,j_n\, c}, \beta\phi_{j_1} \otimes \cdots \otimes \beta\phi_{j_n} \rangle \tag{5.1.136}$$

(for brevity, $\sum_{j_1,\ldots,j_n}$ is suppressed) with

$$\beta\, \phi_j(x) = \phi_j^{\mathfrak{C}}(-x) \quad \forall 1 \le j \le J\;, \tag{5.1.137}$$

where $\phi_j^{\mathfrak{C}}$ is a suitable conjugate of ϕ_j (passing to the adjoint and multiplication by a suitable matrix); in particular it holds that

$$\dim \phi_j^{\mathfrak{C}} = \dim \phi_j\;. \tag{5.1.138}$$

(ii) *β needs not to be an involution; we only assume*

$$\beta^{2N} \equiv \beta \circ \cdots \circ \beta = \mathrm{Id} \quad \textit{for some } N \in \mathbb{N} \setminus \{0\}. \tag{5.1.139}$$

(iii) *β commutes with the star product:*

$$\beta(F \star G) = \beta(F) \star \beta(G) \quad \forall F, G \in \mathcal{F}\;. \tag{5.1.140}$$

(iv) *The interaction $L \in \mathcal{P}$ is PCT-invariant:*

$$\beta L(x) = L(-x)\;. \tag{5.1.141}$$

As immediate consequences of (5.1.136) we note that β is *antilinear* and fulfills

$$\beta(FG) = (\beta F)(\beta G) \quad \text{and} \quad \omega_0(\beta\, F) = f_0^c = \omega_0(F)^c = \omega_0(F^\star) \ . \qquad (5.1.142)$$

As explained above, the assumptions 5.1.23 are satisfied for QED. For a real scalar field the PCT-transformation is simply $\beta\,\varphi(x) = \varphi(-x)$ and for a complex scalar field it reads $\beta\,\phi(x) = \phi^c(-x)$.

We require PCT-symmetry of the T-product as an additional axiom: Since β includes time-reversal, we expect that conjugation of T_n and $\overline{T}_n$ by the PCT-operator β amounts to mutual exchange of these products, in detail, the condition reads:

(xii) ***PCT*-symmetry:** Let $(\overline{T}_n)$ be the antichronological product which is uniquely determined by the time-ordered product (T_n), see (3.3.34) or (3.3.37). With that, the axiom reads

$$\beta \circ T_n = \overline{T}_n \circ \beta^{\otimes n} \quad \text{and} \quad \beta \circ \overline{T}_n = T_n \circ \beta^{\otimes n} \quad \forall n \in \mathbb{N} \ . \qquad (5.1.143)$$

The assumptions 5.1.23 suffice to prove PCT-symmetry of the model.

Theorem 5.1.24 (PCT-Theorem).

(a) *The axiom PCT-symmetry is a renormalization condition; it can be fulfilled and it is compatible with all other renormalization conditions.*

(b) *Assuming that the basic axioms and PCT-symmetry are satisfied, the S-matrix* $\mathbf{S}\big(L(g)\big) = T\big(e_\otimes^{iL(g)/\hbar}\big)$ *is PCT-covariant:*

$$\beta\,\mathbf{S}\big(L(g)\big) = \mathbf{S}\big(L(\tilde g)\big)^{\star -1} \quad \textit{where} \quad \tilde g(x) := g(-x)^c \ . \qquad (5.1.144)$$

Proof. (a) $\Rightarrow$ (b): This follows immediately from (5.1.143) and (5.1.141), namely

$$\beta\,\mathbf{S}\big(L(g)\big) = \beta\,T\big(e_\otimes^{iL(g)/\hbar}\big) = \overline{T}\big(e_\otimes^{-i\,\beta L(g)/\hbar}\big) = \overline{T}\big(e_\otimes^{-i\,L(\tilde g)/\hbar}\big) = \mathbf{S}\big(L(\tilde g)\big)^{\star -1} \ .$$

Ad (a): The proof of (a) goes by induction on n, following the inductive construction of the sequence (T_n) described in Sect. 3.3.2. The case $n = 1$ is a consequence of $T_1(F) = F = \overline{T}_1(F)\ \forall F \in \mathcal{F}_{\text{loc}}$.

Going to the inductive step $(n-1) \to n$, we assume that PCT-covariant T-products T_k^{sym} and the pertinent antichronological products $\overline{T}_k^{\text{sym}}$ (3.3.37) are constructed for $1 \le k < n$. We first prove the assertion for $T_n^0\big(\otimes_{j=1}^n B_j(x_j)\big) \in \mathcal{D}'(\mathbb{M}^n \setminus \Delta_n, \mathcal{F})$, where $B_j \in \mathcal{P}$ arbitrary. Using (3.3.17)–(3.3.18) and that the functions f_I (3.3.19) are real-valued, and applying the relation (5.1.140) and the inductive assumption, we obtain

$$\begin{aligned}
&\beta\, T^0_n(\otimes_{j=1}^n B_j(x_j)) \\
&\quad = \sum_I f_I(x_1,\ldots,x_n)\, \beta\, T^{\mathrm{sym}}_{|I|}(\otimes_{j\in I} B_j(x_j)) \star \beta\, T^{\mathrm{sym}}_{|I^c|}(\otimes_{k\in I^c} B_k(x_k)) \\
&\quad = \sum_I f_I(x_1,\ldots,x_n)\, \overline{T}^{\mathrm{sym}}_{|I|}(\otimes_{j\in I} (\beta B_j)(x_j)) \star \overline{T}^{\mathrm{sym}}_{|I^c|}(\otimes_{k\in I^c} (\beta B_k)(x_k)) \\
&\quad = \overline{T}^0_n(\otimes_{j=1}^n (\beta B_j)(x_j)) \, . \qquad (5.1.145)
\end{aligned}$$

In the last step anticausal factorization of the $\overline{T}$-product (3.3.35) is used; this is allowed since $\beta B_j(x) = B_j^{\mathfrak{C}}(-x)$ and since $\{-x_k \,|\, k \in I^c\} \cap (\{-x_j \,|\, j \in I\} + \overline{V}_-) = \emptyset$ for $(x_1,\ldots,x_n) \in \operatorname{supp} f_I$. In the same way one derives $\beta \circ \overline{T}^0_n = T^0_n \circ \beta^{\otimes n}$.

We conclude that the axiom "*PCT*-symmetry" is indeed a renormalization condition. To obtain a *PCT*-covariant renormalization, we take an arbitrary extension $T_n(\ldots) \in \mathcal{D}'(\mathbb{M}^n, \mathcal{F})$ of $T^0_n(\ldots)$ fulfilling all other renormalization conditions and symmetrize it with respect to the *finite* group generated by β – this is an application of the method (3.2.91):

$$T^{\mathrm{sym}}_n := \frac{1}{2N} \sum_{l=0}^{N-1} \Big(\beta^{2l} \circ T_n \circ (\beta^{-2l})^{\otimes n} + \beta^{2l+1} \circ \overline{T}_n \circ (\beta^{-2l-1})^{\otimes n} \Big) \, , \qquad (5.1.146)$$

where $\overline{T}_n$ is the antichronological product obtained from T_n and the inductively known, *PCT*-covariant $(T^{\mathrm{sym}}_k)_{1\leq k<n}$ by means of (3.3.37). T^{sym}_n is also an extension of T^0_n, since $\beta^{2l} \circ T_n \circ (\beta^{-2l})^{\otimes n}$ and $\beta^{2l+1} \circ \overline{T}_n \circ (\beta^{-2l-1})^{\otimes n}$ are respectively extensions of $\beta^{2l} \circ T^0_n \circ (\beta^{-2l})^{\otimes n}$ and $\beta^{2l+1} \circ \overline{T}^0_n \circ (\beta^{-2l-1})^{\otimes n}$, and since

$$\beta^{2l} \circ T^0_n \circ (\beta^{-2l})^{\otimes n} = T^0_n \, , \quad \text{as well as} \quad \beta^{2l+1} \circ \overline{T}^0_n \circ (\beta^{-2l-1})^{\otimes n} = T^0_n \, ,$$

due to (5.1.145).

Let $\overline{T}^{\mathrm{sym}}_n$ be the antichronological product which is uniquely given by T^{sym}_n (5.1.146) and the inductively known $(T^{\mathrm{sym}}_k)_{1\leq k<n}$. Below we show that $\overline{T}^{\mathrm{sym}}_n$ can be written similarly to (5.1.146):

$$\overline{T}^{\mathrm{sym}}_n = \frac{1}{2N} \sum_{l=0}^{N-1} \Big(\beta^{2l} \circ \overline{T}_n \circ (\beta^{-2l})^{\otimes n} + \beta^{2l+1} \circ T_n \circ (\beta^{-2l-1})^{\otimes n} \Big) \, , \qquad (5.1.147)$$

From (5.1.146) and (5.1.147) we see that T^{sym}_n and $\overline{T}^{\mathrm{sym}}_n$ fulfill indeed the assertion (5.1.143), by using $\beta^{2N} = \mathrm{Id}$.

It remains to prove the claim (5.1.147), we proceed in a somewhat tricky way: From

$$\overline{T}^{\mathrm{sym}}(e_\otimes^{-iF/\hbar}) \star T^{\mathrm{sym}}(e_\otimes^{iF/\hbar}) = 1 = T^{\mathrm{sym}}(e_\otimes^{iF/\hbar}) \star \overline{T}^{\mathrm{sym}}(e_\otimes^{-iF/\hbar})$$

we conclude that any extension T_n (of T_n^0) and the corresponding $\overline{T}_n$ (3.3.37) satisfy

$$\begin{aligned} &\big[T_n+(-1)^n\,\overline{T}_n\big]\big(\otimes_{j=1}^n B_j(x_j)\big) \qquad (5.1.148)\\ &\quad= \sum_{\substack{I\subset\{1,\dots,n\}\\ 1\le|I|<n}} (-1)^{|I|+1}\,\overline{T}^{\mathrm{sym}}_{|I|}\big(\otimes_{j\in I} B_j(x_j)\big)\star T^{\mathrm{sym}}_{|I^c|}\big(\otimes_{k\in I^c} B_k(x_k)\big)\\ &\quad= \sum_{\substack{I\subset\{1,\dots,n\}\\ 1\le|I|<n}} (-1)^{n-|I|+1}\,T^{\mathrm{sym}}_{|I|}\big(\otimes_{j\in I} B_j(x_j)\big)\star \overline{T}^{\mathrm{sym}}_{|I^c|}\big(\otimes_{k\in I^c} B_k(x_k)\big)\;. \end{aligned}$$

Since the two expressions on the right-hand side of (5.1.148) are inductively given, the proof is complete if we succeed to show that the equation

$$T_n^{\mathrm{sym}}+(-1)^n\overline{T}_n^{\mathrm{sym}} \overset{?}{=} T_n+(-1)^n\overline{T}_n\,, \qquad (5.1.149)$$

holds true when we substitute the expression (5.1.147) for $\overline{T}_n^{\mathrm{sym}}$, where, on the r.h.s., T_n is the arbitrary extension used in (5.1.146) and (5.1.147). From (5.1.148) and PCT-symmetry of the lower orders (T_k^{sym}) and $(\overline{T}_k^{\mathrm{sym}})$, $k<n$, we obtain

$$\begin{aligned} &\beta\circ\big[\overline{T}_n+(-1)^n\,T_n\big]\big(\otimes_{j=1}^n B_j(x_j)\big)\\ &\quad= \sum_{\substack{I\subset\{1,\dots,n\}\\ 1\le|I|<n}} (-1)^{n+|I|+1}\,T^{\mathrm{sym}}_{|I|}\big(\otimes_{j\in I} (\beta B_j)(x_j)\big)\star \overline{T}^{\mathrm{sym}}_{|I^c|}\big(\otimes_{k\in I^c} (\beta B_k)(x_k)\big)\\ &\quad= \big[T_n+(-1)^n\,\overline{T}_n\big]\big(\otimes_{j=1}^n (\beta B_j)(x_j)\big)\;. \end{aligned}$$

As well,

$$\beta\circ\big[T_n+(-1)^n\,\overline{T}_n\big]=\big[\overline{T}_n+(-1)^n\,T_n\big]\circ\beta^{\otimes n}\;.$$

Using these relations after inserting (5.1.146) and (5.1.147) into the left-hand side of (5.1.149), we see that (5.1.149) holds indeed true:

$$\begin{aligned} T_n^{\mathrm{sym}}+(-1)^n\,\overline{T}_n^{\mathrm{sym}} &= \frac{1}{2N}\sum_{l=0}^{N-1}\Big(\beta^{2l}\circ\big[T_n+(-1)^n\,\overline{T}_n\big]\circ(\beta^{-2l})^{\otimes n}\\ &\qquad\qquad + \beta^{2l+1}\circ\big[\overline{T}_n+(-1)^n\,T_n\big]\circ(\beta^{-2l-1})^{\otimes n}\Big)\\ &= T_n+(-1)^n\,\overline{T}_n\;. \end{aligned}$$

One verifies straightforwardly that all other renormalization conditions are preserved under the symmetrization (5.1.146). For example, for the *Sm-expansion axiom*, we first note that this axiom holds also for $\overline{T}_n$. The vacuum expectation value $\omega_0(\dots)$ of (5.1.146) can be written as

$$t^{\mathrm{sym}}(A_1,\dots)(X)=\frac{1}{2N}\sum_{l=0}^{N-1}\Big(t(\beta^{-2l}A_1,\dots)(X)+\overline{t}(\beta^{-2l-1}A_1,\dots)^c(X)\Big)$$

by using (5.1.142), where $X := (x_1 - x_n, \ldots)$ and $A_j \in \mathcal{P}_{\mathrm{hom}}$ $\forall 1 \leq j \leq n$. Due to (5.1.138), we have $\dim(\beta^{-2l} A_j) = \dim A_j = \dim(\beta^{-2l-1} A_j)$, and with that the assertion follows from part (0) of Lemma 3.1.26.

Or, the *Field independence* can be checked by proceeding analogously to (5.1.120): We use

$$\frac{\partial}{\partial(\beta\phi_j)} \circ \beta = \beta \circ \frac{\partial}{\partial\phi_j} \,,$$

which follows from (5.1.136); with that we get

$$\begin{aligned}\frac{\partial}{\partial\phi_j} \circ \beta^{2l+1} \circ \overline{T}_n \circ (\beta^{-2l-1})^{\otimes n} &= \beta^{2l+1} \circ \frac{\partial}{\partial(\beta^{-2l-1}\phi_j)} \circ \overline{T}_n \circ (\beta^{-2l-1})^{\otimes n} \\ &= \beta^{2l+1} \circ \overline{T}_n \circ \Big(\Big(\frac{\partial}{\partial(\beta^{-2l-1}\phi_j)} \circ \beta^{-2l-1} \Big) \otimes (\beta^{-2l-1})^{\otimes(n-1)} + \cdots \Big) \\ &= \beta^{2l+1} \circ \overline{T}_n \circ \Big(\Big(\beta^{-2l-1} \circ \frac{\partial}{\partial\phi_j} \Big) \otimes (\beta^{-2l-1})^{\otimes(n-1)} + \cdots \Big)\end{aligned}$$

and similarly for $\beta^{2l} \circ T_n \circ (\beta^{-2l})^{\otimes n}$.

As a last example, *Charge conjugation invariance* (5.1.118) is maintained under the symmetrization (5.1.146), because $\beta_C \circ \beta = \beta \circ \beta_C$ (this is a consequence of part (b) of Lemma 5.1.19) and because $\beta_C \circ T_n = T_n \circ (\beta_C)^{\otimes n}$ implies $\beta_C \circ \overline{T}_n = \overline{T}_n \circ (\beta_C)^{\otimes n}$. □

Remark 5.1.25 (*PCT*-transformation applied to interacting fields)**.** The interacting QFT-fields $R(e_\otimes^{F/\hbar}, G)$ studied so far are *retarded* fields, they have causal support due to the axiom Causality (3.1.9). We can introduce "advanced" interacting fields, by modifying the Bogoliubov formula (3.3.30): We mutually exchange T with $\overline{T}$, that is,

$$A(e_\otimes^{F/\hbar}, G) := T(e_\otimes^{iF/\hbar}) \star \overline{T}(e_\otimes^{-iF/\hbar} \otimes G) \,. \tag{5.1.150}$$

By proceeding analogously to Exer. 3.3.11(a), one verifies that the advanced interacting fields have "anticausal" support:

$$A(e_\otimes^{(F+H)/\hbar}, G) = A(e_\otimes^{F/\hbar}, G) \quad \text{if} \quad (\operatorname{supp} G + \overline{V}_+) \cap \operatorname{supp} H = \emptyset \,.$$

Under the *PCT*-transformation β retarded and advanced interacting fields mutually exchange – this is plausible due to the time reversal. In detail: If the axiom *PCT*-symmetry is fulfilled and the interaction L is *PCT*-invariant (5.1.141), we obtain

$$\begin{aligned}\beta\, R(e_\otimes^{L(g)/\hbar}, B(x)) &= \beta\, \overline{T}(e_\otimes^{-iL(g)/\hbar}) \star \beta\, T(e_\otimes^{iL(g)/\hbar} \otimes B(x)) \\ &= T(e_\otimes^{i\,\beta L(g)/\hbar}) \star \overline{T}(e_\otimes^{-i\,\beta L(g)/\hbar} \otimes \beta\, B(x)) \\ &= A(e_\otimes^{L(\tilde{g})/\hbar}, B^{\mathfrak{C}}(-x))\end{aligned} \tag{5.1.151}$$

for any $B \in \mathcal{P}$, and similarly

$$\beta\, A\big(e_\otimes^{L(g)/\hbar}, B(x)\big) = R\big(e_\otimes^{L(\tilde{g})/\hbar}, B^{\mathfrak{C}}(-x)\big)\ , \tag{5.1.152}$$

where $\tilde{g}$ is as above (5.1.144) and $B^{\mathfrak{C}} \in \mathcal{P}$ is defined by

$$\Big(\sum_n \sum_{j_1,\dots} \sum_{a_1,\dots} b^{j_1,\dots}_{n;a_1,\dots}\, \partial^{a_1}\phi_{j_1} \cdots \partial^{a_n}\phi_{j_n}\Big)^{\mathfrak{C}} := \sum_n \sum_{j_1,\dots} \sum_{a_1,\dots} (b^{j_1,\dots}_{n;a_1,\dots})^{c}\, \partial^{a_1}\phi^{\mathfrak{C}}_{j_1} \cdots \partial^{a_n}\phi^{\mathfrak{C}}_{j_n}$$

with coefficients $b^{j_1,\dots}_{n;a_1,\dots} \in \mathbb{C}$.

Remark 5.1.26. If both renormalization conditions Charge conjugation invariance and PCT-symmetry are satisfied, the T-product is also PT-symmetric. Namely, due to $\beta = \beta_P \circ \beta_C \circ \beta_T$, $\beta_C^2 = \mathrm{Id}$ and $\beta_C \circ \beta_T = \beta_T \circ \beta_C$ we have $\beta_P \circ \beta_T = \beta \circ \beta_C$ and, hence,

$$\beta_P \circ \beta_T \circ T_n = \overline{T}_n \circ (\beta_P \circ \beta_T)^{\otimes n} \quad \text{and} \quad \beta_P \circ \beta_T \circ \overline{T}_n = T_n \circ (\beta_P \circ \beta_T)^{\otimes n} \quad \forall n \in \mathbb{N}\ .$$

In QED one can prove that there exists a renormalization of the T-product respecting C-, P- and T-symmetry *individually* and all further renormalization conditions. This can be done by replacing the symmetrization (5.1.146) by a suitable symmetrization with respect to the finite group G_{16} studied in Exer. 5.1.21, see [148, Sect. 4.4].

5.1.7 Reducing the computation of R- and T-products to scalar field renormalization

In this section we explain in terms of a few examples, how the computation of R- and T-products of QED can be traced back to *scalar* field theory, that is, to the construction given in Sect. 3.2 (or the corresponding construction for the T-product, see Sect. 3.3). In particular, the techniques to renormalize explained in Sect. 3.5 can be applied also to QED. The key for this is formula (A.2.11), which expresses the spinor field propagators in terms of the corresponding propagators for a real scalar field. In addition, an important tool is the following set of formulas for the trace of products of γ-matrices (see, e.g., [109, App. A2]).

Lemma 5.1.27 (Trace of products of γ-matrices). *The trace of an odd product of γ-matrices vanishes:*

$$\mathrm{tr}(\gamma^{\mu_1}\gamma^{\mu_2}\cdots\gamma^{\mu_{2n+1}}) = 0 \qquad \forall n \in \mathbb{N}\ ;$$

and for the lowest even products it holds that

$$\mathrm{tr}(\gamma^\mu\gamma^\nu) = 4g^{\mu\nu}\ ,$$
$$\mathrm{tr}(\gamma^\mu\gamma^\nu\gamma^\rho\gamma^\sigma) = 4(g^{\mu\nu}g^{\rho\sigma} - g^{\mu\rho}g^{\nu\sigma} + g^{\mu\sigma}g^{\nu\rho})\ .$$

Proof. For even products we use the cyclicity of the trace and the anticommutation relation of the γ-matrices (5.1.2):

$$\begin{aligned}\operatorname{tr}(\gamma^\mu\gamma^\nu) &= \frac{1}{2}\operatorname{tr}[\gamma^\mu,\gamma^\nu]_+ = g^{\mu\nu}\operatorname{tr}(1_{4\times 4}) = 4g^{\mu\nu}\ ,\\ 2\operatorname{tr}(\gamma^\mu\gamma^\nu\gamma^\rho\gamma^\sigma) &= \operatorname{tr}(\gamma^\mu\gamma^\nu\gamma^\rho\gamma^\sigma + \gamma^\nu\gamma^\rho\gamma^\sigma\gamma^\mu)\\ &= \operatorname{tr}\big([\gamma^\mu,\gamma^\nu]_+\gamma^\rho\gamma^\sigma - \gamma^\nu[\gamma^\mu,\gamma^\rho]_+\gamma^\sigma + \gamma^\nu\gamma^\rho[\gamma^\mu,\gamma^\sigma]_+\big)\\ &= 2g^{\mu\nu}\operatorname{tr}(\gamma^\rho\gamma^\sigma) - 2g^{\mu\rho}\operatorname{tr}(\gamma^\nu\gamma^\sigma) + 2g^{\mu\sigma}\operatorname{tr}(\gamma^\nu\gamma^\rho)\\ &= 8(g^{\mu\nu}g^{\rho\sigma} - g^{\mu\rho}g^{\nu\sigma} + g^{\mu\sigma}g^{\nu\rho})\ .\end{aligned}$$

The procedure for higher even products is analogous.

To treat the odd products, we use the matrix γ^5 defined in (5.1.122), which fulfills the identities (5.1.123). With the latter, we get

$$\operatorname{tr}(\gamma^{\mu_1}\cdots\gamma^{\mu_{2n+1}}) = \operatorname{tr}(\gamma^{\mu_1}\cdots\gamma^{\mu_{2n+1}}\gamma^5\gamma^5) = (-1)^{2n+1}\operatorname{tr}(\gamma^5\gamma^{\mu_1}\cdots\gamma^{\mu_{2n+1}}\gamma^5)$$

by anticommuting one factor γ^5 to the left-hand side. By means of the cyclicity of the trace, this is equal to

$$= (-1)^{2n+1}\operatorname{tr}(\gamma^5\gamma^5\gamma^{\mu_1}\cdots\gamma^{\mu_{2n+1}}) = -\operatorname{tr}(\gamma^{\mu_1}\cdots\gamma^{\mu_{2n+1}})\ . \quad \square$$

As an illustration we compute here the time-ordered product $T_2\big(L(x_1)\otimes L(x_2)\big)$ (where $L = A^\mu j_\mu$); the retarded product $R_{1,1}\big(L(x_1); j^\mu(x)\big)$ is computed in Exer. 5.2.5. By means of the causal Wick expansion (3.1.24) and the notation (3.2.20) we obtain

$$\begin{aligned}T_2\big(L(x_1)\otimes L(x_2)\big) &= L(x_1)\,L(x_2) + t(A^\mu,A^\nu)(y)\,j_\mu(x_1)j_\nu(x_2)\\ &+ \overline{\psi}(x_1)\not{A}(x_1)\,t(\psi,\overline{\psi})(y)\,\not{A}(x_2)\psi(x_2) + \overline{\psi}(x_2)\not{A}(x_2)\,t(\psi,\overline{\psi})(-y)\,\not{A}(x_1)\psi(x_1)\\ &+ \overline{\psi}(x_1)\gamma_\mu\,t(A^\mu\psi, A^\nu\overline{\psi})(y)\,\gamma_\nu\psi(x_2) + \overline{\psi}(x_2)\gamma_\nu\,t(A^\nu\psi, A^\mu\overline{\psi})(-y)\,\gamma_\mu\psi(x_1)\\ &+ t(j^\mu,j^\nu)(y)\,A_\mu(x_1)A_\nu(x_2) + t(L,L)(y)\ ,\quad y := x_1 - x_2\ . \qquad (5.1.153)\end{aligned}$$

From (5.1.63) and (5.1.89) we know that

$$t(\psi,\overline{\psi})(y) = \hbar\, S^F(y) \quad \text{and} \quad t(A^\mu,A^\nu)(y) = -\hbar g^{\mu\nu}\,D^F(y)\ . \qquad (5.1.154)$$

Similarly to Remk. 3.3.4, the contraction schemes belonging to the various terms in (5.1.153) can be symbolized by Feyman diagrams, see Figure 5.1. The wavy lines stand for photon fields or photon contractions, the non-wavy lines for spinor fields or spinor contractions. The latter are oriented, in order to distinguish between $\psi(x)$ and $\overline{\psi}(x)$, or $S^F(y)$ and $S^F(-y)$, respectively; the arrow indicates the flow of electric charge, its direction is opposite to the order in which the pertinent term is written (in matrix notation). Charge conservation of $L(x_j)$ ($j = 1,2$), i.e., $\theta L(x_j) = 0$, is visualized by the fact that in the pertinent vertex there is precisely one incoming and one outgoing spinor line.

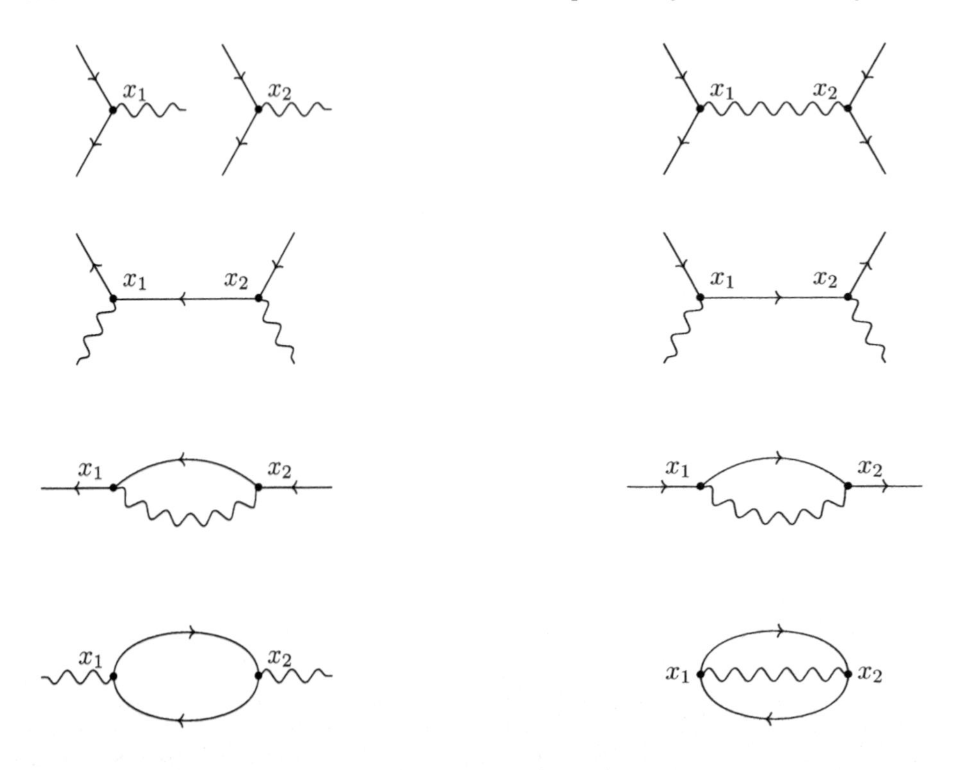

Figure 5.1: The Feynman diagrams are in the same order as the pertinent terms in formula (5.1.153).

Computation of the vacuum polarization term. We are going to compute

$$t^{\mu\nu}(y) := t(j^\mu, j^\nu)(y) = t^{\nu\mu}(-y)$$

which is called the "vacuum polarization tensor", due to its physical interpretation. For the latter and also for the physical relevance of the term $t(A^\mu\psi, A^\nu\overline{\psi})$ (which is called "electron self-energy" and is computed in Exer. 5.1.28), we refer to any book about perturbative QED, e.g., [148]. We put the focus on the peculiarities of our formalism: Renormalization in x-space – we will use the Sm-expansion and differential renormalization – and how to fulfill the condition

$$\partial_\mu^y t^{\mu\nu}(y) = 0 \qquad (5.1.155)$$

in our framework. The requirement (5.1.155) follows from the QED-MWI, as derived in Exer. 5.2.5; it can also be interpreted as the restriction of $t^{\mu\nu}$ coming from gauge invariance (or BRST invariance) of $T_2\big(L(g)^{\otimes 2}\big)$, see, e.g., [148].

To obtain $t^{\mu\nu\,0} \in \mathcal{D}'(\mathbb{M}_4 \setminus \{0\})$ we use causality: For $y \notin \overline{V}_-$ it holds that

$$\begin{aligned} t^{\mu\nu\,0}(y) &= \omega_0\big(j^\mu(x_1) \star j^\nu(x_2)\big) = \hbar^2\, \mathrm{tr}\big(\gamma^\mu S^+(y)\gamma^\nu S^-(-y)\big) \\ &= -\hbar^2\, \mathrm{tr}\big(\gamma^\mu S^F(y)\gamma^\nu S^F(-y)\big)\ ; \end{aligned} \tag{5.1.156}$$

where we have inserted the result of Exap. 5.1.4 and the relations (5.1.66)–(5.1.67). For $y \notin \overline{V}_+$ we have $-y \notin \overline{V}_-$ and, hence, we can apply (5.1.156) in the following way:

$$t^{\mu\nu\,0}(y) = t^{\nu\mu\,0}(-y) = -\hbar^2\, \mathrm{tr}\big(\gamma^\nu S^F(-y)\gamma^\mu S^F(y)\big)\ ,$$

which agrees with the result of (5.1.156) by the cyclicity of the trace. By means of $S^F = (i\partial\!\!\!/ + m)\Delta^F$ and the trace-formulas of Lemma 5.1.27 we obtain

$$\begin{aligned} t^{\mu\nu\,0}(y) &= -4\hbar^2\Big(\big(g^{\mu\alpha}g^{\nu\beta} - g^{\mu\nu}g^{\alpha\beta} + g^{\mu\beta}g^{\nu\alpha}\big)\, t^0_{1\,\alpha\beta}(y) + m^2\, g^{\mu\nu}\, t^0_2(y)\Big) \quad \text{with} \\ t^0_{1\,\alpha\beta}(y) &:= \partial_\alpha\Delta^F(y)\,\partial_\beta\Delta^F(y)\ , \quad t^0_2(y) := \Delta^F(y)\,\Delta^F(y)\ ; \end{aligned} \tag{5.1.157}$$

the diagram belonging to t^0_2 is the fish diagram (3.2.124).

It remains to compute the extension of $t^0_{1\,\alpha\beta}$ and t^0_2, respectively, from $\mathcal{D}'(\mathbb{M}_4 \setminus \{0\})$ to $\mathcal{D}'(\mathbb{M}_4)$; these are scalar field renormalization problems, solved by means of the Sm-expansion and differential renormalization in Exap. 3.5.4 and Exer. 3.5.6. The result for $t_{1\,\alpha\beta}$ is given in (3.5.22) and the relevant Sm-expansion of t_2 reads:

$$t_2(y) = a^2\,\square_y\overline{\Big(\frac{\log(M^2Y)}{4Y}\Big)} + \mathfrak{r}_{1\,m}(y) + C\,\delta(y)\ , \quad Y := -(y^2 - i0)\ , \quad \mathrm{sd}(\mathfrak{r}_{1\,m}) = 2\ , \tag{5.1.158}$$

where $C \in \mathbb{C}$ is arbitrary and $a = \frac{1}{4\pi^2}$. Note that the leading term, i.e., the term $\sim m^0$, can be read off from $u_{2,1}$ given in (3.5.13). Inserting these results into (5.1.157) we end up with

$$t^{\mu\nu}(y) = u_0^{\mu\nu}(y) + m^2\, u_2^{\mu\nu}(y) + \mathfrak{r}_m^{\mu\nu}(y) + \big(C_0\, g^{\mu\nu}\,\square + C_1\,\partial^\mu\partial^\nu + m^2\, C_2\, g^{\mu\nu}\big)\,\delta(y)\ , \tag{5.1.159}$$

$$u_0^{\mu\nu}(y) = -\,4\hbar^2 a^2\,\frac{2y^\mu y^\nu - g^{\mu\nu}\,y^2}{48}\,\square_y\square_y\square_y\overline{\Big(\frac{\log(M^2Y)}{4Y}\Big)}\ , \tag{5.1.160}$$

$$u_2^{\mu\nu}(y) = -\,4\hbar^2 a^2\Big(\frac{2y^\mu y^\nu - g^{\mu\nu}\,y^2}{4}\,\square_y + g^{\mu\nu}\Big)\square_y\overline{\Big(\frac{\log(M^2Y)}{4Y}\Big)}\ , \tag{5.1.161}$$

where $\mathrm{sd}(\mathfrak{r}_m^{\mu\nu}) = 2 < 4$ and C_0, C_1, $C_2 \in \mathbb{C}$ are arbitrary numbers.

Let us turn to the requirement (5.1.155): Since it is a consequence of the QED-MWI and since the MWI is a renormalization condition, it must be satisfied for $t^{\mu\nu\,0}$. (This is a simple way to check formula (5.1.157), by using $\square\Delta^F(y) = -m^2\Delta^F(y)$ for $y \neq 0$.) Therefore, it holds that

$$\mathrm{supp}\,\partial_\mu t^{\mu\nu} \subseteq \{0\}\ .$$

Inserting the Sm-expansion (5.1.159) into this relation, we realize that $\partial_\mu u_0^{\mu\nu}(y)$, $\partial_\mu u_2^{\mu\nu}(y)$ and $\partial_\mu \mathfrak{r}_m^{\mu\nu}(y)$ satisfy this support property individually; hence, we can apply Theorem (A.1.17) to these distributions. Using that $\partial_\mu u_j^{\mu\nu}(y)$ (where $j = 0, 2$) scales almost homogeneously with degree $(7-j)$ and that $\mathrm{sd}(\partial_\mu \mathfrak{r}_m^{\mu\nu}) = 3$, and taking into account additionally Lorentz covariance, we conclude that

$$\partial_\mu u_0^{\mu\nu}(y) = K_0\, \partial^\nu \Box \delta(y)\ , \quad \partial_\mu u_2^{\mu\nu}(y) = K_2\, \partial^\nu \delta(y)\ , \quad \partial_\mu \mathfrak{r}_m^{\mu\nu}(y) = 0\ . \quad (5.1.162)$$

The numbers $K_0, K_2 \in \mathbb{C}$ can explicitly be computed – this is done below. We see that choosing

$$C_0 + C_1 := -K_0 \quad \text{and} \quad C_2 = -K_2\ ,$$

the condition (5.1.155) is fulfilled. Since $(C_0 - C_1)$ is untouched by this condition, it remains the freedom to add a term

$$C\,(g^{\mu\nu}\,\Box - \partial^\mu\partial^\nu)\delta(y) \quad \text{to} \quad t^{\mu\nu}(y)\ , \quad C \in \mathbb{C} \quad \text{arbitrary.}$$

To compute K_0 we consider

$$\partial_\mu u_0^{\mu\nu}(y) = -\frac{\hbar^2 a^2}{12}\left\{2\,y^\nu(5 + y\partial)\Box\Box\Box - \partial^\nu\, y^2\, \Box\Box\Box\right\}\overline{\left(\frac{\log(M^2 Y)}{4Y}\right)}\ ,$$

where $y\partial := y_\alpha \partial_y^\alpha$ and we have inserted (5.1.160). By using the commutation relations

$$[y^\nu\,,\Box] = -2\partial^\nu\ , \quad [y\partial\,,\Box] = -2\Box\ , \quad [y^2\,,\Box] = -8 - 4\,y\partial\ , \qquad (5.1.163)$$

the differential operator in $\{\cdots\}$-brackets can be written as

$$\left\{\cdots\right\} = (2y^\nu\,\Box + 12\partial^\nu)\Box\Box\,(y\partial + 2) - \partial^\nu\Box\Box\Box\,y^2 - 6\,\Box\Box\,y^\nu\,\Box\ .$$

Now we take into account the identities

$$\begin{aligned}
\Box\,y^2\,\overline{\left(\frac{\log(M^2Y)}{4Y}\right)} &= -\frac{1}{4}\,\overline{(\Box\,\log(M^2Y))} = \overline{\left(\frac{1}{Y}\right)} \\
(y\partial + 2)\,\overline{\left(\frac{\log(M^2Y)}{4Y}\right)} &= \overline{(y\partial + 2)\,\frac{\log(M^2Y)}{4Y}} = \frac{1}{2}\,\overline{\left(\frac{1}{Y}\right)} \\
y^\nu\,\Box\,\overline{\left(\frac{\log(M^2Y)}{4Y}\right)} &= \overline{\left(\frac{y^\nu}{Y^2}\right)} = \frac{1}{2}\,\partial^\nu\,\overline{\left(\frac{1}{Y}\right)},
\end{aligned} \qquad (5.1.164)$$

which hold true in $\mathcal{D}'(\mathbb{M}_4)$ for the following reasons: In $\mathcal{D}'(\mathbb{M}_4 \setminus \{0\})$ one easily verifies them by straightforward computation of the derivatives. Considering, e.g., the third identity, the expression on the l.h.s. and the expression on the r.h.s. both are homogeneous extensions[159] of $\frac{y^\nu}{Y^2} \in \mathcal{D}'(\mathbb{M}_4 \setminus \{0\})$; however, the

[159]Note: $\Box\,\overline{\left(\frac{\log(M^2Y)}{4Y}\right)}$ scales only almost homogeneously (Exer. 3.5.5); but multiplication with y^ν removes the inhomogeneity,

$$\begin{aligned}
\rho^3\,(\rho y^\nu)\,\Box_{\rho y}\,\overline{\left(\frac{\log(M^2\rho^2 Y)}{4\,\rho^2 Y}\right)} - y^\nu\,\Box\,\overline{\left(\frac{\log(M^2Y)}{4Y}\right)} &= \log\rho\;\, y^\nu\Box\,\overline{\left(\frac{1}{2Y}\right)} \\
&= -\log\rho\;\, 2i\pi^2\,y^\nu\,\delta(y) = 0\ ,
\end{aligned}$$

as it must be according to Proposition 3.2.16(i).

homogeneous extension of $\frac{y^\nu}{Y^2}$ is unique since the degree of the scaling is $D = 3 < 4$ (Proposition 3.2.16(i)). Therefore, this identity is valid also for the extensions. The reasoning for the first two identities is analogous. Using additionally the identities $\square\overline{\left(\frac{1}{Y}\right)} = -4i\pi^2\,\delta(y)$ (3.5.18) and

$$y^\nu\,\square\square\delta(y) = -4\,\partial^\nu\square\delta(y)\ ,$$

and inserting $a = \frac{1}{4\pi^2}$, we end up with

$$\partial_\mu u_0^{\mu\nu}(y) = -\frac{\hbar^2\, i}{24\,\pi^2}\,\partial^\nu\square\delta(y)\ .$$

The computation of K_2 is analogous: Using (5.1.161) and again the identities (5.1.163) and (5.1.164) we get

$$\begin{aligned}\partial_\mu u_2^{\mu\nu}(y) &= -\hbar^2 a^2\left(\left(2\,y^\nu(5+y\partial) - \partial^\nu\,y^2\right)\square\square + 4\,\partial^\nu\square\right)\overline{\left(\frac{\log(M^2Y)}{4\,Y}\right)}\\ &= -\hbar^2 a^2\left((2\,y^\nu\,\square + 8\,\partial^\nu)\,\square\,(y\partial+2) - \partial^\nu\,\square\square\,y^2 - 2\,\square\,y^\nu\,\square\right)\overline{\left(\frac{\log(M^2Y)}{4\,Y}\right)}\\ &= \hbar^2 a^2\,4i\pi^2\,(y^\nu\,\square + 2\,\partial^\nu)\,\delta(y) = 0\ .\end{aligned}$$

The non-local terms cancel in both $\partial_\mu u_0^{\mu\nu}$ and $\partial_\mu u_2^{\mu\nu}$, as it must be – this is a nontrivial check of the calculations. However, we do not see a deeper reason for the vanishing of K_2.

Exercise 5.1.28 (Electron self-energy). Compute the electron self-energy term

$$t_{rs}^{\mu\nu}(y) := t(A^\mu\psi_r, A^\nu\overline{\psi}_s)(y)$$

by means of the Sm-expansion and differential renormalization.

[*Solution*: To compute $t_{rs}^{\mu\nu\,0} \in \mathcal{D}'(\mathbb{M}\setminus\{0\})$ we use causality and the relations (5.1.66)–(5.1.67) and (A.2.9): For $y \notin \overline{V}_-$ we get

$$t_{rs}^{\mu\nu\,0}(y) = \omega_0\big(A^\mu\psi_r(x_1)\star A^\nu\overline{\psi}_s(x_2)\big) = -\hbar^2 g^{\mu\nu}\,D^+(y)S_{rs}^+(y) = -\hbar^2 g^{\mu\nu}\,D^F(y)S_{rs}^F(y)\ ,$$

where $y = x_1 - x_2$. For $y \notin \overline{V}_+$ we take into account additionally the graded symmetry of the T-product:

$$\begin{aligned}t_{rs}^{\mu\nu\,0}(y) &= -t(A^\nu\overline{\psi}_s, A^\mu\psi_r)(-y) = -\omega_0\big(A^\nu\overline{\psi}_s(x_2)\star A^\mu\psi_r(x_1)\big)\\ &= \hbar^2 g^{\nu\mu}\,D^+(-y)S_{rs}^-(y) = -\hbar^2 g^{\mu\nu}\,D^F(y)S_{rs}^F(y)\ .\end{aligned}$$

Hence, in $\mathcal{D}'(\mathbb{M}\setminus\{0\})$ we may write

$$t^{\mu\nu\,0}(y) = -\hbar^2 g^{\mu\nu}\left(i\,D^F(y)\,\partial\!\!\!/\Delta_m^F(y) + m\,1_{4\times4}\,D^F(y)\,\Delta_m^F(y)\right)\ ,$$

where $S^F = (i\partial\!\!\!/ + m)\Delta^F$ is inserted. Using the Sm-expansion of Δ_m^F in the form

$$\Delta_m^F(y) = D^F(y) + \mathfrak{R}_m(y)\ ,\quad \mathrm{sd}(\mathfrak{R}_m) = 0\ ,\quad D^F(y) = \frac{a}{Y}\ ,\quad Y := -(y^2 - i0)\ ,$$

with $a = \frac{1}{4\pi^2}$, we obtain

$$t^{\mu\nu\,0}(y) = -\hbar^2 g^{\mu\nu}\Big(u_0^0(y) + m\,u_1^0(y) + \mathfrak{r}_m^0(y)\Big)\,, \quad \mathrm{sd}(\mathfrak{r}_m^0) = 3\,, \tag{5.1.165}$$
$$u_0^0(y) := \frac{i}{2}\,\partial\!\!\!/_y\,\big(D^F(y)\big)^2 = \frac{ia^2}{2}\,\partial\!\!\!/_y\,\frac{1}{Y^2}\,, \quad u_1^0(y) := 1_{4\times 4}\,\big(D^F(y)\big)^2 = a^2\,1_{4\times 4}\,\frac{1}{Y^2}\,.$$

This is an example for an Sm-expansion, in which even and odd powers of m appear. To extend $\mathfrak{r}_m^0$ to $\mathcal{D}'(\mathbb{M})$ we use the direct extension. To compute the extensions u_0, u_1 of u_0^0, u_1^0, respectively, we apply differential renormalization: u_1 can be read off from (5.1.158),

$$u_1(y) = a^2\,1_{4\times 4}\,\Box_y\,\overline{\Big(\frac{\log(M^2Y)}{4Y}\Big)} + C_1\,1_{4\times 4}\,\delta(y)\,, \quad C_1 \in \mathbb{C}\,,$$

and by using this formula we get also a result for u_0,

$$u_0(y) = \frac{ia^2}{2}\,\partial\!\!\!/_y\Box_y\,\overline{\Big(\frac{\log(M^2Y)}{4Y}\Big)} + C_0\,\partial\!\!\!/_y\delta(y)\,, \quad C_0 \in \mathbb{C}\,.$$

Inserting these extensions into (5.1.165) we obtain the final result.
Another formula for u_0 can be found by writing

$$u_0^0(y) = 2ia^2\,y\!\!\!/\,\frac{1}{Y^3}\,.$$

Using the version (3.5.21) of differential renormalization and the extension of Y^{-3} given in (3.5.13) we get

$$u_0(y) = 2ia^2\,y\!\!\!/\,\Box_y\Box_y\,\overline{\Big(\frac{-\log(M^2Y)}{32\,Y}\Big)} + \tilde{C}_0\,\partial\!\!\!/_y\delta(y)\,, \quad \tilde{C}_0 \in \mathbb{C}\,.$$

To verify explicitly that these two results for u_0 are equal, we use $[y^\nu\,,\,\Box] = -2\partial^\nu$, the third identity of (5.1.164) and (3.5.18):

$$\begin{aligned}
2ia^2\,y\!\!\!/\,\Box_y\Box_y\,\overline{\Big(\frac{-\log(M^2Y)}{32\,Y}\Big)} &= \frac{ia^2}{4}\,\big(-\Box_y\,y\!\!\!/\,\Box_y + 2\,\partial\!\!\!/_y\Box_y\big)\,\overline{\Big(\frac{\log(M^2Y)}{4Y}\Big)}\\
&= -\frac{ia^2}{8}\,\partial\!\!\!/_y\Box_y\,\overline{\Big(\frac{1}{Y}\Big)} + \frac{ia^2}{2}\,\partial\!\!\!/_y\Box_y\,\overline{\Big(\frac{\log(M^2Y)}{4Y}\Big)}\\
&= -\frac{a^2\pi^2}{2}\,\partial\!\!\!/_y\delta(y) + \frac{ia^2}{2}\,\partial\!\!\!/_y\Box_y\,\overline{\Big(\frac{\log(M^2Y)}{4Y}\Big)}\,.]
\end{aligned}$$

5.2 The relevant Master Ward Identity for Quantum Electrodynamics

On the one hand this section illustrates the MWI formalism, on the other hand we derive a crucial tool for the construction of QED.

5.2.1 Formulation of the QED Master Ward Identity

Since the Faddeev–Popov ghost field do not couple (see (5.1.81)), we may omit them here. The free action $S_0 := S_0^{\text{spinor}} + S_0^{\text{photon}}$, given by (5.1.28) and (5.1.83), is invariant under the global $U(1)$-transformation

$$\psi \longmapsto \psi_\alpha := e^{i\alpha}\psi \ , \quad \overline{\psi} \longmapsto \overline{\psi}_\alpha := e^{-i\alpha}\overline{\psi} \ , \quad A^\mu \longmapsto A^\mu \ , \tag{5.2.1}$$

where $\alpha \in \mathbb{R}$. The pertinent conserved Noether current can be computed by means of (4.2.18), it is the electromagnetic current $j^\mu = \overline{\psi} \wedge \gamma^\mu \psi$ (5.1.27):

$$\partial_\mu j^\mu(x)\Big|_{\mathcal{C}_{S_0}} = 0 \ . \tag{5.2.2}$$

In terms of off-shell fields, this current conservation can be written as

$$\partial_\mu j^\mu(x) = \frac{\delta_r S_0}{\delta\psi(x)} \wedge Q(x) + \overline{Q}(x) \wedge \frac{\delta S_0}{\delta\overline{\psi}(x)} \ , \quad Q := i\psi \ , \ \overline{Q} := Q^\dagger \gamma_0 = -i\overline{\psi} \ , \tag{5.2.3}$$

by using (5.1.29), in accordance with (4.2.17). Let B be a "monomial" in ψ and $\overline{\psi}$, for simplicity we assume that it does not contain any derivatives of ψ or $\overline{\psi}$; more precisely

$$B(y) := \sum_{k_1,\dots,k_n} P_{k_1\dots k_n}(y)\, \overline{\psi}_{k_1}(y) \wedge \cdots \wedge \overline{\psi}_{k_l}(y) \wedge \psi_{k_{l+1}}(y) \wedge \cdots \wedge \psi_{k_n}(y) \ , \tag{5.2.4}$$

where $P_{k_1\dots k_n}$ is a polynomial in the gauge field A^μ. In addition, let

$$F := \int dy\, g(y)\, B(y) \in \mathcal{F}_{\text{QED}}^{\text{loc}} \quad \text{with} \quad g \in \mathcal{D}(\mathbb{M}) \ .$$

To compute the application of the derivation δ_{hQ} (4.1.4) to such a local functional F, we use the definition of the functional derivatives $\frac{\delta}{\delta\overline{\psi}}$ and $\frac{\delta_r}{\delta\psi}$ (given in (5.1.14) and (5.1.33), resp.):

$$\begin{aligned}\delta_{hQ}F &:= i \int dx\ h(x) \Big(\frac{\delta_r F}{\delta\psi(x)} \wedge \psi(x) - \overline{\psi}(x) \wedge \frac{\delta F}{\delta\overline{\psi}(x)} \Big) \\ &= i \int dx\ h(x) \int dy\, g(y)\, \delta(x-y)\, (\theta B)(y) \ ,\end{aligned} \tag{5.2.5}$$

where

$$\theta B := \big((n-l) - l\big)\, B \ ; \tag{5.2.6}$$

that is, θ is the charge number operator for *field monomials* – we use the same notation θ as for the charge number operator $\theta\colon \mathcal{F}_{\rm QED} \longrightarrow \mathcal{F}_{\rm QED}$ (5.1.99) acting on functionals. With that we may write $\theta \int dy\, g(y)\, B(y) = \int dy\, g(y)\, (\theta B)(y)$.

Setting $h(x) = 1\ \forall x$, the operator $\delta_{hQ}|_{h=1} : \mathcal{F}_{\rm QED} \longrightarrow \mathcal{F}_{\rm QED}$ agrees with the charge number operator θ (5.1.99). Note also that we may interpret $i\theta$ as the infinitesimal action of the transformation (5.2.1): $i\theta\, B = \frac{d}{d\alpha}|_{\alpha=0}\, B_\alpha$ where B_α is obtained from B (5.2.4) by replacing ψ and $\overline{\psi}$ by ψ_α and $\overline{\psi}_\alpha$, respectively.

For $B_1,\dots,B_n$ being of the form (5.2.4), we obtain the following result for the MWI belonging to the symmetry $\mathcal{A} = \int dx\, h(x)\, \partial^\mu j_\mu(x)$, by proceeding analogously to Exer. 4.2.6:

$$\begin{aligned}
&\partial_x^\mu R\big(B_1(x_1)\otimes\cdots\otimes B_n(x_n); j_\mu(x)\big)\\
&\qquad + i\hbar\sum_{l=1}^n \delta(x-x_l)\ \mathrm{sgn}_g\pi_l\ R\big(B_1(x_1)\otimes\cdots\widehat{l}\cdots\otimes B_n(x_n); (\theta B_l)(x_l)\big)\\
&\quad = \mathrm{sgn}_g\pi\,\big(\partial_x^\mu\overline{\psi}(x)\gamma_\mu - im\overline{\psi}(x)\big)\wedge R\big(B_1(x_1)\otimes\cdots\otimes B_n(x_n);\psi(x)\big)\\
&\qquad + R\big(B_1(x_1)\otimes\cdots\otimes B_n(x_n);\overline{\psi}(x)\big)\wedge(\not\partial_x + im)\psi(x)\ ,
\end{aligned}\tag{5.2.7}$$

where sgn_g is the graded signature (introduced in (5.1.45)) and the permutations are $\pi_l\colon (B_1,\dots,B_n) \longmapsto (B_1,\dots\widehat{l},\dots,B_n,B_l)$ and $\pi\colon (B_1,\dots,B_n,\overline{\psi}) \longmapsto (\overline{\psi},B_1,\dots,B_n)$. The signs $\mathrm{sgn}_g\pi_l$ and $\mathrm{sgn}_g\pi$, can easily be derived from the purely bosonic case by using the η-trick (5.1.48). We call (5.2.7) the "QED-MWI".

We will prove the QED-MWI for $B_1,\dots,B_n$ being arbitrary submonomials of the interaction $L = j_\mu A^\mu$ (5.1.102), i.e., for

$$B_1,\dots,B_n \in \{\psi,\ \overline{\psi},\ A^\mu,\ j^\mu,\ \overline{\psi}\not{A},\ \not{A}\psi,\ L\}\ ;$$

this suffices for our construction of QED.

The equivalent formulation of the QED-MWI in terms of the T-product reads

$$\begin{aligned}
&\partial_y^\mu T\big(\tilde{B}_1(x_1)\otimes\cdots\otimes\tilde{B}_n(x_n)\otimes j_\mu(y)\big)\\
&\qquad + \hbar\sum_{l=1}^n \delta(y-x_l)\, T\big(\tilde{B}_1(x_1)\otimes\cdots\otimes\widetilde{\theta B_l}(x_l)\otimes\cdots\otimes\tilde{B}_n(x_n)\big)\\
&\quad = \big(\partial_y^\mu\overline{\psi}(y)\gamma_\mu - im\overline{\psi}(y)\big)\wedge T\big(\psi(y)\otimes\tilde{B}_1(x_1)\otimes\cdots\otimes\tilde{B}_n(x_n)\big)\\
&\qquad + T\big(\tilde{B}_1(x_1)\otimes\cdots\otimes\tilde{B}_n(x_n)\otimes\overline{\psi}(y)\big)\wedge(\not\partial_y + im)\psi(y)\ ,
\end{aligned}\tag{5.2.8}$$

where we use the η-trick, that is, by $\tilde{B}_j$ we mean a bosonic field polynomial which is defined in terms of B_j, which is assumed to be of the form (5.2.4), by $\tilde{B}_j(x) := 1\otimes B_j(x)$ if B_j is bosonic, or $\tilde{B}_j(x) := \eta_j\otimes B_j(x)$ if B_j is fermionic, analogously to (5.1.48).

Taking the VEV of the QED-MWI, the r.h.s. of (5.2.7) or (5.2.8), respectively, does not contribute – a trivial but useful remark.

A famous example is the Ward identity which connects the "vertex function" $t(\not{A}\psi, \overline{\psi}\not{A}, L\ldots, L, j^\mu)$ with the electron self-energy $t(\not{A}\psi, \overline{\psi}\not{A}, L\ldots, L)$: The VEV of the QED-MWI (5.2.8) with $B_1 := \not{A}\psi$, $B_2 := \overline{\psi}\not{A}$ and $B_3 := L, \ldots, B_n := L$ reads

$$\partial_\mu^y t(\not{A}\psi, \overline{\psi}\not{A}, L, \ldots, L, j^\mu)(x_1 - y, x_2 - y, x_3 - y, \ldots, x_n - y) \\ = \hbar\big(\delta(y - x_2) - \delta(y - x_1)\big)\, t(\not{A}\psi, \overline{\psi}\not{A}, L, \ldots, L)(x_1 - x_n, x_2 - x_n, x_3 - x_n, \ldots)\,. \tag{5.2.9}$$

Remark 5.2.1 (Derivation of the Ward identity (5.2.9) from gauge invariance of the S-matrix). Usually the Ward identity (5.2.9) is derived as follows (see, e.g., [148, Chap. 4.6]) – for simplicity we study the case of lowest order, that is, $n = 2$ in (5.2.9). One starts with the condition

$$\partial_\mu^y T\big(L(x_1) \otimes L(x_2) \otimes j^\mu(y)\big)\Big|_{\mathcal{C}_{S_0}} = 0\;, \tag{5.2.10}$$

which is motivated by invariance of the on-shell S-matrix in the formal adiabatic limit with respect to a gauge (or BRST) transformation of the asymptotically free fields (see Definition 5.5.1 for the BRST transformation of the free theory). We point out that (5.2.10) is the on-shell version of the QED-MWI (5.2.8) for $n = 2$ and $B_1 = L = B_2$, since $\theta L = 0$.

After expanding $T\big(L(x_1) \otimes L(x_2) \otimes j^\mu(y)\big)$ by the causal Wick expansion, one then selects from (5.2.10) all terms which have precisely one pair $(\overline{\psi}, \psi)$ and no further fields; the following contraction patterns of $T\big(L(x_1) \otimes L(x_2) \otimes j^\mu(y)\big)$ contribute:

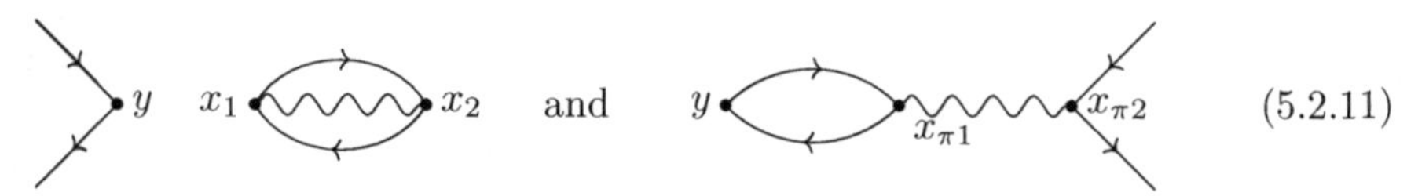

(5.2.11)

as well as

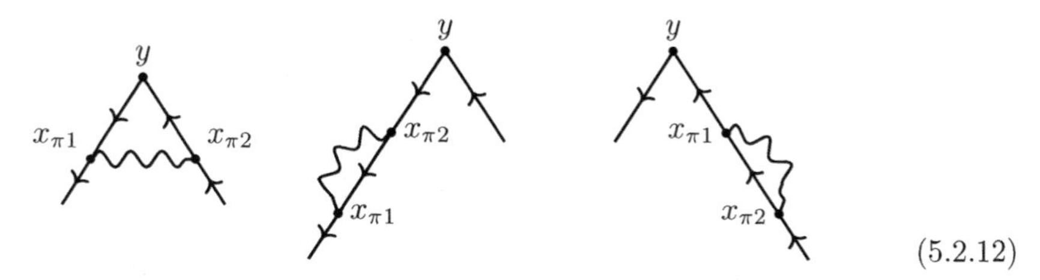

(5.2.12)

where $\pi \in S_2$. Considering the l.h.s. of (5.2.10), the two diagrams in (5.2.11) do not contribute: For the first one this is due to $\partial_\mu j^\mu(y)|_{\mathcal{C}_{S_0}} = 0$ and for the second one this follows from $\partial_\mu^y t(j^\mu, j^\nu)(y - x_{\pi 1}) = 0$ (5.1.155), which is the VEV of the QED-MWI (5.2.8) for $n = 1$ and $B_1 = j^\nu$ (since $\theta j^\nu = 0$) and, when working without the QED-MWI, can also be motivated by gauge invariance of the S-matrix in the formal adiabatic limit.

So there remain the contributions of the three diagrams in (5.2.12):

$$0 = \Big(\overline{\psi}(x_1)\,\partial_\mu^y t(A\!\!\!/\psi,\overline{\psi}A\!\!\!/,j^\mu)(x_1-y,x_2-y)\,\psi(x_2) \\ +\hbar\,\overline{\psi}(x_1)\,t(A\!\!\!/\psi,\overline{\psi}A\!\!\!/)(x_1-x_2)\,\partial_\mu^y[S^F(x_2-y)\gamma^\mu\psi(y)] \\ +\hbar\,\partial_\mu^y[\overline{\psi}(y)\gamma^\mu S^F(y-x_1)]\,t(A\!\!\!/\psi,\overline{\psi}A\!\!\!/)(x_1-x_2)\,\psi(x_2)\Big)\Big|_{\mathcal{C}_{S_0}} + (x_1\leftrightarrow x_2)\,. \tag{5.2.13}$$

Adding $0 = im\,S^F(x_2-y)\psi(y) - im\,S^F(x_2-y)\psi(y)$ to the term in the first $[\cdots]$-bracket and proceeding analogously for the second $[\cdots]$-bracket, we can insert the (adjoint) Dirac equation (5.1.29) and the corresponding relation for S^F (5.1.65). Hence, the identity (5.2.13) simplifies to

$$0 = \overline{\psi}(x_1)\,\Big[\partial_\mu^y t(A\!\!\!/\psi,\overline{\psi}A\!\!\!/,j^\mu)(x_1-y,x_2-y) - \hbar\,t(A\!\!\!/\psi,\overline{\psi}A\!\!\!/)(x_1-x_2)\,\delta(x_2-y) \\ +\hbar\,\delta(y-x_1)\,t(A\!\!\!/\psi,\overline{\psi}A\!\!\!/)(x_1-x_2)\Big]\,\psi(x_2)\Big|_{\mathcal{C}_{S_0}} + (x_1\leftrightarrow x_2)\,. \tag{5.2.14}$$

It follows that the term in $[\cdots]$-brackets must vanish – this is the particular QED-MWI (5.2.9) to order $n=2$.

We point out: To derive the Ward identity (5.2.9) the procedure in (5.2.9) is much faster than the derivation given here; because the QED-MWI (5.2.8) is more general than the identity (5.2.10) (and its generalization to higher orders) expressing gauge invariance of the S-matrix, since in the QED-MWI $B_1,\ldots,B_n$ may be true submonomials of L.

Because the MWI holds always true in classical field theory, it is satisfied for the contribution of all (connected) tree diagrams to the QED-MWI (5.2.7) – see Claim 3.1.28 in Sect. 3.1.7. We illustrate this statement by a particular simple case of the QED-MWI, in which solely tree diagrams appear:

Exercise 5.2.2. Verify explicitly the MWI (5.2.7) for $n=1$ and $B_1 = \overline{\psi}A\!\!\!/$.

[*Solution*: By using

$$R\big(\overline{\psi}A\!\!\!/(x_1);j^\mu(x)\big) = -\hbar\,\overline{\psi}(x)\gamma^\mu S^{\mathrm{ret}}(x-x_1)A\!\!\!/(x_1)$$

and $(\partial\!\!\!/ + im)S^{\mathrm{ret}}(z) = -i\delta(z)$, we get

$$\partial_\mu^x R\big(\overline{\psi}A\!\!\!/(x_1);j^\mu(x)\big) \\ = -\hbar\,\big(\partial_x^\mu\overline{\psi}(x)\gamma_\mu - im\overline{\psi}(x)\big)S^{\mathrm{ret}}(x-x_1)A\!\!\!/(x_1) - \hbar\,\overline{\psi}(x)(\partial\!\!\!/_x + im)S^{\mathrm{ret}}(x-x_1)A\!\!\!/(x_1) \\ = -\big(\partial_x^\mu\overline{\psi}(x)\gamma_\mu - im\overline{\psi}(x)\big)\wedge\Big[\hbar S^{\mathrm{ret}}(x-x_1)A\!\!\!/(x_1)\Big] - i\hbar\,\delta(x-x_1)\,\theta(\overline{\psi}A\!\!\!/)(x_1)\,.$$

Since the $[\ldots]$-bracket is equal to $R\big(\overline{\psi}A\!\!\!/(x_1);\psi(x)\big)$ and since $R\big(\overline{\psi}A\!\!\!/(x_1);\overline{\psi}(x)\big)=0$, this equation agrees with the MWI (5.2.7) for the considered case.]

The result of the following exercise will be used in Sect. 5.3.

Exercise 5.2.3. Let B_j, B be of the form (5.2.4) and let $\tilde{B}_j$, $\tilde{B}$ be the corresponding bosonic field polynomials defined as in (5.2.8). Derive from the QED-MWI (5.2.7) the on-shell Ward identity

$$\begin{aligned} &\partial^\mu_y\, R\big(\tilde{B}_1(x_1)\otimes\cdots\otimes\tilde{B}_{n-1}(x_{n-1})\otimes j_\mu(y);\tilde{B}(x)\big)\Big|_{\mathcal{C}_{S_0}} \qquad (5.2.15)\\ &\quad = -i\hbar\,\delta(y-x)\,R\big(\tilde{B}_1(x_1)\otimes\cdots\otimes\tilde{B}_{n-1}(x_{n-1});\widetilde{\theta B}(x)\big)\Big|_{\mathcal{C}_{S_0}}\\ &\qquad -i\hbar\sum_{l=1}^{n-1}\delta(y-x_l)\,R\big(\tilde{B}_1(x_1)\otimes\cdots\otimes\widetilde{\theta B_l}(x_l)\otimes\cdots\otimes\tilde{B}_{n-1}(x_{n-1});\tilde{B}(x)\big)\Big|_{\mathcal{C}_{S_0}}. \end{aligned}$$

[*Solution*: By using the GLZ relation and the QED-MWI, the l.h.s. of (5.2.15) is equal to

$$\begin{aligned} &\partial^\mu_y\, R\big(\tilde{B}_1(x_1)\otimes\cdots\otimes\tilde{B}_{n-1}(x_{n-1})\otimes\tilde{B}(x);j_\mu(y)\big)\Big|_{\mathcal{C}_{S_0}}\\ &\quad -i\sum_{I\subseteq\{1,\dots,n-1\}}\Big[\partial^\mu_y\,R\big(\tilde{B}_I(x_I);j_\mu(y)\big)\,,\,R\big(\tilde{B}_{I^c}(x_{I^c});\tilde{B}(x)\big)\Big]\Big|_{\mathcal{C}_{S_0}}\\ &= -i\hbar\bigg(\delta(y-x)\,R\big(\tilde{B}_1(x_1)\otimes\cdots\otimes\tilde{B}_{n-1}(x_{n-1});\widetilde{\theta B}(x)\big)\\ &\quad +\sum_{l=1}^{n-1}\delta(y-x_l)\,R\big(\tilde{B}_1(x_1)\otimes\cdots\hat{l}\cdots\otimes\tilde{B}_{n-1}(x_{n-1})\otimes\tilde{B}(x);\widetilde{\theta B_l}(x_l)\big)\\ &\quad -i\sum_{I\subseteq\{1,\dots,n-1\}}\sum_{l\in I}\delta(y-x_l)\Big[R\big(\tilde{B}_{I\setminus l}(x_{I\setminus l});\widetilde{\theta B_l}(x_l)\big)\,,\,R\big(\tilde{B}_{I^c}(x_{I^c});\tilde{B}(x)\big)\Big]\bigg)\Big|_{\mathcal{C}_{S_0}} \end{aligned}$$

where $\tilde{B}_I(x_I) := \otimes_{j\in I}\tilde{B}_j(x_j)$ and $I\setminus l := I\setminus\{l\}$. Reordering the sums the last term (i.e., the commutator term) is equal to

$$-i\sum_{l=1}^{n-1}\delta(y-x_l)\sum_{I\subseteq\{1,\dots\hat{l}\dots,n-1\}}\Big[R\big(\tilde{B}_I(x_I);\widetilde{\theta B_l}(x_l)\big)\,,\,R\big(\tilde{B}_{I^c}(x_{I^c});\tilde{B}(x)\big)\Big]\Big|_{\mathcal{C}_{S_0}}$$

where now $I^c := \{1,\dots\hat{l}\dots,n-1\}\setminus I$. Using again the GLZ relation, we obtain the r.h.s. of (5.2.15).]

5.2.2 Proof of the QED-MWI

Essentially we follow [48, App. B].

For technical reasons we work with the time-ordered product and – as in (5.2.8) and Exer. 5.2.3 – with the bosonic fields $\tilde{B}_j$. We use the axiom "Scaling degree" instead of "Sm-expansion". Proceeding by induction on the order in the fields $\tilde{B}_j$, we assume that the QED-MWI is satisfied to lower orders $k<n$. With

that, there is some simplification in the formula for the anomalous term to order n (4.3.11):

$$\begin{aligned}\frac{\hbar^n}{i^n}\,\Delta\big(\tilde{B}_1(x_1),\ldots,\tilde{B}_n(x_n);y\big) &:= -\partial_y^\mu\, T\big(\tilde{B}_1(x_1)\otimes\cdots\otimes\tilde{B}_n(x_n)\otimes j_\mu(y)\big)\\ &\quad -\hbar\sum_{l=1}^n \delta(y-x_l)\,T\big(\tilde{B}_1(x_1)\otimes\cdots\otimes\widetilde{\theta B_l}(x_l)\otimes\cdots\otimes\tilde{B}_n(x_n)\big)\\ &\quad +\big(\partial_y^\mu\overline{\psi}(y)\gamma_\mu - im\overline{\psi}(y)\big)\wedge T\big(\psi(y)\otimes\tilde{B}_1(x_1)\otimes\cdots\otimes\tilde{B}_n(x_n)\big)\\ &\quad +T\big(\tilde{B}_1(x_1)\otimes\cdots\otimes\tilde{B}_n(x_n)\otimes\overline{\psi}(y)\big)\wedge(\not{\partial}_y+im)\psi(y)\ . \qquad (5.2.16)\end{aligned}$$

We will show that $\Delta(\tilde{B}_1(x_1),\ldots,\tilde{B}_n(x_n);y)$ can be removed by finite renormalizations of the T-product which maintain the other renormalization conditions.

Step 1: Charge number conservation. Let B_j be of the form (5.2.4) with r_j factors ψ and s_j factors $\overline{\psi}$, i.e., $\theta B_j = (r_j - s_j)\,B_j$.

In this first step we will prove that we can easily satisfy Charge number conservation (5.1.101) for $T\big(B_1(x_1)\otimes\cdots\otimes B_n(x_n)\big)$, that is,

$$\theta\, T\big(B_1(x_1)\otimes\cdots\otimes B_n(x_n)\big) = T\big(B_1(x_1)\otimes\cdots\otimes B_n(x_n)\big)\cdot\sum_{j=1}^n (r_j - s_j) \quad (5.2.17)$$

for all $B_1,\ldots,B_n$ of the mentioned form. To obtain a necessary condition for the requirement (5.2.17), we consider its VEV: Using $\omega_0\circ\theta = 0$ we get $t(B_1,\ldots,B_n)\cdot\sum_{j=1}^n (r_j - s_j) = 0$, that is,

$$t(B_1,\ldots,B_n) = 0 \quad\text{if}\quad \sum_{j=1}^n (r_j - s_j) \neq 0\ . \qquad (5.2.18)$$

By means of (5.1.100), one easily verifies that in the inductive construction of the T-product (see Sect. 3.3), the property (5.2.18) can get lost only in the extension to the thin diagonal, i.e., (5.2.18) is a renormalization condition. To fulfill it, we simply extend zero by zero – this is compatible with all other renormalization conditions.

We are now going to show that (5.2.18) is also sufficient for (5.2.17). We use the causal Wick expansion:

$$T\big(B_1(x_1)\otimes\ldots\otimes B_n(x_n)\big) = \sum_{\underline{B}_l\subseteq B_l} t(\underline{B}_1,\ldots,\underline{B}_n)(x_1-x_n,\ldots)\,\overline{B}_1(x_1)\wedge\cdots\wedge\overline{B}_n(x_n)\ .$$

Let $\underline{r}_j$ and $\overline{r}^j$ be the number of factors ψ appearing in $\underline{B}_j$ and $\overline{B}_j$, respectively, and let $\underline{s}_j$ and $\overline{s}^j$ be defined analogously. Obviously it holds that

$$\underline{r}_j + \overline{r}_j = r_j\ ,\qquad \underline{s}_j + \overline{s}_j = s_j\ .$$

From (5.2.18) we know that all non-vanishing terms fulfill $\sum_j(\underline{r}_j - \underline{s}_j) = 0$. Using these relations we obtain

$$\theta\Big(t(\underline{B}_1,\dots,\underline{B}_n)\,\overline{B}_1\wedge\cdots\wedge\overline{B}_n\Big) = t(\underline{B}_1,\dots,\underline{B}_n)\,\overline{B}_1\wedge\cdots\wedge\overline{B}_n\cdot\sum_j(\overline{r}_j-\overline{s}_j)$$
$$= \Big(t(\underline{B}_1,\dots,\underline{B}_n)\,\overline{B}_1\wedge\cdots\wedge\overline{B}_n\Big)\cdot\sum_j(r_j-s_j)\ ,$$

which yields the assertion (5.2.17). In the following we assume that (5.2.17) holds true.

Remark 5.2.4. Comparing (5.2.18) with (3.1.34) we see why Charge number conservation corresponds to the Field parity axiom. Namely, for the unrenormalized R- or T-product, given by formulas (3.1.69) or (3.3.10), respectively, both requirements are automatically satisfied for the same reason: $\omega_0\big(T^{\mathrm{unrenorm}}(B_1,\dots,B_n)\big)$ vanishes, if it is not possible to contract *all* basic fields appearing in $B_1,\dots,B_n$.

Step 2: Vanishing of the integrated anomalous term. In this step we prove

$$\int dy\ \Delta\big(\tilde{B}_1(x_1),\dots,\tilde{B}_n(x_n);y\big)_0 = 0\ , \tag{5.2.19}$$

where we use the notation $F_0 := F|_{C_{S_0}}$ for $F\in\mathcal{F}_{\mathrm{QED}}$.

For a given configuration $(x_1,\dots,x_n)\in\mathbb{M}^n$ let $\mathcal{O}\subset\mathbb{M}$ be an open double cone (3.7.7) which contains the points $x_1,\dots,x_n$ and let g be a test function which is equal to 1 on a neighbourhood of $\overline{\mathcal{O}}$. With that and (5.2.16), and due to the Locality of Δ (4.3.4), we may write

$$\frac{\hbar^n}{i^n}\int dy\ \Delta\big(\tilde{B}_1(x_1),\dots,\tilde{B}_n(x_n);y\big)_0 = \frac{\hbar^n}{i^n}\int dy\ g(y)\,\Delta\big(\tilde{B}_1(x_1),\dots,\tilde{B}_n(x_n);y\big)_0$$
$$= -\int dy\ g(y)\,\partial^\mu_y\,T\big(\tilde{B}_1(x_1)\otimes\cdots\otimes\tilde{B}_n(x_n)\otimes j_\mu(y)\big)_0$$
$$-\hbar\sum_{l=1}^n T\big(\tilde{B}_1(x_1)\otimes\cdots\otimes\widetilde{\theta B_l}(x_l)\otimes\cdots\otimes\tilde{B}_n(x_n)\big)_0\ . \tag{5.2.20}$$

Now we decompose $\partial^\mu g = a^\mu - b^\mu$ such that $\operatorname{supp} a^\mu\cap(\mathcal{O}+\overline{V}_-)=\emptyset$ and $\operatorname{supp} b^\mu\cap(\mathcal{O}+\overline{V}_+)=\emptyset$. By causal factorization of the T-product, the first term on the r.h.s. of (5.2.20) becomes equal to

$$j^\mu(a_\mu)_0\star T\big(\tilde{B}_1(x_1)\otimes\cdots\otimes\tilde{B}_n(x_n)\big)_0 - T\big(\tilde{B}_1(x_1)\otimes\cdots\otimes\tilde{B}_n(x_n)\big)_0\star j^\mu(b_\mu)_0$$
$$= [j^\mu(a_\mu)_0\,,\,T\big(\tilde{B}_1(x_1)\otimes\cdots\otimes\tilde{B}_n(x_n)\big)_0]_\star + T\big(\tilde{B}_1(x_1)\cdots\tilde{B}_n(x_n)\big)_0\star j^\mu(\partial_\mu g)_0.$$

The second term vanishes because $\partial_\mu j^\mu_0 = 0$. From the Field independence of T we know $\operatorname{supp} T\big(\tilde{B}_1(x_1)\otimes\cdots\otimes\tilde{B}_n(x_n)\big)_0\subset\mathcal{O}$; therefore, we may vary a^μ in the set

$$\{\,z\in\mathbb{M}\,|\,(z-x)^2<0\ \forall x\in\mathcal{O}\,\}\quad\text{without affecting}\quad [j^\mu(a_\mu)_0, T\big(\tilde{B}_1(x_1)\otimes\cdots\big)_0]_\star\ ,$$

due to spacelike commutativity (Lemma 2.4.1). Hence, we may choose for a_μ a smooth approximation to $-\delta_{\mu 0}\,\delta(x^0-c)$, where $c\in\mathbb{R}$ is a sufficiently large constant:

$$a_\mu(x) = -\delta_{\mu 0}\, h(x^0) \quad \text{with} \quad \int dx^0\; h(x^0) = 1\;, \quad h\in\mathcal{D}([c-\varepsilon, c+\varepsilon])$$

for some $\varepsilon > 0$. Taking into account additionally that the integral

$$[Q_0^\psi\,,\, T(\ldots)_0]_\star := \int d\vec{x}\; [j^0(x^0,\vec{x})_0\,,\, T(\ldots)_0]_\star$$

exists and does not depend on x^0 – as we know from Exer. 5.1.7 – we obtain

$$\begin{aligned}[j^\mu(a_\mu)_0\,,\, T(\ldots)_0]_\star &= -\int dx^0\; h(x^0) \int d\vec{x}\; [j^0(x^0,\vec{x})_0\,,\, T(\ldots)_0]_\star \\ &= -\,[Q_0^\psi\,,\, T(\ldots)_0]_\star\;.\end{aligned}$$

From that exercise we also know that

$$-[Q_0^\psi\,,\, F_0]_\star = \hbar\,(\theta F)_0 \qquad \forall F\in\mathcal{F}\;. \tag{5.2.21}$$

Summing up we have

$$\begin{aligned}\frac{\hbar^n}{i^n}\int dy\; \Delta\big(\tilde{B}_1(x_1),\ldots,\tilde{B}_n(x_n);y\big)_0 &= \hbar\,\theta\, T\big(\tilde{B}_1(x_1)\otimes\cdots\otimes \tilde{B}_n(x_n)\big)_0 \\ &\quad -\hbar\sum_{l=1}^n T\big(\tilde{B}_1(x_1)\otimes\cdots\otimes \widetilde{\theta B_l}(x_l)\otimes\cdots\otimes\tilde{B}_n(x_n)\big)_0\;.\end{aligned}$$

The r.h.s. vanishes due to (5.2.17).

Step 3: Structure of the anomalous term. From Theorem 4.3.1 we know that the anomalous term (5.2.16) satisfies the causal Wick expansion:

$$\Delta\big(\tilde{B}_1(x_1),\ldots,\tilde{B}_n(x_n);y\big) = \sum_{\underline{B}_l\subseteq B_l} d(\underline{\tilde{B}}_1,\ldots,\underline{\tilde{B}}_n)(x_1-y,\ldots)\,\overline{\tilde{B}}_1(x_1)\wedge\cdots\wedge\overline{\tilde{B}}_n(x_n)\;, \tag{5.2.22}$$

where

$$d(\tilde{B}_1,\ldots,\tilde{B}_n)(x_1-y,\ldots,x_n-y) := \omega_0\Big(\Delta\big(\tilde{B}_1(x_1),\ldots,\tilde{B}_n(x_n);y\big)\Big)\in\mathcal{D}'(\mathbb{R}^{4n})\;. \tag{5.2.23}$$

Here we have used that there is no true submonomial of $Q=i\psi$ (or of $Q^\dagger\gamma^0=-i\overline{\psi}$) contributing to the causal Wick expansion; because the only submonomial of such a kind is the constant field $\underline{Q}(y)=c=$ constant, and because $\Delta(\ldots;c)=0$ (4.3.7) due to the Off-shell field equation axiom.

A further consequence of the latter axiom is worked out in Exer. 4.3.3; this result holds also in QED. Therefore, on the r.h.s. of (5.2.22) the sum is restricted to submonomials $\underline{B_l}$ of B_l which are at least bilinear in the basic fields for all l.

Summing up we conclude: To prove the QED-MWI (5.2.8) for all $B_1, \ldots, B_n \subsetneq L$, we only have to find admissible finite renormalizations which remove the VEVs $d(\tilde{B}_1, \ldots, \tilde{B}_n)$ (5.2.23) for all $B_1, \ldots, B_n \in \{j^\mu, \overline{\psi} A\!\!\!/, A\!\!\!/\psi, L\}$.

By Locality of the anomaly map Δ (4.3.4), the distributions $d(\tilde{B}_1, \ldots, \tilde{B}_n)$ are of the form

$$d(\tilde{B}_1, \ldots, \tilde{B}_n)(x_1 - y, \ldots, x_n - y) = \sum_a C_a(B_1, \ldots, B_n)\, \partial^a \delta(x_1 - y, \ldots, x_n - y) \tag{5.2.24}$$

with coefficients $C_a(B_1, \ldots, B_n) \in \mathbb{C}$. Now, for the vacuum expectation values the restriction to $\mathcal{C}_{S_0}$ in (5.2.19) is irrelevant, hence we know that

$$\int dy\; d(\tilde{B}_1, \ldots, \tilde{B}_n)(x_1 - y, \ldots, x_n - y) = 0 \; .$$

By means of Lemma 4.5.1 we conclude that the sum over a in (5.2.24) is restricted to $|a| \geq 1$. On the other hand, from (4.3.6) we have also an upper bound on $|a|$. Summing up, the sum is restricted to

$$1 \leq |a| \leq \sum_{j=1}^{n} \dim B_j + 4 - 4n =: \omega(B_1, \ldots, B_n) \; . \tag{5.2.25}$$

To be precise, formula (4.3.6) holds only for scalar fields: In general the term $\frac{d+2}{2}$ is replaced by $\dim \frac{\delta S_0}{\delta \phi}$, where ϕ is the relevant basic field. For the QED-MWI we have $\dim Q = \frac{3}{2} = \dim Q^\dagger \gamma^0$ and $\dim \frac{\delta S_0}{\delta \psi} = \frac{5}{2} = \dim \frac{\delta S_0}{\delta \overline{\psi}}$.

From Theorem 4.3.1 we know that $d(\tilde{B}_1, \ldots, \tilde{B}_n)$ is Lorentz covariant and respects the $*$-structure:

$$d(\tilde{B}_1^*, \ldots, \tilde{B}_n^*)(x_1 - y, \ldots, x_n - y) = d(\tilde{B}_1, \ldots, \tilde{B}_n)(x_1 - y, \ldots, x_n - y)^c \; , \tag{5.2.26}$$

where the upper index c denotes complex conjugation. This can be seen also directly from (5.2.16) – or, for the $*$-structure it is simpler to argue with the corresponding formula for the R-product (instead of the T-product).

Step 4: Removing the anomalous term by finite renormalizations. In detail, Lemma 4.5.1 states that we can write $d(\tilde{B}_1, \ldots, \tilde{B}_n)$ as

$$d(\tilde{B}_1, \ldots, \tilde{B}_n)(x_1 - y, \ldots, x_n - y) = \partial^y_\mu u^\mu(\tilde{B}_1, \ldots, \tilde{B}_n)(x_1 - y, \ldots, x_n - y) \tag{5.2.27}$$

where $u^\mu(\tilde{B}_1, \ldots, \tilde{B}_n)$ is of the form

$$u^\mu(\tilde{B}_1, \ldots, \tilde{B}_n)(x_1 - y, \ldots) = \sum_{|a| \leq \omega(B_1, \ldots, B_n) - 1} \tilde{C}_a(B_1, \ldots, B_n)\, \partial^a \delta(x_1 - y, \ldots) \tag{5.2.28}$$

and the upper bound (5.2.25) is also taken into account. Obviously, $u^\mu(\tilde{B}_1,\ldots,\tilde{B}_n)$ can be chosen such that it is Lorentz covariant. With all that, the replacement

$$t(\tilde{B}_1,\ldots,\tilde{B}_n,j^\mu) \;\to\; t(\tilde{B}_1,\ldots,\tilde{B}_n,j^\mu)+\frac{\hbar^n}{i^n}\,u^\mu(\tilde{B}_1,\ldots,\tilde{B}_n) \tag{5.2.29}$$

removes the anomalous term $d(\tilde{B}_1,\ldots,\tilde{B}_n)$, as we see from (5.2.16). Under this replacement nearly all axioms for the T-product remain satisfied, in particular Causality and the Scaling degree condition

$$\operatorname{sd} t(\tilde{B}_1,\ldots,\tilde{B}_n,j^\mu)\le\sum_{j=1}^n \dim B_j+\dim j^\mu=\omega(B_1,\ldots,B_n)-1+4n\;;$$

the latter is due to the upper bound on $|a|$ in (5.2.28). But (5.2.29) is only an admissible finite renormalization if $u^\mu(\tilde{B}_1,\ldots,\tilde{B}_n)$ has the same symmetries as required for $t(\tilde{B}_1,\ldots,\tilde{B}_n,j^\mu)$. Especially if there are factors $j^{\mu_l}(x_l)$ among $B_1(x_1),\ldots,B_n(x_n)$ the permutation symmetry with respect to $(y,\mu)\leftrightarrow(x_l,\mu_l)$ must be maintained (for all l). There is a prominent counterexample where this is impossible: In axial QED, introduced in Exer. 5.1.17, the Ward identity

$$\partial^y_\mu\, t(j_a^{\mu_1},j_a^{\mu_2},j_a^\mu)(x_1-y,x_2-y)\stackrel{?}{=}0 \tag{5.2.30}$$

is part of the relevant MWI (i.e., the MWI corresponding to the QED-MWI), where, for simplicity, we consider the case of massless spinor fields: $m=0$. The condition (5.2.30) cannot be fulfilled by any admissible finite renormalization, because such a renormalization must maintain the invariance under permutations of (x_1,μ_1), (x_2,μ_2) and (y,μ) – this is the "axial anomaly". We point out that the latter is an experimentally measurable phenomenon; the theoretical predictions are in agreement with the experimental results.[160]

Note that in general $u^\mu(\tilde{B}_1,\ldots,\tilde{B}_n)$ (5.2.27) is not uniquely determined by $d(\tilde{B}_1,\ldots,\tilde{B}_n)$. We do not know a general argument (for non-axial QED) that for all possible anomalous terms $d(\tilde{B}_1,\ldots,\tilde{B}_n)$ one can find a $u^\mu(\tilde{B}_1,\ldots,\tilde{B}_n)$ which has the wanted symmetries. The famous Ward identity (5.2.9) has only one factor j; hence, the renormalization (5.2.29) maintains the symmetries in that case. Taking into account the results of Step 3, and that $t(L\ldots,L,j,j,j)=0$ (due to Furry's

[160] For a treatment of the axial anomaly in the framework of causal perturbation theory and its application to the decay of the neutral π^0-meson into two photons, see, e.g., [148, Sect. 5.3] for the case $m>0$ and [41] for $m=0$. To be precise, the π^0-decay concerns the case $m>0$, and it is not precisely the anomaly of (5.2.30) which yields this decay rate; the latter is obtained from the somewhat different anomaly

$$a^{\mu_1\mu_2}(x_1-y,x_2-y):=\partial^y_\mu\, t(j^{\mu_1},j^{\mu_2},j^\mu_a)(x_1-y,x_2-y)-2m\,t(j^{\mu_1},j^{\mu_2},j_\pi)(x_1-y,x_2-y)\,, \tag{5.2.31}$$

where j^μ is the usual electromagnetic current (5.1.27) and $j_\pi(y):=i\,\overline{\psi}(x)\wedge\gamma^5\psi(x)$. The anomaly (5.2.31) is also called "axial anomaly" and one can show that it is related to the anomaly of (5.2.30) by: $a^{\mu_1\mu_2}=3\cdot$ [anomaly of (5.2.30)].

Theorem, see Remk. 5.1.16), there remain only two cases for which it is unclear whether the procedure (5.2.27)–(5.2.29) is applicable: These are the cases with $(n-1)$ factors L and two factors j, or $(n-3)$ factors L and four factors j, that is, the cases

$$\begin{aligned}&\partial_\mu^y t(L\dots,L,j^\nu,j^\mu)(x_{11}-y,\dots,x_{1m}-y,x_2-y)\\&\quad=\sum_{1\le|a|\le 3} C_{1a}^\nu\,\partial^a\delta(x_{11}-y,\dots,x_{1m}-y,x_2-y)\,,\quad m:=n-1\,,\end{aligned}\tag{5.2.32}$$

and

$$\begin{aligned}&\partial_\mu^y t(L\dots,L,j^{\nu_1},j^{\nu_2},j^{\nu_3},j^\mu)(x_{11}-y,\dots,x_{1m}-y,x_{21}-y,\dots,x_{23}-y)\\&\quad=\sum_{|a|=1} C_{2a}^{\nu_1\nu_2\nu_3}\,\partial^a\prod_{l,j}\delta(x_{lj}-y)\,,\quad m:=n-3\,,\end{aligned}\tag{5.2.33}$$

of the anomalous QED-MWI (5.2.16). Due to (5.2.26) and the factor $(-i)^n$ appearing on the l.h.s. of (5.2.16), the coefficients $C_{ra}^{\dots}$, $r=1,2$, of the anomalous term are elements of $i^n\,\mathbb{R}$; in addition they are restricted by Lorentz covariance and the permutation symmetry in $x_{11},\dots,x_{1m}$. A further restriction can be obtained from the symmetry in the factors j of the terms on the l.h.s., e.g.,

$$\partial_\nu^{x_2}\partial_\mu^y t(L\dots,L,j^\nu,j^\mu)(x_{11}-y,\dots,x_{1m}-y,x_2-y)\quad\text{is symmetrical in } y,x_2.$$

By working this out one proves that in both cases (5.2.32) and (5.2.33) one can find a $u^\mu(\dots)$ (5.2.27)–(5.2.28) such that the renormalization (5.2.29) preserves the symmetries of $t(L\dots,L,j^\nu,j^\mu)$ and $t(L\dots,L,j^{\nu_1},j^{\nu_2},j^{\nu_3},j^\mu)$, respectively, see, e.g., [42]. □

5.2.3 Conservation of the interacting current and the corresponding charge

An immediate consequence of the QED-MWI (5.2.7) is that the interacting electromagnetic current $j^\mu_{\mathcal{L}(g)}(x)_0$ is conserved:

$$\partial_\mu^x\, j^\mu_{\mathcal{L}(g)}(x)_0=0\tag{5.2.34}$$

(where the lower index 0 means restriction to $\mathcal{C}_{S_0}$, i.e., $F_0:=F|_{\mathcal{C}_{S_0}}$), since $\theta L=0$. We emphasize that (5.2.34) holds true without performing any kind of adiabatic limit.

Exercise 5.2.5. Show that the current conservation (5.2.34) to first order in e is equivalent to the condition

$$\partial_\mu^y\, r(j^\nu,j^\mu)(y)=0\tag{5.2.35}$$

(where the notation (3.2.20) is used) and that the latter is equivalent to $\partial_\mu^y\, t(j^\mu,j^\nu)(y)=0$ (5.1.155).

[*Solution*: Using the causal Wick expansion and the fact that only connected diagrams contribute to an R-product (Claim 3.1.27 in Sect. 3.1.7) we obtain

$$R\big(L(x_1);j^\mu(x)\big)=r(j^\nu,j^\mu)(x_1-x)\,A_\nu(x_1) \qquad (5.2.36)$$
$$+\,\overline{\psi}(x_1)A\!\!\!/(x_1)r(\psi,\overline{\psi})(x_1-x)\gamma^\mu\psi(x)-\overline{\psi}(x)\gamma^\mu r(\overline{\psi},\psi)(x_1-x)A\!\!\!/(x_1)\psi(x_1)\;,$$

where

$$r(\psi,\overline{\psi})(y)=-\hbar\,S^{\mathrm{av}}(y)\quad\text{and}\quad r(\overline{\psi},\psi)(y)=\hbar\,S^{\mathrm{ret}}(-y)$$

as computed in (5.1.52). The contraction patterns belonging to the three terms appearing on the r.h.s. of (5.2.36) can be visualized by the Feynman diagrams

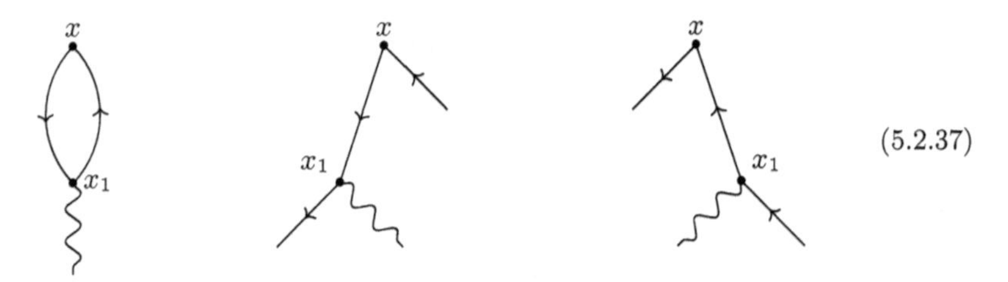

the diagrams are in the same order as the corresponding terms in (5.2.36). The computation of the l.h.s. of (5.2.34) amounts to

$$\partial^x_\mu\, R\big(L(x_1)\otimes j^\mu(x)\big)_0=\partial^x_\mu\, r(j^\nu,j^\mu)(x_1-x)\,A_\nu(x_1)_0$$
$$+\,i\hbar\,\overline{\psi}(x_1)_0A\!\!\!/(x_1)_0\Big(S^{\mathrm{av}}(x_1-x)(i\partial\!\!\!/_x-m)\psi(x)_0$$
$$+\,\big(i\partial^x_\mu\, S^{\mathrm{av}}(x_1-x)\gamma^\mu+m\,S^{\mathrm{av}}(x_1-x)\big)\psi(x)_0\Big)$$
$$+\,i\hbar\,\Big(\big(i\partial^x_\mu\,\overline{\psi}(x)_0\gamma^\mu+m\,\overline{\psi}(x)_0\big)S^{\mathrm{ret}}(x-x_1)$$
$$+\,\overline{\psi}(x)_0(i\partial\!\!\!/_x-m)S^{\mathrm{ret}}(x-x_1)\Big)A\!\!\!/(x_1)_0\psi(x_1)_0\;.$$

Due to the (adjoint) Dirac equation (5.1.29) and the corresponding relations for $S^{\mathrm{ret/av}}$ (5.1.55), all terms except the first one cancel. Hence the condition (5.2.34) to first order in e is equivalent to (5.2.35).
To prove that the latter is equivalent to the corresponding relation (5.1.155) for $t(j^\mu,j^\nu)$, we use Bogoliubov's formula (3.3.41) (or (5.1.73)):

$$i\,r(j^\nu,j^\mu)(x_1-x)+t(j^\nu,j^\mu)(x_1-x)=\omega_0\big(j^\nu(x_1)\star j^\mu(x)\big)\;.$$

The r.h.s. is computed in Exap. 5.1.4, its divergence $\partial^x_\mu\,\mathrm{tr}\big(\gamma^\nu S^+(x_1-x)\gamma^\mu S^+(x-x_1)\big)$ vanishes due to (5.1.31). This implies the assertion, since $t(j^\mu,j^\nu)(y)=t(j^\nu,j^\mu)(-y)$.]

In the following exercise we construct the charge $Q^\psi_{\mathcal{L}(g),0}$ belonging to the current conservation (5.2.34) and derive its most important property.

Exercise 5.2.6. [48] In Exer. 5.1.7 we have shown that the charge $\frac{-1}{\hbar}\,Q^\psi_0$ (5.1.41) implements the charge number operator θ (5.1.99) for non-interacting on-shell fields, that is,

Q_0^ψ fulfills (5.2.21). Prove that the QED-MWI implies the following generalization to interacting fields: Let $\mathcal{O}$ be an open double cone (3.7.7); then it holds that

$$\frac{-1}{\hbar}\,[Q^\psi_{\mathcal{L}(g),0}\,,\,B_{\mathcal{L}(g)}(x)_0]_\star = (\theta B)_{\mathcal{L}(g)}(x)_0\ , \quad \forall x\in\mathcal{O}$$

and for all submonomials B of $L=j^\mu A_\mu$. The charge $Q^\psi_{\mathcal{L}(g),0}$ appearing here is defined by

$$Q^\psi_{\mathcal{L}(g),0} := j^0_{\mathcal{L}(g)}(f)_0\ ,$$

where the test function $f\in\mathcal{D}(\mathbb{R}^4)$ is of the following form: There exists $h\in\mathcal{D}(\mathbb{R})$ such that

$$f(y)=h(y_0)\quad \forall y=(y_0,\vec{y})\in\overline{\mathcal{O}}+(\overline{V}_+\cup\overline{V}_-)$$

and

$$\int dy_0\ h(y_0)=1\ .$$

[*Solution*: Due to spacelike commutativity of the interacting fields (see Remk. 3.1.6) we may write

$$[Q^\psi_{\mathcal{L}(g),0}\,,\,B_{\mathcal{L}(g)}(x)_0]_\star = \int d^4y\ h(y_0)\,[j^0_{\mathcal{L}(g)}(y)_0\,,\,B_{\mathcal{L}(g)}(x)_0]_\star \quad \text{for}\quad x\in\mathcal{O}\ .$$

By using the GLZ relation, this is equal to

$$\begin{aligned}
&=\sum_{n=0}^\infty \frac{i\,e^n}{\hbar^n\,n!}\int dy_1\cdots dy_n\ g(y_1)\cdots g(y_n)\int dy\\
&\quad\cdot\Big\{\Big([h(y_0)-h(y_0-a)]+h(y_0-a)\Big)\,R\big(L(y_1),\ldots,L(y_n),j^0(y);B(x)\big)_0\\
&\quad-\Big([h(y_0)-h(y_0-b)]+h(y_0-b)\Big)\,R\big(L(y_1),\ldots,L(y_n),B(x);j^0(y)\big)_0\Big\}\ .
\end{aligned}$$

Taking into account the causal support of the R-products (axiom (d)), we can choose a and b such that the contributions from $h(y_0-a)$ and $h(y_0-b)$ vanish, i.e.,

$$\text{supp}\big(y\mapsto h(y_0-a)\big)\cap(\overline{\mathcal{O}}+\overline{V}_-)=\emptyset\ ,\quad \text{supp}\big(y\mapsto h(y_0-b)\big)\cap(\overline{\mathcal{O}}+\overline{V}_+)=\emptyset\ . \tag{5.2.38}$$

Setting

$$k(y)\equiv k_0(y_0) := \int_{-\infty}^{y_0} dz\ [h(z)-h(z-a)]\ ,$$

the $[h(y_0)-h(y_0-a)]$-term is equal to

$$\begin{aligned}
&\ldots\int dy\ \partial_0^y k(y)\,R\big(L(y_1),\ldots,j^0(y);B(x)\big)_0\\
&=\ldots\int dy\ \partial_\mu^y k(y)\,R\big(L(y_1),\ldots,j^\mu(y);B(x)\big)_0\\
&=-\ldots\int dy\ k(y)\,\partial_\mu^y R\big(L(y_1),\ldots,j^\mu(y);B(x)\big)_0\\
&=\ldots i\hbar\int dy\ k(y)\,\delta(y-x)\,R\big(L(y_1),\ldots;(\theta B)(x)\big)_0\\
&=-\hbar\,k(x)\,(\theta B)_{\mathcal{L}(g)}(x)_0\ ,
\end{aligned}$$

where we have inserted the version (5.2.15) of the QED-MWI and $\theta L=0$ is taken into account.

To compute the $[h(y_0) - h(y_0 - b)]$-term we proceed analogously: We set

$$l(y) := \int_{-\infty}^{y_0} dz\ [h(z-a) - h(z-b)]\ ;$$

with that, this term is equal to

$$\begin{aligned}
&= \dots \int dy\ \big(k(y) + l(y)\big)\, \partial_\mu^y R\big(L(y_1), \dots, B(x); j^\mu(y)\big)_0 \\
&= - \dots i\hbar \int dy\ \big(k(y) + l(y)\big)\, \delta(y-x)\, R\big(L(y_1), \dots; (\theta B)(x)\big)_0 \\
&= \hbar\, \big(k(x) + l(x)\big)\, (\theta B)_{\mathcal{L}(g)}(x)_0\ ,
\end{aligned}$$

by using the QED-MWI (5.2.7). From (5.2.38) we conclude that for $x \in \mathcal{O}$ we have

$$l(x) = -\int_{-\infty}^{\infty} dz\ h(z-b) = -1\ .$$

Combining these results, the proof is complete.]

5.3 The local algebras of interacting fields

The reference for this section is [48].

We are going to derive algebraic properties of the interacting fields $B_{\mathcal{L}(g)}(x)$, where $B = A^\mu, \psi, \overline{\psi}, j^\mu, \dots$ and $\mathcal{L}(g) \in \mathcal{F}_{\mathrm{QED}}^{\mathrm{loc}}$ is the QED-interaction (5.1.102). Mostly, we will study the corresponding on-shell fields $B_{\mathcal{L}(g)}(x)_0 := B_{\mathcal{L}(g)}(x)|_{\mathcal{C}_{S_0}}$.

$*$-operation and field equations. In order that $\mathcal{L}(g)^* = \mathcal{L}(g)$, we assume that g is real-valued. Then, the $*$-structure axiom implies the relations

$$A^\mu_{\mathcal{L}(g)}(x)^* = A^\mu_{\mathcal{L}(g)}(x)\ , \qquad j^\mu_{\mathcal{L}(g)}(x)^* = j^\mu_{\mathcal{L}(g)}(x), \qquad \psi_{\mathcal{L}(g)}(x)^* \gamma^0 = \overline{\psi}_{\mathcal{L}(g)}(x)\ ,$$

by remembering that $\psi(x)^* := \psi^\dagger(x)$.

The off-shell field equation axiom yields the following on-shell field equations: the interacting Maxwell equation, the interacting Dirac equation and its adjoint, that is,

$$\begin{aligned}
&\Box_x A^\mu_{\mathcal{L}(g)}(x)_0 = -g(x) e\, j^\mu_{\mathcal{L}(g)}(x)_0, \\
&(i\partial\!\!\!/_x - m)\psi_{\mathcal{L}(g)}(x)_0 = -g(x) e\, (A\!\!\!/\psi)_{\mathcal{L}(g)}(x)_0\ , \\
&i\partial_x^\mu \overline{\psi}_{\mathcal{L}(g)}(x)_0 \gamma_\mu + m\overline{\psi}_{\mathcal{L}(g)}(x)_0 = g(x) e\, (\overline{\psi}A\!\!\!/)_{\mathcal{L}(g)}(x)_0\ .
\end{aligned} \tag{5.3.1}$$

In contrast to the conservation of the interacting current, given in (5.2.34), these field equations contain a trace of the adiabatic switching of the interaction: The elementary charge e is multiplied by $g(x)$.

This factor $g(x)$ is replaced by 1 in the algebraic adiabatic limit (introduced in Sect. 3.7): Let $\mathcal{O}$ be an open double cone and let $g \in \mathcal{D}(\mathbb{M})$ with

$$g(x) = 1 \qquad \forall x \text{ in a neighbourhood of } \overline{\mathcal{O}}. \tag{5.3.2}$$

The *algebra of interacting on-shell fields localized in* $\mathcal{O}$ is defined by

$$\tilde{\mathcal{F}}_0(\mathcal{O}) := \bigvee_\star \{ B_{\mathcal{L}(g)}(f)_0 \,|\, f \in \mathcal{D}(\mathcal{O}),\, B = A^\mu,\, \psi,\, \overline{\psi},\, \partial_\mu A^\mu,\, F^{\mu\nu},\, j^\mu,\, L, \dots \} \,, \tag{5.3.3}$$

where $B_{\mathcal{L}(g)}(f) := \int d^4x\; B_{\mathcal{L}(g)}(x) f(x)\,,\ F^{\mu\nu} := \partial^\mu A^\nu - \partial^\nu A^\mu$ and $\bigvee_\star$ means the algebra, under the $\star$-product, generated by the elements of the indicated set. Moreover, the lower index "0" denotes restriction to $\mathcal{C}_{S_0}$. In agreement with the notations used later on in Sect. 5.4.3, we write $\tilde{\mathcal{F}}_0(\mathcal{O})$ instead of $\mathcal{F}_0(\mathcal{O})$ – the tilde means "interacting".

The algebraic adiabatic limit can be summarized as follows: Due to Theorem 3.7.1, the algebraic relations of $\tilde{\mathcal{F}}(\mathcal{O})_0$ are independent of the behaviour of g outside of $\overline{\mathcal{O}}$. Since $\mathcal{O}$ is arbitrary, the full net of local field algebras can be constructed in this way without performing the adiabatic limit $g(x) \to 1$ in the traditional sense (see Remk. 3.7.3 and App. A.6). And, for $x \in \mathcal{O}$ we have the usual field equations, i.e., without the factor $g(x)$.

From the integrated field equation

$$A^\mu_{\mathcal{L}(g)}(x)_0 = A^\mu(x)_0 + e \int dy\; D^{\mathrm{ret}}(x-y)\, g(y)\, j^\mu_{\mathcal{L}(g)}(y)_0$$

and current conservation (5.2.34) we get

$$\partial^x_\mu A^\mu_{\mathcal{L}(g)}(x)_0 = \partial^x_\mu A^\mu(x)_0 + e \int dy\; D^{\mathrm{ret}}(x-y)\, (\partial_\mu g)(y)\, j^\mu_{\mathcal{L}(g)}(y)_0 \,. \tag{5.3.4}$$

Thus, performing formally the limit $g(y) \to 1\ \forall y \in x + \overline{V}_-$ ("partial adiabatic limit"), the interacting field $\partial^x_\mu A^\mu_{\mathcal{L}(g)}(x)_0$ agrees with the corresponding free one. However, note that the algebraic adiabatic limit does not suffice for that, even for $x \in \mathcal{O}$.

Commutators of interacting fields. By means of the GLZ relation we can explicitly compute commutators of interacting fields $B_{\mathcal{L}(g)}(f)_0 \in \tilde{\mathcal{F}}_0(\mathcal{O})$. For these commutators we will use the notions "advanced part" and "retarded part", introduced in Remk. 3.1.6. In the following proposition we compute some commutators of $\partial^x_\mu A^\mu_{\mathcal{L}(g)}(x)_0$ (where $x \in \mathcal{O}$) with elements of $\tilde{\mathcal{F}}_0(\mathcal{O})$.

Proposition 5.3.1 ([48]). *Let $\mathcal{O}$ and g be as described above in* (5.3.2) *and let the QED-MWI* (5.2.7) *be satisfied. For the interacting fields at points $x, y \in \mathcal{O}$ the* *following commutation relations hold true:*

$$\begin{aligned}
&\text{(a)} \quad [\partial_\mu^x A^\mu_{\mathcal{L}(g)}(x)_0\,,\, A^\nu_{\mathcal{L}(g)}(y)_0]_\star = -i\hbar\,\partial^\nu D(x-y) = [\partial_\mu A^\mu(x)\,,\, A^\nu(y)]_\star\,,\\
&\text{(b)} \quad [\partial_\mu^x A^\mu_{\mathcal{L}(g)}(x)_0\,,\, \partial_\nu A^\nu_{\mathcal{L}(g)}(y)_0]_\star = 0\,,\\
&\text{(c)} \quad [\partial_\mu^x A^\mu_{\mathcal{L}(g)}(x)_0\,,\, \psi_{\mathcal{L}(g)}(y)_0]_\star = -e\hbar^2\, D(x-y)\,\psi_{\mathcal{L}(g)}(y)_0\,,\\
&\text{(d)} \quad [\partial_\mu^x A^\mu_{\mathcal{L}(g)}(x)_0\,,\, \overline{\psi}_{\mathcal{L}(g)}(y)_0]_\star = e\hbar^2\, D(x-y)\,\overline{\psi}_{\mathcal{L}(g)}(y)_0\,,\\
&\text{(e)} \quad [\partial_\mu^x A^\mu_{\mathcal{L}(g)}(x)_0\,,\, L_{\mathcal{L}(g)}(y)_0]_\star = -i\hbar\,(\partial_\mu D)(x-y)\,j^\mu_{\mathcal{L}(g)}(y)_0\,,
\end{aligned}$$

where $L := j^\mu A_\mu$.

Proof. We will prove that the advanced part of the commutator on the l.h.s. (denoted by $[\cdot\,,\,\cdot]_\star^{\mathrm{av}}$) is equal to the advanced part of the r.h.s., which is obtained by replacing $D(x-y)$ by $-D^{\mathrm{av}}(x-y) := -D^{\mathrm{ret}}(y-x)$. The equality of the retarded parts can be shown analogously.

(a) and (b): The second equation in (a) is obvious. To compute

$$[A^\mu_{\mathcal{L}(g)}(x)_0\,,\, A^\nu_{\mathcal{L}(g)}(y)_0]_\star^{\mathrm{av}} = i\,R\big(e_\otimes^{\mathcal{L}(g)/\hbar}\otimes A^\mu(x); A^\nu(y)\big)_0\,, \tag{5.3.5}$$

we use the off-shell field equation in the form (3.2.61).[161] The first term on the r.h.s. of (3.2.61) contributes only to zeroth order in $\mathcal{L}(g)$, due to $\frac{\delta A^\nu(y)}{\delta A_\mu(z)} = g^{\mu\nu}\,\delta(y-z)$ and Remk. 3.1.4. With that, the expression (5.3.5) is equal to

$$i\hbar\Big(g^{\mu\nu}\,D^{\mathrm{av}}(x-y) + e\int dz\; D^{\mathrm{av}}(x-z)\,g(z)\,R\big(e_\otimes^{\mathcal{L}(g)/\hbar}\otimes j^\mu(z); A^\nu(y)\big)_0\Big)\,. \tag{5.3.6}$$

Due to the causal support of the R-product and $\operatorname{supp} D^{\mathrm{av}} \subseteq \overline{V}_-$, the z-integration is confined to the double cone

$$z \in \big((x+\overline{V}_+)\cap(y+\overline{V}_-)\big) \subset \mathcal{O}\,, \quad \text{therefore we may replace } g(z) \text{ by } 1. \tag{5.3.7}$$

Let us now consider $[\partial_\mu^x A^\mu_{\mathcal{L}(g)}(x)_0\,,\, A^\nu_{\mathcal{L}(g)}(y)_0]_\star^{\mathrm{av}}$. We want to show that the divergence ∂_μ^x of the second term in (5.3.6) vanishes. In fact, ∂_μ^x can be written as $-\partial_\mu^z D^{\mathrm{av}}(x-z)$. So, after an integration by parts with respect to z, we get $\partial_\mu^z R\big(e_\otimes^{\mathcal{L}(g)/\hbar}\otimes j^\mu(z); A^\nu(y)\big)_0$, which vanishes due to the version (5.2.15) of the QED-MWI. Hence,

$$[\partial_\mu^x A^\mu_{\mathcal{L}(g)}(x)_0\,,\, A^\nu_{\mathcal{L}(g)}(y)_0]_\star^{\mathrm{av}} = i\hbar\,\partial^\nu D^{\mathrm{av}}(x-y), \qquad \forall x,y\in\mathcal{O}\,,$$

which proves (a). Formula (b) is obtained by applying ∂_ν^y to the claim (a) and by using $\square_z D(z) = 0$.

[161] Comparing the two point function of the photon field (5.1.89) with the corresponding formula for the real scalar field (A.2.6), we see that there is an additional global factor (-1) on the r.h.s. of (3.2.61) when this formula is applied to $R(\cdots\otimes A^\mu(x);\cdots)$.

(c): Proceeding analogously to (5.3.5)–(5.3.7), the first term on the r.h.s. of (3.2.61) does not contribute since $\frac{\delta\psi}{\delta A_\mu}=0$. So we obtain

$$
\begin{aligned}
[\partial^x_\mu A^\mu_{\mathcal{L}(g)}(x)_0\,,\psi_{\mathcal{L}(g)}(y)_0]^{\mathrm{av}}_\star &= i\,\partial^x_\mu R\big(e_\otimes^{\mathcal{L}(g)/\hbar}\otimes A^\mu(x);\psi(y)\big)_0\\
&= ie\hbar\int dz\;\partial^x_\mu D^{\mathrm{av}}(x-z)\,R\big(e_\otimes^{\mathcal{L}(g)/\hbar}\otimes j^\mu(z);\psi(y)\big)_0\\
&= ie\hbar\int dz\;D^{\mathrm{av}}(x-z)\,\partial^z_\mu R\big(e_\otimes^{\mathcal{L}(g)/\hbar}\otimes j^\mu(z);\psi(y)\big)_0\\
&= e\hbar^2\int dz\;D^{\mathrm{av}}(x-z)\,\delta(z-y)\,R\big(e_\otimes^{\mathcal{L}(g)/\hbar};\psi(y)\big)_0\\
&= e\,\hbar^2\,D^{\mathrm{av}}(x-y)\,\psi_{\mathcal{L}(g)}(y)_0\;;
\end{aligned}
$$

this yields (c). A main difference to the proof of (a) is that in the application of the QED-MWI (5.2.15) there is a non-vanishing contribution coming from $\theta\psi=\psi$.

(d): This relation follows by applying the $*$-operation to the relation (c). Or, in a direct proof along the lines of the proof of (c), there is an additional minus sign due to $\theta\overline{\psi}=-\overline{\psi}$.

(e): Proceeding similarly to (5.3.5)–(5.3.7), both terms on the r.h.s. of (3.2.61) contribute:

$$
\begin{aligned}
[\partial^x_\mu A^\mu_{\mathcal{L}(g)}(x)_0\,,L_{\mathcal{L}(g)}(y)_0]^{\mathrm{av}}_\star &= i\,\partial^x_\mu R\big(e_\otimes^{\mathcal{L}(g)/\hbar}\otimes A^\mu(x);L(y)\big)_0\\
&= i\hbar\Big(\partial^x_\mu D^{\mathrm{av}}(x-y)\,R\big(e_\otimes^{\mathcal{L}(g)/\hbar};j^\mu(y)\big)_0\\
&\quad +e\int dz\;\partial^x_\mu D^{\mathrm{av}}(x-z)\,R\big(e_\otimes^{\mathcal{L}(g)/\hbar}\otimes j^\mu(z);L(y)\big)_0\Big)\,.
\end{aligned}
$$

After an integration by parts with respect to z, the second term vanishes due to the QED-MWI (5.2.15). The first term is equal to

$$
i\hbar\,(\partial_\mu D^{\mathrm{av}})(x-y)\,j^\mu_{\mathcal{L}(g)}(y)_0\;,
$$

which is the advanced part of the r.h.s. of the assertion. □

5.4 Connection of observable algebras and field algebras

We continue to follow [48].

The construction given in this section is not restricted to QED; it uses some assumptions and applies to all perturbative gauge or massive vector boson theories satisfying these assumptions. The latter are verified for QED in Sect. 5.5.

5.4.1 Local construction of observables in gauge theories

In gauge theories the local algebras $\mathcal{F}_0(\mathcal{O})$ of on-shell fields – for QED these are the algebras $\tilde{\mathcal{F}}_0(\mathcal{O})$ introduced in (5.3.3) – contain unphysical fields like vector potentials and ghosts. How to select the observables? Using the framework of AQFT (see Sect. 3.7), this question can be reformulated as follows: How can we obtain from the net $\mathcal{O} \longrightarrow \mathcal{F}_0(\mathcal{O})$ the net $\mathcal{O} \longrightarrow \mathcal{A}(\mathcal{O})$ of on-shell observables localized in $\mathcal{O}$? We use the BRST transformation for this; basic references for the BRST formalism are [7, 8, 96, 164].[162]

A short outline of BRST symmetry. (Cf., e.g., [58].) The BRST quantization is based on earlier work of Feynman [72], Faddeev and Popov [69], and of Slavnov. The basic idea is that after adding the "gauge fixing term" (5.1.85) to the action, which makes the Cauchy problem well posed but which is not gauge invariant, one enlarges the number of fields by the ghosts u and the anti-ghosts $\tilde{u}$. This amounts to adding a further term to the action: The "ghost action" S^{ghost}. In non-Abelian gauge theories, the latter contains, besides the free part S_0^{ghost} (5.1.77), a coupling of u and $\tilde{u}$ to the gauge field. The purpose of this enlargements is that the total action (i.e., the original action plus the gauge fixing term plus the ghost action) is invariant under the BRST transformation s, which has the following crucial properties:

- On the gauge fields and the fermionic matter fields, s acts as an *infinitesimal gauge transformation* (cf. (5.5.53)); hence, BRST invariance can be interpreted as a generalization of gauge invariance to models containing ghost fields, which overcomes the problem that the gauge fixing term is not gauge invariant.

- If the field equations hold true (in our formalism we ensure this by restricting all functionals to $\mathcal{C}_{S_0}$), the BRST transformation is *nilpotent*:[163] $s^2\, F_0 = 0\ \forall F_0 \in \mathcal{F}_0(\mathcal{O})$. This opens the door for the application of cohomological methods.

We illustrate these statements for the free photon field. We proceed in an intuitive way, a rigorous definition of the BRST transformation s for free and interacting QED is given in Sect. 5.5. For the photon field, s is defined by

$$s(A^\mu)(x) := i\frac{\partial}{\partial\alpha}\Big|_{\alpha=0}(A^\mu + \alpha\partial^\mu u)(x) = i\partial^\mu u(x) \ ,$$

where u is the ghost field. The free action S_0^{photon} (5.1.83) is BRST-invariant, $s\,S_0^{\text{photon}} = 0$, because $A^\mu \longmapsto -i\,sA^\mu$ can formally be understood as an infinitesimal gauge transformation. The gauge fixing term S_0^{gf} (5.1.85) is neither gauge-

[162] The name "BRST" stands for Becchi, Rouet, Stora (the authors of the first two references) and for Tyutin (the author of the last reference, which is unfortunately unpublished).

[163] By introducing an additional, auxiliary field, the "Nakanishi–Lautrup field" [121, 128], which is a real scalar field, one can reach that the BRST transformation is nilpotent for the off-shell fields, i.e., without using any field equation.

nor s-invariant:

$$\begin{aligned} s(S_0^{\mathrm{gf}}) &:= \frac{-i\lambda}{2}\,\frac{\partial}{\partial\alpha}\Big|_{\alpha=0}\int dx\,\Big(\partial_\mu(A^\mu+\alpha\partial^\mu u)(x)\Big)^2 \\ &= \frac{-\lambda}{2}\int dx\,\Big(\partial_\mu s(A^\mu)(x)\,\partial_\nu A^\nu(x)+\partial_\mu A^\mu(x)\,\partial_\nu s(A^\nu)(x)\Big) \\ &= -\,i\lambda\int dx\,\big(\partial_\mu A^\mu(x)\big)\,\Box u(x)\;. \end{aligned} \tag{5.4.1}$$

To give a corresponding definition of the BRST transformation of the free ghost action (5.1.77), we adopt (from the second last expression of (5.4.1)) that s is a derivation with respect to the classical product and commutes with partial derivatives; in addition, the BRST transformation of the basic ghost fields u and $\tilde{u}$ must be defined such that $s(S_0^{\mathrm{ghost}})=-s(S_0^{\mathrm{gf}})$. This can be achieved by setting

$$s(u)(x):=0\;,\quad s(\tilde{u})(x):=-i\lambda\,\partial_\mu A^\mu(x)\;.$$

Namely, with these definitions we obtain[164]

$$\begin{aligned} s(S_0^{\mathrm{ghost}}) &:= \int dx\;\big(\partial^\nu s(\tilde{u})(x)\,\partial_\nu u(x)-\partial^\nu\tilde{u}(x)\,\partial_\nu s(u)(x)\big) \\ &= -\,i\lambda\int dx\,\big(\partial^\nu\partial_\mu A^\mu(x)\big)\,\partial_\nu u(x) \end{aligned} \tag{5.4.2}$$

and, hence,[165]

$$\begin{aligned} s(S_0^{\mathrm{photon}}+S_0^{\mathrm{gf}}+S_0^{\mathrm{ghost}}) &= s(S_0^{\mathrm{gf}})+s(S_0^{\mathrm{ghost}}) \\ &= -\,i\lambda\int dx\;\partial^\nu\Big(\partial_\mu A^\mu(x)\,\partial_\nu u(x)\Big)=0\;. \end{aligned}$$

Finally, we check nilpotency of s on the basic fields:

$$\begin{aligned} &s^2(A^\mu)(x)=i\,\partial^\mu s(u)(x)=0\;,\qquad s^2(u)=s(0)=0\;,\\ &s^2(\tilde{u})(x)\big|_{\mathcal{C}_{S_0}}=-i\lambda\partial_\mu s(A^\mu)(x)\big|_{\mathcal{C}_{S_0}}=\lambda\,\Box u(x)\big|_{\mathcal{C}_{S_0}}=0\;. \end{aligned}$$

Remark 5.4.1 (Massive vector boson fields). If the vector boson field A^μ is massive, a crucial difference between "gauge symmetry" and "BRST symmetry" is the

[164] When applied to ghost fields, the Leibniz rule for s is modified by a sign (which is irrelevant here since $s(u)=0$), i.e., s is a graded derivation, analogously to (5.4.3) below. Otherwise it would, e.g., not hold that $s(0)=0$, as we see from the following calculation:

$$\begin{aligned} s(0) &= s\big(\tilde{u}(x)\wedge\tilde{u}(x)\big)=s(\tilde{u})(x)\,\tilde{u}(x)-\tilde{u}(x)\,s(\tilde{u})(x) \\ &= -i\lambda\big(\partial_\mu A^\mu(x)\,\tilde{u}(x)-\tilde{u}(x)\,\partial_\mu A^\mu(x)\big)=0\;. \end{aligned}$$

[165] Also here, we understand the various free actions $S_0^{\cdots}$ as "generalized Lagrangian" in the sense of Remk. 1.5.1.

following: In contrast to gauge invariance, which is "spontaneously" broken, *BRST invariance holds exactly.* More precisely, the total action of such a model – which contains for each massive vector boson field A^μ a corresponding "Stückelberg field" (which is a bosonic, real, scalar ghost field) and a corresponding pair $(u, \tilde{u})$ of Faddeev–Popov ghost fields, and which contains at least one Higgs field[166] (which is an observable, real scalar field) and possibly spinor fields – this action is BRST-invariant. (For an explicit verification of this well-known statement for the most simple model, see, e.g., [65, Sect. 3].)

Selection of the observables. Returning to the problem of selecting the observables, we make the following *assumptions: Let* $\mathcal{O} \longmapsto \mathcal{F}_0(\mathcal{O})$ *be the net of local on-shell field algebras of a gauge model (for QED we mean the net* $\mathcal{O} \longmapsto \tilde{\mathcal{F}}_0(\mathcal{O})$ *introduced in* (5.3.3)); *the relevant product is the star-product and we will always work on-shell. To simplify the notation, we will omit the lower index "0" of* $F_0 \in \mathcal{F}_0(\mathcal{O})$. *Each* $\mathcal{F}_0(\mathcal{O})$ *is a* $*$*-algebra and is even-odd graded with respect to the ghost number; for QED this grading is given in* (5.1.107). *In addition, we assume that the BRST transformation*[167]

$$s : \mathcal{F}_0 \longrightarrow \mathcal{F}_0 \ , \quad \mathcal{F}_0 := \cup_{\mathcal{O}} \mathcal{F}_0(\mathcal{O}) \ ,$$

is given and has the following properties:

- *s is a graded derivation with respect to the $\star$-product, i.e., it is linear and satisfies the relation*[168]

$$s(F \star G) = s(F) \star G + (-1)^{\delta(F)} F \star s(G) \ ; \tag{5.4.3}$$

- *moreover, s lowers the ghost number by 1 and is nilpotent,*

$$\delta\big(s(F)\big) = \delta(F) - 1 \quad \text{and} \quad s^2 = 0 \ ; \tag{5.4.4}$$

- *in addition, s commutes with the $*$-operation up to the following sign:*

$$s(F^*) = -(-1)^{\delta(F)} s(F)^* \ ; \tag{5.4.5}$$

- *and finally s preserves the localization region,*

$$s\big(\mathcal{F}_0(\mathcal{O})\big) \subseteq \mathcal{F}_0(\mathcal{O}) \ . \tag{5.4.6}$$

[166] There is an exception: Massive QED does not need any Higgs field.

[167] More precisely, by $\mathcal{F}_0 := \cup_{\mathcal{O}} \mathcal{F}_0(\mathcal{O})$ we mean the inductive limit of the algebras $\mathcal{F}_0(\mathcal{O})$, introduced in (3.7.19). Due to (5.4.6), the BRST transformation on $\mathcal{F}_0$ is defined by

$$s\Big(\big[(F, \mathcal{O})\big]\Big) := \Big[\big(s(F), \mathcal{O}\big)\Big] \quad \forall \, [(F, \mathcal{O})] \in \mathcal{F}_0 \ .$$

[168] When writing $\delta(F)$ we implicitly assume that F is an eigenvector of the ghost number operator θ_u (5.1.103); that is, the relations (5.4.3)–(5.4.5) are assumed for all $F, G \in \mathcal{F}_0$ where F is such an eigenvector.

To select the observables, we take into account that they should be s-invariant. Therefore, we consider the kernel of s,

$$\mathcal{A}^0 := s^{-1}(0) \subset \mathcal{F}_0 \ , \quad \text{and} \quad \mathcal{A}^0(\mathcal{O}) := \mathcal{A}^0 \cap \mathcal{F}_0(\mathcal{O}) \ .$$

$\mathcal{A}^0$ is a $*$-algebra, because

$$F, G \in \mathcal{A}^0 \quad \text{implies} \quad s(F \star G) = s(F) \star G + (-1)^{\delta(F)} F \star s(G) = 0$$

and $s(F^*) = 0$, i.e., $F \star G \in \mathcal{A}^0$ and $F^* \in \mathcal{A}^0$. We set

$$\mathcal{A}^{00} := s(\mathcal{F}_0) \quad \text{and} \quad \mathcal{A}^{00}(\mathcal{O}) := \mathcal{A}^{00} \cap \mathcal{F}_0(\mathcal{O}) \ .$$

Because of $s^2 = 0$, the space $\mathcal{A}^{00}$ is a subspace of $\mathcal{A}^0$; it is even a 2-sided ideal in $\mathcal{A}^0$, which means that the relations

$$\mathcal{A}^{00} \star \mathcal{A}^0 \subseteq \mathcal{A}^{00} \quad \text{and} \quad \mathcal{A}^0 \star \mathcal{A}^{00} \subseteq \mathcal{A}^{00} \tag{5.4.7}$$

hold true. Indeed, for $F \in \mathcal{F}_0$ and $G \in \mathcal{A}^0$ we get

$$s(F) \star G = s(F \star G) - (-1)^{\delta(F)}\, F \star s(G) = s(F \star G) \in \mathcal{A}^{00} \ ,$$

and similarly we verify $G \star s(F) \in \mathcal{A}^{00}$.

Because of these facts, the quotient

$$\mathcal{A} := \frac{\mathcal{A}^0}{\mathcal{A}^{00}} \equiv \{[F] := F + \mathcal{A}^{00} \,|\, F \in \mathcal{A}^0\} \tag{5.4.8}$$

with the induced product,

$$[F] \star [G] := [F \star G] \quad \text{for} \quad F, G \in \mathcal{A}^0 \ ,$$

is a well-defined algebra; we interpret it as the "algebra of observables". Due to (5.4.5) it holds that

$$\begin{aligned} &\mathcal{A}^{0*} = \mathcal{A}^0 \ , \quad \mathcal{A}^{00*} = \mathcal{A}^{00} \quad \text{and, hence,} \\ &[F]^* = F^* + \mathcal{A}^{00} = [F^*] \ , \quad \mathcal{A}^* = \mathcal{A} \ , \end{aligned} \tag{5.4.9}$$

that is, $\mathcal{A}$ is also a $*$-algebra. In addition,

$$\mathcal{O} \to \mathcal{A}(\mathcal{O}) := \frac{\mathcal{A}^0(\mathcal{O})}{\mathcal{A}^{00}(\mathcal{O})} \tag{5.4.10}$$

is the net of algebras of local observables.

So, on the level of observables, we identify any two s-invariant fields which differ by a field in the range of s. For example, for $g \in \mathcal{D}(\mathbb{M})$ we have $A^\mu(\partial_\mu g) \in \operatorname{Ker} s$; however, it is well known (e.g., from Gupta–Bleuler quantization) that $[A^\mu(\partial_\mu g)]$ is a trivial observable, it should be identified with the observable $[F] = 0$. This happens indeed, since $A^\mu(\partial_\mu g) = \frac{1}{i\lambda}\, s(\tilde{u})(g) \in \operatorname{Ran} s$.

5.4.2 Construction of physical states on the algebra of observables

The positivity problem in the quantization of gauge fields. We aim to construct physical states on the algebra $\mathcal{A}$ of observables in terms of vector states given by a (pre) Hilbert space representation, as introduced in (2.5.4). Hence, we are searching a nontrivial representation π of $\mathcal{A}$ by linear operators on a pre Hilbert space[169] $\mathcal{H}$, such that

$$\big\langle \pi(A^*)v, w\big\rangle = \big\langle v, \pi(A)w\big\rangle, \qquad \forall A \in \mathcal{A}\ , \quad \forall v, w \in \mathcal{H}\ . \tag{5.4.11}$$

The "positivity problem" in the quantization of spin 1 gauge fields is the following: The field algebra $\mathcal{F}_0$ cannot be nontrivially represented on a pre Hilbert space such that the relation (5.4.11) holds true. We explain this for the representation Φ of the free photon field (restricted to $\mathcal{C}_{S_0}$) in the pertinent Fock space $\mathfrak{F} := \mathfrak{F}^{\rm photon}$, see Theorem 2.6.3. Let Ω be the Fock vacuum and $\langle\,\cdot\,,\,\cdot\,\rangle_{\mathfrak{F}}$ the scalar product. Taking into account $A^\mu(x)^* = A^\mu(x)$, we assume that[170]

$$\big\langle \Phi\big(A^\mu(g)_0\big)v, w\big\rangle_{\mathfrak{F}} = \big\langle v, \Phi\big(A^\mu(\overline{g})_0\big)w\big\rangle_{\mathfrak{F}}\ , \quad g \in \mathcal{D}(\mathbb{M})\ ,$$

for all μ and for all v, w in the domain of $\Phi\big(A^\mu(g)_0\big)$ or $\Phi\big(A^\mu(\overline{g})_0\big)$, respectively; again the lower index "0" of $A^\mu(g)_0$ denotes restriction to $\mathcal{C}_{S_0}$. Then we obtain

$$\begin{aligned} 0 \le \|\Phi\big(A^\mu(g)_0\big)\Omega\|^2 &= \big\langle \Omega, \Phi\big(A^\mu(\overline{g})_0\big)\, \Phi\big(A^\mu(g)_0\big)\Omega\big\rangle_{\mathfrak{F}} \\ &= \big\langle \Omega, \Phi\big(A^\mu(\overline{g})_0 \star A^\mu(g)_0\big)\Omega\big\rangle_{\mathfrak{F}} = \omega_0\big(A^\mu(\overline{g})_0 \star A^\mu(g)_0\big) \\ &= -\hbar g^{\mu\mu}\, \langle \overline{g}, D^+ g\rangle\ , \end{aligned}$$

where we do not sum over μ. Due to positivity of the Wightman two-point function, $\langle \overline{g}, D^+ g\rangle \ge 0\ \forall g \in \mathcal{D}(\mathbb{M})$ (2.2.10), we obtain a contradiction for $\mu = 0$. Therefore, for a representation of the field algebra we have to give up the relation (5.4.11) or the positive definiteness of the scalar product. We will do the latter.

Representation of the field algebra on an inner product space. To use the Kugo–Ojima formalism [118] we make the following *assumptions*: *Let a faithful representation ρ of $(\mathcal{F}_0, \star)$ on an inner product*[171] *space $(\mathcal{K}, \langle\cdot\,,\cdot\rangle)$ be given such that*

$$\big\langle \rho(F^*)v, w\big\rangle = \big\langle v, \rho(F)w\big\rangle, \qquad \forall F \in \mathcal{F}_0\ , \ v, w \in \mathcal{K}\ . \tag{5.4.12}$$

---

[169] A "pre Hilbert space" is a generalization of a Hilbert space: It does not need to be complete with respect to the norm given by the scalar product. The completion of a pre Hilbert space (with respect to the mentioned norm) is a Hilbert space.

[170] In the remaining sections we return to the standard notation "$\overline{z}$" for the complex conjugated number of $z \in \mathbb{C}$.

[171] That $(\mathcal{K}, \langle\cdot\,,\cdot\rangle)$ is an "inner product space" means that $\mathcal{K}$ is a vector space and that the "inner product" $\langle\cdot\,,\cdot\rangle : \mathcal{K} \times \mathcal{K} \longrightarrow \mathbb{C}$ is bilinear and satisfies $\langle w, v\rangle = \overline{\langle v, w\rangle}$ for all $v, w \in \mathcal{K}$ – but, in contrast to a scalar product, an inner product needs not to be positive definite.

To simplify the notation we will frequently write F for $\rho(F)$ and the product of operators on $\mathcal{K}$ is denoted by $\rho(F)\,\rho(G)$, so we have $\rho(F\star G) = \rho(F)\,\rho(G)$. In addition, let an operator Q on $\mathcal{K}$ be given which implements the BRST transformation s, i.e.,

$$\rho\big(s(F)\big) = Q\,\rho(F) - (-1)^{\delta(F)}\,\rho(F)\,Q\ , \tag{5.4.13}$$

and is symmetric (with respect to the indefinite inner product) and nilpotent:

$$\langle Qv, w\rangle = \langle v, Qw\rangle \quad \forall v, w \in \mathcal{K} \quad \textit{and} \quad Q^2 \equiv Q\,Q = 0\ . \tag{5.4.14}$$

Q is called the "BRST charge".

Exercise 5.4.2. Let the assumptions (5.4.12) and (5.4.14) be satisfied and let $Q \in \operatorname{Ran}\rho$ and $\delta\big(\rho^{-1}(Q)\big) = -1$. With that, we can define a linear map $s : \mathcal{F}_0 \longrightarrow \mathcal{F}_0$ by (5.4.13), that is, by

$$s(F) := \rho^{-1}(Q) \star F - (-1)^{\delta(F)}\, F \star \rho^{-1}(Q)\ .$$

Verify that the so-defined s fulfills the properties (5.4.3)–(5.4.5) of the BRST transformation.

[*Solution*: $\delta\big(s(F)\big) = \delta(F) - 1$ follows from (5.1.105). The verifications of the other relations are straightforward calculations; we omit ρ and ρ^{-1}:

$$\begin{aligned} s(F)\star G &+ (-1)^{\delta(F)}\, F \star s(G) \\ &= \big(Q\star F - (-1)^{\delta(F)}\, F\star Q\big)\star G + (-1)^{\delta(F)}\, F\star\big(Q\star G - (-1)^{\delta(G)}\, G\star Q\big) \\ &= Q\star(F\star G) - (-1)^{\delta(F\star G)}\,(F\star G)\star Q = s(F\star G) \end{aligned}$$

by using again (5.1.105);

$$\begin{aligned} s^2(F) &= Q\star\big(Q\star F - (-1)^{\delta(F)}\, F\star Q\big) - (-1)^{\delta(F)-1}\big(Q\star F - (-1)^{\delta(F)}\, F\star Q\big)\star Q \\ &= 0\ ; \\ \big\langle s(F)^* v, w\big\rangle &= \big\langle v, s(F) w\big\rangle = \big\langle v, \big(Q\,F - (-1)^{\delta(F)}\, F\,Q\big) w\big\rangle \\ &= \big\langle \big(F^*\,Q - (-1)^{\delta(F)}\, Q\,F^*\big) v, w\big\rangle = \big\langle -(-1)^{\delta(F)}\, s(F^*) v, w\big\rangle \quad \forall v, w \in \mathcal{K}\ , \end{aligned}$$

where $\delta(F^*) = \delta(F)$ (5.1.106) is used in the last step.]

Note that if the inner product on $\mathcal{K}$ would be positive definite, we would find $\langle Qv, Qv\rangle = \langle v, Q^2 v\rangle = 0$ for all $v \in \mathcal{K}$, hence $Q = 0$ and thus also $s = 0$. Hence for nontrivial s the inner product must necessarily be indefinite.

Since the physical states should be s-invariant, we consider the kernel of Q: $\mathcal{K}^0 := \operatorname{Ker} Q$. Let $\mathcal{K}^{00}$ be the range of Q. Because of $Q^2 = 0$ we have $\mathcal{K}^{00} \subseteq \mathcal{K}^0$. We assume:

"Positivity"

$$\begin{aligned} &\text{(i)}\quad \langle v, v\rangle \geq 0 \qquad \forall v \in \mathcal{K}^0 \quad \text{and} \\ &\text{(ii)}\quad \big(v \in \mathcal{K}^0\ \wedge\ \langle v, v\rangle = 0\big) \quad \Longrightarrow \quad v \in \mathcal{K}^{00}\ . \end{aligned} \tag{5.4.15}$$

Then

$$\mathcal{H} := \frac{\mathcal{K}^0}{\mathcal{K}^{00}} \equiv \big\{[v] := v + \mathcal{K}^{00}\,\big|\, v \in \mathcal{K}^0\big\} \tag{5.4.16}$$

with the scalar product

$$\big\langle [v_1],[v_2]\big\rangle_{\mathcal{H}} := \langle w_1, w_2\rangle_{\mathcal{K}} \quad \text{where} \quad w_j \in [v_j] \;, \tag{5.4.17}$$

is a *pre Hilbert space*. Due to (5.4.14) the definition of $\langle [v_1],[v_2]\rangle_{\mathcal{H}}$ is independent of the choice of the representatives $w_j \in [v_j]$, $j=1,2$. Since part (ii) of the Positivity assumption (5.4.15) can equivalently be written as $(v \in \mathcal{K}^0 \wedge \langle v,v\rangle = 0) \Leftrightarrow v \in \mathcal{K}^{00}$, this assumption is precisely what is needed in order that the inner product (5.4.17) is positive definite.

Now we construct a representation of the algebra of observables $\mathcal{A}$ on the pre Hilbert space $\mathcal{H}$.

Lemma 5.4.3 ([48]). *Let $F \in \mathcal{A}^0$, $v \in \mathcal{K}^0$ and $[F] := F + \mathcal{A}^{00}$. A representation π of $\mathcal{A}$ on $\mathcal{H}$ is well defined by setting*

$$\pi\big([F]\big)[v] := \big[\rho(F)v\big] \;. \tag{5.4.18}$$

Proof. Mostly, we will omit ρ. Let $F+s(G)$ (where $F \in \mathcal{A}^0$, $G \in \mathcal{F}_0$) be a representative of $[F] \in \mathcal{A}$ in $\mathcal{F}_0$, and let $v + Qw$ (with $v \in \mathcal{K}^0$, $w \in \mathcal{K}$) be a representative of $[v] \in \mathcal{H}$ in $\mathcal{K}$. In order that the definition (5.4.18) makes sense, we have to show that

(i) $Fv \in \mathcal{K}^0$,
(ii) $\big(F + s(G)\big)(v+Qw) - Fv \in \mathcal{K}^{00} = Q\mathcal{K}$, and
(iii) we also have to verify that $\pi([F_1])\,\pi([F_2]) = \pi([F_1]\star[F_2])$ for $F_1, F_2 \in \mathcal{A}^0$.

Relation (i) is obtained from

$$QFv = s(F)v + (-1)^{\delta(F)} FQv = 0 \;.$$

To verify (ii) we compute

$$\begin{aligned} &s(G)(v+Qw) + (F\,Q)w \\ &\qquad = \big(Q\,G - (-1)^{\delta(G)} G\,Q\big)(v+Qw) - (-1)^{\delta(F)}\big(s(F) - Q\,F\big)w \\ &\qquad = (Q\,G)(v+Qw) + (-1)^{\delta(F)}(Q\,F)w \in Q\mathcal{K} \;. \end{aligned}$$

Finally, (iii) follows from

$$\begin{aligned} \big(\pi([F_1])\,\pi([F_2])\big)[v] &= \pi([F_1])[\rho(F_2)v] = [\rho(F_1)\rho(F_2)v] \\ &= [\rho(F_1\star F_2)v] = \pi([F_1\star F_2])[v] = \pi([F_1]\star[F_2])[v] \end{aligned}$$

for all $v \in \mathcal{K}^0$. □

Due to this lemma and (2.5.4), every vector $[v] \in \mathcal{H}$ induces a *state* $\omega_{[v]}$ on the algebra $\mathcal{A}$:

$$\omega_{[v]}\big([F]\big) := \big\langle [v], \pi([F])[v]\big\rangle_{\mathcal{H}} = \big\langle v, \rho(F)v\big\rangle_{\mathcal{K}} \;. \tag{5.4.19}$$

Accordingly, we interpret $\mathcal{H}$ as the *space of physical states.*

5.4.3 Stability under deformations

For a *free* spin 1 gauge model, the assumptions made in the previous sections are satisfied; for the free theory underlying QED we will verify this in Sect. 5.5.1. What happens when we turn on the interaction? We will show that, with reasonable assumptions, the described structure is stable under such a deformation. The deformation parameter is the coupling constant κ. Similarly to Sects. 2.5–2.6 we assume that the fields are *polynomials in $\hbar$*, where $\hbar > 0$ is a *fixed* parameter.

Let $\mathcal{O} \longmapsto \mathfrak{F}_0(\mathcal{O})$ be the net of local on-shell field algebras of the *free* theory. In addition let ρ be a fixed representation of $(\mathfrak{F}_0, \star)$ (where $\mathfrak{F}_0 := \cup_{\mathcal{O}} \mathfrak{F}_0(\mathcal{O})$) on an inner product space $(\mathcal{K}, \langle \cdot , \rangle)$ fulfilling the assumptions made in the previous section. To simplify the notation, we omit to write ρ. Now, we replace every generator F of $\mathfrak{F}_0(\mathcal{O})$ by a formal power series

$$\tilde{F} = \sum_{n=0}^{\infty} \kappa^n F_{(n)} \quad \text{with} \quad F_{(0)} := F\,,\;\; F_{(n)} \in \mathfrak{F}_0\,,\;\; \delta(F_{(n)}) = \delta(F) \;\; \forall n\,. \tag{5.4.20}$$

For our construction of QED, we have $\kappa = -e$, the set of generators is

$$\begin{aligned} &\{\, B(f)_0 \,|\, f \in \mathcal{D}(\mathcal{O}),\, B = A^\mu,\, \psi,\, \overline{\psi},\, u,\, \tilde{u},\, \partial_\mu A^\mu,\, F^{\mu\nu},\, j^\mu,\, L, \dots \} \quad \text{and} \\ &\widetilde{B(f)_0} = \big(B(f)_{eL(g)}\big)_0 \equiv R\big(e_\otimes^{eL(g)}, B(f)\big)_0\,; \end{aligned} \tag{5.4.21}$$

note that $L(g)$ and $B(f)$, appearing in the argument of the R-product, are off-shell fields. Under the star product the set of power series $\tilde{F}$ (5.4.20) (obtained from the generators of $\mathfrak{F}_0(\mathcal{O})$) generates a $*$-algebra, which we denote by $\tilde{\mathfrak{F}}_0(\mathcal{O})$ – for QED $\tilde{\mathfrak{F}}_0(\mathcal{O})$ is given by (5.3.3). In addition, let $\tilde{\mathfrak{F}}_0 := \cup_{\mathcal{O}} \tilde{\mathfrak{F}}_0(\mathcal{O})$, where again Footn. 167 applies.

In the same way we assume that the free BRST transformation $s : \mathfrak{F}_0 \longrightarrow \mathfrak{F}_0$ and the pertinent free BRST charge Q (5.4.13) are replaced by formal power series $\tilde{s}$ and $\tilde{Q}$, respectively, and that $\tilde{s}$ and $\tilde{Q}$ satisfy the assumptions made in the previous sections. In detail, we assume

$$\tilde{s} = \sum_{n=0}^{\infty} \kappa^n s_n : \begin{cases} \tilde{\mathfrak{F}}_0 \longrightarrow \tilde{\mathfrak{F}}_0 \\ \tilde{s}\Big(\sum_j \kappa^j F_{(j)}\Big) = \sum_{n,j} \kappa^{n+j} s_n(F_{(j)})\,, \end{cases}$$

where each $s_n : \mathfrak{F}_0 \longrightarrow \mathfrak{F}_0$ is a graded derivation and $\quad s_0 := s$;

and, setting $\tilde{\mathcal{K}} := \mathcal{K}[\![\kappa]\!]$,

$$\tilde{Q} = \sum_{n=0}^{\infty} \kappa^n Q_{(n)} : \begin{cases} \tilde{\mathcal{K}} \longrightarrow \tilde{\mathcal{K}} \\ \tilde{Q}\Big(\sum_j \kappa^j v_j\Big) = \sum_{n,j} \kappa^{n+j} Q_{(n)} v_j\,, \end{cases}$$

where each $Q_{(n)}$ is a linear operator $Q_{(n)} : \mathcal{K} \longrightarrow \mathcal{K}$ and $\quad Q_{(0)} := Q$.

In addition we assume that the following relations hold true:

$$\tilde{s}^2 = 0 \,, \quad \tilde{Q}^2 = 0, \quad \tilde{s}(\tilde{F}) = \tilde{Q}\tilde{F} - (-1)^{\delta(\tilde{F})}\tilde{F}\tilde{Q} \quad \text{and} \quad \langle \tilde{Q}\tilde{v}, \tilde{w}\rangle = \langle \tilde{v}, \tilde{Q}\tilde{w}\rangle \tag{5.4.22}$$

for all $\tilde{v}, \tilde{w} \in \tilde{\mathcal{K}}$.

We can then define the (*local*) *algebra of observables of the interacting model:*

$$\tilde{\mathcal{A}} := \frac{\operatorname{Ker}\tilde{s}}{\operatorname{Ran}\tilde{s}} \quad \text{and} \quad \tilde{\mathcal{A}}(\mathcal{O}) := \frac{\operatorname{Ker}\tilde{s} \cap \tilde{\mathcal{F}}_0(\mathcal{O})}{\operatorname{Ran}\tilde{s} \cap \tilde{\mathcal{F}}_0(\mathcal{O})} \,.$$

$\mathcal{K}^0$ and $\mathcal{K}^{00}$ are replaced by spaces of formal power series with coefficients in $\mathcal{K}$:

$$\tilde{\mathcal{K}}^0 := \operatorname{Ker}\tilde{Q} \quad \text{and} \quad \tilde{\mathcal{K}}^{00} := \operatorname{Ran}\tilde{Q} \,.$$

Due to Lemma 5.4.3, the algebra $\tilde{\mathcal{A}}$ has a natural representation on

$$\tilde{\mathcal{H}} := \frac{\tilde{\mathcal{K}}^0}{\tilde{\mathcal{K}}^{00}} \,. \tag{5.4.23}$$

The inner product on $\mathcal{K}$ induces an inner product on $\tilde{\mathcal{H}}$ which assumes values in $\mathbb{C}[\![\kappa]\!]$: For $\tilde{v} = \sum_n \kappa^n v_n \in \tilde{\mathcal{K}}^0$ and $\tilde{w} = \sum_n \kappa^n w_n \in \tilde{\mathcal{K}}^0$ it is defined by

$$\langle \tilde{v} + \tilde{\mathcal{K}}^{00}, \tilde{w} + \tilde{\mathcal{K}}^{00}\rangle_{\tilde{\mathcal{H}}} := \langle \tilde{v}, \tilde{w}\rangle_{\tilde{\mathcal{K}}} = \sum_{n,j} \kappa^{n+j}\langle v_n, w_j\rangle_{\mathcal{K}} \in \mathbb{C}[\![\kappa]\!] \,.$$

To discuss positivity of this inner product, we work with the following definition, which is used also in [157]:

Definition 5.4.4 (Positivity of formal power series). A formal power series $\tilde{b} = \sum_n \kappa^n b_n \in \mathbb{C}[\![\kappa]\!]$ is called *positive* – we write $\tilde{b} \geq 0$ – if and only if there exists another formal power series $\tilde{c} \in \mathbb{C}[\![\kappa]\!]$ such that[172] $\tilde{b} = \overline{\tilde{c}}\,\tilde{c}$.

An immediate consequence of this Definition is that, for $\tilde{a}, \tilde{b} \in \mathbb{C}[\![\kappa]\!]$ with $\tilde{a} \geq 0$ and $\tilde{b} \geq 0$, it holds that $\tilde{a}\,\tilde{b} \geq 0$. It also holds that $\tilde{a} + \tilde{b} \geq 0$; this follows from the following exercise.

Exercise 5.4.5. Prove that $\tilde{b} = \sum_n \kappa^n b_n \geq 0$ (where $b_n \in \mathbb{C}$ for all n) is equivalent to the condition

$$\begin{aligned} &b_n \in \mathbb{R} \,, \qquad \forall n \in \mathbb{N}, \quad \text{and} \\ &\exists k \in \mathbb{N} \cup \{\infty\} \quad \text{such that} \quad b_l = 0 \quad \forall l < 2k \;\; \text{and, if } k < \infty, \;\; b_{2k} > 0 \,. \end{aligned} \tag{5.4.24}$$

[*Solution*: (i) (5.4.24) *is necessary for* $\tilde{b} \geq 0$: Let $\tilde{b} = \overline{\tilde{c}}\,\tilde{c}$ where $\tilde{c} = (0, \ldots, 0, c_k, \ldots)$ with $c_k \neq 0$ for some $k \in \mathbb{N}$. From

$$b_n = \sum_{j=0}^{n} \overline{c_j}\, c_{n-j}$$

we see that $b_n \in \mathbb{R}$ $\forall n \in \mathbb{N}$ and that $\tilde{b} = (0, \ldots, 0, b_{2k} = \overline{c_k}c_k, \ldots)$ with $b_{2k} > 0$.

[172] For $\tilde{c} = \sum_n c_n \kappa^n \in \mathbb{C}[\![\kappa]\!]$ the definition (A.1.2) says that $\overline{\tilde{c}} := \sum_n \overline{c_n}\, \kappa^n$.

(ii) (5.4.24) *is sufficient for* $\tilde{b} \geq 0$: Let $\tilde{b} = (0,\ldots,0,b_{2k},\ldots)$ be given with $b_{2k} > 0$ and $b_n \in \mathbb{R}\ \forall n \in \mathbb{N}$. There exists a unique real solution $\tilde{c} \in \mathbb{R}[\![\kappa]\!]$ of the equation $\tilde{b} = \overline{\tilde{c}}\,\tilde{c}$. Namely, making the ansatz $\tilde{c} = (0,\ldots,0,c_l,c_{l+1},\ldots)$ we get

$$\tilde{c}^2 = \Big(0,\ldots,0,c_l^2,\, 2c_lc_{l+1},\ldots,\, 2c_lc_{l+j} + \sum_{r=1}^{j-1} c_{l+r}c_{l+j-r}\,,\ldots\Big)\,.$$

We conclude:

$$c_s = 0\ \forall s < k\,,\quad c_k = \sqrt{b_{2k}}\,,\quad c_{k+1} = \frac{b_{2k+1}}{2c_k}\,,\quad c_{k+j} = \frac{1}{2c_k}\Big(b_{2k+j} - \sum_{r=1}^{j-1} c_{k+r}c_{k+j-r}\Big)\,.$$

If we admit complex solutions $\tilde{c} \in \mathbb{C}[\![\kappa]\!]$ of $\tilde{b} = \overline{\tilde{c}}\,\tilde{c}$, then $\tilde{c}$ is non-unique.]

In the following exercise, which is taken from [38, Lemma 4.2.2], we generalize the definition of a state (Definition 2.5.1) to formal power series in κ, that is, from $(\mathcal{F}_\hbar, \star)$ to $(\mathcal{F}_\hbar[\![\kappa]\!], \star)$, where $\mathcal{F}_\hbar$ is defined in (2.1.6) and $\mathcal{F}_\hbar[\![\kappa]\!] \ni 1 = (1,0,0,0,\ldots)$.

Exercise 5.4.6. For the model of a real scalar field, the map

$$\omega_0 : \begin{cases} (\mathcal{F}_\hbar[\![\kappa]\!],\star) \longrightarrow \mathbb{C}[\![\kappa]\!] \\ F = \sum_{n=0}^\infty F_n\,\kappa^n \longmapsto \sum_{n=0}^\infty \omega_0(F_n)\,\kappa^n \end{cases} \qquad \text{where} \quad \omega_0\Big(f_0 + \sum_{j\geq 1}\langle f_j, \varphi^{\otimes j}\rangle\Big) := f_0\,,$$

is a state according to Definition 2.5.1 – the "vacuum state", if positivity is understood in the just introduced sense. The only non-obvious part of this statement is positivity, that is,

$$\omega_0(F^*\star F) \equiv \sum_{n,j=0}^\infty \kappa^{n+j}\,\omega_0(F_n^*\star F_j) \geq 0\,.$$

Prove this relation.

[*Hint*: To verify that the criterion (5.4.24) is fulfilled, use the Fock space representation (Thm. 2.6.3) and the Cauchy–Schwarz inequality.]

[*Solution*: First, the relation

$$\overline{\omega_0(F^*\star F)} = \omega_0\big((F^*\star F)^*\big) = \omega_0(F^*\star F) \quad \text{implies} \quad \omega_0(F^*\star F) \in \mathbb{R}[\![\kappa]\!]\,.$$

Next, let $k \in \mathbb{N}\cup\{\infty\}$ be the smallest number such that $\omega_0(F_k^*\star F_k) \neq 0$. By using Thm. 2.6.3, the relation (2.6.7) and the Cauchy–Schwarz inequality, we obtain

$$\begin{aligned} 0 &\leq |\omega_0(F_n^*\star F_j)|^2 = |\langle\Omega\mid\Phi\big((F_n^*\star F_j)_0\big)\mid\Omega\rangle|^2 = |\langle\Phi(F_{n,0})\Omega, \Phi(F_{j,0})\Omega\rangle|^2 \\ &\leq \|\Phi(F_{n,0})\Omega\|^2 \cdot \|\Phi(F_{j,0})\Omega\|^2 = \omega_0(F_n^*\star F_n)\cdot\omega_0(F_j^*\star F_j)\,, \end{aligned}$$

where the lower index "0" in the respective argument of Φ denotes restriction to $\mathcal{C}_0$. Hence, it holds that

$$\omega_0(F_n^*\star F_j) = 0 \quad \text{if} \quad n < k \quad \text{or} \quad j < k\,.$$

If $k = \infty$, we get $\omega_0(F^*\star F) = 0$. In the case $k < \infty$, we conclude that

$$\omega_0(F^*\star F) = \kappa^{2k}\,\omega_0(F_k^*\star F_k) + \mathcal{O}(\kappa^{2k+1})\,;$$

to complete the proof we mention that

$$0 \neq \omega_0(F_k^*\star F_k) = \|\Phi(F_{k,0})\Omega\|^2 \geq 0\,.]$$

Remark 5.4.7 (Weaker definition of positivity). Bordemann and Waldmann [13] work with a weaker definition of positivity in the case of $\mathbb{R}[\![\lambda]\!]$ (or, more generally, $\mathbb{R}$ may be replaced by an ordered ring R): They only require that the smallest non-vanishing coefficient is positive, it does not need to be an even coefficient. Mathematically this definition has several advantages, in particular the ring $R[\![\lambda]\!]$ is then ordered as well, that is, each non-vanishing $r \in R[\![\lambda]\!]$ is either positive ($r > 0$) or negative ($-r > 0$) [167]. Physically this definition is well suited for the definition of positivity of states on $(\mathcal{F}[\![\hbar]\!], \star)$ (cf. Definition 2.5.1); a main reason for this is that it should hold $(0, 1, 0, 0, \ldots) > 0$, because $\hbar > 0$. But in the present framework, the deformation parameter is the coupling constant κ, which may be negative and, hence, $(0, 1, 0, 0, \ldots)$ should not be positive; therefore, we work with Definition 5.4.4.

Remark 5.4.8 (Positivity of polynomials). Interpreting a polynomial $p : D \longrightarrow \mathbb{C}$, where $D \subset \mathbb{R}$ is an open interval containing $x = 0$, as a formal power series, i.e., $p(x) \in \mathbb{C}[\![x]\!]$, the question arises whether Definition 5.4.4 agrees with the usual definition: $p(x) \geq 0 \ \forall x \in D$. The answer is "yes" *if D is sufficiently small.* We verify this statement for two examples:

- A simple example is $p_1 : (-\varepsilon, \varepsilon) \longrightarrow \mathbb{C}\,;\ p_1(x) = x\ ,\ \varepsilon > 0$ arbitrary, which is neither positive in the usual sense, nor does it fulfill the criterion (5.4.24).
- For $p_2 : D := (-1, 1) \longrightarrow \mathbb{C}\,;\ p_2(x) = 1 - x$ the condition (5.4.24) is satisfied and it holds that $p_2(x) > 0 \ \forall x \in D$. But for $x > 1$ we have $p_2(x) < 0$ – this can be understood as follows: From $p(x) = \overline{\tilde{c}(x)}\, \tilde{c}(x)$, where $\tilde{c}(x) \in \mathbb{C}[\![x]\!]$, we certainly obtain $p(x) \geq 0$ for all x which lie inside the circle of convergence of the power series $\tilde{c}(x)$; but for x outside this region, $p(x)$ may be negative. The latter happens for p_2; namely, the radius of convergence of the power series

$$\sqrt{1-x} = \sum_{n=0}^{\infty} \binom{1/2}{n} (-x)^n = 1 - \tfrac{1}{2}\, x - \tfrac{1}{8}\, x^2 - \tfrac{1}{16}\, x^3 - \cdots$$

  is $r = 1$.

The following theorem is very useful and gratifying: It states that the assumptions concerning the positivity of the inner product are automatically fulfilled for the deformed theory, if they hold true in the initial, undeformed model.

Theorem 5.4.9 (Stability under deformations, [48]). *Let the positivity assumption* (5.4.15) *be fulfilled to zeroth order in* κ. *Then, the following statements about the deformed theory hold true:*[173]

(i) $\langle \tilde{v}, \tilde{v} \rangle_{\tilde{\mathcal{K}}} \geq 0 \qquad \forall \tilde{v} \in \tilde{\mathcal{K}}^0$,

(ii) $(\tilde{v} \in \tilde{\mathcal{K}}^0 \ \wedge\ \langle \tilde{v}, \tilde{v} \rangle_{\tilde{\mathcal{K}}} = 0) \quad \Longrightarrow \quad \tilde{v} \in \tilde{\mathcal{K}}^{00}$.

[173] The kth coefficient of a formal power series $\tilde{A}$ is denoted by $(\tilde{A})_k$.

(iii) *For every* $v \in \mathcal{K}^0$ *there exists a power series* $\tilde{v} \in \tilde{\mathcal{K}}^0$ *with* $(\tilde{v})_0 = v$.

(iv) *Let* π *and* $\tilde{\pi}$ *be the representations* (5.4.18) *of* $\mathcal{A}$, $\tilde{\mathcal{A}}$ *on* $\mathcal{H}$, $\tilde{\mathcal{H}}$, *respectively. Then* $\tilde{\pi}([\tilde{F}]) \neq 0$ *if* $\pi\big([(\tilde{F})_0]\big) \neq 0$.

Proof. (i) and (ii): Let $\tilde{v} = \sum_n \kappa^n v_n \in \tilde{\mathcal{K}}^0$ and

$$b_n := \big(\langle \tilde{v}, \tilde{v} \rangle_{\tilde{\mathcal{K}}}\big)_n = \sum_{k=0}^{n} \langle v_k, v_{n-k} \rangle_{\mathcal{K}} \; .$$

Obviously, b_n is real. $\tilde{Q}\tilde{v} = 0$ implies $Q_{(0)} v_0 = 0$, hence $v_0 \in \mathcal{K}^0$ and $b_0 \geq 0$. If $b_0 > 0$ (i) follows. If $b_0 = 0$ we know that there is some $w_0 \in \mathcal{K}$ with $v_0 = Q_{(0)} w_0$. Let $w_k^{(0)} := w_0 \, \delta_{k,0}$ and $\tilde{w}^{(0)} := \sum_k \kappa^k w_k^{(0)} = w_0$. Then,

$$\tilde{\eta}^{(0)} := \tilde{v} - \tilde{Q}\tilde{w}^{(0)}$$

is a formal power series with vanishing term of zeroth order. We now proceed by induction and assume that $b_0 = b_1 = \cdots = b_{2n} = 0$ and that there exists some formal power series $\tilde{w}^{(n)} = \sum_k \kappa^k w_k^{(n)}$ with coefficients in $\mathcal{K}$ such that

$$\tilde{\eta}^{(n)} \equiv \sum_k \kappa^k \eta_k^{(n)} := \tilde{v} - \tilde{Q}\tilde{w}^{(n)}$$

vanishes up to order n: $\tilde{\eta}^{(n)} = (0, \ldots, 0, \eta_{n+1}^{(n)}, \ldots)$. Note that $\tilde{Q}\tilde{\eta}^{(n)} = \tilde{Q}\tilde{v} = 0$. Then,

$$b_{2n+1} = \big(\langle \tilde{\eta}^{(n)} + \tilde{Q}\tilde{w}^{(n)}, \tilde{\eta}^{(n)} + \tilde{Q}\tilde{w}^{(n)} \rangle_{\tilde{\mathcal{K}}}\big)_{2n+1} = \big(\langle \tilde{\eta}^{(n)}, \tilde{\eta}^{(n)} \rangle_{\tilde{\mathcal{K}}}\big)_{2n+1} = 0$$

and $b_{2n+2} = \langle \eta_{n+1}^{(n)}, \eta_{n+1}^{(n)} \rangle_{\mathcal{K}}$. Since $\tilde{Q}\tilde{\eta}^{(n)} = 0$ we get $Q_{(0)} \eta_{n+1}^{(n)} = 0$, i.e., $\eta_{n+1}^{(n)} \in \mathcal{K}^0$; hence $b_{2n+2} \geq 0$. If $b_{2n+2} > 0$ we obtain (i), otherwise there exists a $w_{n+1} \in \mathcal{K}$ with $\eta_{n+1}^{(n)} = Q_{(0)} w_{n+1}$, and we can define

$$w_k^{(n+1)} := w_k^{(n)} + \delta_{n+1,k} \, w_{n+1} \; .$$

Setting $\tilde{w}^{(n+1)} := \sum_k \kappa^k w_k^{(n+1)}$, we obtain

$$\eta_k^{(n+1)} := (\tilde{v} - \tilde{Q}\tilde{w}^{(n+1)})_k = \begin{cases} (\tilde{v} - \tilde{Q}\tilde{w}^{(n)})_k = 0 & \text{for } 0 \leq k \leq n \\ (\tilde{v} - \tilde{Q}\tilde{w}^{(n)})_{n+1} - Q_{(0)} w_{n+1} = 0 & \text{for } k = n+1 \; . \end{cases}$$

Either the induction stops at some n, then (i) is satisfied; otherwise, we are precisely in the case $\langle \tilde{v}, \tilde{v} \rangle_{\tilde{\mathcal{K}}} = 0$. In the latter case we obtain a

$$\tilde{w} := \lim_{n \to \infty} \tilde{w}^{(n)} \quad \text{with} \quad \tilde{v} = \tilde{Q}\tilde{w} \; ,$$

that is, $\tilde{v} \in \tilde{\mathcal{K}}^{00}$.

(iii): Again we proceed by induction and assume that there exists a power series $\tilde{v}^{(n)}$ such that $\tilde{Q}\tilde{v}^{(n)}$ vanishes up to order n and $(\tilde{v}^{(n)})_0 = v$. This is certainly true for $n = 0$. Then $0 = (\tilde{Q}^2\tilde{v}^{(n)})_{n+1} = Q_{(0)}(\tilde{Q}\tilde{v}^{(n)})_{n+1}$, hence $(\tilde{Q}\tilde{v}^{(n)})_{n+1} \in \mathcal{K}^0$. In addition,

$$\begin{aligned} 0 &= \big(\langle \tilde{v}^{(n)}, \tilde{Q}^2\tilde{v}^{(n)}\rangle_{\tilde{\mathcal{K}}}\big)_{2n+2} = \big(\langle \tilde{Q}\tilde{v}^{(n)}, \tilde{Q}\tilde{v}^{(n)}\rangle_{\tilde{\mathcal{K}}}\big)_{2n+2} \\ &= \big\langle (\tilde{Q}\tilde{v}^{(n)})_{n+1}, (\tilde{Q}\tilde{v}^{(n)})_{n+1}\big\rangle_{\mathcal{K}} , \end{aligned}$$

thus $(\tilde{Q}\tilde{v}^{(n)})_{n+1} \in \mathcal{K}^{00}$ and there exists a $v_{n+1} \in \mathcal{K}$ with $(\tilde{Q}\tilde{v}^{(n)})_{n+1} + Q_{(0)}v_{n+1} = 0$. We then set

$$(\tilde{v}^{(n+1)})_k := (\tilde{v}^{(n)})_k + \delta_{n+1,k}\, v_{n+1}$$

and find

$$(\tilde{Q}\tilde{v}^{(n+1)})_k = \begin{cases} (\tilde{Q}\tilde{v}^{(n)})_k = 0 & \text{for} \quad 0 \le k \le n \\ (\tilde{Q}\tilde{v}^{(n)})_{n+1} + Q_{(0)}v_{n+1} = 0 & \text{for} \quad k = n+1 \,. \end{cases}$$

Therefore,

$$\tilde{v} := \lim_{n\to\infty} \tilde{v}^{(n)} \in \tilde{\mathcal{K}}^0$$

is then the wanted formal power series.

(iv): Let $\tilde{\pi}([\tilde{F}]) = 0$. This means that $\tilde{F} = \sum_k \kappa^k F_{(k)} \in \operatorname{Ker}\tilde{s}$ and $\tilde{F}\tilde{v} \in \tilde{\mathcal{K}}^{00}$, $\forall \tilde{v} \in \tilde{\mathcal{K}}^0$. The former implies $F_{(0)} \in \operatorname{Ker} s_0$, hence $\pi([F_{(0)}])$ is well defined. Let $v_0 \in \mathcal{K}^0$ arbitrary. According to (iii), there exists a $\tilde{v} \in \tilde{\mathcal{K}}^0$ with $(\tilde{v})_0 = v_0$. From $\tilde{F}\tilde{v} = \tilde{Q}\tilde{w}$ for some $\tilde{w} \in \tilde{\mathcal{K}}$, we conclude $F_{(0)}v_0 = Q_{(0)}(\tilde{w})_0$, i.e., $F_{(0)}v_0 \in \mathcal{K}^{00}$. Hence, $\pi([F_{(0)}]) = 0$, which contradicts the assumption. □

With regard to the statement (iii) we point out that in general $v \longmapsto \tilde{v}$ is non-unique and this holds true also for the induced relation $v + \mathcal{K}^{00} \longmapsto \tilde{v} + \tilde{\mathcal{K}}^{00}$ between $\mathcal{H}$ and $\tilde{\mathcal{H}}$. Namely, we may add to $\tilde{v}$ a non-vanishing solution $\tilde{\eta}$ of $\tilde{Q}\tilde{\eta} = 0$ with $(\tilde{\eta})_0 = 0$ and $\tilde{\eta} \notin \tilde{\mathcal{K}}^{00}$.

We understand $\tilde{\mathbb{C}} := \mathbb{C}[\![\kappa]\!]$ as a ring with unit $\tilde{1} = (1, 0, 0, \ldots)$; $\tilde{a} \in \tilde{\mathbb{C}}$ is invertible if and only if $(\tilde{a})_0 \neq 0$, cf. (A.1.3). Note that $\tilde{\mathbb{C}}\,\tilde{\mathcal{K}} = \tilde{\mathcal{K}}$, but

$$\hat{\mathfrak{F}}_0 := \tilde{\mathbb{C}}\,\tilde{\mathfrak{F}}_0 = \{\, \tilde{a}\tilde{F} \,|\, \tilde{a} \in \mathbb{C},\ \tilde{F} \in \tilde{\mathfrak{F}}_0 \,\}$$

is in general larger than $\tilde{\mathfrak{F}}_0$, since the set of generators is larger: $\tilde{a}\,\big(B(f)_{\kappa L(g)}\big)_0 = \big((\tilde{a}B(f))_{\kappa L(g)}\big)_0$, cf. (5.4.21).

In the following we interpret $\tilde{\mathcal{K}}$ and $\hat{\mathfrak{F}}_0$ as $\tilde{\mathbb{C}}$-modules.[174] This is possible because the usual multiplication of formal power series (A.1.2) yields maps

$$\begin{cases} \tilde{\mathbb{C}} \times \hat{\mathfrak{F}}_0 \longrightarrow \hat{\mathfrak{F}}_0 \\ (\tilde{a}, \tilde{F}) \longmapsto \tilde{a}\tilde{F} = \tilde{F}\tilde{a} \end{cases} \quad \text{and} \quad \begin{cases} \tilde{\mathbb{C}} \times \tilde{\mathcal{K}} \longrightarrow \tilde{\mathcal{K}} \\ (\tilde{a}, \tilde{v}) \longmapsto \tilde{a}\tilde{v} = \tilde{v}\tilde{a} \,, \end{cases}$$

[174] A $\tilde{\mathbb{C}}$-module is a generalization of a $\mathbb{C}$-vector space; the difference is that the "scalars" are elements of $\tilde{\mathbb{C}}$ instead of $\mathbb{C}$.

respectively, which fulfill the relations

$$\tilde{F}(\tilde{a}\tilde{v}) = \tilde{a}(\tilde{F}\tilde{v}) = (\tilde{a}\tilde{F})\tilde{v} \ , \qquad (\tilde{a}\tilde{F})^* = \overline{\tilde{a}}\,\tilde{F}^*, \qquad \langle \tilde{a}\tilde{v}, \tilde{b}\tilde{w}\rangle = \overline{\tilde{a}}\,\tilde{b}\,\langle \tilde{v}, \tilde{w}\rangle \tag{5.4.25}$$

and

$$\tilde{s}(\tilde{a}\tilde{F}) = \tilde{a}\,\tilde{s}(\tilde{F}) \ .$$

Also the physical pre Hilbert space $\tilde{\mathcal{H}}$ and the local algebra of observables $\hat{\mathcal{A}}(\mathcal{O}) := \tilde{\mathbb{C}}\tilde{\mathcal{A}}(\mathcal{O})$ are $\tilde{\mathbb{C}}$-modules, and the multiplications by a "scalar",

$$\begin{cases} \tilde{\mathbb{C}} \times \hat{\mathcal{A}}(\mathcal{O}) \longrightarrow \hat{\mathcal{A}}(\mathcal{O}) \\ (\tilde{a}, [\tilde{F}]) \longmapsto \tilde{a}\,[\tilde{F}] = [\tilde{a}\tilde{F}] = [\tilde{F}]\,\tilde{a} \end{cases} \quad \text{and} \quad \begin{cases} \tilde{\mathbb{C}} \times \tilde{\mathcal{H}} \longrightarrow \tilde{\mathcal{H}} \\ (\tilde{a}, [\tilde{v}]) \longmapsto \tilde{a}\,[\tilde{v}] = [\tilde{a}\tilde{v}] = [\tilde{v}]\,\tilde{a} \ , \end{cases}$$

satisfy the relations (5.4.25) (with $\tilde{v}$, $\tilde{w}$ and $\tilde{F}$ replaced by $[\tilde{v}]$, $[\tilde{w}]$ and $[\tilde{F}]$, respectively).

We are now going to prove that every $[\tilde{v}] \in \tilde{\mathcal{H}}$ can be normalized:

Lemma 5.4.10 ([48]). *For every $[\tilde{v}] \in \tilde{\mathcal{H}}$ with $[\tilde{v}] \neq 0$, there exist $[\tilde{w}] \in \tilde{\mathcal{H}}$ and $\tilde{a} \in \tilde{\mathbb{C}}$ such that*

$$[\tilde{v}] = \tilde{a}\,[\tilde{w}] \quad \textit{and} \quad \big\langle [\tilde{w}], [\tilde{w}] \big\rangle_{\tilde{\mathcal{H}}} = \tilde{1} \ . \tag{5.4.26}$$

Proof. We set $\tilde{b} := \langle [\tilde{v}], [\tilde{v}] \rangle_{\tilde{\mathcal{H}}} \in \tilde{\mathbb{C}}$. From Thm. 5.4.9 we know that $\tilde{b} = \sum_{n=2k}^{\infty} b_n \kappa^n$ with $b_n \in \mathbb{R}$, $b_{2k} > 0$.

- *Case* $k = 0$: There exists an $\tilde{a} \in \tilde{\mathbb{C}}$ with $\overline{\tilde{a}}\,\tilde{a} = \tilde{b}$; from $b_0 \neq 0$ we conclude $(\tilde{a})_0 \neq 0$, that is, $\tilde{a}$ is invertible. Then $[\tilde{w}] := \tilde{a}^{-1}[\tilde{v}]$ satisfies the assertion (5.4.26).
- *Case* $k > 0$: To trace this case back to the case $k = 0$, we are searching

  $$\text{a formal power series} \quad \tilde{\tau}_k \in \tilde{\mathcal{K}}^0 \quad \text{with} \quad [\tilde{v}] = \kappa^k\,[\tilde{\tau}_k] \ . \tag{5.4.27}$$

  Having constructed such a $\tilde{\tau}_k$, we obtain

  $$\langle [\tilde{\tau}_k], [\tilde{\tau}_k] \rangle_{\tilde{\mathcal{H}}} = \kappa^{-2k}\,\tilde{b} = (b_{2k}, \kappa b_{2k+1}, \ldots) \ .$$

  Similarly to the case $k = 0$, we then conclude that there exists an invertible $\tilde{c} \in \tilde{\mathbb{C}}$ with $\overline{\tilde{c}}\,\tilde{c} = \kappa^{-2k}\,\tilde{b}$. Then $[\tilde{w}] := \tilde{c}^{-1}\,[\tilde{\tau}_k]$ satisfies $\langle [\tilde{w}], [\tilde{w}] \rangle_{\tilde{\mathcal{H}}} = \tilde{1}$ and $[\tilde{v}] = \kappa^k\,\tilde{c}\,[\tilde{w}]$, that is, the assertion (5.4.26) is satisfied for $\tilde{a} := \kappa^k\,\tilde{c}$.

  To construct $\tilde{\tau}_k$ (5.4.27), we consider a representative $\tilde{v} = \sum_n v_n \kappa^n \in \tilde{\mathcal{K}}^0$ of $[\tilde{v}]$. The problem is that in general $\kappa^{-k}\,\tilde{v}$ contains terms with negative powers of κ. Due to $\langle v_0, v_0 \rangle_{\mathcal{K}} = b_0 = 0$ and $Q_{(0)} v_0 = 0$, there exists $\eta_0 \in \mathcal{K}$ with $Q_{(0)}\eta_0 = v_0$. Then,

  $$\tilde{\tau}_1 := \kappa^{-1}\big(\tilde{v} - \tilde{Q}\eta_0\big)$$

is a formal power series, which fulfills $\tilde{\tau}_1 \in \tilde{\mathcal{K}}^0$ and $[\tilde{v}] = \kappa\,[\tilde{\tau}_1]$. If $k > 1$, we have

$$\big\langle (\tilde{\tau}_1)_0, (\tilde{\tau}_1)_0 \big\rangle_{\mathcal{K}} = \big(\langle [\tilde{\tau}_1], [\tilde{\tau}_1] \rangle_{\tilde{\mathcal{H}}} \big)_0 = b_2 = 0 \quad \text{and} \quad Q_{(0)}(\tilde{\tau}_1)_0 = 0 \ ;$$

hence we can repeat this procedure: There exists $\eta_1 \in \mathcal{K}$ with $Q_{(0)}\eta_1 = (\tilde{\tau}_1)_0$. Thus,

$$\tilde{\tau}_2 := \kappa^{-1}\big(\tilde{\tau}_1 - \tilde{Q}\eta_1\big)$$

is a formal power series satisfying $\tilde{\tau}_2 \in \tilde{\mathcal{K}}^0$ and $\kappa^2\,[\tilde{\tau}_2] = \kappa\,[\tilde{\tau}_1] = [\tilde{v}]$. If $k > 2$ we repeat this procedure again until we obtain (5.4.27). □

The definition of a *state* (Definition 2.5.1) can directly be generalized to the present framework in which the algebra of observables is a $\tilde{\mathbb{C}}$-module. Explicitly, a state ω on the algebra of observables $\hat{\mathcal{A}}(\mathcal{O})$ is defined by

$$\begin{aligned}
&\text{(i)} \quad \omega\colon \hat{\mathcal{A}}(\mathcal{O}) \longrightarrow \tilde{\mathbb{C}} \quad \text{is linear, i.e.,} \quad \omega(\tilde{a}\,[\tilde{F}] + [\tilde{G}]) = \tilde{a}\,\omega([\tilde{F}]) + \omega([\tilde{B}]), \ \ \tilde{a} \in \tilde{\mathbb{C}}, \\
&\text{(ii)} \quad \omega\big([\tilde{F}]^*\big) = \overline{\omega\big([\tilde{F}]\big)} \ , \quad \forall [\tilde{F}] \in \hat{\mathcal{A}}(\mathcal{O}) \ , \\
&\text{(iii)} \quad \omega\big([\tilde{F}]^* \star [\tilde{F}]\big) \geq 0 \ , \quad \forall [\tilde{F}] \in \hat{\mathcal{A}}(\mathcal{O}) \ , \\
&\text{(iv)} \quad \omega\big([\tilde{\mathbf{1}}]\big) = \tilde{1} \ .
\end{aligned} \tag{5.4.28}$$

Since we are dealing with formal power series in the coupling constant κ, positivity in (iii) is meant in the sense of Definition 5.4.4, and not in the Bordemann–Waldmann sense (Remk. 5.4.7). On the l.h.s. of (iv), $\tilde{\mathbf{1}} = (1, 0, 0, \ldots)$ is a formal power series of functionals, the zeroth component is $F(h) = 1 \ \forall h$.

The constructed physical states, i.e., the vector states

$$\omega_{[\tilde{v}]}\big([\tilde{F}]\big) := \big\langle [\tilde{v}], [\tilde{F}]\,[\tilde{v}] \big\rangle_{\tilde{\mathcal{H}}} = \langle \tilde{v}, \tilde{F}\,\tilde{v} \rangle_{\tilde{\mathcal{K}}} \ , \quad \tilde{v} \in \tilde{\mathcal{K}}^0 \ , \tag{5.4.29}$$

satisfy obviously (i) and, if $\langle [\tilde{v}], [\tilde{v}] \rangle_{\tilde{\mathcal{H}}} = \tilde{1}$, also (iv). The conditions (ii) and (iii) (positivity) are also fulfilled:

$$\begin{aligned}
\omega_{[\tilde{v}]}\big([\tilde{F}]^*\big) &= \omega_{[\tilde{v}]}\big([\tilde{F}^*]\big) = \langle \tilde{v}, \tilde{F}^*\,\tilde{v} \rangle_{\tilde{\mathcal{K}}} = \langle \tilde{F}\,\tilde{v}, \tilde{v} \rangle_{\tilde{\mathcal{K}}} = \overline{\langle \tilde{v}, \tilde{F}\,\tilde{v} \rangle_{\tilde{\mathcal{K}}}} = \overline{\omega_{[\tilde{v}]}\big([\tilde{F}]\big)} \ ; \\
\omega_{[\tilde{v}]}\big([\tilde{F}]^* \star [\tilde{F}]\big) &= \omega_{[\tilde{v}]}\big([\tilde{F}^* \star \tilde{F}]\big) = \langle \tilde{v}, \tilde{F}^*\tilde{F}\,\tilde{v} \rangle_{\tilde{\mathcal{K}}} = \langle \tilde{F}\,\tilde{v}, \tilde{F}\,\tilde{v} \rangle_{\tilde{\mathcal{K}}} \geq 0 \ ,
\end{aligned}$$

where in the last step we use Thm. 5.4.9(i) and that $\tilde{Q}\tilde{F}\,\tilde{v} = (-1)^{\delta(\tilde{F})}\,\tilde{F}\tilde{Q}\,\tilde{v} = 0$ since $\tilde{s}(\tilde{F}) = 0$.

Summing up, we may interpret $\tilde{\mathcal{H}}$ (5.4.23) as the *space of physical states of the interacting model.*

Remark 5.4.11 (Simplifications in case that strong adiabatic limit exists)**.** If the strong adiabatic limit $g(x) \to 1 \ \forall x \in \mathbb{M}$ exists and is unique in the sense of Epstein and Glaser (see App. A.6), it holds that

$$\lim_{g \to 1} \tilde{Q} = (Q_{(0)}, 0, 0, 0, \ldots) =: \tilde{Q}_{g=1} \quad \text{where} \quad Q_{(0)} := (\tilde{Q})_0 \ . \tag{5.4.30}$$

We will verify this statement for the formal adiabatic limit of QED in Sect. 5.5.2. For purely massive models the strong adiabatic limit exists [67] and, hence, the relation (5.4.30) can be applied. Then, following the Kugo–Ojima formalism [118], the charge $Q_{(0)}$ obtained by the limit (5.4.30), can be identified with the BRST charge of the incoming free fields. In this case, the physical pre Hilbert space is

$$\tilde{\mathcal{H}} := \frac{\operatorname{Ker} \tilde{Q}_{g=1}}{\operatorname{Ran} \tilde{Q}_{g=1}} \ .$$

From

$$\tilde{Q}_{g=1}\colon \begin{cases} \tilde{\mathcal{K}} \longrightarrow \tilde{\mathcal{K}} \\ \tilde{v} = \sum_n \kappa^n v_n \longmapsto Q_{(0)}\tilde{v} = \sum_n \kappa^n\, Q_{(0)} v_n \ . \end{cases}$$

we see that $\tilde{\mathcal{H}}$ is the space of formal power series with coefficients in $\mathcal{H} = \frac{\mathcal{K}^0}{\mathcal{K}^{00}}$, that is,

$$\tilde{\mathcal{H}} = \Big\{ [\tilde{v}] = \sum_n \kappa^n (v_n + \mathcal{K}^{00}) \,\Big|\, v_n \in \mathcal{K}^0 \Big\} \ ,$$

where $\mathcal{K}^0$ and $\mathcal{K}^{00}$ are the kernel and range, respectively, of $Q_{(0)} : \mathcal{K} \longrightarrow \mathcal{K}$.

Of major interest is the *physical scattering matrix* $\mathbf{S}_{\text{phys}}(\kappa L)$ which is obtained as follows (cf. Definition 3.3.8): The operator

$$\mathbf{S}(\kappa L) := \lim_{g\to 1} \rho\Big(\mathbf{S}\big(\kappa L(g)\big)\Big) = \lim_{g\to 1} \rho\Big(T\big(e_\otimes^{i\kappa\, L(g)/\hbar}\big)\Big) \ : \tilde{\mathcal{K}} \longrightarrow \tilde{\mathcal{K}} \ ,$$

induces a well-defined operator

$$\mathbf{S}_{\text{phys}}(\kappa L)\colon \begin{cases} \tilde{\mathcal{H}} \longrightarrow \tilde{\mathcal{H}} \\ [\tilde{v}] \longmapsto \mathbf{S}_{\text{phys}}(\kappa L)\,[\tilde{v}] := \big[\mathbf{S}(\kappa L)\,\tilde{v}\big] \end{cases}$$

if and only if[175]

$$[\tilde{Q}_{g=1}\,,\,\mathbf{S}(\kappa L)]\Big|_{\operatorname{Ker}\tilde{Q}_{g=1}} \equiv \lim_{\tilde{g}\to 1} \Big[\tilde{Q}_{g=1}\,,\,\rho\Big(\mathbf{S}\big(\kappa L(\tilde{g})\big)\Big)\Big]\Big|_{\operatorname{Ker}\tilde{Q}_{g=1}} = 0 \ , \qquad (5.4.31)$$

where we assume that $\mathbf{S}(\kappa L(\tilde{g}))$ satisfies the $*$-structure axiom (3.3.29); see, e.g., [50, Lemma 1] for a proof of this statement. This is the motivation to require "perturbative gauge invariance" [43, 49, 56, 59, 149], which is a somewhat stronger condition than (5.4.31), but has the advantage that it is well defined independent of the adiabatic limit.

---

[175] $[\,\cdot\,,\,\cdot\,]$ denotes the commutator of operators defined on $\tilde{\mathcal{K}}$.

5.5 Verification of the assumptions for QED

Also this section follows [48].

The construction given in the previous section relies on some assumptions, which we are now going to verify for QED. The deformation is given by going over from the free theory to the interacting fields $\tilde{F}$, see (5.4.21). For the free theory we will first define the BRST transformation s and then we will construct a nilpotent and symmetric operator Q which implements s in a representation space with indefinite inner product. For the interacting theory we will proceed essentially as in Exercise 5.4.2: We first construct a symmetric and nilpotent operator $\tilde{Q}$, with that we define $\tilde{s}(\tilde{F})$ as the graded commutator of $\tilde{Q}$ with $\tilde{F}$. For both, the free and the interacting theory, the local observables (defined by (5.4.10)) are then naturally represented on $\mathcal{H} = \frac{\operatorname{Ker} Q}{\operatorname{Ran} Q}$ or $\tilde{\mathcal{H}} = \frac{\operatorname{Ker} \tilde{Q}}{\operatorname{Ran} \tilde{Q}}$, respectively, by (5.4.18).

It remains to prove the positivity of the inner product induced in $\mathcal{H}$ and $\tilde{\mathcal{H}}$, respectively. For the free theory this can be done by determining explicitly (distinguished) representatives of the equivalence classes in $\mathcal{H}$ and by showing that the inner product is positive definite on the subspace formed by this representatives. Turning to the interacting theory, we want to use Thm. 5.4.9 to obtain positivity for $\tilde{\mathcal{H}}$. For this purpose we need that the 0th coefficient $(\tilde{Q})_0$ satisfies the positivity assumption (5.4.15). It is unclear whether this holds true, because the integrals defining $(\tilde{Q})_0$ and the charge Q of the free theory, differ by a spatial infrared cutoff. By means of a spatial compactification we can reach that $(\tilde{Q})_0 = Q$ and, with that, Thm. 5.4.9 can be applied.

We assume that the reader is familiar with the quantization of the free photon, spinor and ghost field(s) in the Fock spaces $\mathfrak{F}^{\text{photon}}$, $\mathfrak{F}^{\text{spinor}}$ and $\mathfrak{F}^{\text{ghost}}$, respectively; references are given in App. A.5. For the photon field we choose Feynman gauge throughout, that is, we set $\lambda = 1$ in (5.1.85).

5.5.1 The free theory

In this section all fields $F \in \mathcal{F}_{\text{QED}}$ and all field polynomials $B(x)$ are throughout restricted to $\mathcal{C}_{S_0}$. To simplify the notation we nearly always omit the lower index "0" of

$$F_0 \in \mathcal{F}_{\text{QED}}\big|_{\mathcal{C}_{S_0}} =: \mathcal{F}_{\text{QED},0} \;, \tag{5.5.1}$$

which is frequently written in other sections to indicate the restriction to $\mathcal{C}_{S_0}$.

Representation of the field algebra on an inner product space. We consider the field algebra

$$\mathcal{F}_0 := \bigvee_{\star} \{\, B(f) \,\Big|\, f \in \mathcal{D}(\mathbb{M}),\, B = A^\mu, \psi, \overline{\psi}, u, \tilde{u}, \partial_\mu A^\mu, F^{\mu\nu}, j^\mu, \overline{\psi}\not{A}, \not{A}\psi, L, \dots \} \tag{5.5.2}$$

where $B(f) \equiv B(f)_0$ and $F^{\mu\nu} := \partial^\mu A^\nu - \partial^\nu A^\mu$; later on we will specify which field polynomials we mean by the dots. The representation ρ of $\mathcal{F}_0$, which is part of the assumptions made in Sect. 5.4.2, can be obtained as follows: According to Thm. 2.6.3, the algebra $(\mathcal{F}_0, \star)$ has a faithful representation on the Fock space $\mathfrak{F}^{\text{QED}} := \mathfrak{F}^{\text{photon}} \otimes \mathfrak{F}^{\text{spinor}} \otimes \mathfrak{F}^{\text{ghost}}$. There is a dense subspace $\mathcal{K} = \mathcal{K}^{\text{photon}} \otimes \mathcal{K}^{\text{spinor}} \otimes \mathcal{K}^{\text{ghost}}$ of $\mathfrak{F}^{\text{QED}}$ which is defined analogously to the subspace D (2.6.8). And, similarly to that space D, $\mathcal{K}$ is "invariant", i.e., satisfies the relation (2.6.9). This representation of $(\mathcal{F}_0, \star)$ on $\mathcal{K}$ is the representation ρ. For B as in (5.5.2), we define $\rho\big(B(x)\big) \equiv \rho\big(B(x)_0\big)$ in the usual way:

$$\int dx\; f(x)\, \rho\big(B(x)\big) := \rho\big(B(f)\big) \qquad \forall f \in \mathcal{D}(\mathbb{M})\ . \tag{5.5.3}$$

We still have to construct the indefinite product $\langle \cdot\,, \cdot \rangle$ on $\mathcal{K}$: it is defined in terms of the Fock space scalar product $(\cdot\,, \cdot)$ and a "Krein operator" $J\colon \mathcal{K} \longrightarrow \mathcal{K}$:

$$\langle v, w \rangle := (v, J\, w)\ , \quad \forall v, w \in \mathcal{K}\ , \tag{5.5.4}$$

with J satisfying the relations

$$J^2 = \text{Id}\ , \quad J^+ = J\ . \tag{5.5.5}$$

Here and in the following, O^+ denotes the adjoint of the operator $O\colon \mathcal{K} \longrightarrow \mathcal{K}$ with respect to the scalar product $(\cdot\,, \cdot)$. Defining

$$O^K := J\, O^+\, J \quad \text{we obtain} \quad \langle O\, v, w \rangle = (v, O^+\, J\, w) = \langle v, O^K\, w \rangle \quad \forall v, w \in \mathcal{K}\ . \tag{5.5.6}$$

Note also that $(O_1 O_2)^K = O_2^K\, O_1^K$.

The condition determining the Krein operator J is (5.4.12), which can be written as[176]

$$\rho(F^*) = \rho(F)^K\ , \quad \forall F \in \mathcal{F}_0 \tag{5.5.7}$$

by using (5.5.6).

In a first step we determine J such that (5.5.7) holds true for F being a basic field. Working in Feynman gauge, one can construct the photon field in Fock space such that

$$\rho(A^0(x))^+ = -\rho(A^0(x))\ , \quad \rho(A^k(x))^+ = \rho(A^k(x)) \quad \text{for} \quad k = 1, 2, 3\ ,$$

see, e.g., [148, 149]. Setting

$$J := J^{\text{photon}} \otimes J^{\text{spinor}} \otimes J^{\text{ghost}} \quad \text{with} \quad J^{\text{photon}} := (-1)^{N_0}\ , \tag{5.5.8}$$

where N_0 is the particle number operator (see (A.5.7)) for scalar photons (i.e., the particles belonging to the component $\mu = 0$ of the photon field $A^\mu(x)$), we indeed obtain

$$\rho(A^\mu(x))^K = J^{\text{photon}}\, \rho(A^\mu(x))^+\, J^{\text{photon}} = \rho(A^\mu(x)) = \rho(A^\mu(x)^*)$$

[176] In accordance with (3.4.21) we define $(F_0)^* := (F^*)_0$ for all $F \in \mathcal{F}_{\text{QED}}$.

for $\mu = 0,1,2,3$, since $(-1)^{N_0}\,\rho(A^0(x))\,(-1)^{N_0} = -\rho(A^0(x))$. Following widespread conventions, the spinor field in Fock space satisfies

$$\rho(\psi(x))^+ = \rho(\overline{\psi}(x))\,\gamma^0 = \rho(\psi(x)^*)\;, \quad \rho(\overline{\psi}(x))^+ = \gamma^0\,\rho(\psi(x)) = \rho(\overline{\psi}(x)^*)\;,$$

where (5.1.25) is used. Therefore, to fulfill (5.5.7), we set

$$J^{\text{spinor}} := \text{Id}\;. \tag{5.5.9}$$

For brevity, we do not give an explicit definition of J^{ghost}; we only mention that it can be defined such that

$$\rho(u(x))^K = \rho(u(x)^*)(= \rho(u(x)))\;, \quad \rho(\tilde{u}(x))^K = \rho(\tilde{u}(x)^*)(= -\rho(\tilde{u}(x)))\;,$$

where (5.1.76) is inserted; for details we refer to [48, 117, 149].

In a second step we explain that this definition of J implies the required relation (5.5.7) for all $F \in \mathfrak{F}_0$. Let $\phi_1, \phi_2 \in \{A^\mu, \psi, \overline{\psi}, u, \tilde{u}\}$, so we know $\rho(\phi_j^*(x)) = \phi_j(x)^K$. The claim (5.5.7) is quite obvious if F is the *star product* of basic fields, e.g., for $F = \phi_1(x) \star \phi_2(y)$. Namely:

$$\begin{aligned}\rho\Big(\phi_1(x) \star \phi_2(y)\Big)^K &= \Big(\rho(\phi_1(x))\,\rho(\phi_2(y))\Big)^K = \rho(\phi_2(y))^K\,\rho(\phi_1(x))^K \\ &= \rho(\phi_2^*(y))\,\rho(\phi_1^*(x)) = \rho(\phi_2^*(y) \star \phi_1^*(x)) = \rho\Big(\big(\phi_1(x) \star \phi_2(y)\big)^*\Big)\;.\end{aligned} \tag{5.5.10}$$

More complicated is the case in which F is the *classical product* of basic fields, e.g., for $F = (\phi_1 \wedge \phi_2)(x)$. By using (5.5.10), $c^K = J\,\overline{c}\,J = \overline{c}\ \forall c \in \mathbb{C}$, $\overline{\omega_0(F)} = \omega_0(F^*)$ and finally (5.1.26), we get

$$\begin{aligned}\rho((\phi_1 \wedge \phi_2)(x))^K &= \lim_{y \to x} \rho\Big(\phi_1(x) \star \phi_2(y) - \omega_0\big(\phi_1(x) \star \phi_2(y)\big)\Big)^K \\ &= \lim_{y \to x} \rho\Big(\big(\phi_1(x) \star \phi_2(y)\big)^*\Big) - \overline{\omega_0\big(\phi_1(x) \star \phi_2(y)\big)} \\ &= \lim_{y \to x} \rho\Big(\phi_2^*(y) \star \phi_1^*(x) - \omega_0\big(\phi_2^*(y) \star \phi_1^*(x)\big)\Big) \\ &= \rho(\phi_2^*(x) \wedge \phi_1^*(x)) = \rho((\phi_1 \wedge \phi_2)^*(x))\;.\end{aligned} \tag{5.5.11}$$

To generalize the results (5.5.10) and (5.5.11) to *derivated* basic fields, we remark that

$$\rho((\partial^a B)(x)) = \partial_x^a \rho(B(x))\;, \quad a \in \mathbb{N}^d\;, \tag{5.5.12}$$

which follows from (5.5.3). With that, we obtain

$$\rho((\partial^a \phi)(x))^K = \Big(\partial_x^a \rho(\phi(x))\Big)^K = \partial_x^a \Big(\rho(\phi(x))^K\Big) = \partial_x^a \rho(\phi^*(x)) = \rho((\partial^a \phi)^*(x))\;.$$

Due to this relation, the calculations (5.5.10) and (5.5.11) remain true, when we replace ϕ_j by $\partial^{a_j}\phi_j$ for $j = 1, 2$, throughout.

BRST transformation. Motivated by (5.4.1) and (5.4.2), we are going to define $s : \mathcal{F}_{\mathrm{QED},0} \longrightarrow \mathcal{F}_{\mathrm{QED},0}$ such that it is a graded derivation with respect to the *classical* product.

Definition 5.5.1 (BRST transformation of the free theory). For the basic fields we set

$$s(A^\mu) := i\partial^\mu u\,, \quad s(\psi) := 0\,, \quad s(\overline{\psi}) := 0\,, \quad s(u) := 0\,, \quad s(\tilde{u}) := -i\,\partial_\mu A^\mu\,. \tag{5.5.13}$$

We extend this definition to field polynomials (which may be local or non-local) by requiring that s is a graded derivation with respect to the tensor product and the classical product, where the relevant grading is the even-odd grading with respect to the ghost number (5.1.107) (similarly to (5.4.3)). Explicitly, we define

$$\begin{aligned}
&s\Big(A_{\mu_1}(y_1)\cdots A_{\mu_r}(y_r)\otimes\overline{\psi}(x_1)\wedge\cdots\wedge\psi(x_n)\otimes\tilde{u}(z_1)\wedge\cdots\wedge u(z_s)\Big) \qquad (5.5.14)\\
&:= s(A_{\mu_1})(y_1)\cdots A_{\mu_r}(y_r)\otimes\overline{\psi}(x_1)\wedge\cdots\wedge\psi(x_n)\otimes\tilde{u}(z_1)\wedge\cdots\wedge u(z_s)+\cdots\\
&\quad + A_{\mu_1}(y_1)\cdots s(A_{\mu_r})(y_r)\otimes\overline{\psi}(x_1)\wedge\cdots\wedge\psi(x_n)\otimes\tilde{u}(z_1)\wedge\cdots\wedge u(z_s)\\
&\quad + A_{\mu_1}(y_1)\cdots A_{\mu_r}(y_r)\otimes\overline{\psi}(x_1)\wedge\cdots\wedge\psi(x_n)\otimes s(\tilde{u})(z_1)\wedge\cdots\wedge u(z_s)+\cdots\\
&\quad + (-1)^{s-1}\,A_{\mu_1}(y_1)\cdots A_{\mu_r}(y_r)\otimes\overline{\psi}(x_1)\wedge\cdots\wedge\psi(x_n)\otimes\tilde{u}(z_1)\wedge\cdots\wedge s(u)(z_s)\,.
\end{aligned}$$

With that we define $s : \mathcal{F}_{\mathrm{QED},0} \longrightarrow \mathcal{F}_{\mathrm{QED},0}$ by writing $F \in \mathcal{F}_{\mathrm{QED},0}$ as[177] $F = \sum_n \langle f_n, \phi^{\otimes n}\rangle$ and by setting

$$s\Big(\sum_n \langle f_n, \phi^{\otimes n}\rangle\Big) := \sum_n \langle f_n, s(\phi^{\otimes n})\rangle\,. \tag{5.5.15}$$

Indeed, for monomials $F := \langle f_n^{j_1\ldots j_n}, \otimes_{r=1}^n \phi_{j_r}\rangle$ and $G := \langle g_k^{l_1\ldots l_k}, \otimes_{s=1}^k \phi_{l_s}\rangle$ (i.e., we do not sum over $j_1,\ldots,j_n,l_1,\ldots l_k$) this definition yields the derivation property:

$$\begin{aligned}
s(F\wedge G) &= s\Big\langle f_n^{j_1\cdots}\otimes g_k^{l_1\cdots}, (\otimes_r\phi_{j_r})\wedge(\otimes_s\phi_{l_s})\Big\rangle\\
&= \Big\langle f_n^{j_1\cdots}\otimes g_k^{l_1\cdots}, s\Big((\otimes_r\phi_{j_r})\wedge(\otimes_s\phi_{l_s})\Big)\Big\rangle\\
&= \Big\langle f_n^{j_1\cdots}\otimes g_k^{l_1\cdots}, \Big(s\big(\otimes_r\phi_{j_r}\big)\wedge\big(\otimes_s\phi_{l_s}\big) + (-1)^{\sum_{r=1}^n \delta(\phi_{j_r})}\big(\otimes_r\phi_{j_r}\big)\wedge s\big(\otimes_s\phi_{l_s}\big)\Big)\Big\rangle\\
&= s(F)\wedge G + (-1)^{\delta(F)}\,F\wedge s(G)\,,
\end{aligned}\tag{5.5.16}$$

by using (5.1.19) and (5.1.105).

Writing a local functional $F \in \mathcal{F}_{\mathrm{QED}}^{\mathrm{loc}}$ as $F = \sum_k \int dx\; g_k(x)\,B_k(x)$, where $g_k \in \mathcal{D}(\mathbb{M})$ and B_k is a local field polynomial (this is a shorthand notation for

[177] This is a shorthand notation for (5.1.93): ϕ is a vector whose components are the various basic fields restricted to $\mathcal{C}_{S_0}$, f_n is a matrix and $\otimes$ stands for the tensor product or the $\wedge$-product.

(5.1.96)) Definition 5.5.1 yields

$$s\Big(\sum_k \int dx\; g_k(x)\, B_k(x)\Big) = \sum_k \int dx\; g_k(x)\, s\big(B_k(x)\big)\;;$$

this implies

$$s\big(\partial^a B(x)\big) = \partial_x^a s\big(B(x)\big)\;, \qquad \forall a \in \mathbb{N}^d$$

and for all local field polynomials B.

Note also that the values $s(\psi) = 0 = s(\overline{\psi})$ are motivated by gauge invariance of $\psi(x)$ and $\overline{\psi}(x)$ in *free* field theory, cf. (5.5.53).

Proposition 5.5.2. *The BRST transformation $s : \mathcal{F}_{\mathrm{QED},0} \longrightarrow \mathcal{F}_{\mathrm{QED},0}$ is a graded derivation also with respect to the* star product*:*

$$s(F \star G) = s(F) \star G + (-1)^{\delta(F)}\, F \star s(G)\;, \tag{5.5.17}$$

if F is an eigenvector of the ghost number operator θ_u (5.1.103).

This proposition holds only, if the signs of the propagators of the star product for the photon and the ghost fields (which are a matter of conventions) are coordinated with each other, as the following example shows:

$$\begin{aligned} s(A^\mu)(x) \star \tilde{u}(y) + A^\mu(x) \star s(\tilde{u})(y) - s\big(A^\mu(x) \star \tilde{u}(y)\big) \\ = i\,\omega_0\big(\partial^\mu u(x) \star \tilde{u}(y)\big) - i\,\omega_0\big(A^\mu(x) \star \partial_\nu A^\nu(y)\big) = 0\;. \end{aligned}$$

Proof. Using $\mathcal{M} : (\mathcal{F}_{\mathrm{QED},0})^{\otimes 2} \longrightarrow \mathcal{F}_{\mathrm{QED},0}$ (2.1.8) to denote the classical product, the derivation property of s with respect to the classical product (5.5.16) can be written as

$$s \circ \mathcal{M} = \mathcal{M} \circ \hat{s} \quad \text{where} \qquad \hat{s} := s \otimes \mathrm{Id} + (-1)^{\delta_1}\, \mathrm{Id} \otimes s\;, \tag{5.5.18}$$

and the linear operator $(-1)^{\delta_1} : (\mathcal{F}_{\mathrm{QED},0})^{\otimes 2} \longrightarrow (\mathcal{F}_{\mathrm{QED},0})^{\otimes 2}$ is defined as follows: Writing $L \in \mathcal{F}_{\mathrm{QED},0}$ as a sum of eigenvectors of θ_u,

$$L = \sum_j L_j \quad \text{with} \quad \delta(L_j) = j\;,$$

we set

$$(-1)^{\delta_1}(L \otimes K) := \sum_j (-1)^j (L_j \otimes K)\;, \quad \forall L, K \in \mathcal{F}_{\mathrm{QED},0}\;.$$

With (2.1.7) and (5.5.18) the assertion can be written as

$$\mathcal{M} \circ \hat{s} \circ e^{\hbar D_H} = \mathcal{M} \circ e^{\hbar D_H} \circ \hat{s}\;; \tag{5.5.19}$$

that is, we only have to show that the commutator of $\hat{s}$ with

$$D_H = \int dx\,dy\; D^+(x-y)\Big(-\frac{\delta}{\delta A_\nu(x)} \otimes \frac{\delta}{\delta A^\nu(y)} + \frac{\delta_r}{\delta u(x)} \otimes \frac{\delta}{\delta \tilde{u}(y)} - \frac{\delta_r}{\delta \tilde{u}(x)} \otimes \frac{\delta}{\delta u(y)} \Big) + \cdots \tag{5.5.20}$$

vanishes. In the latter formula, the dots stand for the terms generating the spinor-contractions (see (5.1.34)); they do not contribute to this commutator, since $s(\psi) = 0 = s(\overline{\psi})$.

We will use that the BRST transformation can be written as

$$s = i \int dz\, \Big(\partial^\mu u(z)\, \frac{\delta}{\delta A^\mu(z)} - \partial_\mu A^\mu(z)\, \frac{\delta}{\delta \tilde{u}(z)} \Big)\ ; \tag{5.5.21}$$

this follows from the definition of the functional derivative, see (1.3.1) and (5.1.14)

The computation of the commutator $[D_H, \hat{s}]_\circ$ requires a careful treatment of the signs coming from Fermi statistics. Taking into account relations as, e.g.,

$$\frac{\delta^2}{\delta u(y) \delta \tilde{u}(z)} = -\frac{\delta^2}{\delta \tilde{u}(z) \delta u(y)}\ , \quad \text{but} \quad \frac{\delta_r}{\delta u(x)}\, \frac{\delta}{\delta \tilde{u}(z)} = \frac{\delta}{\delta \tilde{u}(z)}\, \frac{\delta_r}{\delta u(x)}\ ,$$

and

$$\frac{\delta}{\delta \tilde{u}(y)} \Big(\partial^\mu u(z)\, \frac{\delta}{\delta A^\mu(z)} \Big) = -\partial^\mu u(z)\, \frac{\delta^2}{\delta A^\mu(z) \delta \tilde{u}(y)}\ ,$$

$$\frac{\delta_r}{\delta u(x)} \Big(\partial^\mu u(z)\, \frac{\delta F}{\delta A^\mu(z)} \Big) = (-1)^{\delta(F)}\, \partial^\mu \delta(z-x)\, \frac{\delta F}{\delta A^\mu(z)} + \partial^\mu u(z)\, \frac{\delta}{\delta A^\mu(z)}\, \frac{\delta_r F}{\delta u(x)}$$

(where $F \in \mathcal{F}_{\mathrm{QED},0}$), we obtain that all terms cancel, except the terms in which the functional derivatives appearing in D_H (5.5.20) act on the fields $\partial^\mu u(z)$ and $\partial_\mu A^\mu(z)$ appearing in $\hat{s}$ (after insertion of (5.5.21) into the expression (5.5.18) for $\hat{s}$). There are four terms of the latter kind, namely:

$$\begin{aligned}
[D_H, \hat{s}]_\circ (F \otimes G) = \int dx\,dy\,dz\; D^+(x-y)\, \Big(&\frac{\delta(\partial A(z))}{\delta A_\nu(x)}\, \frac{\delta F}{\delta \tilde{u}(z)} \otimes \frac{\delta G}{\delta A^\nu(y)} \\
&+ (-1)^{\delta(F)}\, \frac{\delta F}{\delta A_\nu(x)} \otimes \frac{\delta(\partial A(z))}{\delta A^\nu(y)}\, \frac{\delta G}{\delta \tilde{u}(z)} \\
&+ (-1)^{\delta(F)}\, \frac{\delta_r(\partial^\mu u(z))}{\delta u(x)}\, \frac{\delta F}{\delta A^\mu(z)} \otimes \frac{\delta G}{\delta \tilde{u}(y)} \\
&- (-1)^{\delta(F)}\, \frac{\delta_r F}{\delta \tilde{u}(x)} \otimes \frac{\delta(\partial^\mu u(z))}{\delta u(y)}\, \frac{\delta G}{\delta A^\mu(z)} \Big)\ .
\end{aligned}$$

Computing the functional derivative of $\partial A(z)$ and $\partial^\mu u(z)$, respectively, we get

$$[D_H, \hat{s}]_\circ (F \otimes G) = \int dx\,dy\,dz\; D^+(x-y) \Big(\partial^\nu \delta(z-x)\, \frac{\delta F}{\delta \tilde{u}(z)} \otimes \frac{\delta G}{\delta A^\nu(y)}$$

$$+ (-1)^{\delta(F)}\, \frac{\delta F}{\delta A_\nu(x)} \otimes \partial_\nu \delta(z-y)\, \frac{\delta G}{\delta \tilde{u}(z)}$$

$$+ (-1)^{\delta(F)}\, \partial^\mu \delta(z-x)\, \frac{\delta F}{\delta A^\mu(z)} \otimes \frac{\delta G}{\delta \tilde{u}(y)}$$

$$- (-1)^{\delta(F)}\, \frac{\delta_r F}{\delta \tilde{u}(x)} \otimes \partial^\mu \delta(z-y)\, \frac{\delta G}{\delta A^\mu(z)} \Big) = 0\ .$$

So these terms cancel, too – this completes the proof. □

Exercise 5.5.3. Verify that $s : \mathcal{F}_{\mathrm{QED},0} \longrightarrow \mathcal{F}_{\mathrm{QED},0}$ fulfills the properties (5.4.4), (5.4.5) and (5.4.6), that is,

$$\delta\big(s(F)\big) = \delta(F) - 1\ , \quad s^2(F) = 0\ , \quad s(F^*) = -(-1)^{\delta(F)}\, s(F)^* \qquad \forall F \in \mathcal{F}_{\mathrm{QED},0} \tag{5.5.22}$$

and $s\big(\mathcal{F}_{\mathrm{QED},0}(\mathcal{O})\big) \subseteq \mathcal{F}_{\mathrm{QED},0}(\mathcal{O})$ for all open regions $\mathcal{O} \subseteq \mathbb{M}$.

[*Solution*: If we replace F by a basic field, the relations (5.5.22) hold true, as one verifies straightforwardly by using the explicit formulas (5.5.13).
Now, let us assume that the relations (5.5.22) are satisfied for $F = F_1$ and $F = F_2$. Then they hold also for the classical product $F = F_1 \wedge F_2$. This follows from the derivation property (5.5.16), as we are now going to verify: For $\delta(s(F)) = \delta(F) - 1$ this is obvious due to (5.1.105); for the nilpotency this is a straightforward calculation,

$$\begin{aligned} s^2(F_1 \wedge F_2) &= s\big(s(F_1) \wedge F_2 + (-1)^{\delta(F_1)}\, F_1 \wedge s(F_2)\big) \\ &= s^2(F_1) \wedge F_2 + (-1)^{\delta(s(F_1))}\, s(F_1) \wedge s(F_2) + (-1)^{\delta(F_1)}\, s(F_1) \wedge s(F_2) + F_1 \wedge s^2(F_2) \\ &= 0\ ; \end{aligned}$$

the compatibility with the $*$-operation is obtained by using (5.1.26) and (5.1.106):

$$\begin{aligned} s\big((F_1 \wedge F_2)^*\big) &= s(F_2^* \wedge F_1^*) = s(F_2^*) \wedge F_1^* + (-1)^{\delta(F_2)}\, F_2^* \wedge s(F_1^*) \\ &= -(-1)^{\delta(F_2)}\, s(F_2)^* \wedge F_1^* - (-1)^{\delta(F_2)+\delta(F_1)}\, F_2^* \wedge s(F_1)^* \\ &= -(-1)^{\delta(F_1 \wedge F_2)} \Big(\big(s(F_1) \wedge F_2\big)^* + (-1)^{\delta(F_1)} \big(F_1 \wedge s(F_2)\big)^* \Big) \\ &= -(-1)^{\delta(F_1 \wedge F_2)}\, s(F_1 \wedge F_2)^*\ . \end{aligned}$$

Proceeding iteratively, we conclude that the relations (5.5.22) are fulfilled for F being an arbitrary (local or non-local) field polynomial. By means of (5.5.15) we find that they hold true for all $F \in \mathcal{F}_{\mathrm{QED},0}$, e.g.,

$$\begin{aligned} s\big(\langle f, \phi^{\otimes n}\rangle^*\big) &= s\big(\langle \overline{f}, (\phi^{\otimes n})^*\rangle\big) = \langle \overline{f}, s\big((\phi^{\otimes n})^*\big)\rangle \\ &= \pm \langle \overline{f}, s(\phi^{\otimes n})^*\rangle = \pm \langle f, s(\phi^{\otimes n})\rangle^* = \pm s\big(\langle f, \phi^{\otimes n}\rangle\big)^*\ . \end{aligned}$$

Finally, from (5.1.97) and (5.5.21) we conclude that

$$\operatorname{supp} s(F) \subseteq \operatorname{supp} F \qquad \forall F \in \mathcal{F}_{\mathrm{QED},0}\ ;$$

this implies $s\big(\mathcal{F}_{\mathrm{QED},0}(\mathcal{O})\big) \subseteq \mathcal{F}_{\mathrm{QED},0}(\mathcal{O})$ for all $\mathcal{O}$.]

Turning to the subspace $\mathcal{F}_0$ (5.5.2) of $\mathcal{F}_{\mathrm{QED},0}$, we want that $s : \mathcal{F}_{\mathrm{QED},0} \longrightarrow \mathcal{F}_{\mathrm{QED},0}$ induces a well-defined map $s : \mathcal{F}_0 \longrightarrow \mathcal{F}_0$. All we need for that is $s(\mathcal{F}_0) \subseteq \mathcal{F}_0$. To reach the latter we specify the dots in formula (5.5.2): They stand for $s(B)$ where B runs through all the field monomials written in that formula before the dots.

Implementation of the BRST transformation by the Kugo–Ojima charge Q. Due to the free field equations and (5.5.12) the current

$$J_\mu(x) := \hbar^{-1}\, \rho\big((\partial_\nu A^\nu(x))\overset{\leftrightarrow}{\partial}_\mu u(x)\big) \quad \text{is conserved:} \quad \partial^\mu_x J_\mu(x) = 0\;, \tag{5.5.23}$$

where the shorthand notation

$$f(x)\overset{\leftrightarrow}{\partial}_\mu g(x) := f(x)\,\partial_\mu g(x) - (\partial_\mu f)(x)\, g(x) \tag{5.5.24}$$

is used. Proceeding *heuristically*, the pertinent charge reads

$$Q := \int d^3x\; J_0(c, \vec{x})\;. \tag{5.5.25}$$

It does not depend on the choice of $c \in \mathbb{R}$; this follows from $\partial^\mu J_\mu = 0$ and Gauss' integral theorem (A.1.13).

Working *rigorously*, the existence of the integral in (5.5.25) is unclear, since $J_0(x)$ is integrated out with $\delta(x^0 - c)$ instead of an element of $\mathcal{D}(\mathbb{R}^4)$. We point out that this problem does not appear for the graded commutator

$$[Q, \rho(F)]_\mp := Q\,\rho(F) - (-1)^{\delta(F)}\,\rho(F)\,Q = \int d^3x\; [J_0(c,\vec{x}), \rho(F)]_\mp\;, \quad F \in \mathcal{F}_0\;, \tag{5.5.26}$$

because the region of integration is bounded; and, due to the latter, the reasoning by means of Gauss' integral theorem that the commutator (5.5.26) does not depend on c, is rigorous. This is completely analogous to the commutator $[Q_0^\psi\,, F_0]_\star$, studied in detail in Exer. 5.1.7.

Nevertheless, for our construction of physical states in Sect. 5.4.2, we need to give the integral (5.5.25) a rigorous sense, in order that $Q : \mathcal{K} \longrightarrow \mathcal{K}$ is a well-defined linear operator. This can be done by the following method due to Requardt [144]: The integral

$$\int d^4x\; k(x_0)h(\vec{x})\, J_0(x_0, \vec{x}) \quad \text{certainly exists, if} \quad k(x_0)h(\vec{x}) \in \mathcal{D}(\mathbb{R}^4, \mathbb{R})\;.$$

We choose $k \in \mathcal{D}(\mathbb{R}, \mathbb{R})$ and $h \in \mathcal{D}(\mathbb{R}^3, \mathbb{R})$ such that

$$\int dx_0\; k(x_0) = 1 \quad \text{and} \quad h(\vec{x}) = 1 \;\; \forall \vec{x} \in \mathbb{R}^3 \text{ with } |\vec{x}| \le R \tag{5.5.27}$$

for some $R > 0$. The scaled test functions

$$k_\lambda(x_0) := \lambda\, k(\lambda x_0)\;, \quad h_\lambda(\vec{x}) := h(\lambda \vec{x})\;, \quad \lambda > 0\;, \tag{5.5.28}$$

satisfy the properties (5.5.27) with R replaced by $\frac{R}{\lambda}$, and the operator

$$Q_\lambda := \int d^4x \; k_\lambda(x_0) h_\lambda(\vec{x}) \, J_0(x_0, \vec{x})$$

exists. By a straightforward calculation one can prove that

$$\lim_{\lambda \downarrow 0} \|Q_\lambda \Omega\|^2 = \lim_{\lambda \downarrow 0} (Q_\lambda \Omega, Q_\lambda \Omega) = 0 \; , \tag{5.5.29}$$

where $(\cdot\,,\,\cdot)$ denotes the Fock space scalar product and $\Omega \in \mathcal{K}$ is the vacuum vector. In addition, we consider the limit

$$\lim_{\lambda \downarrow 0} [Q_\lambda, \rho(F)]_\mp = \lim_{\lambda \downarrow 0} \int dx_0 \; k_\lambda(x_0) \int d^3x \; h_\lambda(\vec{x}) \, [J_0(x_0, \vec{x}), \rho(F)]_\mp \; . \tag{5.5.30}$$

Let $\operatorname{supp} k \subseteq [a, b]$; then we have $\operatorname{supp} k_\lambda \subseteq [\frac{a}{\lambda}, \frac{b}{\lambda}]$. Choosing $R > 0$ big enough compared to $\max\{|a|, \, |b|\}$, we can reach that

$$\begin{aligned} &\big([\tfrac{a}{\lambda}, \tfrac{b}{\lambda}] \times \mathbb{R}^3\big) \cap \big(\operatorname{supp} F + (\overline{V}_+ \cup \overline{V}_-)\big) \\ &\qquad = \big([\tfrac{a}{\lambda}, \tfrac{b}{\lambda}] \times \{\vec{x} \,\big|\, |\vec{x}| \leq \tfrac{R}{\lambda}\}\big) \cap \big(\operatorname{supp} F + (\overline{V}_+ \cup \overline{V}_-)\big) \end{aligned}$$

for all $\lambda > 0$ sufficiently small. With that we may replace $h_\lambda(\vec{x})$ by 1 in the integral (5.5.30). By using that the commutator (5.5.26) does not depend on c, we get

$$\begin{aligned} \lim_{\lambda \downarrow 0} [Q_\lambda, \rho(F)]_\mp &= \int d^3x \; [J_0(c, \vec{x}), \rho(F)]_\mp \; \lim_{\lambda \downarrow 0} \Big(\int dx_0 \; k_\lambda(x_0) \Big) \\ &= [Q, \rho(F)]_\mp \; . \end{aligned} \tag{5.5.31}$$

With that we can show that the strong limit $\lim_{\lambda \downarrow 0} Q_\lambda$ exists, that is, $\lim_{\lambda \downarrow 0} Q_\lambda w$ exists for all $w \in \mathcal{K}$. Namely, since $\mathcal{K}$ is defined analogously to the subspace D (2.6.8), it suffices to consider

$$\lim_{\lambda \downarrow 0} Q_\lambda \, \rho\big(\phi_1(f_1)\big) \cdots \rho\big(\phi_n(f_n)\big) \, \Omega \; , \tag{5.5.32}$$

where $\phi_1, \ldots \phi_n$ are arbitrary basic fields and $n \in \mathbb{N}$, $f_1, \ldots, f_n \in \mathcal{D}(\mathbb{R}^4)$ are also arbitrary. Using (5.5.29) and (5.5.31), the limit (5.5.32) is equal to

$$= \lim_{\lambda \downarrow 0} [Q_\lambda, \rho\big(\phi_1(f_1) \star \cdots \star \phi_n(f_n)\big)]_\mp \, \Omega = [Q, \rho\big(\phi_1(f_1) \star \cdots \star \phi_n(f_n)\big)]_\mp \, \Omega \; . \tag{5.5.33}$$

Since in the last commutator the heuristic definition of Q (5.5.25) appears (which is well defined in this context, see (5.5.26)), we may give the integral (5.5.25) a rigorous sense by defining

$$Qw := \lim_{\lambda \downarrow 0} Q_\lambda w \qquad \forall w \in \mathcal{K} \; . \tag{5.5.34}$$

This definition does not depend on the choice of the test functions h and k (provided they satisfy the properties (5.5.27)), because the r.h.s. of (5.5.33) is independent of this choice. In particular we get

$$\Omega \in \operatorname{Ker} Q \tag{5.5.35}$$

from (5.5.29).

In view of a particle interpretation of $\mathcal{H} = \frac{\operatorname{Ker} Q}{\operatorname{Ran} Q}$ we introduce the "scalar and longitudinal component of the photon field" $A^\mu(x)$. The former is simply $A^0(x)$. The latter is defined in momentum space by

$$\widehat{\rho(A)_{\|}}(k) := \frac{\vec{k} \cdot \widehat{\rho(\vec{A})}(\vec{k})}{|\vec{k}|} \,. \tag{5.5.36}$$

Note the following: Since the elements of $\mathcal{C}_{S_0}$ (more precisely $\mathcal{C}_{\text{photon},(S_0^{\text{photon}}+S_0^{\text{gf}})}$) may have bad decay properties, it is a priory unclear whether the Fourier transformation of $A(x)$ exists. However, it is well known that the Fourier transformation applied to the free photon field in Fock space yields a well-defined operator-valued distribution. Therefore, we introduce the longitudinal component for $\rho(A)$ instead of A. To introduce the transversal components of the photon field, let $\vec{\varepsilon}(\vec{k}) \in \mathbb{R}^3$ be a polarization vector, i.e.,

$$\vec{\varepsilon}(\vec{k}) \cdot \vec{k} = 0 \,, \quad |\varepsilon(\vec{k})| = 1 \,.$$

Then,

$$\widehat{\rho(A)_{\vec{\varepsilon}}}(k) := \vec{\varepsilon}(\vec{k}) \cdot \widehat{\rho(\vec{A})}(\vec{k})$$

is the "transversal component of $\widehat{\rho(A)}$ with polarization $\vec{\varepsilon}$". Returning to x-space $\rho(A)_{\|}(x)$ and $\rho(A)_{\vec{\varepsilon}}(x)$ are obtained from $\widehat{\rho(A)_{\|}}(k)$ and $\widehat{\rho(A)_{\vec{\varepsilon}}}(k)$ by inverse Fourier transformation.

We introduce the following 1-particle states:

$$\frac{v}{\|v\|} \text{ is a(n)} \begin{cases} \text{scalar photon state for} \quad v = \rho(A^0)(f)\,\Omega \\ \text{longitudinal photon state for} \quad v = \rho(A)_{\|}(f)\,\Omega \\ \text{transversal photon state (with polariz. } \vec{\varepsilon}\text{) for } v = \rho(A)_{\vec{\varepsilon}}(f)\,\Omega \\ \text{ghost state for} \quad v = \rho(u)(f)\,\Omega \\ \text{anti-ghost state for} \quad v = \rho(\tilde{u})(f)\,\Omega \\ \text{positron state for} \quad v = \rho(\psi)(f)\,\Omega \\ \text{electron state for} \quad v = \rho(\overline{\psi})(f)\,\Omega \,, \end{cases} \tag{5.5.37}$$

where $f \in \mathcal{D}(\mathbb{M})$. n-particle states are constructed according to (2.6.11). Of course, for $n > 1$ the free field operators have to be normally ordered; only for $n = 1$ it may be omitted, due to $:\varphi^{\text{op}}(f): = \varphi^{\text{op}}(f)$.

Proposition 5.5.4 (Basic properties of the free BRST charge Q, [48]).

(a) *The free BRST charge* Q (5.5.25) (*or more precisely* (5.5.34)) *fulfills*

$$Q^K = Q\ , \qquad Q^2 = 0$$

and implements the BRST transformation s (*given in Definition* 5.5.1)*:*

$$[Q, \rho(F)]_{\mp} = \rho\big(s(F)\big)\ , \quad \forall F \in \mathcal{F}_0\ , \tag{5.5.38}$$

where $[\cdot\,,\,\cdot]_{\mp}$ *denotes the graded commutator* (5.4.13).

(b) *Let* $Q^+ : \mathcal{K} \longrightarrow \mathcal{K}$ *be the adjoint operator of* $Q : \mathcal{K} \longrightarrow \mathcal{K}$ *with respect to the Fock space scalar product* $(\,\cdot\,,\,\cdot\,)$*. With that, the identity*

$$\operatorname{Ker} Q \cap \operatorname{Ker} Q^+ = \operatorname{Ker}(QQ^+ + Q^+Q)$$

holds true, and the elements of this subspace are distinguished representatives of the equivalence classes in the physical space $\mathcal{H} = \frac{\operatorname{Ker} Q}{\operatorname{Ran} Q}$*. They are states which do not contain any longitudinal or scalar photons, ghosts or anti-ghosts; that is they are built up from electrons, positrons and transversal photons only.*

(c) *The inner product* $\langle\cdot\,,\,\cdot\rangle$ *on* $\mathcal{K}$ *satisfies positivity* (5.4.15)*, that is, it is positive semidefinite on* $\operatorname{Ker} Q$ *and the subspace of null vectors in* $\operatorname{Ker} Q$ *is precisely* $\operatorname{Ran} Q$.

Proof. Essentially we follow [149, Sect. 1.4] and [48, App. A], which are partly based on [44, Sect. 5]; see also [87].

Part (a): The claim $Q^K = Q$ follows immediately from

$$\begin{aligned}\hbar\, J_\mu^K &= \rho\Big((\partial A)\overset{\leftrightarrow}{\partial}_\mu u\Big)^K = \rho\Big(\big((\partial A)\overset{\leftrightarrow}{\partial}_\mu u\big)^*\Big) = \rho\Big((\partial A^*)\overset{\leftrightarrow}{\partial}_\mu u^*\Big) = \rho\Big((\partial A)\overset{\leftrightarrow}{\partial}_\mu u\Big)\\ &= \hbar\, J_\mu\ .\end{aligned}$$

Next we prove the implementation (5.5.38) for $F = \phi(x)$ with ϕ being a basic field. For $\phi \in \{u, \psi, \overline{\psi}\}$ this is obvious. To treat the cases $\phi \in \{A, u\}$, we use (5.5.26), $\Box_x \rho\big(u(x)\big) = 0$ and the result (A.2.16):

$$\begin{aligned}[Q\,, \rho\big(A^\mu(y)\big)]_- &= \hbar^{-1} \int_{x^0=c} d^3x\ [\rho\big(\partial_\nu A^\nu(x)\big), \rho\big(A^\mu(y)\big)]_-\ \overset{\leftrightarrow x}{\partial_0}\, \rho\big(u(x)\big)\\ &= -i \int_{x^0=c} d^3x\ \partial_x^\mu D(x-y)\ \overset{\leftrightarrow x}{\partial_0}\, \rho\big(u(x)\big)\\ &= -i\partial_y^\mu \int_{x^0=c} d^3x\ D(y-x)\ \overset{\leftrightarrow x}{\partial_0}\, \rho\big(u(x)\big) = \rho\big(i\partial^\mu u(y)\big)\ ,\\ [Q\,, \rho\big(\tilde{u}(y)\big)]_+ &= \hbar^{-1} \int_{x^0=c} d^3x\ \rho\big(\partial_\nu A^\nu(x)\big)\ \overset{\leftrightarrow x}{\partial_0}\, [\rho\big(u(x)\big), \rho\big(\tilde{u}(y)\big)]_+\\ &= i \int_{x^0=c} d^3x\ D(y-x)\ \overset{\leftrightarrow x}{\partial_0}\, \rho\big(\partial_\nu A^\nu(x)\big) = \rho\big(-i\partial_\nu A^\nu(y)\big)\ .\end{aligned}$$

Let $[Q, \rho(F_j)]_\mp = s(F_j)$, $j = 1, 2$. Since $X \longmapsto [Q, X]_\mp$ is a graded derivation, the implementation (5.5.38) holds then also for the *star product* $F = F_1 \star F_2$; explicitly:

$$\begin{aligned}[Q\,,\,\rho(F_1 \star F_2)]_\mp &= [Q\,,\,\rho(F_1)]_\mp\,\rho(F_2) \pm \rho(F_1)\,[Q\,,\,\rho(F_2)]_\mp \\ &= \rho\big(s(F_1) \star F_2 \pm F_1 \star s(F_2)\big) = \rho\big(s(F_1 \star F_2)\big)\ .\end{aligned}$$

More complicated is the proof of the corresponding statement for the *classical product*:

$$[Q\,,\,\rho(F_1 \wedge F_2)]_\mp = \rho\big(s(F_1 \wedge F_2)\big)\ . \tag{5.5.39}$$

We only treat the example $F = (\phi_1 \wedge \phi_2)(x)$, where ϕ_1, ϕ_2 are basic fields. Since $s(1) = 0$ – this is a consequence of the derivation property of s – we obtain

$$\begin{aligned}[Q\,,\rho(\phi_1 \wedge \phi_2)(x)]_\mp &= \lim_{y\to x}\Big[Q\,,\,\rho\big(\phi_1(x) \star \phi_2(y)\big) - \omega_0\big(\phi_1(x) \star \phi_2(y)\big)\Big]_\mp \\ &= \lim_{y\to x}\rho\Big(s\big(\phi_1(x) \star \phi_2(y)\big)\Big) \\ &= \lim_{y\to x}\rho\Big(s\Big(\phi_1(x) \star \phi_2(y) - \omega_0\big(\phi_1(x) \star \phi_2(y)\big)\Big)\Big) = \rho\Big(s\big((\phi_1 \wedge \phi_2)(x)\big)\Big)\ .\end{aligned}$$

For an n-fold classical product of basic fields ϕ_j, one can show by induction on n that

$$\begin{aligned}[Q\,,\rho\big(\phi_1(x_1) \wedge \phi_2(x_2) \wedge \cdots\big)]_\mp &= \rho\Big(s\big(\phi_1(x_1) \wedge \phi_2(x_2) \wedge \cdots\big)\Big) \\ &= \rho\big(s(\phi_1)(x_1) \wedge \phi_2(x_2) \wedge \cdots\big) \pm \rho\big(\phi_1(x_1) \wedge s(\phi_2)(x_2) \wedge \cdots\big) \pm \cdots\ ,\end{aligned} \tag{5.5.40}$$

where (5.5.16) is used in the last step. From Thm. 2.6.3 we know that the classical product of on-shell fields corresponds to the normally ordered product of Fock space operators,

$$\rho\big(\phi_1(x_1) \wedge \phi_2(x_2) \wedge \cdots\big) = {:}\phi_1^{\rm op}(x_1)\,\phi_2^{\rm op}(x_2)\cdots{:}\ , \quad \text{where} \quad \phi_j^{\rm op}(x) := \rho\big(\phi_j(x)\big)\ .$$

With that and with $s(\phi_j)^{\rm op}(x) \equiv \rho\big(s(\phi_j)(x)\big) = [Q\,,\,\phi_j^{\rm op}(x)]_\mp$ the relation (5.5.40) can be written as

$$\begin{aligned}&[Q\,,{:}\phi_1^{\rm op}(x_1)\,\phi_2^{\rm op}(x_2)\cdots{:}]_\mp \\ &\quad = {:}[Q\,,\phi_1^{\rm op}(x_1)]_\mp\,\phi_2^{\rm op}(x_2)\cdots{:} \pm {:}\phi_1^{\rm op}(x_1)\,[Q\,,\phi_2^{\rm op}(x_2)]_\mp\cdots{:} \pm \ldots\ ,\end{aligned} \tag{5.5.41}$$

that is, the map $X \longmapsto [Q, X]_\mp$ commutes with normal ordering of Fock space operators. For an alternative, direct proof of (5.5.41) we refer to [149, Lemma 3.1.1].

Since $\mathcal{F}_0$ (5.5.2) can be generated from the basic fields by applying the classical product and the star product, the proof of the implementation (5.5.38) is complete.

Finally, the nilpotency of Q can easily be shown:

$$2\,Q^2 = [Q\,,\,Q]_+ = 2\hbar^{-1}\int_{x^0=c} d^3x\;[Q\,,\,\rho(\partial_\nu A^\nu(x)\overset{\leftrightarrow x}{\partial_0}u(x))]_+$$
$$= 2\hbar^{-1}\int_{x^0=c} d^3x\;\rho\Big(s\big(\partial_\nu A^\nu(x)\overset{\leftrightarrow x}{\partial_0}u(x)\big)\Big) = 0\;,$$

since $s(\partial_\nu A^\nu) = \square u = 0$ and $s(u) = 0$.

Part (b): Let

$$(\mathrm{Ran}\,Q^+)^\perp := \{\,v \in \mathcal{K}\,\big|\,(v,w) = 0\;\;\forall w \in \mathrm{Ran}\,Q^+\,\}$$

and generally we write "$\perp$" for "orthogonal with respect to $(\cdot\,,\,\cdot)$". For $v \in \mathcal{K}$ it holds that

$$Qv = 0 \quad\Leftrightarrow\quad (v, Q^+w) = 0 \quad \forall w \in \mathcal{K}\;,$$

that is, $\mathrm{Ker}\,Q = (\mathrm{Ran}\,Q^+)^\perp$, which implies

$$\mathcal{K} = \mathrm{Ker}\,Q \oplus \mathrm{Ran}\,Q^+ \quad \text{with} \quad \mathrm{Ker}\,Q \perp \mathrm{Ran}\,Q^+\;. \tag{5.5.42}$$

Exchanging $Q \leftrightarrow Q^+$ we also get

$$\mathcal{K} = \mathrm{Ker}\,Q^+ \oplus \mathrm{Ran}\,Q \quad \text{with} \quad \mathrm{Ker}\,Q^+ \perp \mathrm{Ran}\,Q\;. \tag{5.5.43}$$

Due to $Q^2 = 0$ it holds that

$$\mathrm{Ran}\,Q \subseteq \mathrm{Ker}\,Q \quad \text{and, hence,} \quad \mathrm{Ran}\,Q \perp \mathrm{Ran}\,Q^+\;. \tag{5.5.44}$$

Let P_1 and P_2 be the orthogonal projections onto $\mathrm{Ran}\,Q^+$ and $\mathrm{Ran}\,Q$, respectively, that is, $P_j : \mathcal{K} \longrightarrow \mathcal{K}$ satisfies $P_j^2 = P_j$, $P_j^+ = P_j$ and $\mathrm{Ran}\,P_1 = \mathrm{Ran}\,Q^+$, $\mathrm{Ran}\,P_2 = \mathrm{Ran}\,Q$. Because of (5.5.42), the operator $(\mathrm{Id}-P_1)$ is the orthogonal projection onto $\mathrm{Ker}\,Q$ and, similarly, $(\mathrm{Id}-P_2)$ is the orthogonal projection onto $\mathrm{Ker}\,Q^+$. From (5.5.44) we know $P_1P_2 = 0$, hence we obtain

$$\mathrm{Id}-P_1-P_2 = (\mathrm{Id}-P_1)(\mathrm{Id}-P_2) = (\mathrm{Id}-P_2)(\mathrm{Id}-P_1)\;;$$

this operator is the orthogonal projection onto $\mathrm{Ker}\,Q \cap \mathrm{Ker}\,Q^+$. We conclude that

$$\mathcal{K} = P_1\mathcal{K} \oplus P_2\mathcal{K} \oplus (\mathrm{Id}-P_1-P_2)\mathcal{K}$$
$$= \mathrm{Ran}\,Q^+ \oplus \mathrm{Ran}\,Q \oplus (\mathrm{Ker}\,Q \cap \mathrm{Ker}\,Q^+) \tag{5.5.45}$$

and that these three spaces are pairwise orthogonal (w.r.t. $(\cdot\,,\,\cdot)$). Comparing (5.5.42) with (5.5.45) we find

$$\mathrm{Ker}\,Q = \mathrm{Ran}\,Q \oplus (\mathrm{Ker}\,Q \cap \mathrm{Ker}\,Q^+)\;. \tag{5.5.46}$$

Therefore, the vectors $v \in (\mathrm{Ker}\,Q \cap \mathrm{Ker}\,Q^+)$ are distinguished representatives of the equivalence classes $[v] \in \frac{\mathrm{Ker}\,Q}{\mathrm{Ran}\,Q} = \mathcal{H}$.

To analyze $(\mathrm{Ker}\, Q \cap \mathrm{Ker}\, Q^+)$ we use

$$(\mathrm{Ker}\, Q \cap \mathrm{Ker}\, Q^+) = \mathrm{Ker}(QQ^+ + Q^+Q) \tag{5.5.47}$$

which can be seen as follows: The relation "$\subseteq$" is obvious. To verify "$\supseteq$", let $v \in \mathrm{Ker}(QQ^+ + Q^+Q)$. Then

$$0 = \big(v, (QQ^+ + Q^+Q)v\big) = \|Q^+v\|^2 + \|Qv\|^2, \text{ which implies } Q^+v = 0 \wedge Qv = 0.$$

Let N_{unphys} be the particle number operator of unphysical particles; that is, the eigenvalues of N_{unphys} are: [Number of scalar photons] + [number of longitudinal photons] + [number of ghosts] + [number of anti-ghosts]. Inserting the momentum space representation of the basic fields in terms of creation and annihilation operators (see App. A.5) into $(QQ^+ + Q^+Q)$, one obtains that $(QQ^+ + Q^+Q)$ is a modified version of N_{unphys}; however, the kernels agree: $\mathrm{Ker}(QQ^+ + Q^+Q) = \mathrm{Ker}\, N_{\mathrm{unphys}}$, see the mentioned references. This completes the proof of Part (b).

Part (c): From this characterization of $\mathrm{Ker}(QQ^+ + Q^+Q)$ and from the definition of J^{photon} (5.5.8) and J^{spinor} (5.5.9), we see that

$$J\big|_{\mathrm{Ker}(QQ^+ + Q^+Q)} = \mathrm{Id}\ . \tag{5.5.48}$$

Now, let $v \in \mathrm{Ker}\, Q$. From (5.5.46) and (5.5.47) we know that

$$v = v_1 + Qw \quad \text{for some} \quad v_1 \in \mathrm{Ker}(QQ^+ + Q^+Q)\ , \quad w \in \mathcal{K}\ ,$$

and that this splitting into v_1 and Qw is unique. By using $Q = Q^K$, $Qv_1 = 0$, $Q^2 = 0$ and (5.5.48) we get

$$\langle v, v\rangle = \langle v_1, v_1\rangle = (v_1, v_1) \geq 0$$

and

$$\langle v, v\rangle = 0 \ \Leftrightarrow\ (v_1, v_1) = 0 \ \Leftrightarrow\ v_1 = 0 \ \Leftrightarrow\ v \in \mathrm{Ran}\, Q\ .$$

As a byproduct (which is not needed for the proof of the proposition) we note: From $Q^K = Q$ and $J^2 = \mathrm{Id}$ we obtain $JQ = Q^+J$ and $JQ^+ = QJ$; since $J\mathcal{K} = \mathcal{K}$ these relations imply

$$J\ \mathrm{Ran}\, Q = \mathrm{Ran}\, Q^+\ , \quad J\ \mathrm{Ran}\, Q^+ = \mathrm{Ran}\, Q\ . \quad \square$$

Unfortunately, the representation π (5.4.18) of the observables $\mathcal{A} = \frac{\mathrm{Ker}\, s}{\mathrm{Ran}\, s}$ on the physical pre Hilbert space $\mathcal{H} = \frac{\mathrm{Ker}\, Q}{\mathrm{Ran}\, Q}$ is *not faithful*. To give a counterexample first note that

$$u(\partial_\mu h) = i\, s(A_\mu(h)) \in \mathrm{Ran}\, s \quad \text{for} \quad h \in \mathcal{D}(\mathbb{M})\ .$$

However, for $g \in \mathcal{D}(\mathbb{M}, \mathbb{R})$ with $\int d^4x\ g(x) \neq 0$ (i.e., there does not exist any $h \in \mathcal{D}(\mathbb{M})$ with $g = \partial_\mu h$ for some μ) we find

$$u(g) \in \mathrm{Ker}\, s \quad \text{and} \quad u(g) \notin \mathrm{Ran}\, s\ ,$$

that is, $[u(g)]$ is a nontrivial element of $\mathcal{A}$. But, due to

$$\|\pi([u(g)])\,[v]\|^2_{\mathcal{H}} = \langle [\rho(u(g))\,v], [\rho(u(g))\,v]\rangle_{\mathcal{H}} = \langle v, \rho(u(g))^K\,\rho(u(g))\,v\rangle_{\mathcal{K}}$$
$$= \langle v, \rho(u(g)^* \star u(g))\,v\rangle_{\mathcal{K}} = \langle v, \rho(u(g)\star u(g))\,v\rangle_{\mathcal{K}} = \langle v, \rho(0)\,v\rangle_{\mathcal{K}} = 0$$

for all $[v] \in \mathcal{H}$, the observable $[u(g)]$ is represented by zero on $\mathcal{H}$.

To make the representation π faithful, we reduce $\mathcal{A}$ as follows: Let $\mathcal{A}^{\text{phys}}$ be the subalgebra of $\mathcal{A}$ with ghost number zero, explicitly

$$\mathcal{A}^{\text{phys}} := \{\,[F] = F + \operatorname{Ran} s \,\big|\, F \in \operatorname{Ker} s \;\wedge\; \theta_u F = 0\,\}\,.$$

Note that for $[F], [G] \in \mathcal{A}^{\text{phys}}$ we indeed have $[F] \star [G] = [F \star G] \in \mathcal{A}^{\text{phys}}$, due to (5.1.105). The representation π (5.4.18) of $\mathcal{A}^{\text{phys}}$ on $\mathcal{H}$ is in fact *faithful*. To make this plausible we mention that the elements of $\mathcal{A}^{\text{phys}}$ are built up from the (derivated) basic fields $F^{\mu\nu} := \partial^\mu A^\nu - \partial^\nu A^\mu$, ψ and $\overline{\psi}$, more precisely[178]

$$\mathcal{A}^{\text{phys}} = \bigvee\nolimits_{\star} \{\,[P(F^{\mu\nu}, \psi, \overline{\psi})(f)] \,\big|\, f \in \mathcal{D}(\mathbb{M})\,,\; P \,\text{an allowed monomial}\,\}\;;$$

this matches precisely with the particle interpretation of $\mathcal{H}$ given in part (b) of Proposition 5.5.4.

5.5.2 The interacting theory: The interacting Kugo–Ojima charge

In this section we have to distinguish: The arguments of the R-product are off-shell fields; however, the values of the R-product are throughout restricted to $\mathcal{C}_{S_0}$.[179] Hence, we write the lower index "0" indicating the restriction to $\mathcal{C}_{S_0}$, except for

$$F_{\mathcal{L}(g)} := F_{\mathcal{L}(g),0} = R\big(e_\otimes^{\mathcal{L}(g)}, F\big)_0\,, \quad \text{where} \quad F, \mathcal{L}(g) \in \mathcal{F}^{\text{loc}}_{\text{QED}}\,.$$

For a field polynomial B we use the notation (3.4.5): $B_0(x) := B(x)_0 \equiv B(x)\big|_{\mathcal{C}_0}$.

Local algebras of interacting fields. We now replace the fields $F_0 \in \mathcal{F}_0$ (5.5.2) of the free theory by the corresponding interacting fields $F_{\mathcal{L}(g)} = \sum_{n=0}^\infty e^n\,F_{n,0}$, where $\mathcal{L}(g) = eL(g)$ is taken into account. The Fock space representation $\rho : \mathcal{F}_{\text{QED},0} \longrightarrow L(\mathcal{K})$ introduced for the free theory ($L(\mathcal{K})$ denotes the space of linear operators form $\mathcal{K}$ to $\mathcal{K}$), yields a representation of the interacting fields: Since $F_{n,0} \in \mathcal{F}_{\text{QED},0}$, $\rho(F_{n,0})$ is well defined for all $n \in \mathbb{N}$. By using (5.5.7), $\mathcal{L}(g)^* = \mathcal{L}(g)$ and the $*$-structure axiom (axiom (g)), we get

$$\rho(F_{\mathcal{L}(g)})^K = \rho((F_{\mathcal{L}(g)})^*) = \rho\Big(\big(R(e_\otimes^{\mathcal{L}(g)}, F)^*\big)_0\Big) = \rho\Big(R(e_\otimes^{\mathcal{L}(g)}, F^*)_0\Big)$$
$$= \rho\big((F^*)_{\mathcal{L}(g)}\big)\,. \tag{5.5.49}$$

---

[178] P is not an arbitrary monomial, due to the condition $P(F^{\mu\nu}, \psi, \overline{\psi})(f) \in \mathcal{F}_0$, see (5.5.2).
[179] Alternatively, one could throughout work with on-shell fields by using the on-shell R-product R^{on} introduced in Sect. 3.4, or one could work with the R-product of Epstein and Glaser [66], which is a map from local Wick polynomials to Wick polynomials – the latter is used in [48].

Since the interaction $L(g)$ does not contain any ghost fields $u, \tilde{u}$, we may apply the renormalization condition (5.1.81): With P and Q being polynomials of the same kind as in that formula, it holds that

$$\begin{aligned} &\big(P(\psi,\overline{\psi},A)Q(u,\tilde{u})\big)_{\mathcal{L}(g)}(x) = P(\psi,\overline{\psi},A)_{\mathcal{L}(g)}(x)\cdot Q(u,\tilde{u})_0(x) \;, \\ &\text{in particular}\quad u_{\mathcal{L}(g)}(x) = u_0(x)\;,\quad \tilde{u}_{\mathcal{L}(g)}(x) = \tilde{u}_0(x)\;, \end{aligned} \tag{5.5.50}$$

that is, the ghost fields do not couple.

We shall study the algebra $\tilde{\mathcal{F}}_0(\mathcal{O})$ of interacting on-shell fields localized in $\mathcal{O}$, as introduced in (5.3.3) and after (5.4.21). However, note the change of notation: For an element of $\tilde{\mathcal{F}}_0 := \cup_{\mathcal{O}}\tilde{\mathcal{F}}_0(\mathcal{O})$, we now write $\tilde{F}_0 = \sum_{n=0}^{\infty} e^n\, F_{n,0}$ instead of (5.4.20). For simplicity, we choose

$$\mathcal{O} := \big((-r,\vec{0}) + V_+\big) \cap \big((r,\vec{0}) + V_-\big)\;,\quad r>0\;, \tag{5.5.51}$$

and for the switching function $g \in \mathcal{D}(\mathbb{M})$ we assume

$$g(x) = 1 \quad \forall x \in \mathcal{U} \quad \text{for a bounded, open neighbourhood } \mathcal{U} \text{ of } \overline{\mathcal{O}}. \tag{5.5.52}$$

The generalization of the derivations and proofs given in this section to arbitrary open double cones $\mathcal{O}$ is an easy task.

BRST transformation and interacting Kugo–Ojima charge. For QED the BRST transformation is of the form

$$\tilde{s} = s_0 + e\, s_1 \,:\, \tilde{\mathcal{F}}_0(\mathcal{O}) \longrightarrow \tilde{\mathcal{F}}_0(\mathcal{O})\;.$$

For $x \in \mathcal{O}$ (and, hence, $g(x) = 1$) it should have the following values on the basic fields: For $A_{\mathcal{L}(g)}$, $\psi_{\mathcal{L}(g)}$, $\overline{\psi}_{\mathcal{L}(g)}$ it is *formally* an infinitesimal gauge transformation,

$$\begin{aligned} &\tilde{s}(A^\mu_{\mathcal{L}(g)})(x) := i\frac{\partial}{\partial\alpha}\Big|_{\alpha=0}(A^\mu_{\mathcal{L}(g)} + \alpha\partial^\mu u_0)(x) = i\partial^\mu u_0(x) = s_0(A^\mu_{\mathcal{L}(g)})(x)\;, \\ &\tilde{s}(\psi_{\mathcal{L}(g)})(x) \\ &\quad := i\frac{\partial}{\partial\alpha}\Big|_{\alpha=0}\psi_{\mathcal{L}(g)}(x)\, e^{\alpha\, i e\hbar\, u_0(x)} = -e\hbar\, \psi_{\mathcal{L}(g)}(x)\, u_0(x) = e\, s_1(\psi_{\mathcal{L}(g)})(x)\;, \\ &\tilde{s}(\overline{\psi}_{\mathcal{L}(g)})(x) \\ &\quad := i\frac{\partial}{\partial\alpha}\Big|_{\alpha=0}\overline{\psi}_{\mathcal{L}(g)}(x)\, e^{-\alpha\, i e\hbar\, u_0(x)} = e\hbar\,\overline{\psi}_{\mathcal{L}(g)}(x)\, u_0(x) = e\, s_1(\overline{\psi}_{\mathcal{L}(g)})(x)\;; \end{aligned} \tag{5.5.53}$$

for the ghost fields $\tilde{s}$ has the same form as in the free theory:

$$\tilde{s}(u_{\mathcal{L}(g)})(x) = s_0(u_{\mathcal{L}(g)})(x) = 0\;,\quad \tilde{s}(\tilde{u}_{\mathcal{L}(g)})(x) = s_0(\tilde{u}_{\mathcal{L}(g)})(x) = -i\partial^x_\mu A^\mu_{\mathcal{L}(g)}(x)\;, \tag{5.5.54}$$

where we could also write u_0 and $\tilde{u}_0$ in place of $u_{\mathcal{L}(g)}$ and $\tilde{u}_{\mathcal{L}(g)}$, respectively, due to (5.5.50). Note that $\tilde{s}(\tilde{u}_{\mathcal{L}(g)})(x)$ depends on the interaction $\mathcal{L}(g)$, although this does not hold for $\tilde{u}_{\mathcal{L}(g)}(x) = \tilde{u}_0(x)$. We point out: In contrast to the free theory, $\psi_{\mathcal{L}(g)}$ and $\overline{\psi}_{\mathcal{L}(g)}$ are not observables in the sense of our definition (5.4.8). This different behaviour can be understood physically by the fact that an electron and a positron are always accompanied by a soft photon cloud.

In addition, $\tilde{s}$ should be a graded derivation which is nilpotent and compatible with the $*$-operation (properties (5.4.3)–(5.4.5)). These properties will be automatically satisfied by proceeding similarly to Exer. 5.4.2: We construct the BRST charge $Q_{\mathcal{L}} : \mathcal{K} \longrightarrow \mathcal{K}$ for the interacting theory, satisfying $Q_{\mathcal{L}}^K = Q_{\mathcal{L}}$ and $Q_{\mathcal{L}}^2 = 0$, and define $\tilde{s}$ by $\tilde{s}(\tilde{F}_0) := \rho^{-1}([Q_{\mathcal{L}}\, , \rho(\tilde{F}_0)]_\mp)$; then we will have to verify that this definition induces the transformations (5.5.53)–(5.5.54) on the basic fields.

To get $Q_{\mathcal{L}}$ we follow Kugo and Ojima [118]: We replace the current J_μ (5.5.23) underlying the free charge Q (5.5.25) by the corresponding interacting field

$$J^\mu_{\mathcal{L}(g)}(x) := \hbar^{-1}\, \rho\Big(\partial_\nu A^\nu_{\mathcal{L}(g)}(x) \overset{\leftrightarrow}{\partial}{}^\mu_x u_0(x)\Big)\, . \tag{5.5.55}$$

By means of $\Box u_0 = 0$, the field equation for $A^\nu_{\mathcal{L}(g)}$ (5.3.1) and the QED-MWI (5.2.34) we find

$$\begin{aligned} \hbar\, \partial^x_\mu J^\mu_{\mathcal{L}(g)}(x) &= -\rho\Big(\big(\Box\partial_\nu A^\nu_{\mathcal{L}(g)}(x)\big)\, u_0(x)\Big) = e\, \rho\Big(\partial^x_\nu \big(g(x)\, j^\nu_{\mathcal{L}(g)}(x)\big)\, u_0(x)\Big) \\ &= e\, (\partial_\nu g)(x)\, \rho\Big(j^\nu_{\mathcal{L}(g)}(x)\, u_0(x)\Big)\, . \end{aligned} \tag{5.5.56}$$

Taking into account also (5.5.52), we get

$$\partial^x_\mu J^\mu_{\mathcal{L}(g)}(x) = 0 \qquad \forall x \in \mathcal{U}\, . \tag{5.5.57}$$

We may therefore define $\tilde{s}$ on the algebra $\tilde{\mathfrak{F}}_0(\mathcal{O})$ in the following way:

$$\rho\Big(\tilde{s}(\tilde{F}_0)\Big) := \int d\vec{x}\; [J^0_{\mathcal{L}(g)}(0,\vec{x})\, , \rho(\tilde{F}_0)]_\mp\, , \qquad \tilde{F}_0 \in \tilde{\mathfrak{F}}_0(\mathcal{O})\, . \tag{5.5.58}$$

cf. (5.5.26). In order that $\tilde{s}(\tilde{F}_0) \in \tilde{\mathfrak{F}}_0(\mathcal{O})$ it must hold that the integral on the r.h.s. is an element of $\rho\big(\tilde{\mathfrak{F}}_0(\mathcal{O})\big)$. The latter relies on the fact that the region of integration in (5.5.58) is bounded by $|\vec{x}| < r$, that is $(0,\vec{x}) \in \mathcal{O}$; hence $J^0_{\mathcal{L}(g)}(0,\vec{x}) \in \rho\big(\tilde{\mathfrak{F}}_0(\mathcal{O})\big)$, which implies $[J^0_{\mathcal{L}(g)}(0,\vec{x})\, , \rho(\tilde{F}_0)]_\mp \in \rho\big(\tilde{\mathfrak{F}}_0(\mathcal{O})\big)$. The bound $|\vec{x}| < r$ is due to

$$\mathrm{supp}_x\, [J^0_{\mathcal{L}(g)}(x)\, , \rho(\tilde{F}_0)]_\mp \subseteq \mathcal{O} + (\overline{V}_+ \cup \overline{V}_-) \quad \forall \tilde{F}_0 \in \tilde{\mathfrak{F}}_0(\mathcal{O})\, . \tag{5.5.59}$$

This support property can be seen as follows: If $\tilde{F}_0$ is a generator of $\tilde{\mathfrak{F}}_0(\mathcal{O})$, i.e., $\tilde{F}_0 = F_{\mathcal{L}(g)}$ with $\mathrm{supp}\, F \subset \mathcal{O}$, the relation (5.5.59) relies on Remk. 3.1.6. The generalization to an arbitrary $\tilde{F}_0 \in \tilde{\mathfrak{F}}_0(\mathcal{O})$ can be obtained by means of the derivation property of $[J^0_{\mathcal{L}(g)}(x), \cdot\,]_\mp$, i.e.,

$$[J^0_{\mathcal{L}(g)}(x)\,,\rho(\tilde F_{1,0}\star\tilde F_{2,0})]_\mp \\ = [J^0_{\mathcal{L}(g)}(x)\,,\,\rho(\tilde F_{1,0})]_\mp\;\rho(\tilde F_{2,0})\pm\rho(\tilde F_{1,0})\;[J^0_{\mathcal{L}(g)}(x)\,,\,\rho(\tilde F_{2,0})]_\mp\;,$$

and by taking into account the definition of the support (5.1.97).

Now, for a given $r>0$ let $\mathcal{O}$, $\mathcal{U}$ and g as in (5.5.51)–(5.5.52); in addition, let $\varepsilon>0$ such that $r\gg\varepsilon$ and

$$\Big([-\varepsilon,\varepsilon]\times\{\,\vec x\,\Big|\,|\vec x|\le r+3\varepsilon\,\}\Big)\subset\mathcal{U}\,.$$

Then, there exists a function $\tilde h\equiv(\tilde h_\mu)$,

$$\tilde h_\mu(x):=\delta_{\mu 0}\,h(x)\quad\text{with}\quad h(x)=h_{(0)}(x^0)\,h_{(1)}(\vec x)\in\mathcal{D}(\mathcal{U},\mathbb{R})\;,\tag{5.5.60}$$

$$h_{(1)}(\vec x)=\begin{cases}0 & \forall|\vec x|\ge r+2\varepsilon\\ 1 & \forall|\vec x|\le r+\varepsilon\end{cases}\;,\quad h_{(0)}\in\mathcal{D}\big((-\varepsilon,\varepsilon),\mathbb{R}\big)\;,\quad\int dx^0\;h_{(0)}(x^0)=1\;.$$

This implies that

$$(\operatorname{supp}h+\overline{V}_+)\cap(\operatorname{supp}h+\overline{V}_-)\subset\mathcal{U}\;,\tag{5.5.61}$$

see Figure 5.2. We are going to show, that the operator

$$Q_{\mathcal{L}}(g,\tilde h):=\int d^4x\;\tilde h_\mu(x)\,J^\mu_{\mathcal{L}(g)}(x)\in\rho\big(\tilde{\mathcal{F}}_0(\mathcal{U})\big)\tag{5.5.62}$$

implements the transformation $\tilde s$ (5.5.58). Namely, in the expression

$$[Q_{\mathcal{L}}(g,\tilde h)\,,\,\rho(\tilde F_0)]_\mp=\int dx^0\;h_{(0)}(x^0)\int d\vec x\;h_{(1)}(\vec x)\,[J^0_{\mathcal{L}(g)}(x^0,\vec x)\,,\,\rho(\tilde F_0)]_\mp$$

we may replace $h_{(1)}(\vec x)$ by 1, since for $x^0\in\operatorname{supp}h_{(0)}$ it holds that

$$\{\,\vec x\,\Big|\,(x^0,\vec x)\in\operatorname{supp}_x[J^0_{\mathcal{L}(g)}(x)\,,\,\rho(\tilde F_0)]_\mp\,\}\subset\{\,\vec x\,\Big|\,|\vec x|<r+\varepsilon\,\}\;,$$

see again Figure 5.2. Then, because of current conservation (5.5.57) and Gauss' integral theorem (A.1.13), we may replace $J^0_{\mathcal{L}(g)}(x^0,\vec x)$ by $J^0_{\mathcal{L}(g)}(0,\vec x)$ and end up with

$$\begin{aligned}[Q_{\mathcal{L}}(g,\tilde h)\,,\,\rho(\tilde F_0)]_\mp&=\int_{|\vec x|<r}d\vec x\;[J^0_{\mathcal{L}(g)}(0,\vec x)\,,\,\rho(\tilde F_0)]_\mp\cdot\Big(\int dx^0\;h_{(0)}(x^0)\Big)\\&=\rho\big(\tilde s(\tilde F_0)\big)\;.\end{aligned}\tag{5.5.63}$$

Actually, the freedom in the definition of $Q_{\mathcal{L}}(g,\tilde h)$ (5.5.62) is larger than (5.5.60): Due to current conservation (5.5.57) we have

$$Q_{\mathcal{L}}(g,\tilde h)=Q_{\mathcal{L}}(g,k)\quad\text{where}\quad k\equiv(k_\mu)\;,\quad k_\mu:=\tilde h_\mu+\partial_\mu f\quad\text{with}\quad f\in\mathcal{D}(\mathcal{U},\mathbb{R})\tag{5.5.64}$$

arbitrary.

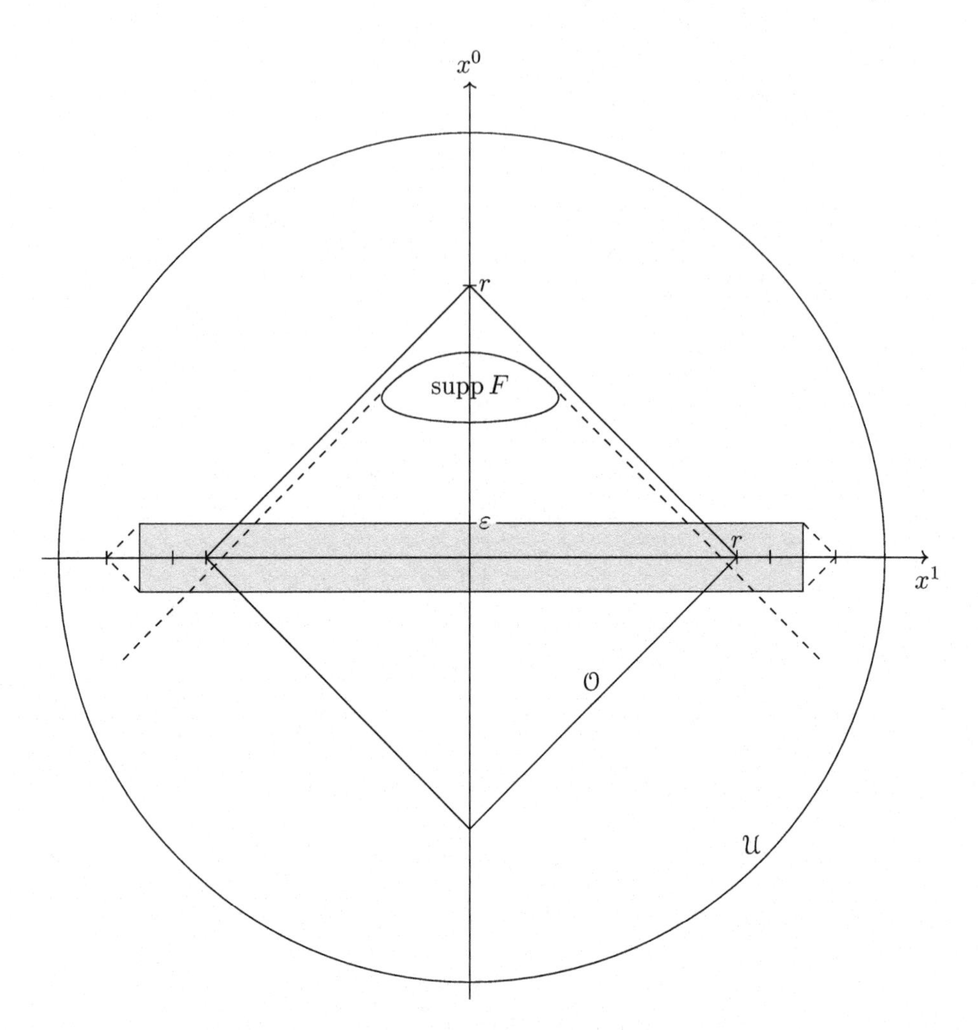

Figure 5.2: Illustration in $(d = 2)$-dimensions for the computation of the commutator $[Q_{\mathcal{L}}(g, \tilde{h}), \rho(F_{\mathcal{L}(g)})]_{\mp}$ with $\operatorname{supp} F \subset \mathcal{O}$. The region $\mathcal{G} := [-\varepsilon, \varepsilon] \times [-r-2\varepsilon, r+2\varepsilon]$ is marked, it contains $\operatorname{supp} h$. The set $(\mathcal{G} + \overline{V}_+) \cap (\mathcal{G} + \overline{V}_-)$ and the past of $\operatorname{supp} F$ are sketched by dashed lines.

The relation

$$Q_{\mathcal{L}}(g, k)^K = Q_{\mathcal{L}}(g, k) \tag{5.5.65}$$

follows immediately from (5.5.49) and $A^{\nu *} = A^\nu$, $u^* = u$. It is also easy to see that $Q_{\mathcal{L}}(g, k)$ has ghost number (-1):

$$\delta\big(\rho^{-1}(Q_{\mathcal{L}}(g, k))\big) = -1 \;, \quad \text{since} \quad \delta\big(A^\nu_{\mathcal{L}(g)}(x)\big) = 0 \;. \tag{5.5.66}$$

The nilpotency can be shown in the following way:

$$2\hbar^2\, Q_{\mathcal{L}}(g,k)^2 = \hbar^2 \left[Q_{\mathcal{L}}(g,k)\,,\, Q_{\mathcal{L}}(g,k)\right]_+ \tag{5.5.67}$$

$$= \int d^4x\; h(x) \int d^4y\; h(y)\, \left[\partial_\nu A^\nu_{\mathcal{L}(g)}(x)\,,\, \partial_\tau A^\tau_{\mathcal{L}(g)}(y)\right]_- \overset{\leftrightarrow 0}{\partial}_x \overset{\leftrightarrow 0}{\partial}_y u_0(x)\, u_0(y)$$

where we omit ρ. With that, $Q_{\mathcal{L}}(g,k)^2 = 0$ is proved, if we can show that

$$[\partial_\nu A^\nu_{\mathcal{L}(g)}(x)\,,\, \partial_\tau A^\tau_{\mathcal{L}(g)}(y)]_- = 0 \qquad \forall x,y \in \operatorname{supp} h\ . \tag{5.5.68}$$

Due to Remk. 3.1.6, the vanishing of this commutator is nontrivial only for $(x-y)^2 \geq 0$. Let $x \in y + \overline{V}_-$. Then we know from (5.5.61) that $g(z) = 1$ for all z in a neighbourhood of $(x+\overline{V}_+)\cap(y+\overline{V}_-)$. Therefore, we can apply Proposition 5.3.1(b) to conclude that the commutator (5.5.68) vanishes indeed. The case $y \in x + \overline{V}_-$ is completely analogous.

With that all assumptions made in Exer. 5.4.2 are satisfied. In addition, from $Q_{\mathcal{L}}(g,\tilde{h}) \in \rho(\tilde{\mathfrak{F}}_0)$ (5.5.62) we conclude that

$$\rho\big(\tilde{s}(\tilde{F}_0)\big) = [Q_{\mathcal{L}}(g,\tilde{h})\,,\, \rho(\tilde{F}_0)]_\mp \in \rho(\tilde{\mathfrak{F}}_0) \qquad \forall \tilde{F}_0 \in \tilde{\mathfrak{F}}_0\ ,$$

that is, $\tilde{s}(\tilde{\mathfrak{F}}_0) \subseteq \tilde{\mathfrak{F}}_0$. Summing up, formula (5.5.58) defines a linear transformation $\tilde{s} : \tilde{\mathfrak{F}}_0 \longrightarrow \tilde{\mathfrak{F}}_0$ fulfilling properties (5.4.3)–(5.4.5).

That the so-defined $\tilde{s}$ preserves the localization region, $\tilde{s}\big(\tilde{\mathfrak{F}}_0(\mathcal{O})\big) \subseteq \tilde{\mathfrak{F}}_0(\mathcal{O})$ (5.4.6), follows from Thm. 5.5.5 below: An interacting field at $y \in \mathcal{O}$ is mapped to an interacting field at the same point y.[180] In addition, Thm. 5.5.5 states that the map $\tilde{s}$ (5.5.58) has the required values (5.5.53)–(5.5.54) on the basic fields; and also for all further interacting fields belonging to sub monomials of the interaction Lagrangian $L = j^\mu A_\mu = \overline{\psi} \not{A} \psi$, the values of $\tilde{s}$ agree with what one expects from the BRST transformation in classical Electrodynamics.

Theorem 5.5.5 (BRST transformation for QED, [48]). *We assume that the interacting fields are renormalized as described in Sects.* 5.1 *and* 5.2, *especially that they fulfill the field equations* (5.3.1) *and the QED-MWI* (5.2.7). *Furthermore, let* $\mathcal{O}$, g *and* k *be as before, see* (5.5.51), (5.5.52) *and* (5.5.64), *respectively. Then, the BRST transformation of interacting fields localized at* $y \in \mathcal{O}$ *has the following values: For the basic interacting fields* (*with derivatives*)

$$[Q_{\mathcal{L}}(g,k)\,,\, A^\mu_{\mathcal{L}(g)}(y)]_- = i\,\partial^\mu u_0(y)\ , \qquad [Q_{\mathcal{L}}(g,k)\,,\, \partial_\mu A^\mu_{\mathcal{L}(g)}(y)]_- = 0\ , \tag{5.5.69}$$

$$[Q_{\mathcal{L}}(g,k)\,,\, \psi_{\mathcal{L}(g)}(y)]_- = -e\hbar\, \psi_{\mathcal{L}(g)}(y)\, u_0(y)\ ,$$

$$[Q_{\mathcal{L}}(g,k)\,,\, \overline{\psi}_{\mathcal{L}(g)}(y)]_- = e\hbar\, \overline{\psi}_{\mathcal{L}(g)}(y)\, u_0(y)\ , \tag{5.5.70}$$

$$[Q_{\mathcal{L}}(g,k)\,,\, u_0(y)]_+ = 0\ , \qquad [Q_{\mathcal{L}}(g,k)\,,\, \tilde{u}_0(y)]_+ = -i\,\partial_\mu A^\mu_{\mathcal{L}(g)}(y)\ , \tag{5.5.71}$$

$$[Q_{\mathcal{L}}(g,k)\,,\, F^{\mu\nu}_{\mathcal{L}(g)}(y)]_- = 0\ , \tag{5.5.72}$$

[180] Note that, e.g., the r.h.s. of (5.5.74) can be written as *one* interacting field, namely $\hbar\left(-e\not{A}\psi\, u + i\,(\not{\partial} u)\,\psi\right)_{\mathcal{L}(g)}(y)$ by using (5.5.50).

and for the composite interacting fields

$$[Q_{\mathcal{L}}(g,k)\,,\,j^{\mu}_{\mathcal{L}(g)}(y)]_- = 0\ , \tag{5.5.73}$$

$$[Q_{\mathcal{L}}(g,k)\,,\,(\not{A}\psi)_{\mathcal{L}(g)}(y)]_- = -e\hbar\,(\not{A}\psi)_{\mathcal{L}(g)}(y)\,u_0(y) + i\hbar\,(\not{\partial}u_0)(y)\ \psi_{\mathcal{L}(g)}(y)\,, \tag{5.5.74}$$

$$[Q_{\mathcal{L}}(g,k)\,,\,(\overline{\psi}\not{A})_{\mathcal{L}(g)}(y)]_- = e\hbar\,(\overline{\psi}\not{A})_{\mathcal{L}(g)}(y)\,u_0(y) + i\hbar\overline{\psi}_{\mathcal{L}(g)}(y)\ \not{\partial}u_0(y)\ , \tag{5.5.75}$$

$$[Q_{\mathcal{L}}(g,k)\,,\,L_{\mathcal{L}(g)}(y)]_- = i\,j^{\mu}_{\mathcal{L}(g)}(y)\,\partial_\mu u_0(y)\ , \tag{5.5.76}$$

where $j^\mu := \overline{\psi}\gamma^\mu\psi$. *Here, for brevity, we omit* ρ, *i.e., we write*

$$[Q_{\mathcal{L}}(g,k)\,,\,B_{\mathcal{L}(g)}(y)]_\mp = C_{\mathcal{L}(g)}(y) \quad \textit{for} \quad [Q_{\mathcal{L}}(g,k)\,,\,\rho\big(B_{\mathcal{L}(g)}(y)\big)]_\mp = \rho\big(C_{\mathcal{L}(g)}(y)\big)$$

with B *and* C *being field polynomials.*

We emphasize that the commutation relations given in this theorem are independent of the choice of g (provided we remain in the class of test functions satisfying (5.5.52)). This is a nice illustration of Thm. 3.7.1, which is the basis of the algebraic adiabatic limit. Namely, we may choose for $\mathcal{U}$ an open double cone (3.7.7); then, Thm. 3.7.1 states that the algebraic relations of the algebra $(\tilde{\mathcal{F}}_0(\mathcal{U}),\star)$ are independent of g, this holds in particular for the commutators (5.5.69)–(5.5.76), since $Q_{\mathcal{L}}(g,k) \in \rho\big(\tilde{\mathcal{F}}_0(\mathcal{U})\big)$.

Proof. The proof goes by explicit computation of the various commutators, it is based on the commutation relations derived in Proposition 5.3.1.

Using (5.5.63), we compute $[J^0_{\mathcal{L}(g)}(x), \tilde{F}_0]_\mp$, that is, we have to compute commutators of type

$$\begin{aligned}&[(\partial^0)\partial A_{\mathcal{L}(g)}(x)\otimes(\partial^0)u_0(x), B_{\mathcal{L}(g)}(y)\otimes 1]_-\\ &\quad = \big((\partial^0_x)[\partial A_{\mathcal{L}(g)}(x), B_{\mathcal{L}(g)}(y)]_-\big)\otimes(\partial^0)u_0(x)\end{aligned} \tag{5.5.77}$$

or

$$\begin{aligned}&[(\partial^0)\partial A_{\mathcal{L}(g)}(x)\otimes(\partial^0)u_0(x), 1\otimes G(y)]_+\\ &\quad = (\partial^0)\partial A_{\mathcal{L}(g)}(x)\otimes(\partial^0_x)[u_0(x),G(y)]_+\ ,\end{aligned} \tag{5.5.78}$$

where $G \in \{u_0,\,\tilde{u}_0\}$ and B is a field monomial which does not contain any ghost fields. The symbol "$\otimes$" stands for the tensor product of $(\mathcal{F}_{\text{photon},0}\otimes\mathcal{F}_{\text{spinor},0})$ with $\mathcal{F}_{\text{ghost},0}$. In addition, (∂^0) means that at most one derivative $\frac{\partial}{\partial x_0}$ is there.

The verification of formulas (5.5.71) amounts to the computation of commutators of type (5.5.78): For $[Q_{\mathcal{L}}(g,k), u_0(y)]_+$ we have $G = u_0$ and, hence, $[(\partial^0)u_0(x), G(y)]_+ = 0$; this yields the assertion. For $[Q_{\mathcal{L}}(g,k), \tilde{u}_0(y)]_+$, formulas (5.5.63), (5.5.78) (with $G = \tilde{u}_0$) and (5.1.79) give

$$[Q_{\mathcal{L}}(g,k)\,,\,\tilde{u}_0(y)]_+ = i\int_{|\vec{x}|<r} d\vec{x}\ \partial A_{\mathcal{L}(g)}(x)\,\overset{\leftrightarrow}{\partial}{}^0_x\,D(x-y)\Big|_{x^0=0}\ . \tag{5.5.79}$$

From (5.5.56) we know that $\Box_x (\partial A)_{\mathcal{L}(g)}(x) = 0$ for $x \in \mathfrak{U}$. Therefore, we may apply (A.2.16) to (5.5.79); this yields $[Q_{\mathcal{L}}(g,k) , \tilde{u}_0(y)]_+ = -i\,\partial A_{\mathcal{L}(g)}(y)$.

The proofs of (5.5.69), (5.5.70) and (5.5.76) amount to the computation of commutators of type (5.5.77):

$$[Q_{\mathcal{L}}(g,k) , B_{\mathcal{L}(g)}(y)]_- = \hbar^{-1} \int_{|\vec{x}|<r} d\vec{x}\ [\partial A_{\mathcal{L}(g)}(x) , B_{\mathcal{L}(g)}(y)]_- \overset{\leftrightarrow 0}{\partial_x} u_0(x)\Big|_{x^0=0} .$$

Since both x and y lie in $\mathcal{O}$, we may insert the formulas derived in Proposition 5.3.1 into $[\partial A_{\mathcal{L}(g)}(x), B_{\mathcal{L}(g)}(y)]_-$. For example, we obtain

$$\begin{aligned}[Q_{\mathcal{L}}(g,k) , \psi_{\mathcal{L}(g)}(y)]_- &= -e\hbar\,\psi_{\mathcal{L}(g)}(y) \int_{|\vec{x}|<r} d\vec{x}\ D(x-y) \overset{\leftrightarrow 0}{\partial_x} u_0(x)\Big|_{x^0=0} \\ &= -e\hbar\,\psi_{\mathcal{L}(g)}(y)\, u_0(y) ,\end{aligned}$$

by taking into account $\Box u_0 = 0$ in order to use (A.2.16).

The claim (5.5.74) can be obtained from the first relation of (5.5.70) by means of the field equation for $\psi_{\mathcal{L}(g)}(y)$ (5.3.1):

$$\begin{aligned}-e\,[Q_{\mathcal{L}}(g,k) , (\not{A}\psi)_{\mathcal{L}(g)}(y)]_- &= (i\not{\partial}_y - m)[Q_{\mathcal{L}}(g,k) , \psi_{\mathcal{L}(g)}(y)]_- \\ &= -e\hbar\,(i\not{\partial}_y - m)\Big(\psi_{\mathcal{L}(g)}(y)\, u_0(y)\Big) \\ &= -e\hbar\Big(-e(\not{A}\psi)_{\mathcal{L}(g)}(y)\, u_0(y) + i\,(\not{\partial} u_0)(y)\ \psi_{\mathcal{L}(g)}(y)\Big) .\end{aligned}$$

Analogously, the second relation of (5.5.70) and the field equation for $\overline{\psi}_{\mathcal{L}(g)}(y)$ imply (5.5.75).

Finally the claims (5.5.72) and (5.5.73) can be obtained by applying the derivatives ∂^ν_y and $\Box_y$, respectively, to the first relation of (5.5.69), and by using $F^{\nu\mu}_{\mathcal{L}(g)} = \partial^\nu A^\mu_{\mathcal{L}(g)} - \partial^\mu A^\nu_{\mathcal{L}(g)}$ in the first case, and the field equation for $A^\mu_{\mathcal{L}(g)}(y)$ (5.3.1) and $\Box u_0(y) = 0$ in the second case. □

In the *formal* adiabatic limit, $\partial_\nu A^\nu_{\mathcal{L}(g)}$ converges to $\partial_\nu A^\nu_0$, as noted in (5.3.4). Therefore, if we perform the adiabatic limit by the scaling $g_\lambda(x) := g(\lambda x)$ (with $\lambda \downarrow 0$, cf. (A.6.1)–(A.6.2)) and if we scale the test function $h(x) = h_{(0)}(x^0)\, h_{(1)}(\vec{x})$ (5.5.60) similarly to (5.5.28), i.e.,

$$\tilde{h}_{\lambda\,\mu} = \delta_{\mu 0}\, h_\lambda(x) , \quad h_\lambda(x) := \lambda\, h_{(0)}(\lambda x^0)\, h_{(1)}(\lambda\vec{x}) , \tag{5.5.80}$$

we expect that

$$\lim_{\lambda\downarrow 0} Q_{\mathcal{L}}(g_\lambda, \tilde{h}_\lambda) \equiv \lim_{\lambda\downarrow 0} \int d^4x\ h_\lambda(x)\, J^0_{\mathcal{L}(g_\lambda)}(x) \overset{?}{=} \lim_{\lambda\downarrow 0} \int d^4x\ h_\lambda(x)\, J^0(x) \equiv Q \tag{5.5.81}$$

with Q being the free Kugo–Ojima charge (5.5.34). However, a proof of this statement requires a rigorous control of the adiabatic limit, which is beyond the scope of this book. Whereas in QED the reasoning (5.5.81) seems to be correct, the corresponding argument does certainly not work in massless non-Abelian gauge theories – perturbation theory seems to be unable to describe the long distance properties of these models, as indicated by experiments showing confinement of gluons and quarks.[181]

We therefore prefer *not* to work in the adiabatic limit. The price to pay is that $Q_{\mathcal{L}}(g,\tilde{h}) = Q_{\mathcal{L}}(g,k)$ does not agree with Q. To construct the physical Hilbert space by means of Thm. 5.4.9, we have to check the assumptions made there. As noted above, $Q_{\mathcal{L}}(g,k)$ is Krein-hermitian (5.5.65) and nilpotent (5.5.67); but we need in addition that the zeroth order term

$$Q_{(0)}(k) := \int d^4x\; k_\mu(x)\, J^\mu(x) = \int d^4x\; h(x)\, J^0(x) =: Q_{(0)}(\tilde{h})$$

of $Q_{\mathcal{L}}(g,k)$ satisfies the positivity assumption (5.4.15). We only know that the strong limit

$$\lim_{\lambda\downarrow 0} Q_{(0)}(\tilde{h}_\lambda) = Q \quad (\equiv \text{free BRST charge (5.5.34)})$$

fulfills positivity. However, it is an open question whether this holds for $Q_{(0)}(\tilde{h}_\lambda)$ for any fixed $\lambda > 0$; or, whether the freedom in the choice of k^μ (5.5.64) is sufficient to prove positivity of $Q_{(0)}(k)$.

Spatial compactification. Taking into account the local character of our construction, this problem can be solved by a *spatial compactification.* Following [48], we may embed our double cone $\mathcal{O}$ (5.5.51) isometrically into the cylinder $\mathbb{R} \times C_{\mathfrak{L}}$, where $C_{\mathfrak{L}} \subset \mathbb{R}^3$ is a cube of length $\mathfrak{L}$, $\mathfrak{L} \gg r$, with suitable boundary conditions, and where the first factor denotes the time axis. If we choose the compactification length $\mathfrak{L}$ so large that any signal emitted from $\overline{\mathcal{O}}$ which runs around the cylinder $\mathbb{R} \times C_{\mathfrak{L}}$ cannot reach $\overline{\mathcal{O}}$, the physical properties of the local algebra $\tilde{\mathfrak{F}}_0(\mathcal{O})$ are not changed.

For the Fock space quantization of the free fields on this cylinder and in particular for the choice of suitable boundary conditions we refer to [48, App. A] and [47]. The inductive construction of the perturbation series for the interacting fields is not changed by the compactification, in particular the QED-MWI, the field equations (5.3.1) and the commutation relations given in Proposition 5.3.1 can be adopted without any modification, as explained in the mentioned references and, to some extent, in the following remark.

Remark 5.5.6. Let $\varphi_{\mathfrak{L}}(x)$ be the free, real scalar field to the mass $m \geq 0$ on $\mathbb{R} \times C_{\mathfrak{L}}$ with Dirichlet boundary conditions, more precisely, $\varphi_{\mathfrak{L}}(x^0, \vec{x}) = 0$ if $\vec{x} \in \partial C_{\mathfrak{L}}$. The commutator function

$$\Delta_{\mathfrak{L}}(x,y) := [\varphi_{\mathfrak{L}}(x)\,,\, \varphi_{\mathfrak{L}}(y)]_\star$$

[181] On the theoretical side, a proof of confinement for QCD does not exist, as far as we know.

is the fundamental solution of the Klein–Gordon equation on $\mathbb{R}\times C_{\mathfrak{L}}$ with Dirichlet boundary conditions, which has the explicit form [48, App. A]

$$\Delta_{\mathfrak{L}}(x^0,\vec{x}\,,\,y^0,\vec{y}) = \sum_{s\in S}(-1)^{n(s)}\,\Delta\big(x^0-y^0\,,\,\vec{x}-s(\vec{y})\big)\ , \tag{5.5.82}$$

where Δ is the Minkowski space commutator function and S is the group generated by the reflections on the planes which bound $C_{\mathfrak{L}}$ and $n(s)$ is the number of reflections occurring in s (which is well defined modulo 2). Note that for $\vec{y}\in\partial C_{\mathfrak{L}}$ there is an $s_0\in S$ such that $s_0(\vec{y})=\vec{y}$ and $n(s_0)=1$; therefore, the r.h.s. of (5.5.82) vanishes in this case, as required by the Dirichlet boundary conditions.

Now let $x,y\in\mathcal{O}$. Due to $\mathrm{supp}\,\Delta\subseteq(\overline{V}_+\cup\overline{V}_-)$ only the summand $s=\mathrm{Id}$ contributes on the r.h.s. of (5.5.82), that is,

$$\Delta_{\mathfrak{L}}(x,y)=\Delta(x-y)\qquad\forall x,y\in\mathcal{O}\ . \tag{5.5.83}$$

Therefore, the Minkowski space commutator function $D\equiv\Delta_{m=0}$ appearing in the commutation relations of Proposition 5.3.1 needs not to be modified when we formulate this proposition on $\mathbb{R}\times C_{\mathfrak{L}}$. This illustrates explicitly that the algebraic relations of $\tilde{\mathcal{F}}_0(\mathcal{O})$ are not changed by the compactification.

The definition of the BRST transformation $\tilde{s}$ (5.5.58) is not touched by the compactification, because the region of integration in (5.5.58) is bounded by $|\vec{x}|<r$. The construction of the interacting Kugo–Ojima charge $Q_{\mathcal{L}}(g,\tilde{h})$ can now be simplified as follows: The open neighbourhood $\mathcal{U}$ of $\overline{\mathcal{O}}$ (on which the switching function g is constant: $g|_{\mathcal{U}}=1$ (5.5.52)) needs to be bounded only in the x^0-direction. More precisely, $\mathcal{U}$ is now an open neighbourhood of $\overline{\mathcal{O}}\cup\big([-\varepsilon,\varepsilon]\times C_{\mathfrak{L}}\big)$ (where still $r\gg\varepsilon>0$), and we may replace $h_{(1)}(\vec{x})$ by 1 in the definition of $\tilde{h}$ (5.5.60), because $(x^0,\vec{x})\longmapsto h_{(0)}(x^0)$ is an admissible test function on $\mathbb{R}\times C_{\mathfrak{L}}$. With these modifications, the definition of the interacting Kugo–Ojima charge (5.5.62) goes over into

$$Q_{\mathcal{L}}(g):\mathcal{K}\longrightarrow\mathcal{K},\quad Q_{\mathcal{L}}(g):=\int dx^0\,h_{(0)}(x^0)\int_{C_{\mathfrak{L}}}d\vec{x}\;J^0_{\mathcal{L}(g)}(x^0,\vec{x})\ .$$

Since the region of integration is a subset of the time-slice $[-\varepsilon,\varepsilon]\times C_{\mathfrak{L}}\subset\mathcal{U}$ and since $\partial^x_\mu J^\mu_{\mathcal{L}(g)}(x)=0\quad\forall x\in\mathcal{U}$ (5.5.57), we may apply Gauss integral theorem (A.1.13) to conclude that the result of the spatial integration over $C_{\mathfrak{L}}$ is independent of x^0 and, hence, the arbitrariness in the choice of $h_{(0)}(x^0)$ drops out:

$$Q_{\mathcal{L}}(g)=\int_{C_{\mathfrak{L}},\,|c|<\varepsilon}d\vec{x}\;J^0_{\mathcal{L}(g)}(c,\vec{x})\ . \tag{5.5.84}$$

The term of zeroth order is obtained by setting $g=0$; it agrees with the free Kugo–Ojima charge Q on $\mathbb{R}\times C_{\mathfrak{L}}$, since

$$Q_{(0)}=Q_{\mathcal{L}}(g=0)=\int_{C_{\mathfrak{L}},\,|c|<\varepsilon}d\vec{x}\;J^0(c,\vec{x})\equiv Q\ ;$$

in contrast to (5.5.25), this integral is unproblematic. The proofs that $Q_{\mathcal{L}}(g,\tilde{h})$ (5.5.62)

- implements the BRST transformation (5.5.63),
- is Krein-hermitian (5.5.65),
- has ghost number (-1) (5.5.66)
- and is nilpotent (5.5.67)–(5.5.68),

can easily be translated into proofs of the same properties for the modified version $Q_{\mathcal{L}}(g)$ (5.5.84) – the compactification simplifies these proofs somewhat. Also Thm. 5.5.5 holds true on $\mathbb{R} \times C_{\mathfrak{L}}$ (with $Q_{\mathcal{L}}(g)$ in place of $Q_{\mathcal{L}}(g,k)$), because the proof of this theorem relies on formula (5.5.63) and on Proposition 5.3.1, which are valid also on $\mathbb{R} \times C_{\mathfrak{L}}$. With that, our construction of QED is complete.

We emphasize that our construction describes locally QED also in the *non-compactified* Minkowski space (this is our main concern) and, therefore, should not depend on the compactification length $\mathfrak{L}$. On the level of the algebras this is evident: The local algebras of interacting fields or observables belonging to different values of $\mathfrak{L}$ are *isomorphic*. We conjecture that also the state space (i.e., the set of expectation functionals induced by vectors in $\frac{\mathrm{Ker}\, Q_{\mathcal{L}}(g)}{\mathrm{Ran}\, Q_{\mathcal{L}}(g)}$) is independent of $\mathfrak{L}$, but this is not yet proved.

5.6 Reasons in favour of a LOCAL construction

From experiments we know that strong interactions have the property of confinement. Therefore, perturbation theory fails in the description of the long distance behaviour. Hence, a *local* construction of the observables (as developed in this chapter for QED) seems to be the best one can do in a perturbative approach to QCD. This is worked out in references [53, 54]; however, with the gap that a proof of the MWI expressing BRST symmetry, which is needed for this construction, is missing in these papers. This gap was closed by S. Hollands [105] (cf. below), by using (among other techniques) our version of the "Quantum Action Principle" (Thm. 4.3.1). Note that also these subsequent developments (in particular the papers [53, 105]) need a spatial compactification, to prove, by means of the stability under deformations given in Thm. 5.4.9, that the space of physical (vector) states $\frac{\mathrm{Ker}\, Q_{\mathcal{L}}}{\mathrm{Ran}\, Q_{\mathcal{L}}}$ is positive definite. A nice application of this formalism would be an explicit computation of the operator product expansion for QCD, by using also the results and techniques of [104].

A *local* construction of a QFT model has also the following crucial advantages, which are not restricted to gauge theories:

- There is a clear separation of the IR-problem (in causal perturbation theory this is the existence and uniqueness of the adiabatic limit) from the UV-problem (extension of retarded or time-ordered products to coinciding points). The former is absent as long as we switch the interaction with a test function $g \in \mathcal{D}(\mathbb{M})$. In particular the MWI, which expresses the symmetries of the model, can be formulated and proved independently of the IR-behaviour. At the level of the net $\mathcal{O} \longmapsto \mathcal{A}_{\mathcal{L}}(\mathcal{O})$ of local observables, the dependence on the IR-cutoff $g \in \mathcal{D}(\mathbb{M})$ can be removed in a well-defined manner by the algebraic adiabatic limit (see Sect. 3.7); in particular this limit can be applied to massless non-Abelian gauge theories. We recall that the algebraic adiabatic limit enables a treatment of the conventional form of the renormalization group, as a renormalization of the coupling *constants*, without meeting any IR-divergences (Sect. 3.8).

Also quantum states are well defined independently of the IR-behaviour; they are functionals on the local algebras $\mathcal{A}_{\mathcal{L}}(\mathcal{O})$ with certain properties (see Sects. 2.5 and 5.4.3). For QED we have constructed the space of physical states on $\mathcal{A}_{\mathcal{L}}(\mathcal{O})$ in terms of the Fock–Krein space representation of $\mathcal{A}_{\mathcal{L}}(\mathcal{O})$ and the interacting Kugo–Ojima charge $Q_{\mathcal{L}}(g)$. But, in the construction of scattering states, we would meet the traditional IR-divergences, cf. App. A.6.

- The generalization of QFT from Minkowski spacetime to a globally hyperbolic curved spacetime encounters serious conceptual problems, in particular the non-existence of a preferred vacuum state on time-dependent spacetimes. In addition, useful flat spacetime techniques, such as "momentum space" and "functional integral", are not applicable. It has turned out that these problems can be overcome by using algebraic QFT [94], that is, by a local construction of the observables. For a perturbative construction of the interacting fields in this framework, causal perturbation theory seems to be the best method; this is worked out in [23–25, 100–103]. This series of papers culminates in a construction of "Renormalized Quantum Yang–Mills fields in Curved Spacetime" by S. Hollands [105], which is based on the local construction of QED given in this chapter.

Appendix

A.1 Some notations and a few mathematical preliminaries

Numbers and permutations. For the set of natural numbers we use the French convention:

$$\mathbb{N} = \{0, 1, 2, 3, \dots\} \quad \text{and} \quad \mathbb{N}^* := \mathbb{N} \setminus \{0\}\ .$$

By S_n we mean the *symmetric group*, i.e., the group of all permutations of the set $\{1, \dots, n\}$. For the signature of a permutation $\pi \in S_n$ we use the notation $\operatorname{sgn} \pi$; for brevity we sometimes write πj instead of $\pi(j)$, where $j \in \{1, \dots, n\}$.

Complex numbers $z \in \mathbb{C}$: By $\overline{z} = \operatorname{Re} z - i \operatorname{Im} z$ we denote the complex conjugate of $z = \operatorname{Re} z + i \operatorname{Im} z$, where $\operatorname{Re} z$, $\operatorname{Im} z \in \mathbb{R}$. In the presence of spinor fields we write z^c instead of $\overline{z}$, to avoid confusion with the overline denoting the conjugate spinor field $\overline{\psi}(x) := \psi^\dagger(x)\gamma^0$.

Algebras. An *algebra* $\mathcal{A} \equiv (\mathcal{A}, \cdot)$ over $\mathbb{K} = \mathbb{R}$ or $\mathbb{K} = \mathbb{C}$ is a $\mathbb{K}$-vector space with a product $\cdot : \mathcal{A} \times \mathcal{A} \longrightarrow \mathcal{A}$ which is

- associative, $(A \cdot B) \cdot C = A \cdot (B \cdot C) \quad \forall A, B, C \in \mathcal{A}$,
- distributive, $A \cdot (B + C) = A \cdot B + A \cdot C$ and $(B + C) \cdot A = B \cdot A + C \cdot A$ for all $A, B, C \in \mathcal{A}$,
- and satisfies $\alpha(A \cdot B) = (\alpha A) \cdot B = A \cdot (\alpha B) \quad \forall A, B \in \mathcal{A},\ \alpha \in \mathbb{K}$.

An algebra $\mathcal{A}$ is *unital* if and only if it has a unit, that is, there exists an element $\mathbf{1} \in \mathcal{A}$ such that $\mathbf{1} \cdot A = A = A \cdot \mathbf{1}$ for all $A \in \mathcal{A}$.

A $*$-*algebra* is a $\mathbb{C}$-algebra with an *involution*, that is, a map $* : \mathcal{A} \longrightarrow \mathcal{A}$; $A \longmapsto A^*$ (called "$*$-operation") satisfying

$$(A + \alpha B)^* = A^* + \overline{\alpha}\, B^*\ , \quad (A \cdot B)^* = B^* \cdot A^* \quad \text{and} \quad (A^*)^* = A\ ,$$

for all $A, B \in \mathcal{A}\ ,\ \alpha \in \mathbb{C}$.

An *algebra isomorphism* is a map $f : (\mathcal{A}_1, \cdot_1) \longrightarrow (\mathcal{A}_2, \cdot_2)$, where $(\mathcal{A}_j, \cdot_j)$, $j = 1, 2$, are algebras,

- which is a vector space isomorphism, that is, f is *bijective* and *linear*,

M. Dütsch, *From Classical Field Theory to Perturbative Quantum Field Theory*,
Progress in Mathematical Physics 74, https://doi.org/10.1007/978-3-030-04738-2

- and which intertwines the products:

$$f(A \cdot_1 B) = f(A) \cdot_2 f(B) \qquad \forall A, B \in \mathcal{A}_1 \ .$$

If f is only linear and intertwines the products, then it is called an *algebra homomorphism*.

Formal power series. Let $\mathcal{V}$ be a $\mathbb{C}$-vector space and $\lambda \in \mathbb{R}\setminus\{0\}$. The set of *formal power series in* λ whose coefficients are elements of $\mathcal{V}$,

$$\mathcal{V}[\![\lambda]\!] := \left\{ V \equiv \sum_{n=0}^{\infty} V_n \lambda^n \equiv (V_n)_{n\in\mathbb{N}} \,\middle|\, V_n \in \mathcal{V} \right\} , \tag{A.1.1}$$

is also a $\mathbb{C}$-vector space: $(V + cW)_n := V_n + cW_n$ for $c \in \mathbb{C}$.

If $\mathcal{V}$ is a $*$-algebra (the product may be non-commutative), then $\mathcal{V}[\![\lambda]\!]$ is also a $*$-algebra:

$$\begin{aligned} \Big(\sum_{r=0}^{\infty} V_r \lambda^r\Big) \cdot \Big(\sum_{s=0}^{\infty} V_s \lambda^s\Big) &:= \sum_{n=0}^{\infty} \Big(\sum_{k=0}^{n} V_k \cdot V_{n-k}\Big) \lambda^n \ , \\ \Big(\sum_{r=0}^{\infty} V_r \lambda^r\Big)^* &:= \sum_{r=0}^{\infty} V_r^* \lambda^r \ . \end{aligned} \tag{A.1.2}$$

If $\mathcal{V}$ has a unit $\mathbf{1}$, i.e., $\mathbf{1} \cdot V = V = V \cdot \mathbf{1}\ \forall V \in \mathcal{V}$, then this holds also for $\mathcal{V}[\![\lambda]\!]$; the unit reads $\mathbf{1}\,\lambda^0 + 0 \equiv (\mathbf{1}, 0, 0, 0, \ldots)$.

One verifies straightforwardly that a formal power series of the form

$$V = c\,\mathbf{1} + \sum_{n=1}^{\infty} V_n \lambda^n \in \mathcal{V}[\![\lambda]\!] \quad \text{(where } c \in \mathbb{C}\text{), is } \textit{invertible} \text{ if and only if} \quad c \neq 0 \ . \tag{A.1.3}$$

The first terms of the inverse read:

$$\begin{aligned} V^{-1} = c^{-1}\,\mathbf{1} - c^{-2}\, V_1\, \lambda + \big(-c^{-2}\, V_2 + c^{-3}\, V_1 \cdot V_1\big)\, \lambda^2 \\ + \big(-c^{-2}\, V_3 + c^{-3}(V_2 \cdot V_1 + V_1 \cdot V_2) - c^{-4}\, V_1 \cdot V_1 \cdot V_1\big)\, \lambda^3 + \cdots \ . \end{aligned} \tag{A.1.4}$$

Completion of the tensor algebra over a vector space and pertinent linear maps. By the completion of the tensor algebra over a $\mathbb{C}$-vector space $\mathcal{V}$ we mean

$$\mathcal{T}(\mathcal{V}) := \overline{\mathbb{C} \oplus \bigoplus_{n=1}^{\infty} \mathcal{V}^{\otimes n}} \ , \tag{A.1.5}$$

where the completion concerns the infinite direct sum:[182] $\mathcal{T}(\mathcal{V})$ is not restricted to $(\otimes, \oplus)$-polynomials, it contains also $\oplus$-sums with infinitely many non-vanishing

[182] Alternatively one may write $\mathcal{T}(\mathcal{V})$ as infinite *cartesian* product: $\mathcal{T}(\mathcal{V}) := \mathbb{C} \times \big(\times_{n=1}^{\infty} \mathcal{V}^{\otimes n}\big)$.

summands; a typical element is

$$e_{\otimes}^{V} := 1 \oplus \bigoplus_{n=1}^{\infty} V^{\otimes n}/n! \equiv (1,\, V,\, \tfrac{1}{2!}V^{\otimes 2}, \ldots) \in \mathcal{T}(\mathcal{V})\ , \quad \text{where} \quad V \in \mathcal{V}\ . \tag{A.1.6}$$

Let $\mathcal{W}$ be a further vector space. A sequence of linear maps

$$f_n \,:\, \mathcal{V}^{\otimes n} \longrightarrow \mathcal{W}\ , \quad n \in \mathbb{N}\ , \quad \mathcal{V}^{\otimes 0} := \mathbb{C}\ ,$$

may be interpreted as a linear map

$$f \equiv (f_n) \,:\, \begin{cases} \mathcal{T}(\lambda\mathcal{V}) \longrightarrow \mathcal{W}[\![\lambda]\!] \\ f\big(c + \sum_{n=1}^{\infty} V_n \lambda^n\big) := f_0(c) + \sum_{n=1}^{\infty} f_n(V_n)\,\lambda^n \end{cases} \tag{A.1.7}$$

for $c \in \mathbb{C}$, $V_n \in \mathcal{V}^{\otimes n}$, in particular

$$f\big(e_{\otimes}^{\lambda V}\big) = f_0(1) + \sum_{n=1}^{\infty} \frac{f_n(V^{\otimes n})}{n!}\,\lambda^n\ . \tag{A.1.8}$$

For vector spaces $\mathcal{V}$, $\mathcal{W}$ note that a linear map $f_n \,:\, \mathcal{V}^{\otimes n} \longrightarrow \mathcal{W}$ may be identified with a multilinear map $\tilde{f}_n \,:\, \mathcal{V}^{\times n} \longrightarrow \mathcal{W}$. For simplicity we sometimes write $f_n(v_1, \ldots, v_n)$ instead of $f_n(v_1 \otimes \cdots \otimes v_n)$. If, in addition, such a map is symmetrical, i.e.,

$$f_n(v_{\pi 1} \otimes \cdots \otimes v_{\pi n}) = f_n(v_1 \otimes \cdots \otimes v_n) \qquad \forall \pi \in S_n\ ,$$

it suffices to know f_n for diagonal entries, that is, $f_n(v \otimes \cdots \otimes v)$ $\forall v \in \mathcal{V}$. From that, the values of f_n for arbitrary entries can be obtained by[183]

$$f_n(v_1 \otimes \cdots \otimes v_n) = \frac{1}{n!}\,\frac{\partial^n}{\partial\lambda_1 \cdots \partial\lambda_n}\, f_n\Big(\big(\sum_{j=1}^{n} \lambda_j v_j\big)^{\otimes n}\Big)\ . \tag{A.1.9}$$

If $f = (f_n)$ is a sequence of linear maps of type (A.1.7) and if each f_n is symmetrical, then we get

$$f_n(v_{j_1} \otimes \cdots \otimes v_{j_n}) = \frac{\partial^n}{\partial\lambda_{j_1} \cdots \partial\lambda_{j_n}}\Big|_{\lambda_1 = \ldots = \lambda_J = 0} f\Big(e_{\otimes}^{\sum_{j=1}^{J} \lambda_j v_j}\Big) \tag{A.1.10}$$

for all $n \geq 1\ ,\ 1 \leq j_k \leq J$.

---

[183] With regard to this formula note that, by linearity, $f_n\Big(\big(\sum_{j=1}^{n} \lambda_j v_j\big)^{\otimes n}\Big)$ is of the form $\sum_k P_k(\lambda_1, \ldots, \lambda_n)\, w_k$, where the sum over k is finite and for each k the map $P_k : \mathbb{R}^n \longrightarrow \mathbb{C}$ is a monomial of degree n and $w_k \in \mathcal{W}$. We use the obvious definition

$$\partial_{\lambda_j}\Big(\sum_k P_k(\lambda_1, \ldots, \lambda_k)\, w_k\Big) := \sum_k \Big(\partial_{\lambda_j} P_k(\lambda_1, \ldots, \lambda_k)\Big)\, w_k\ .$$

The Minkowski space. By $\mathbb{M}_d \equiv \mathbb{M}$ we mean the d-dimensional Minkowski space; we assume $d > 2$. For $x \equiv (x^\mu)_{\mu=0,\dots,d-1} \in \mathbb{M}$ we sometimes write $x =: (x^0, \vec{x}) =: (t, \vec{x})$. The sign convention for the Minkowski metric reads

$$g := \mathrm{diag}(+, -, \dots, -) \ .$$

As usual, the forward and backward light cones are defined by

$$V_+ := \{x \in \mathbb{M} \,|\, x^2 > 0,\, x^0 > 0\} \ , \quad V_- := \{x \in \mathbb{M} \,|\, x^2 > 0,\, x^0 < 0\} \ , \qquad \text{(A.1.11)}$$

and $\overline{V}_+$ and $\overline{V}_-$ are the closures of these open sets. The relevant symmetry groups of the Minkowski space are $\mathcal{L}_+^\uparrow$, which is the proper, orthochronous Lorentz group, and $\mathcal{P}_+^\uparrow$, which is the pertinent Poincaré group. For Epstein–Glaser renormalization, the *thin diagonal* in $\mathbb{M}^n$, which is defined as

$$\Delta_n := \{\, (x_1, \dots, x_n) \in \mathbb{M}^n \,|\, x_1 = x_2 = \dots = x_n \,\} \ , \qquad \text{(A.1.12)}$$

plays an important role.

In four-dimensional Minkowski spacetime the *Integral Theorem of Gauss* takes the following Lorentz covariant form: For a compact subset G of $\mathbb{M}$ with a sufficiently smooth boundary ∂G and a vector field $v = (v^\mu) \in C^1(\mathbb{M}, \mathbb{R}^4)$ it holds that

$$\int_G d^4x \; \partial_\mu v^\mu(x) = \oint_{\partial G} d\sigma_\mu(x)\, v^\mu(x) \ , \qquad \text{(A.1.13)}$$

where the 4-vector $d\sigma \equiv (d\sigma^\mu)$ is pointing to the outside of G and is given by

$$(\pm d\sigma)_\mu(x) := \varepsilon_{\mu\alpha\beta\gamma} \, \frac{\partial x^\alpha}{\partial u} \, \frac{\partial x^\beta}{\partial v} \, \frac{\partial x^\gamma}{\partial w} \, du\, dv\, dw \ ,$$

with a C^1-parametrization $\big(x^\nu(u,v,w)\big)_{\nu=0,\dots,3}$ of the three-dimensional surface ∂G. For a *spacelike* hypersurface Σ of $\mathbb{M}$ (i.e., $(x-y)^2 < 0 \;\; \forall x, y \in \Sigma$ with $x \neq y$), the infinitesimal surface element $d\sigma$ is a timelike vector:

$$d\sigma^\mu(x) = n^\mu(x)\, d\tilde{\sigma}(x) \quad \text{with} \quad n^\mu(x)\, n_\mu(x) = 1$$

and a Euclidean, scalar surface element $d\tilde{\sigma}(x)$. A simple example is the family of surfaces $x^0 =$ constant, for which we get

$$d\sigma = (\pm 1,\, 0,\, 0,\, 0)\; dx^1\, dx^2\, dx^3 \ .$$

Function spaces and distributions. We frequently use the shorthand notation

$$\mathcal{C} := C^\infty(\mathbb{M}, \mathbb{R}) \equiv C^\infty(\mathbb{R}^d, \mathbb{R}) \ . \qquad \text{(A.1.14)}$$

In view of the generalization to curved spacetimes our test function space is usually $\mathcal{D}(\Omega) \equiv \mathcal{D}(\Omega, \mathbb{C})$, where Ω is an open subset of $\mathbb{M}^n$ ($n \in \mathbb{N}^*$), frequently we have

$\Omega = \mathbb{M}^n$. $\mathcal{D}(\Omega)$ is the space of all smooth functions $f \in C^\infty(\Omega, \mathbb{C})$, whose support is compact (with respect to the Euclidean norm of $\mathbb{R}^{dn}$) and is a subset of Ω. For the definition of the topology of the function space $\mathcal{D}(\Omega)$, we refer to the literature, e.g., [141].

The relation $\mathcal{D}(\Omega) \subseteq \mathcal{D}(\mathbb{M}^n)$ (for $\Omega \subseteq \mathbb{M}^n$) is reversed for the corresponding spaces of distributions: $\mathcal{D}'(\mathbb{M}^n) \subseteq \mathcal{D}'(\Omega)$. For the application of a distribution $t \in \mathcal{D}'(\Omega)$ to a test function $g \in \mathcal{D}(\Omega)$ we use several notations

$$t(g) \equiv \langle t, g\rangle \equiv \big\langle t(x), g(x)\big\rangle_x \equiv \int_\Omega d^{dn}x\; t(x)\, g(x) \in \mathbb{C}\ .$$

Convergence of a sequence of distributions is understood with respect to the *weak topology*, that is, for $t_n, t \in \mathcal{D}'(\Omega)$ (or $\mathcal{S}'(\mathbb{M}^n)$, introduced below) we define

$$\lim_{n\to\infty} t_n = t \quad \text{iff} \quad \lim_{n\to\infty} \langle t_n, g\rangle = \langle t, g\rangle \quad \forall g \in \mathcal{D}(\Omega) \text{ (or } g \in \mathcal{S}(\mathbb{M}^n)\,). \tag{A.1.15}$$

The formula of Sokhotski–Plemelj,

$$\frac{1}{x \pm i0} = \mathcal{P}\frac{1}{x} \mp i\pi\, \delta(x) \quad \text{in} \quad \mathcal{D}'(\mathbb{R}) \quad \text{or} \quad \mathcal{S}'(\mathbb{R})\ , \tag{A.1.16}$$

is frequently used, it can be interpreted as a splitting into real- and imaginary part of $\frac{1}{x\pm i0}$.

The *support* $\operatorname{supp} t$ *of a distribution* $t \in \mathcal{D}'(\Omega)$ is defined to be the smallest closed set $K(\subseteq \Omega)$ such that $t\big|_{\mathcal{D}(\Omega\setminus K)} = 0$. For example, $\operatorname{supp} \partial^a \delta_{(n)} = \{0\}$ for any $a \in \mathbb{N}^n$, where $\delta_{(n)}$ denotes the Dirac delta distribution in $\mathbb{R}^n$.

For Epstein–Glaser renormalization, the following theorem is important: Every distribution $t \in \mathcal{D}'(\mathbb{R}^n)$ satisfying

$$\operatorname{supp} t \subseteq \{0\} \quad \text{is a linear combination of derivatives of } \delta_{(n)}\text{:} \quad t = \sum_a C_a\, \partial^a \delta_{(n)} \tag{A.1.17}$$

for some coefficients $C_a \in \mathbb{C}$.

For a mathematician some notations may be sloppy; in particular, for $t \in \mathcal{D}'(\mathbb{R}^n)$ and an open set $\Omega \subset \mathbb{R}^n$, the statement "$t(x) = 0\ \forall x \in \Omega$" should be read as "$t\big|_{\mathcal{D}(\Omega)} = 0$".

Fourier transformation. When using Fourier transformation we work with the space $\mathcal{S}(\mathbb{M}^n) \equiv \mathcal{S}(\mathbb{R}^{dn}, \mathbb{C})$ of smooth and "rapidly decreasing" functions (also called "Schwartz functions") and consider the distributions as elements of $\mathcal{S}'(\mathbb{M}^n)$. For the definition of the topology of $\mathcal{S}(\mathbb{M}^n)$ see any book on Functional Analysis, e.g., [141].

The Fourier transformation can be defined as a map $\mathfrak{F} : \mathcal{S}(\mathbb{M}^n) \longrightarrow \mathcal{S}(\mathbb{M}^n)$; we use the conventions

$$(\mathfrak{F}f)(k) \equiv \hat{f}(k) := \frac{1}{(2\pi)^{dn/2}} \int d^{dn}x\; f(x)\, e^{ikx}\ , \quad \text{hence}$$

$$(\mathfrak{F}^{-1}f)(k) \equiv \check{f}(k) = \hat{f}(-k)\ ,$$

where $kx := \sum_{l=1}^{n}(k_l^0 x_l^0 - \vec{k}_l \vec{x}_l)$ and $f \in \mathcal{S}(\mathbb{M}^n)$. The so-defined map $\mathfrak{F}$ is an automorphism of $\mathcal{S}(\mathbb{M}^n)$, that is, $\mathfrak{F}$ is linear, bijective and both $\mathfrak{F}$ and $\mathfrak{F}^{-1}$ are continuous.

The Fourier transformation of distributions $t \in \mathcal{S}'(\mathbb{M}^n)$ (which we denote also by $\mathfrak{F}$) is defined in terms of the Fourier transformation of test functions:

$$\big\langle \mathfrak{F}(t), g \big\rangle := \big\langle t, \mathfrak{F}(g) \big\rangle \qquad \forall g \in \mathcal{S}(\mathbb{M}^n)\ . \tag{A.1.18}$$

The so-obtained map $\mathfrak{F}\colon \mathcal{S}'(\mathbb{M}^n) \longrightarrow \mathcal{S}'(\mathbb{M}^n)$ is also an automorphism.

Derivatives of a function of functionals. Let H be a map $H : \mathcal{F} \longrightarrow \mathcal{F}$ or $H : \mathcal{F}_{\mathrm{loc}} \longrightarrow \mathcal{F}$ or $H : \mathcal{F}_{\mathrm{loc}} \longrightarrow \mathcal{F}_{\mathrm{loc}}$; moreover, $\mathcal{F}$ and/or $\mathcal{F}_{\mathrm{loc}}$ may be replaced by $\mathcal{F}[\![\hbar, \kappa]\!]$ and/or $\mathcal{F}_{\mathrm{loc}}[\![\hbar, \kappa]\!]$, respectively.[184] Such a map H is N-times differentiable at $F \in \mathcal{F}_{[\mathrm{loc}]}$ if and only if

$$H^{(n)}(F)(G_1 \otimes \cdots \otimes G_n)(h) := \frac{d}{d\lambda}\Big|_{\lambda=0} H^{(n-1)}(F + \lambda G_n)(G_1 \otimes \cdots \otimes G_{n-1})(h) \tag{A.1.19}$$

exists for all $h \in \mathcal{C}$, $G_1 \otimes \cdots \otimes G_n \in \mathcal{F}_{[\mathrm{loc}]}^{\otimes n}$ and for all $1 \le n \le N$. By applying the functional to an arbitrary configuration $h \in \mathcal{C}$, we have traced back differentiability of H to differentiability of a function from $\mathbb{R}$ to $\mathbb{C}$ – in accordance with (1.2.3). This recursive definition of the nth derivative of H at F can be written non-recursively:

$$H^{(n)}(F)(G_1 \otimes \cdots \otimes G_n) = \frac{\partial^n}{\partial\lambda_1 \cdots \partial\lambda_n}\Big|_{\lambda_1=\cdots=\lambda_n=0} H\Big(F + \sum_{j=1}^{n} \lambda_j G_j\Big)\ , \tag{A.1.20}$$

as one verifies easily by induction on n; for brevity the application to $h \in \mathcal{C}$ is omitted. For diagonal entries, this formula simplifies to

$$H^{(n)}(F)(G^{\otimes n}) = \frac{d^n}{d\lambda^n}\Big|_{\lambda=0} H(F + \lambda G)\ . \tag{A.1.21}$$

We are now going to show that

$$H^{(n)}(F) : \mathcal{F}_{[\mathrm{loc}]}^{\otimes n} \longrightarrow \mathcal{F}_{[\mathrm{loc}]} \quad \text{is linear and totally symmetric.} \tag{A.1.22}$$

The latter, i.e., that $H^{(n)}(F)(G_1 \otimes \cdots \otimes G_n)$ is invariant under permutations of $G_1, \ldots, G_n$, follows immediately from (A.1.20). To prove linearity, we proceed by induction on n. With that, we know that the r.h.s. of (A.1.19) is linear in

[184] The spaces of functionals $\mathcal{F}$ and $\mathcal{F}_{\mathrm{loc}}$ are defined in Definitions 1.2.1 and 1.3.4, respectively. In the following we write "$\mathcal{F}_{[\mathrm{loc}]}$" for "$\mathcal{F}$ or $\mathcal{F}_{\mathrm{loc}}$".

$G_1 \otimes \cdots \otimes G_{n-1}$, and it remains to verify linearity in G_n:

$$H^{(n)}(F)\big(G_1 \otimes \cdots \otimes G_{n-1} \otimes (G_n + \alpha\tilde{G}_n)\big) = \lim_{\lambda\to 0} \frac{1}{\lambda}$$
$$\cdot \Big(H^{(n-1)}\big(F + \lambda(G_n + \alpha\,\tilde{G}_n)\big) - H^{(n-1)}(F + \lambda\alpha\,\tilde{G}_n)\Big)(G_1 \otimes \cdots \otimes G_{n-1})$$
$$+ \alpha \lim_{\lambda\to 0} \frac{1}{\lambda\alpha}\Big(H^{(n-1)}(F + \lambda\alpha\,\tilde{G}_n) - H^{(n-1)}(F)\Big)(G_1 \otimes \cdots \otimes G_{n-1})$$
$$= H^{(n)}(F)(G_1 \otimes \cdots \otimes G_{n-1} \otimes G_n) + \alpha\, H^{(n)}(F)(G_1 \otimes \cdots \otimes G_{n-1} \otimes \tilde{G}_n)\,.$$

Shorthand notations for powers and higher partial derivatives. For powers and higher partial derivatives we introduce a multi-index notation: let $a = (a_1, \ldots, a_k) \in \mathbb{N}^k$ and $x = (x_1, \ldots, x_k) \in \mathbb{R}^k$; with that we write

$$|a| := a_1 + \cdots + a_k\,, \quad a! := a_1! \cdot \cdots \cdot a_k!\,, \quad x^a := x_1^{a_1} \cdot \cdots \cdot x_k^{a_k}\,, \quad \partial_x^a := \partial_{x_1}^{a_1} \cdots \partial_{x_k}^{a_k}$$

or shorter $\partial^a := \partial_1^{a_1} \cdots \partial_k^{a_k}$. In particular for $x = (x_1, \ldots, x_n) \in \mathbb{M}^n$ and a function $f \in C^\infty(\mathbb{M}^n, \mathbb{C})$ or a distribution $f \in \mathcal{D}'(\mathbb{M}^n)$, this takes the following form: $a := (a_j^\mu)_{j=1,\ldots,n}^{\mu=0,\ldots,d-1} \in (\mathbb{N}^d)^n$ and

$$|a| := \sum_{j=1}^{n}\sum_{\mu=0}^{d-1} a_j^\mu\,, \quad a! := \prod_{j=1}^{n}\prod_{\mu=0}^{d-1} a_j^\mu!\,, \quad x^a := \prod_{j=1}^{n}\prod_{\mu=0}^{d-1} (x_j^\mu)^{a_j^\mu}\,,$$
$$\partial^a f(x) := \Big(\prod_{j=1}^{n}\prod_{\mu=0}^{d-1} (\partial_{x_j^\mu})^{a_j^\mu}\Big)\, f(x)\,. \tag{A.1.23}$$

For powers of $\log x$ and of the field $\varphi(x)$ we use a notation which differs from the one for powers of distributions as, e.g., the propagators treated in the next appendix or the Hadamard function $H_m^\mu \in \mathcal{D}'(\mathbb{M})$:

$$\log^k x := (\log x)^k\,, \quad \varphi^k(x) := \big(\varphi(x)\big)^k$$

versus

$$\Delta_m^+(x)^k := \big(\Delta_m^+(x)\big)^k\,, \quad H_m^\mu(x)^k := \big(H_m^\mu(x)\big)^k\,, \quad k \in \mathbb{N}\,.$$

A.2 Propagators: Conventions and properties

To a large extent, this appendix is a summary of definitions and results given in the main text. With regard to factors (-1), i and 2π in the definition of propagators one finds various conventions in the literature. We work throughout with the following definitions.

- The *retarded propagator* Δ_m^{ret} is the retarded Green function of the Klein–Gordon operator, i.e., it is defined by

$$(\square + m^2)\Delta_m^{\mathrm{ret}}(x) = -\delta(x) \quad \text{and} \quad \operatorname{supp}\Delta_m^{\mathrm{ret}} \subseteq \overline{V}_+\,; \tag{A.2.1}$$

and can be written as

$$\Delta_m^{\rm ret}(x) = \frac{1}{(2\pi)^d} \int d^dp \, \frac{e^{-ipx}}{p^2 - m^2 + ip^0 0} \,. \tag{A.2.2}$$

$\Delta_m^{\rm ret}$ is the propagator of classical retarded fields:

$$R_{{\rm cl}\,1,1}\big(\varphi(y), \varphi(x)\big) = -\Delta_m^{\rm ret}(x-y) \,.$$

- The *Jordan–Pauli "function"* or *"commutator function"* is defined by

$$\Delta_m(x) := \Delta_m^{\rm ret}(x) - \Delta_m^{\rm ret}(-x) = \frac{-i}{(2\pi)^{d-1}} \int d^dp \; {\rm sgn}\, p^0 \, \delta(p^2 - m^2) \, e^{-ipx} \tag{A.2.3}$$

(the momentum integral representation is obtained by inserting (A.2.2) and using (A.1.16)). Besides the mathematical properties

$$\begin{aligned} &\Delta_m(-x) = -\Delta_m(x) \,, \quad (\Box + m^2)\Delta_m(x) = 0 \,, \\ &{\rm supp}\, \Delta_m \subseteq (\overline{V}_+ \cup \overline{V}_-) \quad \text{and} \quad \Delta_m(x)\, \theta(x^0) = \Delta_m^{\rm ret}(x) \,, \end{aligned} \tag{A.2.4}$$

it is the propagator of the Poisson bracket:

$$\frac{1}{i\hbar}\, [\varphi(x), \varphi(y)]_\star = \{\varphi(x), \varphi(y)\} = \Delta_m(x-y) \,.$$

- The *Wightman two-point function* is the "positive frequency part of $i\Delta_m$", that is,

$$\Delta_m^+(x) := \frac{1}{(2\pi)^{d-1}} \int d^dp \, \theta(p^0) \, \delta(p^2 - m^2) \, e^{-ipx} \,. \tag{A.2.5}$$

It fulfills

$$(\Box + m^2)\Delta_m^+(x) = 0 \,.$$

Δ_m^+ is a possible propagator of the star product:

$$\omega_0\big(\varphi(x) \star \varphi(y)\big) = \hbar \, \Delta_m^+(x-y) \,; \tag{A.2.6}$$

a necessary condition for that is the property

$$-i\big(\Delta_m^+(x) - \Delta_m^+(-x)\big) = \Delta_m(x) \,. \tag{A.2.7}$$

Obviously, if $x^2 < 0$ then $\Delta_m^+(-x) = \Delta_m^+(x)$, but this is wrong for $x^2 \geq 0$.

- The *Feynman propagator* is defined by

$$\Delta_m^F(x) := \theta(x^0)\, \Delta_m^+(x) + \theta(-x^0)\, \Delta_m^+(-x) \,. \tag{A.2.8}$$

It can be written as

$$\Delta_m^F(x) = \frac{i}{(2\pi)^d} \int d^d p \, \frac{e^{-ipx}}{p^2 - m^2 + i0} ,$$

and fulfills

$$\begin{aligned} &\Delta_m^F(-x) = \Delta_m^F(x) , \quad (\Box + m^2)\Delta_m^F(x) = -i\delta(x) , \\ &\Delta_m^F(x) = i\,\Delta_m^{\rm ret}(x) + \Delta_m^+(-x) . \end{aligned} \tag{A.2.9}$$

Δ_m^F is the propagator of the time-ordered product:

$$\omega_0\big(T_2(\varphi(x) \otimes \varphi(y))\big) = \hbar\,\Delta_m^F(x-y) .$$

All propagators $p_m = \Delta_m^{\rm ret}$, Δ_m, Δ_m^+, Δ_m^F are Lorentz-invariant and scale homogeneously with degree $(d-2)$:

$$p_m(\Lambda x) = p_m(x) \quad \text{for all} \quad \Lambda \in \mathcal{L}_+^\uparrow \quad \text{and} \quad \rho^{d-2}\, p_{\rho^{-1}m}(\rho x) = p_m(x) \text{ for all } \rho > 0 .$$

The wave front sets of Δ_m, Δ_m^+, Δ_m^F and $\Delta^{\rm ret}$ are given in (1.8.9), (2.2.7) and (A.3.5)–(A.3.6), respectively.

The corresponding propagators for a *massless* scalar field are obtained by the limit $m \downarrow 0$:

$$p_0(x) = \lim_{m\downarrow 0} p_m(x) , \quad p = \Delta^{\rm ret},\ \Delta,\ \Delta^+,\ \Delta^F.$$

All definitions and properties given in this appendix are maintained in this limit – one may simply set $m = 0$ in these formulas. Instead of $\Delta_0^{\rm ret}$, Δ_0, Δ_0^+ and Δ_0^F one usually writes $D^{\rm ret}$, D, D^+ and D^F, respectively. For $d = 4$ the x-space formulas for the massless propagators read:

$$\begin{aligned} D^{\rm ret}(x) &= \frac{-1}{2\pi}\,\delta(x^2)\,\theta(x^0) , & D(x) &= \frac{-1}{2\pi}\,\delta(x^2)\,\mathrm{sgn}(x^0) , \\ D^+(x) &= \frac{-1}{4\pi^2}\,\frac{1}{x^2 - ix^0 0} , & D^F(x) &= \frac{-1}{4\pi^2}\,\frac{1}{x^2 - i0} . \end{aligned} \tag{A.2.10}$$

Exercise A.2.1. Let $d = 4$. Verify directly the relations

$$D(x) = -i\big(D^+(x) - D^+(-x)\big) \quad \text{and} \quad D^F(x) = i\,D^{\rm ret}(x) + D^+(-x)$$

for the x-space expressions (A.2.10).

[*Solution*: These verifications are straightforward applications of the Sokhotski–Plemelj formula (A.1.16).]

The corresponding propagators for a *Dirac spinor field* in $d = 4$ dimensions are defined by

$$S_m^{\rm ret/\cdot/+/F}(x) := (i\partial\!\!\!/_x + m)\,\Delta_m^{\rm ret/\cdot/+/F}(x) , \tag{A.2.11}$$

where $S_m^{\cdot} := S_m$, $\Delta_m^{\cdot} := \Delta_m$, details are given in Sect. 5.1.1. Here, we only point out some crucial differences between scalar and spinor propagators: For a scalar field the "advanced propagator", $\Delta_m^{\rm av}$, and the "negative frequency part of $i\Delta_m$", Δ_m^-, are defined in terms of $\Delta_m^{\rm ret}$ and Δ_m^+, respectively, by the simple formulas

$$\Delta_m^{\rm av}(x) := \Delta_m^{\rm ret}(-x) \; ; \quad \Delta_m^-(x) := -\Delta_m^+(-x) \; ; \tag{A.2.12}$$

however, for a Dirac spinor field the corresponding definitions read:

$$\begin{aligned} S_m^{\rm av}(x) &:= (i\partial\!\!\!/_x + m)\,\Delta_m^{\rm av}(x) \neq S_m^{\rm ret}(-x) \\ S_m^-(x) &:= (i\partial\!\!\!/_x + m)\Delta_m^-(x) \neq -S_m^+(-x) \; . \end{aligned} \tag{A.2.13}$$

We also emphasize that

$$S_m(-x) \neq -S_m(x) \quad \text{and} \quad S_m^F(-x) \neq S_m^F(x) \; .$$

Exercise A.2.2 (Equal-time commutation relations).

(a) In Exer. A.3.4(d) we show that

$$\partial^a \Delta_m(0,\vec{x}) = -i\Big(\partial^a \Delta_m^+(0,\vec{x}) - (-1)^{|a|}\,\partial^a \Delta_m^+(0,-\vec{x})\Big) \in \mathcal{D}'(\mathbb{R}^{d-1}) \; , \quad a \in \mathbb{N}^d \; ,$$

is well defined. Derive the following time-zero relations of the commutator function,

$$\begin{aligned} &\Delta_m(0,\vec{x}) = 0 \; , &&\partial_k \Delta_m(0,\vec{x}) = 0 \quad (\text{for } k = 1,\dots,d-1) \; , \\ &\partial_0 \Delta_m(0,\vec{x}) = -\delta(\vec{x}) \; , &&(\partial_0)^2 \Delta_m(0,\vec{x}) = 0 \; , \end{aligned} \tag{A.2.14}$$

on a formal level by manipulating the momentum space integral representation (A.2.3).

Hence, interpreting $\pi(t,\vec{x}) := \partial_t \varphi(t,\vec{x})$ as canonical conjugate momentum of $\varphi(t,\vec{x})$, the *equal-time fields satisfy canonical commutation relations*:

$$[\varphi(t,\vec{x}),\varphi(t,\vec{y})]_\star = 0 \; , \quad [\varphi(t,\vec{x}),\pi(t,\vec{y})]_\star = i\hbar\,\delta(\vec{x}-\vec{y}) \; , \quad [\pi(t,\vec{x}),\pi(t,\vec{y})]_\star = 0 \; , \tag{A.2.15}$$

see, e.g., [109, Sect. 3-1].

(b) Prove the following: Any solution $f \in C^2(\mathbb{M},\mathbb{C})$ of $(\Box_y + m^2)f(y) = 0$ can be written as

$$f(y) = -\int_{x^0=c} d^3x \; \Delta_m(y-x)\overset{\leftrightarrow x}{\partial_0} f(x) \qquad \forall c \in \mathbb{R} \; , \tag{A.2.16}$$

where the notation (5.5.24) is used. We point out that the formula (A.2.16) yields an integral representation of the solution of the Cauchy problem for the Klein-Gordon equation (cf. Thm. 1.6.6).

[*Solution*: (a) In the formula for Δ_m (A.2.3) we perform the p_0-integral by using

$$\operatorname{sgn} p^0 \; \delta(p^2 - m^2) = \frac{1}{2\omega_{\vec{p}}}\Big(\delta(p^0 - \omega_{\vec{p}}) - \delta(p^0 + \omega_{\vec{p}})\Big) \; , \quad \omega_{\vec{p}} := \sqrt{\vec{p}^{\,2} + m^2} \; .$$

This yields

$$\Delta_m(x) = \frac{-1}{(2\pi)^{d-1}} \int \frac{d\vec{p}}{\omega_{\vec{p}}} \; \sin(\omega_{\vec{p}} x_0) \; e^{i\,\vec{p}\vec{x}} \; ,$$

which implies

$$\partial_k \Delta_m(x) = \frac{-i}{(2\pi)^{d-1}} \int \frac{d\vec{p}}{\omega_{\vec{p}}} \sin(\omega_{\vec{p}} x_0)\, p_k\, e^{i\,\vec{p}\vec{x}} \;,$$

$$\partial_0 \Delta_m(x) = \frac{-1}{(2\pi)^{d-1}} \int d\vec{p}\, \cos(\omega_{\vec{p}} x_0)\, e^{i\,\vec{p}\vec{x}} \;,$$

$$(\partial_0)^2 \Delta_m(x) = \frac{1}{(2\pi)^{d-1}} \int d\vec{p}\, \omega_{\vec{p}} \sin(\omega_{\vec{p}} x_0)\, e^{i\,\vec{p}\vec{x}} \;,$$

Setting $x^0 := 0$ we get the asserted relations.
(b) First we point out that the region of integration in (A.2.16) is bounded, due to $\operatorname{supp}\Delta_m \subseteq (\overline{V}_+ \cup \overline{V}_-)$. We denote the r.h.s. of (A.2.16) by $g_c(y)$. By means of $(\Box + m^2)\Delta_m = 0$ we find

$$\partial_x^\mu \big(\Delta_m(y-x) \overset{\leftrightarrow x}{\partial}_\mu f(x)\big) = 0 \;.$$

Due to this and Gauss' integral theorem (A.1.13), $g_c(y)$ does not depend on c; hence, we may choose $c = y^0$. With that and by using the relations (A.2.14), we get

$$g_{y^0}(y) = -\int d^3x\; \Delta_m(0, \vec{y}-\vec{x}) \overset{\leftrightarrow x}{\partial}_0 f(y^0, \vec{x}) = \int d^3x\; \delta(\vec{y}-\vec{x}) f(y^0, \vec{x}) = f(y) \;.]$$

A.3 A short introduction to wave front sets

As references we recommend the books [107] and [141, Volume II] and in particular the article [21]. It was Radzikowski [140] who realized the usefulness of wave front sets for the formulation of (Quantum) Field Theory.

The wave front set is a central tool of "microlocal analysis". The wave front set of a distribution $t \in \mathcal{D}'(\mathbb{R}^m)$ is a characterization of the singularities of t; which relies on the connection between "smoothness" of a distribution and "rapid decay" of its Fourier transform in all directions.

To characterize the singularity of a distribution $t(x) \in \mathcal{D}'(\mathbb{R}^m)$ at a given point $x_0 \in \mathbb{R}^m$ we have to select from t the behaviour at x_0. We do this by replacing $t(x)$ by $(ft)(x) := f(x)t(x)$ where f runs through all functions $f \in \mathcal{D}(\mathbb{R}^m)$ with $f(x_0) \neq 0$. Note that the support of f may be an arbitrary small neighbourhood of x_0. Now, if t is singular at x_0 (i.e., there does not exist any open neighbourhood U of x_0 such that t restricted to $\mathcal{D}(U)$ can be identified with a smooth function) and if f belongs to the mentioned class of functions, then ft is still singular at x_0 and its support is compact, so that its Fourier transform $\widehat{ft}$ is a C^∞ function.

Since the Fourier transformation is a bijection $\mathcal{S}(\mathbb{R}^m) \longrightarrow \mathcal{S}(\mathbb{R}^m)$, we may say: Non-smoothness of ft is equivalent to non-rapid decay of $\widehat{ft}$ in at least one direction, for all f of the above kind. Recall that a function g *decays rapidly in the direction* $k \in \mathbb{R}^m \setminus \{0\}$ if and only if there exists an open cone C containing k such that

$$\sup_{y \in C} |y|^N |g(y)| < \infty \quad \text{for all} \quad N \in \mathbb{N}.$$

Motivated by these explanations, we define:

Definition A.3.1 (Wave front set). The *wave front set* $\mathrm{WF}(t)$ of a distribution $t \in \mathcal{D}'(\mathbb{R}^m)$ is the set

$$\mathrm{WF}(t) := \{ (x,k) \in \mathbb{R}^m \times (\mathbb{R}^m \setminus \{0\}) \,\big|\, \widehat{ft} \text{ does not decay rapidly in direction } k, \text{ for all } f \in \mathcal{D}(\mathbb{R}^m) \text{ with } f(x) \neq 0 \}.$$

Obviously, $(x,k) \in \mathrm{WF}(t)$ implies $(x, \lambda k) \in \mathrm{WF}(t)$ for all $\lambda > 0$.

The *singular support of a distribution* $t \in \mathcal{D}'(\mathbb{R}^m)$, denoted by $\operatorname{sing\,supp}(t)$, is the set of points x at which t is singular. As explained above, it is related to the wave front set of t by

$$\operatorname{sing\,supp}(t) = \{ x \in \mathbb{R}^m \,\big|\, \exists k \in \mathbb{R}^m \setminus \{0\} \text{ with } (x,k) \in \mathrm{WF}(t) \} . \tag{A.3.1}$$

Example A.3.2 (Wave front set of the δ-distribution). For $t = \delta \equiv \delta_{(m)} \in \mathcal{D}'(\mathbb{R}^m)$, we know from (A.3.1) that the wave front set is of the form $\mathrm{WF}(\delta) = \{0\} \times (\ldots)$. Since $\widehat{f\delta}(k) \sim \int dx\, e^{ikx} f(x)\, \delta(x) = f(0)$ does not decay rapidly in *any* direction k, we obtain

$$\mathrm{WF}(\delta) = \{0\} \times (\mathbb{R}^m \setminus \{0\}) . \tag{A.3.2}$$

We mention some further *basic properties of the wave front set.* For $t, s \in \mathcal{D}'(\mathbb{R}^m)$ and $g \in C^\infty(\mathbb{R}^m)$ it holds that

(a) $\mathrm{WF}(t+s) \subseteq \mathrm{WF}(t) \cup \mathrm{WF}(s)$,
(b) $\mathrm{WF}(gt) \subseteq \big(\operatorname{supp} g \times (\mathbb{R}^m \setminus \{0\})\big) \cap \mathrm{WF}(t)$.
(c) Let $P(x) = \sum_a g_a(x) \partial^a$ be a differential operator, where $g_a \in C^\infty(\mathbb{M})$ and the sum over a is finite. Then, $\mathrm{WF}(Pt) \subseteq \mathrm{WF}(t)$.

We make these statements plausible by the following explanations:

(a) In the sum $t+s$ there may be cancellations of singularities; therefore, $\mathrm{WF}(t+s)$ may be truly smaller than $\mathrm{WF}(t) \cup \mathrm{WF}(s)$.
(b) Multiplying a distribution t with a smooth function g, the singularities of t lying outside the support of g are suppressed.

For a singularity of t at some x_0 with $g(x_0) \neq 0$, the statement is that

$$\{ k \in (\mathbb{R}^m \setminus \{0\}) \,|\, (x_0, k) \in \mathrm{WF}(gt) \} = \{ k \in (\mathbb{R}^m \setminus \{0\}) \,|\, (x_0, k) \in \mathrm{WF}(t) \} .$$

This follows immediately from Definition A.3.1, since there is a neighbourhood U of x_0 such that $g(x) \neq 0 \;\forall x \in U$ and, hence,

$$\{ ((fg)t)|_{\mathcal{D}(U)} \,|\, f \in \mathcal{D}(\mathbb{M}),\, f(x_0) \neq 0 \} = \{ (ft)|_{\mathcal{D}(U)} \,|\, f \in \mathcal{D}(\mathbb{M}),\, f(x_0) \neq 0 \} .$$

(c) Due to the statements (a) and (b), it suffices to prove $\mathrm{WF}(\partial^\mu t) \subseteq \mathrm{WF}(t)\ \forall t \in \mathcal{D}'(\mathbb{R}^m)$. Now let $(x_0, k_0) \in \mathrm{WF}(\partial^\mu t)$ and $f \in \mathcal{D}(\mathbb{R}^m)$ with $f(x_0) \neq 0$. Writing

$$(-ik^\mu)\,\widehat{ft}\,(k) = \widehat{f\partial^\mu t}(k) + \widehat{(\partial^\mu f)t}(k)\ , \tag{A.3.3}$$

we know that the first term on the r.h.s. does not decay rapidly in direction k_0. For simplicity we assume that x_0 is an isolated singularity, that is, there is some neighbourhood U of x_0 such that t restricted to $\mathcal{D}(U \setminus \{x_0\})$ is a smooth function. To show that $(x_0, k_0) \in \mathrm{WF}(t)$ by means of Definition A.3.1, it suffices to consider all functions $f \in \mathcal{D}(U)$ with $f(x_0) \neq 0$ and $x_0 \notin \mathrm{supp}(\partial^\mu f)$. For such an f statement (b) says $(\partial^\mu f)t \in C^\infty(\mathbb{R}^m)$; hence, the second term on the r.h.s. of (A.3.3) decays rapidly in all directions k. We conclude that $\widehat{ft}\,(k)$ does not decay rapidly in direction k_0, that is, $(x_0, k_0) \in \mathrm{WF}(t)$.

An obvious example for $\mathrm{WF}(Pt) \neq \mathrm{WF}(t)$ is

$$\mathrm{WF}\big((\square + m^2)\Delta_m^+\big) = \mathrm{WF}(0) = \emptyset \neq \mathrm{WF}(\Delta_m^+)\ .$$

As mentioned in (3.1.1), a fundamental technical difficulty in QFT is that in general the *pointwise product* $t(x)s(x)$ of distributions $t, s \in \mathcal{D}'(\mathbb{R}^m)$ does not exist, if t and s have an "overlapping singularity", i.e., if there exists an $x_0 \in \mathbb{R}^m$ such that both t and s are singular at x_0. Nevertheless, $t(x)s(x)$ exists in this case, if the following criterion is satisfied.

Theorem A.3.3 (Hörmander's criterion [107]). *Let* $t, s \in \mathcal{D}'(\mathbb{R}^m)$. *Suppose that*

$$\mathrm{WF}(t) \oplus \mathrm{WF}(s) := \{(x, k_1 + k_2)\,\big|\,(x, k_1) \in \mathrm{WF}(t)\ \wedge\ (x, k_2) \in \mathrm{WF}(s)\}$$

does not contain any element of the form $(x, 0)$. *Then the pointwise product* $ts \in \mathcal{D}'(\mathbb{R}^m)$ *is well defined and*

$$\mathrm{WF}(ts) \subseteq \Big(\mathrm{WF}(t) \cup \mathrm{WF}(s) \cup \Big(\mathrm{WF}(t) \oplus \mathrm{WF}(s)\Big)\Big)\ .$$

Heuristic explanation for the existence of $t(x)s(x)$. Let t and s both be singular at some point x_0. Take $f \in \mathcal{D}(\mathbb{R}^m)$, $f(x_0) \neq 0$ and, for simplicity, assume that ft and fs have only one overlapping singularity, namely at x_0. In general, the product $ft(x)\,fs(x)$ does not exist since the integral $\int dk\ \widehat{ft}\,(p-k)\ \widehat{fs}\,(k)$ diverges. However, if the assumption of the theorem is satisfied, we see that $(x_0, k) \in \mathrm{WF}(s)$ implies $(x_0, -k) \notin \mathrm{WF}(t)$; so there exists an f with the above-mentioned properties such that the integral does converge, hence $(ft)(x)\,(fs)(x) \in \mathcal{D}'(\mathbb{R}^m)$ exists for this f in the distributional sense. □

Exercise A.3.4.

(a) Compute the wave front set of $\theta(x^0) \in \mathcal{D}'(\mathbb{R}^d)$ (where $\mathbb{R}^d \ni x \equiv (x^0, \ldots, x^{d-1})$).

(b) Investigate the Hörmander criterion for $\big(\theta(x^0), \delta(x)\big) \in \mathcal{D}'(\mathbb{R}^d)^{\times 2}$.

(c) By using the result (2.2.7) for $\mathrm{WF}(\Delta_m^+)$, prove existence of the products

$$\theta(x^0)\,\Delta_m^+(x) \quad \text{and} \quad \theta(x^0)\,\Delta_m^+(-x) \quad \text{in } \mathcal{D}'(\mathbb{R}^d).$$

This shows that

$$\theta(x^0)\,\Delta_m(x) = -i\,\theta(x^0)\,\big(\Delta_m^+(x) - \Delta_m^+(-x)\big)$$

(cf. (A.2.4)) and the products appearing in the definition of the Feynman propagator (2.3.8) (or (A.2.8)) are well defined in $\mathcal{D}'(\mathbb{R}^d)$.

(d) Prove that

$$(\partial^a \Delta_m^+)(0, \pm\vec{x}) \in \mathcal{D}'(\mathbb{R}^{d-1})\,, \qquad a \in \mathbb{N}^d\,,$$

is well defined.

[*Solution*: (a) The set of all $x \in \mathbb{R}^d$ at which $t(x) := \theta(x^0)$ is singular is $\{0\} \times \mathbb{R}^{d-1}$. Working with Definition A.3.1 it suffices to consider functions $f(x)$ of the form $f(x) = f_0(x^0)\, f_1(\vec{x})$, $f_0 \in \mathcal{D}([-L, L])$, $f_1 \in \mathcal{D}(\mathbb{R}^{d-1})$ for some $L > 0$. From

$$\widehat{ft}\,(k) = I(k^0) \int d\vec{x}\; f_1(\vec{x})\, e^{-i\vec{k}\vec{x}}\,, \quad I(k^0) := \int_0^L dx^0\; f_0(x^0)\, e^{i\,k^0 x^0}\,,$$

we see that $\widehat{ft}\,(k)$ decays rapidly in the directions $(0, \vec{k})$, $\vec{k} \in \mathbb{R}^{d-1} \setminus \{0\}$. Following [21], we get

$$I(k^0) = \frac{i f_0(0)}{k^0} - \frac{f_0'(0)}{(k^0)^2} - \frac{1}{(k^0)^2} \int_0^L dx^0\; f_0''(x^0)\, e^{i\,k^0 x^0}$$

by two-fold integration by parts. Let $|f_0''(x_0)| \le M$, $\forall x^0 \in [-L, L]$. With that we conclude

$$\Big| I(k^0) - \frac{i f_0(0)}{k^0} \Big| \le \frac{1}{(k^0)^2}\,\big(|f'(0)| + LM\big)\,,$$

that is, $I(k^0)$ does not rapidly decay for $k^0 \to \pm\infty$. Summing up, we obtain

$$\mathrm{WF}\big(\theta(x^0)\big) = \Big(\{0\} \times \mathbb{R}^{d-1}\Big) \times \Big((\mathbb{R} \setminus \{0\}) \times \{\vec{0}\}\Big)\,.$$

(b) $\theta(x^0)$ and $\delta(x)$ have an overlapping singularity at $x = 0$. Since

$$\big(0, (k^0, \vec{0})\big) \in \mathrm{WF}\big(\theta(x^0)\big) \quad \text{and} \quad \big(0, (-k^0, \vec{0})\big) \in \mathrm{WF}\big(\delta(x)\big) \quad \forall k^0 \neq 0\,,$$

the criterion is not satisfied.

(c) $\theta(x^0)$ and $\Delta_m^+(x)$ have an overlapping singularity at $x = 0$ only. The Hörmander criterion is fulfilled, since for $(0, k) \in \mathrm{WF}\big(\theta(x^0)\big)$ and $(0, p) \in \mathrm{WF}\big(\Delta_m^+(x)\big)$ we have $k = (k^0, \vec{0})$, $k^0 \neq 0$, and $p = (|\vec{p}|, \vec{p})$, $\vec{p} \neq \vec{0}$, respectively; hence $k + p \neq 0$.
Turning to $\theta(x^0)\,\Delta_m^+(-x)$, note that

$$\mathrm{WF}\big(t(-x)\big) = -\,\mathrm{WF}\big(t(x)\big)\,, \tag{A.3.4}$$

this follows from Definition A.3.1. With that the verification of the Hörmander criterion is analogous to $\theta(x^0)\,\Delta_m^+(x)$.

(d) The assertion is obtained by means of

$$(\partial^a \Delta_m^+)(0, \pm\vec{x}) = \int dx^0\; \delta(x^0)\,(\partial^a \Delta_m^+)(\pm x)\,,$$

if we know that the pair $\big(\delta(x^0), (\partial^a\Delta_m^+)(\pm x)\big) \in \mathcal{D}'(\mathbb{R}^d)^{\times 2}$ satisfies the Hörmander criterion. The latter follows from the above-given property (c) of the wave front set, that is,

$$\mathrm{WF}\big(\delta(x^0)\big) = \mathrm{WF}\big(\partial_{x^0}\theta(x^0)\big) \subseteq \mathrm{WF}\big(\theta(x^0)\big) \quad \text{and} \quad \mathrm{WF}\big((\partial^a\Delta_m^+)(\pm x)\big) \subseteq \mathrm{WF}\big(\Delta_m^+(\pm x)\big) \ ,$$

and the solution of part (c) of this exercise.]

From the result of part (b) one cannot conclude that the product of $\theta(x^0)$ and $\delta(x)$ does not exist in $\mathcal{D}'(\mathbb{R}^d)$, since the Hörmander criterion is not a necessary condition for the existence of the pointwise product of distributions. Nevertheless, one can show that this product fails to exist.

Exercise A.3.5 (Wave front set of some propagators).

(a) For the Feynman propagator $\Delta_m^F \in \mathcal{D}'(\mathbb{R}^d)$ (defined in (2.3.8) or (A.2.8)) prove that

$$\begin{aligned} \mathrm{WF}(\Delta_m^F) = \Big(\{0\} \times (\mathbb{R}^d \setminus \{0\})\Big) & \\ \cup \big\{(x,k)\,\big|\, x^2 = 0,\ x \neq 0,\ k^2 = 0,\ x = \lambda k \text{ for some } \lambda > 0\big\} \ , & \end{aligned} \tag{A.3.5}$$

by using the results (2.2.7) and (A.3.2) for the wave front sets of Δ_m^+ and $\delta = i(\Box + m^2)\Delta_m^F$, respectively.

(b) Verify the formula

$$\begin{aligned} \mathrm{WF}(\Delta_m^{\mathrm{ret}}) = \Big(\{0\} \times (\mathbb{R}^d \setminus \{0\})\Big) & \\ \cup \big\{(x,k)\,\big|\, x^2 = 0,\ x^0 > 0,\ k^2 = 0,\ k^0 \neq 0,\ x = \lambda k \text{ for some } \lambda \in \mathbb{R}\big\} \ , & \end{aligned} \tag{A.3.6}$$

for the wave front set of the retarded propagator $\Delta_m^{\mathrm{ret}}(x) = \theta(x^0)\,\Delta_m(x)$, by using the result (1.8.9) for $\mathrm{WF}(\Delta_m)$ and $(\Box + m^2)\Delta_m^{\mathrm{ret}} = -\delta$.

[*Solution*: (a) By using (2.3.9) and (A.3.4) we obtain:

$$\begin{aligned} \big((\mathbb{R}^d\backslash\overline{V}_-) \times \mathbb{R}^d\big) \cap \mathrm{WF}(\Delta_m^F) &= \big((\mathbb{R}^d \setminus \overline{V}_-) \times \mathbb{R}^d\big) \cap \mathrm{WF}(\Delta_m^+) \\ &= \{(x,k)\,|\, x^2 = 0,\ x^0 > 0,\ k^2 = 0,\ k^0 > 0,\ x = \lambda k \text{ for some } \lambda > 0\} \ , \\ \big((\mathbb{R}^d\backslash\overline{V}_+) \times \mathbb{R}^d\big) \cap \mathrm{WF}(\Delta_m^F) &= \big((\mathbb{R}^d \setminus \overline{V}_+) \times \mathbb{R}^d\big) \cap \big(-\mathrm{WF}(\Delta_m^+)\big) \\ &= \{(x,k)\,|\, x^2 = 0,\ x^0 < 0,\ k^2 = 0,\ k^0 < 0,\ x = \lambda k \text{ for some } \lambda > 0\} \ . \end{aligned}$$

It remains to verify

$$\big(\{0\} \times (\mathbb{R}^d \setminus \{0\})\big) \subseteq \mathrm{WF}(\Delta_m^F) \ .$$

This follows from (A.3.2) and the above-given property (c) of the wave front set:

$$\big(\{0\} \times (\mathbb{R}^d \setminus \{0\})\big) = \mathrm{WF}(\delta) = \mathrm{WF}\big((\Box + m^2)\Delta_m^F\big) \subseteq \mathrm{WF}(\Delta_m^F) \ . \tag{A.3.7}$$

(b) Proceeding analogously to part (a) we get

$$\begin{aligned} \big((\mathbb{R}^d\backslash\overline{V}_-) \times \mathbb{R}^d\big) \cap \mathrm{WF}(\Delta_m^{\mathrm{ret}}) &= \big((\mathbb{R}^d \setminus \overline{V}_-) \times \mathbb{R}^d\big) \cap \mathrm{WF}(\Delta_m) \\ &= \{(x,k)\,|\, x^2 = 0,\ x^0 > 0,\ k^2 = 0,\ k^0 \neq 0,\ x = \lambda k \text{ for some } \lambda \in \mathbb{R}\} \ , \\ \big((\mathbb{R}^d\backslash\overline{V}_+) \times \mathbb{R}^d\big) \cap \mathrm{WF}(\Delta_m^{\mathrm{ret}}) &= \emptyset \ . \end{aligned}$$

Finally the relation $\big(\{0\} \times (\mathbb{R}^d \setminus \{0\})\big) \subseteq \mathrm{WF}(\Delta_m^{\mathrm{ret}})$ is obtained similarly to (A.3.7).]

A.4 Perturbative QFT based on quantization with a Hadamard function

This appendix is based on reference [55].

For brevity we assume that the spacetime dimension is $d = 4$, in odd dimensions there are crucial differences as explained at the end of Sect. 2.2.

A main difficulty of QFT in presence of an external (i.e., non-quantized) field – this may be a gravitational background (i.e., QFT on curved spacetimes) or, e.g., QED with an external electromagnetic field – is that in general there does not exist a unique vacuum state. Instead one usually works with a family of so-called "Hadamard states", which are uniquely determined by their two-point function – a "Hadamard function".

In this appendix we study perturbative QFT based on quantization with a Hadamard function in Minkowski spacetime. The retarded propagator Δ_m^{ret} (A.2.1) and the commutator function Δ_m (A.2.3) are unchanged; but the Wightman two-point function Δ_m^+ is replace by a Hadamard function H_m^μ (introduced in Sect. 2.2), which is non-unique: The mass parameter $\mu > 0$ is arbitrary. We recall some basic properties of H_m^μ making it possible to use H_m^μ as propagator of the star product:

$$-i\big(H_m^\mu(x) - H_m^\mu(-x)\big) = \Delta_m(x)\ , \quad \mathrm{WF}(H_m^\mu) = \mathrm{WF}(\Delta_m^+) \tag{A.4.1}$$

and

$$(\square + m^2)H_m^\mu = 0\ , \quad H_m^\mu(\Lambda x) = H_m^\mu(x) \quad \forall \Lambda \in \mathcal{L}_+^\uparrow\ , \quad \overline{H_m^\mu(x)} = H_m^\mu(-x)\ . \tag{A.4.2}$$

The equivalence of the corresponding star products $\star_{m,\mu} := \star_{H_m^\mu}$, for different values of μ, is shown in Sect. 2.3.

Axioms for the retarded product. The axioms for perturbative QFT, given in Sect. 3.1 can be modified as follows: Since H_m^μ *is smooth in* $m \geq 0$ (2.2.14), the Sm-expansion axiom is replaced by two axioms requiring smoothness of the retarded products in $m \geq 0$ and almost homogeneous behaviour under the scaling $(X, m) \mapsto (\rho X, m/\rho)$ (where $X := (x_1, \ldots, x_n) \in \mathbb{M}^n$ denotes the spacetime arguments of the R-product, see (3.1.66)). This simplifies the Sm-expansion to the Taylor expansion in m. However, there is the complication that H_m^μ is non-unique; to take this into account we require in an additional axiom that the retarded products transform under a change of μ in the same way as the star product $\star_{m,\mu}$ does.

Let us explain the modifications in detail. For the *basic axioms* there is only a change in the GLZ-relation: The commutator is now meant with respect to the star product $\star_{m,\mu}$. Turning to the *renormalization conditions*, the Sm-expansion axiom is replaced by the following three axioms:

- **Smoothness in the mass $m \geq 0$:** We require that

$$0 \leq m \longmapsto (F_S)^{H_m^\mu} \equiv R^{H_m^\mu}\big(e_\otimes^{S/\hbar}, F\big) \quad \text{be smooth, for each} \quad S, F \in \mathcal{F}_{\mathrm{loc}}\ . \tag{A.4.3}$$

This excludes the Wightman two-point function Δ_m^+. Instead, we must take a Hadamard function H_m^μ with the additional mass parameter $\mu > 0$. We shall write $R_{n-1,1}^{(m,\mu)}$, $r_{n-1,1}^{(m,\mu)} := \omega_0(R_{n-1,1}^{(m,\mu)})$ etc. to indicate that we are using H_m^μ.

Turning to the *scaling behaviour*, we first remark that

$$\rho^2 H_{m/\rho}^{\mu/\rho}(\rho x) = H_m^\mu(x) \ ,$$

as we see from (2.2.13) and (2.2.9); from this relation one derives that

$$\sigma_\rho(\sigma_\rho^{-1} F_1 \star_{m/\rho,\mu/\rho} \cdots \star_{m/\rho,\mu/\rho} \sigma_\rho^{-1} F_n) = F_1 \star_{m,\mu} \cdots \star_{m,\mu} F_n \tag{A.4.4}$$

by proceeding analogously to part (a) of Exer. 3.1.21. However, the relevant scaling transformation leaves μ unchanged; but under $(x,m) \mapsto (\rho x, m/\rho)$ a Hadamard function $H_m^\mu(x)$ behaves only almost homogeneously, see (2.2.15). From that formula we conclude that

$$\sigma_\rho(\sigma_\rho^{-1} F_1 \star_{m/\rho,\mu} \cdots \star_{m/\rho,\mu} \sigma_\rho^{-1} F_n) = F_1 \star_{m,\mu} \cdots \star_{m,\mu} F_n + \sum_{k=1}^{N} \widetilde{L}_k \log^k \rho \tag{A.4.5}$$

for some $\widetilde{L}_k \in \mathcal{F}$, which do not depend on ρ, and some $N < \infty$; this conclusion is analogous to the derivation of (A.4.4).[185] Taking into account additionally that, as a consequence of the basic axioms, the unrenormalized $R^{(m,\mu)}$-product is an iterated, retarded commutator with respect to $\star_{m,\mu}$ (that is, formula (3.1.69) holds true on $\mathcal{D}(\check{\mathbb{M}}^{n+1})$ (3.1.68) for $R^{(m,\mu)}$ on the l.h.s. and $\star_{m,\mu}$ on the r.h.s.), we conclude that the unrenormalized $R^{(m,\mu)}$-product fulfills the scaling relation (A.4.6) given below. Since almost homogeneous scaling can be maintained in the extension of distributions (Proposition 3.2.16), we may require (A.4.6) also for the renormalized $R^{(m,\mu)}$-product.

- **Scaling:** For all $F_1, \ldots, F_n \in \mathcal{F}_{\text{loc}}$, there exist m- and μ-dependent functionals $L_1^{(m,\mu)}, \ldots, L_N^{(m,\mu)} \in \mathcal{F}$, which depend also on $(F_1, \ldots, F_n)$, such that

$$\sigma_\rho R_{n-1,1}^{(m/\rho,\mu)}(\sigma_\rho^{-1} F_1 \otimes \cdots \otimes \sigma_\rho^{-1} F_n) = R_{n-1,1}^{(m,\mu)}(F_1 \otimes \cdots \otimes F_n) + \sum_{k=1}^{N} L_k^{(m,\mu)} \log^k \rho \tag{A.4.6}$$

  for some $N < \infty$.

The three sentences directly after (3.1.67) are appropriate also here.

Finally we study the μ-dependence. Arguing analogously to the scaling behaviour, we conclude from the "μ-covariance" of the $\star_{m,\mu}$-product, given in (2.3.6) (Exer. 2.3.2), and the formula for the unrenormalized $R^{(m,\mu)}$-product (3.1.69), that the latter behaves under a change of μ according to formula (A.4.7) below; hence we require (A.4.7) also for the renormalized $R^{(m,\mu)}$-product.

[185] Alternatively, one can also argue as follows: Due to (A.4.4), the l.h.s. of (A.4.5) is equal to $F_1 \star_{m,\rho\mu} \cdots \star_{m,\rho\mu} F_n$ and then one uses that $H_m^{\rho\mu} - H_m^\mu \sim \log \rho$.

- **μ-covariance:**

$$R^{(m,\mu_2)}_{n-1,1} = \Big(\frac{\mu_2}{\mu_1}\Big)^{\Gamma} \circ R^{(m,\mu_1)}_{n-1,1} \circ \Big(\Big(\frac{\mu_2}{\mu_1}\Big)^{-\Gamma}\Big)^{\otimes n}\,, \tag{A.4.7}$$

where the functional derivative operator Γ is defined in (2.3.5) in terms of the analytic function f (2.2.4), which appears in the definition of the Hadamard function (Eq. (2.2.13)).

Remark A.4.1 (Particular case $\mu = m$). Setting $\mu := m$ we have $H^m_m = \Delta^+_m$. Therefore, from a retarded product $R^{(m,\mu)}$ based on quantization with H^μ_m, we can obtain a retarded product based on the Wightman two-point function Δ^+_m by setting $\mu := m$, or by the transformation formula

$$R^{(m,m)}_{n,1} = \Big(\frac{m}{\mu}\Big)^{\Gamma} \circ R^{(m,\mu)}_{n,1} \circ \Big(\Big(\frac{m}{\mu}\Big)^{-\Gamma}\Big)^{\otimes(n+1)}\,. \tag{A.4.8}$$

By definition, the so-obtained $R^{(m,m)}$ satisfies the axioms given here, i.e., the axioms with respect to the star product $\star_{H^\mu_m}$ (with $m = \mu$), in particular the GLZ relation with respect to $\star_{H^m_m} = \star_{\Delta^+_m}$. Hence, $R^{(m,m)}$ satisfies the axioms for $R^{(m)}$ given in Sect. 3.1 – this is obvious, except for the validity of the Sm-expansion axiom.

To illustrate the relation (A.4.8), we study the example of the setting sun diagram. The unrenormalized retarded product $r^{(m,\mu)\,0}(\varphi^3,\varphi^3) \in \mathcal{D}'(\mathbb{R}^4\setminus\{0\})$ can be computed similarly to Sect. 3.1.2:

$$r^{(m,\mu)\,0}(\varphi^3,\varphi^3)(z) = \frac{6\hbar^3}{i}\,\big(H^\mu_m(z)^3 - H^\mu_m(-z)^3\big)\,\theta(-z^0)\,. \tag{A.4.9}$$

Setting $\mu := m$ and inserting the relation

$$H^m_m(z) = \Delta^+_m(z) = H^\mu_m(z) + d^\mu_m(z)\,, \quad \text{where} \quad d^\mu_m(z) = d^\mu_m(-z) \in C^\infty(\mathbb{R}^4)$$

is given in (2.2.13), we obtain

$$\begin{aligned} r^{(m,m)\,0}(\varphi^3,\varphi^3)(z) &= \frac{6\hbar^3}{i}\Big(\big(H^\mu_m(z)^3 - H^\mu_m(-z)^3\big) \\ &\quad + 3\big(H^\mu_m(z)^2 - H^\mu_m(-z)^2\big)\,d^\mu_m(z) + 3\big(H^\mu_m(z) - H^\mu_m(-z)\big)\,d^\mu_m(z)^2\Big)\,\theta(-z^0) \\ &= r^{(m,\mu)\,0}(\varphi^3,\varphi^3)(z) + 9\hbar\, r^{(m,\mu)\,0}(\varphi^2,\varphi^2)(z)\,d^\mu_m(z) - 18\hbar^3\,\Delta^{\mathrm{ret}}_m(-z)\,d^\mu_m(z)^2. \end{aligned} \tag{A.4.10}$$

In the last step we have taken into account that $r^{(m,\mu)\,0}(\varphi^2,\varphi^2)(z) = \frac{2\hbar^2}{i}\big(H^\mu_m(z)^2 - H^\mu_m(-z)^2\big)\,\theta(-z^0)$, which is obtained analogously to (A.4.9). The μ-covariance axiom requires that the relation (A.4.10) is preserved in the extension to $\mathcal{D}'(\mathbb{R}^4)$, that is, the extensions $r^{(m,m)}(\varphi^3,\varphi^3)$, $r^{(m,\mu)}(\varphi^3,\varphi^3)$, $r^{(m,\mu)}(\varphi^2,\varphi^2) \in \mathcal{D}'(\mathbb{R}^4)$ must also satisfy (A.4.10).

The *axioms for the time-ordered product*, given in Sect. 3.3, are modified accordingly. In particular, in the Causality axiom the causal factorization is now meant with respect to the star product $\star_{m,\mu}$. Hence, the Feynman propagator is replaced by

$$H_m^{\mu,F}(x) := \theta(x^0)\, H_m^{\mu}(x) + \theta(-x^0)\, H_m^{\mu}(-x) \ , \tag{A.4.11}$$

where $\mu > 0$ is arbitrary. Also in the Bogoliubov formula (3.3.30), it is now the star product $\star_{m,\mu}$ which is used.

Construction of the retarded product. The inductive construction given elaborately in Sect. 3.2 has to be modified as follows – we only explain the changes:

- The Sm-expansion is replaced by the *Taylor expansion in* m. In detail, Sect. 3.2.3 is replaced by the following: Let $A_1, \dots, A_n \in \mathcal{P}_{\mathrm{hom}}$. By induction we know that the totally symmetric part $s^{(m)\,0} := s_n^{(m,\mu)\,0}(A_1, \dots, A_n) \in \mathcal{D}'(\mathbb{R}^{4(n-1)} \setminus \{0\})$ of $r_{n-1,1}^{(m,\mu)\,0}(A_1, \dots, A_n)$, which is defined analogously to (3.2.50), is smooth in $m \geq 0$ and scales almost homogeneously:

$$(\rho\, \partial_\rho)^{N+1} \big[\rho^D s^{(m/\rho)\,0}(\rho y)\big] = 0 \quad \text{for some } N \in \mathbb{N}, \text{ where } \quad D := \sum_{j=1}^{n} \dim A_j \tag{A.4.12}$$

is a natural number. We now seek an extension $s^{(m)} \in \mathcal{D}'(\mathbb{R}^{4(n-1)})$ of $s^{(m)\,0}$ fulfilling the same properties, with the same D but N may be replaced by $N+1$.

If $m = 0$, we can apply Proposition 3.2.16 directly, since there is no scaling in the mass.

On the other hand, if $m > 0$, due to the smoothness in m we can make a Taylor expansion:

$$s^{(m)\,0}(y) = \sum_{l=0}^{D-4(n-1)} \frac{m^l}{l!}\, u_l^0(y) + m^{D-4(n-1)+1} s_{\mathrm{red}}^{(m)\,0}(y) \tag{A.4.13}$$

$$\text{with} \quad u_l^0(y) := \frac{\partial^l s^{(m)\,0}(y)}{\partial m^l}\bigg|_{m=0} .$$

Differentiating Equation (A.4.12), we get

$$\begin{aligned} 0 &= \frac{\partial^l}{\partial m^l}\bigg|_{m=0} (\rho\, \partial_\rho)^{N+1} \big[\rho^D s^{(m/\rho)\,0}(\rho y)\big] \\ &= (\rho\, \partial_\rho)^{N+1} \Big[\rho^{D-l} \frac{\partial^l}{\partial (m/\rho)^l}\bigg|_{m=0} s^{(m/\rho)\,0}(\rho y)\Big] \\ &= (\rho\, \partial_\rho)^{N+1} \big[\rho^{D-l}\, u_l^0(\rho y)\big] \ , \quad \text{where} \quad 0 \leq l \leq D - 4(n-1) \ . \end{aligned} \tag{A.4.14}$$

Proposition 3.2.16 now implies that an extension u_l of u_l^0 exists, satisfying

$$(\rho\, \partial_\rho)^{N+2} \big[\rho^{D-l}\, u_l(\rho y)\big] = 0.$$

Combining (A.4.12), (A.4.13) and (A.4.14), we get

$$(\rho\,\partial_\rho)^{N+1}\big[\rho^{4(n-1)-1}\,s_{\rm red}^{(m/\rho)\,0}(\rho y)\big]=0\;; \tag{A.4.15}$$

this is equivalent to

$$\rho^{4(n-1)-1}s_{\rm red}^{(m)\,0}(\rho y)=s_{\rm red}^{(\rho m)\,0}(y)+\sum_{j=1}^{N}l_j^{(\rho m)}(y)\,\log^j\rho \tag{A.4.16}$$

for some $l_j^{(m)}\in\mathcal{D}'(\mathbb{R}^{4(n-1)}\setminus\{0\})$ which are smooth in $m\geq 0$. Since also $s_{\rm red}^{(m)\,0}$ is smooth in $m\geq 0$, we conclude that

$$\lim_{\rho\downarrow 0}\rho^{4(n-1)}s_{\rm red}^{(m)\,0}(\rho y)=0\;,\quad\text{i.e.,}\quad \mathrm{sd}(s_{\rm red}^{(m)\,0})<4(n-1)\;. \tag{A.4.17}$$

Due to that, Theorem 3.2.8(a) provides a unique extension $s_{\rm red}^{(m)}$ with $\mathrm{sd}(s_{\rm red}^{(m)})=\mathrm{sd}(s_{\rm red}^{(m)\,0})$ by the direct extension (3.2.22). The latter maintains smoothness in $m\geq 0$. And, $s_{\rm red}^{(m)}$ also fulfills the scaling relation (A.4.15); this can be seen as follows:

$$t_\rho^{(m)}(y):=(\rho\,\partial_\rho)^{N+1}\big[\rho^{4(n-1)-1}\,s_{\rm red}^{(m/\rho)}(\rho y)\big]\quad\text{has support in } \{0\};$$

hence, according to (A.1.17), it is of the form

$$t_\rho^{(m)}(y)=\sum_a C_{\rho,a}^{(m)}\,\partial^a\delta_{(4(n-1))}(y)\quad\text{for some } C_{\rho,a}^{(m)}\in\mathbb{C}.$$

In addition, for each fixed $\rho>0$, we obtain $\mathrm{sd}(t_\rho^{(m)})<4(n-1)$ since $\mathrm{sd}(s_{\rm red}^{(m)})<4(n-1)$; these results imply $t_\rho^{(m)}=0$.

Putting it all together, we arrive at

$$s_n^{(m,\mu)}(A_1,\ldots,A_n)(y)\equiv s^{(m)}(y):=\sum_{l=0}^{D-4(n-1)}\frac{m^l}{l!}u_l(y)+m^{D-4(n-1)+1}s_{\rm red}^{(m)}(y), \tag{A.4.18}$$

which is an extension of $s_n^{(m,\mu)\,0}(A_1,\ldots,A_n)$ with the desired smoothness and scaling properties.

- To fulfill the *μ-covariance* (A.4.7) we can proceed as follows: First we construct the extension $S_n^{(m,\mu_1)}$ of $S_n^{(m,\mu_1)\,0}$ for a fixed μ_1. For arbitrary μ, the extension $S_n^{(m,\mu)}$ is then obtained by μ-covariance, that is, by the relation (A.4.7) with $S_n^{(m,\mu)}$ in place of $R_{n-1,1}^{(m,\mu)}$. One verifies that the validity of the other renormalization conditions goes over from $S_n^{(m,\mu_1)}$ to the so-constructed $S_n^{(m,\mu)}$.
- The most general solution of the axioms can be described as follows: Let $R_{n-1,1}^{(m,\mu)}$ and $\widehat{R}_{n-1,1}^{(m,\mu)}$ be two solutions of the inductive step $(n-2)\mapsto(n-1)$ and

let $A_1,\dots,A_n \in \mathcal{P}_{\mathrm{hom}}$. Taking into account Theorem 3.2.8 and the axioms Smoothness in m (A.4.3) and Scaling (A.4.6), we conclude that

$$\omega_0\Big(\big(\widehat{R}^{(m,\mu)}_{n-1,1} - R^{(m,\mu)}_{n-1,1}\big)\big(A_1(x_1),\dots,A_n(x_n)\big)\Big) \tag{A.4.19}$$
$$= \mathfrak{S}_n \sum_{\substack{|a|+l=D-4(n-1)\\ l\geq 0}} m^l\, C_{l,a}(A_1,\dots,A_n;\mu/M)\,\partial^a\delta(x_1-x_n,\dots,x_{n-1}-x_n)$$

where $D := \sum_j \dim A_j$; in addition, $\mathfrak{S}_n$ denotes symmetrization w.r.t. permutations of the pairs $(A_1,x_1),\dots,(A_n,x_n)$ and M is a fixed mass parameter, which must be introduced for the renormalization of the massless theory. (Remember that the latter does not contain the scale $\mu > 0$, since $\lim_{m\downarrow 0} H^\mu_m = D^+$; so it is unnatural to take μ as renormalization mass scale for the $m = 0$ theory.) The appearance of M has the consequence that the dimensionless numbers $C_{l,a}$ may depend on μ (via μ/M); this corrects an oversight in [55]. The requirement that $l \geq 0$ comes from smoothness in m (via the factor m^l).

We point out that $C_{l,a}$ is independent of m – in contrast to (3.2.71), powers of $\log(m/\mu)$ or $\log(m/M)$ do not appear in $C_{l,a}$ since they would violate smoothness in $m \geq 0$. Due to this, the r.h.s. of (A.4.19) *scales homogeneously* under $(X,m) \mapsto (\rho X, m/\rho)$, although $\widehat{R}^{(m,\mu)}_{n-1,1}$ and $R^{(m,\mu)}_{n-1,1}$ scale only almost homogeneously.

Stückelberg–Petermann renormalization group. As for quantization with Δ^+_m, the Stückelberg–Petermann renormalization group $\mathcal{R}$ is defined such that it is precisely the set of maps $Z : \mathcal{F}_{\mathrm{loc}} \longrightarrow \mathcal{F}_{\mathrm{loc}}$ relating different $T^{(m,\mu)}$-products by the Main Theorem (Thm. 3.6.3). This gives the following modifications of the explicit definition of $\mathcal{R}$ (Definition 3.6.1): The analytic maps $Z \in \mathcal{R}$ depend now on both parameters m and μ of the underlying star product: $Z \equiv Z_{(m,\mu)}$. The defining property "(3) Locality, Translation covariance and Sm-expansion" is replaced by the following two properties (3a) and (3b):

(3a) *Locality, Translation covariance, Smoothness in $m \geq 0$ and Scaling*: For all $A_1,\dots,A_n \in \mathcal{P}_{\mathrm{hom}}$ the VEV $z^{(n)}_{(m,\mu)}(A_1,\dots,A_n) := \omega_0\big(Z^{(n)}_{(m,\mu)}(A_1\otimes\cdots\otimes A_n)\big)$ depends only on the relative coordinates and is of the form

$$z^{(n)}_{(m,\mu)}(A_1,\dots,A_n)(x_1-x_n,\dots) \tag{A.4.20}$$
$$= \mathfrak{S}_n \sum_{\substack{|a|+l=D-4(n-1)\\ l\geq 0}} m^l\, C_{l,a}(A_1,\dots,A_n;\mu/M)\,\partial^a\delta(x_1-x_n,\dots,x_{n-1}-x_n)$$

with m-independent coefficients $C_{l,a}(A_1,\dots,A_n;\mu/M) \in \mathbb{C}$; in addition, $D := \sum_j \dim A_j$, $\mathfrak{S}_n$ and M are as above in (A.4.19).

(3b) *μ-covariance*: $Z_{(m,\mu_2)} = (\mu_2/\mu_1)^\Gamma \circ Z_{(m,\mu_1)} \circ (\mu_2/\mu_1)^{-\Gamma}$.

The defining property (3a) can be understood as follows: From the proof of the Main Theorem, more precisely from formula (3.6.28), we know that $Z^{(n)}_{(m,\mu)}$ is the difference of two time-ordered products of nth order which agree to lower orders (3.6.29). Therefore, $z^{(n)}_{(m,\mu)}(A_1,\ldots)$ (where $A_1,\ldots\in\mathcal{P}_{\mathrm{hom}}$) is of the from of the r.h.s. of (A.4.19). From the explanation after (A.4.19) it follows that $z^{(n)}_{(m,\mu)}$ may depend on μ and, hence, this holds also for $Z^{(n)}_{(m,\mu)}$ – this corrects [55, Thm. 4.2(iii)].

Remark A.4.2. The defining property (3a) can more elegantly be formulated in terms of $Z_{(m,\mu)}$ by the four properties:

- *Locality*, given in formula (3.6.33);
- *Translation covariance:* $\beta_a\circ Z_{(m,\mu)}=Z_{(m,\mu)}\circ\beta_a\quad\forall\,a\in\mathbb{R}^4$;
- *Smoothness in* $m\geq 0$ *of* $Z\equiv Z_{(m,\mu)}$ in the sense of (A.4.3);
- *Scaling:* $\sigma_\rho\circ Z_{(m/\rho,\mu)}\circ\sigma_\rho^{-1}=Z_{(m,\mu)}$.

We point out: The elements $Z_{(m,\mu)}$ of the Stückelberg–Petermann group appearing here, scale *homogeneously* under $(X,m)\mapsto(\rho X,m/\rho)$. This is in contrast to the elements $Z_{(m)}$ of the Stückelberg–Petermann group relying on Δ^+_m, which scale *only almost homogeneously*; see[186] (3.6.15).

Chap. 4 about the MWI can be translated into quantization based on a Hadamard function H^μ_m without any obstacle, see [16]; a main reason for this is that also H^μ_m is a solution of the Klein–Gordon equation.

A.5 The Fock space

Almost every introductory book to QFT treats the Fock space, however there are different conventions. Essentially we follow [73].

The Fock space is a Hilbert space, on which our on-shell fields $F\big|_{\mathcal{C}_0}$, $F\in\mathcal{F}$ can faithfully be represented, see Thm. 2.6.3. The normalized Fock space vectors are interpreted as vector states in the sense of (2.5.4). A main advantage of the Fock space description of physical states is that it provides a clear particle number interpretation for free field theories, in particular for the asymptotic states in a scattering experiment, see (2.6.11), (5.5.37) and (A.5.25). Hence, the Fock space

[186] By means of the tools introduced in this appendix, the relation (3.6.15) can be derived in an alternative way: Using that $Z_{(m)}$ and $Z_{(m,\mu)}$ are related by $Z_{(m)}:=Z_{(m,m)}$ (see Remk. A.4.1) and applying the properties “Scaling” and “μ-covariance” of $Z_{(m,\mu)}$, we obtain

$$\sigma_\rho\circ Z_{(m/\rho)}\circ\sigma_\rho^{-1}=\sigma_\rho\circ Z_{(m/\rho,m/\rho)}\circ\sigma_\rho^{-1}=Z_{(m,m/\rho)}=\rho^{-\Gamma}\circ Z_{(m,m)}\circ\rho^{\Gamma}=\rho^{-\Gamma}\circ Z_{(m)}\circ\rho^{\Gamma}\,.$$

Now, for $G\in\mathcal{F}$, the term $\rho^\Gamma G$ is a polynomial in $\log\rho$, since by definition G is a polynomial in $(\partial^a)\varphi$, i.e., of finite order in φ. Therefore, $\rho^{-\Gamma}\circ Z^{(n)}_{(m)}\big((\rho^\Gamma F)^{\otimes n}\big)$ is a polynomial in $\log\rho$ for all $F\in\mathcal{F}_{\mathrm{loc}}$, $n\in\mathbb{N}^*$.

is well suited for scattering processes in high energy physics, in which particles are created and annihilated.

We give a short and self-contained introduction to the bosonic Fock space, and at the end we sketch how one has to change this construction to obtain the fermionic Fock space.

Definition of the bosonic Fock space. Let $\mathbb{M}^* \equiv \mathbb{M}^*_d$ be the momentum space belonging to the Minkowski space $\mathbb{M}_d$, that is $\mathbb{M}^*_d := \mathbb{R}^d$ and $p^2 := (p^0)^2 - \vec{p}^{\,2}$ for $p = (p^0, \vec{p}) \in \mathbb{M}^*$. We equip the mass shell

$$\mathcal{H}_m^+ := \{\, p = (p^0, \vec{p}) \in \mathbb{M}^* \,\Big|\, p^2 = m^2 \;,\; p^0 > 0 \,\} \;, \quad m > 0 \;,$$

with the Lorentz invariant (more precisely: $\mathcal{L}_+^\uparrow$ invariant) measure

$$\frac{d^{d-1}\vec{p}}{2\omega_{\vec{p}}} \quad \text{where} \quad \omega_{\vec{p}} := \sqrt{\vec{p}^{\,2} + m^2} \;; \tag{A.5.1}$$

Lorentz invariance follows from

$$\int d^dp \;\theta(p^0)\,\delta(p^2 - m^2)\,f(p^0, \vec{p}) = \int \frac{d^{d-1}\vec{p}}{2\omega_{\vec{p}}}\; f(\omega_{\vec{p}}, \vec{p}) \quad \forall f \in \mathcal{D}(\mathbb{R}^d) \;.$$

For simplicity we study the massive, real scalar field, that is, a relativistic particle with spin 0 and mass $m > 0$. The pertinent 1-*particle space* is

$$\mathfrak{H}_1 := \{\, \phi : \mathcal{H}_m^+ \longrightarrow \mathbb{C} \,\Big|\, \|\phi\|_{\mathfrak{H}_1} < \infty \,\} \;, \quad \|\phi\|^2_{\mathfrak{H}_1} := \hbar \int \frac{d^{d-1}\vec{p}}{2\omega_{\vec{p}}}\; |\phi(\omega_{\vec{p}}, \vec{p})|^2 \;. \tag{A.5.2}$$

Due to Bose statistics, the elements of the *n-particle space* $\mathfrak{H}_n^+$ are totally symmetric wave functions, in detail:

$$\mathfrak{H}_n^+ := \{\, \phi : \mathcal{H}_m^+ \times \cdots \times \mathcal{H}_m^+ \longrightarrow \mathbb{C} \text{ symmetric} \,\Big|\, \|\phi\|_{\mathfrak{H}_n^+} := \sqrt{\langle \phi, \phi \rangle_{\mathfrak{H}_n^+}} < \infty \,\} \;, \tag{A.5.3}$$

$$\langle \phi, \psi \rangle_{\mathfrak{H}_n^+} := \hbar^n \int \frac{d^{d-1}\vec{p}_1}{2\omega_{\vec{p}_1}} \cdots \frac{d^{d-1}\vec{p}_n}{2\omega_{\vec{p}_n}}\; \overline{\phi(\omega_{\vec{p}_1}, \vec{p}_1, \ldots, \omega_{\vec{p}_n}, \vec{p}_n)}\; \psi(\omega_{\vec{p}_1}, \vec{p}_1, \ldots, \omega_{\vec{p}_n}, \vec{p}) \;,$$

where "symmetric" means that

$$\phi(\omega_{\vec{p}_{\pi 1}}, \vec{p}_{\pi 1}, \ldots, \omega_{\vec{p}_{\pi n}}, \vec{p}_{\pi n}) = \phi(\omega_{\vec{p}_1}, \vec{p}_1, \ldots, \omega_{\vec{p}_n}, \vec{p}_n) \quad \forall \pi \in S_n \;.$$

With the given L^2-scalar product, each $\mathfrak{H}_n^+$ is a Hilbert space.

The *bosonic Fock space* is an infinite direct sum of Hilbert spaces, which is again a Hilbert space:

$$\mathfrak{F}^+ := \bigoplus_{n=0}^{\infty} \mathfrak{H}_n^+ \;, \quad \mathfrak{H}_0^+ := \mathbb{C} \;, \quad \mathfrak{H}_1^+ := \mathfrak{H}_1 \;. \tag{A.5.4}$$

Elements $\phi \in \mathfrak{F}^+$ are sequences[187]

$$\phi = (\phi_0, \phi_1, \phi_2, \ldots) \equiv (\phi_n)_{n\in\mathbb{N}} \quad \text{with} \quad \phi_n \in \mathfrak{H}_n^+ \ , \ \|\phi\|^2_{\mathfrak{F}^+} := \langle \phi, \phi \rangle_{\mathfrak{F}^+} < \infty \ ,$$

where the scalar product in $\mathfrak{F}^+$ is defined by

$$\langle \phi, \psi \rangle_{\mathfrak{F}^+} := \sum_{n=0}^{\infty} \langle \phi_n, \psi_n \rangle_{\mathfrak{H}_n^+} \quad \text{for} \quad \phi \equiv (\phi_n) \in \mathfrak{F}^+ \ , \ \psi \equiv (\psi_n) \in \mathfrak{F}^+ \ . \tag{A.5.5}$$

The vector

$$\Omega := (1, 0, 0, 0, \ldots) \in \mathfrak{F}^+ \tag{A.5.6}$$

is the "vacuum vector", it describes a state with zero particles. The *particle number operator* N is defined by[188]

$$(N\phi)_n := n\, \phi_n \quad \text{for} \quad \phi \equiv (\phi_n) \in \mathfrak{F}^+ \ . \tag{A.5.7}$$

Obviously, N is symmetric: $\langle \phi, N\psi \rangle_{\mathfrak{F}^+} = \langle N\phi, \psi \rangle_{\mathfrak{F}^+}$ for all ϕ, ψ in the domain of N. A normalized eigenvector,

$$\phi = (0, \ldots, 0, \phi_n, 0, \ldots) \quad \text{with} \quad \|\phi_n\|_{\mathfrak{H}_n^+} = 1 \ ,$$

describes a state with n particles.

Creation- and annihilation operators; the free, real scalar field. An operator which changes the particle number is the "annihilation operator" $a(f)$, which annihilates a particle with wave function $f \in \mathfrak{H}_1$; it is defined by:

$$\big(a(f)\phi\big)_n(p_1, \ldots, p_n) := \sqrt{n+1}\, \hbar \int \frac{d^{d-1}\vec{p}}{2\omega_{\vec{p}}}\, \overline{f(p)}\, \phi_{n+1}(p, p_1, \ldots, p_n) \ ; \tag{A.5.8}$$

here and in the following we use the notation

$$p := (\omega_{\vec{p}}, \vec{p}) \ , \ p_j := (\omega_{\vec{p}_j}, \vec{p}_j) \ .$$

The adjoint operator, $a(f)^*$ is a "creation operator"; as we will see, it creates a particle with wave function $f \in \mathfrak{H}_1$. By inserting the above definitions into

$$\big\langle \psi \, , \, a(f)^*\phi \big\rangle_{\mathfrak{F}^+} := \big\langle a(f)\psi \, , \, \phi \big\rangle_{\mathfrak{F}^+} \ ,$$

[187] In contrast to the infinite direct sum of vector spaces (cf. the comment to formula (A.1.5)), the infinite direct sum of Hilbert spaces may contain sequences $(\phi_n)_{n\in\mathbb{N}}$ for which infinitely many ϕ_n are non-vanishing; an example is given by the coherent states (treated in Exer. (A.5.3)).

[188] Most of the Fock space operators we introduce are *unbounded* operators, e.g., N, the annihilation and creation operators $a(f)$, $a(f)^*$, (see (A.5.8) and (A.5.9) below) and the free field $\varphi^{\rm op}(g) = \int dx\, \varphi^{\rm op}(x)\, g(x)$ (with $g \in \mathcal{D}(\mathbb{M})$) (A.5.17); we omit a study of the pertinent domains.

we obtain

$$\sum_{n=0}^{\infty} \hbar^n \int \frac{d\vec{p}_1}{2\omega_{\vec{p}_1}} \cdots \int \frac{d\vec{p}_n}{2\omega_{\vec{p}_n}} \overline{\psi_n(p_1,\ldots,p_n)} \left(a(f)^*\phi\right)_n(p_1,\ldots,p_n)$$
$$= \sum_{n=1}^{\infty} \hbar^{n-1} \int \frac{d\vec{p}_1}{2\omega_{\vec{p}_1}} \cdots \frac{d\vec{p}_{n-1}}{2\omega_{\vec{p}_{n-1}}}\, \hbar \int \frac{d\vec{p}_n}{2\omega_{\vec{p}_n}} \overline{\psi_n(p_1,\ldots,p_{n-1},p_n)}$$
$$\cdot \sqrt{n}\, f(p_n)\, \phi_{n-1}(p_1,\ldots,p_{n-1}) ,$$

from which we conclude that

$$(a(f)^*\phi)_n(p_1,\ldots,p_n) = \begin{cases} 0 & \text{for } n=0 \\ \frac{1}{\sqrt{n}} \sum_{k=1}^{n} f(p_k)\, \phi_{n-1}(p_1,\ldots,\widehat{p_k},\ldots,p_n) & \text{for } n \geq 1 \end{cases} \tag{A.5.9}$$

see Footn. 27 for the notation. Note the particular relations

$$a(f)\,\Omega = 0 , \quad a(f)^*\,\Omega = (0, f, 0, 0, \ldots) , \quad \forall f \in \mathfrak{H}_1 . \tag{A.5.10}$$

Obviously it holds that

$$[N\,,\, a(f)] = -a(f) , \quad [N\,,\, a(f)^*] = a(f)^* , \tag{A.5.11}$$

which also shows that $a(f)$ ($a(f)^*$ respectively) lowers (increases resp.) the particle number by 1. By direct calculation we get the important commutation relations

$$[a(f)\,,\, a(g)^*] = \langle f, g\rangle_{\mathfrak{H}_1} = \hbar \int \frac{d\vec{p}}{2\omega_{\vec{p}}}\, \overline{f(p)}\, g(p) ,$$
$$[a(f)\,,\, a(g)] = 0 , \quad [a^*(f)\,,\, a(g)^*] = 0 , \quad \forall f, g \in \mathfrak{H}_1 . \tag{A.5.12}$$

For $\|f\|_{\mathfrak{H}_1} = 1$, the operator $a(f)^*a(f)$ can be interpreted as "number of particles with wave function f", because

$$[a(f)^*a(f)\,,\, a(f)] = -a(f) , \quad [a(f)^*a(f)\,,\, a(f)^*] = a(f)^* \quad \text{and}$$
$$[a(f)^*a(f)\,,\, a(g)] = 0 = [a(f)^*a(f)\,,\, a(g)^*] \quad \text{if} \quad \langle f, g\rangle_{\mathfrak{H}_1} = 0 ; \tag{A.5.13}$$

these relations follow immediately from (A.5.12) by using $[a(f), a(f)^*] = \|f\|^2_{\mathfrak{H}_1} = 1$.

Creation and annihilation operators with sharp momentum can be introduced by

$$\int \frac{d\vec{p}}{2\omega_{\vec{p}}}\, \overline{f(p)}\, a(p)\Big|_{p^0=\omega_{\vec{p}}} := a(f) , \quad \int \frac{d\vec{p}}{2\omega_{\vec{p}}}\, f(p)\, a^*(p)\Big|_{p^0=\omega_{\vec{p}}} := a(f)^* , \quad \forall f \in \mathfrak{H}_1 ;$$

they can be interpreted as operator-valued "distributions" – we put quotation marks, because $a(p) : f \longmapsto a(f)$ is antilinear and the "test functions" f are elements of $\mathfrak{H}_1$. The commutation relations (A.5.12) translate into

$$[a(p)\,,\, a^*(q)] = \hbar\, 2\omega_{\vec{p}}\, \delta_{(d-1)}(\vec{p} - \vec{q}) , \quad [a(p)\,,\, a(q)] = 0 , \quad [a^*(p)\,,\, a^*(q)] = 0 . \tag{A.5.14}$$

We can also define creation and annihilation operators in position space

$$a(x) := (2\pi)^{-(d-1)/2} \int \frac{d\vec{p}}{2\omega_{\vec{p}}}\, e^{-ipx}\, a(p)\Big|_{p^0=\omega_{\vec{p}}} ,$$

$$a^*(x) := (2\pi)^{-(d-1)/2} \int \frac{d\vec{p}}{2\omega_{\vec{p}}}\, e^{ipx}\, a^*(p)\Big|_{p^0=\omega_{\vec{p}}} ; \qquad \text{(A.5.15)}$$

they are operator-valued distributions in a rigorous sense, that is, they are elements of $\mathcal{D}'(\mathbb{M}_d, L(\mathfrak{F}^+))$ where $L(\mathfrak{F}^+)$ denotes the set of linear operators on $\mathfrak{F}^+$. The commutation relations (A.5.14) are equivalent to

$$[a(x)\,,\, a^*(y)] = \hbar\,\Delta_m^+(x-y)\;, \quad [a(x)\,,\, a(y)] = 0\;, \quad [a^*(x)\,,\, a^*(y)] = 0\;, \qquad \text{(A.5.16)}$$

with the Wightman two-point function (2.2.1) or (A.2.5), as we obtain by inserting the definitions (A.5.15).

Obviously, $a(x)$ and $a^*(x)$ solve the Klein–Gordon equation,

$$(\Box_x + m^2)a(x) = 0\;, \quad (\Box_x + m^2)a^*(x) = 0\;;$$

hence this holds also for the *free, real scalar field* on $\mathfrak{F}^+$, which is defined by

$$\varphi^{\mathrm{op}}(x) := a(x) + a^*(x) \in \mathcal{D}'(\mathbb{M}_d, L(\mathfrak{F}^+))\;. \qquad \text{(A.5.17)}$$

By using (A.5.16) and (A.2.7) we get the commutation relation

$$[\varphi^{\mathrm{op}}(x), \varphi^{\mathrm{op}}(y)] = \hbar\big(\Delta_m^+(x-y) - \Delta_m^+(y-x)\big) = i\hbar\,\Delta_m(x-y)\;. \qquad \text{(A.5.18)}$$

Correlations appear also for spacelike separated points – even in the vacuum state, e.g.,

$$\langle \Omega\,,\, \varphi^{\mathrm{op}}(x)\varphi^{\mathrm{op}}(y)\,\Omega\rangle_{\mathfrak{F}^+} = \langle \Omega\,,\, a(x)a^*(y)\,\Omega\rangle_{\mathfrak{F}^+} = [a(x)\,,\, a^*(y)] = \hbar\,\Delta_m^+(x-y)\;. \qquad \text{(A.5.19)}$$

The normally ordered product. Powers of the free field $\varphi^{\mathrm{op}}(x)$ do not exist – this is not a surprise, since $\varphi^{\mathrm{op}}(x)$ is an operator-valued distribution. For example, trying to compute $\varphi^{\mathrm{op}}(x)^2$, we would obtain

$$\langle \Omega\,,\, \varphi^{\mathrm{op}}(x)^2\,\Omega\rangle_{\mathfrak{F}^+} = \hbar\,\Delta_m^+(0) = \infty\;. \qquad \text{(A.5.20)}$$

This is the main motivation to introduce the *normally ordered product* (also called "Wick product") $:\!\varphi^{\mathrm{op}}(x_1)\cdots\varphi^{\mathrm{op}}(x_n)\!:$, with respect to which powers of $\varphi^{\mathrm{op}}(x)$ exist, as we explain below. It is defined by:

$$:\!\varphi^{\mathrm{op}}(x_1)\cdots\varphi^{\mathrm{op}}(x_n)\!: \stackrel{\mathrm{Def}}{=} \sum_{J\subseteq\{1,\ldots,n\}} \prod_{j\in J} a^*(x_j) \prod_{k\in J^c} a(x_k)\;, \quad J^c := \{1,\ldots,n\}\setminus J\;, \qquad \text{(A.5.21)}$$

that is, all factors $\varphi^{\mathrm{op}}(x_j)$ are splitted into creation and annihilation operators and the annihilation operators are put on the right-hand side of the creation operators. For $n = 1$, we get $:\varphi^{\mathrm{op}}(x): = \varphi^{\mathrm{op}}(x)$. The normally ordered product of $:\varphi^{\mathrm{op}}(x_1)\cdots\varphi^{\mathrm{op}}(x_n):$ and $:\varphi^{\mathrm{op}}(x_{n+1})\cdots\varphi^{\mathrm{op}}(x_{n+k}):$ is defined to be

$$:\Big(\big(:\varphi^{\mathrm{op}}(x_1)\cdots\varphi^{\mathrm{op}}(x_n):\big)\big(:\varphi^{\mathrm{op}}(x_{n+1})\cdots\varphi^{\mathrm{op}}(x_{n+k}):\big)\Big): \stackrel{\mathrm{Def}}{=} :\varphi^{\mathrm{op}}(x_1)\cdots\varphi^{\mathrm{op}}(x_{n+k}): . \tag{A.5.22}$$

By requiring bilinearity, the latter definition is extended to "Wick polynomials", that is, linear combinations of terms of the form (A.5.21). In terms of smeared Wick polynomials $\sum_n\langle f_n\,, :\varphi^{\mathrm{op}\,\otimes n}:\rangle$, where the sum is finite and f_n is as in Definition 1.2.1, the definition (A.5.22) reads

$$:\Big(\sum_n\langle f_n\,, :\varphi^{\mathrm{op}\,\otimes n}:\rangle\sum_k\langle g_k\,, :\varphi^{\mathrm{op}\,\otimes k}:\rangle\Big): \stackrel{\mathrm{Def}}{=} \sum_{n,k}\langle f_n\otimes_{\mathrm{sym}} g_k\,, :\varphi^{\mathrm{op}\,\otimes(n+k)}:\rangle\ . \tag{A.5.23}$$

Obviously, the normally ordered product is *commutative* and *associative.*

From $a(x)\Omega = 0$ we conclude

$$:\varphi^{\mathrm{op}}(x_1)\cdots\varphi^{\mathrm{op}}(x_n):\Omega = a^*(x_1)\cdots a^*(x_n)\,\Omega\ ; \tag{A.5.24}$$

taking into account also $[N, a^*(x)] = a^*(x)$ and $N\,\Omega = 0$, we obtain

$$N\,:\varphi^{\mathrm{op}}(f_1)\cdots\varphi^{\mathrm{op}}(f_n):\Omega = n\,:\varphi^{\mathrm{op}}(f_1)\cdots\varphi^{\mathrm{op}}(f_n):\Omega\ , \tag{A.5.25}$$

where $f_1,\ldots,f_n\in\mathcal{D}(\mathbb{M})$ and $\varphi^{\mathrm{op}}(f) := \int dx\ \varphi^{\mathrm{op}}(x)\,f(x)$; that is, the Fock space vector $:\varphi^{\mathrm{op}}(f_1)\cdots\varphi^{\mathrm{op}}(f_n):\Omega$ describes an *n particle state.*

Obviously, the formal power series

$$:e^{i\,\varphi^{\mathrm{op}}(f)}: \stackrel{\mathrm{Def}}{=} 1+\sum_{n=1}^{\infty}\frac{i^n}{n!}\int dx_1\cdots dx_n\ f(x_1)\cdots f(x_n)\,:\varphi^{\mathrm{op}}(x_1)\cdots\varphi^{\mathrm{op}}(x_n): \tag{A.5.26}$$

(where $f\in\mathcal{D}(\mathbb{M})$) is the generating functional of the normally ordered product:

$$i^n\,:\varphi^{\mathrm{op}}(x_1)\cdots\varphi^{\mathrm{op}}(x_n): = \frac{\delta^n}{\delta f(x_1)\cdots\delta f(x_n)}\bigg|_{f=0}:e^{i\,\varphi^{\mathrm{op}}(f)}:\ . \tag{A.5.27}$$

Defining $e^{i\,\varphi^{\mathrm{op}}(f)}$ in analogy to (A.5.26) by replacing the normally ordered product by the operator product, there appears a significant difference: The formal power series $e^{i\,\varphi^{\mathrm{op}}(f)}$ is the generating functional of the *symmetrized* operator product, that is,

$$\frac{i^n}{n!}\sum_{\pi\in S_n}\varphi^{\mathrm{op}}(x_{\pi 1})\cdots\varphi^{\mathrm{op}}(x_{\pi n}) = \frac{\delta^n}{\delta f(x_1)\cdots\delta f(x_n)}\bigg|_{f=0} e^{i\,\varphi^{\mathrm{op}}(f)}\ . \tag{A.5.28}$$

The following theorem is very important for Fock space computations in practice.

Theorem A.5.1 (Wick's Theorem in terms of generating functionals). *The normally ordered product is related to the symmetrized operator product by*

$$:e^{i\,\varphi^{\mathrm{op}}(f)}: = e^{i\,\varphi^{\mathrm{op}}(f)}\; e^{\frac{\hbar}{2}\,\Delta^+_m(f,f)}\;, \tag{A.5.29}$$

where

$$\Delta^+_m(f,g) := \int dx\,dy\;f(x)\,\Delta^+_m(x-y)\,g(y)\;.$$

The operator product of two normally ordered products can be expanded in normally ordered products by means of

$$:e^{i\,\varphi^{\mathrm{op}}(f)}:\;:e^{i\,\varphi^{\mathrm{op}}(g)}: = \;:e^{i\,\varphi^{\mathrm{op}}(f+g)}:\;e^{-\hbar\,\Delta^+_m(f,g)}\;. \tag{A.5.30}$$

Proof. The first assertion (A.5.29) can be obtained from the Baker–Campbell–Hausdorff formula for formal power series,

$$e^A\,e^B = e^{A+B}\;e^{\frac{1}{2}[A,B]}\quad\text{if } [A,B] \text{ is a } C\text{-number}, \tag{A.5.31}$$

by setting $A := i\int dx\;a^*(x)\,f(x)$, $B := i\int dx\;a(x)\,f(x)$ and by using the commutation relation (A.5.16) and

$$:e^{i\,\varphi^{\mathrm{op}}(f)}: = \;:e^{i\int dx\;(a(x)+a^*(x))\,f(x)}: = e^{i\int dx\;a^*(x)\,f(x)}\;e^{i\int dx\;a(x)\,f(x)}\;;$$

the latter formula is an immediate consequence of the definition of the normally ordered product (A.5.21).

To derive the second assertion (A.5.30), we proceed as follows: By using (A.5.29) for both $:e^{i\,\varphi^{\mathrm{op}}(f)}:$ and $:e^{i\,\varphi^{\mathrm{op}}(g)}:$ we get

$$:e^{i\,\varphi^{\mathrm{op}}(f)}:\;:e^{i\,\varphi^{\mathrm{op}}(g)}: = e^{i\,\varphi^{\mathrm{op}}(f)}\;e^{i\,\varphi^{\mathrm{op}}(g)}\;e^{\frac{\hbar}{2}\left(\Delta^+_m(f,f)+\Delta^+_m(g,g)\right)}\;.$$

Now we apply the identity (A.5.31) for $A := i\,\varphi^{\mathrm{op}}(f)$, $B := i\,\varphi^{\mathrm{op}}(g)$ and the commutation relation (A.5.18):

$$= e^{i\,\varphi^{\mathrm{op}}(f+g)}\;e^{\frac{\hbar}{2}\left(-\Delta^+_m(f,g)+\Delta^+_m(g,f)+\Delta^+_m(f,f)+\Delta^+_m(g,g)\right)}\;;$$

next we use again (A.5.29), however now for $:e^{i\,\varphi^{\mathrm{op}}(f+g)}:$, and finally we simplify the result

$$= \;:e^{i\,\varphi^{\mathrm{op}}(f+g)}:\;e^{-\frac{\hbar}{2}\,\Delta^+_m(f+g,f+g)}\;e^{\frac{\hbar}{2}\left(-\Delta^+_m(f,g)+\Delta^+_m(g,f)+\Delta^+_m(f,f)+\Delta^+_m(g,g)\right)}\;,$$

$$= \;:e^{i\,\varphi^{\mathrm{op}}(f+g)}:\;e^{-\hbar\,\Delta^+_m(f,g)}\;.\quad\square$$

To write the version (A.5.30) of Wick's Theorem in a more handy form, we apply the derivatives

$$\frac{(-i)^{l+k}\,\delta^{l+k}}{\delta f(x_1)\cdots\delta f(x_l)\delta g(y_1)\cdots\delta g(y_k)}\Big|_{f=0=g}\;.$$

For $l = 1 = k$ we obtain

$$\varphi^{\rm op}(x_1)\varphi^{\rm op}(x_2) = :\varphi^{\rm op}(x_1)\varphi^{\rm op}(x_2): + \hbar\,\Delta_m^+(x_1 - x_2)\ ; \tag{A.5.32}$$

for $l = 2 = k$ we get

$$\begin{aligned} :\varphi^{\rm op}(x_1)\varphi^{\rm op}(x_2):\ :\varphi^{\rm op}(y_1)\varphi^{\rm op}(y_2): &= :\varphi^{\rm op}(x_1)\varphi^{\rm op}(x_2)\varphi^{\rm op}(y_1)\varphi^{\rm op}(y_2): \\ &+ \hbar\Big(:\varphi^{\rm op}(x_1)\varphi^{\rm op}(y_1):\ \Delta_m^+(x_2 - y_2) + :\varphi^{\rm op}(x_1)\varphi^{\rm op}(y_2):\ \Delta_m^+(x_2 - y_1) \\ &+ :\varphi^{\rm op}(x_2)\varphi^{\rm op}(y_1):\ \Delta_m^+(x_1 - y_2) + :\varphi^{\rm op}(x_2)\varphi^{\rm op}(y_2):\ \Delta_m^+(x_1 - y_1)\Big) \\ &+ \hbar^2\Big(\Delta_m^+(x_1 - y_1)\,\Delta_m^+(x_2 - y_2) + \Delta_m^+(x_1 - y_2)\,\Delta_m^+(x_2 - y_1)\Big)\ . \end{aligned}$$

For general values of $l, k \in \mathbb{N}$, the result can be written as

$$\begin{aligned} :\varphi^{\rm op}(x_1)\cdots\varphi^{\rm op}(x_l):\ :\varphi^{\rm op}(y_1)\cdots\varphi^{\rm op}(y_k): &= \sum_{n=0}^{\min\{l,k\}} \frac{\hbar^n}{n!}\int du_1\cdots du_n\,dv_1\cdots dv_n \\ :\Big(\frac{\delta^n :\varphi^{\rm op}(x_1)\cdots\varphi^{\rm op}(x_l):}{\delta\varphi^{\rm op}(u_1)\cdots\delta\varphi^{\rm op}(u_n)}\ &\frac{\delta^n :\varphi^{\rm op}(y_1)\cdots\varphi^{\rm op}(y_k):}{\delta\varphi^{\rm op}(v_1)\cdots\delta\varphi^{\rm op}(v_n)}\Big):\ \prod_{j=1}^n \Delta_m^+(u_j - v_j)\ , \end{aligned} \tag{A.5.33}$$

where the functional derivative is defined by

$$\frac{\delta :\varphi^{\rm op}(x_1)\cdots\varphi^{\rm op}(x_l):}{\delta\varphi^{\rm op}(u)} \overset{\rm Def}{=} \sum_{j=1}^l \delta(x_j - u)\ :\varphi^{\rm op}(x_1)\cdots\widehat{\varphi^{\rm op}(x_j)}\cdots\varphi^{\rm op}(x_l):\ , \tag{A.5.34}$$

see again Footn. 27 for the notation. We point out the similarity of the formulation (A.5.33) of Wick's Theorem to our definition (2.1.5) of the star product; the precise formulation of this connection is given in Theorem 2.6.3.

Finally, we give a heuristic explanation why the normally ordered product exists for coinciding points, that is, why

$$:\varphi^{\rm op}(x)^n:\quad \text{or more generally}\quad :\prod_{j=1}^n \partial_x^{a_j}\varphi^{\rm op}(x):\quad \text{exists,} \tag{A.5.35}$$

where $n \in \mathbb{N}^*$, $a_j \in \mathbb{N}^d$; more precisely, we explain why divergences of the kind (A.5.20) do not appear for the normally ordered product. For this purpose we study

$$\big\langle\Omega\,,\, :\varphi^{\rm op}(x_{11})\cdots\varphi^{\rm op}(x_{1n_1}):\ \cdots\ :\varphi^{\rm op}(x_{k1})\cdots\varphi^{\rm op}(x_{kn_k}):\Omega\big\rangle_{\mathfrak{F}^+}\ , \tag{A.5.36}$$

which we compare with the Wightman function

$$W_n(x_{11},\ldots,x_{kn_k}) := \big\langle\Omega\,,\,\varphi^{\rm op}(x_{11})\cdots\varphi^{\rm op}(x_{1n_1})\cdots\varphi^{\rm op}(x_{k1})\cdots\varphi^{\rm op}(x_{kn_k})\,\Omega\big\rangle_{\mathfrak{F}^+}\ , \tag{A.5.37}$$

where $n := n_1 + \cdots + n_k$. Due to Theorem 2.6.3 and (2.6.7), this definition of the Wightman function agrees with the one given in Exer. 2.5.4. From that exercise we know that $W_n(x_{11}, \dots, x_{kn_k})$ vanishes if n is odd, and for n even it is the sum over all possibilities to contract *all* the fields $\varphi^{\mathrm{op}}(x_{11}), \dots, \varphi^{\mathrm{op}}(x_{kn_k})$ ($n/2$ contractions in each term). Hence, contractions of both types

(a) $\Delta_m^+(x_{jr} - x_{ls})$ with $1 \le j < l \le k$, and
(b) $\Delta_m^+(x_{jr} - x_{js})$ with $1 \le r < s \le n_j$

appear.

Computing the expression (A.5.36), there is a crucial difference: By iterated use of Wick's Theorem in the form (A.5.33), we obtain the sum over all possible contractions of type (a), but no contractions of type (b). Therefore, restricting the normally ordered products $:\varphi^{\mathrm{op}}(x_{j1}) \cdots \varphi^{\mathrm{op}}(x_{jn_j}):$ to coinciding points, by performing the limit

$$x_{jr} \to x_{j1} \quad \forall\, 2 \le r \le n_j \;, \; 1 \le j \le k \;,$$

divergent "self-contractions" $\Delta_m^+(0)$ ("tadpole diagrams") do not appear.

Remark A.5.2. It is instructive to explain the difference between (A.5.36) and the Wightman function (A.5.37) by *elementary computations.* The Wightman function can be computed by generalizing the procedure in (A.5.19): Inserting $\varphi^{\mathrm{op}}(x_{jr}) = a(x_{jr}) + a^*(x_{jr})$ for all (j,r) and multiplying out, $W_n(x_{11}, \dots, x_{kn_k})$ becomes a sum of terms of the form $\langle \Omega, a(x_{11}) \cdots a^*(x_{kn_k})\,\Omega \rangle$. Only terms with $n/2$ creation- and $n/2$ annihilation operators can contribute. Then, in each term the annihilation operator farthest right is commuted further to the right until it hits Ω, which gives zero. Every commutator $[a(x_{jr}), a^*(x_{ls})]$ is a contraction $\Delta_m^+(x_{jr} - x_{ls})$, both types (a) and (b) appear. This procedure is iterated until all operators $a(x_{jr})$, $a^*(x_{ls})$ are commuted away.

Computing analogously the expression (A.5.36), solely contractions of type (a) appear, because all $a(x_{jr})$ are already to the right of all $a^*(x_{js})$ ($s \neq r$), due to the normal ordering (A.5.21).

Exercise A.5.3 (Coherent states in Fock space). Let $h \in \mathcal{C} = C(\mathbb{M}, \mathbb{R})$ be a solution of the Klein–Gordon equation, that is,[189]

$$\begin{aligned} h(x) &= \frac{1}{(2\pi)^{(d-1)/2}} \int d^d q\; \delta(q^2 - m^2)\, \hat{h}(q)\, e^{-iqx}\, \big(\theta(q^0) + \theta(-q^0)\big) \\ &= \frac{1}{(2\pi)^{(d-1)/2}} \int \frac{d\vec{q}}{2\omega_{\vec{q}}} \Big(\hat{h}(q)\, e^{-iqx}\Big|_{q^0=\omega_{\vec{q}}} + \hat{h}(q)\, e^{-iqx}\Big|_{q^0=-\omega_{\vec{q}}} \Big) \\ &= \frac{1}{(2\pi)^{(d-1)/2}} \int \frac{d\vec{q}}{2\omega_{\vec{q}}}\Big|_{q^0=\omega_{\vec{q}}} \Big(\hat{h}(q)\, e^{-iqx} + \overline{\hat{h}(q)}\, e^{iqx} \Big) \;, \end{aligned} \tag{A.5.38}$$

where we take into account that $\overline{\hat{h}(q)} = \hat{h}(-q)$ since $h(x)$ is $\mathbb{R}$-valued. The Fock space vector

$$\psi_h := \frac{e^{a(h)^*}\,\Omega}{\|e^{a(h)^*}\,\Omega\|_{\mathfrak{F}^+}} \quad \text{with} \quad e^{a(h)^*}\,\Omega := \Omega + \sum_{n=1}^{\infty} \frac{1}{n!} \left(a(h)^*\right)^n \Omega \tag{A.5.39}$$

[189] Here, $\hat{h}$ does not precisely agree with the distribution obtained by Fourier transformation of h.

and

$$a(h)^* := \int \frac{d\vec{p}}{2\omega_{\vec{p}}}\Big|_{p^0=\omega_{\vec{p}}} \hat{h}(p)\, a^*(p)$$

describes a "coherent state"; such states are distinguished by the property

$$\big\langle \psi_h \,,\, {:}\varphi^{\mathrm{op}}(x_1)\cdots\varphi^{\mathrm{op}}(x_n){:}\, \psi_h \big\rangle_{\mathfrak{F}^+} = \hbar^n\, h(x_1)\cdots h(x_n)\ . \tag{A.5.40}$$

(a) Compute $\|e^{a(h)^*}\,\Omega\|_{\mathfrak{F}^+}$.

(b) Prove the property (A.5.40).

[*Hint*: Use the Baker–Campbell–Hausdorff formula (A.5.31) on a formal level, that is, questions of convergence of the series may be ignored.]

[*Solution*: First note the identity

$$e^A\, e^B = e^{[A,B]}\, e^B\, e^A \quad \text{if } [A,B] \text{ is a } C\text{-number}, \tag{A.5.41}$$

which follows from the Baker–Campbell–Hausdorff formula (A.5.31).

(a) By applying (A.5.41) and (A.5.12), we get

$$\|e^{a(h)^*}\,\Omega\|^2 = \big\langle \Omega\,,\, e^{a(h)}\, e^{a(h)^*}\,\Omega\big\rangle = e^{[a(h),a(h)^*]}\,\big\langle \Omega\,,\, e^{a(h)^*}\, e^{a(h)}\,\Omega\big\rangle = e^{\|\hat{h}\|^2_{\mathfrak{H}_1}}$$

since $e^{a(h)}\,\Omega = \Omega$; hence, $\|e^{a(h)^*}\,\Omega\|_{\mathfrak{F}^+} = e^{\frac{1}{2}\,\|\hat{h}\|^2_{\mathfrak{H}_1}}$.

(b) By using

$$[a(q)\,,\, a(h)^*] = \int \frac{d\vec{p}}{2\omega_{\vec{p}}}\Big|_{p^0=\omega_{\vec{p}}} \hat{h}(p)\,[a(q)\,,\, a^*(p)] = \hbar\,\hat{h}(q)$$

and again (A.5.41), we obtain

$$\begin{aligned} a(q)\, e^{a(h)^*}\,\Omega &= \frac{d}{d\lambda}\Big|_{\lambda=0} e^{\lambda\, a(q)}\, e^{a(h)^*}\,\Omega = \frac{d}{d\lambda}\Big|_{\lambda=0} \Big(e^{\lambda\,[a(q),a(h)^*]}\, e^{a(h)^*}\, e^{\lambda\, a(q)}\Big)\Omega \\ &= \hbar\,\hat{h}(q)\, e^{a(h)^*}\,\Omega\ ; \end{aligned}$$

taking into account (A.5.15), this can be written as

$$a(x)\, e^{a(h)^*}\,\Omega = \frac{\hbar}{(2\pi)^{(d-1)/2}} \int \frac{d\vec{q}}{2\omega_{\vec{q}}}\Big|_{q^0=\omega_{\vec{q}}} e^{-iqx}\,\hat{h}(q)\, e^{a(h)^*}\,\Omega\ . \tag{A.5.42}$$

The latter relation implies

$$\begin{aligned} \big\langle \Omega\,,\, e^{a(h)}\, a^*(x)\,\ldots\big\rangle &= \big\langle a(x)\, e^{a(h)^*}\,\Omega\,,\,\ldots\big\rangle \\ &= \frac{\hbar}{(2\pi)^{(d-1)/2}} \int \frac{d\vec{q}}{2\omega_{\vec{q}}}\Big|_{q^0=\omega_{\vec{q}}} e^{iqx}\,\overline{\hat{h}(q)}\,\big\langle \Omega\,,\, e^{a(h)}\,\ldots\big\rangle\ . \end{aligned} \tag{A.5.43}$$

By iterated use of (A.5.42) and (A.5.43) we get

$$\begin{aligned}
&\big\langle e^{a(h)^*}\,\Omega\,,\,:\!\varphi^{\rm op}(x_1)\cdots\varphi^{\rm op}(x_n)\!:\,e^{a(h)^*}\,\Omega\big\rangle\\
&\quad= \sum_{J\subseteq\{1,\ldots,n\}} \big\langle\Omega\,,\,e^{a(h)}\prod_{j\in J}a^*(x_j)\prod_{k\in J^c}a(x_k)\,e^{a(h)^*}\,\Omega\big\rangle\\
&\quad= \big\langle\Omega\,,\,e^{a(h)}\,e^{a(h)^*}\,\Omega\big\rangle\,\frac{\hbar^n}{(2\pi)^{(d-1)n/2}}\\
&\qquad\cdot\int\prod_{r=1}^{n}\frac{d\vec q_r}{2\omega_{\vec q_r}}\Big|_{q_r^0=\omega_{\vec q_r}}\sum_{J\subseteq\{1,\ldots,n\}}\prod_{j\in J}e^{iq_jx_j}\,\overline{\hat h(q_j)}\prod_{k\in J^c}e^{-iq_kx_k}\,\hat h(q_k)\\
&\quad= e^{\|\hat h\|^2_{\mathfrak{H}_1}}\,\hbar^n\prod_{r=1}^{n}\Big[\frac{1}{(2\pi)^{(d-1)/2}}\int\frac{d\vec q_r}{2\omega_{\vec q_r}}\Big|_{q_r^0=\omega_{\vec q_r}}\Big(\hat h(q_r)\,e^{-iq_rx_r}+\overline{\hat h(q_r)}\,e^{iq_rx_r}\Big)\Big]\\
&\quad= e^{\|\hat h\|^2_{\mathfrak{H}_1}}\,\hbar^n\,h(x_1)\cdots h(x_n)\,,
\end{aligned}$$

which yields the assertion (A.5.40).]

The quantization of the free photon field in the bosonic Fock space can be found, e.g., in [109, Sect. 3-2] or [148, Sect. 2.11].

The fermionic Fock space. Fermionic particles (such as all leptons and all quarks or composite particles as, e.g., the proton and the neutron) satisfy Pauli's exclusion principle and obey *Fermi statistics.* Usually their spin has an half-integer value, $s\in\frac{1}{2}(2\mathbb{N}+1)$, an exception are the Faddeev–Popov ghosts introduced in Sect. 5.1.2.

To construct the Fock space describing a certain species of fermionic particles (mostly a particle-antiparticle pair as, e.g., for the Dirac spinor field) with mass $m>0$ and spin s, let $\mathfrak{H}_n^-$ be the pertinent n-particle space. $\mathfrak{H}_n^-$ is a Hilbert space, its elements are *totally antisymmetric* wave functions $\phi\colon(\mathcal{H}_m^+)^{\times n}\longrightarrow\mathcal{H}_n^{(s)}$, where $\mathcal{H}_n^{(s)}$ is a suitable Hilbert space depending on n and s. Similarly to (A.5.3), the scalar product in $\mathfrak{H}_n^-$ ($n\in\mathbb{N}^*$) is a kind of L^2-scalar product, for the explicit definitions we refer, e.g., to [73].

The *fermionic Fock space* is the infinite direct sum of the Hilbert spaces $\mathfrak{H}_n^-$:

$$\mathfrak{F}^-:=\bigoplus_{n=0}^{\infty}\mathfrak{H}_n^-\,,\qquad\mathfrak{H}_0^-:=\mathbb{C}\,. \tag{A.5.44}$$

Defining the scalar product as in (A.5.5), $\mathfrak{F}^-$ is also a Hilbert space. The definition of the particle number operator N (A.5.7) can directly be adopted to the fermionic Fock space and, hence, we have the same particle number interpretation. Again, N is symmetric.

Writing the definition of the bosonic annihilation operator (A.5.8) as a $\mathfrak{H}_1^+$ scalar product, this definition applies also to the fermionic Fock space:

$$\big(a(f)\phi\big)_n(p_1,\ldots,p_n):=\sqrt{n+1}\,\big\langle f\,,\,\phi_{n+1}(\,\cdot\,,p_1,\ldots,p_n)\big\rangle_{\mathfrak{H}_1^-}\,, \tag{A.5.45}$$

where $f\in\mathfrak{H}_1^-$ and $p_j\equiv(\omega_{\vec p_j},\vec p_j)$. In (A.5.45) we use that formally it is possible to interpret $\mathcal{H}_m^+\ni p\longmapsto\phi_{n+1}(p,p_1,\ldots,p_n)$ as an element of $\mathfrak{H}_1^-$, for fixed $(p_1,\ldots,p_n)\in(\mathcal{H}_m^+)^{\times n}$.

Again, the adjoint operator $a(f)^*$ is a creation operator: $a(f)$ and $a(f)^*$ annihilate or create, respectively, a particle with wave function f; in particular they have the properties (A.5.10) and (A.5.11). One verifies the *anti*commutation relations[190]

$$[a(f)\,,\,a(g)^*]_+ = \langle f,g\rangle_{\mathfrak{H}_1^-}\,, \quad [a(f)\,,\,a(g)]_+ = 0\,, \quad [a^*(f)\,,\,a(g)^*]_+ = 0 \tag{A.5.46}$$

for all $f,g \in \mathfrak{H}_1^-$.

For $\|f\|_{\mathfrak{H}_1^-} = 1$, the operator $a(f)^*a(f)$ satisfies again the relations (A.5.13); hence it can again be interpreted as "number of particles with wave function f". Because of Fermi statistics, $a(f)^*a(f)$ has only the eigenvalues 0 and 1; mathematically this result can be derived form the anticommutation relations (A.5.46) in the following way:

$$\begin{aligned}\big(a(f)^*a(f)\big)^2 &= a(f)^*a(f)a(f)^*a(f)\\ &= a(f)^*\,[a(f),a(f)^*]_+\,a(f) - a(f)^*a(f)^*a(f)a(f)\\ &= a(f)^*a(f)\,,\end{aligned}$$

since $[a(f),a(f)^*]_+ = \|f\|^2_{\mathfrak{H}_1^-} = 1$ and $\big(a(f)\big)^2 = \frac{1}{2}\,[a(f),a(f)]_+ = 0$.

The quantization of free Dirac spinors in the fermionic Fock space can be found, e.g., in [148, Sect. 2.2], [109, Sect. 3-3] or [73, Sects. III.3], and for the free Faddeev–Popov ghosts fields we refer to [149, Sect. 1.2].

A.6 Weak adiabatic limit: Wightman- and Green functions

References for the weak adiabatic limit are [66, Sect. 8.2], [9] and [38]; for the strong adiabatic limit, which we do not treat in this book, we refer to [67]. In some parts of this appendix we assume that the reader is familiar with the Wightman axioms [162] and with the Gell-Mann–Low formula for the Green function of interacting fields [81].

Weak adiabatic limit. A main characteristics of causal perturbation theory is that the coupling constant is adiabatically switched off: κ is replaced by $\kappa g(x)$ with $g \in \mathcal{D}(\mathbb{M})$. To make contact to traditional approaches to pQFT, we have to remove this IR-regularization, we have to perform the adiabatic limit $g(x) \to 1$ in some form; cf. Sect. 3.7. In this appendix we study the adiabatic limit in the sense of Epstein and Glaser: We choose a test function

$$g \in \mathcal{D}(\mathbb{M}) \quad \text{with} \quad g(0) = 1 \quad \text{and set} \quad g_\varepsilon(x) := g(\varepsilon x) \tag{A.6.1}$$

[190] The second anticommutation relation follows immediately from (A.5.45), the third one is then obtained by taking the adjoint. To prove the $[a(f)\,,\,a(g)^*]_+$-relation, one first derives an explicit formula for $a(f)^*$ by using the definition of the scalar product in $\mathfrak{H}_n^-$ – similarly to our procedure in the bosonic Fock space.

for $\varepsilon > 0$. With that the adiabatic limit of a distribution $t \in \mathcal{D}'(\mathbb{M}^n)$ is the limit

$$\lim_{\varepsilon \downarrow 0} \int dx_1 \cdots dx_n \; g_\varepsilon(x_1) \cdots g_\varepsilon(x_n) \, t(x_1, \ldots, x_n) \; ; \tag{A.6.2}$$

it *exists* if this limit exists for all such g, and it is *unique* if this limit does not depend on the choice of g.

By the *weak adiabatic limit* we mean the adiabatic limit $\varepsilon \downarrow 0$ of the expectation value in a suitable state of an observable or field $F\big(\mathcal{L}(g_\varepsilon)\big) \in \mathcal{F}$, or of a suitable matrix element of the S-matrix $\mathbf{S}\big(\mathcal{L}(g_\varepsilon)\big)$ (where $\mathcal{L} = \kappa L \in \kappa \mathcal{P}_{\mathrm{bal}}$ is the interaction). We give some major examples:

- Epstein and Glaser define the *Wightman function* of the interacting fields $A_{j,\mathcal{L}}(x_j)$ (where $A_j \in \mathcal{P}_{\mathrm{bal}}$, $1 \le j \le k$) as the weak adiabatic limit

$$W_{\mathcal{L}}\big(A_1(x_1), \ldots, A_k(x_k)\big) := \lim_{\varepsilon \downarrow 0} \omega_0\Big(A_{1,\mathcal{L}(g_\varepsilon)}(x_1) \star \cdots \star A_{k,\mathcal{L}(g_\varepsilon)}(x_k)\Big) \, , \tag{A.6.3}$$

  and the *Green function* of these interacting fields as

$$G_{\mathcal{L}}\big(A_1(x_1), \ldots, A_k(x_k)\big) := \lim_{\varepsilon \downarrow 0} \omega_0\Big(T_{\mathcal{L}(g_\varepsilon)}\big(A_1(x_1) \otimes \cdots \otimes A_k(x_k)\big)\Big) \, , \tag{A.6.4}$$

  where the time-ordered product $T_{\mathcal{L}(g)}(\cdots)$ of the interacting fields $A_{j,\mathcal{L}(g)}(x_j)$ is defined in (4.2.45), see also Remk. 4.2.4.

- In a scattering experiment the *transition probability* from the initial state ω_H to the final state ω_G, where $H, G \in \mathcal{F}_\hbar$ and ω_H, ω_G are given in Exer. 2.5.5, is obtained by the weak adiabatic limit

$$\lim_{\varepsilon \downarrow 0} \Big| \omega_0\big(G^* \star \mathbf{S}\big(\mathcal{L}(g_\varepsilon)\big) \star H\big) \Big|^2 \, .$$

In the literature, there appears also the *strong adiabatic limit*, which is the limit $\varepsilon \downarrow 0$ of the corresponding Fock space operator $F^{\mathrm{op}}\big(\mathcal{L}(g_\varepsilon)\big)$ or $\mathbf{S}^{\mathrm{op}}\big(\mathcal{L}(g_\varepsilon)\big)$, respectively, in the strong operator sense.

Example A.6.1 (Adiabatic limit of the vacuum expectation value of $\mathbf{S}\big(\mathcal{L}(g_\varepsilon)\big)$). In some cases, a necessary and sufficient condition for the existence and uniqueness of the weak adiabatic limit is that the T- or R-products are suitably renormalized. We are going to explain this in terms of a simple example (given in [46, App.]): The vacuum expectation value of $\mathbf{S}\big(\mathcal{L}(g_\varepsilon)\big)$ in the case that all fields are massive. We proceed in $d = 4$ dimensions and assume $\dim \mathcal{L} = 4$. With

$$\widehat{g_\varepsilon}(p) = \varepsilon^{-4} \, \hat{g}(p/\varepsilon)$$

and the notation

$$t_n(x_1 - x_n, \ldots) := \omega_0\Big(T\big(\mathcal{L}(x_1) \otimes \cdots \otimes \mathcal{L}(x_n)\big)\Big) \tag{A.6.5}$$

we obtain

$$\lim_{\varepsilon\downarrow 0}\omega_0\Big(\mathbf{S}\big(\mathcal{L}(g_\varepsilon)\big)\Big) = 1+\sum_{n=2}^{\infty}\lim_{\varepsilon\downarrow 0}\frac{i^n}{n!\hbar^n}\int dx_1\cdots dx_n\; g_\varepsilon(x_1)\cdots g_\varepsilon(x_n)\,t_n(x_1-x_n,\dots)$$
$$= 1+\sum_{n=2}^{\infty}\lim_{\varepsilon\downarrow 0}\frac{(2\pi)^2 i^n}{\varepsilon^4\, n!\hbar^n}\int dp_1\cdots dp_{n-1}\;\hat g(-p_1)\cdots\hat g(-p_{n-1})$$
$$\cdot\,\hat g(p_1+\cdots+p_{n-1})\,\widehat{t_n}(\varepsilon p_1,\dots,\varepsilon p_{n-1})\;. \tag{A.6.6}$$

Working with the axiom "Scaling degree" (3.2.62) (instead of "Sm-expansion"), we inductively know that $t_n^0 := t_n\big|_{\mathcal{D}(\mathbb{R}^{4(n-1)}\setminus\{0\})}$ fulfills

$$\mathrm{sd}(t_n^0)\le n\,\dim\mathcal{L}=4n\;,$$

as proved in Remk. 3.2.25. We assume that $\mathrm{sd}(t_n^0)=4n$, which is almost always satisfied in physically relevant examples. As mentioned in Remk. 3.2.13, the central solution t_n^{c} for t_n exists and it satisfies

$$\partial_p^a\widehat{t_n^{\mathrm{c}}}(0)=0\qquad\forall|a|\le\omega=4\;. \tag{A.6.7}$$

Hence, the most general, Lorentz invariant solution for $\widehat{t_n}$ fulfilling the renormalization condition "Scaling degree" can be written as

$$\widehat{t_n}(p_1,\dots,p_{n-1})=\widehat{t_n^{\mathrm{c}}}(p_1,\dots,p_{n-1})+C_n^0+\sum_{j,k}C_n^{jk}\,(p_jp_k)+\sum_{jkrs}C_n^{jkrs}\,(p_jp_k)(p_rp_s)\;, \tag{A.6.8}$$

where C_n^0, C_n^{jk}, $C_n^{jkrs}\in\mathbb{C}$ are constants, which are restricted by invariance of $t_n(x_1-x_n,\dots)$ under permutations of $x_1,\dots,x_n$, that is, by the axiom "Symmetry" for the T-product. Inserting (A.6.8) into (A.6.6) and taking into account (A.6.7), we conclude: The adiabatic limit (A.6.6) exists iff

$$C_n^0=0\,,\quad C_n^{jk}=0\quad\forall j,k,n\;; \tag{A.6.9}$$

it is unique iff

$$C_n^{jkrs}=0\quad\forall j,k,r,s,n\;. \tag{A.6.10}$$

With this renormalization, which is the central solution t_n^{c}, the "vacuum diagrams" vanish in the adiabatic limit:

$$\lim_{\varepsilon\downarrow 0}\omega_0\Big(\mathbf{S}\big(\mathcal{L}(g_\varepsilon)\big)\Big)=1\;. \tag{A.6.11}$$

Epstein and Glaser's Wightman- and Green functions. If all basic fields are *massive*, the weak adiabatic limits (A.6.3) and (A.6.4) defining the Wightman and Green functions, respectively, exist and are unique for any renormalization respecting our axioms. More precisely, Epstein and Glaser proved the following (see [66, Sect. 8.2] and also [38, Sect. 5.1]):

Theorem A.6.2 (Existence and uniqueness of weak adiabatic limit for massive fields)**.** *Let $d=4$, let all basic fields be massive, scalar fields and let the interacting fields and the time-ordered product of interacting fields appearing on the r.h.s. of* (A.6.3) *and* (A.6.4), *respectively, be constructed in accordance with our axioms* *for the R- (or T-) product. (The axiom "Sm-expansion" may be replaced by the weaker axiom "Scaling degree"* (3.2.62).) *Then, the weak adiabatic limits* (A.6.3) *and* (A.6.4) *exist and are unique.*

The resulting Wightman and Green functions are elements of $\mathcal{D}'(\mathbb{M}^k)$ *and they satisfy all linear properties one expects for them – in the sense of formal power series in* κ. *For the Wightman functions these are the Wightman axioms (formulated in terms of the Wightman functions)* [162, Sect. 3.3] *except the cluster decomposition property, where positivity of the Wightman functions is understood in the sense of Definition* 5.4.4. *For the Green functions these are Poincaré covariance, symmetry, i.e.,*

$$G_{\mathcal{L}}\big(A_{\pi 1}(x_{\pi 1}),\ldots,A_{\pi k}(x_{\pi k})\big)=G_{\mathcal{L}}\big(A_1(x_1),\ldots,A_k(x_k)\big)\quad\forall\pi\in S_k\ ,$$

and Causality in the form of Lemma 3.3.2(a)*:*

$$G_{\mathcal{L}}\big(A_1(x_1),\ldots,A_k(x_k)\big)=W_{\mathcal{L}}\big(A_1(x_1),\ldots,A_k(x_k)\big)$$
$$\textit{if}\quad x_j\notin\big(\{x_{j+1},\ldots,x_k\}+\overline{V}_-\big)\quad\forall 1\le j\le k-1\ .$$

Rough idea of the proof: For the existence and uniqueness of the weak adiabatic limit we argue in terms of the Green function, for the Wightman function analogous arguments hold true. First we integrate out (A.6.4) with an arbitrary test function $f(x_1,\ldots,x_k)\in\mathcal{D}(\mathbb{M}^k)$. Then, to write down the r.h.s. of (A.6.4) to nth order in $\mathcal{L}$, we use the generalized retarded product $R_{n,k}$, introduced in (3.3.49); omitting constant prefactors we get

$$\int dx_1\cdots dx_k\ G_{\mathcal{L}}^{(n)}\big(A_1(x_1),\ldots,A_k(x_k)\big)\,f(x_1,\ldots,x_k)$$
$$\sim\lim_{\varepsilon\downarrow 0}\int dp_1\cdots dp_n\ \widehat{g_\varepsilon}(-p_1)\cdots\widehat{g_\varepsilon}(-p_n)\,\widehat{r_{n,k}}(p_1,\ldots,p_n;f)\ ,\tag{A.6.12}$$

where

$$\widehat{r_{n,k}}(p_1,\ldots,p_n;f):=\int dy_1\cdots dy_n\ e^{i(p_1y_1+\cdots+p_ny_n)}\int dx_1\cdots dx_k\,f(x_1,\ldots,x_k)$$
$$\cdot\,\omega_0\Big(R_{n,k}\big(\mathcal{L}(y_1),\ldots,\mathcal{L}(y_n);A_1(x_1),\ldots,A_k(x_k)\big)\Big)\ .$$

The assumption that all basic fields are massive simplifies the proof crucially: It implies that there is a mass gap in the energy-momentum spectrum of the model, that is, the vacuum state is separated from the rest of the spectrum. By using this gap, one can show that $(p_1,\ldots,p_n)\longmapsto\widehat{r_{n,k}}(p_1,\ldots,p_n;f)$ is *smooth in a neighbourhood of* $(p_1,\ldots,p_n)=0$ – this is the core of the proof. Noting that $\widehat{g_\varepsilon}(-p)$ converges for $\varepsilon\downarrow 0$ to $(2\pi)^2\,\delta_{(4)}(p)$ for all allowed g (A.6.1), explicitly

$$\lim_{\varepsilon\downarrow 0}\int dp\ \widehat{g_\varepsilon}(-p)\,h(p)=\lim_{\varepsilon\downarrow 0}\int d\tilde{p}\ \widehat{g}(-\tilde{p})\,h(\varepsilon\tilde{p})=h(0)\int d\tilde{p}\ \widehat{g}(-\tilde{p})=(2\pi)^2\,h(0)$$

for all $h \in \mathcal{D}(\mathbb{M})$, it is then obvious that the adiabatic limit (A.6.12) exists and is unique.[191]

The asserted properties of $W_{\mathcal{L}}(\cdots)$ and $G_{\mathcal{L}}(\cdots)$ are not difficult to see, apart from the spectral condition [66] and the positivity of the Wightman functions (see below), because corresponding properties hold true before the adiabatic limit is taken. For the Causality of the Green functions this is explained in Remk. 4.2.4.

For example, the Poincaré covariance of the Wightman functions, that is,

$$W_{\mathcal{L}}\big(A_1(Lx_1),\ldots,A_k(Lx_k)\big) = W_{\mathcal{L}}\big(A_1(x_1),\ldots,A_k(x_k)\big) \qquad \forall L \equiv (\Lambda,a) \in \mathcal{P}_+^\uparrow \tag{A.6.13}$$

if all field polynomials $A_j \in \mathcal{P}_{\mathrm{bal}}$ $(1 \le j \le k)$ and the interaction $\mathcal{L} \in \kappa\mathcal{P}_{\mathrm{bal}}$ are Lorentz scalars (cf. Exap. 3.1.13), relies on

$$\begin{aligned} &\omega_0\Big(A_{1,\mathcal{L}(g_\varepsilon)}(x_1)\star\cdots\star A_{k,\mathcal{L}(g_\varepsilon)}(x_k)\Big) = \omega_0\Big(\beta_L\big(A_{1,\mathcal{L}(g_\varepsilon)}(x_1)\star\cdots\star A_{k,\mathcal{L}(g_\varepsilon)}(x_k)\big)\Big) \\ &= \omega_0\Big(A_{1,\mathcal{L}(g^L_\varepsilon)}(Lx_1)\star\cdots\star A_{k,\mathcal{L}(g^L_\varepsilon)}(Lx_k)\big)\Big) \quad \text{with} \quad g^L_\varepsilon(x) := g\circ\Lambda^{-1}\big(\varepsilon(x-a)\big) \end{aligned} \tag{A.6.14}$$

(where (3.1.37) and axiom (h) (3.1.39) are used) and on the uniqueness of the adiabatic limit. The latter is used here in a somewhat more general form: Replacing in (A.6.3) $g_\varepsilon(x)$ by $g^a_\varepsilon(x) := g\big(\varepsilon(x-a)\big)$, where $a \in \mathbb{M}$ and still $g \in \mathcal{D}(\mathbb{M})$ with $g(0) = 1$, the limit $\varepsilon \downarrow 0$ is not only independent of g, it does also not depend on a.

Finally, we give the proof of the positivity of the Wightman functions, which is due to Paweł Duch [39, Sect. 5.2]: Let

$$F_\varepsilon := A_{1,\mathcal{L}(g_\varepsilon)}(h_1)\star\cdots\star A_{k,\mathcal{L}(g_\varepsilon)}(h_k) \quad \text{and} \quad F := F_{\varepsilon=1}\ ,$$

with arbitrary $A_j \in \mathcal{P}_{\mathrm{bal}}$, $h_j \in \mathcal{D}(\mathbb{M})$ and with $\mathcal{L} = \kappa L \in \kappa\mathcal{P}_{\mathrm{bal}}$. We have to prove that

$$\mathbb{C}[\![\kappa]\!] \ni \lim_{\varepsilon\downarrow 0}\omega_0(F^*_\varepsilon\star F_\varepsilon) \ge 0; \tag{A.6.15}$$

we will do this by using the criterion (5.4.24). First, we conclude from

$$\overline{\omega_0(F^*_\varepsilon\star F_\varepsilon)} = \omega_0\big((F^*_\varepsilon\star F_\varepsilon)^*\big) = \omega_0(F^*_\varepsilon\star F_\varepsilon) \quad \text{that} \quad \lim_{\varepsilon\downarrow 0}\omega_0(F^*_\varepsilon\star F_\varepsilon) \in \mathbb{R}[\![\kappa]\!]\ .$$

To verify the further parts of the criterion (5.4.24), let $\mathcal{O}$ be an open double cone containing $\{0\}\cup\operatorname{supp}h_1\cup\cdots\cup\operatorname{supp}h_k$ and let $g \in \mathcal{G}(\mathcal{O})$ (3.7.10). Due to $0 \in \mathcal{O}$

[191] Naively one might think that the result (A.6.9)–(A.6.10) contradicts the statement here that the adiabatic limit (A.6.4) exists and is unique for *all* renormalizations of $r_{n,k}$ respecting our axioms. But the result (A.6.9)–(A.6.10) does not apply here for the following reason: Due to the support property (3.3.50) of $R_{n,k}\big(\mathcal{L}(y_1),\ldots,\mathcal{L}(y_n);A_1(x_1),\ldots,A_k(x_k)\big)$, all contributing diagrams have the property that every interaction vertex y_j $(1 \le j \le n)$ is connected with at least one field vertex x_s $(1 \le s \le k)$ – this argumentation is analogous to the proof of Claim 3.1.27. Therefore, expressing $r_{n,k}$ completely in terms of T-products by means of (3.3.51) and (3.3.37), there is no term containing a vacuum subdiagram, that is, a factor of the form (A.6.5).

it holds that $g_\varepsilon \in \mathcal{G}(\mathcal{O})$ for all $0 < \varepsilon \leq 1$. Using the Fock space representation (Theorem 2.6.3) we show below that, writing

$$\Phi(F_{\varepsilon,0}) = \kappa^n \, F^{\mathrm{op}}_{n,\varepsilon} + \mathcal{O}(\kappa^{n+1}) \quad \text{with} \quad F^{\mathrm{op}}_{n,\varepsilon} \neq 0 \tag{A.6.16}$$

(where the lower index "0" in the argument of Φ denotes restriction to $\mathcal{C}_0$), the first non-vanishing order n does not depend on ε and even the first non-vanishing coefficient does not depend on ε:

$$F^{\mathrm{op}}_{n,\varepsilon} = F^{\mathrm{op}}_{n,\varepsilon=1} =: F^{\mathrm{op}}_n \quad \forall 0 < \varepsilon \leq 1 \, . \tag{A.6.17}$$

From that and by using (2.6.7), we conclude

$$\begin{aligned}\lim_{\varepsilon \downarrow 0} \omega_0(F_\varepsilon^* \star F_\varepsilon) &= \lim_{\varepsilon \downarrow 0} \langle \Omega \mid \Phi(F^*_{\varepsilon,0}) \, \Phi(F_{\varepsilon,0}) \mid \Omega \rangle \\ &= \kappa^{2n} \, \langle \Omega \mid F^{\mathrm{op}\,*}_n \, F^{\mathrm{op}}_n \mid \Omega \rangle + \mathcal{O}(\kappa^{2n+1}) \, .\end{aligned}$$

Since F^{op}_n lies in the range of the map Φ, it is of the form

$$F^{\mathrm{op}}_n = f_0 + \sum_{k=1}^{N} \langle f_k, :\varphi^{\mathrm{op}\,\otimes k}: \rangle \quad \text{with } f_0 \text{ and } f_k \text{ as in Def. 1.2.1 and } N < \infty .$$

In addition, we know from (2.6.11) that such an operator does not annihilate the vacuum: $F^{\mathrm{op}}_n \Omega \neq 0$. Therefore, we obtain

$$\langle \Omega \mid F^{\mathrm{op}\,*}_n \, F^{\mathrm{op}}_n \mid \Omega \rangle = \| F^{\mathrm{op}}_n \Omega \|^2 > 0 \, .$$

It remains to prove (A.6.17). This is a consequence of Theorem 3.7.1:

$$\begin{aligned} F_\varepsilon &= U_\varepsilon \star F \star (U_\varepsilon)^{\star -1} \quad \text{with} \\ U_\varepsilon &:= U_{g, g_\varepsilon} = \mathbf{S}\big(\mathcal{L}(g + b_\varepsilon)\big)^{\star -1} \star \mathbf{S}\big(\mathcal{L}(g)\big) \, , \end{aligned} \tag{A.6.18}$$

where $g_\varepsilon - g = a_\varepsilon + b_\varepsilon$ with supports according to (3.7.13). The crucial observation is that

$$U_\varepsilon = 1 + \mathcal{O}(\kappa) \, .$$

Applying the map Φ to the relation (A.6.18), we obtain:

$$\Phi(F_{\varepsilon,0}) = \Phi\big(U_{\varepsilon,0} \star F_0 \star (U_{\varepsilon,0})^{\star -1}\big) = \Phi(U_{\varepsilon,0}) \, \Phi(F_0) \, \Phi(U_{\varepsilon,0})^{-1}$$

with $\Phi(U_{\varepsilon,0}) = \mathrm{Id} + \mathcal{O}(\kappa)$. Using additionally (A.6.16) for $\varepsilon = 1$, we get the assertion:

$$\Phi(F_{\varepsilon,0}) = \Phi(F_0) \, (1 + \mathcal{O}(\kappa)) = \kappa^n \, F^{\mathrm{op}}_n + \mathcal{O}(\kappa^{n+1}) \, . \qquad \square$$

Remark A.6.3. Naively one might think that the claim (A.6.17) follows from the Initial condition (axiom (c)) for R-products:

$$A_{\kappa L(g_\varepsilon)}(h) = R\big(e_\otimes^{\kappa L(g_\varepsilon)}, \, A(h)\big) = A(h) + \mathcal{O}(\kappa) \, .$$

But for $A(h) \in \mathcal{J}^{(m)}$ (2.6.2), we have $A(h)_0 = 0$ and, hence, $\Phi\big(A(h)_0\big) = 0$. For such an $A(h)$, the statement (A.6.17) says that

$$\Phi\Big(R_{1,1}\big(L(g_\varepsilon), A(h)\big)_0\Big) \quad \text{does not depend on } \varepsilon. \tag{A.6.19}$$

To understand the reason for this on a more explicit level, we give an example: For $A = (\Box + m^2)\varphi$ the Off-shell field equation (axiom (j)) yields

$$R_{1,1}\big(L(g_\varepsilon), (\Box + m^2)\varphi(h)\big) = -\hbar \int dx\, dy\; h(x)(\Box_x + m^2)\Delta^{\mathrm{ret}}(x-y)$$
$$\cdot \sum_a (-1)^{|a|}\, \partial_y^a \Big(g_\varepsilon(y)\, \frac{\partial L}{\partial(\partial^a \varphi)}(y)\Big)\,,$$

where (1.3.11) is used. The point is the following: Due to $(\Box_x + m^2)\Delta^{\mathrm{ret}} = -\delta$ and $g_\varepsilon\big|_{\mathrm{supp}\, h} = 1$, we may replace $g_\varepsilon(y)$ by 1. For more complicated $A(h) \in \mathcal{J}^{(m)}$, loop diagrams contribute to $R_{1,1}\big(L(g_\varepsilon), A(h)\big)$; nevertheless, the core of the mechanism underlying (A.6.19) is the same.

Remark A.6.4 (Generalization of Thm. A.6.2 to massless fields)**.** This theorem can be generalized to models with *massless* basic fields; in addition, fermionic and gauge fields (which may be massive or massless) can also be admitted. Ph. Blanchard and R. Seneor [9] have worked this out for QED and the massless, scalar $\kappa\,\varphi^{2n}$ model, with $n \geq 2$. P. Duch has generalized their result to a larger class of models in his very readable Ph.D. thesis [38] and in [39].

If the model contains massless basic fields, there is a significant difference to the purely massive case: Existence and uniqueness of the weak adiabatic limits (A.6.3) and (A.6.4)needs the validity of a certain renormalization condition! P. Duch has shown the following:

- If *all* basic fields are massless, this renormalization condition follows from almost homogeneous scaling of $r(A_1, \ldots, A_n)$ for all monomials $A_1, \ldots, A_n \in \mathcal{P}$, that is, from our Sm-expansion axiom, see Remk. 3.1.23.
- For models containing massless and massive basic fields, the existence and uniqueness of the weak adiabatic limits (A.6.3) and (A.6.4) follows from the validity of an additional renormalization condition; which is compatible with our list of renormalization conditions, if the "Sm-expansion" is replaced by the weaker renormalization condition "Scaling degree" (3.2.62). (By "compatible" we mean that there exists a common solution.)

Following [38, Theorem 4.2.1], a main consequence of Theorem A.6.2 and its generalizations (Remk. A.6.4) is that we can define a *state* ω *on the algebra* $\mathcal{A}_{\mathcal{L}}^{\mathrm{loc}}$ *of all local observables* (3.7.19), which is the inductive limit of the algebras $\mathcal{A}_{\mathcal{L}}(\mathcal{O})$

(3.7.16) obtained by the algebraic adiabatic limit:

$$\omega\colon \begin{cases} \mathcal{A}^{\mathrm{loc}}_{\mathcal{L}} \longrightarrow \mathbb{C}[\![\kappa]\!] \\ \omega\Big(\big[(A_1(h_1)^{\mathcal{O}}_{\mathcal{L}} \star \cdots \star A_k(h_k)^{\mathcal{O}}_{\mathcal{L}}\,,\,\mathcal{O})\big]\Big) \\ \qquad := \int dx_1 \cdots dx_k\; h_1(x_1)\cdots h_k(x_k)\, W_{\mathcal{L}}\big(A_1(x_1),\ldots,A_k(x_k)\big)\ , \end{cases} \tag{A.6.20}$$

where $A_j \in \mathcal{P}_{\mathrm{bal}}$ and $h_j \in \mathcal{D}(\mathcal{O})\ \forall 1 \le j \le k$ and $\mathcal{O}$ is an open double cone. Since the r.h.s. of (A.6.20) depends only on the pairs $(A_1, h_1), \ldots, (A_k, h_k)$, this definition is consistent, that is, it does not depend on the choice of the representative $(A, \mathcal{O})$ of the equivalence class $[(A, \mathcal{O})]$. In view of the definition of the algebraic adiabatic limit (Definition 3.7.2), it is instructive to write

$$\omega\Big(\big[(A_1(h_1)^{\mathcal{O}}_{\mathcal{L}} \star \cdots \star A_k(h_k)^{\mathcal{O}}_{\mathcal{L}}\,,\,\mathcal{O})\big]\Big) = \lim_{\varepsilon\downarrow 0} \omega_0\Big(A_1(h_1)_{\mathcal{L}(g_\varepsilon)} \star \cdots \star A_k(h_k)_{\mathcal{L}(g_\varepsilon)}\Big)$$
$$= \lim_{\varepsilon\downarrow 0} \omega_0\Big(A_1(h_1)^{\mathcal{O}}_{\mathcal{L}}(G_\varepsilon) \star \cdots \star A_k(h_k)^{\mathcal{O}}_{\mathcal{L}}(G_\varepsilon)\Big) \quad \text{with} \quad G_\varepsilon(x) := g_\varepsilon(x)\mathcal{L}(x)\ ,$$

where $g \in \mathcal{G}(\mathcal{O})$ (3.7.10) and we assume that the representative of $[(A_1(h_1)^{\mathcal{O}}_{\mathcal{L}} \star \cdots\,,\,\mathcal{O})]$ is chosen such that $0 \in \mathcal{O}$, in order that $G_\varepsilon \in \mathcal{G}_{\mathcal{L}}(\mathcal{O})$ for all $0 < \varepsilon \le 1$.

Let us check the defining properties of a state (Definition 2.5.1): Obviously, ω is linear and normalized. That ω is real and positive follows from the corresponding properties of the Wightman function $W_{\mathcal{L}}\big(A_1(x_1), \ldots\big)$ (Theorem A.6.2); again, positivity is meant in the sense of Definition 5.4.4.

One checks straightforwardly that ω is Poincaré invariant:

$$\omega \circ \alpha_L = \omega \qquad \forall L \in \mathcal{P}^{\uparrow}_{+}\ , \tag{A.6.21}$$

where α_L is defined in (3.7.20) – the argumentation is essentially the same as in the verification of the Poincaré covariance of the Wightman functions (A.6.13). Due to (A.6.21), we may interpret ω as *vacuum state*.

Agreement with the Gell-Mann–Low formula. In the traditional literature there is a famous formula for the perturbative computation of Green functions: The Gell-Mann–Low formula [81]. In the presence of massless fields, this formula usually contains IR-divergences. Therefore, we write it down in an IR-improved form [46, App.]:

$$G^{\mathrm{GML}}_{\mathcal{L}}\big(A_1(x_1), \ldots, A_k(x_k)\big) = \lim_{\varepsilon\downarrow 0} \frac{\omega_0\Big(T\big(e_\otimes^{i\mathcal{L}(g_\varepsilon)/\hbar} \otimes A_1(x_1) \otimes \cdots \otimes A_k(x_k)\big)\Big)}{\omega_0\Big(T\big(e_\otimes^{i\mathcal{L}(g_\varepsilon)/\hbar}\big)\Big)}. \tag{A.6.22}$$

One can show that this formula agrees with Epstein and Glaser's definition of Green functions; we give the precise statement and the proof for the case of solely *massive*, scalar fields; for more general fields, in particular models involving massless fields, we refer to [38, App. B].

Proposition A.6.5 (Agreement with Gell-Mann–Low formula for massive fields, [46, App.]). *Let $d = 4$, let all basic fields be massive, scalar fields and let* $\dim \mathcal{L} = 4$. *We assume that*

$$t_n^0(x_1 - x_n, \ldots) := \omega_0\Big(T_n^0(\mathcal{L}(x_1) \otimes \cdots \otimes \mathcal{L}(x_n))\Big) \in \mathcal{D}'(\mathbb{R}^{4(n-1)} \setminus \{0\})$$

has scaling degree $\mathrm{sd}(t_n^0) = 4n$ *and is renormalized by the central solution* (A.6.7) *for all $n \geq 2$. For all other time-ordered or generalized retarded products appearing in $G_{\mathcal{L}}^{\mathrm{GML}}(\cdots)$* (A.6.22) *and $G_{\mathcal{L}}(\cdots)$* (A.6.4)*, respectively, we only assume that they are constructed in accordance with our axioms, where the axiom "Sm-expansion" is replaced by the weaker axiom "Scaling degree"* (3.2.62)*. Then, the weak adiabatic limit* (A.6.22) *exists, is unique and the relations*

$$\begin{aligned} G_{\mathcal{L}}(A_1(x_1), \ldots, A_k(x_k)) &= \lim_{\varepsilon \downarrow 0} \omega_0\Big(T\big(e_{\otimes}^{i\mathcal{L}(g_\varepsilon)/\hbar} \otimes A_1(x_1) \otimes \cdots \otimes A_k(x_k)\big)\Big) \\ &= G_{\mathcal{L}}^{\mathrm{GML}}(A_1(x_1), \ldots, A_k(x_k)) \end{aligned} \tag{A.6.23}$$

hold true for all $A_1, \ldots, A_k \in \mathcal{P}_{\mathrm{bal}}$.

Proof. Computing the nth order in $\mathcal{L}$ of $G_{\mathcal{L}}(\cdots)$ by using (3.3.51) to express $R_{n,k}$ in terms of $\overline{T}$- and T-products, we obtain

$$\begin{aligned} &G_{\mathcal{L}}^{(n)}(A_1(x_1), \ldots, A_k(x_k)) \\ &\quad = \frac{i^n}{n!\hbar^n} \lim_{\varepsilon \downarrow 0} \sum_{I \subseteq \{1,\ldots,n\}} (-1)^{|I|} \int dy_1 \cdots dy_n \; g_\varepsilon(y_1) \cdots g_\varepsilon(y_n) \\ &\quad\quad \cdot \omega_0\Big(\overline{T}\Big(\otimes_{i \in I} \mathcal{L}(y_i)\Big) \star^{\geq 1} T\Big((\otimes_{j \in I^c} \mathcal{L}(y_j)) \otimes (\otimes_{s=1}^k A_s(x_s))\Big)\Big) \,. \end{aligned} \tag{A.6.24}$$

In this formula, we have replaced $\star$ by $\star^{\geq 1}$, which is defined by $F \star^{\geq 1} G := F \star G - F \cdot G$, i.e., $\star^{\geq 1}$ is that part of $\star$ with ≥ 1 contractions. This replacement is allowed, because all diagrams contributing to $R_{n,k}(\mathcal{L}(y_1), \ldots, \mathcal{L}(y_n); A_1(x_1), \ldots, A_k(x_k))$ have the property that every interaction vertex y_j $(1 \leq j \leq n)$ is connected with at least one field vertex x_s $(1 \leq s \leq k)$, as explained in Footn. 191.

It suffices to show that

$$\begin{aligned} &\lim_{\varepsilon \downarrow 0} \int dy_1 \cdots dy_n \; g_\varepsilon(y_1) \cdots g_\varepsilon(y_n) \\ &\quad \cdot \omega_0\Big(\overline{T}\Big(\otimes_{i \in I} \mathcal{L}(y_i)\Big) \star^{\geq 1} T\Big((\otimes_{j \in I^c} \mathcal{L}(y_j)) \otimes (\otimes_{s=1}^k A_s(x_s))\Big)\Big) = 0 \quad \text{if} \quad I \neq \emptyset, \end{aligned} \tag{A.6.25}$$

for the following reasons: From Theorem A.6.2 we know that the limit $\varepsilon \downarrow 0$ on the r.h.s. of (A.6.24) exists and is unique, hence, this holds then for the $(I = \emptyset)$-term separately. Since the latter is precisely the nth order in $\mathcal{L}$ of the expression written in the middle of (A.6.23), we then have proved the first equality in (A.6.23). Finally, the second equality in (A.6.23) follows immediately from (A.6.11), since we renormalize the vacuum diagrams (A.6.5) by the central solution.

Our proof of (A.6.25) relies on the existence of the mass gap, more precisely, on the fact that all Δ^+_m-propagators contributing to $\star^{\geq 1}$ in (A.6.25) have $m > 0$. Inserting the causal Wick expansion (3.1.24) of $\overline{T}(\cdots)$ and $T(\cdots)$ into (A.6.25), we see that all terms are of the form

$$f(y_1,\ldots,y_n;x_1,\ldots,x_k) := \bar{t}(y_1-y_s,\ldots,y_{s-1}-y_s)$$
$$\cdot\Big(\prod_{r=1}^{l}\partial^{a_r}\Delta^+_{m_r}(y_{j_r}-z_{i_r})\Big)\, t(y_{s+1}-x_k,\ldots,y_n-x_k;x_1-x_k,\ldots,x_{k-1}-x_k)\;,$$

where

$$l\geq 1 \quad\text{and}\quad j_r\in\{1,\ldots,s\}\;,\quad z_{i_r}\in\{y_{s+1},\ldots,y_n,x_1,\ldots,x_k\}\quad \forall 1\leq r\leq l\;.$$

In addition, we have $\bar{t}\in\mathcal{D}'(\mathbb{M}^{s-1})$, $t\in\mathcal{D}'(\mathbb{M}^{n-s+k-1})$, $a_r\in\mathbb{N}^4$ and, to simplify the notation, we consider the case $I=\{1,\ldots,s\}$. It may be that $y_{j_r}=y_{j_{r'}}$ or $z_{i_r}=z_{i_{r'}}$ for $r\neq r'$. It suffices to consider the adiabatic limit with respect to the vertices $y_1,\ldots,y_s$:

$$\lim_{\varepsilon\downarrow 0}\int dy_1\cdots dy_s\; g_\varepsilon(y_1)\cdots g_\varepsilon(y_s)f(y_1,\ldots,y_n;x_1,\ldots,x_k) \tag{A.6.26}$$
$$=\lim_{\varepsilon\downarrow 0}\int dk_1\cdots dk_s\;\hat{g}(-k_1)\cdots\hat{g}(-k_s)\,\hat{f}(\varepsilon k_1,\ldots,\varepsilon k_s,y_{s+1},\ldots y_n;x_1,\ldots,x_k),$$

where $\hat{f}(k_1,\ldots,k_s,y_{s+1},\ldots;x_1,\ldots)$ is obtained from $f(y_1,\ldots y_n;x_1,\ldots)$ by Fourier transformation with respect to the variables $y_1,\ldots,y_s$. To simplify the notations, we compute this distributional Fourier transformation (A.1.18) as if f would be a function – the modifications making this computation rigorous can easily be done. Neglecting constant prefactors, we get

$$\hat{f}(k_1,\ldots,k_s,y_{s+1},\ldots;x_1,\ldots)\sim\int dy_1\ldots dy_s\; e^{i(k_1y_1+\cdots+k_sy_s)}\int dq_1\cdots dq_l$$
$$\cdot\, e^{-i(q_1(y_{j_1}-z_{i_1})+\cdots+q_l(y_{j_l}-z_{i_l}))}\int dp_1\ldots dp_{s-1}\; e^{-i(p_1(y_1-y_s)+\cdots+p_{s-1}(y_{s-1}-y_s))}$$
$$\cdot\,\hat{\bar{t}}(p_1,\ldots,p_{s-1})\Big(\prod_{r=1}^{l}q_r^{a_r}\,\widehat{\Delta^+_{m_r}}(q_r)\Big)\, t(y_{s+1}-x_k,\ldots;x_1-x_k,\ldots)\;.$$

Writing

$$y_{j_r}=:\alpha_{r1}y_1+\cdots+\alpha_{rs}y_s\quad\text{with}\quad \alpha_{ru}\in\{0,1\}\;,\quad \sum_{u=1}^{s}\alpha_{ru}=1\;,\quad\forall 1\leq r\leq l\;,$$

the integration $\int dy_v$ yields

$$\begin{cases}\delta(k_v-p_v-\alpha_{1v}q_1-\cdots-\alpha_{lv}q_l) & \text{for}\quad 1\leq v\leq s-1\\ \delta(k_s+p_1+\cdots+p_{s-1}-\alpha_{1s}q_1-\cdots-\alpha_{ls}q_l) & \text{for}\quad v=s\;.\end{cases}$$

In the product of these δ-distributions we may replace $\delta(k_s + p_1 + \cdots + p_{s-1} - \alpha_{1s}q_1 - \cdots - \alpha_{ls}q_l)$ by $\delta(k_1 + \cdots + k_s - q_1 - \cdots - q_l)$. Removing the δ-distributions by performing the integrations $\int dp_1 \ldots dp_{s-1} \int dq_l$, we obtain

$$\hat{f}(\varepsilon k_1, \ldots, \varepsilon k_s, y_{s+1}, \ldots; x_1, \ldots)$$
$$\sim \int dq_1 \cdots dq_{l-1}\; e^{i(q_1 z_{i_1} + \cdots + q_l z_{i_l})}\, \widehat{\bar{t}}(\ldots(\varepsilon k_v; q_r)\ldots)$$
$$\cdot \Big(\prod_{r=1}^{l} q_r^{a_r}\, \widehat{\Delta^+_{m_r}}(q_r)\Big)\, t(y_{s+1} - x_k, \ldots; x_1 - x_k, \ldots)\Big|_{q_l = \varepsilon(k_1 + \cdots + k_s) - q_1 - \cdots - q_{l-1}} .$$

Because of

$$\widehat{\Delta^+_m}(q) \sim \theta(q^0)\,\delta(q^2 - m^2)\ , \text{ the product } \prod_{r=1}^{l} q_r^{a_r}\, \widehat{\Delta^+_{m_r}}(q_r)\Big|_{q_l = \varepsilon(k_1 + \cdots + k_s) - q_1 - \cdots - q_{l-1}}$$

vanishes for $\varepsilon > 0$ sufficiently small if all possible values of $(k_1, \ldots, k_s)$ lie in a bounded subset of $\mathbb{M}^s$. The latter assumption is not satisfied in (A.6.26); but, $\hat{g}$ decays rapidly; namely, due to $g \in \mathcal{D}(\mathbb{M}) \subset \mathcal{S}(\mathbb{M})$ we have $\hat{g} \in \mathcal{S}(\mathbb{M})$. This suffices to conclude that the adiabatic limit (A.6.26) vanishes indeed. □

A.7 Remarks about the connection to traditional approaches

We aim to compare the approach given in this book with traditional approaches to pQFT.[192] This appendix is far from an exhaustive comparison, we only discuss a few topics, for which remarkable differences appear – differences which are not pointed out in the main text (see in particular Sect. 3.9.5 for the comparison with the functional integral approach) or in Appendix A.6. We will not explain the procedures in the traditional approaches to which we refer, we assume that the reader knows them – at least roughly.

Composite interacting fields. We study this topic in terms of the interacting electromagnetic current of QED; for details see [40, Sect. 2.3]. Epstein and Glaser's [66] definition of $j^\mu_{\mathcal{L}(g)}$ (which we use in this book), satisfies the interacting Maxwell equation (i.e., the first equation of (5.3.1)) and, augmented by the relevant MWI (Sect. 5.2), the current conservation (5.2.34); hence, it is physically acceptable. In contrast to conventional definitions of the interacting electromagnetic current, it does not need any regularization.

Let us explain this difference: Mostly one defines $j^\mu_{\mathcal{L}(g)}(x)$ in terms of the interacting spinor fields $\psi_{\mathcal{L}(g)}(x)$ and $\overline{\psi}_{\mathcal{L}(g)}(x)$. The naive attempts

$$\text{(a) } \overline{\psi}_{\mathcal{L}(g)}(x) \star \gamma^\mu \psi_{\mathcal{L}(g)}(x) \quad \text{or} \quad \text{(b) } \overline{\psi}_{\mathcal{L}(g)}(x) \wedge \gamma^\mu \psi_{\mathcal{L}(g)}(x)$$

[192] This appendix is based on discussions with Detlev Buchholz and Klaus Fredenhagen.

fail: The pointwise products of distributions appearing in (a) do not exist. And, (b) is physically wrong – this can be seen already to first order in $\mathcal{L}(g)$: Looking at the Feynman diagrams (i.e., the contraction patterns) of the first-order contribution

- to the Epstein–Glaser expression, that is, $R\big(L(x_1); j^\mu(x)\big)$ (given in (5.2.36), where $\mathcal{L} = eL$),
- to (a) (where we ignore the divergences)
- and to (b), respectively,

we see that the former two contributions have precisely the same diagrams, the three diagrams in (5.2.37). But (b) does not contain the first diagram in (5.2.37), because in (b) there do not appear any contractions between the interacting fields $\overline{\psi}_{\mathcal{L}(g)}(x)$ and $\gamma^\mu\psi_{\mathcal{L}(g)}(x)$.

The usual way out is to regularize (a) by a point splitting:

$$\tilde{j}^\mu_{\mathcal{L}(g)} := \lim_{\zeta\to 0}\Big(\overline{\psi}_{\mathcal{L}(g)}(x)\star\gamma^\mu\psi_{\mathcal{L}(g)}(x+\zeta) - q^\mu(x,\zeta)\Big)\,, \tag{A.7.1}$$

where $q^\mu(x,\zeta)$ is chosen such that it compensates the singularities of $\overline{\psi}_{\mathcal{L}(g)}(x)\star\gamma^\mu\psi_{\mathcal{L}(g)}(x+\zeta)$ appearing in the limit $\zeta\to 0$. Of course this does not fix $q^\mu(x,\zeta)$ uniquely, it may be further restricted by requiring, besides some obvious conditions as Poincaré covariance, that $\tilde{j}^\mu_{\mathcal{L}(g)}$ satisfies the interacting Maxwell equation and current conservation, see [14].

Terms like $\overline{\psi}_{\mathcal{L}(g)}(x)\star\gamma^\mu\psi_{\mathcal{L}(g)}(x+\zeta)$ can be systematically described for $\zeta\to 0$ by the *operator product expansion*, which is due to Wilson [170] and has been established in pQFT by Zimmermann [177]. A treatment in the framework of perturbative AQFT (see Sect. 3.7), which points out the algebraic aspects and can be generalized to curved spacetimes, has been given by Hollands [104].

The role of the interacting field equations. In most traditional textbooks on perturbative QFT, the field equations for the interacting quantum fields play a minor role; a reason for this may be that one is primarily interested in the S-matrix, that is, in the T-product. However, there are a few constructions of pQFT, for which the interacting field equations are a major ingredient; for example, Brandt's construction of QED [14], Steinmann's construction of Wightman functions [156] and parts of the work of Zimmermann, e.g., [177].

The quantization of clFT given in this book amounts to an inductive construction of the R-product. A main justification that $R\big(e_\otimes^{S/\hbar}, F\big)$ can be physically interpreted as perturbative interacting quantum field is that it fulfills the interacting field equation(s). In addition, the field equations play also an important role in our formulation of symmetries: Classically, the MWI is simply the off-shell field equation multiplied pointwise with an arbitrary interacting field.

In our construction of the R-product (and also in [40]), we show that renormalization can be done such that the field equations are maintained: The renormalized fields satisfy precisely the same field equation(s) as the classical fields and the unrenormalized fields, respectively, see (5.3.1). In particular, the mass(es) and the coupling constant(s) have the same value, which is the physical value.

This contrasts with the construction of Brandt [14]: In a first step he constructs the renormalized, interacting basic fields[193] $\psi_{eL}(x), \overline{\psi}_{eL}(x)$ and $A^\mu_{eL}(x)$, which involves mass and coupling constant renormalization. Then, in terms of these fields, he defines interacting composite fields, e.g., the electromagnetic current $\tilde{j}^\mu_{eL}(x)$, by the point splitting method (A.7.1), such that the interacting Dirac and Maxwell equation are satisfied; the parameters m and $(-e)$ appearing in the latter equations, which may be interpreted as the physical electron mass and the physical electron charge, respectively, are required to agree with the respective renormalized quantities obtained in the first step. Needless to say that the resulting formulas for the counter terms (i.e., $q^\mu(x,\zeta)$ in case of $\tilde{j}^\mu_{eL}(x)$) are rather complicated.

About the notions "on-" and "off-shell". In causal perturbation theory the notions "on-" and "off-shell field" are defined as explained in Sect. 1.6. We are now going to explain, how this definition is related to the usage of the words "on-shell" and "off-shell" in the traditional literature.

(a) **Traditional literature:** In this formalism, the notions "on-" and "off-shell" are mostly used for the renormalized *amplitude* belonging to a given Feynman diagram in momentum space. Since the adiabatic limit is performed, this amplitude is a function of the external momenta $p_1, \ldots, p_s$ only, i.e., the momenta belonging to the external lines. Let $\tau(p_1, \ldots, p_s)$ be the "amputated" amplitude, that is, the amplitude without the factors coming from the external lines.

- When computing the S-matrix, each external line represents a free field in Fock space: $\hat{\Phi}^{\mathrm{op}}_k(p_k)$, $k = 1, \ldots, s$. Hence, the contribution to the S-matrix reads

$$\int dp_1 \cdots dp_s \; \tau(p_1, \ldots, p_s)\Big|_{p_j \in H_{m_j}\ \forall j} \cdot :\prod_{k=1}^{s} \hat{\Phi}^{\mathrm{op}}_k(p_k): \,, \qquad \text{(A.7.2)}$$

where H_m denotes the mass shell (1.6.1). Therefore, only the "on-shell amplitude" $\tau(p_1, \ldots, p_s)\big|_{p_j \in H_{m_j}\ \forall j}$ is relevant, because each external momentum p_k appears also as argument of a factor $\hat{\Phi}^{\mathrm{op}}_k$, which is an on-shell field: $\mathrm{supp}\, \hat{\Phi}^{\mathrm{op}}_k \subseteq H_{m_k}$.

- When computing the vacuum expectation value of the time-ordered product of the interacting fields $\hat{\Phi}^{\mathrm{op}}_{1,\kappa L}(p_1), \ldots, \hat{\Phi}^{\mathrm{op}}_{s,\kappa L}(p_s)$, i.e., the Green function, by means of the Gell-Mann–Low formula (cf. App. A.6), each

[193] Where $L := \overline{\psi} \not{A} \psi$. We write $\Phi_{eL}(x)$ instead of $\Phi_{\mathcal{L}(g)}(x)$, since Brandt works throughout in the adiabatic limit. By $(-e)$ we mean the renormalized coupling constant.

external line represents a Feynman propagator. Therefore, the contribution is

$$\tau(p_1,\ldots,p_s)\cdot\prod_{k=1}^{s}\hat{\Delta}^F_{m_k}(p_k)\ . \tag{A.7.3}$$

Since $\mathrm{supp}\,\hat{\Delta}^F_m = \mathbb{M}_d$, the "off-shell amplitude" is needed, that is, $\tau(p_1,\ldots,p_s)$ without restriction of the external momenta to the mass shell.

The contribution to the S-matrix (A.7.2) is obtained from the contribution of the same Feynman diagram to the Green function (A.7.3) by the "LSZ reduction formula" [124].

(b) **On-shell formulation of causal perturbation theory:** In the original formulation of causal perturbation theory given by Epstein and Glaser [66], the arguments and the values of the R- and T-product are Fock space operators, that is, on-shell fields. Such an on-shell R- or T-product can be introduced also in our functional formalism; this is worked out in Sect. 3.4. If the adiabatic limit is performed, the picture is essentially the same as in (a). The main difference is that instead of the expansion in terms of Feynman diagrams in momentum space, one studies the causal Wick expansion (3.1.24) in x-space.

- Let us explain this more explicitly in terms of the contribution[194]

$$\int dx_1\cdots dx_n\; t^{\mathrm{EG}}(x_1-x_n,\ldots,x_{n-1}-x_n)\,{:}\varphi^{\mathrm{op}\,2}(x_1)\,\varphi^{\mathrm{op}\,2}(x_n){:} \tag{A.7.4}$$

to the Fock space S-matrix

$$\lim_{g\to 1}\mathbf{S}^{\mathrm{EG}}\big(\kappa\,{:}\varphi^{\mathrm{op}\,4}{:}\,,g\big)$$

(with $\mathbf{S}^{\mathrm{EG}}$ defined similarly to $\mathbf{S}^{\mathrm{on}}$ (3.4.28)), where

$$t^{\mathrm{EG}}(x_1-x_n,\ldots) \tag{A.7.5}$$

$$:= \langle\Omega\,|\,T^{\mathrm{EG}}_n\Big({:}\varphi^{\mathrm{op}\,2}(x_1){:}\otimes\Big(\bigotimes_{j=2}^{n-1}{:}\varphi^{\mathrm{op}\,4}(x_j){:}\Big)\otimes{:}\varphi^{\mathrm{op}\,2}(x_n){:}\Big)\,|\,\Omega\rangle\ .$$

The integrand of (A.7.4) can be illustrated by the diagram

x_1 $\quad t^{\mathrm{EG}}(x_1-x_n,\ldots)\quad$ x_n

(A.7.6)

[194]The upper index ^{EG} stands for "Epstein and Glaser".

Here and in the following we omit combinatorial factors and powers of 2π, and the adiabatic limit is naively performed by setting $g(x) := 1\ \forall x \in \mathbb{M}_d$.

For example, for $n = 4$ there are four different x-space Feynman diagrams (that is, contraction schemes, see Remk. 3.3.4) contributing to $t^{\mathrm{EG}}(x_1 - x_4, x_2 - x_4, x_3 - x_4)$, namely the connected diagrams

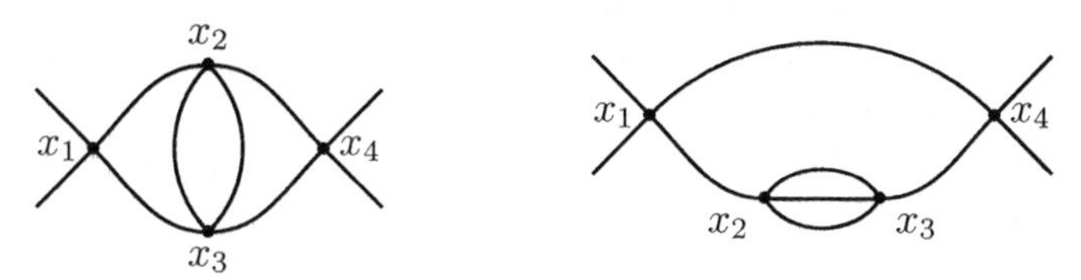

plus the latter diagram with x_2 and x_3 mutually exchanged, plus the disconnected diagram

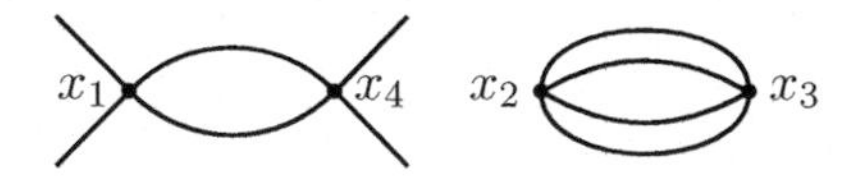

If $m > 0$ and if the adiabatic limit is carried out according to Epstein and Glaser, then the disconnected diagram does not contribute to (A.7.4), because the vacuum subdiagram (i.e., the connected component with vertices x_2 and x_3) vanishes in the adiabatic limit, as derived in Exap. A.6.1.

The term (A.7.4) can be rewritten in momentum space:[195]

$$\text{(A.7.4)} \sim \int dp_1 dp_2 dp_3 dp_4\ \widehat{t^{\mathrm{EG}}}\big(-(p_1 + p_2), 0, \ldots, 0\big) \qquad \text{(A.7.7)}$$
$$\cdot\, \delta(p_1 + p_2 + p_3 + p_4)\ {:}\hat{\varphi}^{\mathrm{op}}(p_1)\hat{\varphi}^{\mathrm{op}}(p_2)\hat{\varphi}^{\mathrm{op}}(p_3)\hat{\varphi}^{\mathrm{op}}(p_4){:}\ .$$

Hence the amputated amplitude reads

$$\tau^{\mathrm{EG}}(p_1, p_2, p_3, p_4) := \widehat{t^{\mathrm{EG}}}\big(-(p_1 + p_2), 0, \ldots, 0\big)\, \delta(p_1 + p_2 + p_3 + p_4)\ . \qquad \text{(A.7.8)}$$

Similarly to (A.7.2), only the on-shell amplitude $\tau^{\mathrm{EG}}\big|_{H_m^{\times 4}}$ contributes to (A.7.7), that is, to the S-matrix.

[195] In (A.7.4) the number of (d-dimensional) integrations is the order n of the T-product, the integrations run over the x-coordinates of the vertices. This is in contrast to (A.7.7), where the number of integrations is the number of external lines, the integrations run over the momenta of the external lines. Since we proceed in the adiabatic limit, the amplitude τ^{EG} contains $\delta(p_1 + p_2 + p_3 + p_4)$ expressing energy-momentum conservation of the scattering process; and the momenta belonging to inner vertices (i.e., vertices which do not carry any external line) are $= 0$, in the example at hand these are the momenta which are conjugate to $(x_2 - x_n), \ldots, (x_{n-1} - x_n)$, respectively.

- Off-shell amplitudes appear also in the Epstein–Glaser on-shell formalism, when one studies Green functions. To explain this by the analogon of the example (A.7.4)–(A.7.5), we investigate

$$G(y_1,\ldots,y_4) := G_{\kappa:\varphi^{\rm op\,4}:}\big(\varphi^{\rm op}(y_1),\ldots,\varphi^{\rm op}(y_4)\big)\ ,$$

where the Fock space version of the definition (A.6.4) is used. This Green function can be computed by means of the Gell-Mann–Low formula, more precisely, the Epstein–Glaser Fock space version of the expression written in the middle of (A.6.23): To nth order we obtain

$$G^{(n)}(y_1,\ldots,y_4) \tag{A.7.9}$$
$$\sim \int dx_1\cdots dx_n\ \langle\Omega\mid T^{\rm EG}_{n+4}\Big(\big(\bigotimes_{r=1}^{4}\varphi^{\rm op}(y_r)\big)\otimes\bigotimes_{j=1}^{n}:\varphi^{\rm op\,4}(x_j):\Big)\mid\Omega\rangle\ ;$$

again we neglect constant prefactors and the adiabatic limit is naively performed. Requiring the Field equation (3.4.46) as a renormalization condition for $T^{\rm EG}$, the contribution of $t^{\rm EG}(x_1-x_n,\ldots)$ (A.7.5) to (A.7.9) is

$$\sim \sum_{\pi\in S_4}\int dx_1\cdots dx_n\ t^{\rm EG}(x_1-x_n,\ldots) \tag{A.7.10}$$
$$\cdot\prod_{r=1,2}\Delta^F_m(y_{\pi r}-x_1)\prod_{s=3,4}\Delta^F_m(y_{\pi s}-x_n)\ ;$$

the integrand of an individual summand can be visualized by the diagram

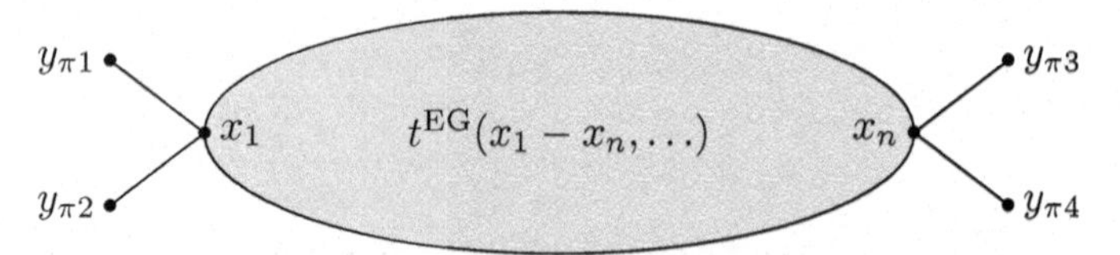

Considering the Green function in momentum space,

$$\hat G^{(n)}(-p_1,\ldots,-p_4)\sim\int dy_1\ldots dy_4\ e^{-i(p_1y_1+\cdots+p_4y_4)}\ G^{(n)}(y_1,\ldots,y_4)\ ,$$

the term (A.7.10) gives the contribution

$$\sim\sum_{\pi\in S_4}\tau^{\rm EG}(p_{\pi 1},p_{\pi 2},p_{\pi 3},p_{\pi 4})\cdot\prod_{r=1}^{4}\hat\Delta^F_m(p_r) \tag{A.7.11}$$

with the same distribution $\tau^{\rm EG}$ as in (A.7.8); the result (A.7.11) is completely analogous to (A.7.3). Similarly to the latter formula, the off-shell amplitude is needed.

(c) **Off-shell formulation of causal perturbation theory:** In the formulation of causal perturbation theory given in this book, both the arguments and the values of the R-/T-product are off-shell fields. This is advantageous for various purposes; examples pointing this out are Remk. 3.1.2 and Footn. 127 for the arguments of the T-product and Remk. 3.6.5 for the values.

For simplicity we assume that all field polynomials $A_j \in \mathcal{P}$ appearing in the argument of the (off-shell) T-product contain only zeroth and first derivatives of φ, that is, they fulfill (3.4.40). Then, we have $\xi\pi(A_j) = A_j$ (3.4.41) and, using the definition of T^{on} (3.4.8), we conclude from the results of Sect. 3.4.2 that T^{EG} is connected to our off-shell T-product by

$$\begin{aligned} T_n^{\mathrm{EG}}\big(:A_1^{\mathrm{op}}(x_1):\otimes\cdots\otimes :A_n^{\mathrm{op}}(x_n):\big) &= \Phi\Big(T_n^{\mathrm{on}}\big(A_{1,0}(x_1)\otimes\cdots\otimes A_{n,0}(x_n)\big)\Big) \\ &= \Phi\Big(T_n\big(A_1(x_1)\otimes\cdots\otimes A_n(x_n)\big)_0\Big)\ , \end{aligned} \tag{A.7.12}$$

where Φ is the algebra isomorphism of Thm. 2.6.3 and

$$:A_j^{\mathrm{op}}(x_j): = \Phi\big(A_{j,0}(x_j)\big) \quad \forall 1\leq j\leq n\ .$$

The relation (2.6.7) yields

$$\langle\Omega\,|\,T_n^{\mathrm{EG}}\big(:A_1^{\mathrm{op}}(x_1):\otimes\cdots\otimes :A_n^{\mathrm{op}}(x_n):\big)\,|\,\Omega\rangle = \omega_0\Big(T_n\big(A_1(x_1)\otimes\cdots\otimes A_n(x_n)\big)\Big)\ . \tag{A.7.13}$$

- We recall from Proposition A.6.5 that in the off-shell formalism of this book Green functions can be computed by the expression written in the middle of (A.6.23). For the Green functions it does not matter whether we work with our off-shell formalism or with the on-shell formalism of Epstein and Glaser: The computation is precisely the same and we obtain precisely the same results, e.g., we also get (A.7.11) for the just considered example. The reason for this is the relation (A.7.13), that is,

$$\begin{aligned} &\omega_0\Big(T_{n+4}\Big(\big(\bigotimes_{r=1}^{4}\varphi(y_r)\big)\otimes\bigotimes_{j=1}^{n}\varphi^4(x_j)\Big)\Big) \\ &\qquad = \langle\Omega\,|\,T_{n+4}^{\mathrm{EG}}\Big(\big(\bigotimes_{r=1}^{4}\varphi^{\mathrm{op}}(y_r)\big)\otimes\bigotimes_{j=1}^{n}:\varphi^{\mathrm{op}\,4}(x_j):\Big)\,|\,\Omega\rangle \end{aligned} \tag{A.7.14}$$

  in case of our example (A.7.9).
- However, considering the off-shell S-matrix

$$\lim_{g\to 1}\mathbf{S}\big(\kappa\varphi^4(g)\big)\ ,$$

  there is an essential difference between the two formalisms: *The entire off-shell amplitudes are needed.* Let us explain this in terms of the above example: The coefficient appearing in the causal Wick expansion, i.e., the distribution

$$t(x_1 - x_n, \ldots) := \omega_0\Big(T_n\Big(\varphi^2(x_1) \otimes \big(\bigotimes_{j=2}^{n-1} \varphi^4(x_j)\big) \otimes \varphi^2(x_n)\Big)\Big) , \quad \text{(A.7.15)}$$

is equal to the corresponding coefficient $t^{\mathrm{EG}}(x_1 - x_n, \ldots)$ (A.7.5) in the Epstein–Glaser framework, again this is an application of the identity (A.7.13). But, the contribution of $t(x_1 - x_n, \ldots)$ (A.7.15) to the off-shell S-matrix reads

$$\int dx_1 \cdots dx_n \; t(x_1 - x_n, \ldots, x_{n-1} - x_n)\, \varphi^2(x_1)\, \varphi^2(x_n) , \quad \text{(A.7.16)}$$

where $\varphi(x)$ is now an off-shell field; the diagram visualizing the integrand of (A.7.16) is again (A.7.6).
To rewrite the term (A.7.16) in momentum space, we introduce the Fourier transformed basic field:

$$\hat{\varphi}(p) := \frac{1}{(2\pi)^{d/2}} \int dx \; e^{ipx} \, \varphi(x) , \quad \text{that is,} \quad \hat{\varphi}(p)(h) := \hat{h}(p) \quad \forall h \in \mathcal{C} ;$$

note that $\hat{h}$ is in general a distribution, because the configurations h do not need to decay rapidly in all directions. With that we obtain

$$\text{(A.7.16)} \sim \int dp_1 dp_2 dp_3 dp_4 \; \tau^{\mathrm{EG}}(p_1, p_2, p_3, p_4)\, \hat{\varphi}(p_1)\hat{\varphi}(p_2)\hat{\varphi}(p_3)\hat{\varphi}(p_4) ,$$

where $\tau^{\mathrm{EG}}(p_1, \ldots)$ is given in terms of $t = t^{\mathrm{EG}}$ by (A.7.8). Since, for the support in momentum space, we have $\mathrm{supp} \int dp\; \hat{\varphi}(p)\, g(p) = \mathrm{supp}\, g$ for any $g \in \mathcal{D}(\mathbb{M})$, it is the off-shell amplitude $\tau^{\mathrm{EG}}(p_1, \ldots)$ which is relevant here – in contrast to the traditional literature (A.7.2) and the Epstein–Glaser formalism (A.7.7).

Glossary of Abbreviations and Symbols

The following *abbreviations* are used in the book:

App.	Appendix
AWI	Action Ward Identity (Sect. 3.1.1)
AQFT	algebraic quantum field theory (Sect. 3.7)
BPH	Bogoliubov, Parasiuk and Hepp (Foreword)
BPHZ	Bogoliubov, Parasiuk, Hepp and Zimmermann (Sect. 3.2.2)
BRST	Becchi, Rouet, Stora and Tyutin (Sect. 5.4.1)
Chap.	Chapter
clFT	classical Field Theory (Chap. 1)
Def.	Definition
Eq.	Equation
Exap.	Example
Exer.	Exercise
Footn.	Footnote
GNS	Gelfand, Naimark and Segal (Foreword)
GLZ	Glaser, Lehmann and Zimmermann (Sects. 1.10 and 3.1.1)
iff	if and only if
IR	infrared
l.h.s.	left-hand side
LSZ	Lehmann, Symanzik and Zimmerman (App. A.7)
MWI	Master Ward Identity (Chap. 4)
MS	minimal subtraction (Prop. 3.5.12)
PCT	parity, charge conjugation and time reversal (Sect. 5.1.6)
1PI	1-particle-irreducible (Footn. 138)
pQFT	perturbative Quantum Field Theory
Prop.	Proposition
QAP	Quantum Action Principle (Thm. 4.3.1)
QCD	Quantum Chromodynamics
QFT	Quantum Field Theory
QED	Quantum Electrodynamics (Chap. 5)
QED-MWI	relevant MWI for QED (Sect. 5.2)
Remk.	Remark
r.h.s.	right-hand side

M. Dütsch, *From Classical Field Theory to Perturbative Quantum Field Theory*,
Progress in Mathematical Physics 74, https://doi.org/10.1007/978-3-030-04738-2

RG	renormalization group (Sect. 3.6)
R-product	retarded product (Sect. 3.1)
Sect.	Section
Sm-expansion	scaling and mass expansion (Def. 3.1.22)
S-matrix	scattering matrix (Def. 3.3.8)
Thm.	Theorem
T-product	time-ordered product (Sect. 3.3)
$\overline{T}$-product	antichronological product (Sect. 3.3.3)
UV	ultraviolet
VEV	vacuum expectation value (Footn. 37)
w.r.t.	with respect to

The following *symbols* are used in the book:

Products and brackets

$\otimes_{\mathrm{sym}}$	symmetrized tensor product
$\otimes_{\mathrm{as}}$	antisymmetrized tensor product (Eq. (5.1.9))
$F \cdot G \equiv FG$	pointwise or "classical" prod. of fields (Eq. (1.2.5))
$\star \equiv \star_\hbar \equiv \star_m$	star product of fields to the mass m (Def. 2.1.1)
$\star_{m,\mu}$	star product with Hadamard function H_m^μ (App. A.4)
$\star_p$	product with propagator p (Exer. 2.3.1)
$\star_\Lambda$	product with propagator p_Λ (Eq. (3.9.4))
$[\cdot,\cdot] \equiv [\cdot,\cdot]_-$	commutator for matrices, linear operators, star prod.
$[\cdot,\cdot]_+$	anticommutator for matrices, linear operators, star prod.
$[\cdot,\cdot]_\star$	graded commutator w.r.t. star product (Eq. (5.1.36))
$\{\cdot,\cdot\}$	Poisson bracket (Def. 1.8.3)
$\langle\cdot,\cdot\rangle_{\mathcal{H}}$, $(\cdot,\cdot)_{\mathcal{H}}$	scalar prod. in Hilbert space $\mathcal{H}$ ($_{\mathcal{H}}$ sometimes omitted)
$\langle\cdot,\cdot\rangle_{\mathcal{K}}$	indefinite inner product in the space $\mathcal{K}$ (Sects. 5.4–5.5)

Conjugations

$\overline{z} \equiv z^c$	complex conjugate of $z \in \mathbb{C}$
$A^\dagger$	adjoint matrix of $A \in \mathbb{C}^{m\times n}$
A^*, A^+	adjoint operator of the linear Hilbert space operator A
$A \mapsto A^*$	$*$-operation in a $*$-algebra (App. A.1)
F^*	$*$-conjugate of the field F (Eq. (1.2.7))

Function spaces and distributions

$\mathcal{C}$	configuration space, e.g., $\mathcal{C} := C^\infty(\mathbb{M},\mathbb{R})$ (Chap. 1)
$\mathcal{D}(\mathbb{M}^n)$	$\equiv \mathcal{D}(\mathbb{R}^{dn},\mathbb{C})$ test funct., compact support, $\mathbb{C}$-valued
$\mathcal{D}(\mathbb{R}^k[\setminus\{0\}])$	$\mathcal{D}(\mathbb{R}^k)$ or $\mathcal{D}(\mathbb{R}^k \setminus \{0\})$, and likewise for $\mathcal{D}'(\mathbb{R}^k[\setminus\{0\}])$
$\mathcal{D}_\omega(\mathbb{R}^k)$	$:= \{\, h \in \mathcal{D}(\mathbb{R}^k) \mid \partial^a h(0) = 0 \ \forall \lvert a\rvert \le \lfloor\omega\rfloor \,\}$ (Eq. (3.2.23))
$\mathcal{S}(\mathbb{M}^n)$	$\equiv \mathcal{S}(\mathbb{R}^{dn},\mathbb{C})$ Schwartz functions (App. A.1)
$\mathcal{D}'(\mathbb{M}^n)$	distributions on $\mathcal{D}(\mathbb{M}^n)$ (App. A.1)

$\mathcal{D}'(\mathbb{M}^n, \mathcal{V})$	distributions on $\mathcal{D}(\mathbb{M}^n)$ with values in vector space $\mathcal{V}$
$\mathcal{S}'(\mathbb{M}^n)$	distributions on $\mathcal{S}(\mathbb{M}^n)$ (App. A.1)
$t(g) = \langle t, g \rangle$	$= \int dx\, t(x)\, g(x)$ application of distrib. t to $g \in \mathcal{D}(\mathbb{M}^n)$
$\mathrm{WF}(t)$	wave front set of the distribution t (App. A.3)
$\mathrm{sing\,supp}(t)$	singular support of the distribution t (Eq. (A.3.1))
$\mathrm{sd}(t)$	scaling degree of the distribution t (Def. 3.2.5)
$\delta_{(k)}$	Delta distribution in $\mathcal{D}'(\mathbb{R}^k)$
$\theta(x)$	Heaviside funct., $\theta(x) = 1$ for $x \geq 0$, $\theta(x) = 0$ for $x < 0$
$\mathrm{sgn}\, x$	$:= \theta(x) - \theta(-x)$ the sign function

Further mathematical symbols

$\mathbb{N}$	natural numbers including zero
$\mathbb{N}^*$	natural numbers without zero
$\mathbb{R}_+$	$:= \{r \in \mathbb{R} \,\vert\, r > 0\}$
$\lfloor r \rfloor$	integer part of $r \in \mathbb{R}$
$\vert I \vert$	number of elements of the finite set $I \subset \mathbb{N}$
$1_{n\times n}$	unit matrix in $\mathbb{C}^{n\times n}$
$\mathrm{Id} \equiv \mathrm{Id}_{\mathcal{M}}$	identity map defined on the set $\mathcal{M}$, i.e., $\mathrm{Id}(x) = x\ ,\ \forall x \in \mathcal{M}$
pp	principle part of a Laurent series (Proposition 3.5.12)
$\mathrm{tr}(\cdot)$	trace of a matrix
S_n	group of all permutations of $\{1, \ldots, n\}$
$\mathfrak{S}_n$	symmetrization w.r.t. the group S_n
$\mathrm{sgn}\, \pi$	signature of the permutation π
$\mathfrak{F}f \equiv \hat{f}$	Fourier transf. of the function/distribution f (App. A.1)
$\mathcal{V}[\![\lambda]\!]$	formal power series in λ with coefficients in $\mathcal{V}$ (Eq. (A.1.1))
$\mathcal{T}(\mathcal{V})$	completion of tensor algebra over $\mathcal{V}$ (Eq. (A.1.5))
$e_{\otimes}^{V}$	typical element of $\mathcal{T}(\mathcal{V})$ (Eq. (A.1.6))

Minkowski spacetime

$\mathbb{M}_d \equiv \mathbb{M}$	d-dimensional Minkowski spacetime
$V_\pm$	open forward/backward light cone (App. A.1)
$\overline{V}_\pm$	closed forward/backward light cone (App. A.1)
Δ_n	thin diagonal in $\mathbb{M}^n$ (defined in (A.1.12))
$\check{\mathbb{M}}^n$	$:= \{(x_1, \ldots, x_n) \in \mathbb{M}^n \,\vert\, x_l \neq x_j\ \forall 1 \leq l < j \leq n\}$ (Eq. (3.1.68))
$\Vert x \Vert$	$:= \sqrt{(x^0)^2 + \cdots + (x^{d-1})^2}$ for $x = (x^0, \ldots, x^{d-1}) \in \mathbb{M}_d$
$\Box_x$	wave operator: $\Box_x := \partial_{x^0}^2 - \partial_{x^1}^2 - \cdots - \partial_{x^{d-1}}^2$
$\mathcal{L}_+^\uparrow$	proper, orthochronous Lorentz group
$\mathcal{P}_+^\uparrow$	proper, orthochronous Poincaré group

Dynamics

κ	coupling constant
$e(>0)$	elementary charge, i.e., the (physical) charge of a positron
S_0, L_0	free action, free Lagrangian (Sect. 1.5)
$S=\kappa\tilde{S}$	interaction (Sect. 1.5)
$\mathcal{L}_{\rm int}$	or $\mathcal{L}=\kappa L=-\kappa L_{\rm int}$, interaction Lagrangian (Sect. 1.5)
r_{S_0+S,S_0}	retarded wave operator (Sect. 1.6)

Fields and Field spaces

$\mathcal{C}_{S_0+S}$	space of smooth solutions to the action S_0+S (Sect. 1.6)
$\varphi(x)$	real scalar field (Sect. 1.1)
$\phi(x)$	complex scalar field (Exap. 1.3.2)
$\psi(x), \overline{\psi}(x)$	Dirac spinors (Sect. 5.1.1)
$A^\mu(x)$	photon field (Sect. 5.1.3)
$u(x)$, $\tilde{u}(x)$	Faddeev–Popov ghosts (Sect. 5.1.2)
$\mathcal{F}$	space of fields (Def. 1.2.1)
$\mathcal{F}_\hbar$	space of polynomials in $\hbar$ with coefficients $\in\mathcal{F}$ (Eq. (2.1.6))
$\mathcal{F}'(\mathbb{M}^n)$	space of allowed distributions f_n in Def. 1.2.1 of $\mathcal{F}$
$\mathcal{F}_0\equiv\mathcal{F}_0^{(m)}$	on-shell fields, $\mathcal{F}$ restricted to $\mathcal{C}_{S_0}$ (Eq. (2.6.1))
$F_0\in\mathcal{F}_0$	lower index "0" denotes restriction to $\mathcal{C}_{S_0}$
$\frac{\delta F}{\delta\varphi(x)}$	functional derivative of the field $F\in\mathcal{F}$ (Def. 1.3.1)
$\frac{\delta F_0}{\delta\varphi_0(x)}$	functional derivative of $F_0\in\mathcal{F}_0$ (Eq. (3.4.36))
$\operatorname{supp} F$	support of the field $F\in\mathcal{F}$ (Def. 1.3.3)
$\mathcal{F}_{\rm loc}$	space of local fields (Def. 1.3.4)
$\mathcal{J}^{(m)}$	ideal of $(\mathcal{F},\cdot\,(\text{or }\star))$ generated by free field eq. (Eq. (2.6.2))
$\mathcal{P}$	space of polynomials in $\partial^a\varphi$ for $a\in\mathbb{N}^d$ (Def. 1.3.4)
$\mathcal{P}_{\rm hom}$	all $A\in\mathcal{P}$ being homogeneous in mass dimens. (Def. 3.1.18)
$\mathcal{P}_j$	all $A\in\mathcal{P}_{\rm hom}$ with mass dimension $\dim A=j$ (Def. 3.1.18)
$\mathcal{P}_{\rm bal}$	space of balanced fields (Def. 1.4.1)
$\mathcal{J}_\mathcal{P}$	ideal of $\mathcal{P}$ generated by free field equation (Eq. (3.4.1))
$\mathcal{P}_0$	$:=\mathcal{P}/\mathcal{J}_\mathcal{P}$ alg. of local on-shell field polynoms (Eq. (3.4.2))
ξ	alg. homomorph. $\mathcal{P}_0\to\mathcal{P}$ choosing a represent. (Eq. (3.4.9))
$A(g)$	$:=\int dx\, A(x)\,g(x)$ for $g\in\mathcal{D}(\mathbb{M})$ and $A\in\mathcal{P}$ (or $A\in\mathcal{P}_{\rm bal}$)
$\varphi(g)$	$:=\int dx\, \varphi(x)\,g(x)$ for $g\in\mathcal{D}(\mathbb{M})$ (particular case of $A(g)$)
$\frac{\partial A}{\partial(\partial^a\varphi)}$	partial derivative of $A\in\mathcal{P}$ w.r.t. $\partial^a\varphi$ (Exer. 1.3.5)
$\underline{A}\subseteq A$	submonomial of the monomial $A\in\mathcal{P}$ (Eq. (3.1.25))
$\overline{A}\subseteq A$	complementary submonomial of $\underline{A}$ (Eq. (3.1.25))
$\dim A$	mass dimension of $A\in\mathcal{P}$ (Def. 3.1.18, Eq. (3.2.63))
$\lvert A\rvert$	order in φ of a monomial $A\in\mathcal{P}$ (Eq. (3.1.31))
α	field parity transformation, $\alpha:\mathcal{F}\to\mathcal{F}$ (Eq. (1.10.6))
β_a	action of the translation $a\in\mathbb{R}^d$ on $\mathcal{F}$ (Eq. (3.1.35))
$\beta_{\Lambda,a}$	action of Poincaré transf. $(\Lambda,a)\in\mathcal{P}_+^\uparrow$ on $\mathcal{F}$ (Eq. (3.1.35))

Propagators

Δ_m^+	Wightman two-point function to the mass m (App. A.2)
Δ_m	Jordan–Pauli or commutator funct. to the mass m (App. A.2)
Δ_m^{ret}	retarded propagator to the mass m (App. A.2)
Δ_m^F	Feynman propagator to the mass m (App. A.2)
p_Λ	smooth approximation for Δ_m^F (Eqs. (3.9.1)–(3.9.3))
Δ_m^{AF}	anti-Feynman propagator to the mass m (Exer. 3.3.13)
$S^\pm$	two-point function for Dirac spinors (Sect. 5.1.1)
$S^{\mathrm{ret/av}}$	retarded/advanced propagator for Dirac spinors (Sect. 5.1.1)
S^F	Feynman propagator for Dirac spinors (Sect. 5.1.1)
H_m^μ	Hadamard function to the mass m (Eq. (2.2.13), App. A.4)

R- and T-products, pertinent generating functionals

$R_{n,1}$	retarded product of QFT to order n (Sect. 3.1)
$r_{n,1}(\cdot)$	$:= \omega_0\big(R_{n,1}(\cdot)\big)$ (Eq. (3.1.41))
$R_{n,1}^0$	restriction of $R_{n,1}$ to $\mathcal{D}(\mathbb{M}^{n+1}\setminus\Delta_{n+1})$ (Sect. 3.2.1)
S_n	totally symmetric part of $R_{n-1,1}$ (Eq. (3.2.1))
S_n^0	totally symmetric part of $R_{n-1,1}^0$
F_S	perturbative quantum field to the interaction S (Eq. (3.1.2))
$R_{\mathrm{cl}\,n,1}$	classical retarded product to order n (Def. 1.7.1)
F_S^{ret}	classical retarded field to the interaction S (Def. 1.6.7)
T_n	time-ordered product to order n (Sect. 3.3)
$t_n(\cdots)$	$:= \omega_0\big(T_n(\cdots)\big)$ (Eq. (3.2.20))
T_n^0	restriction of T_n to $\mathcal{D}(\mathbb{M}^n\setminus\Delta_n)$ (Sect. 3.3.2)
$\overline{T}_n$	antichronological product to order n (Eq. (3.3.34))
$\mathbf{S}(F)$	S-matrix to the interaction F (Def. 3.3.8)
$T_{\Lambda,n}$	regularized T-product to order n, built of p_Λ (Eq. (3.9.4))
$\mathbf{S}_\Lambda$	regularized S-matrix, built of p_Λ (Eq. (3.9.5))
T_n^c	connected part of T_n (Def. 4.4.1)
T_{tree}^c	time-ordered connected tree product (Def. 4.4.4)
T_{tree}	time-ordered tree product (Remk. 4.4.12)
T_F	T-product of interact. fields to the interaction F (Eq. (4.2.45))
$\mathbf{S}_F$	relative S-matrix (Eq. (3.3.31))
$R_{n,1}^{\mathrm{on}}$	on-shell retarded product to order n (Eq. (3.4.49))
T_n^{on}	on-shell time-ordered product to order n (Sect. 3.4)
$\mathbf{S}^{\mathrm{on}}$	on-shell S-matrix (Eq. (3.4.28))
$\overline{T}_n^{\mathrm{on}}$	on-shell antichronological product to order n (Eq. (3.4.30))

Further physical symbols

ω_0	vacuum state (Def. 2.5.2)
σ_ρ	scaling transformation with $\rho > 0$ (Def. 3.1.20)
$\mathcal{R}$	Stückelberg–Petermann renormalization group (Def. 3.6.1)
$\mathcal{R}_0$	larger version of $\mathcal{R}$ (Remark 3.6.6)

$Z \equiv Z_{(m)}$	general element of $\mathcal{R}$ or $\mathcal{R}_0$ (Def. 3.6.1)
$z^{(n)}(\cdots)$	$:= \omega_0\big(Z^{(n)}(\cdots)\big)$ (Eq. (3.6.2))
Z_ρ	(or Z_ρ^m) $\in \mathcal{R}$, Gell-Mann–Low cocycle (Eq. (3.6.44))
V_Λ	effective potential belonging to the propag. p_Λ (Def. 3.9.1)
W_{Λ,Λ_0}	flow of effective potential from Λ_0 to Λ (Eq. (3.9.9))
$\delta_{\vec{h}\cdot\vec{Q}}$	derivation in MWI belonging to symmetry $\vec{Q}$ (Eq. (4.1.4))
$(\Delta^n)_{n\in\mathbb{N}}$	anomaly map in the anomalous MWI (Thm. 4.3.1)
Γ	vertex functional (describes proper vertices) (Prop. 4.4.6)
$\mathfrak{F}^\pm$	bosonic/fermionic Fock space (App. A.5)
Ω	vacuum vector in Fock space (App. A.5)
$a(f)$, $a(f)^*$	annihilation, creation operator in Fock space (App. A.5)
$\varphi^{\mathrm{op}}(x)$	free, real scalar field in Fock space (App. A.5)
$:\varphi^{\mathrm{op}}(x_1)\cdots:$	normally ordered product of $\varphi^{\mathrm{op}}(x_1),\ldots$ (App. A.5)

Symbols used in the treatment of QED

$\mathcal{C}_{\mathrm{spinor}}$	configuration space for spinor fields (Eq. (5.1.15))
$\mathcal{C}_{\mathrm{ghost}}$	configuration space for Faddeev–Popov ghosts (Eq. (5.1.74))
$\mathcal{C}_{\mathrm{photon}}$	configuration space for photon fields (Eq. (5.1.82))
$\mathcal{F}_{\mathrm{spinor}}$	field space generated by Dirac spinors (Def. 5.1.2)
$\mathcal{F}_{\mathrm{ghost}}$	field space generated by Faddeev–Popov ghosts (Sect. 5.1.2)
$\mathcal{F}_{\mathrm{photon}}$	field space generated by photon field (Sect. 5.1.3)
$\mathcal{F}_{\mathrm{QED}}$	field space for QED (Eq. (5.1.91))
$\mathcal{F}_{\mathrm{QED}}^{\mathrm{loc}}$	space of local QED-fields (Eq. (5.1.96))
$\mathcal{F}_{\mathrm{QED},0}$	$:= \mathcal{F}_{\mathrm{QED}}\vert_{\mathcal{C}_{S_0}}$ space of on-shell QED-fields (Eq. (5.5.1))
γ^μ	γ-matrices for the treatment of Dirac spinors (Sect. 5.1.1)
C	charge conjugation matrix (Eq. (5.1.108))
$\not{a}$	$\not{a} := a^\mu\gamma_\mu$ for a Lorentz covariant vector (a^μ)
$\mathrm{sgn}_g\pi$	graded signature of the permutation $\pi \in S_n$ (Eq. (5.1.45))
θ	charge number operator (Eq. (5.1.99))
θ_u	ghost number operator (Eq. (5.1.103))
$\delta(F)$	ghost number of $F \in \mathcal{F}_{\mathrm{QED}}$ (Eq. (5.1.104))
β_C	charge conjugation (Eq. (5.1.113))
β_P	parity transformation or space reflection (Def. 5.1.18)
β_T	time reversal (Def. 5.1.18)
s	BRST transformation (Sect. 5.4.1)
$(\mathcal{K}, \langle\cdot,\cdot\rangle)$	inner product space (Footn. 171)
J	Krein operator in $\mathcal{K}$ (Eq. (5.5.4))

Bibliography

[1] A. Aste, M. Dütsch and G. Scharf, "Gauge Independence of the S-matrix in the causal approach", J. Phys. A **31** (1998), 1563–1579.

[2] A. Aste, M. Dütsch and G. Scharf, "Perturbative gauge invariance: The electroweak theory II", Ann. Phys. (Leipzig) **8 (5)** (1999), 389–404.

[3] D. Bahns and M. Wrochna, "On-shell extension of distributions", Ann. Henri Poincaré **15** (2014), 2045–2067, DOI 10.1007/s00023-013-0288-y.

[4] C. Bär, N. Ginoux and F. Pfäffle, *Wave Equations on Lorentzian Manifolds and Quantization*, EMS Publishing House, Zürich, 2007.

[5] G. Barnich, F. Brandt and M. Henneaux, "Local BRST cohomology in gauge theories", Phys. Rep. **338** (2000), 439–569.

[6] F. Bayen, M. Flato, C. Fronsdal, A. Lichnerowicz and D. Sternheimer, "Deformation theory and quantization I, II", Ann. Phys. **111** (1978), 61–110 and 111–151.

[7] C. Becchi, A. Rouet and R. Stora, "Renormalization of the abelian Higgs-Kibble Model", Commun. Math. Phys. **42** (1975), 127–162.

[8] C. Becchi, A. Rouet and R. Stora, "Renormalization of gauge theories", Ann. Phys. **98** (1976), 287–321.

[9] Ph. Blanchard and R. Sénéor, "Green's functions for theories with massless particles (in perturbation theory)", Ann. Inst. Henri Poincaré A **23** (1975), 147–209.

[10] Ph. Blanchard, J.M. Gracia-Bondía, S. Lazzarini and I. Todorov, "Stora's legacy: Perturbative renormalization", in preparation.

[11] N.N. Bogoliubov and O. Parasiuk, "On the multiplication of propagators in quantum field theory", Acta Math. **97** (1957) 227–326 (in German).

[12] N.N. Bogoliubov and D.V. Shirkov, *Introduction to the Theory of Quantized Fields*, Interscience Publishers, 1959.

[13] M. Bordemann and S. Waldmann, "Formal GNS construction and states in deformation quantization", Commun. Math. Phys. **195** (1998), 549–583.

[14] R.A. Brandt, "Field Equations in Quantum Electrodynamics", Fortschr. Physik **18** (1970), 249–283.

M. Dütsch, *From Classical Field Theory to Perturbative Quantum Field Theory*, Progress in Mathematical Physics 74, https://doi.org/10.1007/978-3-030-04738-2

[15] P. Breitenlohner and D. Maison, "Dimensional renormalization and the action principle", Commun. Math. Phys. **52** (1977), 11–38.

[16] F. Brennecke and M. Dütsch, "Removal of violations of the Master Ward Identity in perturbative QFT", Rev. Math. Phys. **20** (2008), 119–172.

[17] F. Brennecke and M. Dütsch, "The quantum action principle in the framework of causal perturbation theory", in *Quantum Field Theory: Competitive Models*, B. Fauser, J. Tolksdorf and E. Zeidler, eds., Birkhäuser, Basel, 2009; 177–196.

[18] K. Bresser, G. Pinter and D. Prange, "The Lorentz invariant extension of scalar theories", hep-th/9903266;
D. Prange, "Lorentz Covariance in Epstein–Glaser Renormalization", hep-th/9904136.

[19] C. Brouder, B. Fauser, A. Frabetti and R. Oeckel, "Quantum field theory and Hopf algebra cohomology", J. Phys. **A37** (2004), 5895–5927.

[20] Ch. Brouder and M. Dütsch, "Relating on-shell and off-shell formalism in perturbative quantum field theory", J. Math. Phys. **49** (2008), 052303.

[21] Ch. Brouder, N.V. Dang and F. Hélein, "A smooth introduction to the wavefront set", J. Phys. A **47(44)** (2014), 443001, DOI:10.1088/1751-8113/47/44/443001.

[22] Ch. Brouder, N.V. Dang, C. Laurent-Gengoux and K. Rejzner, "Properties of field functionals and characterization of local functionals", J. Math. Phys. **59** (2018), 023508, DOI:10.1063/1.4998323.

[23] R. Brunetti, K. Fredenhagen and Martin Köhler, "The microlocal spectrum condition and Wick polynomials of free fields on curved spacetimes", Commun. Math. Phys. **180** (1996), 633–652.

[24] R. Brunetti and K. Fredenhagen, "Microlocal analysis and interacting quantum field theories: Renormalization on physical backgrounds", Commun. Math. Phys. **208** (2000), 623–661.

[25] R. Brunetti, K. Fredenhagen and R. Verch, "The Generally Covariant Locality Principle – A New Paradigm for Local Quantum Field Theory", Commun. Math. Phys. **237** (2003), 31–68.

[26] R. Brunetti, M. Dütsch and K. Fredenhagen, unpublished notes (2008).

[27] R. Brunetti, M. Dütsch and K. Fredenhagen, "Perturbative algebraic quantum field theory and the renormalization groups", Adv. Theor. Math. Phys. **13** (2009), 1541–1599.

[28] R. Brunetti, K. Fredenhagen and K. Rejzner, "Quantum gravity from the point of view of locally covariant quantum field theory", Commun. Math. Phys. **345 (3)** (2016), 741–779, DOI: 10.1007/s00220-016-2676-x.

[29] R. Brunetti, K. Fredenhagen and P.L. Ribeiro, "Algebraic Structure of Classical Field Theory I: Kinematics and Linearized Dynamics for Real Scalar Fields", submitted to Commun. Math. Phys., arXiv:1209.2148 [math-ph].

[30] D. Buchholz, "Gauss' law and the infraparticle problem", Phys. Lett. B **174** (1986), 331.

[31] D. Buchholz, I. Ojima and H. Roos, "Thermodynamic Properties of Non-Equilibrium States in Quantum Field Theory", Ann. Phys. **297** (2002), 219–242.

[32] A. Deriglazov, *Classical Mechanics: Hamiltonian and Lagrangian Formalism*, Second Edition, Springer Switzerland, 2017, DOI: 10.1007/978-3-319-44147-4.

[33] M. De Wilde and P.B.A. Lecomte, "Existence of star-products and of formal deformations of the Poisson Lie algebra of arbitrary symplectic manifolds", Lett. Math. Phys. **7** (1983), 487–496.

[34] J. Dimock and B.S. Kay, "Classical Wave Operators and Asymptotic Quantum Field Operators on Curved Space-times", Ann. Inst. H. Poincaré Phys. Theor. **37** (1982), 93.

[35] J. Dito, "Star-Product Approach to Quantum Field Theory: The Free Scalar Field", Lett. Math. Phys. **20** (1990), 125.

[36] J. Dito, "Star-products and nonstandard quantization for Klein–Gordon equation", J. Math. Phys. **33** (1992), 791.

[37] N. Drago, T.-P. Hack, N. Pinamonti, "The generalised principle of perturbative agreement and the thermal mass", Ann. Henri Poincaré **18 (3)** (2017), 807–868, DOI: 10.1007/s00023-016-0521-6.

[38] P. Duch, "Massless fields and adiabatic limit in quantum field theory", Ph.D thesis (2017), https://arxiv.org/abs/1709.09907.

[39] P. Duch, "Weak adiabatic limit in quantum field theories with massless particles", Ann. Henri Poincaré **19 (3)** (2018), 875–935, DOI: 10.1007/s00023-018-0652-z.

[40] M. Dütsch, F. Krahe and G. Scharf, "Interacting Fields in Finite QED", N. Cimento A **103 (6)** (1990), 871–901.

[41] M. Dütsch, F. Krahe and G. Scharf, "Axial Anomalies in Massless Finite QED", N. Cimento A **105 (3)** (1992), 399–422.

[42] M. Dütsch, T. Hurth and G. Scharf, "Gauge invariance of massless QED", Phys. Lett. B **327** (1994), 166.

[43] M. Dütsch, T. Hurth, F. Krahe and G. Scharf, "Causal construction of Yang–Mills theories. II", N. Cimento A **107** (1994), 375–406.

[44] M. Dütsch, T. Hurth and G. Scharf, "Causal construction of Yang–Mills theories. IV. Unitarity", N. Cimento A **108** (1995), 737.

[45] M. Dütsch, "Non-uniqueness of quantized Yang–Mills theories", J. Phys. A **29** (1996), 7597–7617.

[46] M. Dütsch, "Slavnov-Taylor identities from the causal point of view", Int. J. Mod. Phys. A **12 (18)** (1997), 3205–3248.

[47] M. Dütsch and K. Fredenhagen, "Deformation stability of BRST-quantization", AIP Conf. Proc. **453** (1998), 324–333.

[48] M. Dütsch and K. Fredenhagen, "A local (perturbative) construction of observables in gauge theories: The example of QED", Commun. Math. Phys. **203** (1999), 71–105.

[49] M. Dütsch and G. Scharf, "Perturbative gauge invariance: The electroweak theory", Ann. Phys. (Leipzig) **8 (5)** (1999), 359–387.

[50] M. Dütsch and B. Schroer, "Massive vector mesons and gauge theory", J. Phys. A **33 (23)** (2000), 4317–4356.

[51] M. Dütsch and K. Fredenhagen, "Algebraic quantum field theory, perturbation theory, and the loop expansion", Commun. Math. Phys. **219** (2001), 5–30.

[52] M. Dütsch and K. Fredenhagen, "Perturbative algebraic field theory and deformation quantization", in *Mathematical Physics in Mathematics and Physics: Quantum and Operator Algebraic Aspects*, R. Longo, ed., Fields Institute Communications **30**, AMS, Providence, RI, (2001); 151–160.

[53] M. Dütsch and F.-M. Boas, "The Master Ward Identity", Rev. Math. Phys. **14** (2002), 977–1049.

[54] M. Dütsch and K. Fredenhagen, "The Master Ward Identity and generalized Schwinger–Dyson equation in classical field theory", Commun. Math. Phys. **243** (2003), 275–314.

[55] M. Dütsch and K. Fredenhagen, "Causal perturbation theory in terms of retarded products, and a proof of the Action Ward Identity", Rev. Math. Phys. **16** (2004), 1291–1348.

[56] M. Dütsch, "Proof of perturbative gauge invariance for tree diagrams to all orders", Ann. Phys. (Leipzig) **14 (7)** (2005), 438–461, DOI: 10.1002/andp.200510145.

[57] M. Dütsch and K. Fredenhagen, "Action Ward identity and the Stückelberg–Petermann renormalization group", in *Rigorous Quantum Field Theory*, A. Boutet de Monvel, D. Buchholz, D. Iagolnitzer and U. Moschella, eds., Birkhäuser, Basel, 2006; 113–123.

[58] M. Dütsch and K. Fredenhagen, "Perturbative renormalization and BRST", Encyclopedia of Mathematical Physics, Elsevier, Academic Press, 2006, hep-th/0411196.

[59] M. Dütsch, J.M. Gracia-Bondía, F. Scheck and J.C. Várilly, "Quantum gauge models without (classical) Higgs mechanism", Eur. Phys. J. C **69** (2010), 599–622.

[60] M. Dütsch and K.-H. Rehren "Protecting the conformal symmetry via bulk renormalization on Anti deSitter space", Commun. Math. Phys. **307** (2011) 315–350, DOI: 10.1007/s00220-011-1311-0.

[61] M. Dütsch, "Connection between the renormalization groups of Stückelberg–Petermann and Wilson", Confluentes Mathematici **4** (2012), 12400014.

[62] M. Dütsch and J.M. Gracia-Bondía, "On the assertion that PCT violation implies Lorentz non-invariance", Phys. Lett. B (2012) PLB28534, DOI: 10.1016/j.physletb.2012.04.038.

[63] M. Dütsch, K. Fredenhagen, K.J. Keller and K. Rejzner, "Dimensional regularization in position space, and a forest formula for Epstein–Glaser renormalization", J. Math. Phys. **55 (12)** (2014), 122303, DOI: 10.1063/1.4902380.

[64] M. Dütsch, "The scaling and mass expansion", Ann. Henri Poincaré **16 (1)** (2015), 163–188, DOI: 10.1007/s00023-014-0324-6.

[65] M. Dütsch, " Massive vector bosons: Is the geometrical interpretation as a spontaneously broken gauge theory possible at all scales?", Rev. Math. Phys. **27 (10)** (2015), 1550024, DOI: 10.1142/S0129055X15500245.

[66] H. Epstein and V. Glaser, "The role of locality in perturbation theory", Ann. Inst. Henri Poincaré **19A** (1973), 211–295.

[67] H. Epstein and V. Glaser, "Adiabatic limit in perturbation theory" in *Renormalization Theory*, G. Velo and A.S. Weightman, eds., D. Reidel Publishing Company, Dordrecht Holland, 1976, 193–254.

[68] F. Faà di Bruno, "Sullo sviluppo delle funzioni", Annali di Scienze Matematiche e Fisiche di Tortoloni **6** (1855), 479–480; "Note sur une nouvelle formule de calcul différentiel", Quart. J. Pure Appl. Math. **1** (1857), 359–360.

[69] L.D. Faddeev and V.N. Popov, "Feynman diagrams for the Yang–Mills field", Phys. Lett. B **25 (1)** (1967), 29.

[70] B.V. Fedosov, "Quantization and the index", Sov. Phys. Dokl. **31 (11)** (1986), 877–878.

[71] B.V. Fedosov, *Deformation Quantization and Index Theory*, Akademie Verlag, Berlin, 1996.

[72] R.P. Feynman, "Quantum Theory of Gravitation", Acta Phys. Polonica **24** (1963), 697–722.

[73] K. Fredenhagen, *Quantum Field Theory*, Lecture notes (2009–2010), https://unith.desy.de/research/aqft/.

[74] K. Fredenhagen and K. Rejzner, "Batalin–Vilkoviski formalism in perturbative algebraic quantum field theory", Commun. Math. Phys. **317** (2013), 697–725.

[75] K. Fredenhagen and F. Lindner, "Construction of KMS States in Perturbative QFT and Renormalized Hamiltonian Dynamics ", Commun. Math. Phys. **332 (3)** (2014), 895–932.

[76] K. Fredenhagen, "An Introduction to Algebraic Quantum Field Theory", in *Advances in Algebraic Quantum Field Theory*, R. Brunneti, C. Dappiaggi, K. Fredenhagen and J. Yngvason, eds., Math. Physics Studies, Springer 2015; 1–30.

[77] K. Fredenhagen and K. Rejzner, “Perturbative Construction of Models of Algebraic Quantum Field Theory”, in *Advances in Algebraic Quantum Field Theory*, R. Brunneti, C. Dappiaggi, K. Fredenhagen and J. Yngvason, eds., Math. Physics Studies, Springer 2015; 31–74.

[78] D.Z. Freedman, K. Johnson and J.I. Latorre, “Differential regularization and renormalization: A new method of calculation in quantum field theory”, Nucl. Phys. **B 371** (1992) 353–414.

[79] G.B. Folland, *Quantum Field Theory: A Tourist Guide for Mathematicians*, Mathematical Surveys and Monographs **149**, American Mathematical Society, 2008.

[80] I.M. Gelfand and G.E. Shilov, *Generalized Functions I*, Academic Press, New York, 1964.

[81] M. Gell-Mann and F.E. Low, “Bound states in quantum field theory”, Phys. Rev. **84 (2)** (1951), 350.

[82] M. Gell-Mann and F.E. Low, “Quantum Electrodynamics at Small Distances”, Phys. Rev. **95 (5)** (1954), 1300–1312.

[83] M. Gerstenhaber, “On the Deformation of Rings and Algebras I-IV”, Ann. Math. **79** (1964), 59–103; Ann. Math. **84** (1966), 1–19; Ann. Math. **88** (1968), 1–34; Ann. Math. **99** (1974), 257–276.

[84] V. Glaser, H. Lehmann and W. Zimmermann, “Field operators and retarded functions”, N. Cimento **6** (1957), 1122–1128.

[85] J.M. Gracia-Bondía, “Improved Epstein–Glaser renormalization in coordinate space I. Euclidean framework”, Math. Phys. Anal. Geom. **6** (2003), 59–88.

[86] J.M. Gracia-Bondía, H. Gutiérrez and J.C. Várilly, “Improved Epstein–Glaser renormalization in x-space versus differential renormalization”, Nucl. Phys. B **886** (2014), 824–869.

[87] D.R. Grigore, “On the uniqueness of the nonabelian gauge theories in Epstein–Glaser approach to renormalisation theory”, Romanian J. Phys. **44** (1999), 853.

[88] D.R. Grigore, “Scale Invariance in the Causal Approach to Renormalization Theory”, Ann. Phys. (Leipzig) **10** (2001), 473.

[89] H.J. Groenewold, “On the principles of elementary quantum mechanics”, Physica **12** (1946), 405–460.

[90] W. Güttinger, “Generalized functions and dispersion relations in physics”, Fortschr. Physik **14** (1966), 483–602.

[91] W. Güttinger, A. Rieckers, “Spectral Representations of Lorentz Invariant Distributions and Scale Transformation”, Commun. Math. Phys. **7** (1968), 190–217.

[92] R. Haag, “Quantum field theories with composite particles and asymptotic conditions”, Phys. Rev. **112** (1958), 669–673.

[93] R. Haag and D. Kastler, "An algebraic approach to quantum field theory", J. Math. Phys. **5** (1964) 848–861.

[94] R. Haag, *Local Quantum Physics: Fields, Particles, Algebras*, 2nd edn., Springer-Verlag, Berlin, 1996.

[95] E. Hawkins and K. Rejzner, "The Star Product in Interacting Quantum Field Theory", arXiv:1612.09157.

[96] M. Henneaux and C. Teitelboim, *Quantization of gauge systems*, Princeton University Press, Princeton, New Jersey, 1992.

[97] K. Hepp, "On the connection between the LSZ and Wightman quantum field theory", Commun. Math. Phys. **1** (1965), 95–111.

[98] K. Hepp, "Proof of the Bogoliubov–Parasuik theorem on renormalization", Commun. Math. Phys. **2** (1966), 301–326.

[99] A.C. Hirshfeld and P. Henselder, "Star Products and Perturbative Quantum Field Theory", Ann. Phys. **298** (2002), 382–393.

[100] S. Hollands and R.M. Wald, "Local Wick polynomials and time-ordered products of quantum fields in curved spacetime", Commun. Math. Phys. **223** (2001), 289–326.

[101] S. Hollands and R.M. Wald, "Existence of local covariant time-ordered products of quantum fields in curved spacetime", Commun. Math. Phys. **231** (2002), 309–345.

[102] S. Hollands and R.M. Wald, "On the renormalization group in curved spacetime", Commun. Math. Phys. **237** (2003), 123–160.

[103] S. Hollands and R.M. Wald, "Conservation of the stress tensor in perturbative interacting quantum field theory in curved spacetimes", Rev. Math. Phys. **17** (2005), 227–312.

[104] S. Hollands, "The operator product expansion for perturbative quantum field theory in curved spacetime ", Commun. Math. Phys. **273 (1)** (2007), 1–36.

[105] S. Hollands, "Renormalized Quantum Yang–Mills Fields in Curved Spacetime", Rev. Math. Phys. **20** (2008), 1033–1172.

[106] G. 't Hooft and M. Veltman, "Regularization and Renormalization of Gauge Fields", Nuclear Physics B **44**, (1972) 189–219.

[107] L. Hörmander, *The Analysis of Linear Partial Differential Operators. I: Distribution Theory and Fourier Analysis*, 2nd edition, Springer, Berlin, 1990.

[108] V.A. Il'in and D.A. Slavnov, "Observable algebras in the S-matrix approach", Theor. Math. Phys. **36** (1978), 32.

[109] C. Itzykson and J.-B. Zuber, *Quantum Field Theory*, McGraw-Hill, 1980.

[110] G. Jona-Lasinio, "Relativistic field theories with symmetry-breaking solutions", N. Cimento **34** (1964), 1790–1795.

[111] G. Källén, *Quantum Electrodynamics*, Springer-Verlag, Berlin, 1972.

[112] G. Keller, C. Kopper and C. Schophaus, "Perturbative renormalization with flow equations in Minkowski space", Helv. Phys. Acta **70** (1997), 247–274.

[113] K.J. Keller, "Euclidean Epstein–Glaser renormalization", J. Math. Phys. **50** (2009), 103503.

[114] K.J. Keller, "Dimensional Regularization in Position Space and a Forest Formula for Regularized Epstein–Glaser Renormalization", Ph.D. thesis, Hamburg University (2010), arXiv:1006.2148.

[115] M. Kontsevich, "Formality conjecture" in *Deformation Theory and Symplectic Geometry,* D. Sternheimer, J. Rawnsley, S. Gutt, eds., Mathematical Physics Studies **20**, Kluwer Academic, Dordrecht/Boston/London (1997), 139–156.

[116] M. Kontsevich, "Deformation quantization of Poisson manifolds", Lett. Math. Phys. **66** (2003), 157–216.

[117] F. Krahe, "On the Algebra of Ghost Fields", arXiv:hep-th/9502097.

[118] T. Kugo and I. Ojima, "Local covariant operator formalism of non-abelian gauge theories and quark confinement problem", Suppl. Progr. Theor. Phys. **66** (1979), 1.

[119] Y.-M.P. Lam, "Perturbation Lagrangian theory for scalar fields: Ward–Takahasi identity and current algebra", Phys. Rev. **D6** (1972), 2145–2161.

[120] J.I. Latorre, C. Manuel and X. Vilasis-Cardona, "Systematic Differential Renormalization to All Orders", Ann. Phys. (N.Y.) **231** (1994) 149.

[121] B. Lautrup, "Canonical quantum electrodynamics in covariant gauges", Kgl. Danske Videnskab. Selskab. Mat.-fys. Medd. **35** (1967), 1.

[122] S. Lazzarini and J.M. Gracia-Bondía, "Improved Epstein–Glaser renormalization II. Lorentz invariant framework", J. Math. Phys. **44** (2003), 3863–3875.

[123] H. Lehmann, "Über Eigenschaften von Ausbreitungsfunktionen und Renormierungskonstanten quantisierter Felder", N. Cimento **11 (4)** (1954), 342–357.

[124] H. Lehmann, K. Symanzik, and W. Zimmerman, "On the formulation of quantized field theories", N. Cimento **1**, (1955), 205-225.

[125] J.H. Lowenstein, "Differential vertex operations in Lagrangian field theory", Commun. Math. Phys. **24** (1971), 1–21.

[126] J.E. Moyal, "Quantum mechanics as a statistical theory", Proc. Cambridge Philos. Soc. **45** (1949), 99–124.

[127] T. Muta, *Foundations of quantum chromodynamics,* Lecture Notes in Physics **57**, World Scientific, 1998, 2nd ed.

[128] N. Nakanishi, "Covariant quantization of the electromagnetic field in the Landau gauge", Prog. Theor. Phys. **35** (1966), 1111–1116.

[129] K.-H. Neeb, *Monastir Summer Scool: Infinite-Dimensional Lie Groups*, TU Darmstadt Preprint **2433** (2006).

[130] J. von Neumann, "Die Eindeutigkeit der Schrödingerschen Operatoren", Math. Ann. **104** (1931), 570–578.

[131] N.M. Nikolov, R. Stora and I. Todorov, "Renormalization of massless Feynman amplitudes in configuration space", Rev. Math. Phys. **26** (2014), 1430002.

[132] N.M. Nikolov, "Renormalization of massive Feynman amplitudes and homogeneity (based on a joint work with Raymond Stora)", Nucl. Phys. B **912** (2016) 38–50.

[133] R.E. Peierls, "The commutation laws of relativistic field theory", Proc. Roy. Soc. London A **214** (1952), 143–157.

[134] M. Peskin and D. Schroeder, *An Introduction to Quantum Field Theory*, Addison-Wesley, Reading, Massachudetts, 1995.

[135] O. Piguet and S.P. Sorella, *Algebraic renormalization: Perturbative renormalization, symmetries and anomalies*, Springer, Lect. Notes Phys. **M28**, 1995.

[136] G. Pinter, "Finite renormalizations in the Epstein–Glaser framework and renormalization of the S-Matrix of ϕ^4-Theory", Ann. Phys. (Leipzig) **10** (2001), 333.

[137] J. Polchinski, " Renormalization and Effective Lagrangians", Nucl. Phys. B **231** (1984), 269–295.

[138] G. Popineau and R. Stora, "A pedagogical remark on the main theorem of perturbative renormalization theory", Nucl. Phys. B **912** (2016), 70–78, preprint: LAPP–TH, Lyon (1982).

[139] D. Prange, "Epstein–Glaser renormalization and differential renormalization", J. Phys. A **32** (1999), 2225.

[140] M.J. Radzikowski, "Micro-local approach to the Hadamard condition in quantum field theory on curved space-time", Commun. Math. Phys. **179** (1996), 529.

[141] M. Reed and B. Simon, *Methods of Modern Mathematical Physics. I: Functional Analysis; and II: Fourier Analysis, Self-Adjointness*, Academic Press, New York, 1972 and 1975.

[142] K. Rejzner, "Fermionic fields in the functional approach to classical field theory", Rev. Math. Phys. **23 (9)** (2011), 1009–1033.

[143] K. Rejzner, *Perturbative Algebraic Quantum Field Theory. An introduction for Mathematicians*, Mathematical Physics Studies, Springer, 2016.

[144] M. Requardt, "Symmetry conservation and integrals over local charge densities in quantum field theory", Commun. Math. Phys. **50** (1976), 259.

[145] D. Ruelle, "On the asymptotic condition in quantum field theory", Helv. Phys. Acta **35** (1962), 147.

[146] L.H. Ryder, *Quantum Field Theory*, Cambridge University Press, 1985.

[147] M. Salmhofer, *Renormalization – An Introduction*, Texts and Monographs in Physics, Springer-Verlag, Berlin, 1999.

[148] G. Scharf, *Finite Quantum Electrodynamics: The Causal Approach*, Texts and Monographs in Physics, Springer, Berlin, 1995.

[149] G. Scharf, *Quantum Gauge Theories: A True Ghost Story*, Wiley, New York, 2001.

[150] G. Scharf, *Gauge Field Theories: Spin One and Spin Two*, Dover Books, Mineola, NY, 2016.

[151] U. Schreiber, "Mathematical Quantum Field Theory", lecture notes, www.physicsforums.com/insights/a-first-idea-of-quantum-field-theory/.

[152] B. Schroer, "Infrateilchen in der Quantenfeldtheorie", Fortschr. Phys. **173** (1963), 1527.

[153] J. Shatah and M. Struwe, *Geometric Wave Equations*, Courant Institute of Mathematical Sciences, New York University, 1998.

[154] E.R. Speer, "On the structure of analytic renormalization", Commun. Math. Phys. **23** (1971), 23–36.

[155] O. Steinmann, *Perturbation Expansions in Axiomatic Field Theory*, Lecture Notes in Physics **11**, Springer, Berlin, 1971.

[156] O. Steinmann, "Perturbation Theory of Wightman Functions", Commun. Math. Phys. **152** (1993), 627–645.

[157] O. Steinmann, *Perturbative Quantum Electrodynamics and Axiomatic Field Theory*, Springer, Berlin, 2000.

[158] R. Stora, "Differential algebras in Lagrangean field theory", Lectures at ETH, Zürich, 1993, unpublished.

[159] R. Stora, "Pedagogical experiments in renormalized perturbation theory", contribution to the conference "Theory of Renormalization and Regularization", Hesselberg in Germany (2002), unpublished.

[160] R. Stora, "Renormalized perturbation theory: A missing chapter", Int. J. Geom. Meth. Mod. Phys. **5** (2008), 1345.

[161] R. Stora, "Causalité et groupes de renormalisation perturbatifs", in "Théorie quantique des champs: Méthodes et applications", T. Boudjedaa, A. Makhlouf and R. Stora, eds., Hermann, Paris, 2008, 67–120.

[162] R.F. Streater and A.S. Wightman, *PCT, Spin and Statistics, and All That*, Princeton Landmarks in Mathematics and Physics, Princeton University Press, 1964.

[163] E.C.G. Stueckelberg and A. Petermann, "La normalisation des constantes dans la theorie des quanta", Helv. Phys. Acta **26** (1953), 499–520.

[164] I.V. Tyutin, "Gauge Invariance in Field Theory and Statistical Physics in Operator Formalism", preprint of P.N. Lebedev Phys. Institute, n° 39 (1975), arXiv:0812.0580.

[165] S. Waldmann, *Poisson-Geometrie und Deformationsquantisierung*, Springer, Heidelberg, 2007.

[166] S. Waldmann, "A nuclear Weyl algebra", J. Geom. Phys. **81** (2014), 10–46.

[167] S. Waldmann, "Representation Theory of $*$-Algebras", Lecture Notes (2016), University of Würzburg.

[168] S. Weinberg, *The Quantum Theory of Fields: Vols. I and II*, Cambridge University Press, Cambridge, 1995.

[169] H. Weyl, *The Theory of Groups and Quantum Mechanics*, Dover, New York, 1931.

[170] K.G. Wilson, "Non-Lagrangian Models of Current Algebra", Phys. Rev. **179** (1969), 1499–1512.

[171] K.G. Wilson and J. Kogut, "The renormalization group and the ε expansion", Phys. Repts. C **12 (2)** (1974), 75–200.

[172] J. Zahn, "Locally covariant charged fields and background independence", Rev. Math. Phys. **27 (07)** (2015), 1550017.

[173] E. Zeidler, *Quantum Field Theory I: Basics in Mathematics and Physics*, Springer, Berlin, Heidelberg, 2006.

[174] E. Zeidler, *Quantum Field Theory II: Quantum Electrodynamics*, Springer, Berlin, Heidelberg, 2009.

[175] E. Zeidler, *Quantum Field Theory III: Gauge Theory*, Springer, Berlin, Heidelberg, 2011.

[176] W. Zimmermann, "Convergence of Bogoliubov's method of renormalization in momentum space", Commun. Math. Phys. **15** (1969), 208–234.

[177] W. Zimmermann, "Normal products and the short distance expansion in the perturbation theory of renormalizable interactions", Ann. Phys. **77** (1973), 570–601.

Index

M. Dütsch, *From Classical Field Theory to Perturbative Quantum Field Theory*, Progress in Mathematical Physics 74, https://doi.org/10.1007/978-3-030-04738-2